Tips & Tricks with Questions

ADVANCED CSIR-NET/JRF Mathematics

Dr. Gajendra Purohit

Published By

Invincible Publishers Pvt. Ltd.

Published by:

Invincible Publication Pvt. Ltd.

1103-A, 11th Floor, SAS Tower, Sector 38, Gurugram, Haryana-122003

Phone: +91-124-4034247, +91 9599066061

Website: www.invinciblepublishers.com

Sales Office-4760-61/23, Basement, Pratap Street, Ansari Road, Daryaganj, New Delhi-110002

Phone: +91-11-40198405

Email: invinciblepublishers@gmail.com

ISBN: 978-81-94548-14-0

Book Name: Advanced CSIR-NIT/JRF Mathematics

Third Edition: October 2023

To the unconditional love of

My family and My Students

PREFAC

Greetings Upcoming Scholars!

We are here to guide you with CSIR-NET / JRF Mathematics with yet another book which can serve the needs of our beloved Students' Community. This book will go a long way in helping students in preparation of their CSIR-NET / JRF and similar exams like GATE, SET, IIT- JAM, CUET, PhD Entrance, NBHM, PGT etc.

This book is provided with well defined theoretical concepts with tips and tricks to solve MCQs & MSQs of the CSIR-NET / JRF exam comprising a large number of solved examples.

We have taken this fact into consideration that the immediate availability of a solution with tips and tricks will serve to reinforce the learning process. At the end of each chapter, a set of unsolved problems has been incorporated as an exercise for the students.

This book contains an exceptional set of questions with tips and tricks. We hope that this book will be of great use for those it's been written.

Any suggestions from the reader for improvement of this book are appreciated.

You can follow Telegram and YouTube Channels of Dr. Gajendra Purohit to stay updated & connected.

Telegram Channel : https://t.me/mathbygpsir

YouTube Channel-1 : https://www.youtube.com/@gajendrapurohit

YouTube Channel-2 : https://www.youtube.com/@gajendrapurohit-GATE-NET-JAM

YouTube Channel-3 : https://www.youtube.com/@gajendrapurohit-ExamPrep

ACKNOWLEDGEMENT

I am indebted to my parents, my wife Swarna, and my beloved boys Manas and Aashay, whose love, support, and care are beyond words & conditions.

I would like to express my sincere gratitude to everyone who has aided me in whichever manner in finishing this book. Additionally, I would like to thank my team members Harish Kumar Lohar & Manish Sharma.

I want to thank my students in particular because they constantly push me to go the additional mile, ask me thought-provoking questions, and inspire me to learn more and more every day.

Dr. Gajendra Purohit

CONTENTS

LINEAR ALGEBRA

REAL ANALYSIS

COMPLEX ANALYSIS

MODERN ALGEBRA

DIFFERENTIAL EQUATION

COV, NA, I.E.

MODEL TEST PAPERS

Chapter 1

Vector Space and Subspace

1. Vector Space :

Let (F,+,.) be a field. The elements of F are called scalars. Let V be a non-empty set whose elements are called vectors. Then V is a vector space over the field F (denoted by V(F)) if

(a) There is an internal composition in V called addition of vectors which is denoted by '+' such that for the composition V is an abelian group, i.e.,

(i) $u+v \in V$ for all $u,v \in V$.

(ii) $u+v = v+u$ for all $u,v \in V$.

(iii) $u+(v+w)=(u+v)+w$ for all $u,v,w \in V$

(iv) $\exists$ an element $\bar{0} \in V$ such that $u+\bar{0}=u=\bar{0}+u$ for all $u \in V$. This element $\bar{0} \in V$ iscalled the zero vector.

(v) To every vector $u \in V$, there exists a vector $-u \in V$ such that $u+(-u)=0=(-u)+u$

(b) There is an external composition in V over F called scalar multiplication and is denoted multiplicatively, i.e. au ∈ V for all a ∈ F and for all u ∈ V. In other words, V is closed with respect to scalar multiplication.

(c) The two compositions, i.e., scalar multiplication and addition of vectors satisfy the following conditions:

(i) $a(u+v)=au+av \quad \forall a \in F$ and $\forall u,v \in V$.

(ii) $(a+b)u = au+bu \quad \forall a,b \in F$ and $\forall u \in V$.

(iii) $(ab)u = a(bu) \quad \forall a,b \in F$ and $\forall u \in V$.

(iv) $1.u = u \quad \forall u \in V$ and 1 is the unity of the field F.

Example (i) R is a vector space over R denoted by R(R).

(ii) C is a vector space over R denoted by C(R).

(iii) R is not vector space over C because $ia \notin R \;;\forall a \in R, i \in C$

Example: $V = \left\{ \begin{bmatrix} a & b \\ c & d \end{bmatrix} ; a,b,c,d \in R \right\} \& F = \left\{ \begin{bmatrix} a & a \\ a & a \end{bmatrix} : a \in R \right\}$ is a field then

Solution : 1 ∈ F, where 1 = unity

$$\Rightarrow \text{unity} = \begin{bmatrix} 1/2 & 1/2 \\ 1/2 & 1/2 \end{bmatrix} \in F$$

$$1.X = \begin{bmatrix} 1/2 & 1/2 \\ 1/2 & 1/2 \end{bmatrix}\begin{bmatrix} 1 & 0 \\ 0 & 1 \end{bmatrix} \neq X = \begin{bmatrix} 1 & 0 \\ 0 & 1 \end{bmatrix} \Rightarrow \text{V(F) is not vector space.}$$

Some important vector spaces :

(i) $V = R^n = \{(a_1, a_2, \ldots, a_n) : a_i \in R\}$ is a vector space over R and denoted by $R^n(R)$

(ii) $V = \{[a_{ij}]_{m\times n} : a_{ij} \in R\}$ be a collection of real matrix of order $m \times n$ is a vector space over R and denoted by V(R)

(iii) $V = p(x) = \{a_0 + a_1x + a_2x^2 \ldots. + a_nx^n \mid a_j \in R\}$ is a vector space over R and denoted by V(R)

(iv) $V = \{f \mid f : R \to R$; f is real valued function$\}$ is a vector space over R and denoted by V(R)

(v) $V = \{<a_n>\ a_n \in R$; be a sequence$\}$ is a vector space over R and denoted by V(R)

(vi) Every field is vector space over its subfield.

Example R(Q), C (R)

2 Vector Subspace :

Subspace :- Let V(F) is a vector space and $W \subset V$,then we say W is subspace of V if W is itself form a vector spaces over same field F.

Example :-{0} and V always subspace of V.

One-step test : The necessary and sufficient condition for a non-empty subset W of a vector space V(F) to be a subspace of V if a, b $\in$ F and u, v$\in$ W $\Rightarrow$ au + bv$\in$ W.

Two – step test : A non-empty subset W of a vector space V(F) to be a subspace of V if a, b $\in$ F and u, v$\in$ W $\Rightarrow$(i) au$\in$ W (ii) u + v$\in$ W.

Example : Let $V = R^2(R)$. Then $W = \{(x ,0) \mid x \in R\}$ is subspace of V

3. Intersection and union of two Subspaces :

1. The intersection of any two subspaces W_1 and W_2 of a Vector space $V(F)$ is also a subspace of $V(F)$.
2. The union of two subspaces is a subspace if and only if one is contained in the other.

Example : In general, the union of two subspaces need not be a subspace.

Let $W_1 = \{(x, 0) \mid x \in R\}$ and $W_2 = \{(0, y) \mid y \in R\}$

Then $W_1 \cup W_2$ is not subspace of $R^2(R)$

4. Quotient Space

Let V be a vector space over F and W be a subspace of V. Then

$\frac{V}{W} = \{W + u \mid u \in V\}$ is called quotient space with the operation

(W + u) + (W + v) = W + (u + v), u,v$\in$V and a(W + u) = W + au, a $\in$ F.

Example : Let V = $R^2(R)$ and W = {(x ,0) | x ∈ R} is subspace of V. Then

$$\frac{V}{W}=\{W+(x,y)\,|\,x,y\in R\}=\{(x,0)+(x,y)\,|\,x,y\in R\}=\{(x,0)+(x,0)+(0,y)\,|\,x,y\in R\}$$

$$\Rightarrow\frac{V}{W}=\{W+(0,y)\,|\,y\in R\}\text{is quotient space}$$

5. Linear Combination

Let $V(F)$ be a vector space. If $u_1,u_2,...,u_n\in V$, then any vector $u=a_1u_1+a_2u_2+...+a_nu_n$, where $a_1,a_2,...,a_n\in F$ is called a linear combination of the vectors $u_1,u_2,...,u_n$.

6. Linear Span

Let $V(F)$ be a vector space and S be any non-empty subset of V. Then the linear span of S is the set of all linear combinations of elements of S and is denoted by $L(S)$.

if S = {u1, u2, ... un} then $span(S)=L(S)=\{a_1u_1+a_2u_2+...+a_nu_n:u_1u_2,...,u_n\in S\ \&\ a_1,a_2,...,a_n\in F\}$.

Example : Let S = {(1 ,0), (0 ,1)} is a subset of R^2. Then L(S) = {a(1 ,0) + b(0 ,1)|a ,b ∈ R}

⇒ L(S) = {(a ,b)|a ,b ∈ R} = $R^2(R)$

Result

(a) Let S be a subset of a vector space V then set span (S) is a subspace of V which contains S.

(b) Span (S) is the intersection of all subspaces of V containing S.

7. Linear Sum of two subspaces

Let W_1 and W_2 be two subspaces of the vector space $V(F)$. Then the linear sum of the subspaces W_1 and W_2, denoted by W_1+W_2, is the set of all sums u_1+u_2 such that $u_1\in W_1,u_2\in W_2$. Thus, $W_1+W_2=\{u_1+u_2:u_1\in W_1,u_2\in W_2\}$.

Theorem If W_1 and W_2 are subspaces of the vector space $V(F)$, then

(i) W_1+W_2 is a subspace of $V(F)$.

(ii) $W_1+W_2=<W_1\cup W_2>$, i.e., $L(W_1\cup W_2)=W_1+W_2$.

Theorem . If S,T are subsets of $V(F)$, then

(i) $S\subseteq T\Rightarrow L(S)\subseteq L(T)$.

(ii) $L(S\cup T)=L(S)+L(T)$.

(iii) S is a subspace of $V\Leftrightarrow L(S)=S$.

(iv) $L(L(S))=L(S)$.

8. Vector Space as a direct sum of subspaces

Let $V(F)$ be a vector space and let $W_1, W_2, ..., W_m$ be subspaces of V. Then V is said to be the direct sum of $W_1, W_2, ..., W_m$ if every element $\alpha \in V$ can be written in one and only one way as $u = u_1 + u_2 + ... + u_m$, where $u_1 \in W_1, u_2 \in W_2, ..., u_m \in W_m$.

i.e. $V = W_1 \oplus W_2 \oplus \ldots \oplus W_m$

Here, $V = W_1 + W_2 + ... + W_m$ & $W_1 \cap W_2 \cap \ldots \cap W_m = \{0\}$.

Examples:

(i) If V is a vector space of all real valued functions over R, then it is a direct sum of its subspaces of even and odd functions.

(ii) A vector space of all $n \times n$ matrices is direct sum of subspaces of Hermitian and Skew Hermitian matrices.

9. Complementary Subspaces

Let V(F) be a vector space and W_1, W_2 be two subspaces of V. Then the subspace W_2 is called the complement of W_1 in V if V is the direct sum of W_1 and W_2. i.e. $V = W_1 \oplus W_2$

Example : In $V_2(R)$, $W_1 = \langle(1, 0)\rangle$ and $W_2 = \langle(0, 1)\rangle$ are subspaces of $V_2(R)$. Any vector $(x, y) \in V_2(R)$ can be written in one and only one way as $(x, y) = (x, 0) + (0, y)$, where $(x, 0) \in W_1$ and $(0, y) \in W_2$.

$\therefore V_2(R) = W_1 \oplus W_2$ and hence W_1 and W_2 are complementary subspaces.

SOLVED QUESTIONS

1. Let $p(x) = \{a_0 + a_1x + a_2x^2 + a_3x^3 \mid a_i \in R, i = 0, 1, 2, 3\}$
Under the standard operation of addition (+) and scalar multiplication (.) then p(x) is
(a) Not a vector space
(b) A vector space of infinite dimension
(c) A vector space of dimension 3
(d) A vector space of dimension 4

Ans. (d)

Explanation :

Given that p(x) is the set of all polynomial of degree at most 3

$\Rightarrow p(x) = R_3[x]$

Then p(x) is vector space of dimension 4.

2. Let $P = \{(x_1, x_2, ..., x_n) \mid x_1 = 0\}$ and $Q = \{(x_1, x_2, ..., x_n \mid x_1x_2 = 0\}$ be subsets of the vector space R^n. Which of the following is True?
(a) P is not subspace of R^n and Q is subspace of R^n
(b) Both P and Q are subspace of R^n
(c) P is subspace of R^n and Q is not subspace of R^n
(d) Neither P nor Q is a subspace of R^n

Ans. (c)

Explanation :

Let $u = (x_1, x_2, ..., x_n) \in P$ and $v = (y_1, y_2, ... y_n) \in P$ then $x_1 = y_1 = 0$

If $au + bv \in P$ then P is subspace

$au + bv = (ax_1 + by_1, ax_2 + by_2 \ldots\ldots ax_n + by_n)$

here $ax_1 + by_1 = 0$ because $x_1 = y_1 = 0$

$\Rightarrow au + bv \in P \Rightarrow$ P is subspace

Now let $u, v \in Q$

So $x_1x_2 = 0$ and $y_1y_2 = 0$ then $au + bv = (ax_1 + by_1, ax_2 + by_2 \ldots\ldots)$

if $au + bv \in Q$ then $(ax_1 + by_1)(ax_2 + by_2) = 0$

but here $(ax_1 + by_1)(ax_2 + by_2) \neq 0$

$\Rightarrow$ Q is not subspace.

3. Which of the following sets of function from R to R is a vector space over R

$S_1 = \{g \mid \lim_{x\to 3} g(x) = 0\}$, $S_2 = \{f \mid \lim_{x\to 3} f(x) = 1\}$, $S_3 = \{g \mid \lim_{x\to 3} g(x)\ exist\}$

(a) Only S_1 (b) Only S_2 (c) S_1 and S_3 but not S_2

(d) All the three are vector space. **CSIR NET JUNE 2015**

Ans. (c)

Explanation :

Let f_1 and $f_2 \in S_1$

Then S_1 is vector space if $\alpha f_1 + \beta f_2 \in S_1$ where $\alpha, \beta \in F$(field)

$\lim_{x\to 3}(\alpha f_1 + \beta f_2)(x) = \alpha \lim_{x\to 3} f_1(x) + \beta \lim_{x\to 3} f_2(x) = 0 \Rightarrow \alpha f_1 + \beta f_2 \in S_1$

S_1 is vector space. For S_2

If $f_1 \in S_2$ and $f_2 \in S_2$ $\lim_{x\to 3} f_1(x) = 1$ and $\lim_{x\to 3} f_2(x) = 1$ $\lim_{x\to 3}(\alpha f_1 + \beta f_2)(x) = \alpha + \beta$

for $\alpha = 1$ and $\beta = 1$ then $\alpha + \beta \neq 1$

S_2 is not vector space. For S_3

If $f_1 \in S_3$ and $f_1 \in S_3$ $\lim_{x\to 3} f_1\ exist$ and $\lim_{x\to 3} f_2\ exist$

$\lim_{x\to 3}(\alpha f_1 + \beta f_2)\ exist$ $\Rightarrow \alpha f_1 + \beta f_2 \in S_3$

S_3 is vector space.

4. Let M_n denote the vector space of all $n \times n$ real matrices. Among the following subsets of M_n decide which are linear subspace. **CSIR NET DEC 2016**

(a) $V_1 = \{A \in M_n \mid A$ is non-singular$\}$

(b) $V_2 = \{A \in M_n \mid |A| = 0\}$ (c) $V_3 = \{A \times M_n \mid Tr(A) = 0\}$

(d) $V_4 = \{BA; A \in M_n$ where B is some fixed matrix in $M_n\}$

Ans. (c), (d)

Explanation :

Option (a) is incorrect.

$V_1 = \{A \in M_n \mid |A| \neq 0\}$

$0 \notin V_1$ $\Rightarrow V_1$ is not subspace.

Option (b) is incorrect.

$V_2 = \{A \in M_n \mid |A| = 0\}$

Let $A = \begin{bmatrix} 1 & 0 \\ 0 & 0 \end{bmatrix}$ & $B = \begin{bmatrix} 0 & 0 \\ 0 & 1 \end{bmatrix}$ Then $A + B = \begin{bmatrix} 1 & 0 \\ 0 & 1 \end{bmatrix}$

$A \in V_2$ & $B \in V_2$

But $|A + B| \neq 0 \Rightarrow A + B \notin V_2$

By one-step test V_2 is not subspace.

Option (c) is correct. $V_3 = \{A \in M_n \mid Tr(A) = 0\}$ Let $A \in V_3 \Rightarrow Tr(A) = 0$

$B \in V_3 \Rightarrow Tr(B) = 0$

Then $Tr(aA + bB) = 0$ and $aA + bB \in V_3$ where $a, b \in F$.

$\Rightarrow V_3$ is subspace. By one-step test V_3 is vector subspace.

Option (d) is correct.

$V_4 = \{BA \mid A \in M_n$ where B is fixed matrix in $M_n\}$ $(\bar{0}) \in V_4$

Let $A_1, A_2 \in M_n$ Then $BA_1, BA_2 \in V_4$ SBy one step test

$aBA_1 + bBA_2 \in V_4$ V_4 is subspace.

5. Let $p_n(x) = x^n$ for $x \in R$ and let $\rho = \text{span}\{p_0, p_1, p_2, ...\}$. Then **CSIR NET JUNE 2015**

(a) ρ is a vector space of all real valued continuous function on R

(b) ρ is a subspace of all real valued continuous function on R

(c) $\{p_0, p_1, p_2.....\}$ is a L.I. set in the vector space of all continuous function on R

(d) Trigonometric function belong to ρ

Ans. **(b), (c)**

Explanation :

Given that $\rho = \text{span}\{p_0, p_1, p_2 .....\}$ $= \text{span}\{1, x, x^2, ...\}$

$\Rightarrow \rho$ is the set of all polynomial $\Rightarrow \rho = R[x]$

$\Rightarrow \rho$ is vector space of all polynomials over R.

$\Rightarrow \rho$ is subspace of all real valued continuous function on R

Option (b) is correct. Option (c) is correct.

Option (a) is incorrect.

ρ is L.I. set of vector

Option (d) is incorrect.

$\sin x \notin \rho$

UNSOLVED QUESTION

1. Which of the following is a subspace of $R^3(R)$? **IIT JAM 2023**
(a) $\{(x, y, z) \in R^3 : (y + z)^2 + (2x - 3y)^2 = 0\}$
(b) $\{(x, y, z) \in R^3 : x + 2y - 3z + 1 = 0\}$
(c) $\{(x, y, z) \in R^3 : yz = 0\}$
(d) $\{(x, y, z) \in R^3; y \in Q\}$

2. Let S and T be non – empty subsets of R^2, and W be a non – zero proper subspace of R^2. Consider the following statements: **IIT JAM 2023**
I. If span(S) = R^2 then span $(S \cap W) = W$
II. span$(S \cup T)$ = span(S) $\cup$ span(T) Then
(a) Both are false
(b) Both are true
(c) Only I is true
(d) Only II is true

3. Consider the following subsets of R^2
$W_1 = \{(x, y) : 2x + y = 0\}$
$W_2 = \{(x, y); xy = 0\}$
$W_3 = \{(x, y); \sin^2x + \sin^2y = 1\}$
$W_4 = \{(x, y); \sin^2x + \cos^2x = 1\}$
Which of the following pair has the property that they are not subspaces of R^2.
(a) W_1, W_2
(b) W_2 only
(c) W_1, W_4
(d) W_2, W_3

4. Which of the following subspace of the vector space R^3
(a) $\{(x, y, z) \in R^3; 2x + 3y - 2z = 3, z = 0\}$
(b) $\{(x, y, z) \in R^3; 2x + 3y + 4z = 0, z = 0\}$
(c) $\{(x, y, z) \in R^3; x = 0, y = 0\}$
(d) $\{(x, y, z) \in R^3; x = 1, y = 0\}$

5. Let $\{V_1, ... V_n\}$ be a linearly independent subset of vector space V where $n \geq 4$. Set $W_{ij} = V_i - V_j$.
Let W be spane of $\{W_{ij} \mid 1 \leq i,j \leq n\}$. Then **CSIR NET DEC 2013**
(a) $\{W_{ij} \mid 1 \leq i < j \leq n\}$ spans W
(b) $\{W_{ij} \mid 1 \leq i < j \leq n\}$ is a linearly independent subset of W
(c) $\{W_{ij} \mid 1 \leq i \leq n\text{-}1, j = i + 1\}$ span W
(d) dim W = n

6. Let V be a real vector space of real-valued functions on a given set. Assume that constant functions are in V and that if $f \in V$, then $f^2 \in V$ and that $|f| \in V$. Pick out the true statements. **NBHM-2011**
(a) If $f, g \in V$, then $fg \in V$
(b) If $f, g \in V$, then max $\{f, g\} \in V$
(c) If $f \in V$ and p is any polynomial in one variable, with real coefficients, then $p(f) \in V$
(d) None of these

7. Let V_1, V_2 be subspace of a vector space V. Which of the following is/are necessarily a subspace of V? **CSIR UGC NET JUNE - 2012**
(a) $V_1 \cap V_2$
(b) $V_1 \cup V_2$
(c) $V_1 + V_2 = \{x + y : x \in V_1, y \in V_2\}$
(d) $V_1/V_2 = \{x \in V_1 \text{ and } x \notin V_2\}$

8. Which of the following are subspaces of the vector space R^3?

CSIR UGC NET JUNE - 2015

(a) $\{(x, y, z) : x + y = 0\}$
(b) $\{(x, y, z) : x - y = 0\}$
(c) $\{(x, y, z) : x + y = 1\}$
(d) $\{(x, y, z) : x - y = 1\}$

9. The set $V = \{(x, y) \in R^2 : xy \geq 0\}$ is **GATE – 2004**

(a) a vector space of R^2
(b) not a vector space of R^2 since every element does not have, an inverse in V.
(c) not a vector space of R^2 since it is not closed under scalar multiplication.
(d) not a vector space of R^2 since it is not closed under vector addition.

10. Let V be the set of 2 × 2 matrices $\begin{bmatrix} a_{11} & a_{12} \\ a_{21} & a_{23} \end{bmatrix}$ with complex entries such that $a_{11} + a_{22} = 0$. Let W be the set of matrices in V with $a_{12} + \overline{a}_{21} = 0$. Then, under usual matrix addition and scalar multiplication, which of the following is(are) true?

(a) V is a vector space over C
(b) W is a vector space over C
(c) V is a vector space over R
(d) W is a vector space over R

11. For the set $W = \{(x, 1, z) \in R^3\}$. Consider the statements :

P : W is not closed under vector addition
Q : W is closed under vector multiplication
R : W has an additive identity
S : W is not subspace of R^3.

Which of the following is true?

(a) P, R (b) Q, R (c) P, S (d) Q, S

12. Let $V = \{ f \mid f : R \to R \}$ over R. Then which of the following is not a subspace of V over R

(a) $W_1 = \{ f \in V : f \text{ is bounded}\}$
(b) $W_2 = \{ f \in V : f \text{ is continuous}\}$
(c) $W_1 = \{ f \in V : f \text{ is Monotonic}\}$
(d) $W_1 = \{ f \in V : f \text{ is of bounded variation}\}$

Answer Key

1	2	3	4	5	6	7	8	9
a	a	d	b,c	a,c	a,b,c	a,c	a,b	d
10	**11**	**12**						
a,c,d	c	c						

Chapter 2

Basis and Dimension

1. Linear Dependence :

Let $V(F)$ be a vector space. A finite set $\{u_1, u_2, ... u_n\}$ of vectors of V is said to be linearly dependent if there exist scalars $a_1, a_2, ..., a_n \in F$ not all of them 0 (some of them may be zero) such that $a_1u_1 + a_2u_2 + a_3u_3 + ... + a_nu_n = \overline{0}$.

Example : (1,0,0) ,(0,1,0) and (1,1,0) are L.D. vectors

2. Linear Independence :

Definition: Let $V(F)$ be a vector space. A finite set $\{u_1, u_2, ... u_n\}$ of vectors of V is said to be linearly independent if every relation of the form $a_1u_1 + a_2u_2 + a_3u_3 + ... + a_nu_n = 0$, $a_i \in F$, $\Rightarrow a_i = 0$ for all each $1 \le i \le n$.

Example : (1,0,0) ,(0,1,0) and (0,0,1) are L.I. vectors

Remarks:

(1) If two vectors are linearly dependent, then one of them is a scalar multiple of the other.

(2) Every superset of a linearly dependent set of vectors is linearly dependent.

(3) Any subset of a linearly independent set of vectors is also linearly independent.

(4) A set consisting of a single non-zero vector is always linearly independent.

(5) A set of vectors which contains zero vector is linearly dependent.

(6) The set of non-zero vectors $u_1, u_2, ..., u_n$ of $V(F)$ is linearly dependent iff one of these vectors is a linearcombination of the remaining $(n-1)$ vectors.

(7) If two of the vectors $v_1, v_2, ..., v_n$ are equal, then the vectors are linearly dependent.

Example . In $R^3(R)$, where R is the field of real numbers, examine each of the following sets of vectors for linear dependence:

(a) $\{(2,1,2), (8,4,8)\}$ (b) $\{(1,2,0), (0,3,1), (-1,0,1)\}$

Solution: For option (a).

$(8,4,8) = 4(2,1,2)$, hence this set is linearly dependent.

For option(b).

Let $\alpha(1,2,0) + \beta(0,3,1) + \gamma(-1,0,1) = 0$ $\Rightarrow \alpha - \gamma = 0$

$2\alpha + 3\beta = 0, \beta + \gamma = 0$

On solving, we get $\alpha = 0, \beta = 0, \gamma = 0$

$\Rightarrow \{(1,2,0), (0,3,1), (-1,0,1)\}$ is linearly independent.

3. Basis of A Vector Space

A subset S of a vector space $V(F)$ is said to be a basis of $V(F)$, if

(i) S consists of linearly independent vectors.

(ii) S generates $V(F)$, i.e., $L(S)=V$, i.e., each vector in V is a linear combination of a finite number of elements of S.

Example : V = R^3 (R) then β = (1, 0, 0), (0, 1, 0), (0, 0, 1) is a standard basis of V.

Dimension of a vector space : Let V(F) be a vector space over F and β is a basis of V(F). Then number of elements in β (basis) is a dimension of vector space.

Example :- Standard basis of F(F) is {1}, where F is a field.

dim of R(R) = dim of C(C) = dim of Q(Q) = dim of Z_p (Z_p) = 1 where p is a prime.

Example : Standard basis of C(R) is = {1, i}

Dim C(R) = 2

Example :- Standard basis of $P_n(R) = \{a_0 + a_1x + \ldots\ldots + a_nx^n \mid a_i \in R\}$ is $\{1, x, x^2, \ldots\ldots x^n\}$

dim $P_n(R) = n + 1$

Example :- dim of $M_n(R)$ is $n \times n = n^2$

Example :- Standard basis and dim of $R^n(R)$ is

{(1, 0,.....0), (0, 1, 0, ...0), (0, 0, 1, ...0) ......(0, 0, ..., 1}

dim $R^n(R) = n$

Finite Dimension (Finitely generated) vector space :If basis of vector space have a finite element then this vector space is called finite dimensional vector space. Otherwise it is called infinitely dimensional vector space.

Examples :

(i) R(Q), C(Q) all are infinite dimensional vector space.

(ii) R[x] be a infinitely dimensional vector space over R.

Dimension of Quotient space :

(1) If V and W both are finite dimensional vector space and subspaces then

$\dim\left(\frac{V}{W}\right)$ = (dim V – dim W) and $\dim\left(\frac{V}{W}\right)$ is finite.

(2) If V is infinite dimension vector space and W is also infinite dimensional subspace then $\dim\left(\frac{V}{W}\right)$ is either finite or infinite.

(3) $\dim\left(\frac{V}{W}\right)$ = {Number of linearly independent restrictions of W}

Dimension of a Direct Sum: If a finite dimensional vector - space $V(F)$ is a direct sum of two subspaces W_1 and W_2, then $\dim V = \dim W_1 + \dim W_2$

Remark: Converse of above is not true, i.e., If - $\dim V = \dim W_1 + \dim W_2$, then we cannot say that V is a direct sum of W_1 and W_2.

For example: Let $V = R^3$ be a vector space over R
Let, $W_1 = \{(x,0,0): x \in R\}$ and $W_2 = \{(x,y,0): x, y \in R\}$. Both are subspaces of V and also $\dim V = \dim W_1 + \dim W_2$, but V is not direct sum of W_1 and W_2.

Some Important Results :

(1) If V(F) be a vector space and W is a subspace of V then dim W ≤ dim V
(2) If V(F) be a vector space and W be its subspace of V
Then dim of W = dim V – number of linearly independent restrictions of W.

Example :-Consider $R^3(R)$ then find dimension of subspace of W_1.
$W_1 = \{(a, b, c) \in R^3 \mid a + 2b + 3c = 0$ isL.I. restriction/ condition$\}$
$\dim W_1 = \dim R^3(R)$ – number of linearly independent condition.
$\dim W_1 = 3 - 1 = 2$

(3) If W_1, W_2 are two subspaces of a finite dimensional vector space $V(F)$, then
$\dim(W_1 + W_2) = \dim W_1 + \dim W_2 - \dim(W_1 \cap W_2)$.
(4) $\dim(W_1 \cap W_2) \le \min\{\dim(W_1), \dim(W_2)\}$
(5) $\dim(W_1 \cap W_2) \ge \max\{0, \dim(W_1) + \dim(W_2) - \dim V\}$
(6) $\max\{\dim(W_1), \dim(W_2)\} \le \dim\{W_1 + W_2\} \le \min\{\dim(W_1) + \dim(W_2), \dim V\}$
(7) The only subspaces of R are R and the zero subspace.
(8) A subspace of R^2 is R^2 , or the zero subspace or the space consisting of all scalar multiples of some fixed vector in R^2 (R^2 or the zero subspace or the straight line through the origin).
(9) Let V be a finite dimensional vector space andlet n=dimV. Then
(a) Any subset of V which contains more than n vectors is linearly dependent.
(b) No subset of V which contains fewer than n vectors can span V.
(10) If W is a subspace of a finite dimensional vector space V, then every linearly independent subset of W is finite and is part of a (finite) basis for W.
(11) C is a 2-dimensionl vector space over R and is an infinite dimensional space over Q.
(12) If V is an n-dimensional vector space over field F, where $|F| = p$ (prime), then the number of elements in $V = p^n$, $i.e. |V| = p^n$.

Results on dimension:

Let V be a vector space of all matrices with real entries over R
(i) Dimension of vector space containing $n \times n$ symmetric matrices $= (n^2 + n)/2$
(ii) Dimension of vector space containing $n \times n$ skew symmetric matrices $= (n^2 - n)/2$
(iii) Dimension of vector space containing $n \times n$ upper or lower triangular matrices $= (n^2 + n)/2$

4. Counting of basis and subspaces :

Let V be a vector space of dimension n over finite field $F = Z_p, p$ is Prime.

(i) The number of basis of $V = \dfrac{(p^n-1)(p^n-p)..(p^n-p^{n-1})}{n!}$

(ii) The number of ordered basis of $V = (p^n-1)(p^n-p)..(p^n-p^{n-1})$

(iii) The number of subspaces of dimension r $= \dfrac{(p^n-1)(p^n-p)..(p^n-p^{r-1})}{(p^r-1)(p^r-p)..(p^r-p^{r-1})}$

SOLVED QUESTIONS

1. Let $W_1 = \left\{\begin{bmatrix} a & b \\ -b & 2a \end{bmatrix} : a,b \in R\right\}$ and $W_2 = \left\{\begin{bmatrix} a & b \\ -b & -a \end{bmatrix} : a,b \in R\right\}$ be two subspaces of $M_2(R)$. Which of the following statements are true? **CSIR NET NOV 2020**

(a) $W_1 \cap W_2 = \left\{\begin{bmatrix} 0 & 0 \\ 0 & 0 \end{bmatrix}\right\}$ (b) $\left\{\begin{bmatrix} 0 & 0 \\ 0 & 0 \end{bmatrix}\right\}$ is a proper subset $W_1 \cap W_2$

(c) $W_1 + W_2 = M_2(R)$ (d) $W_1 + W_2$ is a proper subset of $M_2(R)$

Ans. (b), (d)

Explanation :

Any matrix of the form $\begin{bmatrix} 0 & b \\ -b & 0 \end{bmatrix} \in W_1 \cap W_2$ So, $W_1 \cap W_2 \neq \left\{\begin{bmatrix} 0 & 0 \\ 0 & 0 \end{bmatrix}\right\}$

So, option (a) is false and option (b) is true.

$W_1 \cap W_2 = \left\{\begin{bmatrix} 0 & b \\ -b & 0 \end{bmatrix} : b \in R\right\}$

So, $\dim(W_1 \cap W_2) = 1$ and $\dim(W_1) = 2$, $\dim(W_2) = 2$

$\dim(W_1 + W_2) = \dim(W_1) + \dim(W_2) - \dim(W_1 \cap W_2) = 2 + 2 - 1$

$\dim(W_1 + W_2) = 3$

So, $W_1 + W_2 \neq M_2(R)$

Thus option (c) is false and option (d) is true.

2. Let $l \geq 1$. What is dimension of R-vector space of all polynomial in two variable over R having degree atmost l? **CSIR NET JUNE 2023**

(a) $l+1$ (b) $l(l-1)$ (c) $\dfrac{l(l+1)}{2}$ (d) $\dfrac{(l+1)(l+2)}{2}$

Ans. (d)

Explanation :

For $l = 1$

Then $f(x, y) = a + bx + cy$

For $l = 2$

$f(x, y) = a + bx + cy + dxy + ex^2 + fy^2$

Option (d) is correct.

3. Let $S = \{u_1, ..., u_k\}$ be a subset of non-zero vectors from R^n. Now, consider the two statements given below :

I : If S is linearly dependent in R^n then u_k is a linear combination of $u_1, ..., u_{k-1}$.

II. : If S is linearly independent set in R^n then $k < n$.

Which of the following statements is true? **CSIR NET NOV 2020**

(a) Statement I is false and Statement II is true.

(b) Statement I is true and Statement II is false.

(c) Each Statement I and Statement II are false.

(d) Both Statement I and Statement II are true.

Ans. (c)

Explanation :

For statement (I) :

Let $S = \{u_1 = (1, 1, 0), u_2 = (2, 2, 0), u_3 = (0, 0, 1)\}$

S is linearly dependent but here u_3 is not a linear combination of u_1 and u_2.

Hence statement (I) is false.

If $S = \{(1, 0, 0), (0, 1, 0), (0, 0, 1)\}$, then S is linearly independent but $|S| = k = 3 = n$, in R^3.

Hence statement (II) is false.

$\therefore$ Option (c) is correct.

4. Let $p(x) = a_0 + a_1x + ... + a_nx^n$ be a non-constant polynomial of degree $n \geq 1$.

Consider the polynomial $q(x) = \int_0^x p(t)dt, r(x) = \frac{d}{dx} p(x)$. Let V denote the real vector space of all polynomials in x. Then which of the followings are true?

CSIR NET JUNE 2019

(a) q and r are linearly independent in V (b) q and r are linearly dependent in V

(c) x^n belongs to the linear space of q and r (d) x^{n+1} belongs to the linear span of q and r

Ans. (a)

Explanation :

Let $p(x) = a_0 + a_1x + ... + a_nx^n$

Thus Let $p(x) = x$; for $n = 1$

So, $r(x) = 1$ and $q(x) = x^2/2$

So, r(x) and q(x) are always linearly independent.

So, option (a) is true, and (b) is false.

(c) $\text{Span}(q, r) = (\alpha.1 + \beta.x^2/2 \mid \alpha, \beta \in R)$

Hence x $\notin$ Span(q ,r)

Option (c) is false

(d) Let p(x) = 3x + 5

Then $q(x)=\int_0^x (3t+5)dt \qquad =\frac{3x^2}{2}+5x$ and r(x) = 3

So, Span(q ,r) = {(α.q(x) + β.r(x) : α, β∈ R} = $\left\{\frac{3\alpha}{2}x^2+5\alpha x+3\beta : \alpha,\beta\in R\right\}$

Now if $x^2 \in$ L(S)

Then $x^2=\frac{3\alpha}{2}x^2+5\alpha x+3\beta$ for some α, β∈ R.

Comparing coefficients we get, $\frac{3\alpha}{2}=1, 5\alpha=0, 3\beta=0$.

α = 0 and $\alpha=\frac{1}{5}$ not possible.

Hence $x^2 \notin$ L(S)

Thus option (c) and (d) are false.

5. Let $M_4(R)$ be the space of all (4 × 4) matrices over R. Let $W=\{a_{ij}\in M_4(R) \mid \sum_{i+j=k} a_{ij}=0 \text{ for k} = 2, 3, 4, 5, 6, 7, 8\}$. Then dim W is

CSIR NET DEC 2019

(a) 7 (b)8 (c)9 (d) 10

Ans. (c)

Explanation :

Given that $W=\{a_{ij}\in M_4(R) \mid \sum_{i+j=k} a_{ij}=0 \text{ for k} = 2, 3, 4, 5, 6, 7, 8\}$.

Let
k = 2 ⇒ $a_{11} = 0$
k = 3 ⇒ $a_{12} + a_{21} = 0 \Rightarrow a_{12} = -a_{21}$
k = 4 ⇒ $a_{13} + a_{22} + a_{31} = 0 \Rightarrow a_{13} = -a_{22} - a_{31}$
k = 5 ⇒ $a_{14} + a_{23} + a_{32} + a_{41} = 0 \Rightarrow a_{14} = -a_{23} - a_{32} - a_{41}$
k = 6 ⇒ $a_{24} + a_{33} + a_{42} = 0 \Rightarrow a_{24} = -a_{33} - a_{42}$
k = 7 ⇒ $a_{34} + a_{43} = 0 \Rightarrow a_{34} = -a_{43}$
k = 8 ⇒ $a_{44} = 0$

$$W=\begin{bmatrix} a_{11} & a_{12} & a_{13} & a_{14} \\ a_{21} & a_{22} & a_{23} & a_{24} \\ a_{31} & a_{32} & a_{33} & a_{34} \\ a_{41} & a_{42} & a_{43} & a_{44} \end{bmatrix}$$

1 + 1 + 2 + 2 + 3 dim W = 9

6. Let $W_1 = \{(u, v, w, x) \in R^4 \mid u + v + w = 0, 2v + x = 0, \quad 2u + 2w - x = 0\}$ and $W_2 = \{(u, v, w, x) \in R^4 \mid u + w + x = 0, u + w - 2x = 0, v - x = 0\}$ Then which of the following is TRUE? **CSIR NET DEC 2018**

(a) $\dim W_1 = 1$ (b) $\dim W_2 = 2$ (c) $\dim (W_1 \cap W_2) = 1$ (d) $\dim (W_1 + W_2) = 3$

Ans. **(c)**

Explanation :

$\dim W_1 = \dim R^4 (R)$ – number of linearly independent condition in W_1

Find number of linearly independent in W_1

$u + v + w = 0, \quad 2v + x = 0, \; 2u + 2w - x = 0$

$(u, v, w, x) = (1, 1, 1, 0), (0, 2, 0, 1), (2, 0, 2, -1)$

$\alpha(1, 1, 1, 0) + \beta(0, 2, 0, 1) + \gamma (2, 0, 2, -1) = 0$

$= (\alpha + 2\gamma, \alpha + 2\beta, \alpha + 2\gamma, \beta - \gamma) = (0, 0, 0, 0)$ where α,β,γ are scalars.

$= \alpha + 2\gamma = 0, \quad \alpha + 2\beta = 0 \quad \& \quad \beta - \gamma = 0$

$\alpha = -2\beta \quad \beta = \gamma$

Then $\alpha = -2, \beta = 1, \gamma = 1$

Then at least one vector is linearly dependent.

$\Rightarrow$ only two linearly independent vector.

$\dim W_1 = 4 - 2 = 2$

Option (b) is incorrect.

$\dim W_2 = \dim V$ – number of linearly independent condition in W_2

Find number of linearly independent condition in W_2.

$x_1 = (1, 0, 1, 1), x_2 = (1, 0, 1, -2), x_3 = (0, 1, 0, -1)$

$\alpha x_1 + \beta x_2 + \gamma x_3 = 0$

$(\alpha + \beta, \gamma, \alpha + \beta, \alpha - 2\beta - \gamma) = (0, 0, 0, 0)$

$\Rightarrow \alpha + \beta = 0, \; \gamma = 0, \quad \alpha - \gamma - 2\beta = 0$

$\alpha - 2\beta = 0$

$\alpha = 0, \beta = 0, \gamma = 0$

$\Rightarrow$ All three vectors are linearly independent.

$\Rightarrow \dim W_2 = 4 - 3 = 1$

Option (c) is correct.

$$W_1 \cap W_2 = \left\{ \begin{array}{c} (u,v,w,x) \in R^4 \mid u + v + w = 0, 2v + x = 0 \\ 2u + 2w - x = 0 \\ u + w + x = 0, u + w - 2x = 0 \\ v - x = 0 \end{array} \right\}$$

$\dim(W_1 \cap W_2) = \dim V$ – number of linearly independent condition in $W_1 \cap W_2$.

Find number of linearly independent condition.

$$\begin{bmatrix}1&1&1&0\\0&2&0&1\\2&0&2&-1\\1&0&1&1\\1&0&1&-2\\0&1&0&-1\end{bmatrix}\sim\begin{bmatrix}1&1&1&0\\0&2&0&1\\0&-2&0&-1\\0&-1&0&1\\0&-1&0&-2\\-1&0&-1&-1\end{bmatrix}\sim\begin{bmatrix}1&1&1&0\\0&2&0&1\\0&0&0&0\\0&-1&0&1\\0&-1&0&-2\\0&1&0&-1\end{bmatrix}$$

$$\sim\begin{bmatrix}1&1&1&0\\0&2&0&1\\0&0&0&0\\0&0&0&3\\0&0&0&-3\\0&0&0&-3\end{bmatrix}\sim\begin{bmatrix}1&1&1&0\\0&2&0&1\\0&0&0&0\\0&0&0&3\\0&0&0&0\\0&0&0&0\end{bmatrix}$$

$\Rightarrow$ Only three vectors are linearly independent.

$\Rightarrow \dim(W_1 \cap W_2) = 4 - 3 = 1$

Option (d) is incorrect.

$$\dim(W_1 + W_2) = \dim W_1 + \dim W_2 - \dim(W_1 \cap W_2)$$
$$= 2 + 1 - 1$$
$$= 2$$

7. Which of the following subsets of R^4 is basis of R^4. **CSIR NET DEC 2016**

$B_1 = \{(1, 0, 0, 0), (1, 1, 0, 0), (1, 1, 1, 0), (1, 1, 1, 1)\}$

$B_2 = \{(1, 0, 0, 0), (1, 2, 0, 0), (1, 2, 3, 0), (1, 2, 3, 4)\}$

$B_3 = \{(1, 2, 0, 0), (0, 0, 1, 1), (2, 1, 0, 0), (-5, 5, 0, 0)\}$

(a) B_1 and B_2 but not B_3 (b) B_1, B_2 and B_3

(c) B_1 and B_3 but not B_2 (d) Only B_1

Ans. **(a)**

Explanation :

For B_1

Linear combination of each vector of B_1

$\alpha(1, 0, 0, 0) + \beta(1, 1, 0, 0) + \gamma(1, 1, 1, 0) + \delta(1, 1, 1, 1) = 0$ where $\alpha,\beta,\gamma,\delta$ are scalars.

$= (\alpha + \beta + \gamma + \delta, \beta + \gamma + \delta, \gamma + \delta, \delta) = (0, 0, 0, 0)$

$\Rightarrow \delta = 0, \gamma = 0, \beta = 0, \alpha = 0$

$\Rightarrow$ each vector of B_1 is linearly independent

B_1 is basis.

For B_2

$\alpha(1, 0, 0, 0) + \beta(1, 2, 0, 0) + \gamma(1, 2, 3, 0) + \delta(1, 2, 3, 4) = 0$

$\Rightarrow \alpha + \beta + \gamma + \delta, 2\beta + 2\gamma + 2\delta, 3\gamma + 3\delta, 4\delta = 0$

$\Rightarrow \alpha = \beta = \delta = \gamma = 0$

$\Rightarrow B_2$ is linearly independent.

For B_3

$\alpha(1, 2, 0, 0) + \beta(0, 0, 1, 1) + \gamma(2, 1, 0, 0) + \delta(-5, 5, 0, 0) = 0$

$= (\alpha + 2\gamma - 5\delta, 2\alpha + \gamma + 5\delta, \beta, \beta) = (0, 0, 0, 0)$

$\alpha + 2\gamma - 5\delta = 0$ .....(1)

$2\alpha + \gamma + 5\delta = 0$ ..... (2)

$\beta = 0$

Solve and we get

$\alpha = -5,\ \beta = 0, \gamma = 5, \delta = 1$

$\Rightarrow B_3$ is linearly dependent $\Rightarrow B_3$ is not a basis.

Option (a) is correct.

8. The dimension of the vector space of all symmetric matrices $A = [a_{ij}]$ of order n ×n (n ≥ 2) with real entries $a_{11} = 0$ and trace zero is - **CSIR NET JUNE 2012**

(a) $\frac{(n^2+n-4)}{2}$ (b) $\frac{n^2-n+4}{2}$ (c) $\frac{n^2+n-3}{2}$ (d) $\frac{n^2-n+3}{2}$

Ans. (a)

Explanation :

For n = 3

$$A = \begin{bmatrix} a & b & c \\ b & d & e \\ c & e & f \end{bmatrix}$$

Given that $a_{11} = 0$ $\Rightarrow a = 0$ and trace of A = zero

$a + d + f = 0$ $\Rightarrow d + f = 0$ $\Rightarrow d = -f$

Then $A = \begin{bmatrix} 0 & b & c \\ b & d & e \\ c & e & -d \end{bmatrix}$ Number of arbitrary vectors are 4.

Then dim of V = 4

Option (a) is correct. Put n = 3

$$\frac{3^2+3-4}{2} = \frac{8}{2} = 4$$

Option (b) is incorrect.

Option (c) is incorrect.

Option (d) is incorrect.

9. Let V be a 3-dimensional vector space over the field $F_3 = Z_3$ of 3 elements. The number of distinct 1-dimensional subspace of V is **CSIR NET DEC 2012**

(a) 13 (b) 26 (c) 9 (d) 15

Ans. (a)

Explanation :

Let dim (V) = n and |F| = p

Then number of subspace of dimension – k are

$$\frac{(p^n-1)(p^n-p).....(p^n-p^{k-1})}{(p^k-1)(p^k-p)......(p^k-p^{k-1})} \Rightarrow \text{Given that dim (V) = 3 and |F| = 3 and k = 1}$$

Then $\frac{(3^3-1)}{(3-1)} = \frac{26}{2} = 13$

UNSOLVED QUESTIONS

1. Let $M_n(R)$ be the real vector space of all n×n matrices with real entries $n \geq 2$
Let $A \in M_n(R)$. Consider the suspace W of M_n (R) spanned by $\{I_n, A, A^2, ...\}$. Then the dimension of W over R is necessarily
(a) ∞ (b) at most n. (c) n^2 (d) n

2. Let V be the vector space of all polynomials with real coefficient of degree atmost 2. If W is the vector subspace of V generated by $1 - x$, $x^2 - x$, $x^2 - 1$ and $x^2 - 3x + 2$ then the dimension of W is
(a) 1 (b) 2 (c) 3 (d) 4

3. Let W_1, W_2, W_3 be three distinct subspaces of R^{10}(R) such that each W_i has dimension 9. Let $W = W_1 \cap W_2 \cap W_3$. Then we can conclude that
(a) W may not be subspace of R^{10} (b) dim $W \leq 8$
(c) dim $W \geq 7$ (d) dim $W \leq 3$

4. Let W = {p(B) : p is a polynomial with real coefficient} where $B = \begin{bmatrix} 0 & 1 & 0 \\ 0 & 0 & 1 \\ 1 & 0 & 0 \end{bmatrix}$.
The dimension d of all vector space W satisfies.
(a) $4 \leq d \leq 6$ (b) $6 \leq d \leq 9$ (c) $3 \leq d \leq 8$ (d) $3 \leq d \leq 4$

5. Let W be the subspace of R^4 given by $W = \{(x, y, z, w) \in R^4 : x + z + w = 0, y + z + w = 0\}$. Then the dimension of W is –
(a) 4 (b) 3 (c) 2 (d) 1

6. Let $x_0 \in R^3$ be the column vector such that $x_0^T = (1, 1, 1)$.
Let $V = \{A \in M_3(R) \mid Ax_0 = 0\}$. Then the dimension of V is
(a) 6 (b) 5 (c) 7 (d) 9

7. Suppose V is a finite dimensional vector space over R. If W_1, W_2 and W_3 are subspaces of V, then which of the following statements is TRUE? **GATE 2020**
(a) If $W_1 + W_2 + W_3 = V$ then span $(W_1 \cup W_2) \cup$ span $(W_2 \cup W_3) \cup$ span $(W_3 \cup W_1) = V$

(b) If $W_1 \cap W_2 = \{0\}$ and $W_1 \cap W_3 = \{0\}$, then $W_1 \cap (W_2 + W_3) = \{0\}$

(c) If $W_1 + W_2 = W_1 + W_3$, then $W_2 = W_3$ (d) If $W_1 \neq V$, then span $(V \setminus W_1) = V$

8. Consider R^3 as a vector space with the usual operations of vector addition and scalar multiplication Let $x \in R^3$ be denoted by $x = (x_1, x_2, x_3)$. Define subspaces W_1 and W_2 by $W_1 := \{x \in R^3 : x_1 + 2x_2 - x_3 = 0\}$ and $W_2 := \{x \in R^3 : 2x_1 + 3x_3 = 0\}$. **GATE 2022**

Let dim(U) denote the dimension of the subspace U. Which of the following statements are TRUE?

(a) dim (W_1) = dim (W_2) (b) dim (W_1) + dim (W_2) – dim $(R^3) = 1$

(c) dim $(W_1 + W_2) = 2$ (d) dim $(W_1 \cap W_2) = 1$

9. Let V be a vector space of 6 × 6 real matrices over the field R. Then the dimension of the subspace of V consisting of all symmetric matrices is –

(a) 15 (b) 18 (c) 21 (d) 35

10. Let $W = \{A \in M_3(R) \mid A^T = -A \ and \sum_{j=1}^{3} a_{ij} = 0\}$. Then the dimension of W is

(a) 1 (b) 2 (c) 3 (d) 4

11. Let C be a n ×n matrix. Let W be the vector space spanned by $\{I, C, C^2, \ldots, C^{2n}\}$ the dimension of the vector space W is

(a) 2n (b) atmost n (c) n^2 (d) atmost 2n

12. The dimension of the subspace $W = \{(x, y, z, w) \in R^4 \mid x + y + z + w = 0, x + y + 2z = 0, x + 3y = 0\}$.

(a) 1 (b) 2 (c) 3 (d) 4

13. Which of the following basis of subspace of $W = \left\{ \begin{bmatrix} x & y \\ 0 & t \end{bmatrix} \middle| \begin{matrix} x + 2y = 0 \\ y + t = 0 \end{matrix} \right\}$

(a) $\begin{bmatrix} 1 & 0 \\ 0 & 0 \end{bmatrix}, \begin{bmatrix} 0 & 1 \\ 0 & 0 \end{bmatrix}, \begin{bmatrix} 0 & 0 \\ 0 & 1 \end{bmatrix}$ (b) $\begin{bmatrix} 2 & 1 \\ 0 & -1 \end{bmatrix}, \begin{bmatrix} 1 & -1 \\ 0 & 1 \end{bmatrix}$

(c) $\begin{bmatrix} -2 & 1 \\ 0 & -1 \end{bmatrix}$ (d) $\begin{bmatrix} 1 & -1 \\ 0 & 1 \end{bmatrix}$

14. Let {X, Y, Z} be a basis of R^3. Consider the following statements P and Q.

P : {X + Y, Y + Z, X – Z} is a basis of R^3

Q : {X + Y + Z, X + 2Y – Z, X – 3Z} is a basis of R^3.

Which of the above statements hold TRUE? **GATE – 2016**

(a) Both P and Q (b) Only P (c) Only Q (d) Neither P nor Q

15. If U and V are the null spaces of $\begin{bmatrix} 1 & 1 & 0 & 0 \\ 0 & 0 & 1 & 1 \end{bmatrix}$ and $\begin{bmatrix} 1 & 2 & 3 & 2 \\ 0 & 1 & 2 & 1 \end{bmatrix}$ respectively, then the dimension of the subspace U + V equals **GATE – 2017**

(a) 1 (b) 2 (c) 3 (d) 4

16. Consider the subspaces $W_1 = \{(x_1, x_2, x_3) \in R^3 : x_1 = x_2 + 2x_3\}$, $W_2 = \{(x_1, x_2, x_3) \in R^3 : x_1 = 3x_2 + 2x_3\}$ of R^3. Then the dimension of $W_1 + W_2$ equals **GATE –2018**

(a) 1 (b) 2 (c) 3 (d) 4

17. Let $v_i = (v_i^{(1)}, v_i^{(2)}, v_i^{(3)}, v_i^{(4)})$, for i = 1, 2, 3, 4, be four vectors in R^4 such that $\sum_{i=1}^{4} v_i^{(j)} = 0$, for each j = 1, 2, 3, 4. Let W be the subspace of R^4 spanned by $\{v_1, v_2, v_3, v_4\}$. Then the dimension of W over R is always. **TIFR – 2016**

(a) either equal to 1 or equal to 4
(b) less than or equal to 3
(c) greater than or equal to 2
(d) either equal to 0 or equal to 4.

18. Let n be an integer, $n \geq 3$ and let $u_1, u_2, \ldots, u_n$ be n linearly independent elements in a vector space over R. Set $u_0 = 0$ and $u_{n+1} = u_1$. Define $v_1 = u_i + u_{i+1}$ and $w_i = u_{i-1} + u_i$ for i = 1, 2, ….., n. Then

(a) $v_1, v_2, \ldots, v_n$ are linearly independent, if n = 2010.
(b) $v_1, v_2, \ldots, v_n$ are linearly independent, if n = 2011.
(c) $w_1, w_2, \ldots, w_n$ are linearly independent, if n = 2010
(d) $w_1, w_2, \ldots, w_n$ are linearly independent, if n = 2011 **CSIR NET DEC – 2012**

19. Let V be the real vector space of all polynomials in one variable with real coefficients and having degree at most 20. Define the subspaces

$$W_1 = \left\{ p \in V : p(1) = 0, p\left(\frac{1}{2}\right) = 0, p(5) = 0, p(7) = 0 \right\}$$

$$W_1 = \left\{ p \in V : p\left(\frac{1}{2}\right) = 0, p(3) = 0, p(4) = 0, p(7) = 0 \right\}$$

Then the dimension of $W_1 \cap W_2$ is

(a) 14 (b) 15 (c) 16 (d) 17

20. Let V denote the vector space $C^5[a,b]$ over R and $W = \left\{ f \in V : \frac{d^4 f}{dt^4} + 2\frac{d^2 f}{dt^2} - f = 0 \right\}$. Then

(a) dim (V) = ∞ and dim (W) = ∞
(b) dim (V) = ∞ and dim (W) = 4
(c) dim (V) = 6 and dim (W) = 5
(d) dim (V) = 5 and dim (W) = 4

ANSWER KEY

1	2	3	4	5	6	7	8	9	10
d	b	b,c	c,d	a	a	d	a,b,d	c	A
11	**12**	**13**	**14**	**15**	**16**	**17**	**18**	**19**	**20**
b,d	a	c	c	c	c	b	b,c,d	b	b

Chapter 3

Linear Transformations

Definition : Let U(F) and V(F) be two vector spaces over the same field F. Then a mapping T : $U \to V$ is called a linear transformation of U to V if
$T(au + bv) = aT(u) + bT(v)\ \ \forall u, v \in U$.and $a, b \in F$

Result :

(i) If a transformation do not keep the variable in linear form then such transformation will not be linear transformation.

(ii) Zero vector maps to zero vector in linear transformation.

Example: Consider $T : R^3(R) \to M_2(R)$ given by the following which are linear transformation.

(a) $T(x,y,z) = \begin{bmatrix} x+y & y+z \\ x+z & 0 \end{bmatrix}$

(b) $T(x,y,z) = \begin{bmatrix} y+1 & z+1 \\ x+3 & x+y+z \end{bmatrix}$

(c) $T(x,y,z) = \begin{bmatrix} |x| & x \\ y & z \end{bmatrix}$

Solution : (a) is linear transformation but (b) and (c) are non-linear.

Example: Consider $T : R^3(R) \to R^2(R)$ be a transformation given by $T(x, y, z) = (x^2, y + z)$

Let $u = (x_1, y_1, z_1)$ and $v = (x_2, y_2, z_2)$

$\alpha u + \beta v = (\alpha x_1 + \beta x_2, \alpha y_1 + \beta y_2, \alpha z_1 + \beta z_2)$

$\Rightarrow T[\alpha u + \beta v] \neq \alpha T(u) + \beta T(v)$ where α, β are scalars.

T is not linear.

1. Kernal of Linear Transformation :

Let $T : V(F \to V'(F)$ be a Linear Transformation then $K = \{x \in V \mid T(x) = 0\}$ is called kernal of T and it is denoted by ker(T)

Example :$T : R^3 \to R^3$ s.t. $T(x, y, z) = (x, 0, 0)$

Then $\ker(T) = \{(x, y, z) \in R^3 \mid T(x, y, z) = 0\} = \{(x, y, z) \in R^3 \mid (x, 0, 0) = 0\}$

$\ker(T) = \{(x, y, z) \in R^3 \mid x = 0\} = \{(0, y, z) \mid y, z \in R\}$

Results :

(1) ker(T) is subspace of V

(2) Dimension of ker(T) is called $\eta(T)$ i.e. nullity(T).

2. Range of Linear Transformation :

Let $T : V(F) \to V'(F)$ be a linear transformation then $R(T) = \{T(x) \mid x \in V\}$ is called Range of Linear Transformation.

(1) dim R(T) = ρ(T) = Rank (T)

(2) R(T) is a subspace of V'(F)

Sylvester's Law :

Let T : V(F) → V'(F) be a linear transformation then ρ(T) + η(T) = n, where dim V(F) = n

3. Non-Singular Transformation :

Definition : Let T be a linear transformation from a vector space U(F) to a vector space V(F). T is said to be non-singular if the null space of Tconsists of the zero vector alone i.e. if u∈U and T(u) = 0 then u = 0 ⇒ker (T) = {0}.

4. Singular Transformation :

Definition : Let T be a linear transformation from a vector space U(F) to a vector space V(F). Then T is said to be singular, if there exists a vector 0 ≠u∈U such that T(u) = 0 ⇒ker (T) ≠ {0}.

5. Matrix representation of Linear transformation :

Let T : U → V be a Linear Transformation then the basis of U & V as

$\beta_U = \{\alpha_1, \alpha_2, \ldots\ldots \alpha_m\}$ & $\beta_V = \{\beta_1, \beta_2, \ldots\ldots \beta_n\}$

Then as $T(\alpha_i) \in V$, each $T(\alpha_i)$ is a linear a combination of the elements of basis $\beta_V = \{\beta_1, \beta_2, \ldots\ldots \beta_n\}$

$$T(\alpha_i) = \sum_{j=1}^{n} a_{ij}\beta_j$$

The matrix representation of linear transformation, $[T:\alpha_i, \beta_j] = \begin{bmatrix} a_{11} & a_{12} & \cdots & a_{1m} \\ a_{21} & a_{22} & \cdots & a_{2m} \\ \vdots & \vdots & & \vdots \\ a_{n1} & a_{n2} & \cdots & a_{nm} \end{bmatrix}$

Example: Consider T : $P_3(R) \to P_3(R)$ given by $T(P(x)] = P(x + 1) + 3P'(x)$

Find transformation matrix w.r.t. β : (1, x, x^2, x^3)

Given that the basis of $P_3(R)$

$\beta = \{1, \quad x, \quad x^2, \quad x^3\}$

$T(x_1) = T(1) = 1 + 0 = 1$; $T(x_2) = T(x) = (x + 1) + 3 = x + 4$

$T(x_3) = T(x^2) = (x + 1)^2 + 3(2x) = x^2 + 2x + 1 + 6x = x^2 + 8x + 1$

$T(x_4) = T(x^3) = (x + 1)^3 + 3(3x^2)$

$= x^3 + 12x^2 + 3x + 1$

T(1) T(x) T(x^2) T(x^3)

$$T[\beta',\beta]=\begin{matrix}1\\x\\x^2\\x^3\end{matrix}\begin{bmatrix}1&4&1&1\\0&1&8&3\\0&0&1&12\\0&0&0&1\end{bmatrix}$$

Here $T(1) = 1 = 1.1 + 0.x + 0.x^2 + 0.x^3$

$T(x) = x + 4 = 4.1 + 1.x + 0.x^2 + 0.x^3$

$T(x^2) = x^2 + 8x + 1 = 1.1 + 8.x + 1.x^2 + 0.x^3$

$T(x^3) = 1 + 3x + 12x^2 + x^3$

6. Invariant Subspaces :

Definition : Let $T : V \to V$ be linear operator and let W be a subspace of V. Then W is called Invariant subspace if $T(W) \subseteq W$. Also it is called T – Invariant subspace, if W is T – invariant. We can restrict T to W to get a new map $T/_w : W \to W$.

Remarks :

(1) If T is any linear operator on a vector space V, then the range T and the null space of T are both invariant subspace under T.

(2) If the set $S = \{W_i\}$ is the collection of subspaces of a vector space V which are invariant under T, then $W = \bigcap_i W_i$ is also invariant under T.

(3) V and {0} are also invarient subspaces.

7. Linear Functional :

Definition : Let V(F) be a vector space over a field F. Then, a mapping f from V to F (i.e. $f : V \to F$) is called a linear functional on V if $f(ax + by) = af(x) + bf(y)$, $\forall$ $x, y \in V$ and $a, b \in F$.

Example: $V = R^3(R)$ be a vector space over R than $T : R^3(R) \to R(R)$ s.t. $T(x, y, z) = x + y + z$.

(i) **Dual Space :** The set V^* of all linear functionals f defined on a vector space V(F), is a vector space over F with addition and scalar multiplication defined by $(f_1 + f_2)(x) = f_1(x) + f_2(x)$, $\forall x \in V$ and $(af_1)(x) = af_1(x)$, $a \in F$ $\forall x \in V$, where f_1, f_2 are linear functionals on V.

This vector space V^* (also denoted by V' or $\hat{V}$) is called the dual space or conjugate space or algebraic dual or the algebraic conjugate of V over F.

Basis of dual space $\Rightarrow$ Let β is basis of V as $\beta = \{x_1, x_2, \ldots, x_k)$ & β^* is basis of V^* then $\beta^* = \{T_j\}$

$$\text{Where } T_j(x_k) = \begin{cases}1 & j = k\\0 & j \neq k\end{cases}$$

Example. Consider the basis $\{u_1, u_2, u_3\}$ of R^3 where $u_1 = (1, 0, 0)$, $u_2 = (1, 1, 0)$,

$u_3 = (1, 1, 1)$. Let $\{f_1, f_2, f_3\}$ be the dual basis of $\{u_1, u_2, u_3\}$ and f be the linear functional defined by $f(a, b, c) = a + b + c$. If $f = \alpha_1 f_1 + \alpha_2 f_2 + \alpha_3 f_3$ then $\{\alpha_1, \alpha_2, \alpha_3\}$ is

(a) (1, 2, 3) (b) (1, 3, 2) (c) (2, 3, 1) (d) (3, 2, 1)

Solution : (a) Given that (f_1, f_2, f_3) are dual basis

$\Rightarrow f_j(u_k) = \begin{cases} 1 & j = k \\ 0 & j \neq k \end{cases}$ & $f = \alpha_1 f_1 + \alpha_2 f_2 + \alpha_3 f_3$

$\Rightarrow f(u_1) = \alpha_1 f_1(u_1) + \alpha_2 f_2(u_1) + \alpha_3 f_3(u_1) \quad = \alpha_1(1) + \alpha_2.0 + \alpha_3.0$

$f(u_1) = \alpha_1$

$f(1, 0, 0) = \alpha_1 \quad [\because f(a, b, c) = a + b + c]$

$= 1 + 0 + 0 = \alpha_1 \Rightarrow \alpha_1 = 1$

Now $f(u_2) = \alpha_2$

$f(1, 1, 0) = \alpha_2 \Rightarrow \alpha_2 = 2$

$f(u_3) = \alpha_1.0 + \alpha_2.0 + \alpha_3.1$

$f(1, 1, 1) = \alpha_3$

$\Rightarrow \alpha_3 = 3$

Option (a) is correct

SOLVED QUESTIONS

1. Let $V = P_{10}(R)$ be a vector space of all polynomial of degree atmost 10. Suppose $T : V \to V$ be a linear operator s.t. $T(P) = P`$, then **CSIR NET JUNE 2023**

(a) $\rho(T) = 10$ (b) $|T| = 0$ (c) $Tr(T) = 0$ (d) All E.V. of T are equal

Ans. (All)

Explanation :

Suppose $V = P_{10}(R)$

$T(1) = 0; \quad T(x) = 1; \quad T(x^2) = 2x; \quad T(x^3) = 3x^2 \; \ldots . T(x^{10}) = 10x^9$

$$\begin{bmatrix} 0 & 1 & 0 & 0 \\ 0 & 0 & 2 \ddots & 0 \\ 0 & 0 & \cdots & 9 \\ 0 & 0 & 0 & 0 \end{bmatrix}$$

2. Let A be a matrix of order n and let V be the vector space of all real n ×n matrix X such that $AX = 0$. what is dimension of V. **CSIR NET JUNE 2022**

(a) nr (b) n^2r (c) $n^2 - nr$ (d) n

Ans. (c)

Explanation :

$V = \{ M_{n\times n} : AX = 0 \}$

We know that $T : M_{n\times n} \to M_{n\times n}$ s.t. $T(X) = AX$

then Rank (T) = n.Rank (A) = nr

So , Nullity(T) = $n^2 - nr$

3. Consider P_3 as the set of all polynomials whose degree is less than or equal to 3. Let T : $P_3 \to P_3$ be the map given by $T[P(x)] = \int_1^x P`(t)dt$. If the matrix of T relative to the standard bases $B_1 = B_2 = \{1, x,, x^2, x^3\}$ is M and M` denotes the transpose of Matrix M, then M + M` is **RSET – 2023**

(a) $\begin{bmatrix} 0 & -1 & -1 & -1 \\ -1 & 2 & 0 & 0 \\ -1 & 0 & 2 & 0 \\ -1 & 0 & 0 & 2 \end{bmatrix}$ (b) $\begin{bmatrix} -1 & 0 & 0 & 2 \\ 0 & -1 & 1 & 0 \\ 0 & 1 & -1 & 0 \\ 2 & 0 & 2 & -1 \end{bmatrix}$

(c) $\begin{bmatrix} 2 & 0 & 0 & -1 \\ 0 & 2 & 1 & 0 \\ 0 & 1 & 2 & -1 \\ -1 & 0 & -1 & 0 \end{bmatrix}$ (d) $\begin{bmatrix} 0 & 2 & 2 & 2 \\ 2 & -1 & 0 & 0 \\ 2 & 0 & -1 & 0 \\ 2 & 0 & 0 & -1 \end{bmatrix}$

Ans. **(a)**

Explanation :

$T(1) = \int_1^x 0.dt = 0$ $T(x) = \int_1^x dt = x - 1$ $T(x^2) = 2\int_1^x xdx = \int_1^x x^2 = x^2 - 1$

$T(x^3) = x^3 - 1$

$$\begin{bmatrix} 0 & -1 & -1 & -1 \\ 0 & 1 & 0 & 0 \\ 0 & 0 & 1 & 0 \\ 0 & 0 & 0 & 1 \end{bmatrix} + \begin{bmatrix} 0 & 0 & 0 & 0 \\ -1 & 1 & 0 & 0 \\ -1 & 0 & 1 & 0 \\ -1 & 0 & 0 & 1 \end{bmatrix} = \begin{bmatrix} 0 & -1 & -1 & -1 \\ -1 & 2 & 0 & 0 \\ -1 & 0 & 2 & 0 \\ -1 & 0 & 0 & 2 \end{bmatrix}$$

4. Let V be a vector space of dim 3 over R let T : V →V be a Linear Transformation given by the matrix $A = \begin{bmatrix} 1 & -1 & 0 \\ 1 & -4 & 3 \\ -2 & 5 & -3 \end{bmatrix}$ with ordered basis $\{V_1, V_2, V_3\}$ of V. Then which of the following is TRUE? **CSIR NET DEC 2019**

(a) $T(V_2) = 0$ (b) $T(V_1 + V_2) = 0$ (c) $T(V_1 + V_2 + V_3) = 0$
(d) $T(V_1 + V_3) = T(V_2)$

Ans. **(c)**

Explanation :

Given that the matrix $A = \begin{bmatrix} 1 & -1 & 0 \\ 1 & -4 & 3 \\ -2 & 5 & -3 \end{bmatrix}$ w.r.t. basis $\{V_1, V_2, V_3\}$

Here $T(V_1) = V_1 + V_2 - 2V_3$ ....(1)

$T(V_2) = -V_1 - 4V_2 + 5V_3$ ....(2)

$T(V_3) = 3V_2 - 3V_3$ .... (3)

Option (a) is incorrect. $T(V_2) \neq 0$

Option (b) is incorrect.

$T(V_1 + V_2) = T(V_1) + T(V_2) = V_1 + V_2 + 2V_3 - V_1 - 4V_2 + 5V_3 = -3V_2 + 3V_3$

Option (c) is correct.

$T(V_1 + V_2 + V_3) = T(V_1) + T(V_2) + T(V_3)$

$= V_1 + V_2 - 2V_3 - V_1 - 4V_2 + 5V_3 + 3V_2 - 3V_3 = 0$

Option (d) is incorrect.

$T(V_1 + V_3) = T(V_1) + T(V_3)$

$= V_1 + V_2 - 2V_3 + 3V_2 - 3V_3$

$= V_1 + 4V_2 - 5V_3 \neq T(V_2)$

Where $T(V_2) = -V_1 - 4V_2 + 5V_3$

$T(V_1 + V_3) = -T(V_2)$

Trick : Let $T : U \to V$ be a Linear Transformation and dim V = m & dim V = n then order of matrix representation of this linear Transformation is $n \times m$

5. Consider the vector space P_n of all real polynomials in x of degree less than or equal to n. Define $T : P_2 \to P_3$ by $T(f(x)) = \int_0^x f(t)dt + f'(x)$ then the matrix representation of T with respect. to basis $\{1, x, x^2\}$ and $\{1, x, x^2, x^3\}$ is **CSIR NET JUNE 2019**

(a) $\begin{bmatrix} 0 & 1 & 0 & 0 \\ 1 & 0 & 1/2 & 0 \\ 0 & 2 & 0 & 1/3 \end{bmatrix}$

(b) $\begin{bmatrix} 0 & 1 & 0 \\ 1 & 0 & 2 \\ 0 & 1/2 & 0 \\ 0 & 0 & 1/3 \end{bmatrix}$

(c) $\begin{bmatrix} 0 & 1 & 0 & 0 \\ 1 & 0 & 2 & 0 \\ 0 & 1/2 & 0 & 1/3 \end{bmatrix}$

(d) $\begin{bmatrix} 0 & 1 & 0 \\ 1 & 0 & 1/2 \\ 0 & 2 & 0 \\ 0 & 0 & 1/3 \end{bmatrix}$

Ans. (b)

Explanation :

Given that T : $P_2 \rightarrow P_3$

dim $P_2 = 2 + 1 = 3$

dim $P_3 = 3 + 1 = 4$ $\Rightarrow$ Order of this matrix will be 4 × 3.

Option (a) & (c) discarded. Now

$$T(p(x)) = \int_0^x p(t)dt + p'(t) \qquad T(1) = \int_0^x dt + 0 = x = 0.1 + 1.x + 0.x^2 + 0.x^3$$

$$T(x) = \int_0^x t\,dt + 1 = 1 + \frac{x^2}{2} = 1 + 0.x + \frac{x^2}{2} + 0.x^3 \qquad T(x^2) = \int_0^x t^2 dt + 2x = 2x + \frac{x^3}{3}$$

$$A = \begin{matrix} 1 \\ x \\ x^2 \\ x^3 \end{matrix}\begin{bmatrix} 0 & 1 & 0 \\ 1 & 0 & 2 \\ 0 & 1/2 & 0 \\ 0 & 0 & 1/3 \end{bmatrix}$$

Option (d) is incorrect. Option (b) is correct.

6. Let (R^n) be the space of R-linear maps from R^n to R^n. If Ker(T) denotes the kernel (null space) of T then which of the following are true? **CSIR NET JUNE 2019**

(a) There exists $T \in L(R^5) \setminus \{0\}$ such that Range (T) = Ker(T)

(b) There does not exist $T \in L(R^5) \setminus \{0\}$ such that Range (T) = Ker (T)

(c) There exist $T \in L(R^6) \setminus \{0\}$ such that Range (T) = Ker(T)

(d) There does not exist $T \in L(R^6) \setminus \{0\}$ such that Range (T) = Ker(T)

Ans. (b), (c)

Explanation :

Since T : $R^n \rightarrow R^n$ is a linear transformation.

So, by Rank Nullity Theorem $\rho(T) + \eta(T) = n$

Now, $\rho(T) = \eta(T)$ gives n is even.

Thus if n is odd, ker T ≠ Range T

So, option (a) is false and option (b) is true.

Now define T : $R^6 \rightarrow R^6$ as $T(x_1, x_2, x_3, x_4, x_5, x_6) = (x_4, x_5, x_6, 0, 0, 0)$

Then $\ker(T) = \{(x_1, x_2, x_3, 0, 0, 0) : x_1, x_2, x_3 \in R\}$

Also Range $T = \{(x_1, x_2, x_3, 0, 0, 0) : x_1, x_2, x_3 \in R\}$

Thus ker T = Range T

Thus option (c) is true and option (d) is false.

7. Let A be a 4 × 4 matrix. Suppose that the null space N(A) of A is $\{(x, y, z, w) \in R^4 \mid x + y + z = 0, x + y + w = 0\}$. Then **CSIR NETJUNE 2017**

(a) dim (column space (A)) = 1 (b) dim (column space (A)) = 2

(c) Rank (A) = 1 (d) S = {(1, 1, 1, 0), (1, 1, 0, 1)} is a basis of N(A)

Ans. (b)

Explanation :

Given that N(A) = {(x, y, z, w) ∈ R^4 | x + y + z = 0, x + y + w = 0}

Here x + y + z = 0 x + y + w = 0 ⇒ z = w and x + y + z = 0

then x = -(y + z) ⇒ N(A) = {(-(y+z), (y), (z), (z)}

Then basis of N(A) = {(-1, 1, 0, 0) , (-1, 0, 1, 1)}

⇒η(A) = 2 ⇒ ρ(A) = 2

Option (d) & (c) incorrect.

We know that ρ(A) = Dim[Column space (A)] = 2

Option (b) is correct.

8. For t ∈R define $M(t)=\begin{bmatrix}1 & t & 0\\ 1 & 1 & t^2\\ 0 & 1 & 1\end{bmatrix}$. Then which of the following are TRUE?

CSIR NET DEC 2019

(a) det M(t) is a polynomial of degree 3 in t (b) det M(t) = 0 ∀ t ∈ R

(c) det (M(t) is zero for infinitely may t ∈ R (d) det M(t) is zero for exactly two t ∈ R

Ans. (d)

Explanation :

det M(t) = 1[1 – t^2] – t[1] + 0 = 1 – t^2 – t = -t^2 – t + 1

Option (a) is incorrect. This is polynomial degree 2 in t

Option (b) is incorrect. Option (c) is incorrect. Option (d) is correct.

det M(t) = t^2 + t – 1 = 0

$$t=\frac{-1\pm\sqrt{1+4}}{2}$$

det M(t) is zero for exactly two t ∈ R

9. Let T : C → M_2(R) be the map given by $T(z)=T(x+iy)=\begin{bmatrix}x & y\\ -y & x\end{bmatrix}$.

Then which of the following statements is false? **CSIR NET DEC 2019**

(a) $T(z_1z_2) = T(z_1)T(z_2)$ for all $z_1, z_2 \in C$ (b) T(z) is singular if and only if z = 0

(c) There does not exist non-zero A ∈ M_2(R) such that the trace of T(z) A is zero for all z ∈ C

(d) $T(z_1 + z_2) = T(z_1) + T(z_2)$ for all $z_1, z_2 \in C$

Ans. (c)

Explanation :

Let $z_1 = x_1 + iy_1$ and $z_2 = x_2 + iy_2$ ⇒$z_1z_2 = (x_1x_2 - y_1y_2) + i(x_1y_2 + y_1x_2)$

$$\Rightarrow \quad T(z_1z_2)=\begin{bmatrix}x_1x_2-y_1y_2 & x_1y_2+y_1x_2\\ -(x_1y_2+y_1x_2) & x_1x_2-y_1y_2\end{bmatrix}=\begin{bmatrix}x_1 & y_1\\ -y_1 & x_1\end{bmatrix}\begin{bmatrix}x_2 & y_2\\ -y_2 & x_2\end{bmatrix}$$

= $T(z_1)T(z_2)$

(b) T(z) is singular when $\begin{vmatrix} x & y \\ -y & x \end{vmatrix} = 0$. $\Rightarrow x^2 + y^2 = 0 \Rightarrow x = 0$ and $y = 0 \Rightarrow z = 0$

If z = 0 then $T(z) = \begin{vmatrix} 0 & 0 \\ 0 & 0 \end{vmatrix} = 0$. So, T(z) is singular iff z = 0

(c) $A = \begin{bmatrix} 1 & 0 \\ 0 & -1 \end{bmatrix}$ So, (c) is false.

(d) Option (d) is true by linearly of T(z)

10. Let n be a positive integer and V be an (n + 1) dimensional vector space over R. If $\{e_1, e_2, \ldots e_{n+1}\}$ is a basis of V and $T : V \to V$ is the linear transformation satisfying $T(e_i) = e_{i+1}$ for i = 1, 2, ….., n and $T(e_{n+1}) = 0$. Then **CSIRNET DEC 2012**

(a) trace of T is non-zero (b) rank of T is n (c) nullity of T is 1

(d) $T^n = T \circ T \circ \ldots \circ T$ (n times) is the zero map.

Ans. (b), (c)

Explanation :

V be n + 1 dimensional vector space over R and $B = \{e_1, e_2, \ldots, e_{n+1}\}$ be basis of V.

$T : V \to V$ be linear transformation given by $T(e_i) = e_{i+1}$ for i = 1, 2, …., n and $T(e_{n+1}) = 0$

$\Rightarrow T(e_1) = e_2, T(e_2) = e_3, \ldots, T(e_n) = e_{n+1}$

$T(e_{n+1}) = 0$

The matrix of T with respect to basis B is given by $\begin{bmatrix} 0 & 1 & 0 & 0 & \cdots & 0 \\ 0 & 0 & 1 & 0 & \cdots & 0 \\ 0 & 0 & 0 & 1 & \cdots & 0 \\ \vdots & \vdots & \vdots & & \ddots & \\ 0 & 0 & 0 & \cdots & 0 & 1 \\ 0 & 0 & 0 & \cdots & 0 & 0 \end{bmatrix}_{(n+1)\times(n+1)}$

$\begin{bmatrix} 0 & 0 & 0 & \cdots & 0 & 0 \\ 1 & 0 & 0 & \cdots & 0 & 0 \\ 0 & 1 & 0 & \cdots & 0 & 0 \\ 0 & 0 & 1 & & 0 & 0 \\ \vdots & \vdots & \vdots & \ddots & & \vdots \\ 0 & 0 & 0 & & 1 & 0 \end{bmatrix}_{(n+1)\times(n+1)}$ $\Rightarrow$ trace of T = zero

So, option (a) is incorrect.
rank (T) = n and nullity = (n + 1) – n = 1
So, options (b) and (c) are correct.
Here, $T^{n+1} = O$ (zero map) but $T^n \neq O$
So, option (d) is incorrect.

11. Consider the vector space C[0, 1] over R. Consider the following statements :
P : If the set $\{t f_1, t^2 f_2, t^3 f_3\}$ is linearly independent, then the set $\{f_1, f_2, f_3\}$ is linearly independent, where $f_1, f_2, f_3 \in C[0, 1]$ and t^n represents the polynomial function $t \mapsto t^n$, $n \in N$

Q : If $F : C[0, 1] \to R$ is given by $F(x) = \int_0^1 x(t^2)dt$ for each $x \in C[0, 1]$, then F is a linear map. Which of the above statements hold TRUE? **GATE-2014**

(a) Only P (b) Only Q (c) Both P and Q (d) Neither P nor Q

Ans. (b)

Explanation :

Let $x, y \in C[0, 1]$ and α, β be scalars. Then

$$F(\alpha x + \beta y) = \int_0^1 (\alpha x + \beta y)(t^2)dt = \int_0^1 (\alpha x)(t^2) + (\beta y)(t)^2]dt = \int_0^1 \alpha x(t^2)dt + \int_0^1 \beta y(t^2)dt$$

$$= \alpha \int_0^1 x(t^2)dt + \beta \int_0^1 y(t^2)dt = \alpha F(x) + \beta F(x)$$

Thus, F is a linear map. So, the statement Q is true.
Let $f_1(x) = 1, f_2(x) = 1, f_3(x) = 1 \;\; \forall x \in [0, 1]$
Then the set $\{tf_1, t^2f_2, t^3f_3\}$ is linearly independent, but the set $\{f_1, f_2, f_3\}$ is linearly dependent.
Hence P statement is false. Hence, option (b) is correct.

12. A linear transformation T rotates each vector in R^2 clockwise through 90°. The matrix T relative to the standard ordered basis $\left(\begin{bmatrix}1\\0\end{bmatrix}, \begin{bmatrix}0\\1\end{bmatrix}\right)$ is **CSIR NET JUNE 2013**

(a) $\begin{bmatrix}0 & -1\\-1 & 0\end{bmatrix}$ (b) $\begin{bmatrix}0 & 1\\-1 & 0\end{bmatrix}$ (c) $\begin{bmatrix}0 & 1\\1 & 0\end{bmatrix}$ (d) $\begin{bmatrix}0 & -1\\1 & 0\end{bmatrix}$

Ans. (b)

Explanation :

T is linear transformation that rotates each vector in R^2 clockwise through 90°.
B = {(1, 0), (0, 1)} is ordered basis of R^2.

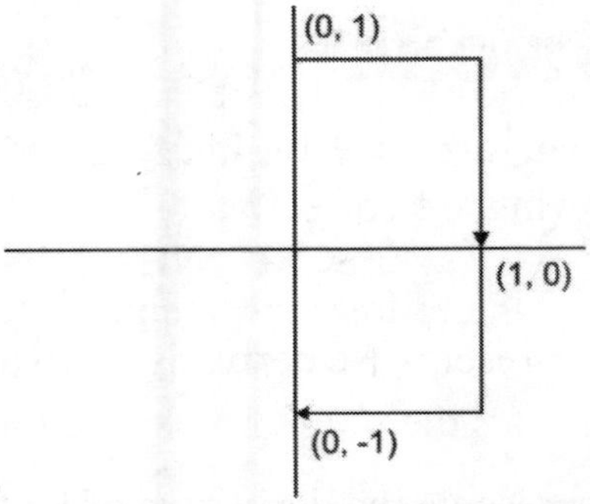

T(1, 0) = (0, -1) = 0(1, 0) – 1(0, 1)
T(0, 1) = (1, 0) = 1(1, 0) + 0(0, 1)

Matrix of T relative to basis B = {(1, 0), (0, 1)} is $\begin{bmatrix} 0 & -1 \\ 1 & 0 \end{bmatrix}^t = \begin{bmatrix} 0 & 1 \\ -1 & 0 \end{bmatrix}$

So, option (b) is correct.

13. Let V be a finite dimensional vector space and T : V →V be a linear transformation. Let R(T) denote the range of T and N(T) denote the null space {v ∈V : Tv = 0} of T. If rank(T) = rank(T^2), then which of the following is/are necessarily true?

(a) N(T) = {0} (b) R(T) = R(T^2) (c) N(T) = N(T^2) (d) N(T)∩R(T) = {0}

Ans. (b), (c), (d) **IIT JAM – 2021**

Explanation :

Suppose that T is idempotent linear operator

Let $A = \begin{bmatrix} 1 & 0 \\ 0 & 0 \end{bmatrix}$

Here η(A) = 1 ⇒ N(T) ≠ {0}

Option (a) is incorrect.

(b) Given that rank (T) = rank(T^2)
⇒ n – nullity(T) = n – nullity(T^2) ⇒ nullity(T) = nullity(T^2)
⇒ Range(T) = Range(T^2) and N(T) = N(T^2)
Option (b) & (c) are correct.

(d) Let x ∈ N(T) ∩ R(T)
⇒ x ∈ N(T) and x ∈ R(T) ⇒ T(x) = 0 and ∃y ∈ T s.t. T(y) = x
Here T^2(y) = T(x)
⇒ T^2(y) = 0 ⇒ y ∈ N(T^2)
⇒ y ∈ N(T) ⇒ T(y) = 0
⇒ x = 0 Option (d) is correct.

UNSOLVED QUESTIONS

1. Let $\varphi : R^2 \to C$ be the map defined by $\varphi(x, y) = z$, where $z = x + iy$. Let $f : C \to C$ be the function $f(z) = z^2$ and $F = \varphi^{-1} f \varphi$. Which of the following are correct?

CSIR NET JUNE 2014

(a) The linear transformation $T(x,y) = 2\begin{pmatrix} x & -y \\ y & x \end{pmatrix}$ represents the derivative of F at (x, y).

(b) The linear transformation $T(x,y) = 2\begin{pmatrix} x & y \\ y & x \end{pmatrix}$ represents the derivative of F at (x, y).

(c) The linear transformation $T(z) = 2z$ represents the derivative of f at $z \in C$.

(d) The linear transformation $T(z) = 2z$ represents the derivative of f only at 0.

2. Let R[x] denotes the vector space of all real polynomials. Let $D : R[x] \to R[x]$ denote the map $Df = \frac{df}{dx}, \forall f$. Then, **CSIR UGC DEC-2014**

(a) D is one-one (b) D is onto

(c) There exists $E : R[x] \to R[x]$ so that $D(E(f)) = f, \forall f$.

(d) There exists $E : R[x] \to R[x]$ so that $E(D(f)) = f, \forall f$.

3. Let $T : R^3 \to R^3$ be defined by $T(x_1, x_2, x_3) = (x_1 - x_2, x_1 - x_2, 0)$. **JAM MA-2006**

If N(T) and R(T) denote the null space and the range space of T respectively, then

(a) dim N(T) = 2(b) dim R(T) = 2 (c) R(T) = N(T) (d) $N(T) \subset R(T)$

4. For $n \neq m$, let $T_1 : R^n \to R^m$ and $T_2 : R^m \to R^n$ be linear transformations such that T_1T_2 is bijective. Then **JAM MA-2010**

(a) rank $T_1 = n$ and rank $T_2 = m$ (b) rank $T_1 = m$ and rank $T_2 = n$

(c) rank $T_1 = n$ and rank $T_2 = n$ (d) rank $T_1 = m$ and rank $T_2 = m$

5. Let $U = \{(a, b, c) \mid a + 2b - 3c = 0, 2a + 5b + 2c = 0, 3a - b - 4c = 0\} \subset R^3$. Let $A = \begin{bmatrix} 1 & 2 & -3 \\ 2 & 5 & 2 \\ 3 & -1 & -4 \end{bmatrix}$. Then **CMI – 2013**

(a) dim U = 0, rank A = 3 (b) dim U = 1, rank A = 2

(c) dim U = 1, rank A = 1 (d) dim U = 3, rank A = 3

6. Let $P_2(R)$ be the vector space of polynomials in x of degree atmost 2 with real coefficients. Let $M_2(R)$ be the vector space of 2 × 2 real matrices. If a linear transformation $T : P_2(R) \to M_2(R)$ is defined as $T(f) = \begin{bmatrix} f(0) - f(2) & 0 \\ 0 & f(1) \end{bmatrix}$. Then

JAM MA-2015

(a) T is one-one but not onto (b) T is onto but not one-one

(c) Range $(T) = span\left\{\begin{bmatrix} 0 & 0 \\ 0 & 1 \end{bmatrix}, \begin{bmatrix} -2 & 0 \\ 0 & 1 \end{bmatrix}\right\}$ (d) Null (T) = span $\{x^2 - 2x, 1 - x\}$

7. Let $T : R^4 \to R^4$ be the linear map satisfying $T(e_1) = e_2$, $T(e_2) = e_3$, $T(e_3) = 0$, $T(e_4) = e_3$, where $\{e_1, e_2, e_3, e_4\}$ is the standard basis of R^4. Then **GATE-2008**

(a) T is idempotent (b) T is invertible (c) Rank T = 3 (d) T is nilpotent

8. For an integer $k \geq 0$, let P_k denote the vector space of all real polynomials in one variable of degree less than or equal to k. Define a linear transformation $T: P_2 \to P_3$ by $Tf(x) = f''(x) + xf(x)$

Which one of the following polynomials is not in the range of T? **IIT JAM 2021**

(a) $x + x^2$ (b) $x^2 + x^3 + 2$ (c) $x + 1$ (d) $x + x^3 + 2$

9. Let P be a fixed 3×3 matrix with entries in R. Which of the following maps from $M_3(R)$ to $M_3(R)$ is/are linear? **IIT JAM - 2022**

(a) $T_1 : M_3(R) \to M_3(R)$ given by $T_1(M) = MP - PM$for $M \in M_3(R)$

(b) $T_2 : M_3(R) \to M_3(R)$ given by $T_2(M) = M^2P - P^2M$ for $M \in M_3(R)$

(c) $T_3 : M_3(R) \to M_3(R)$ given by $T_3(M) = MP^2 - P^2M$ for $M \in M_3(R)$

(d) $T_4 : M_3(R) \to M_3(R)$ given by $T_4(M) = MP^2 - PM^2$ for $M \in M_3(R)$

10. Consider the following statements :

I : There exists a linear transformation from R^3 to R^3 s.t. its range space and null space are same.

II : There exists a linear transformation from R^2 to R^2 s.t. its range space and null space are same.

Then **IIT JAM 2023**

(a) Both are false (b) Only I is true (c) Only II is true (d) Both are true

11. Let V be the real vector space of all polynomials of degree less than or equal to 2 with real coefficients. Let $T : V \to V$ be the linear transformation given by $T(p) = 2p + p'$ for $p \in V$, where p' is the derivative of p. Then the number of non-zero entries in the Jordan canonical form of a matrix of T equals **GATE - 2018**

(a) 3 (b) 4 (c) 5 (d) 6

12. Let $T : P_2(R) \to P_4(R)$ be a linear transformation defined as $T(p(x)) = p(x^2)$ then $\rho(T)$ is **IIT JAM 2023**

(a) 1 (b) 2 (c) 3 (d) 4

13. Let $T : R^3 \to R^3$ be a linear transformation satisfying $T(1, 0, 0) = (0, 1, 1)$, $T(1, 1, 0) = (1, 0, 1)$ and $T(1, 1, 1) = (1, 1, 2)$. Then **GATE 2023**

(a) T is one-one but T is not onto (b) T is one-one and onto

(c) T is neither one-one nor onto (d) T is not one-one but T is onto

14. If V and W are vector spaces over the field F and T is a linear transformation from V into W. If V is finite dimensional, then

(a) Rank (T) + nullity (T) = dim (V) (b) Rank (T) = dim (V) + nullity (T)
(c) Rank (T) + dim (V) = nullity (T) (d) Rank (T) = nullity (T)

15. Let T : V →V be a linear operator, then which of the following is/are not correct?
(a) {0} is invariant under T (b) V is invariant under T
(c) kernel of T is invariant under T (d) None of these

16. Let C be the vector space of all complex numbers over complex field C and T : C →C be defined by $T(z)=\bar{z}, z \in C$. Then
(a) T is a linear mapping (b) T is not linear mapping but a well defined map
(c) T is not well defined (d) None of the above

17. The transformation T : $R^3 \to R^2$ as defined by T(x, y, z) = (x + y, y + z), is
(a) linear and has zero kernel (b) linear and has a proper subspace as kernel
(c) neither linear nor one-one (d) neither linear nor onto

18. Let D : $P_n(R) \to P_{n-1}(R)$ and I : $P_{n-1}(R) \to P_n(R)$ be defined by differentiation and integration, i.e. D(f(x)) = f'(x), I(f(x)) = $\int_0^x f(t)dt$, then
(a) D is linear while I is not linear transformation
(b) D is not linear while I is linear transformation
(c) D and I both are not linear transformations (d) D and I both are linear transformations

19. Which of the following transformations are not linear
(a) T : C(R)→ C(R) given by T(z) = iz (b) T : C(R)→ C(R) given by $T(z)=\bar{z}$
(c) T : R →R given by T(x) = x^2
(d) T : $R^2 \to R^3$ given by T(x, y) = (x, y, xy), where C, R, R^2, R^3 are vector spaces over R.

20. Let dim [hom(V, U)] = 12. Then, there may exists V and U such that
(a) dim V = 3 and dim U = 4 (b) dim V = 6 and dim U = 2
(c) dim V = 1 and dim U = 12 (d) dim V = 4 and dim U = 3

21. Let V be the space of all n ×n matrices and T : V →V be a linear operator defined by $T(A)=\dfrac{A+A^T}{2}$, then nullity of T is not
(a) $2n$ (b) $\dfrac{n^2}{2}$ (c) $\dfrac{n(n+1)}{2}$ (d) $\dfrac{n(n-1)}{2}$

22. Let V be the vector space of 2 × 2 matrices over R. Which of the following is/are linear transformation(s)?

(a) $T([a_{ij}]) = a_{11}$ (b) $T(A) = A + I$ (c) $T(A) = \text{trace } A$ (d) $T([a_{ij}]) = a_{11} + a_{12} + a_{21} + a_{22}$

23. Let X and Y be vector spaces over a field K and let $f : X \to Y$ be an invertible linear map. Then,

(a) $f^{-1} : Y \to X$ is linear (b) $f^{-1} : Y \to X$ is not linear
(c) $f(x_1 + x_2) = f(x_1) + f(x_2)$ (d) $f(x_1 + x_2) \neq f(x_1) + f(x_2)$

24. Let $T : R^4 \to R^4$ be a linear map defined by $T(x, y, z, w) = (x + z, 2x + y + 3z, 2y + 2z, w)$. Then the rank of T is equal to **GATE – 2015**

(a) 4 (b) 3 (c) 5 (d) 2

25. Consider the vector space $V = \{a_0 + a_1x + a_2x^2 : a_i \in R \text{ for } i = 0, 1, 2\}$ of polynomials of degree at most 2. Let $f : V \to R$ be a linear functional such that $f(1 + x) = 0$, $f(1 - x^2) = 0$ and $f(x^2 - x) = 2$. Then $f(1 + x + x^2)$ equals **GATE – 2017**

(a) 4 (b) 3 (c) 2 (d) 1

26. Let P_3 denote the real vector space of all polynomials with real coefficients of degree at most 3. Consider the map $T : P_3 \to P_3$ given by $T(p(x)) = p``(x) + p(x)$. Then **IIT JAM – 2017**

(a) T is neither one-one nor onto (b) T is both one-one and onto
(c) T is one-one but not onto (d) T is onto but not one-one

27. Consider the vector space V over R of polynomial functions of degree less than or equal to 3 defined on R. Let $T : V \to V$ be defined by $(Tf)(x) = f(x) - xf`(x)$. Then the rank T is **IIT JAM – 2018**

(a) 1 (b) 2 (c) 3 (d) 4

28. Let $T \in M_{m\times n}(R)$. Let V be the subspace of $M_{n\times p}(R)$ defined by $V = \{X \in M_{n\times p}(R) : TX = 0\}$. Then the dimension of V is **IIT JAM – 2019**

(a) pn – rank (T) (b) mn – p rank (T) (c) p(m – rank (T)) (d) p(n – rank (T))

Answer Key

1	2	3	4	5	6	7	8	9	10
a,c	b,c	a	d	a	c	d	c	a,c	c
11	**12**	**13**	**14**	**15**	**16**	**17**	**18**	**19**	**20**
c	c	c	a	d	b	b	d	c,d	a,b,c,d
21	**22**	**23**	**24**	**25**	**26**	**27**	**28**		
a,b,c	a,c,d	a,c	b	d	b	c	d		

Chapter 4

System of Linear Equations

1. Consistency and Inconsistency of Non-Homogeneous and Homogeneous System of Linear Equations

By using Rank Method

(i) Non-homogeneous system of linear equations :

Consider a system of m linear equations in n unknowns $x_1, x_2, ..., x_n$ given by

$$a_{11}x_1 + a_{12}x_2 + ... + a_{1n}x_n = b_1$$
$$a_{21}x_1 + a_{22}x_2 + ... + a_{2n}x_n = b_2$$
$$\vdots \quad \vdots \quad \vdots \quad \vdots$$
$$a_{m1}x_1 + a_{m2}x_2 + ... + a_{mn}x_n = b_m$$

This system can be written as AX = B, where

$$A = \begin{bmatrix} a_{11} & a_{12} & \cdots & a_{1n} \\ a_{21} & a_{22} & \cdots & a_{2n} \\ \vdots & \vdots & & \vdots \\ a_{m_1} & a_{m_2} & \cdots & a_{mn} \end{bmatrix}_{m\times n}, X = \begin{bmatrix} x_1 \\ x_2 \\ \vdots \\ x_n \end{bmatrix}_{n\times 1}, B = \begin{bmatrix} b_1 \\ b_2 \\ \vdots \\ b_m \end{bmatrix}_{m\times 1}$$

The matrix A is called the **coefficient matrix** and the matrix

$$[A;B] = \begin{bmatrix} a_{11} & a_{12} & \cdots & a_{1n} & \vdots & b_1 \\ a_{21} & a_{22} & \cdots & a_{2n} & \vdots & b_2 \\ \vdots & \vdots & & \vdots & \vdots & \vdots \\ a_{m_1} & a_{m_2} & \cdots & a_{mn} & \vdots & b_m \end{bmatrix}$$ is called the augmented matrix of the system.

The system of equations AX = B

(i) is consistent iff$\rho[A : B] = \rho(A)$.

(a) has unique solution, if $\rho[A : B] = \rho(A) = n$.

(b) has an infinite number of solutions if $\rho[A ; B] = \rho(A) = r < n$. Giving arbitrary values to n – r unknowns, we may express the other r unknowns in terms of these.

(ii) has no solution iff$\rho[A : B] \neq \rho(A)$ (inconsistent)

(ii) Homogeneous linear equations :

If B = 0, Then system becomes Homogeneous system of linear equations AX = 0

It is always consistent.

The system of equations AX = 0

(a) has unique solution if $\rho[A:0] = \rho(A) = n$ (i.e. $|A| \neq 0$)

(b) has an infinite number of solutions, iff $\rho[A:0] = \rho(A) < n$ (i.e. $|A| = 0$)

NOTE: (1) Let AX = b be a system of linear equation. Then dimension of solution space is $1 + \eta(A)$

(2) Let AX = 0 be a system of linear equation. Then dimension of solution space is $\eta(A)$

Example:

$x + y + z = 1$

$x + y + 2z = 1$

$2x + 2y + 3z = 2$

Augmented matrix

$$\begin{bmatrix} 1 & 1 & 1 & 1 \\ 1 & 1 & 2 & 1 \\ 2 & 2 & 3 & 2 \end{bmatrix} \sim \begin{bmatrix} 1 & 1 & 1 & 1 \\ 0 & 0 & 1 & 0 \\ 0 & 0 & 1 & 0 \end{bmatrix} \sim \begin{bmatrix} 1 & 1 & 1 & 1 \\ 0 & 0 & 1 & 0 \\ 0 & 0 & 0 & 0 \end{bmatrix}$$

Here $\rho(A:b) = \rho(A)$

$\Rightarrow$ This system has a solution and $\eta(A) = 1 > 0$

$\Rightarrow$ This system has infinite solution.

Important conclusions of system of linear equations :

(i) Non-Homogeneous System of linear equations :

Consider the system of non-homogeneous equations AX = b of m-equations and n-variables, where

$$X = \begin{bmatrix} x_1 \\ x_2 \\ \vdots \\ x_n \end{bmatrix}_{n\times 1} \text{ and } b = \begin{bmatrix} b_1 \\ b_2 \\ \vdots \\ b_m \end{bmatrix}_{m\times 1}$$

$A = (a_1, a_2, \ldots, a_n)_{m\times n}$, where a_i is the ith column of matrix.

AX = b has a solution iff

(i) b is linear combination of columns of A

(ii) rank (A) = rank [A : b]

(iii) $b \in F_m$ is linearly dependent with columns of A

(iv) number of linearly independent vectors in $S = \{a_1, a_2, \ldots, a_n\}$ and number of linearly independent vectors in $S_1 = S \cup \{b\}$ are same

(v) $\dim(\langle S\rangle) = \dim(\langle S_1\rangle)$

(ii) Homogeneous System of Equations :

Consider the homogeneous system of equations AX = O of m equations and n – variables.

[It is always consistent (either trivial solution or non-trivial solutions)

$$\because O = \begin{bmatrix} 0 \\ 0 \\ \vdots \\ 0 \end{bmatrix}_{n\times 1}$$ always lie in column space of A]

We are interested to find that when AX = O has non-trivial solutions.

AX = 0 has non-trivial solution iff

(i) $\text{Ker}(A) \neq 0$ (ii) Nullity $(A) \geq 1$ (iii) rank $(A) < n$

(iv) A is not one-one (v) If $m < n$ (i.e. number of equations < number of variables)

Let A be a $m \times n$ matrix with entries from $a_{ij} \in F$ (Field) then $\exists$ a linear transformation $T : F^n \rightarrow F^m$

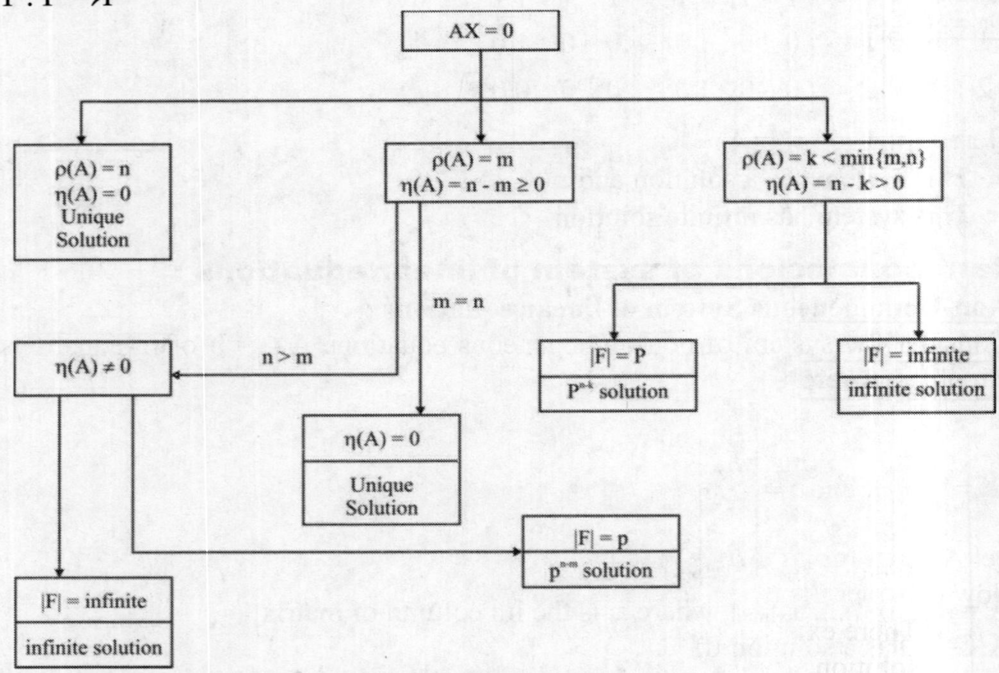

Let A be a $m \times n$ matrix with entries $a_{ij} \in F$(Field)

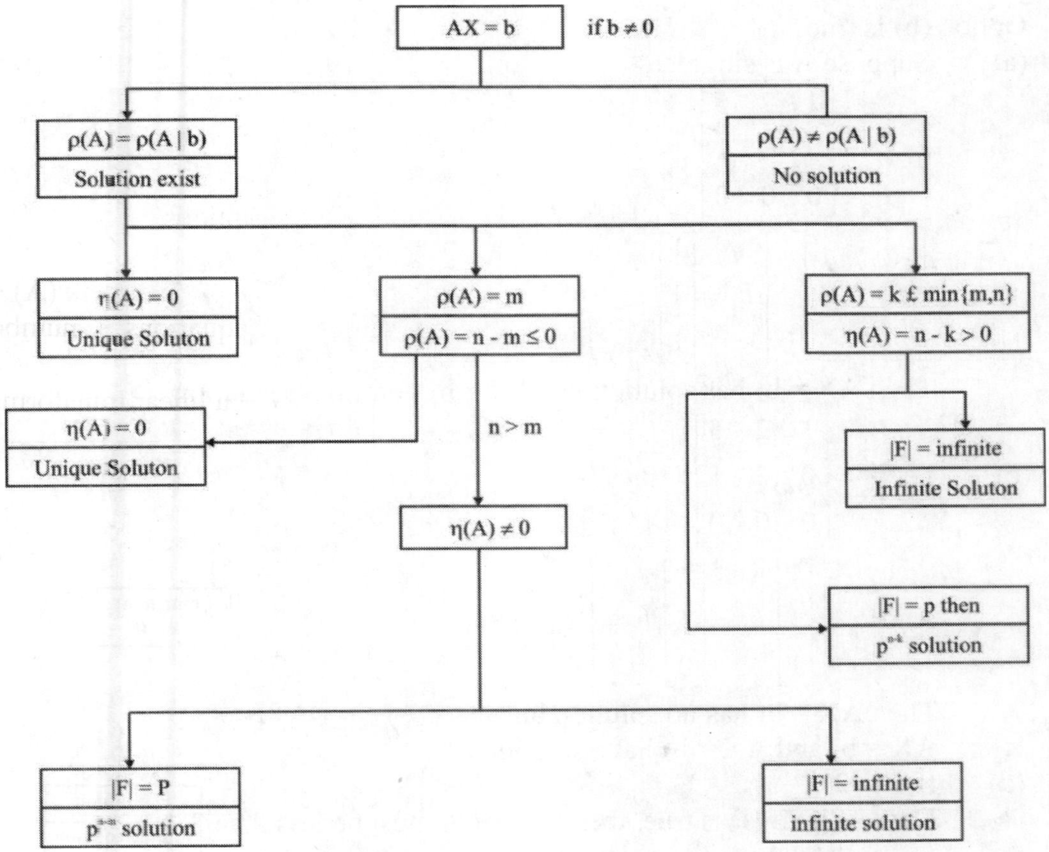

SOLVED QUESTIONS

1. Let A be an m ×m matrix with real entries and let x be an m × 1 vector of unknowns. Now consider the two statements given below :

I There exists non-zero vector $b_1 \in R^m$ such that the linear system $Ax = b_1$ has no solution.

II There exist non-zero vectors $b_2, b_3 \in R^m$, with $b_2 \neq cb_3$ for any $c \in R$, such that the linear systems $Ax = b_2$ and $Ax = b_3$ have solutions.

Which of the following statements are true? **CSIR NET FEB 2022**

(a) II is true whenever A is singular (b) I is true whenever A is singular

(c) Both I and II can be true simultaneously

(d) If m = 2, then at least one of I and II is false.

Ans. (b), (c), (d)

Explanation :

If AX = b has no solution iff A must be singular

Option (b) is true.

(a) Suppose A is singular

i.e. $A = \begin{bmatrix} 1 & 2 & 5 \\ 0 & 3 & 1 \\ 0 & 0 & 0 \end{bmatrix}$

And $b_2 = \begin{bmatrix} 1 \\ 1 \\ 0 \end{bmatrix}, b_3 = \begin{bmatrix} 1 \\ 1 \\ 1 \end{bmatrix}$

Then $AX = b_2$ has solution but $AX = b_3$ don't have solution

(c) i.e. $A = \begin{bmatrix} 1 & 2 & 5 \\ 0 & 3 & 1 \\ 0 & 0 & 0 \end{bmatrix}$

And $b_1 = \begin{bmatrix} 1 \\ 1 \\ 1 \end{bmatrix}, b_2 = \begin{bmatrix} 1 \\ 1 \\ 0 \end{bmatrix}, b_3 = \begin{bmatrix} 1 \\ 2 \\ 0 \end{bmatrix}$

Then $AX = b_1$ has no solution but
$AX = b_2$ and $AX = b_3$ have solution

(d) If $m = 2$

Case – (i) : If (I) is true, then rank of A must be less than 2

i.e. $A = \begin{bmatrix} 1 & 2 \\ 0 & 0 \end{bmatrix}$

Then there doesnot exist two LI vectors b_1 and b_2 such that $Ax = b_2$ and $Ax = b_3$ have solutions.

Case – (ii) : If (II) is true, then rank of A must be 2,

i.e. $|A| \neq 0$

So, option (I) is false.

2. Consider the system **CSIR NET FEB 2022**

$$2x + ky = 2 - k$$
$$kx + 2y = k$$
$$ky + kz = k - 1$$

in three unknowns and one real parameter k. For which of the following values of k is the system of linear equation consistent?

(a) 1 (b) 2 (c) -1 (d) -2

Ans. (a), (c)

Explanation :

As we know a system $Ax = b$ is consistent

If and only if ρ(A) = ρ(A : b)

Here given system is Ax = b, where $A=\begin{bmatrix}2 & k & 0\\ k & 2 & 0\\ 0 & k & k\end{bmatrix}, X=\begin{bmatrix}x\\ y\\ z\end{bmatrix}, b=\begin{bmatrix}2-k\\ k\\ k-1\end{bmatrix}$ and

argument matrix.

$$[A:b]=\left[\begin{array}{ccc|c}2 & k & 0 & 2-k\\ k & 2 & 0 & k\\ 0 & k & k & k-1\end{array}\right] R_2 \to R_2 - R_1 \sim \left[\begin{array}{ccc|c}2 & k & 0 & 2-k\\ k-2 & 2-k & 0 & 2k-2\\ 0 & k & k & k-1\end{array}\right]$$

For k = 2;

$$[A:b]=\left[\begin{array}{ccc|c}2 & 2 & 0 & 0\\ 0 & 0 & 0 & 2\\ 0 & 2 & 2 & 1\end{array}\right]$$

ρ(A) ≠ρ(A : b)

So, it is not consistent.

For k = -2;

$$[A:b]=\left[\begin{array}{ccc|c}2 & -2 & 0 & 4\\ -4 & 4 & 0 & -6\\ 0 & -2 & -2 & -3\end{array}\right], R_2 \to R_2 + 2R_1 \sim \left[\begin{array}{ccc|c}2 & -2 & 0 & 4\\ 0 & 0 & 0 & 2\\ 0 & -2 & -2 & 3\end{array}\right]$$

ρ(A) ≠ρ(A : b)

So, it is not consistent

For k = 1;

$$[A:b]=\left[\begin{array}{ccc|c}2 & 1 & 0 & 1\\ -1 & 1 & 0 & 0\\ 0 & 1 & 1 & 0\end{array}\right], R_2 \to 2R_2 + R_1 \sim \left[\begin{array}{ccc|c}2 & 1 & 0 & 1\\ 0 & 3 & 0 & 1\\ 0 & 1 & 1 & 0\end{array}\right]$$

ρ(A) = ρ(A : B)

So, consistent.

(d) Similarly for k = -1, we can prove, it is consistent.

3. Let M be a 5 × 5 matrix with real entries such that rank of M is 3. consider the linear system Mx = b. let the row reduced echelon form of the augmented matrix [M : b] be R and let R[i , :] denote the ith row of R.Suppose that linear system has admits a solution. which of the following is true **CSIR NET NOV.2020**

(a) R[3 , :] = [0 1 0 ***] (b) R[5 , :] = [0 0 1 ***]

(c) R[4 , :] = [0 0 0 1 **] (d) R[4 , :] = [0 0 0 0 0 0]

Ans. (d)

Explanation:

$$[M:b]\begin{bmatrix}1&0&0&*&*&\vert&*\\0&1&0&*&*&\vert&*\\0&0&1&*&*&\vert&*\\0&0&0&0&0&\vert&0\\0&0&0&0&0&\vert&0\end{bmatrix}$$

Given that Rank of M is 3 and system Mx = b is consistence
Hence (d) is correct

4. Let A be an n ×n real matrix. Let b be an n × 1 vector. Suppose Ax = b has no solution. Which of the following statements are true?
(a) There exists an n × 1 vector c such that Ax = c has a unique solution
(b) There exist infinitely many vectors c such that Ax = c has no solution
(c) If y is the first column of A then Ax = y has a unique solution
(d) det A = 0

Ans. **(b), (d)** **CSIR NET JUNE 2020**

Explanation :

Given, A be an n ×n real matrix, b be an n × 1 vector and Ax = b has no solution. We have, a non- homogeneous system Ax = b has a solution iff r(A) = r(A; b).
If r(A) = n, then r(A) = r(A; b) = n, which is not possible here as our system has no solution, so r(A) < n which imply detA = 0, Option (d) is correct.

(a) We have, Ax = c has unique solution iff r(A) = n, but here r(A) < n, so, Option (b) is incorrect.

(b) We have given Ax = b has no solution, so, for any multiple of b,c = kb, Ax = c has no solution. i.e, there exist infinitely many vectors c such that Ax = c has no solution,
Option (b) is correct.

(c) For unique solution, r(A) must be equal to n. ie, r(A) = r(A: y) = n but here we have already proved that r(A) < n. So, Ax = y has not a unique solution even if y is the first column of A. Option (c) is incorrect.

5. Let $A=\begin{bmatrix}2&0&5\\1&2&3\\-1&5&1\end{bmatrix}$. The system of linear equation AX = Y has a solution. **CSIR NET DEC 2019**

(a) only for $Y=\begin{pmatrix}x\\0\\0\end{pmatrix}$ (b) only for $Y=\begin{pmatrix}0\\y\\0\end{pmatrix}$ (c) only for $Y=\begin{pmatrix}0\\y\\2\end{pmatrix}$ (d) for all Y $\in R^3$

Ans. **(d)**

Explanation :

We find Rank (A) by elementary transformation $\begin{bmatrix} 2 & 0 & 5 \\ 1 & 2 & 3 \\ -1 & 5 & 1 \end{bmatrix} \sim \begin{bmatrix} 2 & 0 & 5 \\ 0 & 4 & 1 \\ 0 & 10 & 7 \end{bmatrix} \sim \begin{bmatrix} 2 & 0 & 5 \\ 0 & 4 & 1 \\ 0 & 0 & 9 \end{bmatrix}$

$\Rightarrow \rho(A) = 3$ $\Rightarrow$ for any value $Y \in R^3$

$\rho(A : Y) = 3$ $\Rightarrow \rho(A) = \rho(A : Y)$ $\Rightarrow$ has a solution

For any $Y \in R^3$

6. If A is 5 × 5 matrix and the dimension of the solution space of Ax = 0 is at least 2 then

CSIR NET JUNE 2018

(a) $\rho(A^2) \le 3$ (b) $\rho(A^2) \ge 3$ (c) $\rho(A^2) = 3$ (d) Det $(A^2) = 0$

Ans. (a), (d)

Explanation :

Solution space of Ax = 0 is ker(A) given that

dim $(\ker(A)) \ge 2 \Rightarrow \eta(A) \ge 2$, and A is 5 × 5 matrix

$\Rightarrow \exists$ a Linear Transformation.T : $R^5 \to R^5$

$\rho(A) + \eta(A) = 5$

$\Rightarrow \rho(A) \le 3$

Let $\eta(A) = 2 \Rightarrow \rho(A) = 3$

$$A = \begin{bmatrix} 0 & 0 & 1 & 0 & 0 \\ 0 & 0 & 0 & 1 & 0 \\ 0 & 0 & 0 & 0 & 1 \\ 0 & 0 & 0 & 0 & 0 \\ 0 & 0 & 0 & 0 & 0 \end{bmatrix} \Rightarrow A^2 = \begin{bmatrix} 0 & 0 & 0 & 0 & 1 \\ 0 & 0 & 0 & 0 & 0 \\ 0 & 0 & 0 & 0 & 0 \\ 0 & 0 & 0 & 0 & 0 \\ 0 & 0 & 0 & 0 & 0 \end{bmatrix}$$

$\Rightarrow$ $\rho(A^2) = 1$

Option (b) and (c) incorrect.

Option (a) is correct. $\Rightarrow \rho(A^2) \le 3$

Option (d) is correct.

$\Rightarrow$ Det $(A^2) = 0$

7. The system of equation **CSIR NET JUNE 2018**

$1x + 2x^2 + 3xy + 0.y = 6$

$2x + x^2 + 3xy + 1.y = 5$

$x - x^2 + 0.xy + y = 7$

(a) has solution in rational form (b) has solution in real number

(c) has solution in complex number (d) has no solution

Ans. (d)

Explanation :

Given that $\begin{bmatrix} 1 & 2 & 3 & 0 \\ 2 & 1 & 3 & 1 \\ 1 & -1 & 0 & 1 \end{bmatrix}\begin{bmatrix} 1 \\ x^2 \\ xy \\ y \end{bmatrix} = \begin{bmatrix} 6 \\ 5 \\ 7 \end{bmatrix}$

then $\rho[A : b] = \left[\begin{array}{cccc|c} 1 & 2 & 3 & 0 & 6 \\ 2 & 1 & 3 & 1 & 5 \\ 1 & -1 & 0 & 1 & 7 \end{array}\right]$ Using elementary transformation

$$\left[\begin{array}{cccc|c} 1 & 2 & 3 & 0 & 6 \\ 0 & -3 & -3 & 1 & -7 \\ 0 & -3 & -3 & 1 & 1 \end{array}\right] \sim \left[\begin{array}{cccc|c} 1 & 2 & 3 & 0 & 6 \\ 0 & -3 & -3 & 1 & -7 \\ 0 & 0 & 0 & 0 & 8 \end{array}\right]$$

$\rho(A) = 2$ and $\rho(A : b) = 3$ $\Rightarrow \rho(A) \neq \rho(A : b)$ $\Rightarrow$ no solution

UNSOLVED QUESTIONS

1. Let $v_1, v_2, \ldots, v_9$ be the column vectors of non-zero 9×9 real matrix A. Let $a_1, a_2, \ldots, a_9 \subset R$ not all zero, be such that $\sum_{i=1}^{9} a_i v_i = 0$. Then the system $AX = \sum_{i=1}^{9} v_i$ has **IIT JAM 2023**

(a) A unique solution
(b) more than one but finite solution
(c) No solution
(d) Infinitely many solution

2. The system of linear equation $\begin{bmatrix} 1 & 1 & 1 \\ 0 & -1 & 1 \\ 2 & 3 & \alpha \end{bmatrix}\begin{bmatrix} x \\ y \\ z \end{bmatrix} = \begin{bmatrix} 3 \\ 1 \\ \beta \end{bmatrix}$ where $\alpha, \beta \in R$. **IIT JAM 2023**

(a) Infinitely many solution for any α when $\beta = 5$
(b) Atleast one solution for any $\alpha \& \beta$
(c) No solution for any α when $\beta \neq 5$
(d) A unique solution for any β when $\alpha \neq 1$.

3. Let m> 1 and n > 1 be integers. Let A be an m×n matrix such thath for some m×1 matrix b_1, the equation $Ax = b_1$ has infinitely many solutions. Let b_2 denote an m×1 matrix different from b_1, then $Ax = b_2$ has **IIT JAM 2021**

(a) Finitely many solutions for some b_2.
(b) No solution for some b_2.
(c) infinitely many solutions for some b_2.
(d) A unique solution for some b_2.

4. The system of linear equations **IIT JAM - 2014**

$x - y + 2z = b_1$
$x + 2y - z = b_2$

$2y - 2z = b_3$ is inconsistent when (b_1, b_2, b_3) equals

(a) (2, 2, 0) (b) (0, 3, 2) (c) (2, 2, 1) (d) (2, -1, -2)

5. Suppose $\alpha, \beta, \gamma \in R$. Consider the following system of linear equations. **IIT JAM - 2018**

$$x + y + z = \alpha,$$
$$x + \beta y + z = \gamma,$$
$$x + y + \alpha z = \beta,$$

If this system has at least one solution, then which of the following statements is(are) TRUE?

(a) If $\alpha = 1$ then $\gamma = 1$ (b) If $\beta = 1$ then $\gamma = \alpha$

(c) If $\beta \neq 1$ then $\alpha = 1$ (d) If $\gamma = 1$ then $\alpha = 1$

6. Let $M = \begin{bmatrix} \alpha & 1 & 1 \\ 1 & \beta & 1 \\ 1 & 1 & \gamma \end{bmatrix}$, $\alpha\beta\gamma = 1$, $\alpha, \beta, \gamma \in R$ and $x = \begin{bmatrix} x_1 \\ x_2 \\ x_3 \end{bmatrix} \in R^3$. Then $Mx = 0$ has infinitely many solutions if trace (M) is **IIT JAM – 2016**

(a) 1 (b) 2 (c) 3 (d) 4

7. Consider the following system of three linear equations in four unknowns x_1, x_2, x_3 and x_4 **IITJAM - 2019**

$$x_1 + x_2 + x_3 + x_4 = 4,$$
$$x_1 + 2x_2 + 3x_3 + 4x_4 = 5,$$
$$x_1 + 3x_2 + 5x_3 + kx_4 = 5$$

If the system has no solutions, then k is

(a) 5 (b) 6 (c) 7 (d) 8

8. Let $m > 1$ and $n > 1$ be integers. Let A be an $m \times n$ matrix such that for some $m \times 1$ matrix b_1, the equation $AX = b_1$ has infinitely many solutions. Let b_2 denote an $m \times 1$ different from b_1. Then $AX = b_2$ has **IIT JAM - 2021**

(a) infinitely many solutions for some b_2. (b) a unique solution for some b_2.

(c) finitely many solutions for some b_2. (d) no solution for some b_2.

9. Let $M = \begin{bmatrix} 2 & 0 & 3 & 2 & 0 & -2 \\ 0 & 1 & 0 & -1 & 3 & 4 \\ 0 & 0 & 1 & 0 & 4 & 4 \\ 1 & 1 & 1 & 0 & 1 & 1 \end{bmatrix}$, $b_1 = \begin{bmatrix} 5 \\ 1 \\ 1 \\ 4 \end{bmatrix}$ & $b_2 = \begin{bmatrix} 5 \\ 1 \\ 3 \\ 3 \end{bmatrix}$

Then which of the following are TRUE? **CSIR NET DEC 2018**

(a) both $MX = b_1$ & $MX = b_2$ are inconsistent.

(b) both system $MX = b_1$ & $MX = b_2$ are consistent.

(c) then system $MX = b_1 - b_2$ is consistent (d) $MX = b_1 - b_2$ is inconsistent.

10. Consider a non-homogeneous system of linear equations representing mathematically an over determined system. Such a system will be **GATE-2005**

(a) consistent having a unique solution (b) consistent having many solutions
(c) inconsistent having a unique solution (d) inconsistent having no solution

11. Solution for the system defined by the set of equations $4y + 3z = 8$; $2x - z = 2$ and $3x + 2y = 5$ is
(a) $x = 0$; $y = 1$; $z = 4/3$ (b) $x = 0$; $y = 1/2$; $z = 2$
(c) $x = 1$; $y = 1/2$; $z = 2$ (d) non-existent **GATE-2006**

12. For what values of α and β, the following simultaneous equations have an infinite number of solutions? **GATE – 2007**
$x + y + z = 5$; $x + 3y + 3z = 9$; $x + 2y + \alpha z = \beta$
(a) 2, 7 (b) 3, 8 (c) 8, 3 (d) 7, 2

13. The following simultaneous equations
$x + y + z = 3$
$x + 2y + 3z = 4$
$x + 4y + kz = 6$
will NOT have a unique solution for k equal to **GATE-2008**
(a) 0 (b) 5 (c) 6 (d) 7

14. Consider the system of simultaneous equations
$x + 2y + z = 6$; $2x + y + 2z = 6$; $x + y + z = 6$ **GATE - 2003**
This system has
(a) unique solution (b) infinite number of solutions (c) no solution (d) exactly two solutions

15. A is a 3×4 real matrix and $AX = b$ is an inconsistent system of equations. The highest possible rank of A is **GATE - 2005**
(a) 1 (b) 2 (c) 3 (d) 4

16. For what value of a, if any, will the following system of equations in x, y and z have a solution?
$2x + 3y = 4$ **GATE-2008**
$x + y + z = 4$
$x + 2y - z = a$
(a) Any real number (b) 0 (c) 1 (d) There is no such value

17. $x + 2y + z = 4$
$2x + y + 2z = 5$
$x - y + z = 1$
The system of algebraic equation given above has **GATE-2012**
(a) A unique solution of $x = 1$, $y = 1$ and $z = 1$
(b) only the two solutions of ($x = 1$, $y = 1$, $z = 1$) and ($x = 2$, $y = 1$, $z = 0$)
(c) infinite number of solutions (d) no feasible solution

18. The equation $\begin{bmatrix} 2 & -2 \\ 1 & -1 \end{bmatrix}\begin{bmatrix} x_1 \\ x_2 \end{bmatrix} = \begin{bmatrix} 0 \\ 0 \end{bmatrix}$ has

(a) no solution

(b) only one solution $\begin{bmatrix} x_1 \\ x_2 \end{bmatrix} = \begin{bmatrix} 0 \\ 0 \end{bmatrix}$

(c) non-zero unique solution

(d) multiple solutions. **GATE-2013**

19. For the set of equations

$x_1 + 2x_2 + x_3 + 4x_4 = 2$

$3x_1 + 6x_2 + 3x_3 + 12x_4 = 6$

The following statement is true **GATE-2015**

(a) Only the trivial solution $x_1 = x_2 = x_3 = x_4 = 0$ exists.

(b) There are no solution

(c) A unique non-trivial solution exists

(d) Multiple non-trivial solution exist

20. Let A be 3 × 3 matrix and consider the system of equations, $AX = \begin{bmatrix} 1 \\ 0 \\ -1 \end{bmatrix}$ then

(a) If the system is consistent, then it has a unique solution.

(b) If A is singular, then the system has infinitely many solutions.

(c) If the system is consistent, then $|A| \neq 0$

(d) If the system has a unique solution, then A is non-singular

21. Let P be a 3 × 3 matrix such that for some C, the linear system PX = C has infinite number of solutions. Which one of the following is true?

(a) The linear system PX = B has infinite number of solutions for all B

(b) Rank (P) = 3

(c) Rank (P) ≠ 1

(d) Rank (P) ≤ 2

22. Let A be a 4 × 3 matrix whose columns form a linearly independent set, which conclusion is justified i.e., which are true?

(a) The set of rows in A is linearly independent.

(b) The equation Ax = b is consistent for every b in R^4.

(c) The equation Ax = 0 has a non-trivial solution.

(d) There is a matrix B s.t. $AB = I_4$

Answer Key

1	2	3	4	5	6	7	8	9	10
d	d	b,c	c	a,b	c	c	a	a,c	a,b,d
11	12	13	14	15	16	17	18	19	20
d	a	d	c	b	b	c	d	d	d
21	22								
d	a								

Chapter 5

Eigenvalues & Eigenvectors

1. Eigen value :

Let V be a vector space over a field(F) and T : V(F) → V(F) be a linear operator Then $\lambda \in F$ is an eigen value.

Iff (i) $\exists\, 0 \neq x \in V$ s.t. $T(x) = \lambda x$

(ii) $(T - \lambda I)x = 0$

(iii) $x \in \ker(T - \lambda I)$, $x \neq 0$

(iv) $\eta(T - \lambda I) \geq 1$

(v) $(T - \lambda I)$ is not 1 – 1

(vi) $(T - \lambda I)$ is singular.

(vii) $\rho(T - \lambda I) < n$

2. Eigen Vector :

Let A be a matrix of order n ×n and λ be a eigen value of A then $\exists\, 0 \neq x \in V$ s.t. $Ax = \lambda x$

This non-zero vector $x \in V$ is called eigen vector corresponding to λ.

Example : Let $A = \begin{bmatrix} 1 & 2 \\ 0 & 1 \end{bmatrix}$

$$C_A(x) = \begin{vmatrix} 1-x & 2 \\ 0 & 1-x \end{vmatrix} = (1-x)^2 = 0$$

$\Rightarrow x = 1, 1$ $\Rightarrow x = 1$ is an eigen value.

Now $(A - I)x = 0$

$$\Rightarrow \begin{bmatrix} 0 & 2 \\ 0 & 0 \end{bmatrix}\begin{bmatrix} x_1 \\ x_2 \end{bmatrix} = \begin{bmatrix} 0 \\ 0 \end{bmatrix} \qquad \Rightarrow x_2 = 0$$

So, $X = \begin{bmatrix} k \\ 0 \end{bmatrix}$ is eigen vector corresponding to $\lambda = 1$. Where $k \neq 0$

3. Characteristic polynomial :

Let A be square matrix of order n then $C_A(x) = |A - xI|$ is a polynomial of degree n called the characteristic polynomial of A.

Note : Let V be a finite dimension Vector space and T : V(F) → V(F) be Linear operator if A be matrix representation of T corresponding to order basis β of V(F) then $\lambda \in F$ is an eigen value of A iff λ is root of characteristic polynomial.

4. Caley-Hamilton Theorem :

Every square matrix satisfies its own characteristic equation.

Example : Characteristic equation of $A = \begin{bmatrix} 2 & -1 & 1 \\ -1 & 2 & -1 \\ 1 & -1 & 2 \end{bmatrix}$

$$|A - \lambda I| = \begin{vmatrix} 2-\lambda & -1 & 1 \\ -1 & 2-\lambda & -1 \\ 1 & -1 & 2-\lambda \end{vmatrix} = \lambda^3 - 6\lambda^2 - 9\lambda - 4 = 0$$

$$\Rightarrow A^3 - 6A^2 - 9A - 4I = 0$$

By CaleyHemilton theorem

5. Algebraic multiplicity(A.M.)& Geometric multiplicity (G.M) :

(i) Algebraic multiplicity :

Let T : V(F) → V(F) be a Linear operator. If $\lambda \in F$ be an eigen value.

Then A.M. of λ is = multiplicity of λ in characteristic equation $C_T(x) = 0$.

Example : Let $A = \begin{bmatrix} 3 & 2 \\ 0 & 3 \end{bmatrix}$

The characteristic equation of A is

$$|\lambda I - A| = 0 \Rightarrow \begin{vmatrix} \lambda - 3 & -2 \\ 0 & \lambda - 3 \end{vmatrix} = 0 \Rightarrow (\lambda - 3)^2 = 0 \Rightarrow \lambda = 3, 3$$

∴ Algebraic multiplicity of eigenvalues '3' is 2.

(ii) Geometric multiplicity :

Number of L.I. eigen vector corresponding to any eigen value is called a geometric multiplicity of this eigen value.

6. Eigen Space :

The set of all eigen vectors corresponding to λ is called eigen space corresponding to λ and denoted by E_λ

$E_\lambda = \{x \in R^n \mid Ax = \lambda x\}$

$\Rightarrow E_\lambda = \{x \in R^n \mid (A - \lambda I)x = 0\}$

$\Rightarrow E_\lambda = \ker(A - \lambda I)$

$\dim(E_\lambda) = \eta(A - \lambda I) = \text{nullity}(A - \lambda I)$ = geometric multiplicity of λ.

Example: Let $A = \begin{bmatrix} 1 & 2 & 7 \\ 0 & 4 & 6 \\ 0 & 0 & 5 \end{bmatrix}$ then eigen value of A are 1 , 4 ,5

(i) Eigen space corresponding to 1 is

$E_1 = \{X \in R^3 \mid (A - I)X = 0\}$

$$(A-I)X = \begin{bmatrix} 0 & 2 & 7 \\ 0 & 3 & 6 \\ 0 & 0 & 4 \end{bmatrix}\begin{bmatrix} x \\ y \\ z \end{bmatrix} = 0 \qquad \Rightarrow z = 0,\ 2y + 7z = 0 \text{ and } 3y + 6z = 0$$

$\Rightarrow y = 0$ and $z = 0$ $\qquad \Rightarrow E_1 = \{(x, y, z) \in R^3 \mid y = 0,\ z = 0\}$

$\text{Dim}E_1 = 1 = \eta(A - I)$

(ii) Eigen space corresponding to 4 is

$E_4 = \{X \in R^3 \mid (A - 4I)X = 0\}$

$$(A-4I)X = \begin{bmatrix} -3 & 2 & 7 \\ 0 & 0 & 6 \\ 0 & 0 & 1 \end{bmatrix}\begin{bmatrix} x \\ y \\ z \end{bmatrix} = 0 \qquad \Rightarrow z = 0,\ -3x + 2y + 7z = 0$$

$\Rightarrow z = 0$ and $-3x + 2y = 0$ $\qquad \Rightarrow E_4 = \{(x, y, z) \in R^3 \mid -3x + 2y = 0,\ z = 0\}$

$\text{Dim}E_4 = 1 = \eta(A - 4I)$

(iii) Eigen space corresponding to 5 is

$E_5 = \{X \in R^3 \mid (A - 5I)X = 0\}$

$$(A-5I)X = \begin{bmatrix} -4 & 2 & 7 \\ 0 & -1 & 6 \\ 0 & 0 & 0 \end{bmatrix}\begin{bmatrix} x \\ y \\ z \end{bmatrix} = 0 \qquad \Rightarrow -y + 6z = 0,\ -4x + 2y + 7z = 0$$

$\Rightarrow E_5 = \{(x, y, z) \in R^3 \mid -y + 6z = 0,\ -4x + 2y + 7z = 0\}$

$\text{Dim}E_5 = 1 = \eta(A - 5I)$

Preposition on eigen values and eigen vectors

Let $T : V(F) \to V(F)$ be a linear operator and $\lambda \& \mu$ are distinct eigen values

(i) $E_\lambda \cap E_\mu = \{0\}$
i.e. an eigen vector cannot corresponding to two distinct eigen values.

(ii) Let x & y be two eigen vectors corresponding to distinct eigen values $\lambda \& \mu$. Then $\alpha x + \beta y$ never eigen vector corresponding to any eigen value.

(iii) Eigen vector corresponding to distinct eigen value are Linear independent

(iv) If every $0 \neq x \in V$ is an eigen vector i.e. if every non-zero vector is an eigen vector then (a) T has unique eigen value (b) $T = \lambda I$

(v) If λ is eigen value of Linear operator T. then $f(\lambda)$ is eigen value of $f(T)$ for any polynomial $f(x)$ over f

(vi) Sum of all eigen values of A = Trace (A)

(vii) Product of all eigen values of A = det (A)

(viii) If λ is eigen value of A then Eigen value of A^{-1} is λ^{-1}

(ix) If λ is eigen value of A then Eigen value of αA is $\alpha\lambda$, where α is scalar

(x) If λ is eigen value of A then Eigen value of A^n is λ^n, $n \in N$

(xi) Eigen value of $adj(A) = \frac{|A|}{\lambda}$.

Example : Let $A = \begin{bmatrix} 2 & 1 \\ -1 & 4 \end{bmatrix}$ then eigen values are 3 and 3

$|A| = 9$

Example : Let $A = \begin{bmatrix} 2 & 1 \\ -1 & 4 \end{bmatrix}$ then eigen values are 3 and 3

7. Minimal Polynomial

Annihilating Polynomial : If f(x) is a polynomial in an indeterminate x and A is a square matrix of order n. If f(A) = 0, then we say that the polynomial f(x) annihilates the matrix A.

Monic Polynomial : A polynomial n'x' in which the coefficient of the highest power of x is unity is called a monic polynomial.

Minimal polynomial : The monic polynomial of lowest degree that annihilates a matrix A is called the minimal polynomial of A. and it is denoted by m(x). Also if f(x) is the minimal polynomial of A, the equation f(x) = 0 is called the minimal equation of the matrix A.

Example : Let $A = \begin{bmatrix} 1 & 0 & 0 \\ 0 & 1 & 0 \\ 0 & 0 & 1 \end{bmatrix}$ then $ch(x) = (x - 1)^3$

By Caley – Hamilton theorem $ch(A) = (A - I)^3 = 0$

But we know that A – I = 0

So , m(x) = (x – 1) is minimal polynomial

Remarks :

(i) The minimal polynomial of a matrix is a divisor of every polynomial that annihilates this matrix.

(ii) The minimal polynomial of a matrix is a divisor of the characteristic polynomial of that matrix.

(iii) Every root of the minimal polynomial of a matrix is also a characteristic root of the matrix.

(iv) Suppose A is lower (upper) triangular block matrix with diagonal block $A_1, A_2, \ldots, A_n$, then $\det A = \det A_1 . \det A_2 . \det A_3 \ldots\ldots \det A_n$.

(v) Suppose A is block diagonal matrix with diagonal blocks $A_1, A_2, \ldots, A_n$ then characteristic polynomial of $A = C_{A_1}(x) C_{A_2}(x) \ldots\ldots\ldots\ldots C_{A_n}(x)$, where $C_{A_i}(x)$ is the characteristic polynomial of A_i and minimal polynomial of $A = lcm(m_{A_1}(x), m_{A_2}(x), \ldots., m_{A_n}(x))$, where $m_{A_i}(x)$ is theminimalpolynomial of A_i.

Example. Let $A \in M_3(R)$ be s.t. $A^8 = I_{3\times 3}$ then **CSIR NET JUNE 2018**

(a) Minimal polynomial of A can only be degree 2

(b) Minimal polynomial of A can only be of degree 3

(c) Either A = I or A = -I (d) There are uncountably many A satisfying the above

Solution : (d)

Given that $A^8 = I$ then annihilating polynomial is $x^8 - 1 = 0$ satisfy by A

$(x^4 - 1)(x^4 + 1)$

$(x^2 - 1)(x^2 + 1)(x^4 + 1) = 0$

$\Rightarrow x^8 - 1 = (x - 1)(x + 1)(x^2 + 1)(x^4 + 1)$

We know that the minimal polynomial is divisor of annihilating polynomial.

Option (a) is incorrect.

If $m(x) = (x - 1)(x^2 + 1)$

Option (b) is incorrect If

$m(x) = x - 1$

Option (c) is incorrect.

$$\text{if } A = \begin{bmatrix} 1 & 0 & 0 \\ 0 & -1 & 0 \\ 0 & 0 & 1 \end{bmatrix}$$

Option (d) is correct.

8. Different Matrices and Their Eigenvalues

1.	Real symmetric	$A^T = A$	All real λ's
2.	Real skew-symmetric	$A^T = -A$	Zero or purely imaginary
3.	Hermitian	$A^\theta = A$	All real λ's
4.	Skew-Hermitian	$A^\theta = -A$	Zero or purely imaginary
5.	Nilpotent	$A^n = 0$ for some $n \in N$	All λ, λ = 0
6.	Real orthogonal matrix	$A^T = A^{-1}$	All λ, |λ| = 1
7.	Involutory	$A^2 = I$	All λ = -1 or 1
8.	Idempotent	$A^2 = A$	All λ = 0 or 1
9.	Positive definite matrix	$x^TAx > 0$	All λ > 0
10.	Positive semi definite matrix	$x^TAx \geq 0$	All λ ≥ 0
11.	Negative definite matrix	$x^TAx < 0$	All λ < 0
12.	Negative semi definite matrix	$x^TAx \leq 0$	All λ ≤ 0
13.	Rank-one matrix	$A = uv^t$	$\lambda = v^tu$(trace A), 0, 0, ..., 0

14.	Inverse	A^{-1}	$\frac{1}{\lambda}$, where 'λ' is E.V. of A
15.	Shift	$A + cI$	$\lambda + c$
16.	Similar	$B = P^{-1}AP$	$\lambda(B) = \lambda(A)$ i.e. Eigenvalues of A and B are same.
17.	Unitary matrix	$AA^{\theta} = I = A^{\theta}A$ or $A^{\theta} = A^{-1}$	All $\|\lambda\| = 1$
18.	Polynomial matrix of A	P(A)	P(λ), where 'λ' is an Eigenvalue of A.

9. Similarity of Matrices

A square matrix 'A' of order n is similar to a square matrix 'B' of order n if $A = P^{-1}BP$ for some non-singular n ×n matrix P.

Note 1 : If the matrix 'A' is similar to the matrix B, then A has the same eigenvalues as B. If X is an eigenvector of B, then $Y = P^{-1}X$ is an eigenvector of A corresponding to the same eigenvector.

Note 2 : If two matrices A and B are similar matrices, then both A and B have same rank, determinant, trace, eigenvalues, characteristic polynomials, minimal polynomial and moreover Jordan form of A and B are also same (discussed later on)

Some Important Matrix :

(1) Idempotent matrix : A square matrix A is called idempotent iff$A^2 = A$

(a) Annhilating polynomial of $A = x^2 - x = x(x - 1)$

(b) Minimal polynomial can be x or (x – 1) or x(x – 1) in each case it contains non-repeated linear factor.

(c) It is diagonalizable matrix over any field.

(d) Eigen value of A are either 0 or 1 or both.

(e) A = diag [0, 0, ..., 0, 1, 1, ...1]

(f) $Tr(A) = \rho(A)$ and $\eta(A) = n - Tr(A)$

(g) $C_A(x) = x^{n-Tr(A)}(x-1)^{Tr(A)}$

(2) Invalutory Matrix :A square matrix A is called invalutoryiff $A^2 = I$

(a) Annhilating polynomial of $A = (x^2 - 1) = (x - 1)(x + 1)$

(b) Minimal polynomial of A can be (x – 1) or (x + 1) or (x – 1)(x + 1) in each case it contains non-repeated linear factor.

(c) It is diagonalizable matrix over any field.

(d) A = diag [-1, -1, ..., -1, 1, 1, ..., 1]

(e) Eigen value of A either 1 or -1 or both.

i.e. A = diag[-1,-1...1,1,1]

(f) $C_A(x) = (x-1)^{\frac{n+Tr(A)}{2}}(x+1)^{\frac{n-Tr(A)}{2}}$

(3) $A = a^Tb$ where $a = (a_1, a_2, \ldots, a_n) \in R^n$ and $b = (b_1, b_2, \ldots, b_n) \in R^n$
Then
(a) $\rho(A) = 1$ and $\eta(A) = n - 1$
(b) $\eta(A) = n - 1 = \text{G.M. of } (\lambda = 0) = \text{A.M. of } (\lambda = 0)$
(c) Eigen values either 0 or Tr(A)
(d) $C_A(x) = (x)^{n-1} (x - Tr(A))$
(e) $M_A(x) = x(x - Tr(A))$

(4) **Nilpotent Matrix :** Let A be a matrix and A is said to be a nilpotent matrix if $\exists$ $k \in N$ such that $A^k = 0$ and the smallest k is said to be index of matrix.
(a) All eigen value are 0.
(b) $C_A(x) = x^n$ and $m_A(x) = x^k = x^{index}$
(c) Nilpotent matrix A is diagonalizable iff A = 0
(d) If $A \neq 0$ and nilpotent then it never diagonalizable.

10. Permutation Matrix

$A = [a_{ij}] : \sum_{i=1}^{n} a_{ij} = 1 \ \forall j = 1,2,\ldots,n$ where $a_{ij} \in \{0, 1\}$ such that $\sum_{i=1}^{n} a_{ij} = 1 \ \forall j = 1,2,\ldots,n$

and $\sum_{j=1}^{n} a_{ij} = 1 \ \forall i = 1,2,\ldots,n$

That is a permutation matrix

Example : $A = \begin{bmatrix} 0 & 1 & 0 \\ 1 & 0 & 0 \\ 0 & 0 & 1 \end{bmatrix}_{3\times3}$

(a) Each permutation matrices of order n correspondence to $\sigma \in S_n$. i.e. A is corresponding to $\sigma = (12) \in S_3$.
(b) Number of permutation matrices of order n are n!.
(c) $\det(A) = \text{sig.}(\sigma)$ where $\sigma \in S_n$ is permutation corresponding to A.
i.e. in above examples
$\sigma = (12)$
$\Rightarrow \quad sig(\sigma) = (-1)^{\text{Number of transposition.}}$
$= (-1)^1 = (-1)$
$= \det(A) = -1$
(d) Tr(A) = Number of symbols which are fixed (self-image in σ)
$\sigma = (12) \in S_3 \Rightarrow \sigma = (12)(3) \in S_3$
Number of self image symbol is 1.
$\Rightarrow Tr(A) = 1$
(e) A is orthogonal/unitary matrix.
(f) A is diagonalizable over complex field.

(h) If $0(\sigma) = k$ then $A^k = I$
in above example.
$\sigma = (12) \in S_3$
$0(\sigma) = 2 \quad\quad \Rightarrow A^2 = I$

(i) $\sigma = c_1.c_2. \ldots\ldots c_k$ product of disjoint cycles such that $l(c_i) = r_i$ where $l(c_i)$ = length of c_i.
Then characteristic of A is $c(x) = \prod_{ri}(x^{r_i} - 1)$

i.e. $\sigma = (12)(3) \in S_3$
$c_1 = (12)$ and $l(c_1) = 2 = r_1$
$c_2 = (3)$ and $l(c_2) = 1 = r_2$ then $c(x) = (x^2 - 1)(x - 1)$

(j) The A.M. of each eigen value is same as G.M.

(k) A is diagonalizable over real field iffσ is either or order one or of order 2.

e.g. $$A = \begin{bmatrix} 0 & 0 & 0 & 1 \\ 0 & 0 & 1 & 0 \\ 0 & 1 & 0 & 0 \\ 1 & 0 & 0 & 0 \end{bmatrix}_{4\times 4}$$

then $\sigma = (14)(23) \in S_4$
$0(\sigma) = 2$
Then A is diagonalizable over R
(1) $A^2 = I$
(2) $\det A = 1 = (-1)^2$
(3) $c(x) = (x^2 - 1)(x^2 - 1)$
(4) $Tr(A) = 0$

SOLVED QUESTIONS

1. Let $A = [a_{ij}]_{3\times 3}$, $a_{ij} \in R$. Then which of the following is false? **CSIR NET JUNE 2023**
(a) A must have real eigen value (b) If $|A| = 0$ then 0 is an eigen value.
(c) If $|A|$ is negative & 3 is an eigen value then A must have three real eigen value
(d) If $|A|$ is positive & 3 is an eigen value then A must have three real eigen value

Ans. **(d)**

Explanation :
(a) Odd degree polynomial has atleast one real root
(b) If $|A| = -$ then 0 must be eigen value
(c) Given that 3 is eigen value.
Suppose $a \pm ib$ are another complex eigen value then $|A| = 3(a^2 + b^2)$ which is always positive.
Hence our assumption is wrong.

(d) Let $A = \begin{bmatrix} 0 & 1 & 0 \\ -1 & 0 & 0 \\ 0 & 0 & 3 \end{bmatrix}$

We know that $|A| > 0$ and 3 is eigen value but eigen value are 3, $\pm i$.
Hence (d) is false.

2. Let T be a linear operator on R and $f(x) \in R[x]$ is its characteristic polynomial. Consider two statements.

I. Suppose T is non-zero & 0 is an eigen value of T.
If we write $f(x) = x\, g(x)$ in $R[x]$, then the linear operator $g(T)$ is zero.

II. Suppose 0 is an eigen value of T with atleast 2 L.I. eigen vector.
If we write $f(x) = x\, g(x)$ in $R[x]$, then linear operator $g(T)$ is zero.

Which of the following is true? **CSIR NET JUNE 2023**

(a) Only I (b) Only II (c) Both I & II (d) Neither I nor II

Ans. (d)

Explanation :

Statement – I : Given that 0 is eigen value of T

Suppose $C_A(x) = f(x) = x(x+1)^2$

Hence $A = \begin{bmatrix} 0 & 0 & 0 \\ 0 & -1 & 0 \\ 0 & 0 & -1 \end{bmatrix}$ and $g(x) = (x+1)^2$

$g(T) = (T + I)^2 \neq 0$

Statement – II : Suppose 0 is an eigen value of T with atleast 2 L.I. eigen vector.

Let $\begin{bmatrix} 0 & 0 & 0 \\ 0 & 0 & 0 \\ 0 & 0 & 2 \end{bmatrix}$. Then $f(x) = x^2(x-2)$ and $g(x) = x(x-2)$

$g(T) = T(T - 2I)$

$= \begin{bmatrix} 0 & 0 & 0 \\ 0 & 0 & 0 \\ 0 & 0 & 2 \end{bmatrix} \begin{bmatrix} -2 & 0 & 0 \\ 0 & -2 & 0 \\ 0 & 0 & 0 \end{bmatrix}$

$= \begin{bmatrix} 0 & 0 & 0 \\ 0 & 0 & 0 \\ 0 & 0 & 0 \end{bmatrix}$

3. Let A be a 4×4 matrix such that -1, 1, 1, -2 are its eigenvalues. If $B = A^4 - 5A^2 + 5I$, then trace$(A + B)$ equals **CSIR NET FEB 2022**

(a) 0 (b) -12 (c) 3 (d) 9

Ans. (c)

Explanation :

Here B = $A^4 - 5A^2 + 5I$

So, A + B = $A^4 - 5A^2 + A + 5I$

And given that eigen values of A are -1, 1, 1, -2.

Then p(A) = A + B and eigen values of p(A) are p(-1), p(1), p(1) and p(-2) and

p(-1) = 1 – 5 – 1 + 5 = 0

p(1) = 1 – 5 + 1 + 5 = 2,

p(-1) = 16 – 20 – 2 + 5 = -1

Thus eigen values of A + B are 0, 2, 2, -1 and sum of eigen values is equal to trace of a matrix. Hence trace (A + B) = 0 + 2 + 2 – 1 = 3.

So, option (c) is correct answer.

4. Let A and B be n ×n matrices. Suppose the sum of the elements in any row of A is 2 and the sum of the elements in any column of B is 2. Which of the following matrices is necessarily singular? **CSIR NET FEB 2022**

(a) $I-\frac{1}{2}BA^T$ (b) $I-\frac{1}{2}AB$ (c) $I-\frac{1}{4}AB$ (d) $I-\frac{1}{4}BA^T$

Ans. (d)

Explanation :

Sum of elements in any row of A is 2.

Sum of elements in any columns of B is 2.

then A and B has two eigen values as 2

Suppose another eigen value of A and B are 0 and 2

then eigen value of A are 0 and 2

eigen value of B are 2 and 2

(a) eigen value of $I-\frac{1}{2}BA^T$ are 1 -0 and 1 – 2 $\Rightarrow I-\frac{1}{2}BA^T$ isnon singular

(b) eigen value of $I-\frac{1}{2}AB$ are 1 -0 and 1 – 2 $\Rightarrow I-\frac{1}{2}AB$ isnon singular

Another method :

Let $A=\begin{bmatrix}1 & 1\\ 1 & 1\end{bmatrix}$ and $B=\begin{bmatrix}2 & 0\\ 0 & 2\end{bmatrix}$.

A has eigen values 2 and 0.

B has eigen values 2 and 2.

$AB=\begin{bmatrix}2 & 2\\ 2 & 2\end{bmatrix}$ and $BA^T=\begin{bmatrix}2 & 2\\ 2 & 2\end{bmatrix}$.

$I-\frac{1}{2}BA^T=\begin{bmatrix}0 & -1\\ -1 & 0\end{bmatrix}$ & $I-\frac{1}{2}AB=\begin{bmatrix}0 & -1\\ -1 & 0\end{bmatrix}$

and both are non-singular.

So, option (a) and (b) are incorrect.
Now, for option (c) :

Let $A=\begin{bmatrix}2 & 0\\ 1 & 1\end{bmatrix}$ and $B=\begin{bmatrix}1 & 0\\ 1 & 2\end{bmatrix}$.

$$I-\frac{1}{4}AB=\begin{bmatrix}1/2 & 0\\ -1/2 & 1/2\end{bmatrix}$$

$\therefore$ $I-\frac{1}{4}AB$ is not singular.

$\therefore$ Option (c) is incorrect.

(d) Given that sum of the elements in any row of A is 2 and the sum of the elements in any column of B is 2
then sum of the elements in any column of A^T is 2
Hence $A^TX = 2X$ and $BX = 2X$

Now $$\left(I-\frac{1}{4}BA^T\right)X = IX-\frac{1}{4}(BA^TX) = X-\frac{1}{4}B(2X) = X-\frac{1}{2}BX$$
$$= X - X = 0$$

So, 0 is eigen value of $I-\frac{1}{4}BA^T$.

Hence, $I-\frac{1}{4}BA^T$ is always singular and option (d) is correct.

5. Let A be an n ×n matrix of rank 1. Let $\alpha = \det(I + A)$, where I is the identity matrix and let β = trace A. Which of the following is true? **CSIR NET NOV 2020**

(a) $\beta - \alpha = 1$ (b) $\alpha - \beta = 1$ (c) $\alpha<\beta + 1$ (d) $\alpha>\beta + 1$

Ans. **(b)**

Explanation :

For (a), (c) and (d)

Let $A=\begin{bmatrix}0 & 1\\ 0 & 0\end{bmatrix}$, then $I+A=\begin{bmatrix}1 & 1\\ 0 & 1\end{bmatrix}$

$\Rightarrow$ $\alpha = 1, \beta = 0$

And we can see all the options (a), (c) and (d) are incorrect.

For option (b) :

Given that A is a n ×n matrix with rank 1 and we know that rank of a matrix is atleast equals to number of non-zero eigenvalues.

Thus here A has only one non-zero eigenvalues say λ.

Thus spectrum of A is $\lambda, \underbrace{0,0,\ldots,0}_{\text{n-1 times}}$ here λ can also be zero.

Now eigenvalues of I + A are $1+\lambda,\underbrace{1,1,....,1}_{\text{n-1 times}}$, and

det(I + A) = (1 + λ), as det(I + A) is product of eigen values of I + A.
Now trace A = λ, as trace A is sum of eigen values of A.
So, here α = 1 + λ and β = λ⇒α - β = 1, and hence option (b) is true.

6. Let T : $C^n \to C^n$ be Linear Transformation$n \geq 2$. Suppose 1 the is only eigen value of T. Which of the following are TRUE? **CSIRNET DEC 2019**

(a) $T^k \neq I$ for any $k \in N$ (b) $(T-I)^{n-1} = 0$ (c) $(T-I)^n = 0$ (d) $(T-I)^{n+1} = 0$

Ans. **(c), (d)**

Explanation :

Let n = 2

Given that 1 is the only eigen value of T

Option (a) is incorrect.

$$T = \begin{bmatrix} 1 & 0 \\ 0 & 1 \end{bmatrix} = I \text{ for k = 1}$$

Option (b) is incorrect.

$$T = \begin{bmatrix} 1 & 1 \\ 0 & 1 \end{bmatrix}$$

$$(T-I) = \begin{bmatrix} 0 & 1 \\ 0 & 0 \end{bmatrix} \neq 0 \quad [n = 2]$$

Option (c) is correct.

Given that 1 is only eigen value of T then B = T – I has eigen value 0

$C_T(B) = (B-0)(B-0) = 0$

$= B^2 = 0$

$\Rightarrow (T-I)^2 = 0$ [for n = 2]

Option (d) is correct.

7. Let A be an n ×n matrix (with n > 1) satisfying $A^2 - 7A + 12I = 0$. Then which of the following are TRUE? **CSIR NET JUNE 2018**

(a) A is invertible (b) $t^2 - 7t + 12 = 0$ where t = Tr(A)

(c) $d^2 - 7d + 12 = 0$ where d = det(A) (d) $\lambda^2 - 7\lambda + 12 = 0$ where λ is eigen value of A

Ans. **(a), (d)**

Explanation :

By Caleyhemilton theorem , the option (d) is correct.

Given that $A^2 - 7A + 12I = 0$

$Ch_T(x) = x^2 - 7x + 12 = 0$

$$x = \frac{7 \pm \sqrt{49-48}}{2}$$

$x = \frac{7 \pm 1}{2} = 4,3$

Eigen value of A are 4 and 3 $\Rightarrow \det(A) = 4 \times 3 = 12$

$\det(A) \neq 0$

Option (a) is correct.

A is invertible.

Option (b) is incorrect.

$t = Tr(A) = 4 + 3 = 7$

then $t^2 - 7t + 12 = 49 - 49 + 12 \neq 0$

Option (c) is incorrect.

$\det(A) = d = 12$

then $d^2 - 7d + 18 = 144 - 84 + 18 \neq 0$

8. Let $T : R^n \to R^n$ be a linear map that satisfies $T^2 - T + I = 0$. Then which of the following are TRUE? **CSIR NET DEC 2018**

(a) T is invertible. (b) $T - I$ is not invertible (c) T has a real eigen value (d) $T^3 = -I$

Ans. (a),(d)

Explanation :

Given that $T^2 = T - I$

Let A be the matrix of $T : R^n \to R^n$

Then $A^2 = A - I$

Annihilating polynomial is $x^2 - x + 1 = 0$

$x = \frac{1 \pm \sqrt{1-4}}{2} = \frac{1 \pm \sqrt{3}i}{2}$ $\Rightarrow$ A has non-zero eigen value.

Option (a) is correct. $\Rightarrow |T| \neq 0$

Option (b) is incorrect.

Since $T - I$ is also invertible.

Option (c) is incorrect.

T has no real eigen value.

Option (d) is correct.

$T^2 = T - I$

$T^3 = T(T - I)$

$= T^2 - T$ $[T^2 = T - I]$

$= T - I - T = -I$

9. Let $n \geq 1$ and $\alpha, \beta \in R$ with $\alpha \neq \beta$. Suppose $A_n(\alpha, \beta) = [a_{ij}]$ is an n ×n matrix s.t.$a_{ij} = \alpha$ and $\alpha_{ij} = \beta$ for $i \neq j$, $1 \leq i, j \leq n$. Let D_n be the det $A_n(\alpha, \beta)$. Which of the following are TRUE? **CSIR NET DEC 2019**

(a) $D_n = (\alpha - \beta)D_{n-1} + \beta$ for $n \geq 2$ (b) $\dfrac{D_n}{(\alpha-\beta)^{n-1}} = \dfrac{D_{n-1}}{(\alpha-\beta)^{n-2}} + \beta$

for $n \geq 2$

(c) $D_n = (\alpha + (n-1)\beta)^{n-1}(\alpha - \beta)$ for $n \geq 2$ (d) $D_n = (\alpha + (n-1)\beta)(\alpha - \beta)^{n-1}$

for $n \geq 2$

Ans. (b), (d)

Explanation :

For n = 2

$$A_2(\alpha,\beta) = \begin{bmatrix} \alpha & \beta \\ \beta & \alpha \end{bmatrix} \& \ D_2 = \alpha^2 - \beta^2$$

$$A_1(\alpha,\beta) = [\alpha] \ \& \ D_1 = \alpha$$

Option (a) is incorrect.

$(\alpha - \beta)\alpha + \beta \quad = \alpha^2 - \alpha\beta + \beta \neq D^2 = \alpha^2 - \beta^2$

Take n = 3

$$\text{Then } A_3(\alpha,\beta) = \begin{bmatrix} \alpha & \beta & \beta \\ \beta & \alpha & \beta \\ \beta & \beta & \alpha \end{bmatrix} \& \ D_3 = \alpha^3 - 3\alpha\beta^2 + 2\beta^3$$

For (c)

$D^3 = (\alpha + 2\beta)^2(\alpha - \beta) = (\alpha^2 + 4\beta^2 + 4\alpha\beta)(\alpha - \beta)$

$= \alpha^3 + 4\alpha\beta^2 + 4\alpha^2\beta - \alpha^2\beta - 4\beta^3 + 4\alpha\beta^2 = \alpha^3 + 3\alpha^2\beta - 4\beta^3$

Option (c) is discarded,

Proof : (d)

$$\text{Let} \quad A_n(\alpha,\beta) = \begin{bmatrix} \alpha & \beta & \beta & \cdots & \beta \\ \beta & \alpha & \beta & \cdots & \beta \\ \vdots & & & & \vdots \\ \vdots & & & & \vdots \\ \beta & \beta & \beta & \cdots & \alpha \end{bmatrix}$$

$$D_n = \begin{bmatrix} \alpha & \beta & \beta & \cdots & \beta \\ \vdots & & & & \\ \vdots & & & & \\ \vdots & & & & \\ \beta & \beta & \beta & \vdots & \alpha \end{bmatrix}$$

$R_1 \to R_1$ + Remaining, R_{n-1} then

$$\begin{bmatrix} \alpha+(n-1)\beta & \alpha+(n-1)\beta & \cdots & \alpha+(n-1)\beta \\ \beta & \alpha & & \beta \\ \beta & \beta & & \vdots \\ \beta & \beta & \cdots & \alpha \end{bmatrix} = \alpha+(n-1)\beta \begin{vmatrix} 1 & 1 & \cdots & \cdots & 1 \\ \beta & \alpha & & & \beta \\ \beta & \beta & & & \beta \\ \\ \beta & \beta & \cdots & \cdots & \alpha \end{vmatrix}$$

$$= \alpha+(n-1)\beta \begin{vmatrix} 1 & 0 & 0 & 0 \\ \beta & \alpha-\beta & 0 & 0 \\ \beta & 0 & \alpha-\beta & 0 \\ \vdots & \vdots & \vdots & \vdots \\ \beta & 0 & 0 & \alpha-\beta \end{vmatrix}$$

$[\alpha+(n-1)\beta]\,[\alpha-\beta]^{n-1} = D_n$

Option (d) is correct.,

For (b)

We know that $D_n = [\alpha + (n-1)\beta]\,[\alpha - \beta]^{n-1}$

$\Rightarrow D_{n-1} = [\alpha + (n-2)\beta]\,[\alpha - \beta]^{n-2}$

$$\frac{D_{n-1}}{(\alpha-\beta)^{n-2}} = (\alpha+(n-2)\beta)$$

$$\Rightarrow \beta + \frac{D_{n-1}}{(\alpha-\beta)^{n-2}} = \beta + (\alpha+(n-2)\beta) \qquad = \beta+\alpha+n\beta-2\beta = [\alpha+(n-1)\beta]$$

$$\because \frac{D_n}{(\alpha-\beta)^{n-1}} = \alpha+(n-1)\beta$$

$$\Rightarrow \frac{D_{n-1}}{(\alpha-\beta)^{n-2}} + \beta = \frac{D_n}{(\alpha-\beta)^{n-1}}$$

Option (b) and (d) is correct.

10. Let $A = [a_{ij}]_{3\times3}$ complex matrix then which of the following are correct.

CSIR NET JUNE 2019

(a) $\det((-1)^{i+j}a_{ij}) = \det A$ (b) $\det((-1)^{i+j}a_{ij}) = -\det A$

(c) $\det((\sqrt{-1})^{i+j}a_{ij}) = \det A$ (d) $\det((\sqrt{-1})^{i+j}a_{ij}) = -\det A$

Ans. (a), (c)

Explanation :

Let $A = \begin{bmatrix} a & b & c \\ d & e & f \\ g & h & k \end{bmatrix}_{3\times3}$

Then $|A| = a(ek - hf) - b(dk - gf) + c(dh - eg)$

$$\Rightarrow (-1)^{i+j} a_{ij} = \begin{bmatrix} a & -b & c \\ -d & e & -f \\ g & -h & k \end{bmatrix}$$

then $\det((-1)^{i+j}a_{ij}) =$

$a(ek - hf) + b(-dk + gf) + c(dh - eg)$

$= a(ek - hf) - b(dk - gf) + c(dh - eg)$

$= |A|$

Option (a) is correct.

& $$\left(\left(\sqrt{-1}\right)^{i+j} a_{ij}\right) = \begin{bmatrix} -a & -ib & c \\ -id & e & if \\ g & ih & -k \end{bmatrix}$$

Then $\det\left(\left(\sqrt{-1)^{i+j}}\, a_{ij}\right)\right)$

$= -a\,[-ek - (i)^2hf] + ib\,[idk - igf] + c[dh - gi]$

$= -a[-ek + hf] + (i)^2b\,[dk - gt] + c(dh - eg)$

$= a(ek - hf] - b[dk - gf] + c[dh - eg]$

$= \det A$

11. Which of the following eigen values of the matrix. **CSIR NET DEC 2014**

$$\begin{bmatrix} 0 & 0 & 0 & 1 & 0 & 0 \\ 0 & 0 & 0 & 0 & 1 & 0 \\ 0 & 0 & 0 & 0 & 0 & 1 \\ 1 & 0 & 0 & 0 & 0 & 0 \\ 0 & 1 & 0 & 0 & 0 & 0 \\ 0 & 0 & 1 & 0 & 0 & 0 \end{bmatrix}$$

(a) 1 (b) -1 (c) i (d) –i

Ans. (a), (b)

Explanation :

This is a permutation matrix $\sigma = (14)(25)(36)$ then $c(x) = (x^2 - 1)(x^2 - 1)(x^2 - 1)$

Eigen value of A are +1, -1

12. Given that the permutation $\sigma = \begin{pmatrix} 1 & 2 & 3 & 4 & 5 \\ 3 & 1 & 2 & 5 & 4 \end{pmatrix}$. Matrix A is denoted to be the one whose i^{th} column is the $\sigma(i)^{th}$ column of the identity matrix I. Which of the following is correct? **CSIR NET JUNE 2014**

(a) $A^2 = A$ (b) $A^{-4} = A$ (c) $A = A^{-5}$ (d) $A = A^{-1}$

Ans. (c)

Explanation :

Given that $\sigma = \begin{pmatrix} 1 & 2 & 3 & 4 & 5 \\ 3 & 1 & 2 & 5 & 4 \end{pmatrix}$

$\sigma = (1\ 3\ 2)\ (4\ 5)$

$O(\sigma) = \text{lcm}\ \{0(1\ 3\ 2), 0(4\ 5)\} = \text{lcm}\ \{3, 2\} = 6$ i.e. $0(\sigma) = 6 \Rightarrow \quad A^6 = I$

Then $\quad A = A^{-5}$

Let $\sigma = (1\ 3\ 5)\ (2\ 6)\ (4) \in S_6$

Then $A = \left[C_{\sigma(1)} \quad C_{\sigma(2)} \quad C_{\sigma(3)} \quad C_{\sigma(4)} \quad C_{\sigma(5)} \quad C_{\sigma(6)}\right]$

$$A = \left[C_3 \quad C_6 \quad C_5 \quad C_4 \quad C_1 \quad C_2\right]$$

i.e. $$A = \begin{bmatrix} 0 & 0 & 0 & 0 & 1 & 0 \\ 0 & 0 & 0 & 0 & 0 & 1 \\ 1 & 0 & 0 & 0 & 0 & 0 \\ 0 & 0 & 0 & 1 & 0 & 0 \\ 0 & 0 & 1 & 0 & 0 & 0 \\ 0 & 1 & 0 & 0 & 0 & 0 \end{bmatrix}_{6\times 6}$$

13. Let A be a 3 × 4 matrix and B be a 4 × 3 matrix with real entries such that AB is non-singular. Consider the following statements :

P : Nullity of A is 0.

Q : BA is a non-singular matrix.

Then

(a) both P and Q are true
(b) P is true and Q is false
(c) P is false and Q is true
(d) Both P and Q are false

Ans. (d) **GATE – 2021**

Explanation :

We know that $\rho(AB) \le \min\ \{\rho(A), \rho(B)\}$

Given that AB is non-singular.

$\Rightarrow \quad \rho(AB) = 3 \quad \Rightarrow \quad \rho(A) = \rho(B) = 3 \quad \Rightarrow \quad \eta(A) = 1\ \&\eta(B) = 0$

Now $\quad \rho(AB) = \rho(BA)$

$\Rightarrow \quad \rho(BA) = 3$ but BA is a matrix of 4 order.

$\Rightarrow \quad$ BA is singular matrix.

14. If $A = \begin{bmatrix} 3 & -2 \\ 2 & -1 \end{bmatrix}$, then A^{20} equals.

(a) $\begin{bmatrix} 41 & 40 \\ -40 & -39 \end{bmatrix}$ (b) $\begin{bmatrix} 41 & -40 \\ 40 & -39 \end{bmatrix}$ (c) $\begin{bmatrix} 41 & -40 \\ -40 & -39 \end{bmatrix}$ (d) $\begin{bmatrix} 41 & 40 \\ 40 & -39 \end{bmatrix}$

Ans. (b) **CSIRNET JUNE 2020**

Explanation :
Here each row sum of A is 1, so, 1 is an eigen value of A and each row sum of A^{20} will be $1^{20} = 1$. So, we can eliminate option (a), (c) and (d) as the row sum not equal to one. So, option (b) is correct.

15. Let A be a 2 × 2 real matrix with det A = 1 and trace (A) = 3. What is the value of trace (A^2)?

(a) 2 (b) 10 (c) 9 (d) 7

Ans. (d) **CSIR NET JUNE 2020**

Explanation :
Consider $A = \begin{bmatrix} 2 & 1 \\ 1 & 1 \end{bmatrix}$, clearly det A = 1 and trace (A) = 3, $A^2 = \begin{bmatrix} 5 & 3 \\ 3 & 2 \end{bmatrix}$, trace ($A^2$) = 7, we can eliminate option (a), (b) and (c). So, option (d) is correct.

Or Characteristic equation of A is $\lambda^2 - tr(A)\lambda + det(A) = 0 \Rightarrow \lambda^2 - 3\lambda + 1 = 0$

$\Rightarrow \lambda_1 = \dfrac{3+\sqrt{5}}{2}$, $\lambda_2 = \dfrac{3-\sqrt{5}}{2}$. So, eigenvalues of A^2 will be λ_1^2 and λ_2^2, so, trace (A^2) =

$\lambda_1^2 + \lambda_2^2 = \left(\dfrac{3+\sqrt{5}}{2}\right)^2 + \left(\dfrac{3-\sqrt{5}}{2}\right)^2 = 7$, option (d) is correct.

Or we have, $(\lambda_1 + \lambda_2)^2 = \lambda_1^2 + \lambda_2^2 + 2\lambda_1\lambda_2 \Rightarrow 3^2 = \lambda_1^2 + \lambda_2^2 + 2 \times 1 \Rightarrow \lambda_1^2 + \lambda_2^2 = 9 - 2 =$
7,

option (d) is correct.

16. What is the rank of the matrix $\begin{pmatrix} 1 & 1 & 1 & 1 & 1 \\ 1 & 2 & 2 & 2 & 2 \\ 1 & 2 & 3 & 3 & 3 \\ 1 & 2 & 3 & 4 & 4 \\ 1 & 2 & 3 & 4 & 5 \end{pmatrix}$. **CSIR NETJUNE – 2019**

(a) 2 (b) 3 (c) 4 (d) 5

Ans. (d)

Explanation :

$$\begin{pmatrix} 1 & 1 & 1 & 1 & 1 \\ 1 & 2 & 2 & 2 & 2 \\ 1 & 2 & 3 & 3 & 3 \\ 1 & 2 & 3 & 4 & 4 \\ 1 & 2 & 3 & 4 & 5 \end{pmatrix} \sim \begin{pmatrix} 1 & 1 & 1 & 1 & 1 \\ 0 & 1 & 1 & 1 & 1 \\ 0 & 0 & 1 & 1 & 1 \\ 0 & 0 & 0 & 1 & 1 \\ 0 & 0 & 0 & 0 & 1 \end{pmatrix}$$

Rank of this matrix is 5.

UNSOLVED QUESTIONS

1. Let $A=\begin{bmatrix}1 & -1 & 0\\ 0 & 0 & 0\\ -2 & -2 & 2\end{bmatrix}$ and $B = A^5 + A^4 + I_3$ are two matrices. Then which of the following is not eigen value of B. **IIT JAM 2023**

(a) 1 (b) 2 (c) 3 (d) 49

2. Let P, M, N be n ×n matrices such that M and N are non-singular. If x is an eigenvector of P corresponding to the eigenvalue λ, then an eigenvector of $N^{-1}MPM^{-1}N$ corresponding to the eigenvalue λ is **JAM CA-2009**

(a) $MN^{-1}x$ (b) $M^{-1}Nx$ (c) $NM^{-1}x$ (d) $N^{-1}Mx$

3. Let $T : R^3 \to R^3$ be the linear transformation whose matrix with respect to standard basis $\{e_1, e_2, e_3\}$ of R^3. Then T **JAM MA-2010**

(a) maps the subspace spanned by e_1 and e_2 into itself.

(b) has distinct eigenvalues. (c) has eigenvectors that span R^3.

(d) has a non-zero null space.

4. Let n > 1 be an integer. Consider the following two statements for an arbitrary n ×n matrix A with complex entries.

I. If $A^k = I_n$ for some integer k ≥ 1, all the eigenvalues of A are k^{th} roots of unity.

II. If, for some integer k ≥ 1, all the eigenvalues of A are k^{th} roots of unity, then $A^k = I_n$. Then **IIT JAM - 2021**

(a) I is false but II is true (b) neither I nor II is true.

(c) I is true but II is false (c) both I and II are true.

5. Let n ≥ 2 be an integer. Let $A : C^n \to C^n$ be the linear transformation defined by

$A(z_1, z_2, \ldots\ldots, z_n) = (z_n, z_1, z_2, \ldots\ldots, z_{n-1})$.

Which one of the following statements is true for every n ≥ 2? **IIT JAM - 2021**

(a) A is nilpotent (b) All eigenvalues of A are of modulus 1.

(c) Every eigenvalue of A is either 0 or 1. (d) A is singular.

6. Let $A=\begin{pmatrix}2 & -1 & 3\\ 2 & -1 & 3\\ 3 & 2 & -1\end{pmatrix}$. Then the largest eigenvalue of A is **IIT JAM - 2021**

(a) 1 (b) 2 (c) 3 (d) 4

7. Let V and W be finite dimensional vector spaces over R and let $T_1 : V \to V$ and $T_2 : W \to W$ be linear transformations whose minimal polynomials are given by $f_1(x) = x^3 + x^2 + x + 1$ and $f_2(x) = x^4 - x^2 - 2$. Let $T : V \oplus W \to V \oplus W$ be the linear transformation defined by $T((v, w)) = (T_1(v), T_2(w))$ for $(v, w) \in V \oplus W$ and let f(x) be the minimal polynomial of T. Then **CSIRNET DEC 2012**

(a) deg f(x) = 7 (b) deg f(x) = 5 (c) nullity (T) = 1 (d) nullity (T) = 0

8. Let $A=\begin{bmatrix} a & -1 & 4 \\ 0 & b & 7 \\ 0 & 0 & 3 \end{bmatrix}$ be a matrix with real entries. If the sum and the product of all the eigenvalues of A are 10 and 30 respectively, then $a^2 + b^2$ equals

(a) 29 (b) 40 (c) 58 (d) 65

9. Let A and B invertible matrix s.t. AB = -BA then **CSIR NET JUNE 2017**

(a) Trace A = Tr(B) = 0 (b) Tr(A) = Tr(B) = 1
(c) Tr(A) = 0, Tr(B) = 1 (d) Tr(A) = 1, Tr(B) = 0

10. Let $A = [a_{ij}]$ be an n ×n matrix such that a_{ij} is an integer for all i,j. Let AB = I with $B = [b_{ij}]$ (where I is the identity matrix). For a square matrix C, det C denotes its determinant. Which of the following statements is true? **CSIR NET DEC 2016**

(a)If det A = 1 then det B = 1
(b)A sufficient condition for each b_{ij} to be an integer is that det A is an integer
(c)B is always an integer matrix
(d)A necessary condition for each b_{ij} to be an integer is det A ∈ {-1, +1}

11. Let S be the set of 3 × 3 real matrices A with $A^T A = \begin{pmatrix} 1 & 0 & 0 \\ 0 & 0 & 0 \\ 0 & 0 & 0 \end{pmatrix}$. Then the set S contains

CSIR NET JUNE 2015

(a)a nilpoint matrix (b)a matrix of rank one (c)a matrix of rank two
(d)a non-zero skew-symmetric matrix

12. Let A be a 5 × 5 skew-symmetric matrix with entries in R and B be the 5 × 5 symmetric matrix whose (i, j)th entry is the binomial coefficient $\begin{pmatrix} i \\ j \end{pmatrix}$ for 1 ≤ i ≤ j ≤ 5. Consider the 10 × 10 matrix, given in block form by $C = \begin{pmatrix} A & A+B \\ 0 & B \end{pmatrix}$. Then **CSIR NET DEC 2012**

(a)det C = 1 or -1 (b)det C = 0 (c)trace of C is 0 (d)trace of C is 5

13. If A is a 5 × 5 real matrix with trace 15 and if 2 & 3 are eigen value of A, each with algebraic multiplicity is 2 & then the determinant of A is equal to

(a) 0 (b) 24 (c) 120 (d) 180

14. Let $A=\begin{bmatrix} 3 & 0 & 0 \\ 0 & 6 & 2 \\ 0 & 2 & 6 \end{bmatrix}$ and let $\lambda_1 \geq \lambda_2 \geq \lambda_3$ be the eigen value of A. The triple $(\lambda_1, \lambda_2, \lambda_3)$ equals.

(a) (9, 4, 2) (b) (8, 4, 3) (c) (9, 3, 3) (d) (7, 5, 3)

15. The minimal polynomial associated with the matrix $\begin{bmatrix} 0 & 0 & 3 \\ 1 & 0 & 2 \\ 0 & 1 & 1 \end{bmatrix}$ is

(a) $x^3 - x^2 - 2x - 3$ (b) $x^3 - x^2 + 2x - 3$
(c) $x^3 - x^2 - 2x + 3$ (d) $x^3 - x^2 + 3x - 3$

16. Let A be 3 × 3 matrix with eigenvalues 1, -1 and 3. Then

(a) $A^2 + A$ is non-singular (b) $A^2 - A$ is non-singular
(c) $A^2 + 3A$ is non-singular (d) $A^2 - 3A$ is non-singular

17. Let A be a 3 × 3 matrix with trace (A) = 3 and det(A) = 2. If 1 is an eigenvalue of A, then the eigenvalues of the matrix $A^2 - 2I$ are

(a) 1, 2 (i – 1), -2(i + 1) (b) -1, 2(i – 1), 2(i + 1)
(c) 1, 2(i + 1), -2(i + 1) (d) -1, 2(i – 1), -2(i + 1)

18. Let a, b, c, d be distinct non-zero real numbers with a + b = c + d. Then an eigenvalue of the matrix $\begin{bmatrix} a & b & 1 \\ c & d & 1 \\ 1 & -1 & 0 \end{bmatrix}$ is

(a) a + c (b) a + b (c) a – b (d) b – d

19. Let $A = \begin{bmatrix} 0 & 1-i \\ -1-i & i \end{bmatrix}$ and $B = A^T.\overline{A}$. Then

(a) An eigenvalue of B is purely imaginary (b) An eigenvalue of A is zero
(c) All eigenvalues of B are real (d) A has a non-zero real eigenvalue

Answer Key

1	2	3	4	5	6	7	8	9	10
b	d	c	d	b	d	b,d	a	a	a,d
11	**12**	**13**	**14**	**15**	**16**	**17**	**18**	**19**	
a,b	b,d	d	b	a	c	d	b	c	

Chapter 6

Diagonalizable Matrix and Jordan Canonical Form

1. Diagonalization of Matrix

Definition :

(1) A matrix is said to be diagonalizable matrix if it is similar to a diagonal matrix. i.e. A is diagonalizable if $\exists$ $D = diag(d_1, d_2, \ldots, d_n)$ and a non-singular matrix P such that $A = PDP^{-1}$ or $P^{-1}AP = D$.

(2) Let V be a finite dimensional vector space and $T : V \rightarrow V$ be a linear transformation. We say T is diagonalizable if there exist a basis β of V such that matrix $[T]_\beta$ is a diagonal matrix.

Important Results :

(1) A square matrix 'A' of order $n \times n$ is diagonalizable iff
   - (i) It has 'n' linearly independent eigenvectors.
   - (ii) Geometric multiplicity of each eigenvalue is equal to its Algebraic multiplicity.
   - (iii) Minimal polynomial has distinct monic linear factors.
   - (iv) Sum of dimensions of eigen spaces corresponding to each eigen value is equal to 'n'.
   - (v) Number of Jordan blocks in Jordan Canonical form of A (Discussed later on) is equal to order of matrix.

(2) If A is diagonalizable, then A^{-1} is also diagonalizable.

(3) If A is diagonalizable, then A^T is also diagonalizable.

(4) If A is diagonalizable, then p(A) is also diagonalizable, where p(A) is polynomial matrix.

(5) If A is diagonalizable, then A + cI is also diagonalizable, where 'c' is a parameter.

(6) If all eigen value of A are distinct then A is diagonalizable.

(7) If A and B are two $n \times n$ square matrices & both are diagonalizable, then their sum and difference may not be diagonalizable.

Example : $A = \begin{bmatrix} 1 & 1 \\ 0 & 0 \end{bmatrix}$ and $B = \begin{bmatrix} -1 & 0 \\ 0 & 0 \end{bmatrix}$ both are diagonalizable matrix but

$A + B = \begin{bmatrix} 0 & 1 \\ 0 & 0 \end{bmatrix}$ is not diagonalizable matrix

(8) Product of two diagonalizable matrices may not be diagonalizable.

2. Jordan Block :

A matrix of ordern ×nin which λ on diagonal ,1 on super diagonal and other entriesare zero is called Jordan block.

Example : $A=\begin{bmatrix} \lambda & 1 & 0 \\ 0 & \lambda & 1 \\ 0 & 0 & \lambda \end{bmatrix}$ is a 3 × 3 Jordan block corresponding to λ.

Jordan Canonical Form :

Jordan Block : A n ×n matrix with λ on the diagonal and 1 on super diagonal, other entries zero is called Jordan block of order n with eigen value λ.

Example : $\begin{bmatrix} \lambda & 1 \\ 0 & \lambda \end{bmatrix}$ is 2 × 2 Jordan block with eigen value λ.

Example : $\begin{bmatrix} \lambda & 1 & 0 \\ 0 & \lambda & 1 \\ 0 & 0 & \lambda \end{bmatrix}$ is 3 × 3 Jordan block with eigen value λ.

Remark :

Let A be an n ×n matrix with characteristic and minimal polynomial.

$C_A(x)=(x-\lambda_1)^{n_1}(x-\lambda_2)^{n_2}.....(x-\lambda_k)^{n_k}$ and $m_A(x)=(x-\lambda_1)^{m_1}(x-\lambda_2)^{m_2}.....(x-\lambda_k)^{m_k}$ respectively.

Rules for number of Jordan Blocks :

(i) There is atleast one Jordan block of order $m_i$corresponding to λ_i and other blocks corresponding to λ_i are of order m_i or less.

(ii) The sum of orders of all Jordan block$M_{ij} = n_i$.

(iii) The number of blocks corresponding to particular λ_i = geometric multiplicity of λ_i.

(iv) The number of blocks of each possible order is uniquely determined by transformation.

(v) If ρ(A), η(A) and index of A are known, then Jordan canonical form of A can be determined uniquely.

(vi) Nullity of a matrix = Number of Jordan blocks corresponding to eigenvalue '0'.

Example : Let A be 3×3 matrix with characteristic and minimal polynomial

$C_A(x) = (x-1)^2 (x-2)$

$m_A(x) = (x-1)^2(x-2)$

Then $A=\begin{bmatrix} 1 & 1 & 0 \\ 0 & 1 & 0 \\ 0 & 0 & 2 \end{bmatrix}_{3\times 3}$

Number of J.C. Form :

(1) Let $C_A(x)=\prod_{i=1}^{k}(x-\alpha_i)^{n_i}$ be a characteristic polynomial.

Then number of possible J.C. form is $\prod_{i=1}^{k}p(n_i)$ where $p(n_i)$ denotes the number of partition of n_i.

Example : Let $C_A(x) = (x-2)^4 (x-3)^3$ then number of J.C. form is $p(4) \times p(3) = 5\times 3 = 15$.

(2) If $Ch_A(x)=\prod_{i=1}^{k}(x-\alpha_i)^{n_i}$ and $M_A(x)=\prod_{i=1}^{k}(x-\alpha_i)^{m_i}$ and let $t_i = n_i - m_i$ then number of possible J.C. form is $\prod_{i=1}^{k}S(t_i)$, where $S(t_i)$ denotes then number of partition of t_i in which no part is greater than m_i.

Example : $C_A(x) = (x-2)^7$ and $m_A(x) = (x-2)^2$ then number of J.C. form is

$n - m = 7 - 2 = 5$

$S(5) = 3$

i.e. $2+1+1+1+1+1$ $\quad 2+2+1+1+1$ $\quad 2+2+2+1$

SOLVED QUESTIONS

1. Let A be a matrix of order 3 × 3 whose entries are real and a polynomial p(T) is characteristic poltnomial $C_A(x)$ which is divisible by T^2, then

CSIR NET JUNE 2023

(a) The eigen space of A for the eigen value 0 is 2 dimensional.
(b) All eigen values of A are real
(c) $A^3 = 0$
(d) A is diagonalizable matrix

Ans. (b)

Explanation :

(a) Suppose a = 0, then $A=\begin{bmatrix}0&0&0\\0&0&0\\0&0&0\end{bmatrix}$

G.M. of ($\lambda = 0$) is 3.

(b) Given that $a_{ij}\in$ R, then coefficient of characteristic polynomial are real. Hence a is real.

(c) For a ≠ 0, then $A=\begin{bmatrix}0&0&0\\0&0&0\\0&0&-a\end{bmatrix}$.

(d) Suppose $A=\begin{bmatrix}0&1&0\\0&0&0\\0&0&-a\end{bmatrix}$

A is not diagonalizable.

2. Let A be a block diagonal matrix of order 5 ×5 s.t. $A=\begin{bmatrix}e&1&0&0&0\\0&e&0&0&0\\0&0&e&1&0\\0&0&0&e&0\\0&0&0&0&e\end{bmatrix}_{5\times5}$.

CSIR NET JUNE 2023

(a) Algebraic multiplicity of E.V.(e) is 5
(b) A is not diagonalizable
(c) Geometric multiplicity of e is 3
(d) Geometric multiplicity of e is 2

Ans. **(a), (b), (c)**

Explanation :

(a) A.M. of e is 5

(b) A is not diagonalizable as A.M. ≠ G.M.

(c) G.M. of e is 3

3. Which of the following is not diagonalizable over R. **CSIR NET JUNE 2019**

(a) $\begin{bmatrix}2&0&1\\0&3&0\\0&0&2\end{bmatrix}$ (b) $\begin{bmatrix}1&1\\1&1\end{bmatrix}$ (c) $\begin{bmatrix}2&1&0\\0&3&0\\0&0&3\end{bmatrix}$ (d) $\begin{bmatrix}1&-1\\2&4\end{bmatrix}$

Ans. **(a)**

Explanation :

Option (a) is correct.

Let $M=\begin{bmatrix}2&0&1\\0&3&0\\0&0&2\end{bmatrix}$

A.M. of $\lambda = 2$ is 2

G.M. of $\lambda = 2$ is $\eta(A-2I)$

$$A-2I=\begin{bmatrix}0&0&1\\0&1&0\\0&0&0\end{bmatrix} \Rightarrow \rho(A-2I)=2 \Rightarrow \eta(A-2I)=1 \Rightarrow \text{A.M. of } 2 \neq \text{G.M. of } 2$$

Not diagonal.

Option (b) is incorrect.

$M=\begin{bmatrix}1&1\\1&1\end{bmatrix}$

Eigen value of this are 0 & 2
Both are distinct ⇒ diagonal
Option (c) is incorrect.

$$M = \begin{bmatrix} 2 & 1 & 0 \\ 0 & 3 & 0 \\ 0 & 0 & 3 \end{bmatrix}$$

A.M. of $\lambda(=2) = 1$
G.M. of $\lambda(=2) = 1$
A.M. of $3 = 2$
G.M. of $3 = 2$ ⇒ diagonal
Option (d) is incorrect.

$$M = \begin{bmatrix} 1 & -1 \\ 2 & 4 \end{bmatrix}$$

Eigen value of these

$C_T(x) = x^2 - 5x + 6 = 0 \quad \Rightarrow x = \dfrac{5 \pm \sqrt{25-24}}{2} \quad \Rightarrow x = \dfrac{5 \pm 1}{2}$

$x = 3, 2$
eigen value are 3 & 2 both are distinct ⇒ diagonalizable.

4. How many non-singular 10 × 10 real matrices exist. Which has the characteristic polynomial $\lambda^3(\lambda - 1)^4 (\lambda + 1)^3$ and the minimal polynomial $\lambda^2(\lambda - 1)^2(\lambda + 1)$.

RSET – 2023

(a) 1 (b) 2 (c) 3 (d) 4

Ans. (d)

Explanation :

2 + 1   2 + 2   1 + 1 + 1
2 + 1 + 1

5. Let A be 7 × 7 matrix such that $2A^2 - A^4 = I$ if A has two distinct eigen values and each eigen value has G.M. 3, then the total number of non-zero entries in the J.C. form

GATE-2017

(a) 5 (b) 6 (c) 8 (d) 10

Ans. (c)

Explanation :

Given that $2A^2 - A^4 = I \quad \Rightarrow \quad x^4 - 2x^2 + 1 = 0$
$(x^2 - 1)^2 = 0$
$x = \pm 1$

⇒ 1 & -1 are possible eigen value of A
Both has G.M. 3   We know that G.M. ≤ A.M. $\forall \lambda$

$\Rightarrow$ Two possibility of A.M.

1, 1, 1, 1, -1, -1, -1   1, 1, 1, -1, -1, -1, -1

$$A=\begin{bmatrix}1&1&0&0&0&0&0\\0&1&0&0&0&0&0\\0&0&1&0&0&0&0\\0&0&0&1&0&0&0\\0&0&0&0&-1&0&0\\0&0&0&0&0&-1&0\\0&0&0&0&0&0&-1\end{bmatrix}_{7\times7}$$

Number of non-zero entries are 8.

6. If the characteristic polynomial and minimal polynomial of a square matrix A are $(\lambda - 1)(\lambda + 1)^4 (\lambda - 2)^5$ and $(\lambda - 1)(\lambda + 1)(\lambda - 2)$, respectively, then the rank of the matrix A + I is **GATE - 2019**

(a) 6   (b) 5   (c) 4   (d) 3

Ans. (a)

Explanation :

Given characteristic is $(\lambda - 1)(\lambda + 1)^4 (\lambda - 2)^5$

and minimal is $(\lambda - 1)(\lambda + 1)(\lambda - 2)$

Since minimal polynomial is of distinct linear factor. So A is diagonalizable.

Eigen value of A are 1, -1,-1,-1,-1,2,2,2,2,2 and all are diagonal entries

So, rank A + I is some as D' which is equal to number of non zero eigenvalues of A + I (being diagonalizable), and these are 1 + 1, –1 + 1, –1 + 1, –1 + 1, 2 + 1, 2 + 1, 2 + 1, 2 + 1, 2 + 1.

So, 2, 0(4 times), 3(5 times)

So rank of A + I is 6.

7. Let $A = [a_{ij}]$ be a 3 × 3 real matrix such that $A\begin{bmatrix}1\\2\\1\end{bmatrix}=2\begin{bmatrix}1\\2\\1\end{bmatrix}, A\begin{bmatrix}0\\1\\1\end{bmatrix}=2\begin{bmatrix}0\\1\\1\end{bmatrix}$ and

$A\begin{bmatrix}-1\\1\\0\end{bmatrix}=4\begin{bmatrix}-1\\1\\0\end{bmatrix}$. If m is the degree of the minimal polynomial of A, then $a_{11} + a_{21} + a_{31} + m$ equals **GATE 2023**

(a) 1   (b) 2   (c) 3   (d) 4

Ans. (d)

Explanation :

Given that $A\begin{bmatrix}1\\2\\1\end{bmatrix}=2\begin{bmatrix}1\\2\\1\end{bmatrix}, A\begin{bmatrix}0\\1\\1\end{bmatrix}=2\begin{bmatrix}0\\1\\1\end{bmatrix}$

Here (1, 2, 1) and (0, 1, 1) are two LI eigen vectors corresponding to 2.

$\Rightarrow$ G.M. of ($\lambda = 2$) is 2.

$\Rightarrow$ Number of Jordan block corresponding to $\lambda = 2$ are 2

Then $A=\begin{bmatrix}2&0&0\\0&2&0\\0&0&4\end{bmatrix}$ $\Rightarrow$ Minimal polynomial m(x) = (x – 2)(x – 4)

Here $a_{11} = 2, a_{21} = 0, a_{31} = 0$ & m = 2

Then $a_{11} + a_{21} + a_{31} + m = 4$

8. Let $M=\begin{bmatrix}1&-1&1\\2&1&4\\-2&1&-4\end{bmatrix}$ given that 1 is eigen value of M. Then which of the following are correct. **CSIR NET DEC 2018**

(a) The minimal polynomial of M = $(x-1)(x+4)$

(b) The minimal polynomial of M = $(x-1)^2(x+4)$

(c) M is not diagonalizable

(d) $M^{-1}=\frac{1}{4}(M+3I)$

Ans. (b), (c)

Explanation :

Eigen value of M are 1, 1, -4

A.M. of ($\lambda = 1$) = 2

And G.M. of 1 = η (A – I)

$$A-I=\begin{bmatrix}0&-1&1\\2&0&4\\-2&1&-5\end{bmatrix}$$

$\rho(A-I)=2$

$\eta(A-I)=1$

$\Rightarrow$ G.M. of 1 = 1 $\Rightarrow$ G.M. of 1 $\neq$ A.M. of 1 $\Rightarrow$ M is not diagonalizable.

Option (c) is correct.

$C_T(x)=(x-1)^2(x+4)$

And we know that

$\rho(M-I)=2$ $\Rightarrow M_T(x)=(x-1)^2(x+4)$

Option (b) is correct.

Option (d) is incorrect.

$$M^{-1}=\frac{1}{4}(M+3I)$$

Suppose it is true.

Then $4M^{-1} = M + 3I$

$4I = M^2 + 3M \Rightarrow \quad M^2 + 3M - 4I = 0 \Rightarrow (M - I)(M + 4I) = 0$

$\Rightarrow \quad M_T(x) = (x - 1)(x + 4)$ contradiction.

9. Let A be a real matrix with characteristic polynomial $(x-1)^3$. Pick the correct statement from below **CSIR NET DEC 2018**

(a) A is necessary diagonalizable

(b) If the minimal polynomial of A is $(x-1)^3$ then A is diagonalizable.

(c) characteristic polynomial of A^2 is $(x-1)^3$

(d) If A has exactly two Jordan block then $(A-I)^2$ is diagonalizable.

Ans. (c), (d)

Explanation :

Given that $C_A(x) = (x-1)^3$

Option (a) is incorrect.

Let $M_A(x) = (x-1)^3$

Option (b) is incorrect.

Then $A=\begin{bmatrix}1&1&0\\0&1&1\\0&0&1\end{bmatrix}$ is not diagonalizable.

Option (c) is correct.

We know that eigen value of A are 1, 1, 1

Then eigen value of A^2 are 1, 1, 1

$C_{A^2}(x) = (x-1)^3$

Option (d) is correct.

Given that if A has two Jordan block

$A=\begin{bmatrix}1&1&0\\0&1&0\\0&0&1\end{bmatrix}$ Then $A-I=\begin{bmatrix}0&1&0\\0&0&0\\0&0&0\end{bmatrix}$ $\quad (A-I)^2=\begin{bmatrix}0&0&0\\0&0&0\\0&0&0\end{bmatrix}$

$(A-I)^2$ is zero matrix.

$\Rightarrow (A-I)^2$ is diagonalizable.

10. Let $T : R^4 \to R^4$ be a linear transformation with characteristic polynomial $(x-2)^4$ and minimal polynomial $(x-2)^2$. Jordan canonical form of T can be

(a) $\begin{bmatrix} 2 & 0 & 0 & 0 \\ 1 & 2 & 0 & 0 \\ 0 & 0 & 2 & 0 \\ 0 & 0 & 1 & 2 \end{bmatrix}$ (b) $\begin{bmatrix} 2 & 0 & 0 & 0 \\ 0 & 2 & 0 & 0 \\ 0 & 0 & 2 & 0 \\ 0 & 0 & 1 & 2 \end{bmatrix}$ (c) $\begin{bmatrix} 2 & 0 & 0 & 0 \\ 0 & 2 & 0 & 0 \\ 0 & 0 & 2 & 0 \\ 0 & 0 & 0 & 2 \end{bmatrix}$ (d) $\begin{bmatrix} 2 & 0 & 0 & 0 \\ 1 & 2 & 0 & 0 \\ 0 & 1 & 2 & 0 \\ 0 & 0 & 0 & 2 \end{bmatrix}$

CSIR NET DEC 2019

Ans. (a), (b)

Explanation :

Given that $c(x) = (x-2)^4$

and $m(x) = (x-2)^2$

$\Rightarrow$ $\exists$ a Jordan block of order (2×2)

and remaining Jordan block are of order less than or equal to (2×2)

Option (a), (b) are correct.

11. Let A be (6×6) matrix over R with characteristic polynomial $= (x-3)^2 (x-2)^4$ and minimal polynomial $= (x-3)(x-2)^2$. Then Jordan canonical form of A can be

CSIR NET JUNE 2018

(a) $\begin{pmatrix} 3 & 0 & 0 & 0 & 0 & 0 \\ 0 & 3 & 0 & 0 & 0 & 0 \\ 0 & 0 & 2 & 1 & 0 & 0 \\ 0 & 0 & 0 & 2 & 1 & 0 \\ 0 & 0 & 0 & 0 & 2 & 1 \\ 0 & 0 & 0 & 0 & 0 & 2 \end{pmatrix}$ (b) $\begin{pmatrix} 3 & 0 & 0 & 0 & 0 & 0 \\ 0 & 3 & 0 & 0 & 0 & 0 \\ 0 & 0 & 2 & 1 & 0 & 0 \\ 0 & 0 & 0 & 2 & 0 & 0 \\ 0 & 0 & 0 & 0 & 2 & 0 \\ 0 & 0 & 0 & 0 & 0 & 2 \end{pmatrix}$

(c) $\begin{pmatrix} 3 & 0 & 0 & 0 & 0 & 0 \\ 0 & 3 & 0 & 0 & 0 & 0 \\ 0 & 0 & 2 & 1 & 0 & 0 \\ 0 & 0 & 0 & 2 & 0 & 0 \\ 0 & 0 & 0 & 0 & 2 & 1 \\ 0 & 0 & 0 & 0 & 0 & 2 \end{pmatrix}$ (d) $\begin{pmatrix} 3 & 1 & 0 & 0 & 0 & 0 \\ 0 & 3 & 0 & 0 & 0 & 0 \\ 0 & 0 & 2 & 1 & 0 & 0 \\ 0 & 0 & 0 & 2 & 0 & 0 \\ 0 & 0 & 0 & 0 & 2 & 1 \\ 0 & 0 & 0 & 0 & 0 & 2 \end{pmatrix}$

Ans. (b), (c)

Exp. In minimal polynomial degree of $(x-3) = 1$ and degree of $(x-2) = 2$

$\Rightarrow$ Atleast one block of 1×1 for eigenvalue 3

$\Rightarrow$ and atleast one block of 2×2 for eigenvalue 2

$\Rightarrow$ Hence, (a), (d) are incorrect.

UNSOLVED QUESTIONS

1. Let A and B are two similar matrices, then
(a) A, B have same characteristic polynomial
(b) A, B may have different characteristic polynomials.
(c) A, B have same determinants.
(d) A, B have different determinants.

2. Let $A=\begin{bmatrix}7&1&0&0\\0&7&1&0\\0&0&7&1\\0&0&0&7\end{bmatrix}$ be the jordan block of order 4, then
(a) Characteristic polynomial = $(t-7)^4$ (b) Minimal polynomial = $(t-7)^4$
(c) Characteristic and minimal polynomials are not equal
(d) 7 is the only eigenvalue.

3. Let N be a 3 × 3 non-zero matrix with the property $N^3 = O$. Which of the following is/are true?
(a) N is not similar to a diagonal matrix
(b) N is similar to a diagonal matrix
(c) N has one non-zero eigenvector
(d) N has three linearly independent eigen vectors.

4. $M=\begin{pmatrix}1&a&b\\0&2&c\\0&0&1\end{pmatrix}$ a, b, c ∈ R. Then, M is diagonalizable, if and only if **GATE-2006**
(a) a = bc (b) b = ac (c) c = ab (d) a = b = c

5. Which of the following matrices is not diagonalizable? **GATE – 2009**
(a) $\begin{pmatrix}1&1\\1&2\end{pmatrix}$ (b) $\begin{pmatrix}1&0\\3&2\end{pmatrix}$ (c) $\begin{pmatrix}0&-1\\1&0\end{pmatrix}$ (d) $\begin{pmatrix}1&1\\0&1\end{pmatrix}$

6. Let $M_2(R)$ denote the set of 2 × 2 real matrices. Let $A \in M_2(R)$ be of trace 2 and determinant -3. Identifying $M_2(R)$ with R^4, consider the linear transformation $T : M_2(R) \to M_2(R)$ defined by T(B) = AB. Then which of the following statement are true?
CSIR NET DEC 2011
(a) T is diagonalizable
(b) 2 is an eigenvalue of T
(c) T is invertible
(d) T(B) = B for some 0 ≠ B in $M_2(R)$

7. For any n ×n matrix B, let $N(B) = \{x \in R^n : BX = 0\}$ be the null space of B. Let A be a 4 × 4 matrix with dim (N(A – 2I)) = 2, dim (N(A – 4I)) = 1 and rank (A) = 3. Then
CSIRNET JUNE 2017
(a) 0, 2 and 4 are eigenvalues of A
(b) determinant (A) = 0
(c) A is not diagonalizable
(d) trace (A) = 8

8. Let A be a real 4 × 4 matrix with characteristic polynomial $p(t) = (t^2 +1)^2$. Which of the following is true?

(a) A is diagonalizable over complex numbers but not over real numbers.
(b) A is nilpotent (c) A is invertible (d) There is no such matrix A

9. Let A be a 4 × 4 matrix over R such that rank (A) = 2 and $A^3 = A^2 \neq 0$. Suppose that A is not diagonalizable. Then **CSIR NET JUNE 2014**

(a) One of the Jordan block of the Jordan canonical form of A is $\begin{bmatrix} 0 & 1 \\ 0 & 0 \end{bmatrix}$

(b) $A^2 = A \neq 0$ (c) There exists a vector v such that $Av \neq 0$ but $A^2v = 0$
(d) The characteristic polynomial of A is $x^4 - x^3$.

10. Which of the following matrices have Jordan canonical form equal to $\begin{pmatrix} 0 & 1 & 0 \\ 0 & 0 & 0 \\ 0 & 0 & 0 \end{pmatrix}$

CSIR NET JUNE 2014

(a) $\begin{pmatrix} 0 & 0 & 1 \\ 0 & 0 & 0 \\ 0 & 0 & 0 \end{pmatrix}$ (b) $\begin{pmatrix} 0 & 0 & 1 \\ 0 & 0 & 1 \\ 0 & 0 & 0 \end{pmatrix}$ (c) $\begin{pmatrix} 0 & 1 & 1 \\ 0 & 0 & 0 \\ 0 & 0 & 0 \end{pmatrix}$ (d) $\begin{pmatrix} 0 & 1 & 1 \\ 0 & 0 & 1 \\ 0 & 0 & 0 \end{pmatrix}$

11. Consider the matrix $A = \begin{bmatrix} 2 & 2 & 1 \\ 0 & 2 & -1 \\ 0 & 0 & 3 \end{bmatrix}$ and $B = \begin{bmatrix} 2 & 1 & 0 \\ 0 & 2 & 0 \\ 0 & 0 & 3 \end{bmatrix}$. Then **CSIRNET DEC 2015**

(a) A and B are similar over the field of rational numbers Q
(b) A is diagonalizable over the field of rational numbers Q.
(c) B is the Jordan canonical form of A.
(d) The minimal polynomial and the characteristic polynomial of A are the same

12. $T : C^n \to C^n$ be a linear operator having n distinct eigen value. Then which of the following may not be true?
(a) T is invertible (b) T is diagonalizable
(c) T is not diagonalizable (d) None of these

13. Which of the following matrices have Jordan canonical form not equal to $\begin{bmatrix} 0 & 1 & 0 \\ 0 & 0 & 0 \\ 0 & 0 & 0 \end{bmatrix}$.

(a) $\begin{bmatrix} 0 & 0 & 1 \\ 0 & 0 & 0 \\ 0 & 0 & 0 \end{bmatrix}$ (b) $\begin{bmatrix} 0 & 0 & 1 \\ 0 & 0 & 1 \\ 0 & 0 & 0 \end{bmatrix}$ (c) $\begin{bmatrix} 0 & 1 & 1 \\ 0 & 0 & 0 \\ 0 & 0 & 0 \end{bmatrix}$ (d) $\begin{bmatrix} 0 & 1 & 1 \\ 0 & 0 & 1 \\ 0 & 0 & 0 \end{bmatrix}$

14. (a) T is invertible (b) T is invertible as well as diagonalizable
(c) T is not diagonalizable (d) T is diagonalizable

15. Let $M = \begin{bmatrix} 1 & 2 & b \\ 0 & 2 & 1 \\ 0 & 0 & 1 \end{bmatrix}$; b ∈ R. Then M is diagonalizable iff b is

(a) 1 (b) 2 (c) 3 (d) 4

16. Which of the following matrices is diagonalizable over R?

(a) $\begin{bmatrix} 2 & 0 & 1 \\ 0 & 3 & 0 \\ 0 & 0 & 2 \end{bmatrix}$ (b) $\begin{bmatrix} 1 & 1 \\ 1 & 1 \end{bmatrix}$ (c) $\begin{bmatrix} 2 & 1 & 0 \\ 0 & 3 & 0 \\ 0 & 0 & 3 \end{bmatrix}$ (d) $\begin{bmatrix} 1 & -1 \\ 2 & 4 \end{bmatrix}$

17. Let A be 4 × 4 matrix over R such that $\rho(A) = 2$ and $A^3 = A^2 \neq 0$. Suppose that A is not diagonalizable. Then

(a) One of the Jordan block of the Jordan canonical form of A is $\begin{bmatrix} 0 & 1 \\ 0 & 0 \end{bmatrix}$.

(b) $A^2 = A \neq 0$ (c) There exist a vector v such that $Av \neq 0$ but $A^2v = 0$

(d) The characteristic polynomial of A is $x^4 - x^3$.

18. Which of the following matrix have J.C. form equal to $\begin{bmatrix} 0 & 1 & 0 \\ 0 & 0 & 0 \\ 0 & 0 & 0 \end{bmatrix}$.

(a) $\begin{bmatrix} 0 & 0 & 1 \\ 0 & 0 & 0 \\ 0 & 0 & 0 \end{bmatrix}$ (b) $\begin{bmatrix} 0 & 0 & 1 \\ 0 & 0 & 1 \\ 0 & 0 & 0 \end{bmatrix}$ (c) $\begin{bmatrix} 0 & 1 & 1 \\ 0 & 0 & 0 \\ 0 & 0 & 0 \end{bmatrix}$ (d) $\begin{bmatrix} 0 & 1 & 1 \\ 0 & 0 & 1 \\ 0 & 0 & 0 \end{bmatrix}$

19. Consider the matrix $A = \begin{bmatrix} 2 & 2 & 1 \\ 0 & 2 & -1 \\ 0 & 0 & 3 \end{bmatrix}$ and $B = \begin{bmatrix} 2 & 1 & 0 \\ 0 & 2 & 0 \\ 0 & 0 & 3 \end{bmatrix}$. Then

(a) A and B are similar over Q (b) A is diagonalizable over Q
(c) B is J.C. form of A (d) The $m_A(x)$ and $c_A(x)$ are equal

Answer Key

1	2	3	4	5	6	7
a,c	a,b,d	a,c	b	d	a,c	a,b,d
8	**9**	**10**	**11**	**12**	**13**	**14**
c	a,c,d	a,b,c	a,c,d	a,c,d	d	d
15	**16**	**17**	**18**	**19**		
b	b,c,d	a,c,d	a,b,c	a,c,d		

Chapter 7

Diagonalizable Matrix and Jordan Canonical Form

1. Quadratic form :

An expression of the form $\sum_{i=1}^{n}\sum_{j=1}^{n} a_{ij}x_ix_j$ .Where a_{ij}'s are elements of a field F, is called a quadratic form in n variables $x_1, x_2, x_3......x_n$ over a field F

Note : Each quadratic form can be written as symmetric matrix.

Example: $\phi(x, y, z) = ax^2 + by^2 + cz^2 + 2hxy + 2gxz + 2fyz = 0$ then

$$\phi(x,y,z) = [xyz]\begin{bmatrix} a & h & g \\ h & b & f \\ g & f & c \end{bmatrix}\begin{bmatrix} x \\ y \\ z \end{bmatrix}.$$

Results :

Let $\phi = x^TAx$ be a real quadratic form where $A = [a_{ij}]$ be a matrix of quadratic form then

(1) Positive definite : If $\phi > 0$ for all real variables other than $x_1 = x_2 = ... = x_n = 0$

Example: $\phi = x_1^2 + x_2^2$ in two variables Or All eigen values of A are positive.

(2) Positive semi definite : If $\phi \geq 0$ for all real values of variables otherthan $x_1 = x_2 ..... x_n = 0$

Example : $\phi = x_1^2 + x_2^2$ is a positive definite of three variables $\phi(0, 0, -1) = 0$

Or All eigen values of A are either positive or zero.

(3) Negative definite : If $\phi < 0$ $\forall$ real values of variables other than $x_1 = ..., x_n = 0$

Example: $\phi(x_1, x_2) = -x_1^2 - x_2^2$ in two variables

Or All eigen value of A are negative.

(4) Negative semi definite : If $\phi \leq 0$ $\forall$ real value of variable other than $x_1 = ... x_n = 0$

Example : $\phi = -x_1^2 - x_3^2$ in three variables

$\phi(0, 1, 0) = 0$ Or All eigen value of A are either negative or zero.

(5) Rank of Quadratic form : Let $\phi = x^TAx$ be a quadratic form then number of non-zero eigen value of A is called rank and denoted by 'r'

(6) Index of Quadratic form :

Number of positive eigen value of A is called index and denoted by 'p'.

(7) Signature : Difference between the number of positive eigen values and negative eigen valuesit is calculated by ' 2p – r' Where p = index and r = rank

Equivalent Quadratic form :

(1) If two quadratic form are equivalent over IR then they have same Rank and signature.

(2) If two quadratic form are equivalent over C then they have same rank.

2. Bilnear form :

Let $f : V \times V \to F$ is called bilinear form. If

(1) $f(u, \alpha v + \beta w) = \alpha f(u, v) + \beta f(u, w)$ (2) $f(\alpha u + \beta v, w) = \alpha f(u, w) + \beta f(v, w)$

Rank of a Bilinear Form

The rank of a bilinear form f on V, written as rank (f) is defined to be the rank of any matrix representation.

Example: Let $V = R^n$. and $f : R^n \times R^n \to R$ s.t. $f(u, v) = a_1b_1 + a_2b_2 + \ldots + a_nb_n$

Here (1) $f(\alpha u + \beta v, w) = \alpha f(u, w) + \beta f(v, w)$

(2) $f(u, \alpha v + \beta w) = \alpha f(u, v) + \beta f(u, w)$

where $u = (a_1, a_2, \ldots, a_n) \in R^n$

$v = (b_1, b_2, \ldots, b_n) \in R^n$

$w = (c_1, c_2, \ldots, c_n) \in R^n$

Hence f(u v) is bilinear form

Hermitian Forms :

Let V be a vector space of finite dimension over the complex field C. A Hermitian form on V is a mapping ff: $V \times V \to C$ such that for all $a, b \in C$ and all $u_i, v_i \in V$,

(i) $f(au_1 + bu_2, v) = af(u_1, v) + bf(u_2, v)$ (ii) $f(u,v) = \overline{f(v,u)}$

(iii) $f(u, av_1 + bv_2) = \bar{a}\ f(u,v_1) + \bar{b}\ f(u,v_2)$

3. Inner Product Space

Let V(F) be a vector space over a field (F). Then $< , > :\ V \times V \to F$ is said to be a inner product if it is satisfy following properties.

(a) Linearly property :

$<\alpha u + \beta v, w> = \alpha<u, w> + \beta<v, w> \forall\ u, v, w \in V \quad \alpha, \beta \in F.$

(b) Conjugate Symmetric property :

$<u,v> = <\overline{v,u}>$ or $<u,v>=<v,u>$

(c) Positive definite property :

$<u, u> \geq 0$ and $<u, u> = 0$ iff $u = 0$

Inner product space : A vector space V with an inner product is called inner product space.

Notation : f(u, v) is generally denoted by any one of the following symbols :

$(u, v) : u.v : <u \mid v>; <u, v>$

Note: $V = R^n(R)$

(1) Then $<x, y> = xy^T = y^Tx$ is called standard innerproduct on V.

(2) $<u, v> = u^TAv$ is inner productiff A is symmetric & positive definite.

Example :

(1) V = Function space C[a, b] = {f | f: [a ,b] → R}

Then $<f,g> = \int_a^b f(t)g(t)dt$; where $f, g \in C[a, b]$

(2) V = polynomial space with real coefficient

then $< p,q >= \int_a^b p(t)q(t)dt$

(3) Matrix space ($M_{m\times n}$) .
Then <A, B> = $Tr(B^TA)$

(4) Let V = $R^n(R)$
Then <x, y> = $x^Ty = x_1y_1+x_2y_2+x_3y_3.....+x_ny_n$ (standard inner product)
where $x = (x_1,x_2,x_3,.....x_n)^T$ and $y = (y_1,y_2,y_3,.....y_n)^T$

4. Orthogonal Vector, Orthogonal Set &Orthogonal Complement

Orthogonal vector : Two vector u and v are said to be a orthogonal if <u, v> = 0

Example : For u = (1, 0) $\in R^2$, v = (0, 1) $\in R^2$s.t.<u, v> = 0
$\Rightarrow$ u and v are orthogonal.

Orthogonal set : Every elements of a set are orthogonal to each other then this set is called orthogonal set.

Example : S = { (1, 0), (0, 2)}then S is orthogonal.

Orthogonal complement : Let S be a subset of inner product space V then orthogonal of S denoted $S^\perp$ is defined as

$S^\perp = \{u \in V \mid <u, v> = 0, \quad \forall\ v \in S\}$

Example : Let V = R^2 and S = {(a, 0 ,0) | a $\in$ R}

then $S^- = \{u \in V \mid <u, v> = 0, \quad \forall\ v \in S\}$

$S^\perp = \{(u ,v ,w) \in V \mid <(u , v , w), (a , 0 , 0) > = 0, \quad u = 0\}$

$S^\perp = \{(0 ,v ,w) \mid v \in R , w \in R\}$

Orthonormal vector : A vector u and v are said to be orthonormal vector if
<u, v> = 0 and ||u|| = 1 = ||v||

Example : u = (1, 0) and v = (0, 1) are orthogonormal vectors
but (1, 0) and (0, 2) are only orthogonal vector because both are not unit vectors

Orthonormal Basis :
A basis of an inner product space that consists only orthonormal vectors is an orthonormal basis.

Results :
(i) Any orthogonal set of non-zero vectors is linearly independent.
(ii) Any orthogonal set of non-zero vectors in V with $v \perp w$, let $\alpha, \beta \in R$ be such that $\alpha v + \beta w = 0$. Then $\alpha = 0 = \beta$ i.e. two non-zero vectors orthogonal to each other are linearly independent.
(iii) The standard basis of R^n is an orthonormal basis.
(iv) If $\{v_i\}$ is an orthonormal subset of V, then the vectors v_i's are Linearly independent. over F. Further if $w = \sum_{i=1}^{k} \alpha_i v_i$, then for each i = 1, 2, ..., k, $\alpha_i = <w, v_i>$

(v) For each subspace W of a finite – dimensional inner product space V, $V = W \oplus W^{\perp}$.
(vi) If W is a subspace of a finite – dimensional inner product space V, then $(W^{\perp})^{\perp} = W$.
(vii) If S, S_1, S_2 are subsets of an inner product space V, then
(a) $S \cap S^{\perp} = \{0\}$ (b) $S_1 \subseteq S_2 \Rightarrow S_2^{\perp} \subseteq S_1^{\perp}$ (c) $S \subseteq S^{\perp\perp}$
(viii) If W_1 and W_2 are subspaces of a finite dimensional inner product space V, then
(a) $(W_1 + W_2)^{\perp} = W_1^{\perp} \cap W_2^{\perp}$ (b) $(W_1 \cap W_2)^{\perp} = W_1^{\perp} + W_2^{\perp}$

5. The Adjoint of Linear Transformation

Let V and W be finite dimensional linear product spaces and let $T \in L(V, W)$. Then there exist a unique linear mapping $T^* : W \to V$ such that for all $v \in V$ and $w \in W$, $\langle Tv, w\rangle = \langle v, T^*w\rangle$.
The mapping T^* is called adjoint of T.

Adjoint Operator

Let V(K) be an inner product space and $T : V \to V$ be linear operator. Let $\exists$ a unique operator T* on V such that $\langle T(u), v\rangle = \langle u, T^*(v)\rangle \forall u, v \in V$.
Then T* is called adjoint of T.
T* is linear.

Self-Adjoint Operator

A linear operator T on an inner product space V(K) is called self adjoint operator iff $T = T^*$, i.e., iff $\langle T(u), v\rangle = \langle u, T(v)\rangle \forall u, v \in V$.

Theorem : Let T be a self adjoint operator on a finite dimensional inner product space V. The following statements are equivalent.
(i) T is positive definite.
(ii) All eigenvalues of T are positive.
(iii) There exists a unique positive definite operator A on V such that $T = A^2$. We denote this operator by $\sqrt{T}$.
(iv) There exists an invertible linear operator S on V such that $T = S^*S$.

6. Unitary Operator :

A linear operator or an inner product space V(F) is said to be unitary operator if
(i) T is one-one and onto.
(ii) $\langle T(u), T(v)\rangle = \langle u, v\rangle \forall u,v \in V$.
A unitary operator on an Euclidean space V(R) is called orthogonal operator.

7. Normal Operator :

A linear operator T on an inner product space V is called Normal if it commutes with its adjoin '*' that is, $TT^* = T^*T$. Unitary and Self adjoint operator are normal operators. $A \in F^{n\times n}$ is called Normal if $AA^* = A^*A$. Thus, a linear operator T on a finite dimensional inner product space is normal iff for any ordered orthonormal basis B, $[T]_n$ is a Normal Matrix.

Results :

(a) Let T be a linear transformation from an inner product space V(F) into an inner product space U(F). Then,

(i) T preserves norms if $||T(u)|| = ||u|| \ \forall u \in V$.

(ii) T preserves inner product if $<T(u), T(v)> \ = <u,v> \forall u,v \in V$

(iii) T is an isometry if T preserves distances i.e., if $||T(u) - T(v)|| = ||u - v|| \ \forall u,v \in V$.

(b) Let V(F) and U(F) be inner product spaces. Let T be a linear transformation from V into U. Then T is called an inner product space isomorism if

(i) T is invertible, i.e., if T is one-one and onto.

(ii) T preserves inner products, i.e. $<T(u), T(v)> = <u,v> \forall u, v \in V$.

In this case V(F) and U(F) are said to be isomorphic.

Remaks : T is unitary iff $TT^* = T^*T = I$.

(c) Let T be a linear operator on an inner product space V(F). Then T is called positive (non-negative) semi-definite operator. i.e. $T \geq 0$ iff T is self-adjoint and $<T(u), u> \geq 0 \ \forall u \in V$.

(d) A linear operator on an inner product space V(F) is called strictly positive (or positive definite operator i.e. $T > 0$ iff T is self adjoint and $\langle T(u), u\rangle > 0 \ \forall u \in V$.

SOLVED QUESTIONS

1. Consider three quadratic form over R. **CSIR NET JUNE 2023**

I. $6x^2 - 13xy + 6y^2$

II. $x^2 - xy + 2y^2$

III. $x^2 - xy - 2y^2$

Then which of the following is true?

(a) I ~ II (b) I ~ III (c) II is positive definite (d) III is positive definite

Ans. (b), (c)

Explanation :

For I $A = \begin{bmatrix} 6 & -13/2 \\ -13/2 & 6 \end{bmatrix}$; E.V. are -1/2, 25/2

For II : $A = \begin{bmatrix} 1 & -1/2 \\ -1/2 & 2 \end{bmatrix}$; E.V. are $\frac{+3 \pm \sqrt{2}}{2}$.

For III : $A = \begin{bmatrix} 1 & -1/2 \\ -1/2 & -2 \end{bmatrix}$; E.V. are $\frac{-1 \pm \sqrt{10}}{2}$.

Hence I is equivalent to III and II is positive definite.

2. Which of the following are inner products on R^2? **CSIR NET FEB 2022**

(a) $\left\langle \begin{pmatrix} x_1 \\ x_2 \end{pmatrix}, \begin{pmatrix} y_1 \\ y_2 \end{pmatrix} \right\rangle = x_1y_1 + 2x_1y_2 + 2x_2y_1 + x_2y_2$

(b) $\left\langle \begin{pmatrix} x_1 \\ x_2 \end{pmatrix}, \begin{pmatrix} y_1 \\ y_2 \end{pmatrix} \right\rangle = x_1y_1 + x_1y_2 + x_2y_1 + 2x_2y_2$

(c) $\left\langle \begin{pmatrix} x_1 \\ x_2 \end{pmatrix}, \begin{pmatrix} y_1 \\ y_2 \end{pmatrix} \right\rangle = x_1y_1 + x_1y_2 + x_2y_1 + x_2y_2$

(d) $\left\langle \begin{pmatrix} x_1 \\ x_2 \end{pmatrix}, \begin{pmatrix} y_1 \\ y_2 \end{pmatrix} \right\rangle = x_1y_1 + 2x_1y_2 + 2x_2y_1 + x_2y_2$

Ans. (b), (d)

Explanation :

(a) Given $\left\langle \begin{pmatrix} x_1 \\ x_2 \end{pmatrix}, \begin{pmatrix} y_1 \\ y_2 \end{pmatrix} \right\rangle = x_1y_1 + 2x_1y_2 + 2x_2y_1 + x_2y_2$

$$A = \begin{bmatrix} 1 & 2 \\ 2 & 1 \end{bmatrix}$$

Here A is symmetric but eigen value of A are 3 and -1

A is indefinite matrix

$\therefore$ This is not an inner product space.

(b) Given $\left\langle \begin{pmatrix} x_1 \\ x_2 \end{pmatrix}, \begin{pmatrix} y_1 \\ y_2 \end{pmatrix} \right\rangle = x_1y_1 + x_1y_2 + x_2y_1 + 2x_2y_2$

Now, $A = \begin{bmatrix} 1 & 1 \\ 1 & 2 \end{bmatrix}$

A is symmetric and positive definite

So, it is an inner product space.

(c) $\left\langle \begin{pmatrix} x_1 \\ x_2 \end{pmatrix}, \begin{pmatrix} y_1 \\ y_2 \end{pmatrix} \right\rangle = x_1y_1 + x_1y_2 + x_2y_1 + x_2y_2$

Now, $A = \begin{bmatrix} 1 & 1 \\ 1 & 1 \end{bmatrix}$

A is symmetric and positive semi definite

So, it is not an inner product space.

(d) $\left\langle \begin{pmatrix} x_1 \\ x_2 \end{pmatrix}, \begin{pmatrix} y_1 \\ y_2 \end{pmatrix} \right\rangle = x_1y_1 + 2x_1y_2 + 2x_2y_1 + x_2y_2$

Now, $A = \begin{bmatrix} 1 & 2 \\ 2 & 1 \end{bmatrix}$ A is symmetric and indefinite

So, it is not an inner product space.

3. Let $v \in R^3$ be a non-zero vector. Define a linear transformation $T : R^3 \to R^3$ by $T(x) = x - 2\frac{x.v}{v.v}v$, where x.y denotes the standard inner product in R^3.
Which of the following statements are true? **CSIR NET NOV 2020**

(a) The eigenvalues of T are +1, -1
(b) The determinant of T is -1
(c) The trace of T is +1
(d) T is distance preserving

Ans. (a), (b), (c), (d)

Explanation :

$T : R^3 \to R^3$ by $T(x) = x - 2\frac{x.v}{v.v}v$.

T(x), option (d) is correct.
Fix v = (1, 0, 0), then

$$T(e_1) = T(1,0,0) = (1,0,0) - 2\frac{(1,0,0).(1,0,0)}{(1,0,0).(1,0,0)}(1,0,0)$$

(1, 0, 0) - (2, 0, 0) = (-1, 0, 0)

$$T(e_2) = T(0,1,0) = (0,1,0) - 2\frac{(0,1,0).(1,0,0)}{(1,0,0).(1,0,0)}(1,0,0) = (0, 1, 0)$$

$$T(e_3) = T(0,0,1) = (0,0,1) - 2\frac{(0,0,1).(1,0,0)}{(1,0,0).(1,0,0)}(1,0,0) = (0, 0, 1)$$

So, $[T] = \begin{bmatrix} -1 & 0 & 0 \\ 0 & 1 & 0 \\ 0 & 0 & 1 \end{bmatrix}$ eigenvalues are nothing but diagonal entries -1, 1, 1 and

trace (T) = -1 + 1 + 1 = 1 and det T = -1 × 1 × 1 = -1, all options are correct.

4. Which of the following is an inner product on the vector space of all real valued continuous functions on [0, 1]? **CSIR NET NOV 2020**

(a) $\langle f, g \rangle = \left| \int_0^1 f(t)g(t)dt \right|$

(b) $\langle f, g \rangle = \int_0^1 | f(t)g(t) | dt$

(c) $\langle f, g \rangle = f(0)g(0) + f(1)g(1)$

(d) $\langle f, g \rangle = \int_0^1 f(t)g(t)dt$

Ans. (a)

Explanation :

(a) $\langle f_1 + f_2, g \rangle$

$$= \left| \int_0^1 (f_1(t) + f_2(t))g(t)dt \right| = \left| \int_0^1 f_1(t)g(t)dt + \int_0^1 f_2(t)g(t)dt \right|$$

$$\leq \left|\int_0^1 f_1(t)g(t)dt\right| + \left|\int_0^1 f_2(t)g(t)dt\right| \qquad = \langle f_1, g\rangle + \langle f_2, g\rangle$$

Thus $\langle f_1 + f_2, g\rangle \leq \langle f_1, g\rangle + \langle f_2, g\rangle$ and $\langle f, g\rangle$ is not inner product.

(b) $|f_1(t) + f_2(t)|$ need not be equals to $|f_1(t)| + |f_2(t)|$, so same as option (a) it is also not an inner product.

(c) Consider f(x) = x(x – 1)
Then $\langle f, f\rangle = f(0)\, f(0) + f(1)\, f(1) = 0.0 + 0.0 = 0$
$\Rightarrow$ $\|f\| = 0$ for non-zero f(x) = x(x – 1) hence it is not an inner product.

(d) For the vector space of all real valued continuous functions defined on [a, b], $-\infty < a < b < \infty$, the product $\langle f, g\rangle = \int_a^b f(t)dt$ is always an inner product, so option (d) is true.

5. A quadratic form Q(x, y, z) over R represents 0 non-trivially if there exists $(a, b, c) \in R^3 \setminus \{(0, 0, 0)\}$ such that Q(a, b, c) = 0. Which of the following quadratic forms Q(x, y, z) over R represents 0 non-trivially? **CSIR NET NOV 2020**

(a) $Q(x, y, z) = xy + z^2$ (b) $Q(x, y, z) = x^2 + 3y^2 - 2z^2$
(c) $Q(x, y, z) = x^2 - xy + y^2 + z^2$ (d) $Q(x, y, z) = x^2 + xy + z^2$

Ans. **(a), (b), (d)**

Explanation :

(a) Q(1, -1, 1) = (1)(-1) + 1 = -1 + 1 = 0, So, option (a) is correct.

(b) As $Q\left(1, 0, \frac{1}{\sqrt{2}}\right) = 1^2 + 3(0)^2 - 2\left(\frac{1}{\sqrt{2}}\right)^2 = 1 + 0 - 2\frac{1}{2} = 0$ So, option (b) is correct.

(d) Q(1, -1, 0) = 1 – 1 = 0
So, option (d) is correct.

(c) For $Q(x, y, z) = x^2 - xy + y^2 + z^2$

We have $A = \begin{bmatrix} 1 & -1/2 & 0 \\ -1/2 & 1 & 0 \\ 0 & 0 & 1 \end{bmatrix}$ Such that $Q(X) = X^TAX$

Where X = [x y z] and eigenvalues of A are $1, \frac{1}{2}, \frac{3}{2}$.

Hence Q(X) > 0 for all $x \neq 0$, because all the eigen values of A are positive. So, as 0 is not an eigen values, so it will never represents a non-zero element trivially.

6. Consider R^3 with the usual inner product. If d is the distance from (1, 1, 1) to the subspace span {(1, 1, 0), (0, 1, 1)} of R^3, then $3d^2$ is **GATE 2018**

(a) 3 (b) 4 (c) 2 (d) 1

Ans. **(d)**

Explanation:
Let W be a subspace of R^3 spanned by {(1, 1, 0), (0, 1, 1)}
Then W = {(a, a + b, b): a, b ∈ R^3}
Let distance of (a, a + b, b) from (1, 1, 1) be d
Then, $d^2 = (a-1)^2 + (a+b-1)^2 + (b-1)^2$

$3d^2 = 3(a-1)^2 + 3(a+b-1)^2 + 3(b-1)^2$

Let $f(x, y) = 3(x-1)^2 + 3(x+y-1)^2 + 3(y-1)^2$
For minima $f_x = 0$ and $f_y = 0$ ⇒ $6(x-1) + 6(x+y-1) = 0$
and $6(x+y-1) + 6(y-1) = 0$ ⇒ $2x + y = 2$ and $x + 2y = 2$
Solving these equation we get

$$x = \frac{2}{3} \quad \text{and} \quad y = \frac{2}{3}$$

$f_{xx} = 2 = r$; $f_{xy} = 1 = s$; $f_{yy} = 2 = t$
$rt - s^2 = 4 - 1 = 3 > 0$
∵ $rt - s^2 > 0$ and $r > 0$
∵ At $r > 0$ we get minima

Now, $3d^2 = f\left(\frac{2}{3}, \frac{2}{3}\right) = 3\left(\frac{2}{3}-1\right)^2 + 3\left(\frac{2}{3}+\frac{2}{3}-1\right)^2 + 3\left(\frac{2}{3}-1\right)^2$

$$3d^2 = 3\times\frac{1}{9} + 3\times\frac{1}{9} + 3\times\frac{1}{9} = 1 \text{ unit}$$

7. Consider R^4 with the inner product $<x, y> = \sum_{i=1}^{4} x_i y_i$ for x = (x_1, x_2, x_3, x_4) and y = (y_1, y_2, y_3, y_4). Let M = {$(x_1, x_2, x_3, x_4) \in R^4 : x_1 = x_3$} and $M^\perp$ denote the orthogonal complement of M. The dimension of $M^\perp$ is equal to **GATE 2023**
(a) 1 (b) 2 (c) 3 (d) 4
Ans. (a)
Explanation :
Dim M = 3
Hence Dim $M^\perp = 1$

8. Let {v_1, v_2, v_3} be an orthogonal basis of R^3. Let V be the 3 × 3 matrix whose columns are v_1, v_2, v_3. Which of the following statements are necessarily true? **CSIR NET NOV 2020**
(a) $VV^T = I$ (b) $V^TV = I$ (c) $V = V^T$ (d) Determinant of V is not zero
Ans. (a), (b), (d)
Explanation :
We know that
Each column vectors of orthogonal matrix are orthonormal basis

9. Let {y_1, y_2, y_3, y_4} be an orthonormal basis of R^4. Which of the following are orthonormal bases? **CSIR NET NOV 2020**

(a) $\{y_1 + y_2, y_1 - y_2, y_3, y_4\}$ (b) $\{y_3, y_4, y_1, y_2\}$

(c) $\left\{\frac{y_1+y_2}{2}, \frac{y_1-y_2}{2}, y_3, y_4\right\}$ (d) $\left\{\frac{3y_1+4y_2}{5}, \frac{4y_1-3y_2}{5}, y_3, y_4\right\}$

Ans. (b), (d)

Explanation :

Given that $\{y_1, y_2, y_3, y_4\}$ be an orthonormal basis of R^4

Hence $\{y_1, y_2, y_3, y_4\}$ are column vectors of orthogonal matrix

Suppose $y_1 = (1,0,0,0)$; $y_2 = (0,1,0,0)$; $y_3 = (0,0,1,0)$; $y_4 = (0,0,0,1)$;

For option (a) := $A = \begin{bmatrix} 1 & 1 & 0 & 0 \\ 1 & -1 & 0 & 0 \\ 0 & 0 & 1 & 0 \\ 0 & 0 & 0 & 1 \end{bmatrix}$

Which is not orthogonal so option (a) is false.

For option (b) : $A = \begin{bmatrix} 0 & 0 & 1 & 0 \\ 0 & 0 & 0 & 1 \\ 1 & 0 & 0 & 0 \\ 0 & 1 & 0 & 0 \end{bmatrix}$

Which is orthogonal so option (b) is true.

For option (c) : $A = \begin{bmatrix} 1/2 & 1/2 & 0 & 0 \\ 1/2 & -1/2 & 0 & 0 \\ 0 & 0 & 1 & 0 \\ 0 & 0 & 0 & 1 \end{bmatrix}$

Which is not orthogonal so option (c) is false.

For option (d) : $A = \begin{bmatrix} 3/5 & 4/5 & 0 & 0 \\ 4/5 & -3/5 & 0 & 0 \\ 0 & 0 & 1 & 0 \\ 0 & 0 & 0 & 1 \end{bmatrix}$

Which is orthogonal so option (d) is true.

So, option (b) and (d) are correct answer.

Another method :

Given that $\{y_1, y_2, y_3, y_4\}$ is an orthonormal basis of R^4.

$\Rightarrow$ All vectors of are orthogonal and unit vector

(a) $S = \{y_1 + y_2, y_1 - y_2, y_3, y_4\}$

Here $y_1 + y_2$ is not unit vector

$$\| y_1+y_2 \| = \sqrt{< y_1+y_2, y_1+y_2 >} = \sqrt{< y_1,y_1 > + < y_2,y_1 > + < y_1,y_2 > + < y_2,y_2 >} = 2 \neq 1$$

(b) $\left\{\frac{y_1+y_2}{2}, \frac{y_1-y_2}{2}, y_3, y_4\right\}$ Here $\frac{y_1+y_2}{2}$ is not unit vector

$$\left\| \frac{y_1+y_2}{2} \right\| = \sqrt{< \frac{y_1+y_2}{2}, \frac{y_1+y_2}{2} >} = \sqrt{< \frac{y_1}{2}, \frac{y_1}{2} > + < \frac{y_2}{2}, \frac{y_1}{2} > + < \frac{y_1}{2}, \frac{y_2}{2} > + < \frac{y_2}{2}, \frac{y_2}{2} >} = 2 \neq 1$$

Because $< \frac{y_1}{2}, \frac{y_1}{2} > = \frac{1}{4} < y_1, y_1 >, < \frac{y_2}{2}, \frac{y_1}{2} > = 0, < \frac{y_1}{2}, \frac{y_2}{2} > = 0, < \frac{y_2}{2}, \frac{y_2}{2} > = \frac{1}{4} < y_2, y_2 >$

(d) $\left\{\frac{3y_1+4y_2}{5}, \frac{4y_1-3y_2}{5}, y_3, y_4\right\}$ $\qquad \left\|\frac{3y_1+4y_2}{5}\right\|^2 = \left\langle \frac{3y_1+4y_2}{5}, \frac{3y_1+4y_2}{5} \right\rangle$

$$\left\langle \frac{3}{5}y_1, \frac{3}{5}y_1 \right\rangle + \left\langle \frac{3}{5}y_1, \frac{4}{5}y_2 \right\rangle + \left\langle \frac{4}{5}y_2, \frac{3}{5}y_1 \right\rangle + \left\langle \frac{4}{5}y_2, \frac{4}{5}y_2 \right\rangle$$

$$\frac{9}{16}\langle y_1, y_1 \rangle + \frac{12}{25}\langle y_1, y_2 \rangle + \frac{12}{25}\langle y_2, y_1 \rangle + \frac{16}{25}\langle y_2, y_2 \rangle = \frac{9}{25} + \frac{16}{25} = 1$$

$\Rightarrow \frac{3y_1+4y_2}{5}$ is unit vector

Now for

$$\left\|\frac{3y_1-4y_2}{5}\right\|^2 = \left\langle \frac{3y_1-4y_2}{5}, \frac{3y_1-4y_2}{5} \right\rangle$$

$$\left\langle \frac{3}{5}y_1, \frac{3}{5}y_1 \right\rangle - \left\langle \frac{3}{5}y_1, \frac{4}{5}y_2 \right\rangle - \left\langle \frac{4}{5}y_2, \frac{3}{5}y_1 \right\rangle + \left\langle \frac{4}{5}y_2, \frac{4}{5}y_2 \right\rangle$$

$$\frac{9}{16}\langle y_1, y_1 \rangle - \frac{12}{25}\langle y_1, y_2 \rangle - \frac{12}{25}\langle y_2, y_1 \rangle + \frac{16}{25}\langle y_2, y_2 \rangle = \frac{9}{25} + \frac{16}{25} = 1$$

$\Rightarrow \frac{3y_1-4y_2}{5}$ is unit vector

and All vectors of (d) are orthogonal

So, (d) is true

10. Consider a inner product space P_2 of all polynomial of degree at most 2 over the field of real number with inner product $< f, g > = \int_0^1 f(t)g(t)dt$. Let $\{f_0, f_1, f_2\}$ be an orthogonal set in P_2, where $f_0 = 1$, $f_1 = t + c_1$, $f_2 = t^2 + c_2 f_1 + c_2$

Then the value of $2c_1 + c_2 + 3c_3$ is equal to – **GATE 2019**

(a) 3 (b) -3 (c) 2 (d) -2

Ans. **(b)**

Explanation :

Given that $\{f_0, f_1, f_2\}$ is orthogonal set then $<f_0, f_1> = <f_1, f_2> = <f_0, f_2> = 0$

Find $< f_0, f_1 > = < 1, t + c_1 > = \int_0^1 1(t + c_1)dt = \left[\frac{t^2}{2} + c_1 t\right]_0^1 = 0 \quad = \frac{1}{2} + c_1 = 0 \Rightarrow c_1 = -\frac{1}{2}$

Now $<f_1, f_2> = 0$

We know that $f_2 = t^2 + c_2 f_1 + c_3 = t^2 + c_2 (t + c_1) + c_3$

$$= t^2 + c_2 t - \frac{c_2}{2} + c_3 \qquad \left[c_1 = -\frac{1}{2}\right]$$

$$< f_1, f_2 > = \int_0^1 \left(t - \frac{1}{2}\right)\left(t^2 + c_2 t - \frac{c_2}{2} + c_3\right)dt = 0$$

$$= \int_0^1 \left(t^3 - \frac{t^2}{2} + c_2 t^2 - \frac{c_2 t}{2} - \frac{c_2 t}{2} + \frac{c_2}{4} + c_3 t - \frac{c_3}{2}\right)dt = 0$$

$$\left[\frac{t^4}{4} - \frac{t^3}{6} + \frac{c_2 t^3}{3} - \frac{c_2 t^2}{4} - \frac{c_2 t^2}{4} + \frac{c_2 t}{4} + \frac{c_3 t^2}{2} - \frac{c_3 t}{2}\right]_0^1 = 0$$

$$= \frac{1}{4} - \frac{1}{6} + \frac{c_2}{3} - \frac{c_2}{4} - \frac{c_2}{4} + \frac{c_2}{4} + \frac{c_3}{2} - \frac{c_3}{2} = 0 = \frac{2}{24} + \frac{c_2}{12} = 0 \Rightarrow c_2 = -1$$

Now $<f_0, f_2> = 0$

$$\int_0^1 1\left(t^2 - t + \frac{1}{2} + c_3\right)dt = 0 = \left[\frac{t^3}{3} - \frac{t^2}{2} + \frac{t}{2} + c_3 t\right]_0^1 = 0 = \frac{1}{3} - \frac{1}{2} + \frac{1}{2} + c_3 = 0$$

$$c_3 = \frac{-1}{3}$$

Then $2c_1 + c_2 + 3c_3$

$$2\left(-\frac{1}{2}\right) + (1) + 3\left(\frac{-1}{3}\right) = -3$$

11. Which of the following statements regarding quadratic form in 3 variable are true?

CSIR NET DEC 2019

(a) Any two quadratic form of rank 3 are isomorphic over R.

(b) Any two quadratic form of rank 3 are isomorphic over C

(c) There are exactly three non-zero quadratic form of rank ≤ 3 upto isomorphic over C.

(d) There are exactly three non-zero quadratic form of rank 2 upto isomorphic over R and C.

Ans. (b), (c)

Explanation :

Option (a) is incorrect.

Two quadratic form are isomorphic over R if they have same rank, index & signature.

Example. $Q_1(x, y, z) = x^2 + y^2 - z^2$
$Q_2(x, y, z) = x^2 + y^2 + z^2$

Coefficient matrix of Q_1 & Q_2

$$A_1 = \begin{bmatrix} 1 & 0 & 0 \\ 0 & 1 & 0 \\ 0 & 0 & -1 \end{bmatrix}, A_2 = \begin{bmatrix} 1 & 0 & 0 \\ 0 & 1 & 0 \\ 0 & 0 & 1 \end{bmatrix}$$

Both have rank = 3 but not equivalent.
Option (a) is incorrect.
Option (b) is correct.
Two quadratic form are isomorphic over C if they have same rank.
Option (c) is correct.
Rank = 1, Rank = 2, Rank = 3} each have exactly one quadratic form upto isomorphic over C
$\Rightarrow$ exactly three non-zero quadratic form upto isomorphic over C.
Option (d) is incorrect.
Exactly one quadratic form of rank 2 upto isomorphic over C.

12. Consider the quadratic form
$Q_1(x, y) = xy$ $\quad$ $Q_2(x, y) = x^2 + 2xy + y^2$ $\quad$ $Q_3(x, y) = x^2 + 3xy + 2y^2$

CSIR NET DEC 2018

(a) Q_1 and Q_2 are equivalent
(b) Q_1 & Q_3 are equivalent
(b) Q_2 and Q_3 are equivalent
(d) All are equivalent

Ans. (b)

Explanation :

Coefficient matrix of $Q_1 = \begin{bmatrix} 0 & 1/2 \\ 1/2 & 0 \end{bmatrix}$ of $Q_2 = \begin{bmatrix} 1 & 1 \\ 1 & 1 \end{bmatrix}$ and of $Q_3 = \begin{bmatrix} 1 & 3/2 \\ 3/2 & 2 \end{bmatrix}$

Q_1 and Q_3 have same rank and index
$\Rightarrow$ Both are equivalent.

13. Let A be nxnnon singular matrix with real entries. Let $B=A^T$ denote the transpose of A. Which of the following matrices are positive definite ? **CSIR NET DEC 2016**

(a) $A-B$ $\quad$ (b) $A^{-1} + B^{-1}$ $\quad$ (c) AB $\quad$ (d) ABA

Ans. (c)

Explanation :

Let A be n ×n non-singular matrix with real entries.
Let $B = A^T$

Let $A = \begin{bmatrix} 1 & 0 \\ 0 & -1 \end{bmatrix}$ $\quad$ Then $B = \begin{bmatrix} 1 & 0 \\ 0 & -1 \end{bmatrix}$

(a) $A + B = \begin{bmatrix} 2 & 0 \\ 0 & -2 \end{bmatrix}$ which is indefinite

(b) $A^{-1} = A$, $B^{-1} = B$ from the taken matrix $A^{-1}+B^{-1}=\begin{bmatrix}2 & 0\\ 0 & -2\end{bmatrix}$

Which is also indefinite
Similarly, ABA indefinite
Hence option (a), (b) & (d) are incorrect.
If A is non-singular then $A.A^T$ is always positive definite.
Therefore, AB is positive definite.
Hence, option (c) is correct.

14. Let C[0, 1] be the space of all real valued continuous function on [0, 1]. Define $<f,g>=\int_0^1 f(t)g(t)^2\,dt$ for all f, g ∈ [0, 1]. Then which of the following are TRUE?

(a) < , > is an inner product
(b) < , > is bilinear form but not inner product
(c) < , > is not bilinear form
(d) <f, f>≥ 0 ∀ f ∈ C[0, 1]

Ans. (c), (d) **CSIR NET DEC 2019**

Explanation :

Let f, g, h ∈ C[0, 1]

(1) $<\alpha f+\beta g,h>=\int_0^1[\alpha f(t)+\beta\ g(t)]\,[h(t)]^2\ dt$

$=\alpha\int_0^1 f\ h^2 dt+\beta\int_0^1 g\ h^2 dt\ =\alpha<f,h>+\beta<g,h>$

(2) $<f,\alpha g+\beta h>=\int_0^1 f(t)[\alpha g+\beta h]^2\,dt$

$\neq\alpha<f,g>+\beta<f,h>\ \Rightarrow$ not bilinear form.

and for inner product.

$<f,g>=\int_0^1 f(t)(g(t))^2\,dt\neq<g,f>\ \Rightarrow$ not I.P.

Option (a), (b) is incorrect.
Option (c) is correct.
Option (d) is correct.

$<f,f>=\int_0^1(f(t))^3\,dt\geq 0$

15. Let V be the vector space of polynomials in the variable t of degree atmost 2 over R.

An inner product on V is defined by $<f,g>=\int_0^1 f(t)g(t)dt$ **CSIR NET JUNE – 2019**

For f, g ∈ V. Let W = span{$1-t^2$, $1+t^2$} and $W^{\perp}$ be the orthogonal complement of W in V. Which of the following condition is satisfied for all h $\in W^{\perp}$

(a) h is even function i.e. h(t) = h(-t)
(b) h is odd function i.e.h(-t) = -h(t)
(c) h(t) = 0 has a real solution
(d) h(0) = 0

Ans. **(c)**

Explanation :

Definition :

(1) $W^{\perp} = \{u \in V \,|\, <u, w> = 0,, w \in W\}$

(2) u& w are called orthogonal if <u, w> = 0

Given that $V = \{a_0 + a_1t + a_2t^2 \mid a_0, a_1, a_2 \in R\}$

$w = \text{span}\{1 - t^2, 1 + t^2\}$

$W^{\perp}$ be the orthogonal complement of W

$w^{\perp} = \{u \in v \,|\, (u, w) = 0,\ w \in w\}$

Find the elements of $W^{\perp}$

$<(1 - t^2), (a_0 + a_1t + a_2t^2)> = 0 = \int_0^1 (1-t^2)(a_0 + a_1t + a_2t^2)dt = 0$

$$= \frac{2a_0}{3} + \frac{a_1}{4} + \frac{2a_2}{15} = 0 \quad \text{.... (1)}$$

$\& < (1+t^2), (a_0 + a_1t + a_2t^2) >= 0$

$$= \int_0^1 (1+t^2)(a_0 + a_1t + a_2t^2)dt = 0 = \frac{2a_0}{3} + \frac{a_1}{4} + \frac{2a_2}{15} = 0 \quad \text{.... (2)}$$

$\& \qquad < (1+t^2), (a_0 + a_1t + a_2t^2) >= 0$

$$= \int_0^1 (1+t^2)(a_0 + a_1t + a_2t^2)dt = 0 \quad = \frac{4a_0}{3} + \frac{3a_1}{4} + \frac{8a_2}{15} = 0 \quad \text{..... (3)}$$

..... (4)

Then $W^{\perp} = \left\{a_0(1 - \frac{16}{3}t + 5t^2) \mid a_0 \in R\right\}$

Let $h = 1 - \frac{16t}{3} + 5t^2 \in W^{\perp}$ for $a_0 = 1$

Option (a) is incorrect. h is not even

Option (b) is incorrect. h is not odd

Option (d) is incorrect. $h(0) = 1 \neq 0$

Option (c) is correct. $h = 15t^2 - 16t + 3 = 0$

$t = \frac{16 \pm \sqrt{256 - 180}}{30} \Rightarrow$ h(t) = 0 has real solution.

16. Let $A = \begin{bmatrix} 3 & 1 & 2 \\ 1 & 2 & 3 \\ 2 & 3 & 1 \end{bmatrix}$ and $Q(X) = X^T AX$ for $X \in R^3$. Then **CSIRNET JUNE 2017**

(a) A has exactly two positive eigenvalues
(b) All the eigenvalues of A are positive
(c) $Q(X) \geq 0$ for all $X \in R^3$.
(d) $Q(X) < 0$ for some $X \in R^3$.

Ans. **(a), (d)**

Explanation :

Given that $A = \begin{bmatrix} 3 & 1 & 2 \\ 1 & 2 & 3 \\ 2 & 3 & 1 \end{bmatrix}$

Eigen values of A are 6, 3i, -3i. A has exactly two positive eigen values

Option (A) is correct

We know that one eigen values is negative
So, option (d) is correct.

17. Which of the following real quadratic forms on R^2 is positive definite?

(a) $Q(X, Y) = XY$
(b) $Q(X, Y) = X^2 - 2XY + Y^2$
(c) $Q(X, Y) = X^2 + 2XY + Y^2$
(d) $Q(X, y) = X^2 + XY$

Ans. **(b)** **CSIR NET JUNE 2020**

Explanation :

General form of a real quadratic form on R^2 be $Q(x, y) = ax^2 + dy^2 + 2cxy$ and the corresponding matrix is $\begin{bmatrix} a & c \\ c & d \end{bmatrix}$, which is symmetric and positive definite,

ie, $D_1 = a > 0$ and $D_2 = ad - c^2 > 0$. We can check options one by one.

(a) $Q(X, Y) = XY$, the corresponding matrix is $\begin{bmatrix} 0 & \frac{1}{2} \\ \frac{1}{2} & 0 \end{bmatrix}$.

$D_1 = 0 > 0$, so Q is not positive definite.

(b) $Q(X,Y) = X^2 - XY + Y^2$, the corresponding matrix is $\begin{bmatrix} 0 & -\frac{1}{2} \\ -\frac{1}{2} & 1 \end{bmatrix}$, $D_1 = 1 > 0$

and $D_2 = \frac{3}{4} > 0$, Q is positive definite.

(c) $Q(X,Y) = X^2 + 2XY + Y^2$, the corresponding matrix is $\begin{bmatrix} 1 & 1 \\ 1 & 1 \end{bmatrix}$, $D_1 = 1 > 0$ and

$D_2 = 0 > 0$. Q is not positive definite.

(d) $Q(X,Y) = X^2 + XY$, the corresponding matrix is $\begin{bmatrix} 1 & \frac{1}{2} \\ \frac{1}{2} & 0 \end{bmatrix}$, $D_1 = 1 > 0$ and $D_2 =$

$-\frac{1}{4} > 0$. Q is not positive definite. So option (b) is correct.

18. for a, b $\in$ R, Let $p(x, y) = a^2x_1y_1 + abx_2y_1 + abx_1y_2 + b^2x_2y_2$, $x = (x_1, x_2)$, $y = (y_1, y_2)$ $\in R^2$. For what values of a and b does $p : R^2 \times R^2 \rightarrow R$ define an inner product?

(a) $a > 0, b > 0$ (b) $ab > 0$ (c) $a = 0, b = 0$ (d) For no values of a, b

Ans. (d) **CSIR NET JUNE 2020**

Explanation :

For a, b $\in$ R, let $p(x, y) = a^2x_1y_1 + abx_2y_1 + abx_1y_2 + b^2x_2y_2$, $x = (x_1,x_2)$, $y = (y_1,y_2) \in R^2$. We have, $\langle x, y \rangle = x^TAy$, where A is symmetric and positive definite matrix, ie, $D_1, D_2 > 0$.

Here the matrix corresponding to p(x, y) is $\begin{bmatrix} a^2 & ab \\ ab & b^2 \end{bmatrix}$, $D_1 = a^2 > 0$, $D_2 = a^2b^2 - a^2b^2 = 0$ > 0, therefore, for no values of a and b, $p: R^2 \times R^2 \rightarrow R$ define an innver product . Option (d) is correct.

UNSOLVED QUESTIONS

1. Which of the following are TRUE? **CSIR NET DEC 2019**

(a) Any two quadratic form of same rank in n-variables over R are isomorphic.
(b) Any two quadratic form of same rank in n-variables over C are isomorphic.
(c) Any two quadratic form in n-variables are isomorphic over C.
(d) A quadratic form in 4 variables may be isomorphic to a quadratic form in 10 variables.

2. For a quadratic form in 3 variable over R. Let r be the rank and s be the signature. The number of possible pair (r, s) is **CSIR NET DEC 2019**
(a) 13 (b) 9 (c) 10 (d) 16

3. Consider a matrix $A = (a_{ij})_{5\times5}$;$1 \le i, j \le 5$ s.t. $a_{ij} = \dfrac{1}{n_i + n_j + 1}$ where $n_i, n_j \in N$

CSIR NET JUNE 2019

Then in which of the following cases A is a positive definite matrix.
(a) $n_i = i$ for all i = 1, 2, 3, 4, 5 (b) $n_1 < n_2 < n_3 \ldots\ldots < n_5$
(c) $n_1 = n_2 = \ldots\ldots = n_5$ (d) $n_1 > n_2 \ldots\ldots > n_5$

4. Let $A = \begin{bmatrix} 1 & 2 & 0 \\ 0 & 0 & -2 \\ 0 & 0 & 1 \end{bmatrix}$ and define for x, y, z ∈ R. Which of the following are TRUE?

CSIR NET JUNE 2018

(a) The matrix of second order partial derivative of quadratic form Q is 2A
(b) The rank of quadratic form Q is 2 (c) The signature of quadratic form Q is (+ + 0)
(d) The quadratic form Q takes the value 0 for some non-zero vector (x, y, z)

5. Consider the quadratic forms q and p given by $q(x, y, z, w) = x^2 + y^2 + z^2 + bw^2$ and $p(x, y, z, w) = x^2 + y^2 + cwz$, Which of the following statements are true?

CSIR NET JUNE 2013

(a) p and q are equivalent over C if b and c are non-zero complex numbers.
(b) p and q are equivalent over R if b and c are non-zero real numbers.
(c) p and q are equivalent over R if b and c are non-zero real numbers with b negative
(d) p and q are NOT equivalent over R if c = 0

6. Let S = {(0,1,1), (1, 0, 1), (-1, 2, 1)} ⊆ R. Suppose R^3 is endowed with the standard inner product <,>. Define M = {x ∈ R^3 : <x, y> = 0 for all y ∈ S}. Then the dimension of M equals **GATE – 2007**
(a) 0 (b) 1 (c) 2 (d) 3

7. Let T : $R^3 \to R^3$ be a linear transformation defined by T(x, y, z) = (x + y, y + z, z – x). Then an orthonormal basis for the range of T is **GATE-2010**

(a) $\left\{\left(\frac{1}{\sqrt{2}},\frac{1}{\sqrt{2}},0\right),\left(\frac{1}{\sqrt{3}},-\frac{1}{\sqrt{3}},.\frac{1}{\sqrt{3}}\right)\right\}$ (b) $\left\{\left(\frac{1}{\sqrt{2}},-\frac{1}{\sqrt{2}},0\right),\left(\frac{1}{\sqrt{6}},\frac{1}{\sqrt{6}},\frac{2}{\sqrt{6}}\right)\right\}$

(c) $\left\{\left(\frac{1}{\sqrt{2}},\frac{1}{\sqrt{2}},0\right),\left(\frac{1}{\sqrt{6}},-\frac{1}{\sqrt{6}},-\frac{2}{\sqrt{6}}\right)\right\}$ (d) $\left\{\left(\frac{1}{\sqrt{2}},\frac{1}{\sqrt{2}},0\right),\left(\frac{1}{\sqrt{3}},-\frac{1}{\sqrt{3}},-\frac{1}{\sqrt{3}}\right)\right\}$

8. Let T : $C^3 \to C^3$ be defined by $T\begin{pmatrix} z_1 \\ z_2 \\ z_3 \end{pmatrix} = \begin{pmatrix} z_1 - iz_2 \\ iz_1 + z_2 \\ z_1 + z_2 + iz_3 \end{pmatrix}$. Then, the adjoint T* of T is given by $T^*\begin{pmatrix} z_1 \\ z_2 \\ z_3 \end{pmatrix} =$

(a) $\begin{pmatrix} z_1 - iz_2 \\ -iz_1 + z_2 \\ z_1 + z_2 - iz_3 \end{pmatrix}$ (b) $\begin{pmatrix} z_1 - iz_2 + z_3 \\ -iz_1 + z_2 + z_3 \\ iz_3 \end{pmatrix}$ (c) $\begin{pmatrix} z_1 - iz_2 + z_3 \\ iz_1 + z_2 + z_3 \\ -iz_3 \end{pmatrix}$ (d) $\begin{pmatrix} iz_1 + z_2 \\ z_1 - iz_2 \\ z_1 - z_2 - iz_3 \end{pmatrix}$

9. Let V be the inner product space consisting of linear polynomials, p : [0, 1] → R(i.e., V consists of polynomials p of the form p(x) = ax + b; a, b ∈ R), with the inner product defined by $\langle p,q \rangle = \int_0^1 p(x)q(x)dx \ for \ p,q \in V$. An orthonormal basis of V is

CSIR NET DEC – 2012

(a) {1, x} (b) $\{1, x\sqrt{3}\}$ (c) $\{1, (2x-1)\sqrt{3}\}$ (d) $\left\{1, x - \frac{1}{2}\right\}$

10. Let $A = \begin{bmatrix} 1 & 0 \\ 0 & -1 \end{bmatrix}$. Let f : $R^2 \times R^2 \to R$ be defined by f(v, w) = w^TAv. Pick the correct statement from below :

CSIR NET JUNE 2016

(a) There exists an eigenvector v of A such that Av is perpendicular to v.
(b) The set $\{v \in R^2 \mid f(v, v) = 0\}$ is a non-zero subspace of R^2.
(c) If v, w ∈ R^2 are non-zero vectors such that f(v, v) = 0 = f(w, w), then v is a scalar multiple of w.
(d) For every v ∈ R^2, there exists a non zero w ∈ R^2 such that f(v, w) = 0

11. Consider the quadratic form Q(v) = v^tAv, where $A = \begin{bmatrix} 1 & 0 & 0 & 0 \\ 0 & 1 & 0 & 0 \\ 0 & 0 & 0 & 1 \\ 0 & 0 & 1 & 0 \end{bmatrix}, v = (x, y, z, w)$.

Then,

CSIR NET DEC 2015

(a) Q has rank 3.
(b) $xy + z^2$ = Q(Pv) for some invertible 4 × 4 real matrix P.
(c) $xy + y^2 + z^2$ = Q(Pv) for some invertible 4 × 4 real matrix P.
(d) $x^2 + y^2 - zw$ = Q(Pv) for some invertible 4 × 4 real matrix P.

12. Let $a_{ij} = a_i a_j$, $1 \le i, j \le n$, where $a_1, \ldots, a_n$ are real numbers. Let $A = ((a_{ij}))$ be the $n \times n$ matrix $((a_{ij}))$. Then **CSIR NET JUNE 2011**

(a) It is possible to choose $a_1, \ldots, a_n$ so as to make the matrix A non-singular.
(b) The matrix A is positive definite if $(a_1, \ldots, a_n)$ is a non-zero vector.
(c) The matrix A is positive semidefinite for all $(a_1, \ldots a_n)$.
(d) for all $(a_1, \ldots, a_n)$, zero is an eigenvalue of A.

13. Let $V = \{f : [0, 1] \to \mathbb{R} \mid f \text{ is a polynomial of degree less than or equal to } n\}$. Let $f_j(x) = x^j$ for $0 \le j \le n$ and let A be the $(n + 1) \times (n + 1)$ matrix given by

$$a_{ij} = \int_0^1 f_i(x) f_j(x) dx.$$

Then which of the following is/are true?

(a) dim V = n
(b) dim V > n.
(c) A is nonnegative definite, i.e. for all $v \in \mathbb{R}^n$, $\langle Av, v \rangle \ge 0$.
(d) det A > 0.

14. Let A be an $n \times n$ matrix with real entries. Define $\langle x, y \rangle_A = \langle Ax, Ay \rangle$, $x, y \in \mathbb{R}^n$. Then $\langle x, y \rangle_A$ defines an inner product if and only if **CSIR NET JUNE 2016**

(a) ker A = {0}
(b) rank A = n
(c) All eigenvalues of A are positive
(d) All eigenvalues of A are non-negative

15. For any real square matrix M let $\lambda^+(M)$ be the number of positive eigenvalues of M counting multiplicities. Let A be an $n \times n$ real symmetric matrix and Q be an $n \times n$ real invertible matrix. Then **CSIR NET DEC 2013**

(a) Rank A = Rank $Q^T AQ$
(b) Rank A = Rank $Q^{-1}AQ$
(c) $\lambda^+(A) = \lambda^+(Q^T AQ)$
(d) $\lambda^+(A) = \lambda^+(Q^{-1}AQ)$

16. Let ζ be a primitive fifth root of unity. Define $A = \begin{pmatrix} \xi^{-2} & 0 & 0 & 0 & 0 \\ 0 & \xi^{-1} & 0 & 0 & 0 \\ 0 & 0 & 1 & 0 & 0 \\ 0 & 0 & 0 & \xi & 0 \\ 0 & 0 & 0 & 0 & \xi^2 \end{pmatrix}$

For a vector $v = (v_1, v_2, v_3, v_4, v_5) \in \mathbb{R}^5$, define $|v|_A = \sqrt{|vAv^T|}$ where v^T is transpose of v. If w = (1, -1, 1, 1, -1), then $|w|_A$ equals

(a) 0
(b) 1
(c) -1
(d) 2

17. Let V be the inner product space consisting of linear polynomials, $p : [0, 1] \to \mathbb{R}$ (i.e., V consists of polynomials p of the form $p(x) = ax + b$, $a, b \in \mathbb{R}$), with the inner product defined by $\langle p, q \rangle = \int_0^1 p(x)q(x)dx$ for $p, q \in V$. An orthonormal basis of V is

(a) $\{1, x\}$ (b) $\{1, x\sqrt{3}\}$ (c) $\{1, (2x-1)\sqrt{3}\}$ (d) $\left\{1, x-\frac{1}{2}\right\}$

18. Consider R^3 with the standard inner product. Let W be the subspace of R^3 generated by (1, 0, -1). Which of the following is a basis for orthogonal complement of W?

(a) {(1, 0, 1), (0, 1, 0)} (b) {(1, 2, 1), (0, 1, 1)}

(c) {(2, 1, 2), (4, 2, 4)} (d) {(2, -1, 2), (1, 3, 1), (-1, -1, -1)}

19. Let $A = \begin{bmatrix} 1 & 0 \\ 0 & -1 \end{bmatrix}$. Let f : $R^2 \times R^2 \rightarrow R$ be defined by $f(v, w) = w^T Av$.

Pick the correct statement from below :

(a) There exists an eigenvector v of A such that Av is perpendicular to v.

(b) The set $\{v \in R^2 \mid f(v, v) = 0\}$ is a non-zero subspace of R^2.

(c) If v, w $\in R^2$ are non-zero vectors such that f(v, v) = 0 = f(w, w) than v is a scalar multiple of w.

(d) For every v $\in R^2$, there exists a non-zero w $\in R^2$ such that f(v, w) = 0

20. Let A be an n ×n self-adjoint matrix with eigenvalues $\lambda_1, ..., \lambda_n$.

Let $\| X \|_2 = \sqrt{|x_1|^2 + ...+ |x_n|^2}$ forX = $(x_1, ..., x_n) \in C^n$.

IfP(A) = $a_0 I + a_1 A + ... + a_n A^n$. Then $\sup_{\|X\|_2 = 1} \|p(A)X\|_2$ is equal to

(a) max $\{a_0 + a_1\lambda_j + ... + a_n\lambda_j^n : 1 \le j \le n\}$

(b) max $\{|a_0 + a_1\lambda_j + ... + a_n\lambda_j^n| : 1 \le j \le n\}$

(c) min $\{a_0 + a_1\lambda_j + ... + a_n\lambda_j^n : 1 \le j \le n\}$

(d) min $\{|a_0 + a_1\lambda_j + ... + a_n\lambda_j^n| : 1 \le j \le n\}$

21. Let R^n, $n \ge 2$ be equipped with standard inner product. Let $\{v_1, v_2, ..., v_n\}$ be n-column vectors forming an orthonormal basis of R^n. Let A be the n ×n matrix formed by the column vectors $v_1, ..., v_n$. Then

(a) $A = A^{-1}$ (b) $A = A^T$ (c) $A^{-1} = A^T$ (d) Det(A) = 1

22. Consider the vector space V of real polynomials p of degree less than or equal to n. Fix distinct real numbers $a_0, a_1, ..., a_k$. For p $\in$ V max $\{|p(a_j)| : 0 \le j \le k\}$ defines a norm on V

CSIR NET JUNE 2017

(a) Only if k < n (b) Only if $k \ge n$ (c) If $k + 1 \le n$ (d) If $k \ge n + 1$

23. Let y be non-zero vector in an inner product space V. Then which of the following are subspaces of V?

CSIR NET DEC 2013

(a) $\{x \in V \mid \langle x, y\rangle = 0\}$ (b) $\{x \in V \mid \langle x, y\rangle = 1\}$

(c) {x ∈ V | <x, z> = 0 for all z such that <z, y> = 0}
(d) {x ∈ V | <x, z> = 1 for all z such that <z, y> = 1}.

24. Let u, v, w be vectors in an inner product space V, satisfying ||u|| = ||v|| = ||w|| = 2 and <u, v> = 0, <u, w> = 1, <v, w> = -1. Then which of the following are true?

CSIR NET JUNE 2014

(a) $\| w+v+u \| = 2\sqrt{2}$

(b) $\left\{\frac{1}{2}u, \frac{1}{2}v\right\}$ forms an orthonormal basis of a two dimensional subspace of V.

(c) w and 4u-w are orthogonal to each other. (d) u, v, w are necessarily linearly independent.

25. Let f be a non-zero symmetric bilinear form on R^3. Suppose that there exists linear transformations $T_i : R^3 \to R$, i = 1, 2 such that form all α, β∈ R^3, f(α, β) = T_1(α), T_2(β). Then

CSIR NET DEC 2014

(a) rank f = 1 (b) dim {β∈ R^3 : f(α, β) = 0 for all α = R^3} = 2
(c) f is positive semi-definite or negative semi-definite.
(d) {α : f (α, α) = 0} is a linear subspace of dimension 2.

26. For the vectors α, β, γ∈ V(F) and non-zero scalars, the inner product <α, cβ + γ> is not equal to
(a) $\bar{c} < \alpha, \beta > + < \alpha, \gamma >$ (b) $\bar{c} < \beta, \alpha > + < \gamma, \alpha >$
(c) $\bar{c} < \alpha > + \beta$ (d) $\bar{c}\alpha\gamma + \alpha\beta$

27. Let V be the vector space of polynomials f(t) with inner product $< f.g >= \int_{-1}^{1} f(t)g(t)dt$. According to Gram-Schmidthorthogonalization process on {1, t, t^2, t^3}
(a) {1, t, $3t^2 - 1$, $5t^3 - 3t$} is orthogonal basis.
(b) {1, t, $3t^2 + 1$, $5t^3 + 3t$} is orthogonal basis.
(c) $\left\{1, t, \frac{1}{2}(3t^2 - 1), \frac{1}{2}(5t^3 - 3t)\right\}$ is orthogonal basis (d) None of the above.

28. Eigenvalues of the matrix with the quadratic form $x_1^2 - 4x_1x_2 + x_2^2 + 2x_1 + 4x_2 + 5$ equals
(a) 3 (b) -1 (c) 2 (d) -3

29. A symmetric matrix in $M_n(R)$ is said to be non-negative definite if $x^T Ax \geq 0$ for all (column vectors) x $\in R^n$. Which of the following statements are true? **NBHM – 2013**
(a) If a real symmetric n × n matrix is non-negative definite, then all of its eigenvalues are non-negative.

(b) If a real symmetric n × n matrix has all its eigenvalues non-negative, then it is non-negative definite.

(c) If A $\in M_n(R)$ then AA^T is non-negative definite. (d) None of these

30. Only one of the following matrices is non-negative definite. Find it **NBHM – 2013**

(a) $\begin{bmatrix} 5 & -3 \\ -3 & 5 \end{bmatrix}$ (b) $\begin{bmatrix} 1 & -3 \\ -3 & 5 \end{bmatrix}$ (c) $\begin{bmatrix} 1 & 3 \\ 3 & 5 \end{bmatrix}$ (d) None of these

31. Which of the following statements are true? **NBHM – 2018**

(a) If A $\in M_n(R)$ such that (Ax, x) = 0 for all x $\in R^n$, then A = 0

(b) If A $\in M_n(R)$ such that (Ax, x) = 0 for all x $\in C^n$, then A = 0

(c) If A $\in M_n(R)$ such that (Ax, x) $\geq$ 0 for all x $\in C^n$, then $A = A^*$

(d) None of these

32. Let v be a non-zero vector in R^3. Consider the linear transformation T : $R^3 \to R^3$ defined by $T(w) = w - 2\dfrac{(w.v)}{(v.v)}v$. Which of the following is false?

(a) $T^2 = 1$ (b) det T = 1 (c) T(v) = -v

(d) There is a non-zero vector w such that T(w) = w.

33. Let T_1, T_2 be two linear transformations from R^n to R^n. Let $\{x_1, x_2, \ldots, x_n\}$ be a basis of R^n. Suppose that $T_1(x_i) \neq 0$ for every i = 1, 2, …, n and $x_i \perp \ker T_2$ for every i = 1, 2, …, n. Which of the following is/are necessarily true?

(a) T_1 is invertible (b) T_2 is invertible

(c) Both T_1, T_2 are invertible (d) Neither T_1 nor T_2 is invertible.

Answer Key

1	2	3	4	5	6	7	8	9	10
b,d	c	a,b,d	d	a,c,d	b	c	c	d	d
11	**12**	**13**	**14**	**15**	**16**	**17**	**18**	**19**	**20**
d	c,d	b,c,d	a,b	a,b,c,d	a	c	a	d	b
21	**22**	**23**	**24**	**25**	**26**	**27**	**28**	**29**	**30**
c	b,d	a,c	a,b,c,d	a,b,c,d	b,c,d	a,c	a,b	a,b,c	a
31	**32**	**33**							
b,c	b	b							

Chapter 1

Elementary Set Theory, Countability and Point Set

REAL ANALYSIS

1. Vector Space :

Let (F,+,.) be a field. The elements of F are called scalars. Let V be a non-empty set whose elements are called vectors. Then V is a vector space over the field F (denoted by V(F)) if

1. SET

A collection of well defined and distinct objects is called a set.

Example : (1) N = set of natural numbers. (2) Z = set of integers.

Subset: A set 'A' is said to be subset of a set B, if A is contained in B i.e. if all the elements of A are also elements of B and is denoted as '$A \subseteq B$'.

Example : (1) A = {1, 2} is a subset of B = {1, 2, 3}.

Note : If B is a set containing 'n' elements, then the total number of subsets ofB is 2^n.

Proper subset: A set 'A' is said to be proper subset of a set 'B' if A is properly contained in B i.e. if all elements of A are also in B and there exists an element of B which is not in A and we write $A \subset B$. If B is a set containing n elements, then number of proper subsets of B is $2^n - 1$.

Cardinality of a set: Number of elements in a set A is called Cardinality of a set.

CARTESIAN PRODUCT OF SETS

For any two sets A and B, Cartesian product of A and B is defined as

$A \times B = \{(a,b) : a \in A, b \in B\}$

In general, $A_1 \times A_2 \times A_3 \times \ldots \times A_n = \{(a_1, a_2, \ldots, a_n) : a_i \in A_i \text{ where } i = 1, 2, \ldots, n\}$

EQUIVALENT SETS / SIMILAR SETS

Two sets A and B are said to be equivalent if $\exists$ a bijection, i.e., a one-one and onto function, from the setA to B and is written as $A \sim B$. Equivalent sets are also called equipotential, equinumeral, or similar sets. A and B are similar sets iff $|A| = |B|$.

For example: The sets N and E ={2, 4, 6, ...}, set of all even natural numbers are equivalent because the function $F : N \to E$ defined by f(n) = 2n, $n \in N$ is one-one from N onto E.

Examples:

(1) $(a,b) \sim (a,b] \sim [a,b] \sim [a,b)$, where a, b $\in$ R

(2) $N \sim Z \sim Q \sim N \times N \sim Q \times Q \times Q$

(3) $R \sim Q'$

(4) $(a,b) \sim [c,d]$, where a, b, c, d $\in$ R

FINITE SET

Let N_m = (1,2,3,...,m}. A set which is either empty or equivalent to N_m is said to be a finite set. Thus a set S is finite if $S = \phi$ or S ~ N_m for some natural number m.

Note. IfS ~ $N_m$then m is called the cardinal number of S.

INFINITE SET

A Set S which is not finite is called an infinite set. Thus S is an infinite set, then S is not equivalent to N_m for any $m \in N$.

Results on infinite set:

(1) Every infinite set is equivalent to a proper subset of itself.

(2) Sis infinite and $S \subset A \Rightarrow A$ is infinite (Every super-set of an infinite set is infinite)

DENUMERABLE (OR ENUMERABLE) SET / COUNTABLY INFINITE SETS

A set S is said to be denumerable if it is equivalent to N(the set of natural numbers). Thus S denumerable $\Rightarrow S \sim N$.

Example : Z (The set of all integers) is a denumerable set.

COUNTABLE SET

A set S is said to be countable if either S is finite or S is denumerable, i.e., if either S is finite or $S \sim N$ the set of all natural numbers.

Examples of Countable Set:

(i) The set of all rational numbers is countable i.e. Q is countable.

(ii) The set of all positive rational numbers is countable i.e., Q^+ is countable.

(iii) The set $N \times N$ is countable.

(iv) The set of all polynomials of degree less than or equal to n with rational coefficients is countable.i.e. $S = \{p(x) = a_0 + a_1x + a_2x^2 + ... + a_nx^n : a_i \in Q, \text{ where } i = 1, 2, ..., n\}$. S is countable.

(v) The set of all the intervals with rational end points is countable. Let $S = \{(a, b) : a, b \in Q\}$, then, $|S| = |Q \times Q|$. S is countable.

(vi) Let S be a subset of R, then the set of isolated points of S is countable.

(vii) The set of all functions from Z_2 to N is countable.

(viii) The set of disjoint intervals in R is countable.

(ix) The set of all finite subsets of N is countable.

(x) The set of algebraic numbers is countable.

UNCOUNTABLE SET

A set which is not countable is said to be an uncountable set. Thus a set A is uncountable if A is not finite and A is not equivalent to N.

Results on Uncountable set:

(1) Every super-set of an uncountable set is uncountable.

(2) If A is uncountable, B is countable, then A - B is uncountable.

(3) Infinite product of denumerable sets is uncountable.

Examples of Uncountable set:

(1) The set of real numbers x such that $0 \le x \le 1$ is not countable, i.e., the unit interval [0, 1] is uncountable

(2) The set of real numbers is uncountable.

(3) The set of irrational numbers is uncountable.

(4) The set of all sequences with entries 0 and 1 is uncountable, i.e., if $S = \{f \mid f : N \to \{0, 1\}\}$ then S is uncountable.

(5) The set of all functions from R to R is uncountable.

(6) The set $S = \{f \mid f : Z \to Z\}$ is uncountable.

(7) The derived set of an uncountable set is uncountable.

Properties of Sets :

(1) Every countable set is similar to subset of N (Set of natural number).
(2) A set is infinite iff it contains a countably infinite subset.
(3) Every infinite set is similar to atleast one of its proper subset.
(4) Countable union of countable sets is countable .

Example : $\bigcup_{i=1}^{\infty} A_i$ is countable if each A_i is countable.

(5) Family of all finite subsets of a countable set is countable.
(6) Family of all subsets of countable infinite sets is uncountable.
(7) P(N), P(Q), P(Z) all are uncountable. Where P(X) is power set of X.
(8) Finite cartisian product of countable sets is countable.
Example : $N \times N \times \ldots\ldots \times N$ is countable set.
(9) Infinite cartisian product of countable sets $A_1, A_2, \ldots\ldots\ldots A_i$, $|A_i| \geq 2$ is uncountable.
(10) Set of all polynomials with rational co-efficient is countable set.

Cardinality of infinite sets :

We usually denote the cardinality of denumerable set as χ_0 (aleph naught) and the cardinality of uncountable set as $\cent$ (continum)

Result

(1) $\chi_0 < \cent$ (2) $2^{\chi_0} = \cent$ (3) $\chi_0^{\chi_0} = \cent$ (4) $\cent^{\cent} = 2^{\cent}$ (5) $n\chi_0 = \chi_0$; $n \in N$ (6) $\chi_0^n = \chi_0$; $n \in N$

Example : $|Q| = \chi_0$& $|R| = \cent$

2. Important definition of point set of topology

NEIGHBOURHOOD OF A POINT

A subset S of Ris called a neighbourhood of a point 'a' if there exists an open interval I containing a and itself is contained in S, i.e., a subset S of Ris called a neighbourhood of a point a if $\exists$ an open interval I containing a such that$I \subset S$.

DELETED NEIGHBOURHOOD OF A POINT

If from the neighbourhood of a point, the point itself is excluded, then we get the deleted neighbourhood of that point. Thus, if S is a neighbourhood of a point a, then the set S – {a} is called a deleted neighbourhood of a.

INTERIOR POINT OF A SET

A real number ais called an interior point of a set S if S is a neighbourhood of a, i.e., $\exists \in > 0$ such that $(a - \in, a + \in) \subset S$

Examples

(i) Every point of the open interval (a, b) is its interior point.
(ii) a and b are not interior points of the closed interval [a, b] since, [a, b] is not a neighbourhood of a and b
(iii) S = N then no real number is interior point of S.

INTERIOR OF A SET

The set of all interior points of a set S is called the interior of S and is denoted by S^0.

Examples

(i) If A = (a, b), then $A^0 = A$, because every point of A is an interior point of A.

(ii) If = [a, b], then $A^0 = (a, b)$, because every point of A is an interior point of A except the end points a and b.

(iii) $R^0 = R$ because R is a neighbourhood of each of its points and so, every point of R is an interior point of R.

(iv) If A is a non-empty finite set, then $A^0 = \phi$

(v) $N^0 = \phi$, $Z^0 = \phi$, $Q^0 = \phi$ and $(Q^c)^o = \phi$ because N, Z, Q & Q^c are not neighbourhoods of any of its points and therefore, no point is an interior point of N, Z, Q.

OPEN SET

Let S be a subset of R, then S is said to be open if it is a neighbourhood of each of its points. Equivalently, S is open if for each $x \in S, \exists \in > 0$ such that $(x - \in, x + \in) \subset S$ and if $S^o =$ S then S is open set

Results on open set

(1) The union of two open sets is an open set.
(2) The union of an arbitrary family of open sets is an open set.
(3) The intersection of two open sets is an open set.
(4) The intersection of a finite number of open sets is an open set.
(5) The intersection of an infinite family of open sets may or may not be an open set

Example : $S = \left(1 - \frac{1}{n}, 1 + \frac{1}{n}\right)$ then $\cap S = \{1\}$ which is not open.

Exterior point:

Let $a \in R$ and $S \subseteq R$, we say a is an exterior point of S if it is an interior point of S^c in R (where S^c means complement of S in R). Set of all exterior point of S is denoted by Ext S.

Examples :

(a) S = Q then no real number which is Exterior point of Q.

(b) S = [0, 1] then set of all exterior point of S is $(-\infty, 0) \cup (1, \infty)$.

LIMIT POINT OF A SET

Definition I A point $a \in R$ is said to be a limit point of a subset S of R if every neighbourhood of 'a' has a point of S other than 'a'. In symbols, a point $a \in R$ is said to be a limit point of a subset S of R if for every $\in > 0$ such that $(a - \in, a + \in) \cap S - \{a\} \neq \phi$.

Definition 2 A point $a \in R$ is said to be a limit point of a subset S of R if every neighbourhood of a contains infinitely many points of S.

Example : (a) $S = \left\{\frac{1}{n} : n \in N\right\}$ then 0 is only limit point of S.

(b) S = [0, 1) then set of all limit points of S is [0, 1].

DERIVED SET

The set of all limit points of a set S is called the derived set of S and is denoted by S'. Thus

$$S' = \{x : x \text{ is a limit point of } S\}$$

CLOSED SET

Let S be a subset of R, then S is said to be a closed if its complement, A^c is an open set.

Another definition : Let S be a subset of R, then S is said to be a closed if it contain all its limit points.

Results on closed set:

The union of an infinite family of closed sets need not be a closed set.

Counter example:

Let $A_n = \left[\frac{1}{n},1\right] \forall n \in N$, then $\{A_n\}_{n\in N}$ is an infinite family of closed sets.

$A_1 = \{1\}, A_2 = \left[\frac{1}{2},1\right], A_3 = \left[\frac{1}{3},1\right],..., \bigcup_{n=1}^{\infty} A_n = \{1\} \cup \left[\frac{1}{2},1\right] \cup \left[\frac{1}{3},1\right] \cup ... = (0,1]$, which is not a closed set.

(4) The intersection of an arbitrary family of closed sets is a closed set.

Examples

(i) $R^c = R / R = \phi$, which is open $\Rightarrow$ R is closed.

(ii) $\phi^c = R/\phi = R$, which is open $\Rightarrow \phi$ is closed.

(iii) R and ϕ are the only two sets which are both open and closed in R.

(iv) The sets [a, b) and (a, b] are neither open nor closed in R.

(v) The set Z of all integers is a closed set.

(vi) Every countable set is closed set.

Example Which of the following sets are closed, open, neither closed nor open?

(i) $\{x : 0 \le x \le 1\}$ (ii) $[0,1] \cup [2,3]$ (iii) $\{x : 1 < x < 8\}$ (iv) $\{x : 4 \le x < 8\}$

Solution

(i) $\{x : 0 \le x \le 1\} = [0,1]$ being a closed interval is a closed set.

(ii) $[0,1]$ and $[2,3]$ being closed interval are closed set and so their union $[0,1] \cup [2,3]$ is also a closed set

(iii) $\{x : 1 < x < 8\} = (1,8)$ being an open interval is an open set.

(iv) $\{x : 4 \le x < 8\} = [4,8)$ is neither closed nor open set.

CLOSURE OF A SET

Let S be a set then its closure is define as $\bar{S} = S \cup S'$, where S` is derived set of S.

Results on closure of a set:

(1) For any set A, $\bar{A}$ is the smallest closed set containing A.

(2) A set A is said to be closed if $A = \bar{A}$, i.e., if $A' = \phi$

FRONTIER POINT :

a $\in$ R is said to be a frontier point of S $\subseteq$ R if it is neither an interior point of S nor an exterior point of S. Set of all frontier points of S is denoted by Fr S.

Example : S = R

$S^\circ = R \Rightarrow (S^c)^0 = \phi$ Ext S = $\phi \Rightarrow$ (Ext S)^c = R

Now, Fr(S) = $(S^c)^\circ \cap (\text{Ext } S)^c = \phi \cap R = \phi$

BOUNDARY POINT :

Frontier point of S which are members of S are referred as boundary points. Set of all boundary point of S is denoted by ∂S.

Example : If S = R, then ∂S = Fr S $\cap$ S = $\phi \cap$ S = ϕ.

DENSE SET

Let A and B be two subsets of Rsuch that A$\subset$ B. The set A is said to dense (or everywhere dense) in B if $B \subset \bar{A}$. A set S is said to be nowhere dense (in R) if complement of the closure of the set S is dense in R i.e., S is nowhere dense if closure of the complement of the set is dense in R.

A set 'A' is said to be **dense in itself** if its every point is a limit point of the set A, i.e., $A \subseteq A'$

COMPACT SET

A subset $S \subset R$ is said to be compact if it is both closed and bounded.

CONNECTED SET

A non-empty subset $S \subset R$ is said to be connected if and only if it is an interval or it is singleton.

ADHERENT POINT :

Let a $\in$ S and S $\subseteq$ R, then a is called Adherent point of S, If for every $\delta > 0$, $(a - \delta, a + \delta) \cap S \neq \phi$.

Set	**Supremum**	**Infimum**	**Set of adherent points**
N	does not exist	1	N
Z	does not exist	does not exist	Z
Q	does not exist	does not exist	R
Q^C	does not exist	does not exist	R
R	does not exist	does not exist	R
$\left\{\frac{1}{n}; n \in N\right\}$	1	0	$\left\{\frac{1}{n}; n \in N\right\} \cup \{0\}$
[a,b); a $\neq$ b	b	a	[a, b]
$\left\{x + \frac{1}{n}; n \in N\right\}$	x + 1	x	$\left\{x + \frac{1}{n}; n \in N\right\}$
{cos n; n $\in$ N}	1	-1	[-1, 1]

Set	**Set of limit points**	**Frontier points**	**Boundary points**	**Nbd. Of**
N	ϕ	N	N	None of its points
Z	ϕ	Z	Z	None of its points
Q	R	R	Q	None of its points
Q^C	R	R	Q^C	None of its points
R	R	ϕ	ϕ	Each of its points
$\left\{\frac{1}{n}; n \in N\right\}$	{0}	$\left\{\frac{1}{n}; n \in N\right\}$	$\left\{\frac{1}{n}; n \in N\right\}$	None of its points
[a, b); a $\neq$ b	[a, b]	{a, b}	{a}	(a, b)
$\left\{x + \frac{1}{n}; n \in N\right\}$	{x}	$\left\{x + \frac{1}{n}; n \in N\right\}$	$\left\{x + \frac{1}{n}; n \in N\right\}$	None of its points
{cosn; n $\in$ N}	[-1, 1]	{cosn; n $\in$ N}	{cosn; n $\in$ N}	None of its points

SOLVED QUESTIONS

1. An infinite set S have **CSIR NET JUNE 2023**

(a) Bijection with set of rational number (b) Bijection with set of real number
(c) Bijection with set of $S \times S$ (d) Bijection with power set of S

Ans. **(c)**

Explanation :

(a) Take S = R. Then R doesnot have bijection with set of rational number
(b) Take S = Q.Then Q doesnot have bijection with set of real number
(d) Take S = N. Then N doesnot have bijection with P(N)

2. Which of the following is countable? **CSIR NET JUNE 2022**

(a) Set of all polynomials with real coefficient
(b) Set of all polynomials with rational coefficient
(c) Set of all polynomials with real coefficient and rational root.
(d) Set of all 2×2 matrices with rational eigen values.

Ans. **(b)**

Explanation :

(a) Set of all polynomials with real coefficient is uncountable
(b) Set of all polynomials with rational coefficient is countable
(c) Set of all polynomials with real coefficient and rational root.is uncountable
(d) $\begin{bmatrix} 1 & a \\ 0 & 1 \end{bmatrix}$; $a \in R$ its eigen value are rational but its uncountable choice

3. Which of the following set is uncountable set? **CSIR N ET JUNE 2019**

(a) $\{x \in R \mid \log(x) = p/q$ for some $p, q \in N\}$
(b) $\{x \in R \mid (\cos x)^n + (\sin x)^n = 1$ for some $n \in N\}$
(c) $\{x \in R \mid x = \log(p/q)$ for some $p, q \in N\}$
(d) $\{x \in R \mid \cos x = p/q$ for some $p, q \in N\}$

Ans. **(b)**

Explanation :

We know that rational number is countable.

$Q = \left\{ \frac{p}{q} \mid p \in Z, q \in N \right\}$ In (a), (c), (d) is countable set. Option (b) is correct.

Let n = 2 Then $S = \{x \in R \mid \cos^2 x + \sin^2 x = 1\}$

$\Rightarrow$ S = R $\Rightarrow$ S is uncountable.

4. Which of the following sets are in bijection with R?

(a) Set of all maps from {0, 1} to N (b) Set of all maps from N to {0, 1}
(c) Set of all subsets of N (d) Set of all subsets of R

Ans. **(b), (c)** **CSIR NET JUNE 2020**

Explanation :

The cardinality of set of all functions from $A \to B$ is $|B|^{|A|}$.

We have, $|N| = \lambda_0$ and $|R| = c$, continuum, $2^{\lambda_0} = c$ and $2^c > c$.

(a) Cardinality of set of all maps from {0, 1} to $N = \lambda_0^2 \neq c$.

(b) Cardinality of set of all maps from N to {0,1} = $2^{\aleph_0}$ = c

(c) Cardinality of set of all subsets of N = $2^{\aleph_0}$ = c

(d) Cardinality of set of all subsets of R = $2^c \neq c$

The sets having the same cardinality as R(= c) are in bijection with R, Option (b) and (c) are correct.

5. Which of the following statement is true?

(a) There are at most countably many continuous maps from R^2 to R

(b) There are at most finitely many continuous surjective maps from R^2 to R

(c) There are infinitely many continuous injective maps from R^2 to R

(d) There are no continuous bijective maps from R^2 to R

Ans. (b), (d) **CSIR NET JUNE 2020**

Explanation :

(a) There are at most countably many continuous maps from R^2 to R
Consider f(x, y) = x + a, for each a ∈ R, f(x, y) is continuous, there are infinitely many continuous maps from R^2 to R. Option (a) is false.

(b) There are at most finitely many continuous surjective maps from R^2 to R
From the same example, f(x, y) = x + a, for each a ∈ R. f(x, y) is continuous surjective map, there are infinitely many continuous surjective maps from R^2 to R. Option (b) is correct.

(c) If f: $R^2 \to$ R is continuous, since R^2 is connected, its image under f is connected, ie, an interval in R. However, if you remove a point in R^2, it remains connected, however, if we remove a point whose image is in the interior of the interval then the image cannot be still connected if the function is injective. So, there is no such map, there are no continuous injective maps from R^2 to R, option (c) is incorrect.

(d) From option (c), we already proved that there are no continuous injective maps from R^2 to R, then clearly there are no continuous bijective maps from R^2 to R, option (d) is correct.

Trick :

1 Let $f : A \to B$ & $|A| \le |B|$
Then it is never onto function but may or may not be 1 – 1.

2. If $|A| \ge |B|$ then it is never 1 – 1 but may or may not be onto.

6. Let S be infinite set. Which of the following are TRUE?

(a) If there is an injection from S to N then S is countable.

(b) If there is a surjection from S to N then S is countable

(c) If there is an injection from N to S then S is countable.

(d) If there is a surjection from N to S then S is countable. **CSIR NET DEC 2018**

Ans. (a), (d)

Explanation :

Option (a) is correct.

$$f : S \xrightarrow{1-1} N$$

REAL ANALYSIS

then $|S| \le |N| \Rightarrow |S| \le \aleph_0$
We know that N is countable set $\Rightarrow$ S is also countable set.
Option (b) is incorrect.
Given that $f: S \xrightarrow{onto} N$ then $|S| \ge |N|$ $\Rightarrow$ S may or may not be countable.
Option (c) is incorrect.
Given that $f: N \xrightarrow{1-1} S$ then $|N| \le |S|$ $\Rightarrow$ S may or may not be countable.
Option (d) is correct.
Given $f: N \xrightarrow{onto} S$
$|N| \ge |S| \Rightarrow$ S is countable.

7. Let $S = \{x \in [-1, 4] \mid \sin x > 0\}$. Which of the following is TRUE?

CSIR NET DEC 2017

(a) inf (S) < 0 (b) sup(S) doesn't exist (c) sup S = π (d) inf (S) = π/2

Ans. (c)

Explanation :

Given that $S = \{x \in [-1, 4] \mid \sin x > 0$

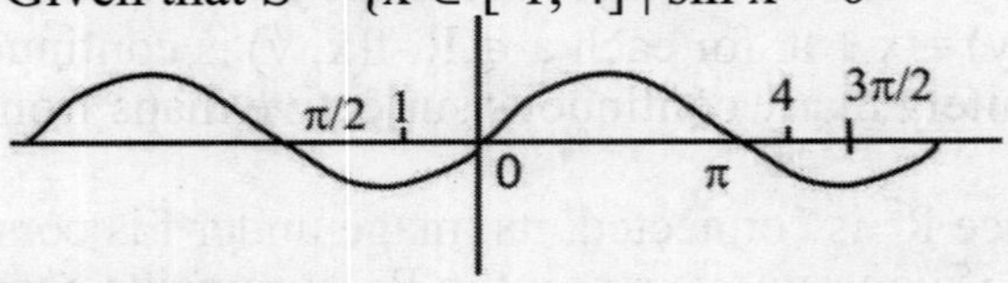

i.e. $S = [0, \pi]$
By option (3) sup s = π

8. Let A be a closed subset of R, $A \neq \phi$, $A \neq R$. Then A is **CSIR NET JUNE 2015**
(a) the closure of the interior of A (b) a countable set
(c) a compact set (d) not open

Ans. (d)

Explanation :

For option (a) and (c), take A = Z
$(\overline{Z^\circ}) = \overline{\phi} = \phi \neq A$ So option (a) is incorrect.
Also Z is not compact. So option (c) is incorrect
For option (b), Take A = [0, 1]
As A is closed but uncountable, so option (b) is incorrect. Hence option (d) is correct.
[Moreover in R, the only sets which are open as well as closed are ϕ and R, itself. since A is proper closed subset of R, so A cannot be open]

9. For each j=1,2,3..., let A be a finite set containing at least two distinct elements. Then

(a) $\bigcup_{j=1}^{\infty} A_j$, is countable set. (b) $\bigcup_{n=1}^{\infty} \prod_{j=1}^{n} A_j$ is uncountable

(c) $\prod_{j=1}^{\infty} A_j$, is uncountable. (d) $\bigcup_{j=1}^{\infty} A_j$, is uncountable.

CSIR NET JUNE 2012

Ans. (a), (c)

Explanation :

Since (i) Countable union of countable sets is countable,
(ii) finite product of countable sets is countable and
(iii) infinite product of countable sets is uncountable

$\Rightarrow \bigcup_{j=1}^{\infty} A_j$ is countable set [By (i)]

$\therefore$ option (a) is correct & (d) is incorrect

Also $\prod_{j=1}^{n} A_j$, is countable [By (ii)]

$\Rightarrow \bigcup_{n=1}^{\infty} \prod_{j=1}^{n} A_j$ is countable [By (i) and (iii)]

$\therefore$ option (b) is incorrect

Further, By (iii) $\prod_{j=1}^{\infty} A_j$, is uncountable

$\therefore$ option (c) is correct

10. Consider the two sets A={1,2,3} and B={1,2,3,4,5}. Choose the correct statements.
(a) The total number of functions from A to B is 125
(b) The total number of functions from A to B is 243
(c) The total number of one-to-one functions from A to B is 60
(d) The total number of one-to-one functions from A to B is 120

CSIR NET JUNE 2013

Ans. (a), (c)

Explanation :

If $f: A \to B$ is a map such that A = {x_1, ..., x_m) and {y_1, y_2,..., y_n), then number of functions from A to B is n^m and if $n \geq m$, then number of one-one functions from A to B is ${}^nP_m$.

Hence, as |A| = 3 and |B|=5

$\Rightarrow$ number of functions from A to B are 5^3=125

And number of one-one functions from A to B are ${}^5P_3$ =60

11. Which of the following subsets of R^2 are uncountable? **CSIR NET DEC 2013**
(a) $\{(a, b) \in R^2 \mid a \leq b\}$
(b) $\{(a, b) \in R^2 \mid a + b \in Q\}$
(c) $\{(a, b) \in R^2 \mid ab \in Z\}$
(d) $\{(a, b) \in R^2 \mid a,b \in Q\}$

Ans. (a), (b), (c)

Explanation :

$Q \times Q = \{(a, b) \in R^2 \mid a, b \in Q\}$
is countable as Q is countable.
Thus, option (d) is incorrect
For Option (a):
Since, $\{(a, a) \in R^2 \mid a \in R\} \subseteq \{(a, b) \in R^2 \mid a \leq b\}$ and $\{(a, a) \in R^2 \mid a \in R\}$ is uncountable
$\Rightarrow \{(a, b) \in R^2 \mid a \leq b\}$ is also uncountable hence (a) is correct.

Option (b) : Let $A = \{(\alpha, -\alpha) \in R \mid \alpha \in Q^c\}$
$A \subseteq \{(a, b) \in R^2 \mid a + b \in Q\}$

A is uncountable then (b) is correct.
$B = \{(\alpha, \alpha) \in R^2 \mid \alpha \in Q^C\}$
B is uncountable then (c) is correct.

Option (c) :
Then $B \subseteq \{(a, b) \in R^2 \mid ab \in Z\}$
Option (a) is correct.

REAL ANALYSIS

12. Which of the following sets of functions are uncountable? (N stands for the set of natural numbers) **CSIR NET JUNE 2015**

(a) $\{f / f : N \to \{1, 2\}\}$
(b) $\{f / f : \{1, 2\} \to N\}$
(c) $\{f / f : \{1, 2\} \to N, f(1) \le f(2)\}$
(d) $\{f / f : N \to \{1, 2\}, f(1) \le f(2)\}$

Ans. (a), (d)

Explanation :

For option (a)
We have to find f such that $f : N \to \{1, 2\}$ is a function
It is equivalent to fill N places using the elements 1 and 2.
Infinite product of countable sets is uncountable
So, option (a) is correct
For option (b)
We have to find f such that $f : \{1, 2\} \to N$ is a function
It is equivalent to set $N \times N$ which is countable as N is countable
For option (c)
Clearly $\{f \backslash f : \{1, 2\} \to N, f(1) \le f(2)\}$ is a subset of $\{f \backslash f : \{1, 2\} \to N\}$ and subset of countable set is also countable
For option (d)
It is equivalent to fill N places in which first place must be 1 and other places are either 1 or 2 which is again infinite product of countable sets.
And infinite product of countable sets is uncountable

13. Let $A_1 = \{f \mid f : A \xrightarrow[1-1]{} B\}$ and $A_2 = \{f \mid f : A \to B, f \text{ is onto function}\}$, Where A & B are non-empty sets then.

(a) $A = N$ & B is finite $\Rightarrow A_1$ countable
(b) A finite & $B = N \Rightarrow A_1$ countable
(c) $A = N$ & $B =$ finite $\Rightarrow A_2$ countable
(d) $A =$ finite & $B = N \Rightarrow A_2$ countable

Ans. (b)

Explanation :

Let A & B are two set and $f : A \to B$ be a one one function then $|A| \le |B|$
and if $f : A \to B$ be onto function then $|B| \le |A|$

Option (a) and (d) discard.
We know that $|A_1| = |B|^{|A|}$
Then $|A_1| = \chi_0^n = \chi_0$
Option (c) is incorrect.
Given that $A = IN$ & $B =$ finite
Let $B = \{0, 1\}$
i.e. A_2 is uncountable.

Option (b) is correct.
Given that A is finite & $B = N$
$\Rightarrow A_1$ is countable
$|A_2| = |B||A|$
$2^{\chi_0} = C$
$|B| = 2$

UNSOLVED QUESTIONS

1. The set of all limit point of the set $\left\{\frac{2}{x+1} : x \in (-1,1)\right\}$ in $\mathbb{R}$ is **IIT-JAM 2014**

(a) $[1,\infty)$ (b) $(1,\infty)$ (c) $[-1,1]$ (d) $[-1,\infty)$

2. Let S be a nonempty subset of R. If S is a finite union of disjoint bounded intervals, then which one of the following is true? **IIT-JAM 2015**

(a) If S is not compact, then sup S $\notin$ S and inf S $\notin$ S

(b) Even if sup S $\in$ S and inf S $\in$ S, S need not be compact

(c) If sup S $\in$ S and inf S $\in$ S, then S is compact

(d) Even if S is compact, it is not necessary that sup S $\in$ S and inf S $\in$ S

3. Let A be non-empty subset of R. Let I(A) denote the set of interior points of A. Then I(A) can be **IIT-JAM 2015**

(a) empty (b) singleton

(c) a finite set containing more than one element (d) countable but not finite

4. Let X be a proper closed subset of [0, 1]. Which of the following statements are always true? **TIFR 2015**

(a) the set X is countable

(b) there exists $x \in X$ such that $X \setminus \{x\}$ is closed

(c) the set X contains an open interval (d) none of the above

5. Let f be a one to one function on the closed interval [0, 1] to the set of real numbers $\mathbb{R}$, then **TIFR 2010**

(a) f must be onto (b) range of f must contain a rational number

(c) range of f must contain an irrational number

(d) range of f must contain both rational and irrational numbers

6. Let $S = \left\{(x, y) \mid x^2 + y^2 = \frac{1}{n^2}\right.$, where n $\in$ N & either x $\in$ Q or y $\in$ Q. Then which of the following are correct. **CSIR NET JUNE 2018**

(a) S is a finite non-empty set (b) S is countable

(c) S is uncountable (d) S is empty.

7. If A & B are non-empty. Define $B^A = \{f \mid f : A \to B\}$ and $A^B = \{f \mid f : B \to A\}$. Then

(a) If B^A & A^B countable $\Rightarrow$ A & B both countable

(b) If A^3 countable $\Rightarrow$ A finite & B countable

(c) If A countable & B finite $\Rightarrow$ A^B is countable. (d) None of these

8. If A and B are two non empty subset of $\mathbb{R}$. If $C = \{x + y : x \in A, y \in B\}$, then which of the following is incorrect?

(a) inf C = inf A + inf B (b) $\inf C \neq \inf A + \inf B$

(c) $\inf C < \inf A + \inf B$ (d) $\inf C > \inf A + \inf B$

9. Which of the following subsets of R^2 are compact?

(a) $\{(x,y):xy=1\}$ (b) $\left\{(x,y):x^{\frac{2}{3}}+y^{\frac{2}{3}}=1\right\}$ (c) $\{(x,y):x^2+y^2<1\}$ (d) None of these

10. Consider the following subsets of R^2, where a, b ∈ R

$$\left\{A=(x,y)\in R^2:\frac{x^2}{a^2}+\frac{y^2}{b^2}=1, a\neq b\right\}$$

$$\left\{B=(x,y)\in R^2:\frac{x^2}{a^2}+\frac{y^2}{b^2}\leq 1, a\neq b\right\}$$

$$C=\{(x,y)\in R^2:ax+by+5=0\}$$

$$D=\{(x,y)\in R^2:ax=by^2\}$$

$$E=\{(x,y)\in R^2:x^3+y^3=1\}$$

Then which of the following is correct?

(a) C and D are compact, but A, B, E are not compact.
(b) A and B are compact, but C, D, E are not compact.
(c) A, B and E are compact, but C, D are not compact.
(d) A and E are compact, but B, C, D are not compact.

11. Let A = $\left\{t\sin\left(\frac{1}{t}\right)\middle| t\in\left(0,\frac{2}{\pi}\right)\right\}$. Which of the following statements are true?

(a) sup (A) $<\frac{2}{\pi}+\frac{1}{n\pi}$ for all n ≥ 1 (b) inf (A) $>\frac{-2}{\pi}-\frac{1}{n\pi}$ for all n ≥ 1
(c) sup (A) =1 (d) inf (A) = -1

12. LetZ denote the set of integers and $Z_{\geq 0}$ denote the set {0, 1, 2, 3, …}. Consider the mapf : $Z_{\geq 0}\times Z\to Z$given by $f(m,n)=2^m.(2n+1)$. Then the map f is

(a) onto (surjective) but not one-one (injective)
(b) one-one (injective) but not onto (surjective)
(c) both one one and onto (d) neither one-one nor onto

13. A set that is neither open or closed in Ris given by

(a) $\left\{1,\frac{1}{2},\frac{1}{3},\frac{1}{4},...\right\}$ (b) $\{1,2,3,...\}$ (c) $(2,\infty)$ (d) $[2,\infty)$

14. If the set A and B are defined as A = {(x, y) : y = ex, x ∈ R}and B = {(x, y) : y = x, x ∈ R}then

(a) $B\subseteq A$ (b) $A\subseteq B$ (c) $A\cap B=\phi$ (d) $A\cup B=A$

15. Let Q be the set of rational numbers. Then

(a) there is no one-one function from [0, 1] to Q
(b) there is no onto function from [0, 1] to Q
(c) there is no one one function from Q to [0, 1]
(d) there is a bijection between Q and [0, 1]

16. If $A=\{1,2,3,4\}$, then which of the following is not function from A to itself?

(a) $f_1 = \{(x,y): x-y=5\}$ (b) $f_2 = \{(x,y): x-y=6\}$

(c) $f_3 = \{(x,y): x+y=5\}$ (d) $f_4 = \{(x,y): x-y>4\}$

17. Let $f,g:[0,1]\to[0,1]$ be function. Let R(f) and R(g) be the ranges of f and g respectively. Which of the following statement is (are) true?

(a) If $f(x)\le g(x)$ for all $x\in[0,1]$, then $\sup R(f)\le \inf R(g)$

(b) If $f(x)\le g(x)$ for some $x\in[0,1]$, then $\inf R(f)\le \sup R(g)$

(c) If $f(x)\le g(y)$ for some $x,y\in[0,1]$, then $\inf R(f)\le \sup R(g)$

(d) If $f(x)\le g(y)$ for all $x,y\in[0,1]$, then $\sup R(f)\le \inf R(g)$

18. Let S be the set of all rational numbers in (0, 1). Then which of the following statement is/are true?

(a) S is a closed subset of R (b) S is not a closed subset of R

(c) S is a open subset of R (d) every $x\in(0,1)\setminus S$ is a limit point of S

19. Let $S\subset R$ and ∂s denote the set of points in Rsuch that every neighborhood of x contain some points of S as well as some points of complement of S. Further, let $\bar{S}$ denote the closure of S. Then which one of the following is false?

(a) $\partial Q = R$ (b) $\partial(R\setminus T) = \partial R,\ T\subset R$

(c) $\partial(\overline{T\cup V}) = \partial T\cup\partial V, T,V\subset R, T\cap V\neq\phi$ (d) $\partial T = T\cap(R\setminus T),\ T\subset R$

20. Let $F_n = \left[-\frac{1}{n},\frac{1}{n}\right]$ be a given interval, then $\bigcap_{n=1}^{\infty} F_n$ is not

(a) closed set (b) open set (c) null set (d) semi open

21. Consider the following subset of R :

$E=\left\{\frac{n}{n+1}: n\in N\right\}, F=\left\{\frac{1}{1-x}: 0\le x<1\right\}$. Then

(a) Both E and F are closed (b) E closed and F is not closed

(c) E is not closed and F is closed (d) neither E nor F is closed.

22. The set of limit points of the set $\left\{\frac{5-x}{x}: x\in(2,3)\right\}$ in R (set of real numbers) is

(a) open (b) bounded but not closed (c) connected (d) compact

ANSWER KEY

1	2	3	4	5	6	7	8	9	10
a	b	a	d	c	b	a,c	b,c,d	b	b
11	**12**	**13**	**14**	**15**	**16**	**17**	**18**	**19**	**20**
a,b	b	a	c	a	a,b,d	b,c,d	b,d	c	b,c,d
21	**22**								
C	c,d								

Chapter 2

Sequence of Real NumbersTopology

1. Sequence

A sequence of real number is a function 'f' whose domain is the set N of all natural numbers and range is a subset of R. In other words, a sequence on a set R is a rule which assigns to each natural number, a unique element of R. A sequence is usually denoted by $\{a_n\}$ or $<a_n>$ where $f(n) = a_n$, a_n is called n^{th} term of the sequence.

Note: unless otherwise stated, by a sequence we shall mean a sequence on the set of real numbers.

Some Important Example of Sequence :

(1) $<a_n> = <n>$ (2) $<a_n> = <-n>$ (3) $<a_n> = (-1)^n n$

(4) $<a_n> = (-1)^n$ (5) $<a_n> = \frac{1}{n}$ (6) $a_n = 1 - \frac{(-1)^n}{n}$

(7) $a_n = \begin{cases} 2 & n = 1 \text{ or prime} \\ n & \text{else} \end{cases}$ (8) $a_n = \begin{cases} 2 & n = 1 \text{ or prime} \\ p & p/n \,\&\, p \text{ is the least prime} \end{cases}$

(9) $a_{n+2} = a_n + a_{n+1}$, $a_1 = 1, a_2 = 1$

(10) $a_{n+2} = \frac{1}{2}(a_n + a_{n+1})$ where a_1& a_2 given

(11) $a_{n+1} = \sqrt{2 + a_n}$ where $a_1 = 0$ (12) $<a_n> = 2012 \ \forall\, n \in N$

(13) $<a_n> = (-1)^n + \sin\left(\frac{n\pi}{6}\right)$ (14) $<a_n> = \begin{cases} -1 - \frac{1}{n} & n \text{ is even} \\ 1 + \frac{1}{n} & n \text{ is odd} \end{cases}$

(15) $<a_n> = \cos n$ (16) $<a_n> = \sin\left(\frac{1}{n}\right)$

(17) $<a_n> =$ Last digit of 3^n. (18) $<a_n> = \{-2, -3, -1, 0, 1, 2, 1, 0, 1, 0, ...\}$

(19) $a_n = 1 + \frac{1}{2} + \frac{1}{3} + ... + \frac{1}{n} - \log n$ (20) $a_n = 1 + \frac{1}{2} + \frac{1}{3} + ... + \frac{1}{n}$

Range of a sequence: The set of all distinct terms of a sequence is called its range.

Example : $a_n = <(-1)^n>$then Range set $= \{-1, 1\}$.

Constant sequence: A sequence $\{a_n\}$on a set S, defined by an $= c \in S$, $\forall\, n \in N$ is called a constant sequence. Thus $\{a_n\} = \{c, c, c, ...\}$ is a constant sequence with range $=\{c\}$, a singleton.

Bounded Sequence

A sequence $<a_n>$ is defined as bounded if its range set is bounded. Hence $<a_n>$ is bounded if there exist real numbers k' and k such that $k' \le a_n \le k \ \forall \ n \in N$.

Note :

(a) Supremum and infimum of the range set is the supremum and infimum of the sequence.

(b) If the range of the sequence $<a_n>$ is finite then there exist $\alpha \in R$ such that $a_n = \alpha$ for infinitely many values of n however converse need not be true.

Example : $a_n = \begin{cases} 4 & n = 1 \text{ or prime} \\ n & \text{else} \end{cases}$

Then $a_n = 4$ for infinitely many values of n but range of the sequence $<a_n>$ is not finite.

Cluster Point (or limit point) of a sequence:

A real number l is said to be a cluster point of a sequence $\langle a_n \rangle$ if for given $\in > 0$, however small and for a positive integer m, there exists a positive integer k > m such that $|a_k - l| < \in$, , i.e., $l - \in < a_k < l + \in$.

Thus $l \in R$ is the cluster point of the sequence $\langle a_n \rangle \Leftrightarrow$ neighbourhood of l contains infinitely many terms of the sequence.

Note: Limit point of a sequence is different from the limit of a sequence.

Examples:

(1) The sequence $\langle (-1)^n \rangle$ has two limit points as 1 and -1.

(2) The sequence $\langle n \rangle$ has no limit point.

Limit of a sequence: Let $\{a_n\}$ be a sequence and $l \in$ R. l is said to be limit of sequence $\{a_n\}$ if for each $\in > 0$, $\exists m \in N$ (m depending $\in$)such that $|a_n - l| < \in \forall n \ge m$.

Algebra of limits: If $\lim_{n\to\infty} a_n = a$ and $\lim_{n\to\infty} b_n = b$, then

(i) $\lim_{n\to\infty} ka_n = ka$, where k is any constant (ii) $\lim_{n\to\infty} |a_n| = |a|$

(iii) $\lim_{n\to\infty} (a_n + b_n) = a + b$ (iv) $\lim_{n\to\infty} (a_n - b_n) = a - b$

(v) $\lim_{n\to\infty} (a_n b_n) = ab$ (vi) $\lim_{n\to\infty} \frac{1}{a_n} = \frac{1}{a}$ provided $a \ne 0$ (vii) $\lim_{n\to\infty} \frac{a_n}{b_n} = \frac{a}{b}$, provided $b \ne 0$.

Some important limits:

(a) If p> 0, then $\lim_{n\to\infty} \frac{1}{n^p} = 0$

(b) If p > 0, then $\lim_{n\to\infty} \sqrt[n]{p} = 1$

(c) $\lim_{n\to\infty} \sqrt[n]{n} = 1$

(d) If p > 0, then $\lim_{n\to\infty} \frac{n^2}{(1+p^n)} = 0$

(e) If $|x| < 1$, then $\lim_{n\to\infty} x^n = 0$

(f) $\lim_{n\to\infty}\left(1+\frac{1}{n}\right)^n = e$

Monotonic Sequences:

(i) A sequence $\{a_n\}$ is said to be monotonically increasing if $a_{n+1} \ge a_n \ \forall n \in N$, i.e., if $a_1 \le a_2 \le \ldots \le a_n \le a_{n+1} \le \ldots$

(ii) A sequence $\{a_n\}$ is said to be monotonically decreasing if $a_{n+1} \le a_n \ \forall n \in N$, i.e., if $a_1 \ge a_2 \ge \ldots \ge a_n \ge a_{n+1} \ge \ldots$.

(iii) A sequence $\{a_n\}$ is said to be monotonic if it is either monotonically increasing or monotonically decreasing

(iv) A sequence $\{a_n\}$ is said to be strictly monotonically increasing if $a_{n+1} > a_n \forall n \in N$

(v) A sequence $\{a_n\}$ is said to be strictly monotonically decreasing if $a_{n+1} < a_n \forall n \in N$.

(vi) A sequence $\{a_n\}$ is said to be strictly monotonic if it is either strictly monotonically increasing or strictly monotonically decreasing

Example :

(1) $a_n = n$ is a strictly increasing sequence.

(2) $a_n = -n$ is strictly decreasing sequence.

(3) $(-1)^n$ is a non-monotonic sequence.

2. Convergent Sequence:

A sequence is called a convergent sequence if limit of this sequence exist.

Note : If $\lim_{n\to\infty} a_n = l$ then we say that this sequence converge to l.

Example : $\lim_{n\to\infty}\left(\frac{1}{n}\right) = 0$, we say that sequence {1/n} converges to 0.

Results on convergent sequence:

(1) Every convergent sequence has a unique limit point

(2) A sequence cannot converge to more than one limit.

(3) Every convergent sequence is bounded. The converse is not true, i.e., a bounded sequence is not necessarily convergent.

Example : The sequence $\{a_n\}$ defined by $a_n = (-1)^n$. Clearly sequence is bounded. -1 is its g.l.b. and 1is its l.u.b., but it is not convergent.

Null Sequence: A sequence $\{a_n\}$ is said to be null sequences if it converges to zero, i.e., if $\lim_{n\to\infty} a_n = 0$

For Example, the sequence $\left\{\frac{1}{n}\right\}, \left\{\frac{1}{n^2}\right\}, \left\{\frac{1}{2^n}\right\}$ are null sequences.

3. Cauchy sequence:

A sequence $< a_n >$ is said to be a Cauchy sequence for given $\in > 0$, however small, there exists a positive integer m (depending on $\in$) such that $|a_n - a_m| < \in \forall n \geq m$

Another Definition : A sequence $< a_n >$ is said to be a Cauchy sequence if for given $\in > 0$, however small, there exists a positive integer m (depending on $\in$) such that $| a_{m+p} - a_m | < \in \forall p > 0, p \in N$.

Results on Cauchy sequence:

(1) A sequence is convergent if and only if it is a Cauchy sequence.
(2) A Cauchy sequence is always bounded.

4. Divergent Sequence:

A monotonic and unbounded sequence is called a divergent sequence.
Example : $a_n = <n>$.

5. Oscillatory Sequences:

If a sequence $\{a_n\}$ neither converges to a finite number nor diverges to $+\infty$ or $-\infty$, then it is called an oscillatory sequence.

Oscillatory sequences are of two types:

(i) A bounded sequence which does not converge is said to oscillate finitely.
Example . $\{(-1)^n\}$

(ii) An unbounded sequence which does not diverge is said to oscillate infinitely
Example $\{(-1)^n n\}$

6. Subsequence:

Let $\langle a_n \rangle$ be a given sequence. If $\langle n_k \rangle$ is a strictly increasing sequence of natural numbers (i.e., $n_1 < n_2 < n_3 \ldots$), then $\langle a_{n_k} \rangle$ is called a subsequence of $\langle a_n \rangle$

Example : Let $a_n = (-1)^n$, then $a^{2n} = 1$ and $a^{3n} = -1$ are subsequence of this sequence.

Results on subsequence:

(1) Every sequence is a subsequence of itself.
(2) If a sequence $\langle a_n \rangle$ converges to l then every subsequence of $\langle a_n \rangle$ also converges to l.

(3) If all subsequences of a sequence $\langle a_n \rangle$ converge to the same limit l, then $\langle a_n \rangle$ converges to l.

(4) If the subsequences $\langle a_{2n-1} \rangle$ and $\langle a_{2n} \rangle$ of a sequence $\langle a_n \rangle$ converge to same limit l, then the sequence $\langle a_n \rangle$ converges to l.

(5) If a sequence $\langle a_n \rangle$ diverges to ∞, then every subsequence of $\langle a_n \rangle$ also diverges to ∞.

(6) If a sequence $\langle a_n \rangle$ diverges to $-\infty$, then every subsequence of $\langle a_n \rangle$ also diverges to $-\infty$.

(7) Every sequence contains a monotonic subsequence.

(8) Every bounded sequence in R contains a convergent subsequence.

Results of a sequence:

(1) A monotonic sequence is convergent $\Leftrightarrow$ it is bounded.

(2) A convergent sequence need not be monotonic. e.g., $< a_n >= \frac{(-1)^n}{n}$

(3) Monotonic sequence can have atmost one limit point.

7. Limit superior and limit inferior :

Limit superior: Let $\langle a_n \rangle$ be any sequence, define $b_n = \sup\{a_n, a_{n+1}, a_{n+2} ...\}$, i.e.,

$b_1 = \sup \{a_1, a_2, a_3, ...\}$

$b_2 = \sup \{a_2, a_3, a_4, ...\}$

$b_3 = \sup \{a_3, a_4, a_5, ...\}$ and so on.

Here, $b_1 \geq b_2 \geq b_3 \geq ...$, i.e., $\langle b_n \rangle$ is monotonically decreasing, then

$\overline{\lim}\, a_n = \sup\{b_1, b_2, b_3, ...\}$

i.e., infimum of $\langle b_n \rangle$ is called limit superior of $\langle a_n \rangle$.

Limit inferior: Let $\langle a_n \rangle$ be any sequence, define $c_n = \inf\{a_n, a_{n+1}, a_{n+2} ...\}$, i.e.,

$c_1 = \inf \{a_1, a_2, a_3, ...\}$

$c_2 = \inf \{a_2, a_3, a_4, ...\}$

$c_3 = \inf \{a_3, a_4, a_5, ...\}$ and so on.

Here, $c_1 \geq c_2 \geq c_3 \geq ...$, i.e., $\langle c_n \rangle$ is monotonically increasing, then

$\underline{\lim}\, a_n = \inf\{c_1, c_2, c_3, ...\}$ i.e., supremum of $\langle c_n \rangle$ is called limit inferior of $\langle a_n \rangle$.

Remark: If $\langle a_n \rangle$ is unbounded above, then we define $\limsup_{n\to\infty} a_n = \infty$ and if $\langle a_n \rangle$ is unbounded below, then we define $\liminf_{n\to\infty} a_n = -\infty$

Limit Superior and limit Inferior of sequence $\Rightarrow$

(1) Convergent sequence $\Rightarrow$ let $a_n \to l$ be a convergent sequence then

$$\underline{\lim}\, a_n = \overline{\lim}\, a_n = l$$

(2) Divergent sequence $\Rightarrow$ let a_n be divergent sequence which diverge to either ∞ or $-\infty$

(3) Oscillatory sequence $\Rightarrow$ let a_n be a oscillatory sequence then $\underline{\lim}\, a_n \neq \overline{\lim}\, a_n$

Remark: Let A be a set of limit point of sequence than sup(A) is limit superior and inf(A) is limit inferior of the sequence.

8. Important results on sequence:

Cauchy's First Theorem on Limits:

If $\{a_n\}$ converges to l, then the sequence $\{x_n\}$,where $x_n = \dfrac{a_1 + a_2 + ... + a_n}{n}$ also converges to l or $\lim\limits_{n\to\infty} a_n = l \Rightarrow \lim\limits_{n\to\infty} \dfrac{a_1 + a_2 + ... + a_n}{n} = l$.

Example Prove that $\lim\limits_{n\to\infty} \dfrac{1}{n}\left(1 + \dfrac{1}{2} + \dfrac{1}{3} + ... + \dfrac{1}{n}\right) = 0$

Solution Let $a_n = \dfrac{1}{n}$, than $\lim\limits_{n\to\infty} a_n = \lim\limits_{n\to\infty} \dfrac{1}{n} = 0$

$\therefore$ By Cauchy's first theorem on limits, we have,

$$\lim_{n\to\infty} \frac{a_1 + a_2 + ... + a_n}{n} = 0 \text{ or } \lim_{n\to\infty} \frac{1}{n}\left(1 + \frac{1}{2} + \frac{1}{3} + ... + \frac{1}{n}\right) = 0$$

Note: Converse of Cauchy's first theorem on limits is not true, i.e.,

If $\lim\limits_{n\to\infty} \dfrac{a_1 + a_2 + ... + a_n}{n} = l$ then $\lim\limits_{n\to\infty} a_n$ may not be l

For example Let $a_n = (-1)^n$, then $\dfrac{a_1 + a_2 + ... + a_n}{n} = \begin{cases} 0, & \text{when } n \text{ is even} \\ -\dfrac{1}{n}, & \text{when } n \text{ is odd} \end{cases}$

$$\Rightarrow \quad \lim_{n\to\infty} \left(\frac{a_1 + a_2 + ... + a_n}{n}\right) = 0$$

but $\{a_n\}$ is not convergent

Cauchy's Second Theorem on Limits:

If $\{a_n\}$ is a sequence of positive terms, then $\lim\limits_{n\to\infty} (a_n)^{1/n} = \lim\limits_{n\to\infty} \dfrac{a_{n+1}}{a_n}$ provided that the limit on the right hand side exists, whether finite or infinite

REAL ANALYSIS

Example prove that $\lim_{n\to\infty}\left[\left(\frac{2}{1}\right)^1\left(\frac{3}{2}\right)^2\left(\frac{4}{3}\right)^3\cdots\left(\frac{n+1}{n}\right)^n\right]^{1/n}=e$

Solution Let $a_n=\left(\frac{2}{1}\right)^1\left(\frac{3}{2}\right)^2\left(\frac{4}{3}\right)^3\cdots\left(\frac{n+1}{n}\right)^n$

$$\therefore a_{n+1}=\left(\frac{2}{1}\right)^1\left(\frac{3}{2}\right)^2\left(\frac{4}{3}\right)^3\cdots\left(\frac{n+1}{n}\right)^n\left(\frac{n+2}{n+1}\right)^{n+1}$$

$$\Rightarrow \frac{a_{n+1}}{a_n}=\left(\frac{n+2}{n+1}\right)^{n+1}=\left(1+\frac{1}{n+1}\right)^{n+1}$$

$$\Rightarrow \lim_{n\to\infty}\frac{a_{n+1}}{a_n}=\lim_{n\to\infty}\left(1+\frac{1}{n+1}\right)^{n+1}=e$$

Also $a_n>0\ \forall n\in N$ by Cauchy's second theorem on limits, we have,

$$\lim_{n\to\infty}(a_n)^{1/n}=e \text{ or } \lim_{n\to\infty}\left[\left(\frac{2}{1}\right)^1\left(\frac{3}{2}\right)^2\left(\frac{4}{3}\right)^3\cdots\left(\frac{n+1}{n}\right)^n\right]^{1/n}=e$$

Converse of Cauchy's second theorem on limits may not be true :

For example

Let $a_n=2^{-n+(-1)^n}\Rightarrow (a_n)^{1/n}=2^{-1+\frac{(-1)^n}{n}}$ $\lim_{n\to\infty}(a_n)^{1/n}=\frac{1}{2}$

but $\frac{a_{n+1}}{a_n}=\frac{2^{-(n+1)+(-1)^{n+1}}}{2^{-n+(-1)^n}}=2^{-1+(-1)^{n+1}-(-1)^n}$

$$\frac{a_{n+1}}{a_n}=\begin{cases}\frac{1}{8}, & \text{when } n \text{ is even}\\ 2, & \text{when } n \text{ is odd}\end{cases}\Rightarrow \lim_{n\to\infty}\frac{a_{n+1}}{a_n} \text{ does not exist}$$

Therefore, $\lim\frac{a_{n+1}}{a_n}$ may not exist even if $\lim(a_n)^{1/n}$ exist

Note1: If $a_n>0$ and $\lim(a_n)=l$ then $\lim(a_1a_2\ldots a_n)^{1/n}=l$

Note2: If $\lim\frac{a_{n+1}}{a_n}=l$ where $|l|<1$, then $\lim(a_n)=0$

Note3: If $a_n>0$ and $\lim\frac{a_{n+1}}{a_n}=l>1$, then $\lim a_n=+\infty$

Ceasaro's Theorem: If $\lim_{n\to\infty}a_n=a$ and $\lim_{n\to\infty}b_n=b$, then $\lim_{n\to\infty}\frac{a_1b_n+a_2b_{n-1}+\ldots+a_nb_1}{n}=ab$

Note: If a sequence $\langle a_n\rangle$ oscillate finitely and $\lim_{n\to\infty}b_n=0$, then $\lim_{n\to\infty}(a_nb_n)=0$

SOLVED QUESTIONS

1. Let $a_n = \frac{1}{n+1} + \frac{1}{n+2} + ... \frac{1}{n+n}$ and $b_n = \frac{1}{n}$ are two sequence. Then

CSIR NET JUNE 2023

(a) $<a_n> \to$ log 2 and both have same rate of convergence.
(b) $<a_n> \to$ log 4 and both have same rate of convergence.
(c) $<a_n> \to$ log2 but both does not have same rate of convergence.
(d) $<a_n>$ is not convergent.

Ans. (a)

Explanation : $a_n = \sum_{k=1}^{\infty}\left(\frac{1}{n+k}\right) = \frac{1}{n}\sum_{k=1}^{\infty}\frac{1}{1+\frac{k}{n}}$ Take $\frac{k}{n} = x \;\&\; \frac{1}{n} = dx$

Then $\int_0^1 \frac{1}{1+x}dx = [log(1+x)]_0^1 = log 2$

Now, we know that $a_n \to a$ and b_n are two sequence s.t. $|a_n - a| \le k|b_n|^r$ for large n

Then rate of convergence of a_n is k

So, $\left|\frac{1}{n+1} + \frac{1}{n+2} + ... \frac{1}{n+n} - \log 2\right| \le \left|\frac{1}{n+1} + \frac{1}{n+2} + ... \frac{1}{n+n}\right| \le \frac{n}{n} \le 1$

For large n i.e. rate of convergence is 1

Now $|b_n - 0| \le \left|\frac{1}{n}\right| \le 1$ for large n.

Hence rate of convergence of both sequences are same.

2. Which of the following implies that given sequence $<x_n>$ is convergent?

CSIR NET JUNE 2023

(a) Subsequence $<x_{2n}>$, $<x_{2n-1}>$ & $<x_{3n}>$ are converges to same limit
(b) $<x_{2n}>$ & $<x_{2n-1}>$ are convergent (c) $<x_{kn}>$ is convergent for all $k \ge 2$ (d) $\lim_{n\to\infty} |x_{n+1} - x_n| = 0$

Ans. (a), (c)

Explanation :

(a) Two complementary subsequence are converges to same limit then sequence converges.

(b) Let $x_n = (-1)^n$

(c) $<x_{2n}>$, $<x_{3n}>$, $<x_{4n}>$, ….. all are convergent.
Then sequence convergent.
Suppose $x_n = \sqrt{n}$ then $\sqrt{n+1} - \sqrt{n}$
$\lim_{n\to\infty} |\sqrt{n+1} - \sqrt{n}| = 0$
But x_n is not convergent.

REAL ANALYSIS

3. Consider the sequence < a_n> , where $a_n = 3+5\left(-\frac{1}{2}\right)^n + (-1)^n\left(\frac{1}{4}+(-1)^n\frac{2}{n}\right)$

then the interval $\left(\liminf_{n\to\infty} a_n, \limsup_{n\to\infty} a_n\right)$

CSIR NET 2022

(a) (-2 , 8) (b) $\left(\frac{11}{4},\frac{13}{4}\right)$ (c) (3 , 5) (d) $\left(\frac{1}{4},\frac{7}{4}\right)$

Ans. (b)

Explanation :

Given that $a_n = 3+5\left(-\frac{1}{2}\right)^n + (-1)^n\left(\frac{1}{4}+(-1)^n\frac{2}{n}\right)$

$$a_n = \begin{cases} 3+5\left(\frac{1}{2}\right)^n + \left(\frac{1}{4}+\frac{2}{n}\right) & n \text{ is even} \\ 3-5\left(\frac{1}{2}\right)^n - \left(\frac{1}{4}-\frac{2}{n}\right) & n \text{ is odd} \end{cases}$$

so , $a_{2n} \to 3 + 1/4 = 13/4$

and $a_{2n+1} \to 3 - 1/4 = 11/4$

Hence $\liminf_{n\to\infty} a_n = \frac{11}{4}$ *and* $\limsup_{n\to\infty} a_n = \frac{13}{4}$.

4. $\lim_{n\to\infty}\left(\frac{1+2^{1/2}+3^{1/3}+......+n^{1/n}}{n}\right)$ is equal to **CSIR NET 2022**

(a) 0 (b) e (c) 1 (d) none of these

Ans. (c)

Explanation :

By Cauchy's first theorem on limits

Result : $\lim_{n\to\infty} a_n = l$ then $\lim_{n\to\infty}(a_1.a_2......a_n)^{1/n} = l$.

5. Let $\{x_n\}$ be a sequence of positive real numbers. Which of the following statements are true?

(a) If the two subsequences $\{x_{2n}\}$ and $\{x_{2n+1}\}$ converge, then the sequence $\{x_n\}$ converges.

(b) If $\{(-1)^n x_n\}$ converges, then the sequence $\{x_n\}$ converges.

(c) If $\lim_{n\to\infty}\frac{x_{n+1}}{x_n}$ exists then $\{(x_n)^{1/n}\}$ is bounded.

(d) If the sequence $\{x_n\}$ is unbounded then every subsequence is unbounded.

Ans. (c) **CSIR NET NOV 2020**

Explanation :

(a) Take $x_n = (-1)^n$

$x_{2n} = 1$ and $x_{2n+1} = -1$ but $(-1)^n$ is not convergent.

(b) $x_n = (-1)^n$

$\Rightarrow$ $(-1)^n x_n = (-1)^n (-1)^n = 1$ $\Rightarrow$ Option (b) is false.

(d) Take $x_n = \begin{cases} n & if \quad n \text{ is even} \\ 1 & if \quad n \text{ is odd} \end{cases}$

$\Rightarrow$ x_{2n+1} is subsequence which is bounded. $\Rightarrow$ Option (d) is false.

(c) Now let $\lim\limits_{n\to\infty} \frac{x_{n+1}}{x_n} = l$ then by using result $\lim\limits_{n\to\infty} (a_n)^{1/n} = \lim\limits_{n\to\infty} \frac{a_{n+1}}{a_n}$ provided later limit exist and $a_n > 0$ for all n. $\Rightarrow$ $\lim\limits_{n\to\infty} (a_n)^{1/n} = l$

$\Rightarrow$ $(a_n)^{1/n}$ is convergent. $\Rightarrow$ $(a_n)^{1/n}$ is bounded.

$\Rightarrow$ Option (c) is true.

6. f : N → N be a bounded function. Which of the following statements is NOT true?

(a) $\lim\limits_{n\to\infty} \sup f(n) \in N$ (b) $\lim\limits_{n\to\infty} \inf f(n) \in N$

(c) $\lim\limits_{n\to\infty} \inf(f(n) + n) \in N$ (d) $\lim\limits_{n\to\infty} \sup(f(n) + n) \notin N$

Ans. (c) **CSIR NET NOV 2020**

Explanation :

Consider f(n) = 1 for all n ∈ N $\Rightarrow$ $\lim\limits_{n\to\infty} \inf(f(n) + n) = \lim\limits_{n\to\infty} \inf(1 + n) = \infty \notin N$

Option (c) is correct.

7. Let $<x_n>$ be a sequence of non-negative real number then which of the following is TRUE? **CSIR NET DEC 2019**

(a) $\underline{\lim}_{n\to\infty}\, x_n = 0 \Rightarrow \lim\limits_{n\to\infty} x_n^{\,2} = 0$ (b) $\overline{\lim}_{n\to\infty}\, x_n = 0 \Rightarrow \lim\limits_{n\to\infty} x_n^{\,2} = 0$

(c) $\underline{\lim}_{n\to\infty}\, x_n = 0 \Rightarrow <x_n>$ is bounded (d) $\underline{\lim}_{n\to\infty}\, x_n^{\,2} > 4 \Rightarrow \overline{\lim}_{n\to\infty}\, x_n > 4$

Ans. (b)

Explanation :

Let x_n = {0, 1, 0, 2, 0, 3, 0, 4, 0, 5, …..}

$\underline{\lim}_{n\to\infty}\, x_n = 0$, but $\lim\limits_{n\to\infty} x_n = \infty \Rightarrow \lim\limits_{n\to\infty} x_n^2 = \infty$

Option (a), (c) is incorrect.

Option (d) is incorrect. Let x_n = {3, 3, 3, ……, 3, …..}

x_n^2 = {9, 9, 9, …..9, …..} $\underline{\lim}_{n\to\infty}\, x_n^2 = 9 > 4$

but $\lim\limits_{n\to\infty} x_n = 3 > 4$

Option (b) is correct.

We know that $\lim\limits_{n\to\infty} |x_n| = 0$ iff $\lim\limits_{n\to\infty} x_n = 0$

Given that $x_n \geq 0$

$\Rightarrow \lim\limits_{n\to\infty} x_n = 0 \Rightarrow \lim\limits_{n\to\infty} x_n = 0 \Rightarrow \lim\limits_{n\to\infty} x_n^2 = 0$

8. Consider a sequence $a_n = (-1)^n\left(\frac{1}{2} - \frac{1}{n}\right)$. Let $b_n = \sum\limits_{k=1}^{n} a_k \ \forall\ n \in N$. Then which of the following are TRUE? **CSIR NET JUNE 2019**

(a) $\lim\limits_{n\to\infty} b_n = 0$ (b) $\overline{\lim\limits_{n\to\infty}}\, b_n < -1/2$ (c) $\overline{\lim\limits_{n\to\infty}}\, b_n > 1/2$

(d) $0 \leq \underline{\lim\limits_{n\to\infty}}\, b_n \leq \overline{\lim\limits_{n\to\infty}}\, b_n \leq 1/2$

Ans. **(c)**

Explanation :

Given that

$$a_1 = -1\left(\frac{1}{2} - 1\right), \qquad a_2 = \left(\frac{1}{2} - \frac{1}{2}\right), \qquad a_3 = -1\left(\frac{1}{2} - \frac{1}{3}\right)$$

$$a_4 = \left(\frac{1}{2} - \frac{1}{4}\right), \qquad a_5 = -\left(\frac{1}{2} - \frac{1}{5}\right), \qquad a_6 = \left(\frac{1}{2} - \frac{1}{6}\right)$$

Then $b_1 = a_1 = (1) - \frac{1}{2}$

$$b_2 = a_1 + a_2 = 1 - \frac{1}{2}$$

$$b_3 = a_1 + a_2 + a_3 = \left(1 - \frac{1}{2} + \frac{1}{3}\right) - \frac{1}{2}$$

$$b_4 = a_1 + a_2 + a_3 + a_4 = \left(1 - \frac{1}{2} + \frac{1}{3} - \frac{1}{4}\right)$$

Then $$b_n = \begin{cases} \left(1 - \frac{1}{2} + \frac{1}{3} - \frac{1}{4} \ldots\ldots \frac{1}{n}\right) - \frac{1}{2} & n \text{ is odd} \\ \left(1 - \frac{1}{2} + \frac{1}{3} - \frac{1}{4} + \ldots\ldots \frac{1}{n}\right) & n \text{ is even} \end{cases}$$

We know that

$$\log(1+x) = x - \frac{x^2}{2} + \frac{x^3}{3} + ...$$

$$\log(2) = \log(1+1) = 1 - \frac{1}{2} + \frac{1}{3} - \frac{1}{4} + ...$$ $$b_n = \begin{cases} \log 2 - \frac{1}{2} & n \text{ is odd} \\ \log 2 & n \text{ is even} \end{cases}$$

$\underline{\lim}_{n\to\infty} b_n = \log 2 = 0.6931$ & $\overline{\lim}_{n\to\infty} b_n = \log 2 - 1/2 = +0.1989$

Option (a), (c), (d) is incorrect. Option (b) is correct.

9. For each $n \in N$, $n \geq 1$ let $a_n = \frac{n}{10^{|\log_{10} n|}}$. where | | = smallest integer greater than or equal to x. Which of the following are TRUE? **CSIR NET DEC 2019**

(a) $\lim_{n\to\infty}$ inf $a_n = 0$ (b) $\lim_{n\to\infty}$ inf a_n doesn't exist

(c) $\lim_{n\to\infty}$ inf $a_n = 0.15$ (d) $\lim_{n\to\infty}$ sup $a_n = 1$

Ans. (d)

Explanation :

Let $n = 10^k$

Then $a_n = \frac{10^k}{10^{\log_{10} 10^k}} = \frac{10^k}{10^k} = 1$ IInd $n \in (10^k, 10^{k+1})$

Then $|\log_{10} n| = k + 1$ e.g. Let $n \in (10^2, 10^3)$ $\therefore$ let $n = 999$

then $|\log_{10} 999| = |\log_{10} 10^{2.(.....)}| = |2.(\ )| = 3$

Then $a_n = \frac{n}{10^{(k+1)}}$ where $n \in (10^k, 10^{k+1})$

Let $n = 10^k + r$ $a_n = \frac{10^k + r}{10^{(k+1)}} = \frac{10^k}{10^k.10} + \frac{r}{10^{k+1}}$

$a_n = \frac{1}{10} + \frac{r}{10^{k+1}}$ Where k = 1, 2, 3, ... 9999

$$\text{Now} \quad a_n = \begin{cases} 1 & n = 10^k \\ \frac{1}{10} + \frac{r}{10^{k+1}} & n \neq 10^k, n \in (10^k, 10^{k+1}) \end{cases}$$

A be the set of limit points Then $A = \left\{1, \frac{1}{10}\right\}$

$\overline{\lim}_{n\to\infty} a_n = 1$ $\underline{\lim}_{n\to\infty} a_n = \frac{1}{10}$

REAL ANALYSIS

10. Define the sequence $\{a_n\}$ as follows : $a_1 = 1$ and for $n \geq 1, a_{n+1} = (-1)^n\left(\frac{1}{2}\right)\left(|a_n|+\frac{2}{|a_n|}\right)$ Which of the following is true?

CSIR NET JUNE 2018

(a) $\lim_{n\to\infty} \sup a_n = \sqrt{2}$ (b) $\lim_{n\to\infty} \inf a_n = -\infty$ (c) $\lim_{n\to\infty} a_n = \sqrt{2}$ (d) $\sup a_n = \sqrt{2}$

Ans. (a)

Explanation :

Given that

$$a_{n+1} = (-1)^n\left(\frac{1}{2}\right)\left(|a_n|+\frac{2}{|a_n|}\right) = (-1)^n\left(\frac{1}{2}\right)|a_n| + (-1)^n\frac{1}{2}\left(\frac{1}{|a_n|}\right)$$

$$a_{2n+1} = (-1)^{2n}\left(\frac{1}{2}\right)\left(|a_{2n}|+\frac{2}{|a_{2n}|}\right)$$

Let $a_n \to l$, then

$$l = \frac{1}{2}\left(l+\frac{2}{l}\right) \Rightarrow l = \pm\sqrt{2}$$

11. If $\{x_n\}$ is a convergent sequence in R and $\{y_n\}$ is a bounded sequence in R, then we can conclude that

CSIR NET JUNE 2018

(a) $\{x_n + y_n\}$ is convergent (b) $\{x_n + y_n\}$ is bounded
(c) $\{x_n + y_n\}$ has no convergent subsequence (d) $\{x_n + y_n\}$ has no bounded subsequence

Ans. (b)

Explanation :

Let $x_n = \frac{1}{n}$ be the convergent sequence and $y_n = (-1)^n$ bounded sequence.

But $\frac{1}{n} + (-1)^n$ is not convergent.

Since it has two limit point 1 & -1.

So, option (a) is incorrect. Also, it has a convergent & bounded subsequence.

Therefore option (c) & (d) are incorrect.

As every convergent sequence is bounded, and the sum of 2 bounded sequence is again a bounded sequence. Hence option (b) is correct.

12. $\lim_{n\to\infty}\left(1+\frac{1}{2^n}+\frac{1}{3^n}+....+\frac{1}{2023^n}\right)^{1/n}$

IIT JAM 2023

(a) 1 (b) 2 (c) 3 (d) 4

Ans. (a)

Explanation :

We know that $\lim_{n\to\infty}\frac{u_{n+1}}{u_n}=\lim_{n\to\infty}(u_n)^{1/n}$

Here $u_n=1+\frac{1}{2^n}+\frac{1}{3^n}+....+\frac{1}{2023^n}$

Then $\lim_{n\to\infty}\frac{1+\frac{1}{2^n}+.....+\frac{1}{2023^n}+\frac{1}{2023^{n+1}}}{1+\frac{1}{2^n}+....+\frac{1}{2023^n}}=1$

UNSOLVED QUESTIONS

1. Let P_n – denote the nth prime number when we enumerate the prime number in increasing order for example $p_1 = 2$, $p_2 = 3$, $p_3 = 5$, ….. and so on.
$S=\{S_n=P_{n+1}-P_n \mid n\in N, n\geq 1\}$ Then which of the following are TRUE?

(a) sup S = ∞ (b) lim sup S_n = ∞
(c) inf S <∞& inf S = 1 (d) lim inf S_n≥ 2

2. Let $a_1=1$ and $a_n=a_{n-1}+4,\ n\geq 2$. Then, $\lim_{n\to\infty}\left[\frac{1}{a_1a_2}+\frac{1}{a_2a_3}+...+\frac{1}{a_{n-1}a_n}\right]$ is equal to

(a) 1/4 (b) 1/2 (c) 1/6 (d) 1/8

3. Let $x_n=\left(1-\frac{1}{3}\right)^2\left(1-\frac{1}{6}\right)^2\left(1-\frac{1}{10}\right)^2...\left(1-\frac{1}{\frac{n(n+1)}{2}}\right)^2$, $n\geq 2$. Then $\lim_{n\to\infty}x_n$ is

(a) $\frac{1}{3}$ (b) $\frac{1}{9}$ (c) $\frac{1}{81}$ (d) 0

4. The sequence $\left\{\cos\left(\frac{1}{2}\tan^{-1}\left(-\frac{n}{2}\right)^n\right)\right\}$ is

(a) Monotone and convergent (b) Monotone but not convergent
(c) Convergent but not monotone (d) Neither monotone nor convergent

5. Let $a_n=\begin{cases}1+\frac{1}{n}, & if\ n\, is\ even\\ -1-\frac{1}{n}, & if\ n\, is\ even\end{cases}$. Then choose incorrect option

(a) $\limsup_{n\to\infty}a_n=\liminf_{n\to\infty}a_n=1$ (b) $\limsup_{n\to\infty}a_n=\liminf_{n\to\infty}a_n=-1$

(c) $\limsup_{n\to\infty} a_n = 1, \liminf_{n\to\infty} a_n = -1$ (d) $\limsup_{n\to\infty} a_n = -1, \liminf_{n\to\infty} a_n = 1$

6. Let $\{a_n\}$ be the sequence of real numbers such that $a_1 = 1$ and $a_{n+1} = a_n + a_n^2$ for all $n \geq 1$. Then

(a) $a_4 = a_1(1 + a_1)(1 + a_2)(1 + a_3)$ (b) $\lim_{n\to\infty} \frac{1}{a_n} = 0$ **IIT-JAM 2019**

(c) $\lim_{n\to\infty} \frac{1}{a_n} = 1$ (d) $\lim_{n\to\infty} a_n = 0$

7. For a, b ∈ N, consider the sequence $d_n = \frac{\binom{n}{a}}{\binom{n}{b}}$ for n > a,b. Which of the following statements are true? As $n \to \infty$ **CSIR NET JUNE 2017**

(a) $\{d_n\}$ converges for all values of a and b

(b) $\{d_n\}$ converges if a < b (c) $\{d_n\}$ converges if a = b (d) $\{d_n\}$ converges if a > b

8. $L = \lim_{n\to\infty} \frac{1}{\sqrt[n]{n!}}$. Then **CSIR NET JUNE 2017**

(a) L = 0 (b) L = 1 (c) $0 < L < \infty$ (d) $L = \infty$

9. Let $\langle a_n \rangle$ be a sequence of real numbers defined by $a_n = \begin{cases} 1 & \text{if } n \text{ is prime} \\ -1 & \text{if } n \text{ is not prime} \end{cases}$.

Let $b_n = \frac{a_n}{n}$ for n ∈ N. Then **IIT JAM 2023**

(a) Both $\langle a_n \rangle$&$\langle b_n \rangle$ are not convergent (b) $\langle a_n \rangle$ is only convergent

(c) Both are convergent (d) Only $\langle b_n \rangle$ is convergent

10. Let $a_n = \left(1 + \frac{1}{n}\right)^n$ and $b_n = n\cos\left(\frac{n!\pi}{2^{10}}\right)$ for n ∈ N. Then **IIT JAM 2023**

(a) $\langle a_n \rangle$ is convergent and $\langle b_n \rangle$ is bounded

(b) $\langle a_n \rangle$ is convergent and $\langle b_n \rangle$ is unbounded.

(c) $\langle a_n \rangle$ is not convergent and $\langle b_n \rangle$ is bounded

(d) $\langle a_n \rangle$ is not convergent and $\langle b_n \rangle$ is unbounded.

11. Let $x_0 = 0$. Define $x_{n+1} = \cos x_n$ for every $n \geq 0$. Then **GATE-2013**

(a) $\{x_n\}$ is increasing and convergent (b) $\{x_n\}$ is decreasing and convergent.

(c) $\{x_n\}$ is convergent and $x_{2n} < \lim_{m\to\infty} x_m < x_{2n+1}$ for every n ∈ N.

(d) $\{x_n\}$ is not convergent.

12. Let $\{a_n\}$ be a sequence of real numbers such that $\lim_{n\to\infty}\left|a_n+3\left(\frac{n-2}{n}\right)^n\right|^{\frac{1}{n}}=\frac{3}{5}$ Then $\lim_{n\to\infty} a_n$ is

(a) $-3e^2$ (b) $3e^{-2}$ (c) $-3e^{-2}$ (d) $3e^2$

13. The sequence $\sqrt{7},\sqrt{7+\sqrt{7}},\sqrt{7+\sqrt{7+\sqrt{7}}},...$ converges to

(a) $\frac{1+\sqrt{33}}{2}$ (b) $\frac{1+\sqrt{32}}{2}$ (c) $\frac{1+\sqrt{30}}{2}$ (d) $\frac{1+\sqrt{29}}{2}$

14. Consider the sequence $a_n=\left(1+(-1)^n\frac{1}{n}\right)^n$. Then

(a) $\limsup_{n\to\infty} a_n=\liminf_{n\to\infty} a_n=1$ (b) $\limsup_{n\to\infty} a_n=\liminf_{n\to\infty} a_n=e$

(c) $\limsup_{n\to\infty} a_n=\liminf_{n\to\infty} a_n=\frac{1}{e}$ (d) $\limsup_{n\to\infty} a_n=e$, $\liminf_{n\to\infty} a_n=\frac{1}{e}$

15. If $a=\lim_{n\to\infty}\left(1+\frac{1}{2}+...+\frac{1}{n}\right)$ and $b=\lim_{n\to\infty}\frac{1}{n}\left(1+\frac{1}{2}+...+\frac{1}{n}\right)$, then

(a) $a=\infty$ (b) $b=\infty$ (c) $a=0$ (d) $b=0$

16. Let $\{a_n\}$ be a sequence of positive real numbers satisfying

$\frac{4}{a_{n+1}}=\frac{3}{a_n}+\frac{a_n^3}{81}$, $n\geq 1$, $a_1=1$. Then all the terms of the sequence lie in

(a) $\left[\frac{1}{2},\frac{3}{2}\right]$ (b) $[0,1]$ (c) $[1,2]$ (d) $[1,3]$

17. Choose the correct statement (s)

(a) Every convergent sequence is a Cauchy sequence

(b) Every Cauchy sequence is a convergent sequence

(c) (a) and (b) both are true (d)(b) is false

18. Let $\{a_n\}$ be the sequence given by $a_n=\max\left\{\sin\left(\frac{n\pi}{3}\right),\cos\left(\frac{n\pi}{3}\right)\right\}, n\geq 1$.

Then which one of the following statements is/are TRUE about the subsequences $\{a_{6n-1}\}$ and $\{a_{6n+4}\}$? **IIT-JAM 2019**

(a) Both the subsequences are convergent (b) Only one of the subsequences is convergent

(c) $\{a_{6n-1}\}$ converges to -1/2 (d) $\{a_{6n+4}\}$ converges to 1/2.

19. Let $a_n = \frac{b_{n+1}}{b_n}$, where $b_1 = 1$, $b_2 = 1$ and $b_{n+2} = b_n + b_{n+1}$, $n \in N$. Then $\lim_{n\to\infty} a_n$ is

IIT-JAM 2018

(a) $\frac{1-\sqrt{5}}{2}$ (b) $\frac{1-\sqrt{3}}{2}$ (c) $\frac{1+\sqrt{3}}{2}$ (d) $\frac{1+\sqrt{5}}{2}$

20. The least upper bound of the sequence $\left\langle 1-\frac{1}{n} \right\rangle$ is

(a) 0 (b) -1 (c) 1 (d) None of these

21. Let $\{x_n\}$ be a real sequence. If the sequence of even terms of $\{x_n\}$ converges to 1 and sequence of odd terms converges to -1. Then the sequence $\{x_n\}$ will

(a) Converge to 0 (b) Converge to 1 (c) Converge to -1 (d) None of these

22. Let $\{a_n\}$, $\{b_n\}$ and $\{c_n\}$ be sequences of real numbers such that $b_n = a_{2n}$ and $c_n = a_{2n+1}$. Then $\{a_n\}$ is convergent

(a) $\{b_n\}$ is convergent but $\{c_n\}$ need not be convergent.

(b) $\{c_n\}$ is convergent but $\{b_n\}$ need not be convergent.

(c) both $\{b_n\}$ and $\{c_n\}$ are convergent.

(d) both $\{b_n\}$ and $\{c_n\}$ are convergent.

Answer Key

1	2	3	4	5	6	7	8	9	10
a,b,d	a	b	a	a,b,d	a,b	b,c	a	d	b
11	**12**	**13**	**14**	**15**	**16**	**17**	**18**	**19**	**20**
c	c	d	d	a,d	d	a,b,c	a	d	c
21	**22**								
d	c								

Chapter 3

Series Of Real Numbers

1. Series

Let <u_n> be a given sequence then a symbol of the form $u_1 + u_2 + ... + u_n + ...$ is called infinite series. It is denoted by $\sum_{n=1}^{\infty} u_n$.

Sequence of partial sums of the series :

Let <u_n> be a sequence and $\sum u_n$ be a series. Then the sequene <S_n> defined by $S_1 = u_1$, $S_2 = u_1 + u_2$, ..., $S_n = u_1 + u_2 + ... + u_n$

The elements of the sequence <S_n> are called the partial sums of the series and the sequence <S_n> is called sequence of partial sums (SOPS).

Sum of Series : Let $\sum u_n$ be a series and S_n is sequence of partial sum of the series then sum of the series = $\lim_{n\to\infty} S_n$.

Example: $\sum u_n = \frac{1}{n(n+1)}$

$$S_1 = \frac{1}{1.2} = 1 - \frac{1}{2}, \qquad S_2 = \frac{1}{2} + \frac{1}{2.3} = 1 - \frac{1}{3}, \quad S_3 = \frac{1}{1.2} + \frac{1}{2.3} + \frac{1}{3.4} = 1 - \frac{1}{4}$$

$$...., \qquad S_n = \frac{1}{1.2} + \frac{1}{2.3} + ... + \frac{1}{n(n+1)} = 1 - \frac{1}{n+1}$$

$\Rightarrow$ Sum of series = $\lim_{n\to\infty} a_n = 1$

Necessary conditon for convergence of series:

(1) If $\sum a_n$ converges, then $\lim_{n\to\infty} a_n = 0$

(2) $a_n \to 0$ is not sufficient to ensure convergence of $\sum a_n$. For example, the series $\sum_{k=1}^{\infty} \frac{1}{n}$ diverges.

(3) If $\lim_{n\to\infty} a_n \neq 0$ then $\sum a_n$ cannot converge

(4) A series of non-negative terms converges if and only if its partial sums form a bounded sequence.

Geometric series:

The geometric series $1 + x + x^2 + ...$

(i) converges if $|x| < 1$ (ii) diverges if $x \geq 1$

(iii) oscillates finitely if x = – 1 (iv) oscillates infinitely if x < – 1

Telescopic Series : Let <an> be a seqence of real numbers, then the series $\sum_{n=1}^{\infty}(a_n - a_{n+1})$ and $\sum_{n=1}^{\infty}(a_{n+1} - a_n)$ are convergent iff the sequence <an> is convergent.

$\sum_{n=1}^{\infty}(a_n - a_{n+1}) = a_1 - \lim_{n\to\infty} a_n$ and $\sum_{n=1}^{\infty}(a_{n+1} - a_n) = \lim_{n\to\infty} a_n - a_1$ = Sum of the Series

Example : $\sum \frac{1}{n(n+1)}$.

2. Test for convergence :

p-series Test:

A positive term series $\sum \frac{1}{n^p}$ is convergent iff p>1. $\sum \frac{1}{n^p}$ is called Hyper Harmonic series or p-series, i.e., the series $\sum \frac{1}{n^p} = \frac{1}{1^p} + \frac{1}{2^p} + \frac{1}{3^p} + ... + \frac{1}{n^p} + ...$converges if p > 1 and diverges if p ≤ 1. We can easily observe that the series $\sum_{n=2}^{\infty} \frac{1}{n(\log n)^p}$ converges if p > 1 and diverges if p ≤ 1.

Comparison test:

Let $\sum a_n$, be a positive term series.

(a) If $a_n \le c_n$, for $n \le N_0$, where No is some fixed integer and if $\sum c_n$ converges, then $\sum a_n$ converges.

(b) If $a_n \le c_n$ for $n \le N_0$ and if $\sum a_n$ diverges, then $\sum c_n$ diverges.

Note that (b) applies only to series of non-negative terms a_n .

Example : Show that $\sum e^{-n^2}$ is convergent.

Solution: We have $e^{n^2} > n^2, \forall n \in N \Rightarrow \frac{1}{e^{n^2}} < \frac{1}{n^2} \Rightarrow e^{-n^2} < \frac{1}{n^2}$

Hence, $\sum e^{-n^2}$ is convergent, because $\sum \frac{1}{n^2}$ is convergent (By p-test)

Limit form test: If $\sum a_n$ and $\sum b_n$ are two positive term series such that $\lim_{n\to\infty}\left(\frac{a_n}{b_n}\right) = l$

[finite and non-zero], then both the series behaves alike, i.e., both converge and diverge together.

Example : Check convergence or divergence of the series $\sum \frac{1}{n^{1+\frac{1}{n}}} = \sum a_n$

Solution: Let $\sum b_n = \frac{1}{n}$, then $\lim_{n\to\infty}\left(\frac{a_n}{b_n}\right) = \lim_{n\to\infty}\left(\frac{\frac{1}{n^{1+\frac{1}{n}}}}{\frac{1}{n}}\right)$

Thus $\lim_{n\to\infty}\left(\frac{a_n}{b_n}\right) = \lim_{n\to\infty}\left(\frac{1}{n^{1+\frac{1}{n}-1}}\right) = \lim_{n\to\infty}\left(\frac{1}{n^{\frac{1}{n}}}\right) = 1 \neq 0, \infty$

$\therefore$ by above comparison test $\sum b_n$ and $\sum a_n$ converge or diverge together, but $\sum b_n$ is divergent.

$\therefore \sum a_n$ is also divergent.

Root Test: Let $\sum a_n$ be a positive terms series and $\alpha = \limsup_{n\to\infty} \sqrt[n]{a_n}$. Then

(a) $\sum a_n$ converges, for $\alpha < 1$,

(b) $\sum a_n$ diverges for $\alpha >$ I and

(c) Test fails for $\alpha = 1$

Example : Check convergence or divergence of the series $\sum a_n = \sum \left(1+\frac{1}{\sqrt{n}}\right)^{-n^{\frac{3}{2}}}$

Solution: Let $a_n = \left(1+\frac{1}{\sqrt{n}}\right)^{-n^{\frac{3}{2}}} = \frac{1}{\left(1+\frac{1}{\sqrt{n}}\right)^{n^{\frac{3}{2}}}} = \frac{1}{\left(1+\frac{1}{\sqrt{n}}\right)^{n^{1+\frac{1}{2}}}}$ i.e. $a_n = \frac{1}{\left(1+\frac{1}{\sqrt{n}}\right)^{n.n^{\frac{1}{2}}}}$

Now, $\limsup_{n\to\infty}(a_n)^{\frac{1}{n}} = \limsup_{n\to\infty} \frac{1}{\left(1+\frac{1}{\sqrt{n}}\right)^{\sqrt{n}}} = \frac{1}{e} < 1$

Here, $l = \frac{1}{e} < 1$ Hence by Root Test, $\sum a_n$ is convergent

Ratio Test: The positive terms series $\sum a_n$

(a) converges if $\limsup\limits_{n\to\infty}\left(\dfrac{a_{n+1}}{a_n}\right)<1$, (b) diverges if $\limsup\limits_{n\to\infty}\left(\dfrac{a_{n+1}}{a_n}\right)>1$

Note: If $\limsup\limits_{n\to\infty}\left(\dfrac{a_{n+1}}{a_n}\right)=1$. Then the series of positive terms may or may not convege.

For Example: $\sum a_n=\sum \dfrac{n^2-1}{n^2+1}.x^n; x>0$

Solution: $a_n=\dfrac{n^2-1}{n^2+1}x^n, a_{n+1}=\dfrac{(n+1)^2-1}{(n+1)^2+1}x^{(n+1)}$

$$\limsup_{n\to\infty}\left(\frac{a_{n+1}}{a_n}\right)=\limsup_{n\to\infty}\left(\frac{(n+1)^2-1}{(n+1)^2+1}.\frac{n^2+1}{n^2-1}.\frac{x^{(n+1)}}{x^n}\right)$$

$$=\limsup_{n\to\infty}\left(\frac{n^2+2n}{n^2+2n+2}.\frac{\left(1+\frac{1}{n^2}\right)}{\left(1-\frac{1}{n^2}\right)}.x\right)=x$$

Hence, by Ratio test series converges if x < 1 and diverges if x > 1. Now, for x = 1

$$a_n=\frac{n^2-1}{n^2+1}.1 \text{ and} =\limsup_{n\to\infty}\left(\frac{\left(1-\frac{1}{n^2}\right)}{\left(1+\frac{1}{n^2}\right)}\right)=1\neq 0$$

By necessary condition of series, this series is not convergent.

Rabbe's Test: If $\sum u_n$ is a positive term series and if $\lim\limits_{n\to\infty} n\left(\dfrac{u_n}{u_{n+1}}-1\right)=l$, then

(i) for $l>1$, series is convergent, (ii) for $l<1$, series is divergent, (iii) for $l=1$, test fails

For Example: By Rabbe's test, verify the series $\sum\dfrac{1}{n^2}$ is convergent

Solution: $u_n=\dfrac{1}{n^2}, u_{n+1}=\dfrac{1}{(n+1)^2}$ Now, $\left(\dfrac{u_n}{u_{n+1}}-1\right)=\dfrac{(n+1)^2}{n^2}-1=\dfrac{2n+1}{n^2}$

$\lim\limits_{n\to\infty} n\left(\dfrac{u_n}{u_{n+1}}-1\right)=\lim\limits_{n\to\infty}\dfrac{2n+1}{n}=2>1$ So, by Rabbe's test, $\sum\dfrac{1}{n^2}$ is convergent

Cauchy Condensation Test: If $\langle a_n\rangle$ is a decreasing sequence of positive real numbers, then the two series $\sum a_n$ and $\sum a_{2^n}2^n$ converge or diverge together.

Example : Check whether the series $\sum \frac{\log n}{n}$, is convergent or divergent

Solution: $\sum 2^n a_{2^n} = \sum 2^n \frac{\log 2^n}{2^n} = \sum 2^n \frac{n \log 2}{2^n} = \log 2 \sum n$

Here $\sum 2^n a_{2^n} = \log 2 \sum n$ is divergent $\Rightarrow \sum \frac{\log n}{n}$ is divergent

Results on series:

(1) If $\sum a_n$ is convergent series of positive terms, then $\sum a_n^2$ is also convergent but converse is not true.

Example : $\sum \frac{1}{n}$ is not convergent but $\sum \frac{1}{n^2}$ is convergent.

(2) If $\sum a_n^2$ and $\sum b_n^2$ both are convergent, where $\sum a_n$ and $\sum b_n$ are positive term series, then $\sum a_n b_n$ is also convergent.

As, $\sqrt{a_n^2 b_n^2} \le \frac{a_n^2 + b_n^2}{2} [\because G.M. \le A.M.] \Rightarrow a_n b_n \le \left(\frac{a_n^2 + b_n^2}{2} \right)$

As, $\sum a_n^2$ and $\sum b_n^2$ are convergent. Therefore, $\sum \left(\frac{a_n^2 + b_n^2}{2} \right)$ is also convergent

Hence, by comparison test $\sum a_n b_n$ is convergent.

(4) If $\sum a_n$ is convergent and $\sum b_n$ is divergent, then $\sum (a_n \pm b_n)$ is divergent

(5) Let $\sum a_n$ be the positive series and $\sum a_n$ be convergent, then $\sum \frac{a_n}{a_{n+1}}$ is convergent.

As, $\frac{a_n}{a_{n-1}} < a_n$ therefore, by comparison test $\sum \frac{a_n}{a_{n+1}}$ is convergent.

3. Alternating series:

A series whose terms are alternatively positive and negative is referred to as an alternating series. It is a series of the form $u_1 - u_2 + u_3 - u_7 + ...$ where $u_n > 0$ for all $n \in N$ and it is denoted by $\sum_{n=1}^{\infty} (-1)^n . u_n$.

Test for nvergence of alternating series :

Leibnitz Test : The alternating series $\sum_{n=1}^{\infty} (-1)^{n-1} u_n = u_1 - u_2 + u_3 ......, u_n > 0 \forall n$, converges if

(i) $u_n \geq u_{n+1}$ for all n i.e. u_n be a monotonic decreasing sequence of positive real numbers.
(ii) $\lim_{n\to\infty} u_n = 0$.

Note : It is the sufficient condition for alternating series.
Result : An alternating series is not convergent if either u_n is not decreasing or $\lim_{n\to\infty} u_n \neq 0$.

Example : Consider the series $\sum_{n=1}^{\infty} \frac{(-1)^{n-1}}{n} = 1 - \frac{1}{2} + \frac{1}{3} - \frac{1}{4} + ...$

Here $a_n = \frac{1}{n}$ clearly $a_n > 0$ and $a_n > a_{n+1}$ for all n and $\lim_{n\to\infty} a_n = 0$.

By Leibnitz test, $\sum_{n=1}^{\infty} \frac{(-1)^{n-1}}{n}$ is convergent.

Absolutely and conditionally convergence of series:

(1) **Absolutely convergent series:** $\sum u_n$ is absolutely convergent if $\sum |u_n|$ is convergent.

Example : $\sum \frac{(-1)^n}{n^2}$

(2) **Conditionally convergent series:** Let $\sum a_n$ is a convergent series but $\sum |a_n|$ is divergent, then $\sum a_n$ is said to be conditionally convergent series or

A series which is convergent but not absolutely is called conditionally convergent series.

Examples:

(1) $\sum (-1)^n \sin \frac{1}{n}$ is conditionally convergent

Solution :

Here, $\sum (-1)^n \sin \frac{1}{n}$ is convergent but $\sum \left|(-1)^n \sin \frac{1}{n}\right| = \sum \sin \frac{1}{n}$ is not convergent

(2) Consider the series $\sum_{n=1}^{\infty} \frac{(-1)^{n-1}}{n^p}$ ...(1)

Find the values of p for which this series is absolutely convergent and also find the value of p for which it is conditionally convergent

Solution: Since, for absolute convergence, we take $\sum_{n=1}^{\infty} \left|\frac{(-1)^{n-1}}{n^p}\right|$

Now, $\sum_{n=1}^{\infty} \frac{1}{n^p}$ is convergent if $p > 1$ and is divergent if $p \leq 1$

But (1) is convergent if p > 0, by Leibnitz test.

Thus, $\sum \frac{(-1)^{n-1}}{n^p}$ is absolutely convergent for p > 1 and conditionally convergent if $0 < p \le 1$

Note : Every absolutely convergent series is convergent.

Tests for series of arbitrary terms:

(1) Abel's Test: If $\{b_n\}$ is positive monotonically decreasing sequence and $\sum u_n$ is convergent, then $\sum b_n u_n$ is convergent

For Example: Test the convergence of the series $\sum \frac{(n^3+1)^{1/3} - n}{\log n}$

Solution: Let $u_n = \left\{(n^3+1)^{1/3} - n\right\}$ and $b_n = \frac{1}{\log n}$. Then the given series can be written as $\sum b_n u_n$

Since $\sum u_n$ converges and $\{b_n\}$ is a positive monotonic decreasing sequence, therefore, by Abel's test the given series converges

(2) Dirichlet's Test: If $\{b_n\}$ is positive monotonically decreasing sequence tending to zero and $\sum u_n$ is a series whose partial sums are bounded, then $\sum b_n u_n$ convergent.

For Example Consider $\sum \frac{\sin n\theta}{n^p}$, discuss about its convergence for different values of p.

Solution: Let $S_n = \sum_{r=1}^{n} \sin r\theta \Rightarrow S_n = \sin\theta + \sin 2\theta + \ldots + \sin n\theta$

$$= \frac{2\sin\frac{\theta}{2}[\sin\theta + \sin 2\theta + \ldots + \sin n\theta]}{2\sin\frac{\theta}{2}}$$

$$= \frac{\left(\cos\frac{\theta}{2} - \cos\frac{3\theta}{2}\right) + \left(\cos\frac{3\theta}{2} - \cos\frac{5\theta}{2}\right) + \ldots + \left(\cos\left(n-\frac{1}{2}\right)\theta - \cos\left(n+\frac{1}{2}\right)\theta\right)}{2\sin\frac{\theta}{2}}$$

$$= \frac{\cos\frac{\theta}{2} - \cos\left(n+\frac{1}{2}\right)\theta}{2\sin\frac{\theta}{2}}$$

REAL ANALYSIS

Taking modulus on both side, we get $|S_n| \le \dfrac{2}{\left|2\sin\frac{\theta}{2}\right|}$

$$|S_n| \le \frac{1}{\left|\sin\frac{\theta}{2}\right|} \text{ for every } \theta \qquad \text{If } \theta \ne 0, 2\pi$$

Then $\sum sin\, n\theta$ is such that sequence of partial sums is bounded, and $\left\{\dfrac{1}{n^p}\right\}$ is positive monotonically decreasing sequence tending to zero, for, $p > 0$

So, by Dirichlet's test, $\sum \dfrac{sin\, n\theta}{n^p}$ is convergent if $p > 0$, for $\theta \ne 2m\pi$ and $\sum \dfrac{sin\, n\theta}{n^p}$ converges to zero $\forall p if\ \theta = 2m\pi, m \in Z$

4. Power Series:

A series of the type $\sum_{n=0}^{\infty} a_n (x-x_0)^n = a_0 + a_1(x-x_0) + a_2(x-x_0)^2 + ...$, is called a power series about x_0. In particular if $x_0 = 0$, then the series $\sum_{n=0}^{\infty} a_n x^n$ is called a power series about 0.

Radius of Convergence and Interval of Convergence:

A power series $a_0 + a_1 + a_2 x^2 + ...$ i.e., $\sum a_n x^n$ is said to have R as radius of convergence if

$\dfrac{1}{R} = \lim\limits_{n\to\infty} (a_n)^{1/n} = \dfrac{a_{n+1}}{a_n}$ & $|x| = R$ is the circle of convergence of the power series.

The set of values of x for which the power series converges is called interval of convergence or its region of convergence. So (- R, R) is called the interval of convergence.

For Example : Find the radius convergence of the power series $\sum\left(1+\dfrac{1}{n}\right)^{n^2} x^n$

Solution: The given power series is $\sum\left(1+\dfrac{1}{n}\right)^{n^2} x^n$

$$\therefore a_n = \left(1+\frac{1}{n}\right)^{n^2} \qquad \therefore |a_n|^{\frac{1}{n}} = \left(1+\frac{1}{n}\right)^{n}$$

$$\therefore \lim_{n\to\infty} |a_n|^{\frac{1}{n}} = \lim_{n\to\infty}\left(1+\frac{1}{n}\right)^{n} = e$$

$$R = \frac{1}{\lim\limits_{n\to\infty} |a_n|^{\frac{1}{n}}} = \frac{1}{e}$$

Note:

(i) If R is the radius of convergence of the power series $\sum_{n=1}^{\infty} a_n x^n$, then the series converges absolutely for $|x| < R$ and diverges for $|x| > R$.

(ii) The power series $\sum_{n=1}^{\infty} a_n x^n$ converges uniformly for $|x| \le \rho < R$, where R is the radius of convergence.

SOLVED QUESTIONS

1. Consider a series $\sum_{n=1}^{\infty} (-1)^{n+1}[\sqrt{n+1} - \sqrt{n}]$, then which of the following is/are true?

CSIR NET JUNE 2023

(a) Divergent (b) Convergent (c) Conditionally convergent (d) Absolutely convergent

Ans. (b), (c)

Explanation :

Using Leibnitz rule $\lim_{n\to\infty} [\sqrt{n+1} - \sqrt{n}] = 0$

$\Rightarrow$ Given series is convergent.

Now $\sum_{n=1}^{\infty} |(-1)^{n+1}[\sqrt{n+1} - \sqrt{n}]$ $= \sum_{n=1}^{\infty} [\sqrt{n+1} - \sqrt{n}]$

Which is divergent. Option (b) & (c) are true.

2. Given $(a_n)_{n\ge 1}$ a sequence of real numbers, which of the following statements is true?

(a) $\sum_{n\ge 1} (-1)^n \dfrac{a_n}{1+|a_n|}$ converges

(b) There is a subsequence $(a_{n_k})_{k\ge 1}$ such that $\sum_{k\ge 1} \dfrac{a_{n_k}}{1+|a_{n_k}|}$ converges

(c) There is a number b such that $\sum_{n\ge 1} \left| b - \dfrac{a_n}{1+|a_n|}(-1)^n \right|$ converges

(d) There is a number b and a subsequence $(a_{n_k})_{k\ge 1}$ such that $\sum_{k\ge 1} \left| b - \dfrac{a_{n_k}}{1+|a_{n_k}|} \right|$ converges

Ans. (d) **CSIR NET JUNE 2020**

Explanation :

(a) If $\lim_{n\to\infty} a_n = \infty$, then $\lim_{n\to\infty} \left| (-1)^n \dfrac{a_n}{1+|a_n|} \right| = \lim_{n\to\infty} \dfrac{1}{\left(\dfrac{1}{|a_n|} + 1\right)} = 1 \to 0,$

so $\sum_{n\geq1}(-1)^n \frac{a_n}{1+|a_n|}$ does not converge, option (a) is incorrect.

Or If $a_n = 1$, then $\sum_{n\geq1}(-1)^n \frac{a_n}{1+|a_n|}$

$=\sum_{n\geq1}\frac{(-1)^n}{2}$, diverges, option (a) is incorrect.

(b) There is a subsequence $(a_{n_k})_{k\geq1}$ such that $\sum_{k\geq1}\frac{a_{n_k}}{1+|a_{n_k}|}$ converges (?)

If $a_n = n^2$, then for any subsequence $(a_{n_k})_{k\geq1}$ of $(a_n)_{n\geq1}$. $\lim_{n\to\infty} a_{n_k} = \infty$, then

$\lim_{n\to\infty}\frac{a_{n_k}}{1+|a_{n_k}|} = 1 \to 0$, so, $\sum_{k\geq1}\frac{a_{n_k}}{1+|a_{n_k}|}$ does not converge,

Option (b) is incorrect.

Or If $a_n = 1$, $(a_{n_k})_{k\geq1} = 1$, then $\sum_{k\geq1}\frac{a_{n_k}}{1+|a_{n_k}|} = \sum_{n\geq1}\frac{1}{2}$, diverges, option (b) is incorrect.

(c) $a_n = (-1)^n$, $\sum_{n\geq1}\left|b-\frac{a_n}{1+|a_n|}\right|(-1)^n$ $=\sum_{n\geq1}\left|b-\frac{(-1)^n}{1+1}\right|(-1)^n$

$=\sum_{n\geq1}\left|b-\frac{(-1)^n}{2}\right|(-1)^n$, there is no number b such that $\sum_{n\geq1}\left|b-\frac{a_n}{1+|a_n|}\right|(-1)^n$ converges, option (3) is incorrect.

Option (a), (b) and (c) are incorrect, so, option (d) is correct.

3. Which of the following statements are true?

(a) The series $\sum_{n\geq1}\frac{(-1)^n}{\sqrt{n}}$ is convergent

(b) The series $\sum_{n\geq1}\frac{(-1)^n}{\sqrt{n}+n}$ is absolutely convergent

(c) The series $\sum_{n\geq1}\frac{[1+(-1)^n]\sqrt{n}+\log n}{n^{3/2}}$. is convergent

(d) The series $\sum_{n\geq1}\frac{((-1)^n\sqrt{n}+1)}{n^{3/2}}$. is convergent

Ans. **(a), (d)** **CSIR NET JUNE 2020**

Explanation :

Take $u_n = \frac{1}{\sqrt{n}}$, $0 < u_{n+1} \le u_n$ and $\lim_{n\to\infty} u_n = \lim_{n\to\infty} \frac{1}{\sqrt{n}} = 0$ by Leibnitz test

$\sum_{n\ge1} \frac{(-1)^n}{\sqrt{n}} < \infty$, the series $\sum_{n\ge1} \frac{(-1)^n}{\sqrt{n}}$ is convergent, Option (a) is correct.

(b) $\sqrt{n} + n \le n + n$ for all $n \ge 1$.

$$\frac{1}{\sqrt{n}+n} \ge \frac{1}{2n} \Rightarrow \sum_{n\ge1} \frac{1}{\sqrt{n}+n} \ge \sum_{n\ge1} \frac{1}{2n}.$$

$\sum_{n\ge1} \frac{1}{2n}$ is divergent series, so is $\sum_{n\ge1} \frac{1}{\sqrt{n}+n}$.

ie, the series $\sum_{n\ge1} \frac{(-1)^n}{\sqrt{n}+n}$ is not absolutely convergent,

Option (b) is not correct.

(c) $$\sum_{n\ge1} \frac{[1+(-1)^n]\sqrt{n}+\log n}{n^{3/2}} = \sum_{n\ge1} \frac{[1+(-1)^n]\sqrt{n}}{n^{3/2}} + \frac{\log n}{n^{3/2}} = \sum_{n\ge1} \frac{1}{n} + \sum_{n\ge1} \frac{(-1)^n}{n} + \sum_{n\ge1} \frac{\log n}{n^{3/2}}$$

Since $\sum_{n\ge1} \frac{1}{n}$ is divergent, whole sum is divergent, Option (c) is incorrect.

(d) $$\sum_{n\ge1} \frac{((-1)^n\sqrt{n}+1)}{n^{3/2}} = \sum_{n\ge1} \frac{(-1)^n\sqrt{n}}{n^{3/2}} + \sum_{n\ge1} \frac{1}{n^{3/2}}$$

$= \sum_{n\ge1} \frac{(-1)^n}{n} + \sum_{n\ge1} \frac{1}{n^{3/2}}$. By Leibnitz test, $\sum_{n\ge1} \frac{(-1)^n}{n}$ converge and

$\sum_{n\ge1} \frac{1}{n^{3/2}}$ converge by p – test. So, the series $\sum_{n\ge1} \frac{((-1)^n\sqrt{n}+1)}{n^{3/2}}$

converges,

Option (d) is correct. Option (a) and (d) are correct.

Necessary condition for convergent of a series.

If Σa_n is convergent then $\lim a_n = 0$ but if $\lim a_n = 0$ then Σa_n may not be convergent.

e.g. $\sum \frac{1}{n}$ where $a_n = \frac{1}{n}$.

$\lim a_n = 0$ but $\sum \frac{1}{n}$ is not convergent.

REAL ANALYSIS

4. Let n be a natural number, then the series $\sum_{m\geq n}(-1)^m \frac{1}{m}$ is **CSIR NET DEC 2019**

(a) Absolutely convergent
(b) divergent
(c) Absolutely convergent if n > 100
(d) convergent

Ans. (d)

Explanation :

Let n = 1

Then $\sum_{m\geq 1}\frac{(-1)^m}{m}$ $\quad a_m = \frac{(-1)^m}{m}$ $\quad$ Then $\sum|a_n| = \sum_{m\geq 1}\frac{1}{m}$ is not convergent.

$\Rightarrow \Sigma a_n$ is not absolutely convergent.

By Leibnitz test given series is convergent.

5. What is the sum of series $\left(\frac{1}{2.3}+\frac{1}{2^2.3}\right)+\left(\frac{1}{2^2.3^2}+\frac{1}{2^3.3^2}\right)+....\left(\frac{1}{2^a.3^a}+\frac{1}{2^{a+1}.3^a}\right)+...$

CSIR NET JUNE 2019

(a) $\frac{3}{8}$
(b) $\frac{3}{10}$
(c) $\frac{3}{14}$
(d) $\frac{3}{16}$

Ans. (b)

Explanation :

Given that $\left(\frac{1}{2.3}+\frac{1}{2^2.3}\right)+\left(\frac{1}{2^2.3^2}+\frac{1}{2^3.3^2}\right)+....\left(\frac{1}{2^a.3^a}+\frac{1}{2^{a+1}.3^a}\right)+...$

We can write this series by $\left(\frac{1}{2.3}+\frac{1}{2^2.3^2}+...+\frac{1}{2^a.3^a}\right)+\frac{1}{2}\left(\frac{1}{2.3}+\frac{1}{2^2.3^2}+...+\frac{1}{2^a.3^a}\right)$

$\frac{1}{6}\left(1+\frac{1}{6}+\frac{1}{6^2}...\right)+\frac{1}{12}\left(1+\frac{1}{6}+\frac{1}{6^2}...\right)$

$\Rightarrow\left(1+\frac{1}{6}+\frac{1}{6^2}...\right)\left(\frac{1}{6}+\frac{1}{12}\right)$ $\quad \Rightarrow\left(\frac{1}{1-\frac{1}{6}}\right)\left(\frac{3}{12}\right)\Rightarrow\left(\frac{6}{5}\right)\left(\frac{3}{12}\right)=\frac{3}{10}$

6. The difference $\log(2)-\sum_{n=1}^{100}\frac{1}{2^n.n}$ is **CSIR NET JUNE – 2018**

(a) Less than 0
(b) less than $\frac{1}{2^{100}.101}$.
(c) greater than 1
(d) greater then $\frac{1}{2^{100}.101}$.

Ans. (b)

Explanation :

We know that $\log(1-x) = -\sum_{n=1}^{\infty}\frac{x^n}{n}$ Put x = 1/2

Then $\log\left(\frac{1}{2}\right) = -\sum_{n=1}^{\infty}\frac{1}{2^n.n}$ $= \log 1 - log(2) = -\sum_{n=1}^{\infty}\frac{1}{2^n.n}$

$$= 0 - \log(2) = -\sum_{n=1}^{100}\frac{1}{2^n.n} - \sum_{n=101}^{\infty}\frac{1}{2^n.n} \qquad = \log 2 - \sum_{n=1}^{100}\frac{1}{2^n.n} = \sum_{n=101}^{\infty}\frac{1}{2^n.n}$$

$$= \frac{1}{2^{101}.101} + \frac{1}{2^{102}.102} + ...$$

$$< \frac{1}{2^{101}.101} + \frac{1}{2^{102}.101} + \frac{1}{2^{103}.101} \qquad \left[\because 102 > 101 \quad \therefore \frac{1}{102} < \frac{1}{101}\right]$$

$$< \frac{1}{101}\left[\frac{1}{2^{101}} + \frac{1}{2^{102}} + \frac{1}{2^{103}} + ...\right] \text{ is G.P. series.}$$

$$< \frac{1}{101}\left[\frac{\frac{1}{2^{101}}}{1-\frac{1}{2}}\right] \qquad \left[\therefore r = \frac{1}{2}\right]$$

$$< \frac{1}{101}\left[\frac{1}{2^{101}} \times 2\right] < \frac{1}{2^{100}101} \qquad = \log 2 - \sum_{n-1}^{100}\frac{1}{2^n.n} < \frac{1}{2^{100}101}$$

7. Given $\{a_n\}$, $\{b_n\}$ two monotone sequences of real numbers and that $\sum a_n b_n$ is convergent, which of the following is true? **CSIR NET JUNE 2018**

(a) $\sum a_n$ is convergent and $\sum b_n$ is convergent.

(b) At least $\sum a_n, \sum b_n$ one is convergent.

(c) $\{a_n\}$ is bounded and $\{b_n\}$ is bounded.

(d) At least one of $\{a_n\}$, $\{b_n\}$ is bounded.

Ans. (d)

Explanation :

(a) Take $a_n = \frac{1}{n}$ and $b_n = \frac{1}{n}$ Option (a) is incorrect.

(b) Take $a_n = \frac{1}{n}$ and $b_n = \frac{1}{n}$ Option (b) is incorrect.

(c) Take $a_n = n$ & $b_n = \frac{1}{n^3}$

Option (c) is incorrect. Option (d) is correct.

REAL ANALYSIS

8. For a > 0, the series $\sum_{n=1}^{\infty} a^{\ln n}$ is convergent if and only if **CSIR NET JUNE 2017**

(a) $0 < a < e$ (b) $0 < a \le e$ (c) $0 < a < \frac{1}{e}$ (d) $0 < a \le \frac{1}{e}$

Ans. (c)

Explanation :

We know that $a^{\log n} = n^{\log a}$ $\Rightarrow \sum a^{\log n} = \sum n^{\log a} = \sum \frac{1}{n^{-\log a}}$

By p-test $\sum \frac{1}{n^{-\log a}}$ is convergent iff $-\log a > 1$

$\Rightarrow \log a < -1 \Rightarrow a < e^{-1}$ $\Rightarrow 0 < a < \frac{1}{e}$ [Given a > 0]

9. Let a_n be a sequence of real number satisfying $\sum_{n=1}^{\infty} |a_n - a_{n-1}| < \infty$. Then the series $\sum a_n x^n$ $x \in R$ is convergent. **CSIR NETJUNE- 2017**

(a) no where on R (b) everywhere on R

(c) on some set containing (-1, 1) (d) only on (-1, 1)

Ans. (c)

Explanation :

Let if we take $a_n = \frac{1}{n^2}$

then $\sum \left| \frac{1}{n^2} - \frac{1}{(n-1)^2} \right| = \sum \left| \frac{n^2 + 1 - 2n - n^2}{n^2 (n-1)^2} \right| = \sum \left| \frac{1-2n}{n^2 (n-1)^2} \right|$

and $\sum \left| \frac{1-2n}{n^2 (n-1)^2} \right| < \infty$ so we can take this example $a_n = \frac{1}{n^2}$

Now given that $\sum \frac{x^n}{n^2}$ is convergent. Let $b_n = \frac{x^n}{n^2}$.

i.e. $\left| \frac{b_{n+1}}{b_n} \right| = \left| \frac{x^{n+1}}{(n+1)^2} \times \frac{n^2}{x^n} \right|$ $\lim_{n \to \infty} \left| \frac{n^2 x}{(n+1)^2} \right| < 1$

$\Rightarrow |x| < 1 \Rightarrow x \in (-1, 1)$

Option (a) & (b) are discard. Option (d) is incorrect. If we take x = 1

Then $\sum \frac{1}{n^2}(1)^2 = \sum \frac{1}{n^2}$ is also convergent.

10. Let u_n be a sequence of real number satisfying the condition :

(1) $(-1)^n u_n \ge 0$ $\forall n \ge 1$

(2) $\left|u_{n+1}\right| < \frac{|u_n|}{2} \qquad \forall n \geq 13$

Which of the following are TRUE? **CSIR NET JUNE - 2018**

(a) $\sum_{n\geq 1} u_n$ doesn't converge in R. (b) $\sum_{n\geq 13} u_n$ converge to zero.

(c) $\sum_{n\geq 13} u_n$ convergent to a non-zero real number.

(d) If $|u_{n+1}| < \frac{|u_n|}{2}$ then $\sum_{n\geq 1} u_n$ is negative real number. $\forall\ 2 \leq n \leq 13$

Ans. **(c), (d)**

Explanation :

Let $u_n = (-1)^n a_n \qquad a_n \geq 0 \quad \& \qquad a_{n+1} < \frac{a_n}{2}$

Now we have to check that $\sum u_n$ is convergent.

Given that $|u_{n+1}| < \frac{|u_n|}{2} \qquad \Rightarrow \left|\frac{u_{n+1}}{u_n}\right| < \frac{1}{2} \qquad \Rightarrow \lim_{n\to\infty} \left|\frac{u_{n+1}}{u_n}\right| < \frac{1}{2} < 1$

By Ratio Test $\sum u_n$ is convergent.

Option (a) is incorrect. Check (b) and (c)

$$\sum_{n\geq 13} u_n = u_{13} + u_{14} + u_{15} + u_{16} + ... = (-a_{13} + a_{14}) + (-a_{15} + a_{16}) + (-a_{17} + a_{18}) + ...$$

Given that $a_{n+1} < \frac{a_n}{2} \Rightarrow a_{n+1} < a_n = \underset{-ve}{(a_{14} - a_{13})} + \underset{-ve}{(a_{16} - a_{15})} + \underset{-ve}{(a_{18} - a_{17})} + ...$

$\Rightarrow \sum_{n\geq 13} u_n < 0$ Option (b) is incorrect. $\Rightarrow \sum_{n\geq 13} u_n$ is convergent to non-zero real number.

Option (c) is correct. Option (d) is correct.

$$\sum_{n\geq 1} u_n = u_1 + u_2 + u_3 + u_4 + u_5 + ...$$

$$= -a_1 + a_2 - a_3 + a_4 - a_5 + a_6 - a_7 ..... - a_{13} + a_{14} - a_{15} .....$$

$$= -a_1 + \underset{-ve}{(a_2 - a_3)} + \underset{-ve}{(a_4 - a_5)} + ...$$

$$+ \underset{-ve}{\left(a_{12} - \frac{a_{13}}{2}\right)} + \underset{-ve}{\left(-\frac{a_{13}}{2} + a_{14}\right)} + \underset{-ve\Theta}{\left(-\frac{a_{15}}{2} + a_{16}\right)} + ...$$

Given that $\left|a_{n-1}\right| < \frac{a_n}{2} \qquad 2 \leq n \leq 13$

& $\quad a_{n+1} < \frac{a_n}{2} \qquad n \geq 13$ Then $\sum u_n < 0$.

Option (d) is correct.

REAL ANALYSIS

REAL ANALYSIS

UNSOLVED QUESTIONS

1. The radius of convergence of power series $\sum_{n=0}^{\infty} 2^{2n} x^{n^2}$ is

(a) $\frac{1}{4}$ (b) 1 (c) 2 (d) 4

2. The sequence $\{s_n\}$ of real numbers given by $s_n = \frac{\sin\frac{\pi}{2}}{1.2} + \frac{\sin\frac{\pi}{2^2}}{2.3} + \ldots + \frac{\sin\frac{\pi}{2^n}}{n.(n+1)}$ is

(a) a divergent sequence (b) an oscillatory sequence
(c) not a Cauchy sequence (d) a Cauchy sequence

3. Let $\{a_0, a_1, \ldots\}$ be a sequence of real numbers. For any $k \geq 1$, let $S_n = \sum_{k=0}^{n} a_{2k}$. Which of the following statements are correct?

(a) If $\lim_{n\to\infty} S_n$ exists, then $\sum_{m=0}^{\infty} a_m$ exists.

(b) If $\lim_{n\to\infty} S_n$ exists, then $\sum_{m=0}^{\infty} a_m$ need not exists.

(c) If $\sum_{m=0}^{\infty} a_m$ exists, then $\lim_{n\to\infty} S_n$ exists.

(d) If $\sum_{m=0}^{\infty} a_m$ exists, then $\lim_{n\to\infty} S_n$ need not exists.

4. For $n \geq 2$, let $a_n = \frac{1}{n\log n}$. Then **CSIR NET JUNE 2015**

(a) The sequence $\{a_n\}_{n=2}^{\infty}$ is convergent (b) The series is $\sum_{n=2}^{\infty} a_n$ convergent

(c) The series $\sum_{n=2}^{\infty} a_n^{\,2}$ is convergent. (d) The series $\sum_{n=2}^{\infty} (-1)^n a_n$ is convergent.

5. Let $\{a_n\}_{n=1}^{\infty}$ be a sequence of positive real numbers such that $\sum_{n=1}^{\infty} a_n$ is divergent.

Which of the following series are convergent?

(a) $\sum_{n=1}^{\infty} \frac{a_n}{1+na_n}$ (b) $\sum_{n=1}^{\infty} \frac{a_n}{1+n^2 a_n}$ (c) $\sum_{n=1}^{\infty} \frac{a_n}{1+a_n}$ (d) none of these

6. Let $\{a_n\}$ be a sequence of real numbers such that the series $\sum_{n=1}^{\infty} |a_n|^2$ is convergent. Which of the following statement are true?

(a) The series $\sum_{n=1}^{\infty} \frac{a_n}{n}$ is convergent

(b) The series $\sum_{n=1}^{\infty} |a_n|^p$ is convergent for all $2 < p < \infty$

(c) The series $\sum_{n=1}^{\infty}|a_n|^p$ is convergent for all $1<p<2$ (d) None of these

7. Let $\{a_n\}_{n=1}^{\infty}$ and $\{b_n\}_{n=1}^{\infty}$ be two sequence of real numbers such that the series $\sum_{n=1}^{\infty}a_n^2$ and $\sum_{n=1}^{\infty}b_n^2$ converges. Then the series $\sum_{n=1}^{\infty}a_nb_n$

(a) is absolutely convergent (b) may not converge

(c) is always convergent, but may not converge absolutely (d) converges to 0

8. Let $\{a_n\}$ be a sequence of real numbers. Which of the following is true?

(a) If $\sum a_n$ converges, then so does $\sum a_n^4$ (b) If $\sum|a_n|$ converges, then so does $\sum a_n^2$

(c) If $\sum a_n$ diverges, then so does $\sum a_n^3$ (d) If $\sum|a_n|$ diverges, then so does $\sum a_n^2$

9. The series $\frac{x+.2}{1}+\frac{(x+.2)^2}{2}+...+\frac{(x+.2)^n}{n}+...$ converges for x =

(a) 1 (b) ∞ (c) zero (d) $\frac{1}{2}$

10. In interval of convergence of the power series $\sum_{n=1}^{\infty}\frac{1}{(-3)^{n+2}}\frac{(4x-12)^n}{n^2+1}$ is **IIT-JAM 2017**

(a) $\frac{10}{4}\le x<\frac{14}{4}$ (b) $\frac{9}{4}\le x<\frac{15}{4}$ (c) $\frac{10}{4}\le x\le\frac{14}{4}$ (d) $\frac{9}{4}\le x\le\frac{15}{4}$

11. Let $a_n=\frac{(-1)^n}{\sqrt{1+n}}$ and let $c_n=\sum_{k=0}^{n}a_{n-k}a_k$, where $n\in N\cup\{0\}$. Then which one of the following is TRUE? **IIT-JAM 2018**

(a) Both $\sum_{n=0}^{\infty}a_n$ and $\sum_{n=1}^{\infty}c_n$ are convegent.

(b) $\sum_{n=0}^{\infty}a_n$ is convergent but $\sum_{n=1}^{\infty}c_n$ is not convergent.

(c) $\sum_{n=1}^{\infty}c_n$ is convergent but $\sum_{n=0}^{\infty}a_n$ is not convergent.

(d) Neither $\sum_{n=0}^{\infty}a_n$ nor $\sum_{n=1}^{\infty}c_n$ is convergent.

12. Let $a_1=b_1=0$ and for each $n\ge 2$, let a_n and b_n be a real numbers given by

$a_n=\sum_{m=2}^{n}\frac{(-1)^m m}{(\log(m))^m}$ and $b_n=\sum_{m=2}^{n}\frac{1}{(\log(m))^m}$. Then which one of the following is TRUE about the sequence $\{a_n\}$ and $\{b_n\}$? **IIT-JAM 2019**

(a) Both {a_n} and {b_n} are divergent (b) {a_n} is convergent and {b_n} is divergent.
(c) {a_n} is divergent and {b_n} is convergent (d) Both {a_n} and {b_n} are convergent.

13. Let $\{a_n\}_{n=0}^{\infty}$ and $\{b_n\}_{n=0}^{\infty}$ be sequences of positive real numberss such that $na_n < b_n < n^2a_n$ for all $n \geq 2$. If the radius of convergence of the power series $\sum_{n=0}^{\infty} a_n x^n$ is 4, then the power series $\sum_{n=0}^{\infty} b_n x^n$. **IIT-JAM 2019**

(a) converges for all x with $|x| < 2$ (b) converges for all x with $|x| > 2$
(c) does not converge for any x with $|x| > 2$ (d) does not converge for any x with $|x| < 2$

14. For $-1 < x < 1$, the sum of the power series $1+\sum_{n=2}^{\infty}(-1)^{n-1} n^2 x^{n-1}$ is **IIT-JAM 2019**

(a) $\frac{1-x}{(1+x)^3}$ (b) $\frac{1+x^2}{(1+x)^4}$ (c) $\frac{1-x}{(1+x)^2}$ (d) $\frac{1+x^2}{(1+x)^3}$

15. Which one of the following is divergent? **IIT-JAM 2019**

(a) $\sum_{n=1}^{\infty} \frac{1}{n} \sin^2 \frac{1}{n}$ (b) $\sum_{n=1}^{\infty} \frac{1}{n} \log n$ (c) $\sum_{n=1}^{\infty} \frac{1}{n^2} \sin \frac{1}{n}$ (d) $\sum_{n=1}^{\infty} \frac{1}{n} \tan \frac{1}{n}$

16. Let {a_n} be a sequence of positive real numbers such that

$a_1 = 1, a_{n+1}^2 - 2a_n a_{n+1} - a_n = 0$ for all $n \geq 1$. Then the sum of the series $\sum_{n=1}^{\infty} \frac{a_n}{3^n}$ lies in the interval **IIT-JAM 2019**

(a) (1, 2] (b) (2, 3] (c) (3, 4] (d) (4, 5]

ANSWER KEY

1	2	3	4	5	6	7	8	9	10
b	d	b,d	a,c,d	b	a,b	b	a	c,d	d
11	**12**	**13**	**14**	**15**	**16**				
b	d	a	a	b	a				

Chapter 4

Continuity and Differentiability

1. Function :

Let $A \subseteq R$ and $B \subseteq R$, then a rule in which assign every element of A to unique element of B is called a function from A to B and denoted by $f : A \to B$, where A is called a domain and B is called a co-domain.

Range : A subset of B which contain image of each element of A is called a Range of the function. i.e., $f(A) = \{f(x) \mid x \in A\}$

Domain of Definition (DOD) : If S be a subset of R and $f : S \to R$ be a real valued function then S is called domain of definition

Example : (1) $f(x) = \frac{1}{x}$, DOD = R – {0}

(2) f(x) = log x, DOD = $(0, \infty)$

(3) $f(x) = \log_2(\log_2 \log_2 x)$

We know that domain of $\log_2 x$ is $x > 0$

$\Rightarrow \log_2 \log_2 x > 0$

$\Rightarrow \log_e x > 2^\circ$

$\Rightarrow \log_2 x > 1$

$\Rightarrow x > 2$

DOD = $(2, \infty)$

One-one function : A function is called a one-one function if image of all distinct element are distinct. i.e., If $f : A \to B$ and f(x) is one –one then $x_1 \neq x_2 \Rightarrow f(x_1) \neq f(x_2)$ for every $x_1, x_2 \in A$.

Onto-function : A function is called a onto if range set is equal to co-domain.

Example : $f : R \to R$ such that $f(x) = x^3$. Which is one-one and onto both.

Limit of function : Let $f : S \to R$, we say $l \in R$ is a limit of f

$\Leftrightarrow$ for any $\varepsilon > 0$, $\exists \delta > 0$ such that $x_1, x_2 \in \{x : 0 < |x - \alpha| < \delta\}$

$\Rightarrow |f(x_1) - f(x_2)| < \varepsilon$

And it is denoted by $\lim_{x \to a} f(x) = l$

One-side limit :

Right hand limit (RHL) : A function f(x) is said to approach L as $x \to a$ from the right or from above if $\lim_{h \to 0} f(a+h) = L$.

Left hand limit (LHL) : A function f(x) is said to approach L as $x \to a$ from the left or from below if $\lim_{h \to 0} f(a-h) = L$.

Note : Limit of f(x) exist at x = a, iff RHL = LHL at x = a.

2. Continuity

Definitions of Continuity:

First Definition: A function $f : A \to R$ is said to be continuous at point x = a ∈ A\` (where a is a limit point of A), if $\lim_{x\to a^+} f(x) = \lim_{x\to a^-} f(x) = f(a)$

Example : $f(x) = \begin{cases} e^x & x > 0 \\ e^{-x} & x \le 0 \end{cases}$

Solution : Here

RHL $\quad f(0 + 0) = e^0 = 1$

LHL $\quad f(0 - 0) = e^{-0} = 1$

So, $\lim_{x\to 0} f(x) = 1$ and $f(0) = e^{-0} = 1$

$\Rightarrow \lim_{x\to 0} f(x) = f(0) \quad \Rightarrow$ This f(x) is continuous.

Sequential definition : Let f : A → R be a function and f is continuous at 'c' iff for every sequence $<x_n>$ converging to c, then $<f(x_n)>$ converging to f(c).

i.e. if ∃ two sequence $<x_n>$ and $<y_n>$ s.t. $<x_n> \to c$ and $<y_n> \to c$ but if $<f(x_n)>$ and $<f(y_n)>$ converging to distinct limit then f(x) is not continuous at c.

Example : $f(x) = \sin\left(\frac{1}{x}\right)$

We have check at x = 0

$$< x_n >= \frac{1}{n\pi} \to 0 \text{ and } < y_n >= \frac{1}{(2n-1)\frac{\pi}{2}} \to 0$$

So, $f(x_n) = \sin(n\pi) = 0$ but $f(y_n) = \sin\left((2n-1)\frac{\pi}{2}\right) \to 1$

Both are converge to distinct limit.

So, $\sin\left(\frac{1}{x}\right)$ is not continuous at x = 0.

Discontinuity of a Function:

A function which is not continuous at a point 'a' is said to be discontinuous at the point 'a' and

'a' is called point of discontinuity of f

Types of Discontinuity

Let f be a function defined on an Interval S. Let f be discontinuous at a point $a \in S$

(1) Removable Discontinuity:

Let 'a' be a limit point of S such that $\lim_{x \to a} f(x)$ exists then f has removable discontinuity at x = a.

If (a) $a \notin S$ or (b) if $a \in S$ but $\lim_{x \to a} f(x) \neq f(a)$

This type of discontinuity can be removed by defining a new function g as

$$g(x) = \begin{cases} f(x) & \text{if } x \neq a \\ \lim_{x \to a} f(x) & \text{if } x = a \end{cases}$$

Then g is continuous at 'a'

Example: $f(x) = \begin{cases} \dfrac{\sin x}{x}, & x \neq 0 \\ 0, & x = 0 \end{cases}$

Domain of definition = R $\lim_{x \to 0} f(x) = 1 \neq f(0)$

Hence removable discontinuity at $x = 0$

We can remove this discontinuity by redefining function $f(x) = \begin{cases} \dfrac{\sin x}{x}, & x \neq 0 \\ 1, & x = 0 \end{cases}$

Domain of definition = R

This re-defined function is continuous at '0'

(2) Discontinuity of First Kind (or Jump Discontinuity):

If $\lim_{x \to a^-} f(x)$ and $\lim_{x \to a^+} f(x)$ both exist, but are unequal, then f is said to have discontinuity of first kind at 'a' or jump discontinuity at 'a'.

Example : Let f : R→ R defined by f(x) = $[x]$ $\forall x \in R$

Then f(x) has jump kind of discontinuity at every $\alpha \in Z$

(3) Discontinuity of Second Kind:

If atleast one of $\lim_{x \to a^-} f(x)$ or $\lim_{x \to a^+} f(x)$ fails to exist, then f is said to have a discontinuity of second kind at 'a'.

Example : $f(x) = \dfrac{1}{x-1}$ has a discontinuity of the second kind at x = 1

(4) Mixed Discontinuity:

If a function f has a discontinuity of the second kind on one side of 'a' and on the other side, f has a discontinuity of the first kind, then f is said to have a mixed discontinuity at 'a'.

Thus f has a mixed discontinuity at 'a' if either

(i) $\lim_{x\to a^-} f(x)$ does not exist and $\lim_{x\to a^+} f(x)$ exists, however $\lim_{x\to a^+} f(x)$ may or may not be equal to $f(a)$.

(ii) $\lim_{x\to a^+} f(x)$ does not exist and $\lim_{x\to a^-} f(x)$ exists, however $\lim_{x\to a^-} f(x)$ may or may not be equal to $f(a)$.

REAL ANALYSIS

Results on discontinuity:

(i) Every monotonic function defined on I has only jump discontinuity

(ii) The set of discontinuities of monotonic function is countable.

3. Some important theorems on continuous function :

1. Boundedness theorem :

If f is continuous function on close interval I then f is bounded on I.

2. The location of roots theorem :

Let I = [a, b] and f : I → R be continuous function on I and if f(a) < 0 < f(b) or f(a) >0 > f(b) then there exist a number c ∈ (a,b) such that f(c) = 0.

3. Intermediate Value Theorem: If a function f is continuous on a closed interval [a, b] and $f(a) \neq f(b)$, then f assumes every value between $f(a)$ and $f(b)$ at least once.

Note: Converse of above theorem is not true, i.e., a function which satisfy the intermediate value property on [a, b] need not be continuous

For example: $f(x)=\begin{cases}\sin\frac{1}{x}, & x\neq 0\\ 0, & x=0\end{cases}$ *defined on* $\left[\frac{-2}{\pi},\frac{2}{\pi}\right]$

Satisfies intermediate value property but not continuous at x = 0

4. If a function f is one to one and satisfy the intermediate value property on [a, b], then function f is continuous on (a, b).

5. The image of a closed interval under a continuous function is a closed interval. If $f : A \to R$ is continuous, then inverse image of an open set is open and inverse image of a closed set is closed.

6. If a one-to-one function f is continuous and strictly monotonic on a closed interval [a, b], then f^{-1} is also continuous on $[f(a), f(b)]$ or $[f(a), f(b)]$ according as f is increasing or decreasing.

7. Let $f(x)=\begin{cases}p(x), & x\in Q\\ q(x) & x\in Q^c\end{cases}$ .Then $f(x)$ is continuous at $x=\alpha$ if $p(\alpha)=q(\alpha)$ .i.e., if α is roots p(x) - q (x) = 0, then $f(x)$ is continuous at $x=\alpha$ and these are the only points at which function will be continuous.

Example: (i)The Dirichlet's function f defined by $f(x)=\begin{cases}1, & if\, x\, is\, rational\\ -1, & f\, x\, is\, irrational\end{cases}$ is discontinuous for every real x.

8. **Fixed Point Theorem:** Let S⊆R and f : A→ R be any function. Then c∈A is said to be a fixed point of f(x) if f(c) = c

Sufficient condition for fixed point:

If $f:[a,b]\to[a,b]$ is continuous, then $\exists c\in[a,b]$ such that f(c) = c.

Example : Let f : R → R be a function defined as f(x) = e^x∀ x ∈ R then f has no fixed point.

Reason : Since the line y = x and y = f(x) does not intersect anywhere.

9. Composite of two functions $(f\circ g)(x)$ may be continuous even if one of f or g is discontinuous or even both are discontinuous

For example:

$$f(x)=\begin{cases}-1 & x\in Q\\ 1 & x\in Q^c\end{cases}$$ Clearly f is discontinuous for all x.

But $(f\circ f)(x)\forall x$, is everywhere continuous, and $g(x)=\begin{cases}x & x\in Q\\ 1-x & x\in Q^c\end{cases}$

Thus, g(x) is continuous at $x=\frac{1}{2}$ only

When $x\in Q,(g\circ g)(x)=g(x)=x$ and when

$x\in Q^c,(g\circ g)(x)=g(1-x)=1-(1-x)=x$

Hence, $(g\circ g)(x)=x$ is continuous function.

Example. Find the conditions on p and q such that for $f(x)=\begin{cases}x^p\sin\frac{1}{x^q}, & x\neq 0\\ 0 & x=0\end{cases}$

(a) Limit of $f(x)$ does not exist at 0 (b) $f(x)$ is continuous in R

Solution: $f(x)$ is continuous everywhere except at x = 0, now at x = 0

$$\lim_{x\to 0}x^p.\sin\frac{1}{x^q}=\lim_{x\to 0}\frac{\sin\frac{1}{x^q}}{\frac{1}{x^p}}$$ Here $x\to 0\Rightarrow\frac{1}{x}\to\infty$ Let $\frac{1}{x}=t$

$$\therefore\lim_{x\to 0}f(x)=\lim_{t\to\infty}\frac{\sin t^q}{t^p}=\lim_{t\to\infty}t^{q-p}\frac{\sin t^q}{t^q}=\lim_{t\to\infty}t^{q-p}$$

So limit does not exist, when q – p > 0 or when p < q and limit will exist and equal to 0, when q – p < 0 or when p > q. So, when p > q function will be continuous at x = 0.

3. Uniform continuity

A function f defined on an interval I is said to be uniformly continuous on I if for given $\in > 0, \exists \delta > 0$ depending upon $\in$ such that $|f(x_1) - f(x_2)| < \in$, whenever $|x_1 - x_2| < \delta$ where $x_1, x_2 \in I$.

REAL ANALYSIS

Examples

(1) $f(x) = \sin x$ is uniformly continuous on R

Proof :

$$|f(x_1) - f(x_2)| = |\sin x_1 - \sin x_2|$$

$$= \left|2\cos\frac{x_1 + x_2}{2}.\sin\frac{x_1 - x_2}{2}\right|$$

$$= 2\left|\cos\frac{x_1 + x_2}{2}\right|.\left|\sin\frac{x_1 - x_2}{2}\right|.$$

$$\leq 2\left|\sin\frac{x_1 - x_2}{2}\right| \qquad \left[\because \left|\cos\frac{x_1 + x_2}{2}\right| \leq 1\right]$$

$$\leq 2\frac{|x_1 - x_2|}{2} \qquad \left[\because |\sin(x_1 - x_2)| \leq |x_1 - x_2|\right]$$

$$= |x_1 - x_2| < \in = \delta$$

Hence $\forall \in > 0, \exists \delta = \in$ such that $|f(x_1) - f(x_2)| < \in \forall x_1, x_2 \in R$ whenever $|x_1 - x_2| < \delta$ $\therefore f(x) = \sin x$ is uniformly continuous on R

(2) $f(x) = x$ is uniformly continuous on R $|f(x_1) - f(x_2)| = |x_1 - x_2| < \delta = \in$

Hence $\forall \in > 0, \exists \delta = \in$ such that $|f(x_1) - f(x_2)| < \in$ whenever $|x_1 - x_2| < \delta$

(3) The function f(x) = sin x^2 is not uniformly continuous on [0, ∞[.

(4) The function $f(x) = \sqrt{x}$ is uniformly continuous in [0, 1].

(5) The function $f(x) = \sin\frac{1}{x}$ is not uniformly continuous on]0, ∞[.

(6) The function f(x) = 1/x is not uniformly continuous on]0, 1[, but it is uniformly continuous on [a, ∞[, where a > 0.

Results on uniform continuity:

(1) If a function f is uniformly continuous on an interval I, then it is continuous on I

(2) If a function f is continuous on closed interval [a, b], then it is uniformly continuous on [a, b]

(3) If a function f is continuous on open interval (a, b), then f is uniformly continuous on (a, b)iff $\lim_{x\to a^+} f(x)$ and $\lim_{x\to b^-} f(x)$ both exists

(4) If a function f is continuous on [a, ∞), (or (- ∞, b]) and $\lim_{x\to\infty} f(x)$ (or $\lim_{x\to-\infty} f(x)$) exists, then f is uniformly continuous on [a, ∞), (or (- ∞, b]).

(5) If a function f is continuous on R and $\lim_{x\to\infty} f(x)$ and $\lim_{x\to-\infty} f(x)$ both exists, then f is uniformly continuous on R

(6) Let f and g be two real functions such that $f:(\phi\neq)S\to R$ and $g:(\phi\neq)T\to S$
Also, let f be uniformly continuous on S and g is uniformly continuous on T
Then, $h(x)=f(g(x))$ is uniformly continuous on T

Example. $f(x)=\begin{cases} x^2, & 0\le x\le 1 \\ \sqrt{x}, & x>1 \end{cases}$, $f(x)$ is uniformly continuous on [0, ∞)

Solution: Let $h:[0,1]\to R$ defined as $h(x)=x^2$ Clearly, $h(x)=x^2$ is uniformly continuous on [0, 1] as $h(x)$ is continuous on [0, 1]. Now take $g:[1,\infty]\to R$, defined as g(x) = $\sqrt{x}$ then, g(x) is uniformly continuous on [1, ∞) as

(i) g(x) is continuous on [1, ∞) and (ii) g'(x)=. $\frac{1}{2\sqrt{x}}$ is a bounded function $\forall x\in[1,\infty)$

And also, h(1) = g(1) = 1

Hence, $f(x)=\begin{cases} h(x), & 0\le x\le 1 \\ g(x), & 1\le x\le\infty \end{cases}=\begin{cases} x^2, & 0\le x\le 1 \\ \sqrt{x}, & x>1 \end{cases}$ is uniformly continuous function on [0, ∞)

(7) A uniformly continuous function may not be bounded. Example : f(x) = x, x ∈ R

(8) A continuous bounded function may not be uniformly continuous.

Example : f(x) = sin (x^2)

Example : Let f(x) = log x ; x ∈ (1, ∞) $f`(x)=\frac{1}{x}$; x ∈ (1, ∞)

f`(x) is bounded on (1, ∞) ⇒ log x is uniformly continuous on (1, ∞)

4. Differentiablity

Derivative of a function at a point:

Let f : [a, b] → R be a function and $c\in(a,b)$, then 'f' is said to be derivable or differentiable at 'c', if $\lim_{x\to c}\frac{f(x)-f(c)}{x-c}$ exists.

Note:

(1) If $\lim_{x\to c^-}\frac{f(x)-f(c)}{x-c}$ or $\lim_{h\to 0}\frac{f(c-h)-f(c)}{-h}$ exist, then limit is called left hand derivative of 'f' at 'c' and is denoted by $f'(c-0)$ or $f'(-c)$ or L $f'(c)$.

(2) If $\lim_{x\to c^+}\frac{f(x)-f(c)}{x-c}$ or $\lim_{h\to 0}\frac{f(c+h)-f(c)}{h}$ exist, then limit is called right hand derivative of 'f' at 'c' and is denoted by $f'(c+0)$ or $f'(+c)$ or R $f'(c)$.

(3) The derivative $f'(c)$ exists iff L $f'(c)=$ R $f'(c)$

(4) 'f' is said to be derivable in (a, b) if $f'(c)$ exists $\forall$ c such that $a<c<b$

Examples:

(1) $f(x)=|x|, \forall x\in R$

$$\lim_{h\to 0}\frac{f(0+h)-f(0)}{h}=\lim_{h\to 0}\frac{|h|-0}{h}=1=R.H.D.$$

$$\lim_{h\to 0}\frac{f(0-h)-f(0)}{-h}=\lim_{h\to 0}\frac{|-h|-0}{-h}=-1=L.H.D.$$

So f is not differentiable at x = 0 because R $f'(0)\neq Lf'(0)$. So function is not differentiable at x = 0.

(2) $f(x)=\begin{cases} x^2\sin\frac{1}{x}, & x\neq 0 \\ 0, & x=0 \end{cases}$ is differentiable at x = 0

As $\lim_{x\to 0}\frac{f(x)-f(0)}{x-0}=\lim_{x\to 0}\frac{x^2\sin\frac{1}{x}-0}{x}=\lim_{x\to 0}x\sin\frac{1}{x}=0$

the required limit exist at x = 0. So f(x) is differentiable at x = 0.

Result :

(a) Let f : [0 ,1] → R be a function s.t. $f(x)=\begin{cases} x^\alpha \sin\frac{1}{x}; & x\neq 0 \\ 0; & x=0 \end{cases}$. Then

(i) f(x) is continuous at x = 0 for α> 0

(ii) f(x) is differentiable at x = 0 for α> 1

(b) Let f : R → R be defined by $f(x)=\begin{cases} \phi_1(x) & x\in Q \\ \phi_2(x) & x\in Q^c \end{cases}$ then f(x) is differentiable at the double root of $\phi_1(x)$ - $\phi_2(x) = 0$

Example : f : R → R be a function s.t. $f(x)=\begin{cases} x^2+1 & x\in Q \\ 2x & x\in Q^c \end{cases}$

Now $x^2 + 1 - 2x = 0 \Rightarrow x = 1, 1$

1 is double root. So, f(x) is differentiable at x = 1.

Local Maxima: Let f be defined on I and $c\in$ I, then f has local maxima at x = c, if $\exists\delta>0$ Such that $f(x)\le f(c)\,\forall x\in(c-\delta,c+\delta)$

Local Minima: Let f be defined on I and $c\in$ I, then f has local minima at x = c, if $\exists\delta>0$ Such that $f(x)\ge f(c)\,\forall x\in(c-\delta,c+\delta)$

Note: (1) x = a is the point of absolute /global maxima when f(a) ≥ f(x) $\forall x\in$ domain.

(2) x = a is the point of absolute /global minima, f(a) ≤ f(x) $\forall x\in$ domain.

Convex Function: Let I ⊆ Rbe an interval. A function f : I → Ris said to be convex on I, if for any t satisfying $0\le t\le 1$ and for any points x_1, x_2 in I, we have $f((1-t)x_1+tx_2)\le(1-t)f(x_1)+tf(x_2)$ or $f(\lambda_1 x_1+\lambda_2 x_2)\le\lambda_1 f(x_1)+\lambda_2 f(x_2)$, where $\lambda_1+\lambda_2=1$ and $\lambda_i\ge 0$.

or

If 'f' is convex on I and if $x_1, x_2\in$ I, then the chord joining any two points ($x_1, f(x_1)$) and ($x_2, f(x_2)$) on the graph of f lies above the graph of f.

A function f is said to be convex on I if at any point of interval I, the tangent lies below the curve.

Concave Function: If f is concave on I and if $x_1, x_2\in$ I, then the chord joining any two-points ($x_1, f(x_1)$) and ($x_2, f(x_2)$) on the graph of f lies below the graph of y = f(x) or f is said to be concave on I if at any point of interval I, the tangent lies above the curve

Results on Differentiability of a Function:

(1) If f is twice differentiable function on an interval I, then f is convex function if and only if $f''(x)\ge 0\ \forall x\in I$ or if $f''(x)>0\ \forall x\in I$

(2) If f is twice differentiable function on an interval I, then f is concave function if and only $f''(x)\le 0\ \forall x\in I$ or if $f''(x)<0\ \forall x\in I$

(3) The points of the curve at which the function is neither convex nor concave are points of inflaxion, i.e., If f is twice differentiable and $f''(\alpha)=0$, then α is called the point of inflexion.

(4) If f is differentiable at x = c and 'c' is the point of extrema (minima or maxima) of y = $f(x)$ then $f'(c)=0$. Converse is not true in general, i.e., this is only necessary condition but not sufficient.

(5) If $f(x)$ is differentiable at $x=a$ and $g(x)$ is differentiable at $f(a)$, then $h(x)=(gof)(x)$ is differentiable at 'a' and have the derivative $h'(x)=g'[f(a)]f'(a)$.

(6) Every differentiable function is continuous also, but converse is not true, i.e., continuous function may or may not be differentiable.

Example : f(x) = |x|

Darboux's Theorem: Let f be differentiable function on I and $a,b \in$ I. If k $\in$ Rsuch that $f'(a)<k<f'(b)$, then $\exists c \in (a,b)$ such that $f'(c)=k$. or

If f has derivative at every point of $[a,b]$, then f takes every value between $f'(a)$ and $f'(b)$,i.e., the image of interval under $f'(x)$ is either an interval or a singleton.

Applications of derivative

Rolle's Theorem: Let f be continuous function on $[a,b]$ and differentiable on (a,b) and $f(a)=f(b)$ then $\exists c \in (a,b)$ such that $f'(c)=0$.

Note: Converse of Rolle'stheorem is not true, i.e., $f'(x)$ may vanish at a point 'c' $\in (a,b)$ without $f(x)$ satisfying all the three conditions of Rolle's theorem

Example Consider $f(x)=\begin{cases} 0, & when\, 0 \le x \le 1 \\ x+1, & when\, 1 < x \le 2 \end{cases}$

Here $f(x)$ is not continuous and derivable at $x=1$, i.e., $f(x)$ is not continuous and derivable on

[0, 2] and (0, 2) respectively.Also $f(0) \ne f(2)$. But $f'(x)=0 \forall x \in (0,1) \subset (0,1)$

Hence, conditions are only sufficient not necessary.

(2) If f is non-negative differentiable function, then every zero of f is also zero of f'.

(3) This theorem is useful to produce the zero of derivative of the function.

(4) A non-constant function satisfy the condition of Rolle's theorem cannot be monotonic.

(5) This theorem tell us that between two distinct zeros of a differentiable function f their lies a zero of f'.

Lagrange's Mean Value Theorem [L.M.V.T]:

If f is a function defined on $[a,b]$ such that f continuous on closed interval $[a,b]$ and derivable on open interval (a,b), then there exists atleast one real numbers $c \in (a,b)$ such that $\dfrac{f(b)-f(a)}{b-a}=f'(c)$

Note: When $f(a)=f(b)$, then L.M.V.T reduces to Rolle's theorem,

Standard form of L.M.V.T

If f be continuous on $[a,b]$ and differentiable on (a,b), then by L.M.V.T.

$$\frac{f(b)-f(a)}{b-a}=f'(c) \text{ for some } c\in(a,b)$$

If $b-a=h \Rightarrow b=a+h$ and $c=a+\theta h, 0<\theta<1$ $\quad \dfrac{f(a+h)-f(a)}{h}=f'(a+\theta h)$

Example : $f(x) = 2x^2 + 3x + 4;\ x \in [1, 2]$

This is continuous & differentiable on [1, 2]

So, $\exists$ c $\in$ (1, 2) s.t. $f'(c)=\dfrac{f(2)-f(1)}{2-1}$

$= 4c + 3 = 8 + 6 + 4 - 2 - 3 - 4$

$\Rightarrow 4c + 3 = 9 \Rightarrow 4c = 6 \qquad \Rightarrow c=\dfrac{3}{2}\in(1,2)$

Some useful deduction from the Mean Value Theorem :

(i) If f is defined and continuous on [a, b] and is derivable on]a, b[, and if $f'(x) = 0$ for all x in]a, b[, then f(x) has a constant value throughout [a, b].

(ii) If f(x) and g(x) are both defined and continuous on [a, b], and are derivable on]a, b[, and if $f'(x) = g'(x)\ \forall\ x \in\ ]a, b[$, then f(x) and g(x) differ by a constant on [a, b].

(iii) If f is continuous on [a, b], and $f'(x) \ge 0$ in]a, b[, then f is increasing in [a, b].

(iv) If f is continuous on [a, b] and $f'(x) > 0$ in]a, b[, then f is strictly increasing in [a, b].

Example : Let a function f be defined on R such that $|f(x) - f(y)| <|x - y|^2$ for all x and y. Then f must be constant function.

Solution : Given $|f(x) - f(y)| <|x - y|^2 \forall\ x, y \in R$ ......(1)

Let c be any real number. Then, for $x \neq c$, (1) gives

$$\left|\frac{f(x)-f(c)}{x-c}\right|\le|x-c| \qquad ..... (2)$$

Let $\varepsilon > 0$ be given. Then, we choose $\delta = \varepsilon$ and so (2) gives

$$\left|\frac{f(x)-f(c)}{x-c}\right|<\varepsilon \text{ , whenever } |x-c|<\delta$$

$$\Rightarrow \lim_{x\to c}\frac{f(x)-f(c)}{x-c}=0\Rightarrow f'(c)=0 \qquad ..... (3)$$

Since c is any point on R, (3) shows that $f'(x) = 0\ \forall\ x\in R$.

Hence, f must be constant function.

Cauchy's Mean Value Theorem [C.M.V.T]:

If f and g are continuous on $[a,b]$ and differentiable on (a,b), then $\exists c \in (a,b)$ such that $\frac{f(b)-f(a)}{g(b)-g(a)} = \frac{f'(c)}{g'(c)}$, $g'(x) \neq 0 \text{ in } (a,b)$.

Example : $f(x) = x^2$, $g(x) = x^3$; $x \in [1, 2]$ Both are continuous & differentiable in [1, 2]

$$\frac{f(2)-f(1)}{g(2)-g(1)} = \frac{f'(c)}{g'(c)} \Rightarrow \frac{4-1}{8-1} = \frac{2c}{3c^2}$$

$$\Rightarrow \frac{3}{7} = \frac{2}{3c} \quad \Rightarrow \quad 9c = 14 \Rightarrow \quad c = \frac{14}{9} \in (1,2)$$

REAL ANALYSIS

SOLVED QUESTIONS

1. Number of real roots of polynomial $f(x) = x^3 + 3x - 2023$ **CSIR NET JUNE 2023**

(a) 1 (b) 2 (c) 3 (d) None of these

Ans. (a)

Explanation :

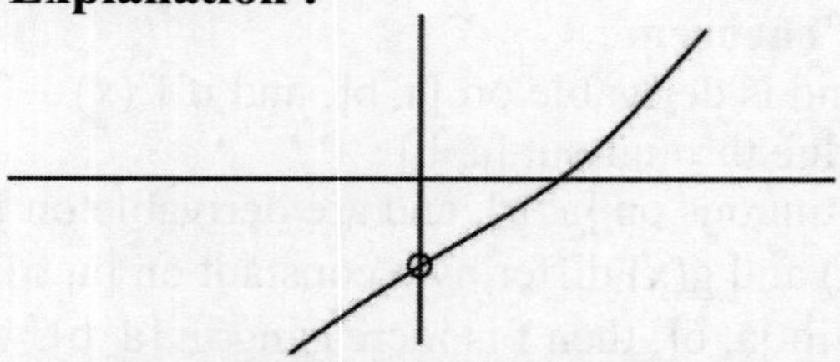

2. Which of the following function are uniformly continuous on (0, 1)?

CSIR NET JUNE 2023

(a) sin(1/x) (b) e^{-1/x^2} (c) $e^x \sin\left(\frac{\pi}{x}\right)$ (d) $\cos x \cos\left(\frac{\pi}{x}\right)$

Ans. (b)

Explanation :

(a) $\lim_{x\to 0} \sin\frac{1}{x}$ does not exist

(b) $\lim_{x\to 0} e^{-1/x^2} = 0 \;\&\; \lim_{x\to 1} e^{-1/x^2} = e^{-1}$

(c) $\lim_{x\to 0} e^x \sin\left(\frac{\pi}{x}\right)$ does not exist

(d) $\lim_{x\to 0} \cos x \cos\left(\frac{\pi}{x}\right)$ does not exist

3. Let $f(x) = \begin{cases} \sin(\pi/x), & x \neq 0 \\ 0 & x = 0 \end{cases}$ on which set the restriction of f is continuous.

CSIR NET JUNE 2023

(a) [-1, 1] (b) (0, 1) (c) $\{0\} \cup \left\{\frac{1}{n}\right\}$ (d) $\left\{\frac{1}{2^n}, n \in N\right\}$

Ans. (b), (c), (d)

Explanation :

(b) We know that f(x) is discontinuous at x = 0

Hence (b) is true.

By sequential definition Let $f(x) = \begin{cases} g(x) & x \neq a \\ b & x = a \end{cases}$

$\exists$ a sequence $\langle a_n \rangle \to b$ and $g(a_n) \to b$, then f(x) is continuous.

(c) $a_n = \frac{1}{n} \to 0$ $\quad sin\left(\frac{\pi}{1/n}\right) = sin\,n\pi = 0$

Hence f(x) is continuous.

(d) $a_n = \frac{1}{2^n} \to 0$ $\quad sin(\pi 2^n) = 0$

Hence f(x) is continuous.

4. In which of the following cases does there exist a continuous and onto function f : X → Y?

(a) X = (0, 1), Y = (0, 1]

(b) X = [0, 1], Y = (0, 1]

(c) X = (0, 1), Y = R

(d) X = (0, 2), Y = {0, 1}

Ans. (a), (c) **CSIR NET FEB 2022**

Explanation :

(a) Consider $f(x) = \begin{cases} 3x, & x \in \left(0, \frac{1}{3}\right) \\ 1, & x \in \left(\frac{1}{3}, 1\right) \end{cases}$ $\Rightarrow$ Option (a) is true.

(b) Image of compact set should be compact under continuous function.

$\Rightarrow$ Option (b) is false.

(d) Image of connected set must be connected under continuous function.

$\Rightarrow$ Option (d) is false.

(c) Consider f : (0, 1) → R

$$f(x) = \tan\left(\left(x - \frac{1}{2}\right)\pi\right)$$

$\Rightarrow$ Option (c) is true.

5. Let f : [0, 1] → (0,1) be a function. Which of the following statements are false?

(a) If f is onto, then f is continuous

(b) If f is continuous then f is not onto

(c) If f is one-to-one then f is continuous

(d) If f is continuous, then f is not one-to-one

Ans. (a), (c), (d) **CSIR NET NOV 2020**

REAL ANALYSIS

Explanation :

(a) We know that Image of compact set under continuous function is also compact set

If f : [0, 1] → (0,1) is onto and continuous then f{[0 ,1]} = (0 ,1) which is not possible

⇒ Option (a) is false hence correct choice.

(b) Suppose f is continuous ⇒ f{[0, 1]} is compact ⇒ f{[0, 1]} ≠ (0, 1)

⇒ f is not onto. ⇒ Option (b) is not false hence incorrect choice.

(c) Take $f = \begin{cases} \frac{1}{2}x & if \quad x \in (0,1) \\ \frac{3}{5} & if \quad x = 0 \\ \frac{4}{5} & if \quad x = 1 \end{cases}$

This is one-one but not continuous. ⇒ Option (c) is false hence correct choice.

(d) Consider $f(x) = \frac{1}{8}x + \frac{1}{8}$ ⇒ f is one-one and continuous.

⇒ Option (d) is false hence correct choice. ⇒ Option (a), (c), (d) are true.

6. Let A ⊆ R and let f : R → R be continuous. Which of the following statements are true?

(a) If A is closed then f(A) is closed.

(b) If A is bounded then $f^{-1}(A)$ is bounded

(c) If A is closed and bounded then f(A) is closed and bounded.

(d) If A is bounded then f(A) is bounded

Ans. **(c), (d)** **CSIR NET FEB 2022**

Explanation :

(a) A is closed then f(A) need not be closed.

Take f : R → R

$f(x) = \tan^{-1}x$

R is closed but $f(R) = \left(-\frac{\pi}{2}, \frac{\pi}{2}\right)$ is not closed.

(c) Result : Continuous function take compact set to compact set.

Since closed and bounded set in R is compact by Heine-Borel theorem.

⇒ Option (c) is true.

(b) For option (b) : Take f(x) = c for all x ∈ R

Then $f^{-1}\{c\} = R$

Here {c} is bounded but R is not bounded.

⇒ Option (b) is false.

(d) If A is bounded then f(A) is bounded

Result : f is uniformly continuous function iff $|f(a_n) - f(b_n)| \to 0$ for all a_n and b_n such that $|a_n - b_n| \to 0$

7. Let f, g : R → R be given by and $f(x) = x^2$ and g(x) = sin x.
Which of the following functions is uniformly continuous on R? **CSIR NET FEB 2022**
(a) h(x) = g(f(x)) (b) h(x) = g(x) f(x) (c) h(x) = f(g(x)) (d) h(x) = f(x) + g(x)
Ans. **(c)**
Explanation :
For option (a) :
$h(x) = g(f(x)) = \sin x^2$
Consider $a_n = \dfrac{1}{\sqrt{n\pi}} \to 0 \;\&\; b_n = \dfrac{1}{\sqrt{(2n+1)\dfrac{\pi}{2}}} \to 0$

but $f(a_n) \to 0$ and $f(b_n)| \to 1$
⇒ $\text{Sin}(x^2)$ is not uniformly continuous. ⇒ h(x) is not uniformly continuous.
Hence option (a) is false.
For option (b)
$h(x) = x^2 \sin x$
We know that product of two UC function is also UC
but $f(x) = x^2$ is not UC in R
h(x) is not uniformly continuous.
Hence option (b) is false.
For option (d) :
f + g is not uniformly continuous since x^2 is not uniformly continuous.
Hence option (d) is false.
For option (c) :
Note that $f(g(x)) = (\sin x)^2$ is uniformly continuous because

$$\left|\frac{d}{dx}(f(g(x)))\right| = |2\sin x.\cos x| = |\sin 2x| \le 1$$

⇒ f(g(x)) has bounded derivative.
⇒ f(g(x)) is Lipschitz function.
⇒ f(g(x)) is uniformly continuous function.
Hence option (c) is true.

8. Let $f(x) = e^{-x}$ and $g(x) = e^{-x^2}$. **CSIR NET NOV 2020**
Which of the following statements are true?
(a) Both f and g are uniformly continuous on R
(b) f is uniformly continuous on every interval of the form [a, +∞), a ∈ R
(c) g is uniformly continuous on R (d) f(x) g(x) is uniformly continuous on R
Ans. **(b), (c), (d)**

Explanation :

Result : If f(x) is continuous in an interval and limit at end point of the interval (a, b) exists then f(x) is uniformly continuous in that interval. Here a and b can be infinite as well.

Note that for any a ∈ R

$\lim_{x\to a} e^{-x} = e^{-a}$ & $\lim_{x\to\infty} e^{-x} = 0$ but $\lim_{x\to-\infty} e^{-x}$ = doesnot exist

and e^{-x} is not uniformly continuous on R.

By using above result e^{-x} is uniformly continuous on [a, ∞), a ∈ R.

⇒ Option (b) and (c) are true.

Now, $f(x).g(x) = e^{x^2}.e^{-x}.e^{-x} = e^{-x^2-x}$

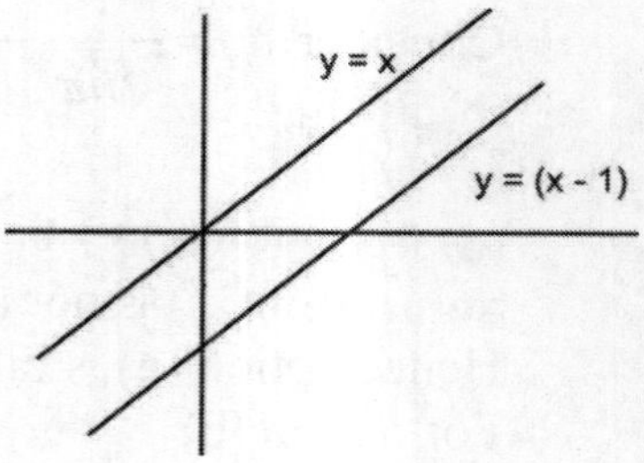

Note that e^{-x^2-x} is continuous on whole R.

Now, $\lim_{x\to\infty} e^{-x^2-x} = \lim_{x\to\infty} e^{-x^2\left(1+\frac{1}{x}\right)} = 0$

And $\lim_{x\to\infty} e^{-x^2-x}$ exist.

By using above result e^{-x^2-x} is uniformly continuous on R.

⇒ Option (b), (c), (d) are true.

9. Which of the following functions are uniformly continuous on (0, 1)?

(a) $\frac{1}{x}$ (b) $\sin\frac{1}{x}$ (c) $x\sin\frac{1}{x}$ (d) $\frac{\sin x}{x}$

Ans. (c), (d) **CSIR NET NOV 2020**

Explanation :

We know that option (c) and (d) are true

Trick : Fixed point of y= f(x) are the point of intersection of the curve y = f(x) and y = x

10. Let p(x) be a polynomial of odd degree & g(x) be a continuous function from R to R then which of the following are TRUE?

CSIR NET DEC 2019

(a) ∃ a point x_0 ∈ R s.t. $p(x_0) = g(x_0)$

(b) If g is polynomial function then ∃ x_0 ∈ R s.t. $p(x_0) = g(x_0)$

(c) If g is bounded function ∃ x_0 ∈ R s.t. $p(x_0) = g(x_0)$

(d) There is unique x_0 ∈ R s.t. $p(x_0) = g(x_0)$

Solution : (c)

(a), (b), (d) are discarded by only one example like

p(x) = x & g(x) = x – 1

∃ any point x_0 ∈ R s.t. $p(x_0) = g(x_0)$

if g(x) is polynomial.

Option (c) :

If g : R → R and g is continuous & bounded function then $|g(x)| \le M \quad \forall x \in R$

$-M \le g(x) \le M \quad \forall x \in R$

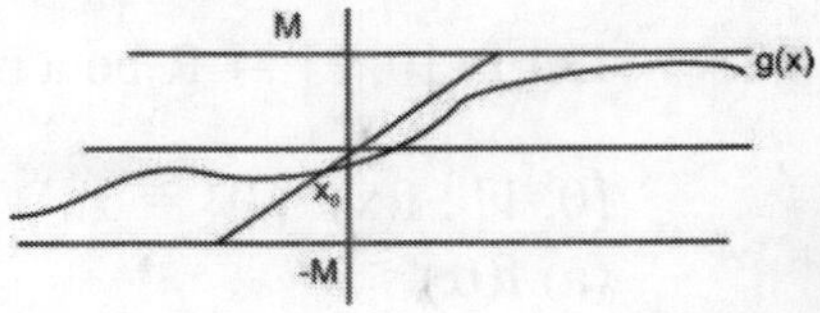

Then we can choose any polynomial function f(x) of odd degree then $\exists x_0 \in R$ s.t. $p(x_0) = g(x_0)$

11. Let f : R → R be given by $f(x) = \dfrac{a}{1+bx^2}$, a, b ∈ R b ≥ 0. Which of the following are TRUE?

CSIR NET JUNE 2018

(a) f is uniformly continuous on compact intervals of R for all value of a & b.
(b) f is uniformly continuous on R and is bounded ∀ a, b ∈ R
(c) f is uniformly continuous on R only if b = 0
(d) f is uniformly continuous on R & unbounded if a ≠ 0, b ≠ 0

Solution : (a), (b)

$f(x) = \dfrac{a}{1+bx^2}$ where $1 + bx^2 \neq 0$

We know that $f = \dfrac{g}{h}$

If g & h are continuous & h ≠ 0 then f is continuous.

Then $f(x) = \dfrac{a}{1+bx^2}$ is continuous on R.

Option (a) is correct.

By properties (2) if f(x) is continuous on [a, b] then it is uniformly continuous on [a, b]

So f(x) is continuous on every compact (closed + bounded) interval then it is uniformly continuous on interval.

Option (b) $\quad f(x) = \dfrac{a}{1+bx^2}$

Given that $\quad b \ge 0 \;\&\; x^2 \ge 0 \Rightarrow \quad bx^2 \ge 0$

$1 + bx^2 \ge 1 \Rightarrow \dfrac{1}{1+bx^2} \le 1 \quad \Rightarrow \dfrac{a}{1+bx^2} \le a$

a is upper bounded.

$\& \lim\limits_{x\to\infty} \dfrac{a}{1+bx^2} = 0 \quad \& \lim\limits_{x\to-\infty} \dfrac{a}{1+bx^2} = 0$

⇒ given f(x) is bounded & uniformly continuity on R = (-∞, ∞)

Option (c) is incorrect.

f(x) is uniformly continuous on R for every a, b ∈ R

Option (d) is incorrect.
f(x) is bounded.

12. Let f : [0, 1] → R be a monotonic function with $f\left(\frac{1}{4}\right).f\left(\frac{3}{4}\right)<0$ & suppose sup {x ∈ [0, 1] : f(x) < 0} = α. Which of the following are correct? **CSIR NET DEC 2019**

(a) f(α) < 0

(b) If f is increasing then f(α) ≤ 0

(c) If f is continuous & $\frac{1}{4}<\alpha<\frac{3}{4}$ then f(α) = 0

(d) If f is decreasing then f(α) < 0

Ans. **(c), (d)**

Explanation :

Given that $f\left(\frac{1}{4}\right).f\left(\frac{3}{4}\right)<0$

then either $f\left(\frac{1}{4}\right)>0\,\&\,f\left(\frac{3}{4}\right)<0$ or $f\left(\frac{1}{4}\right)<0\,\&\,f\left(\frac{3}{4}\right)>0$

Option (a) is incorrect.

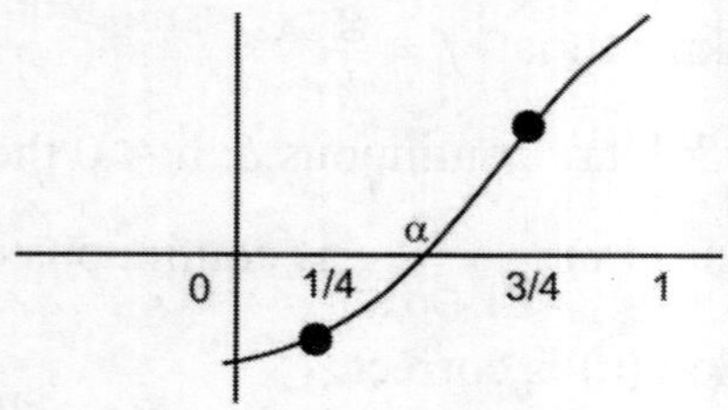

Given that A = {x ∈ [0, 1] | f(x) < 0}
A = [0, α)
& sup A = α and f(α) = 0
hence option (a) is incorrect

Option (b) is incorrect.

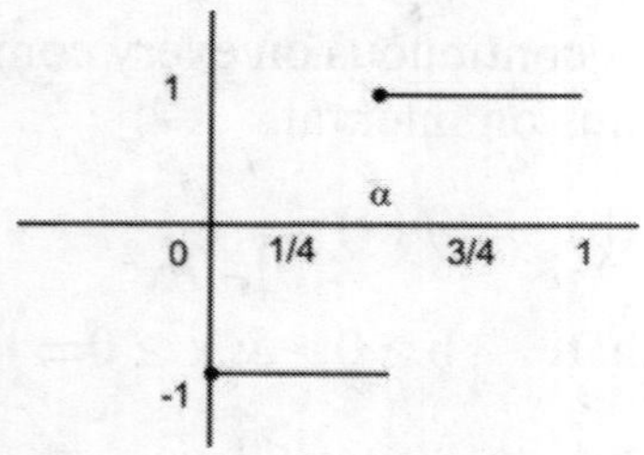

Here A = [0, α)
Sup A = α f(α) = 1
hence (b) is incorrect.

Option (c) is correct

By intermedite value property. ⇒
Let f : [a, b] → R
∃α,β∈ (a, b)
s.t. f(α) . f(β) < 0
then ∃ c∈ (α, β) s.t. f(c) = 0
Given that f : [0, 1] → R
$\alpha=\frac{1}{4}$ & $\beta=\frac{3}{4}$ s.t. f(α) . f(β) < 0
then ∃α∈ (α, β) = (1/4,3/4)
s.t. f(α) = 0
hence (c) is correct.

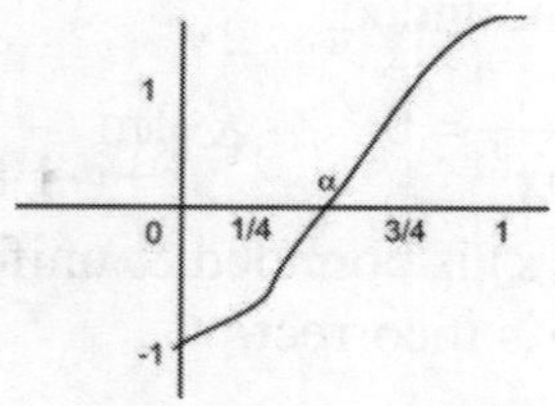

Option (d) is correct.
If f(x) is decreasing.

A = (c, 1), Sup A = 1 = α
f(α) = f(1) < 0
hence (d) is correct.

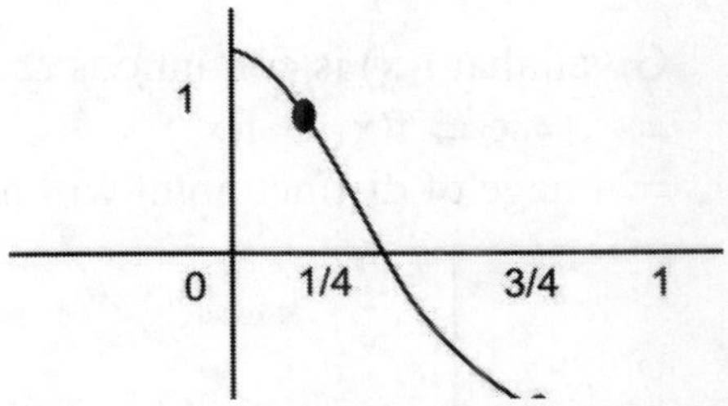

Trick : If a function f : R → R be a continuous and one-one then f is either strictly decreasing or strictly increasing.

13. Let f : R → R be a continuous and one - one function. Then which of the following is TRUE?

CSIR NET JUNE 2019

(a) f is onto
(b) f is either strictly decreasing or strictly increasing
(c) ∃ x∈ R s.t. f(x) = 1
(d) f is unbounded.

Ans. (b)

Explanation :

Given that f : R → R be continuous & one – one

.

Option (a) is incorrect.
$f(x) = e^x$

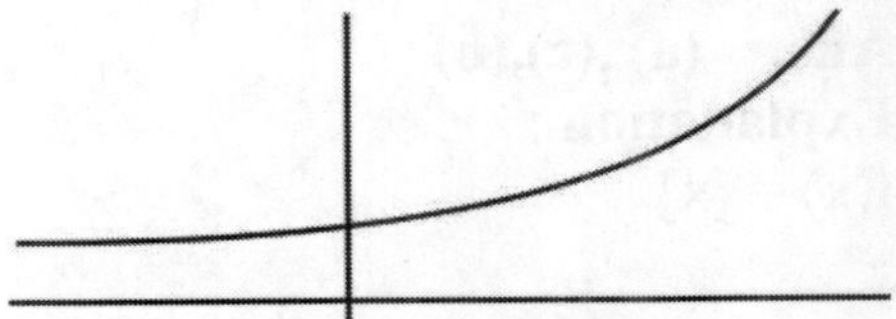

Range of f(x) = (0, ∞)
Not onto.
Option (c) is incorrect.
$f(x) = 1 + e^x$

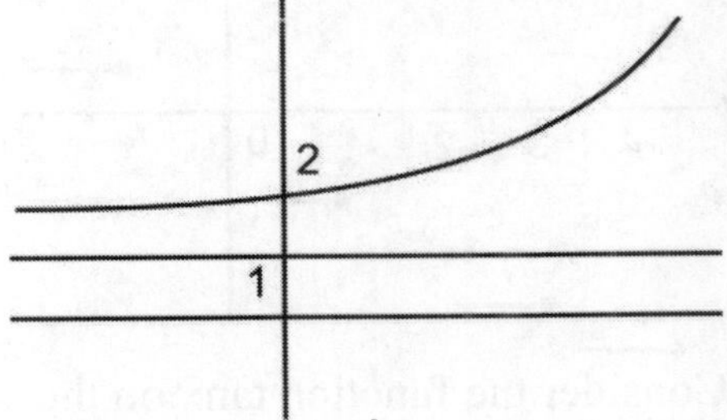

∃ x∈R s.t. f(x) = 1
Option (d) is incorrect.
$f(x) = \tan^{-1}x$

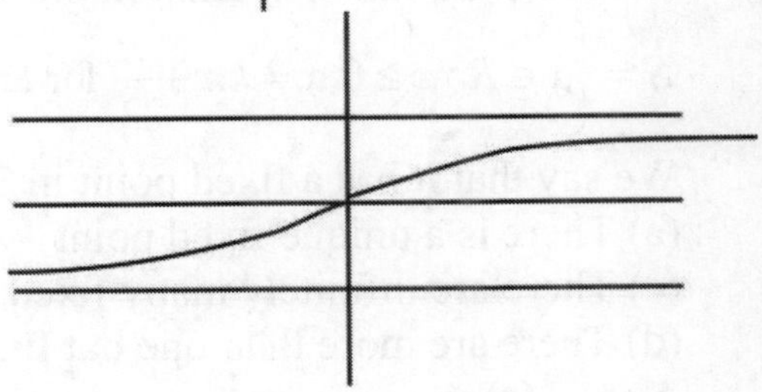

Range $f(x) = \left(-\frac{\pi}{2}, \frac{\pi}{2}\right)$

f(x) is bounded.
Option (b) is correct.

Given that f(x) is continuous & 1 – 1

$\Rightarrow x_1 \neq x_2 \Rightarrow f(x_1) \neq f(x_2)$

$\Rightarrow$ image of distinct point will be distinct.

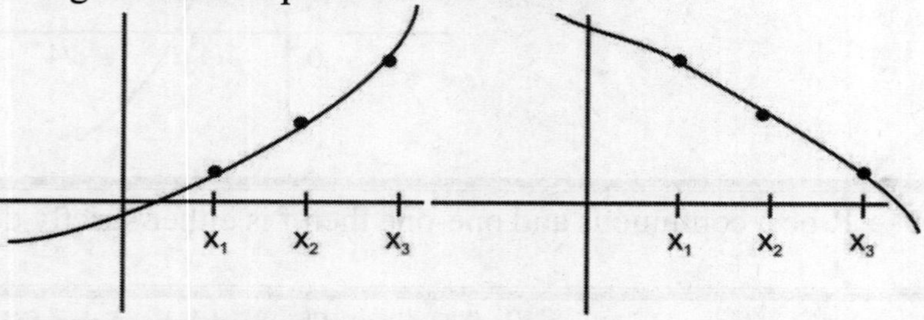

Then either increase or decerase.

> **Trick** : Let f : R→R be a increasing function then the set of point at which f(x) is discontineous is atmost countable

14. Let f : R → R be a non-decreasing function. Which of the following can be set of discontinuous of f. **CSIR NET JUNE 2019**

(a) Z (b) Q (c) N (d) R/Q

Ans. **(a) ,(c),(b)**

Explanation :

f(x) = [x]

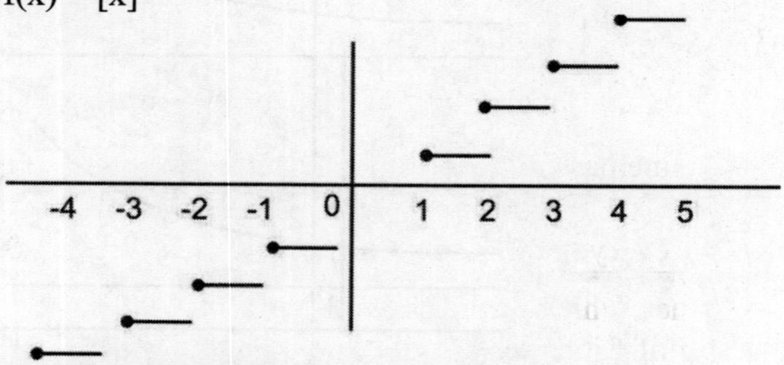

15. Consider the function tanx on the set

$$S = \left\{ x \in R : x \geq 0, x \neq k\pi + \frac{\pi}{2} \text{ for any } k \in N \cup \{0\} \right\}$$

We say that it has a fixed point in S if ∃ x∈S,s.t.tan x = x then**CSIR NET DEC 2018**

(a) There is a unique fixed point (b) There is no fixed point

(c) There are infinitely many fixed point

(d) There are more than one but finitely many fixed point.

Ans. **(c)**

Explanation :

Reason : The line y = x and tan x intersect at infinitely many point.

16. Let $\alpha = \int_0^\infty \frac{dt}{1+t^2}$. Which of the following are TRUE? **CSIR NET JUNE 2018**

(a) $\frac{d\alpha}{dt} = \frac{1}{1+t^2}$ (b) $\log(\alpha) = 1$ (c) α is a rational number (d) $\sin(\alpha) = 1$

Ans. **(d)**

Explanation : $\alpha = \left[\tan^{-1} t\right]_0^\infty = \tan^{-1}\infty - \tan^{-1} 0$

$\alpha = \frac{\pi}{2}$

Option (d) : $\sin\alpha = \sin\frac{\pi}{2} = 1$

Trick :

(1) If $f : A \to B$ and f(x) is one -one

Then $x_1 \neq x_2 \Rightarrow f(x_1) \neq f(x_2)$ for every $x_1, x_2 \in A$

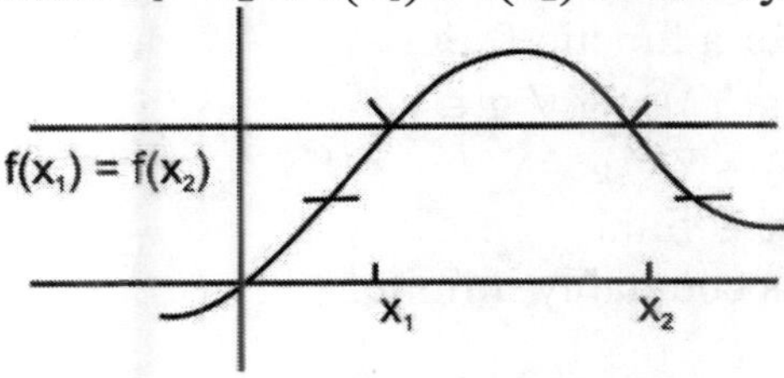

Not one-one

i.e. graph of f(x) intersect some line parallel to x-axis in atmost one point.

(2) f(x) is onto & $f : A \to B$

Then graph of f intersect every parallel to x-axis.

17. Consider a function $f : R \to R$ then which of the following are TRUE?

(a) f is not one-one if the graph of f intersects some line parallel to x-axis in at least two points.

(b) f is one-one if graph of f intersects any line parallel to x-axis in atmost one point.

(c) f is onto if the graph of f intersects every line parallel to x-axis.

(d) f is not onto if graph of f doesn't intersects at least one line parallel to x-axis.

Ans. **(a), (b), (c), (d)** **CSIR NET JUNE 2019**

Explanation :

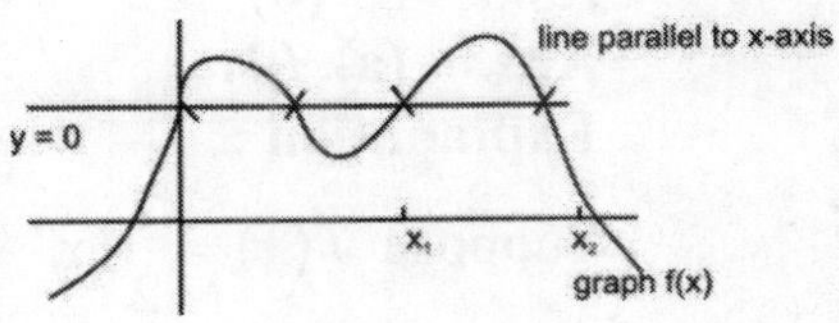

Option (a) is correct.

$x_1 \neq x_2$ but $f(x_1) = f(x_2) = a$

So, not one - one

Option (b) is correct.

Intersect atmost one point $\Rightarrow$ f(x) is one - one

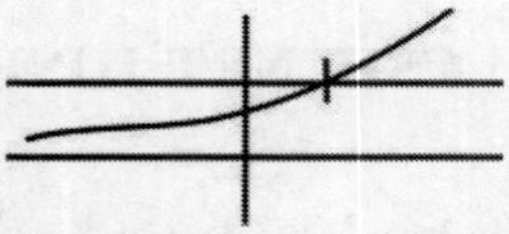

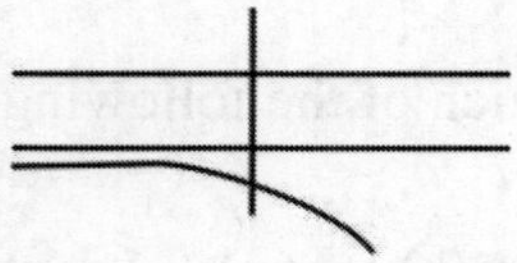

Option (c) is correct.

$f : R \rightarrow R$ and if graph of f intersect every line parallel to x-axis.$\Rightarrow$ Range of f(x) = codomain of f(x)

$\Rightarrow$ f(x) is onto.

Option (d) is correct.

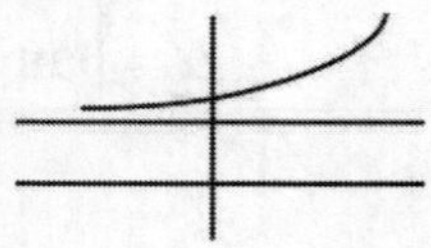

$f : R \rightarrow R$

$f(x) = e^x$

$\Rightarrow$ Range of f(x) $\neq$ codomain of f(x) Not onto.

Trick : Let f(x) & g(x) be functions and if fog is one - one & onto then g(x) is one – one & f(x) is onto function, where fog(x) = f(g(x))

18. Which of the following is necessary true for a functio $f : X \rightarrow Y$.

(a) If f is one - one then $\exists$ $g : Y \rightarrow X$ s.t. $f(g(Y)) = Y \forall g \in Y$

(b) If f is onto then $\exists$ $g : Y \rightarrow X$ s.t. $f(g(Y)) = Y \forall g \in Y$

(c) If f is one - one &Y is countable then X is finite

(d) if f is onto &X is uncountable then Y is countably infinite.

Ans. (b)

CSIR NET DEC 2017

Explanation :

We know that if fog(Y) = Y (identity function) then f is onto & g is one - one

Option (a) is incorrect. Option (b) is correct. Option (c) is incorrect.

f is one - one $\Rightarrow |X| \leq |Y|\backslash$ & given that Y is countable. $\Rightarrow |X| \leq \chi_0$

But X need not be finite. Option (d) is incorrect. f is onto $\Rightarrow |X| \geq |Y|$

given that X is uncountable $\Rightarrow$ |X| = c(continum)

$|Y| \leq c$ (continum) but Y need not be countably infinite.

19. Let f(x) be a real polynomial of degree 4. Suppose f(-1) = 0, f(0) = 0, f(1) = 1 & f'(0) = 0. Which of the following are TRUE? **CSIR NET DEC 2019**

(a) $\exists\ a \in (-1, 1)$ s.t. $f^{(3)}(a) \geq 3$

(b) $f^{(3)}(a) \geq 3 \ \forall a \in (-1, 1)$

(c) $0 < f^{(3)}(0) \leq 2$

(d) $f^{(3)}(0) \geq 3$

Ans. (a), (d)

Explanation :

Suppose $f(x) = \frac{1}{2}(x^4 + x^3)$ f(-1) = 0 = f(0) = f'(0) & f(1) = 1

Option (a) is correct.

$f' = \frac{1}{2}\left[4x^3 + 3x^2\right]$ $\quad f'' = \frac{1}{2}\left[12x^2 + 6x\right]$ $\quad f''' = \frac{1}{2}\left[24x + 6\right]$

Take a = 0 then $f^{(3)}(0) = 3$

Option (b) is incorrect.

$f^{(3)}(a) \geq 3$

$\frac{1}{2}\left[24a + 6\right] \geq 3 \quad \Rightarrow 24a + 6 \geq 6 \quad \Rightarrow 24a \geq 0 \quad \Rightarrow a \geq 0$

i.e. $f^{(3)}(a) \geq 3$ only if $a \in (0, 1)$ Option (c) is incorrect.

$f^{(3)}(0) = 3$ Option (d) is correct.

$f^{(3)}(0) = 3$

20. Let a < c < b, f : (a, b) → R be continuous. Assume that f is differentiable at every point of (a, b) / (c, c) and f' has a limit at c. Then which of the following are TRUE?

(a) f is differentiable at c (b) f need not be differentiable at c

(c) f is differentiable at c & $\lim_{x\to c} f'(x) = f'(c)$.

(d) f is differentiable at c but f'(c) is not necessary $\lim_{x\to c} f'(x)$.

Ans. **(a), (c)** **CSIR NET JUNE 2019**

Explanation :

Given that f(x) is continuous on (a, b)

⇒ f(x) is continuous at x = c a < c < b

$\lim_{x\to c} f(x) - f(c)$ ....(1)

& f' has limit at x = c

$\lim_{x\to c} f'(x) = l$ ..... (2)

Now $f'(c) = \lim_{x\to c} \frac{f(x) - f(c)}{x - c}$ $\left[\frac{0}{0} \text{ form}\right]$

By using D.L. Hospital Rule

$f'(c) = \lim_{x\to c} f'(x)$ (c) is correct.

From (2)

$\lim_{x\to c} f'(x) = l \Rightarrow$ $f'(c) = l$

⇒f(x) is differentiable at x = c

(a) is correct.

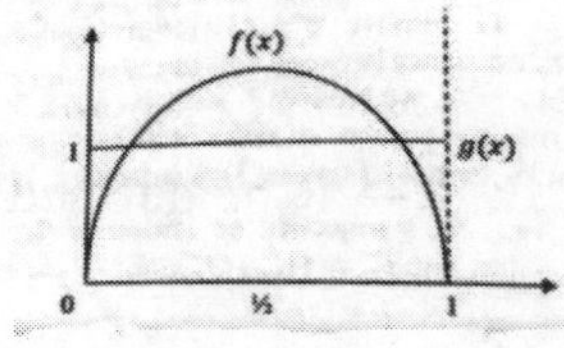

21. Given f, g are continuous functions on [0, 1] such that f(0) = f(1) = 0; g(0) = g(1) = 1 and $f\left(\frac{1}{2}\right) > g\left(\frac{1}{2}\right)$. Which of the following statements is true?

(a) There is no t ∈ [0, 1] such taht f(t) = g(t)

(b) There is exactly one t ∈ [0, 1] such that f(t) = g(t)

(c) There are at least two $t \in [0, 1]$ such that $f(t) = g(t)$
(d) There are always infinitely many $t \in [0, 1]$ such that $f(t) = g(t)$

Ans. (c) **CSIR NET JUNE 2020**

Explanation :

Consider f(x) and g(x) as shown in the figure, satisfies all conditions given in the question, there is two $t \in [0, 1]$ such that $f(t) = g(t)$, we can simply eliminate option (a), (b) and (d), so option (c) is correct.

REAL ANALYSIS

UNSOLVED QUESTIONS

1. Let $f : R \to R$ be a function satisfying $f(x + y) = f(x) . f(y)\ \forall\ x, y \in R\ \&\ \lim_{x\to 0} f(x) = 1$.

Which of the following are TRUE?

(a) f is strictly increasing
(b) f is either constant or bounded
(c) $f(rx) = f(x)^r$ for every rational number $r \in Q$
(d) $f(x) \geq 0\ \forall\ x \in R$

2. Which of the following real – valued function on (0, 1) is uniformly continuous?

(a) $f(x) = \frac{1}{x}$
(b) $f(x) = \frac{\sin x}{x}$
(c) $f(x) = \sin\frac{1}{x}$
(d) $f(x) = \frac{\cos x}{x}$

CSIR NET DEC 2012

3. Which of the following functions is/are uniformly continuous on the interval (0, 1)?

(a) $\frac{1}{x}$
(b) $\sin\frac{1}{x}$
(c) $x\sin\frac{1}{x}$
(d) $\frac{\sin x}{x}$

CSIR NET 2016

4. Let $f : (0, \infty) \to R$ be uniformly continuous. Then **CSIR NET DEC 2017**

(a) $\lim_{x\to 0+} f(x)$ and $\lim_{x\to\infty} f(x)$ exist.
(b) $\lim_{x\to 0+} f(x)$ exists but $\lim_{x\to\infty} f(x)$ need not exist.
(c) $\lim_{x\to 0+} f(x)$ need not exists but $\lim_{x\to\infty} f(x)$ exists.
(d) Neither $\lim_{x\to 0+} f(x)$ nor $\lim_{x\to\infty} f(x)$ exist.

5. Let $f : R \to R$ be defined by $f(x) = \begin{cases} \frac{\sin x}{x} & x \neq 0 \\ 1 & x = 0 \end{cases}$. Then **CSIR NET JUNE 2017**

(a) f is not continuous
(b) f is continuous but not differential
(c) f is differentiable
(d) f is not bounded.

6. Let $f : R \to R$ be a continuous function with $f(1) = 5$ and $f(3) = 11$.

If $g(x) = \int_1^3 f(x+t)\,dt$, then $g'(0)$ is equal to

(a) 4
(b) 5
(c) 6
(d) 7

7. Let $S=[0,1)\cup[2,3]$ and $f: S \to R$ be a strictly increasing function such that $f(S)$ is connected. Which of the following statement is TRUE?

(a) f has exactly one discontinuity (b) f has exactly two discontinuity

(c) f has infinitely many discontinuities (d) f is continuous

8. Let $f(x)=\frac{e^{-\frac{1}{x}}}{x}$, where $x\in(0,1)$. Then on $(0,1)$

(a) f is uniformly continuous (b) f is continuous but not uniformly continuous

(c) f is bounded (d) f is not continuous

9. For a real number $t>0$, let $\sqrt{t}$ denote the positive square root of t. For a real number $x>0$, let $F(x)=\int_{x^2}^{4x^2}\sin\sqrt{t}\,dt$. If F' is the derivative of F, then

(a) $F'\left(\frac{\pi}{2}\right)=0$ (b) $F'\left(\frac{\pi}{2}\right)=\pi$ (c) $F'\left(\frac{\pi}{2}\right)=-\pi$ (d) $F'\left(\frac{\pi}{2}\right)=2\pi$

10. Which of the following functions are uniformly continuous?

(a) $f(x)=x\sin\frac{1}{x}$ on (0, 1) (b) $f(x)=\sin^2 x$ on $(0, \infty)$

(c) $f(x)=\sin(x\sin x)$ on $(0, \infty)$ (d) None of these

11. Let $I=\{1\}\cup\{2\}\subset R$.. For $x\in R$, let $\varphi(x)=dist(x,I)=\inf\{|x-y|: y\in I\}$. Then

(a) φ is continuous on R but not differentiable only at $x=1,3/2$ and 2

(b) φ is continuous on R but not differentiable only at $x=1$

(c) φ s continuous on R but not differentiable only at $x=1$ and 2.

(d) φ is discontinuous somewhere on R.

12. Let f be a twice differentiable function on R. Given that $f''(x)>0$ for all $x\in R$, then

(a) $f(x)=0$ has exactly two solutions in R

(b) $f(x)=0$ has positive solution if $f(0)=0$ and $f'(0)=0$

(c) $f(x)=0$ has no positive solution if $f(0)=0$ and $f'(0)>0$

(d) $f(x)=0$ has no positive solution if $f(0)=0$ and $f'(0)<0$

13. Let $f: A\cup E\to R^2$ be differentiable, where $A=\left\{(x,y)\in R^2 : \frac{1}{2}<x^2+y^2<1\right\}$ and $E=\left\{(x,y)\in R^2 : (x-2)^2+(y-2)^2<\frac{1}{2}\right\}$. Let Df be the derivative of the function f.

Which of the following are necessarily correct?

(a) If $(Df)(x,y)=0$ for all $(x,y)\in A\cup E$, then f is constant

(b) If $(Df)(x,y)=0$ for all $(x,y)\in A$, then f is constant on A

(c) If $(Df)(x,y)=0$ for all $(x,y)\in E$, then f is constant on E

(d) If $(Df)(x,y)=0$ for all $(x,y)\in A\cup E$, then for some $(x_0,y_0),(x_1,y_1)\in R^2$, f$(x,y)=(x_0,y_0)$ for all $(x,y)\in A$ and f$(x,y)=(x_1,y_1)$ for all $(x,y)\in E$

REAL ANALYSIS

14. Let f : R → Rbe a continuous function with period p >0. Then g(x) = $\int_x^{x+p} f(t)$ is a

(a) constant function (b)continuous function
(c) continuous function but not differentiable (d) neither continuous nor differentiable

15. If $f:[0,1]\to[0,1]$ is a continuous mapping, then which of the following is NOT true?

(a) F⊆ [0,1] is a closed set implies f (F) is closed in [0, 1].

(b) If $f(0)<f(1)$, then f ([0,1]) must be equal to $[f(0),f(1)]$.

(c) There must exist $x\in$ (0, 1) such that $f(x)=x$. (d) $f([0,1])\neq(0,1)$

16. Let f : R → R be a differentiable function such that f'(x) > f(x) for all x ∈ R and f(0) = 1. Then f(1) lies in the interval. **IIT-JAM 2019**

(a) (0, e^{-1}) (b) $(e^{-1},\sqrt{e})$ (c) $(\sqrt{e},e)$ (d) (e, ∞)

17. Suppose that f, g : R → R are differentiable functions such that f is strictly increasing and g is strictly decreasing. Define p(x) = f(g(x)) and q(x) = g(f(x)), ∀ x∈R.

Then, for t > 0, the sign of $\int_0^t p'(x)(q'(x)-3)dx$ is **IIT-JAM 2018**

(a) positive (b) negative (c) dependent on t (d) dependent on f and g

18. Let f : R → R be a differentiable function such that f(2) = 2 and |f(x) – f(y)| ≤ 5(|x – y|)$^{3/2}$ for all x ∈ R, y ∈ R. Let g(x) = x^3f(x). Then g'(2) = **IIT-JAM 2017**

(a) 5 (b) 15/2 (c) 12 (d) 24

19. Let f : [a, b] → R be a monotonic function. Then

(a) f is continuous
(b) f is discontinuous at atmost two points
(c) f is discontinuous at finitely many points
(d) f is discontinuous at atmost countable points

20. Let f : R → R be defined by $f(x)=\begin{cases}x^6-1, & x\in Q\\ 1-x^6, & x\notin Q\end{cases}$. Then number of points at which f is continuous is **IIT-JAM 2015**

(a) 2 to 2 (b) 2 to 3 (c) 2to 5 (d) 2 to 8

21. Let S = [0, 1] ∪ [2, 3) and let f : S → R be defined by $f(x)=\begin{cases}2x & if \quad x\in[0,1]\\ 8-2x & if \quad x\in[2,3)\end{cases}$.

If T = {f(x) : x ∈ S}, then the inverse function f^{-1} : T → S **IIT-JAM 2014**

(a) does NOT exist (b) exists and is continuous

(c) exists and is NOT continuous (d) exists and is monotonic

22. Let the function f(x) be defined by $f(x)=\begin{cases}e^x, & x \text{ is rational}\\ e^{1-x}, & x \text{ is irrational}\end{cases}$ for x in (0, 1). Then

(a) f is continuous at every point in (0, 1) **IIT-JAM 2013**

(b) f is discontinuous at every point in (0, 1)

(c) f is discontinuous only at one point in (0, 1)

(d) f is continuous only at one point in (0, 1)

23. Which of the following statements is/are true? \ **CSIRNET DEC 2015**

(a) There exists a continuous map f : R → R such that f(R) = Q.

(b) There exists a continuous map f : R → R such that f(R) = Z

(c) There exists a continuous map f : R → R^2 such that f(R)={(x, y)∈R^2 : $x^2 + y^2 = 1$}

(d) There exists a continuous map f : [0, 1] ∪ [2, 3] → {0, 1}

24. Let f : R → R be a continuous function and f(x + 1) = f(x) for all x ∈ R. Then

(a) f is bounded above, but not bounded below. **CSIR NET DEC 2016**

(b) f is bounded above and below, but may not attain its bounds.

(c) f is bounded above and below and f attains its bounds.

(d) f is uniformly continuous

25. For a continuous function f : R → Rlet $Z(f)=\{x\in R: f(x)=0\}$. Then $Z(f)$ is always

(a) compact (b) open (c) connected (d) closed

26. Let $f:[0,\infty)\to[0,\infty)$ be a continuous function. Which of the following is correct?

(a) There is $x_0\in[0,\infty)$, such that $f(x_0)=x_0$

(b) If $f(x)\le M$ for all $x\in[0,\infty)$ for some M > 0, then there exists $x_0\in[0,\infty)$ such that $f(x_0)=x_0$

(c) If f has a fixed point, then it must be unique

(d) f does not have a fixed point unless it is differentiable on $(0,\infty)$

27. Let f : R → Rbe a function with continuous derivative such that $f(\sqrt{2})=2$ and

$f(x)=\lim_{t\to 0}\frac{1}{2t}\int_{x-t}^{x+t} sf'(s)\,ds$ for all x ∈ R. Then $f(3)$ equals

(a) $\sqrt{3}$ (b) $3\sqrt{2}$ (c) $3\sqrt{3}$ (d) 9

28. The function f(x) = |tanx | is

(a) continuous at x = 0 but not differentiable at x = 0
(b) differentiable at x = 0 (c) discontinuous at x = 0 (d) twice differentiable at x = 0

29. For m, n ∈ N, define $f_{m,n}(x)=\begin{cases} x^m \sin\left(\frac{1}{x^n}\right), & x \neq 0 \\ 0, & x=0 \end{cases}$. Then at $x=0, f_{m,n}$ is

(a) differentiable for each pair m, n with m > n
(b) differentiable for each pair m, n with m < n
(c) not differentiable for each pair m, n with m > n
(d) not differentiable for each pair m, n with m < n

30. Let X=(0,1) ∪ (2,3) is an open subset of R. Let f be a continuous function on X such that thederivative $f'(x)=0$ for all x. Then the range of f has not

(a) uncountable number of points (b) countably infinite number of points
(c) at most 2 points (d) at most 1 point

31. Let f(x) = |sin πx|, x ∈ R.Then,

(a) f is continuous nowhere
(b) f is continuous everywhere and differentiable nowhere
(c) f is continuous everywhere and differentiable everywhere except at integral values of x
(d) f is differentiable everywhere but not continuous

32. Let f:[0, 1] → R be a discontinuous monotonic function. Then the range of f is

(a) a bounded set that is not an interval (b) an unbounded connected set
(c) a bounded interval (d) a set that is neither bounded nor connected

33. Let f : R → R defined by $f(x)=\left[x^2\right]$. The points of discontinuity of f are

(a) only the integral points (b) all rational numbers
(c) { $\pm\sqrt{n}$: n is a non negative integer} (d) all real numbers

Answer Key

1	2	3	4	5	6	7	8	9	10
c,d	b	c,d	b	c	c	d	a	c	a,b
11	**12**	**13**	**14**	**15**	**16**	**17**	**18**	**19**	**20**
a	c	b,c,d	a,b	b	d	a	c	d	a
21	**22**	**23**	**24**	**25**	**26**	**27**	**28**	**29**	**30**
c	d	c,d	c,d	d	b	b	a	a	a,b,d
31	**32**	**33**							
c	a	c							

Chapter 5

Riemann Integration, Bounded Variation And Improper Integral

1. Riemann Integration

Some definitions:

(1) **Partition:** P = $\{x_0, x_1, ..., x_n : a = x_0 < x_1 < ... < x_n = b\}$ is called a partition of closed interval $[a,b]$

(2) **Norm of a partition:** Let P = $\{x_0, x_1, ..., x_n : a = x_0 < x_1 < ... < x_n = b\}$ be a partition of $[a,b]$. For each i = 1, 2, 3,…, n, let $\Delta x_i = x_i - x_{i-1}$ is the length of the sub interval $[x_{i-1}, x_i]$.

Norm or mesh of a partition P is denoted by ||P|| and defined as ||P||= $\max_i \Delta x_i$.

(3) **Refinement of partition:** Let P be a partition of $[a,b]$, then a partition P^* is called a refinement of P if $P \subset P^*$

(4) **Common refinement of P_1 and P_2:** If P_1 and P_2 are partitions of $[a,b]$, then P = $P_1 \cup P_2$ is called a common refinement of P_1 and P_2.

(5) Let f be a bounded real valued function defined on closed interval $[a,b]$ and let P = $\{x_0, x_1, ..., x_n : a = x_0 < x_1 < ... < x_n = b\}$ be a partition of $[a,b]$.

For each i = 1, 2, 3, ..., n, let $M_i = \sup_{x_{i-1} \le x \le x_i} f(x), m_i = \inf_{x_{i-1} \le x \le x_i} f(x)$

(6) **Upper Riemann sum:** $U(P, f) = \sum_{i=1}^{n} M_i \Delta x_i$ is called upper Riemann sum of f with respect to the partition P of $[a,b]$

(7) **Lower Riemann sum:** $L(P, f) = \sum_{i=1}^{n} m_i \Delta x_i$ is called lower Riemann sum of f with respect to the partition P of $[a,b]$

(8) **Oscillatory sum:** $W(P, f) = U(P, f) - L(P, f)$ is called the oscillatory sum with respect to the partition P of $[a,b]$

REAL ANALYSIS

(9) **Upper Riemann integral:** $\inf_P U(P, f)$ is denoted by $\int_a^{\bar{b}} f(x)dx$ and is called upper Riemann integral of f on $[a,b]$ i.e. $\int_a^{\bar{b}} f(x)dx = \inf_P U(P,f)$

(10) **Lower Riemann integral:** $\sup_P L(P, f)$ is denoted by $\int_{\underline{a}}^{b} f(x)dx$ and called lower Riemann integral of f on $[a,b]$ i.e. $\int_{\underline{a}}^{b} f(x)dx = \sup_P L(P,f)$

(11) **Riemann integrable:** A bounded real valued function f defined on $[a,b]$ is said to be Riemann integrable if $\int_{\underline{a}}^{b} f(x)dx = \int_a^{\bar{b}} f(x)dx$. The common value is denoted by $\int_a^b f(x)dx$ and is called Riemann integral of f on $[a,b]$ If f is Riemann integrable on $[a,b]$, we write $f \in R[a,b]$

Example: f(x) = x^3 is defined on [0, a]

Solution:

Let$P = \left\{0, \frac{a}{n}, \frac{2a}{n} \ldots. \frac{(r-1)a}{n}, \frac{ra}{n}, \frac{na}{n} = a\right\}$ is a partition of [0, a]

We consider a subinterval$\left[\frac{(r-1)a}{n}, \frac{ra}{n}\right]$

Then $M_r = \left(\frac{ra}{n}\right)^3$ & $m_r = \left(\frac{(r-1)a}{n}\right)^3$

$U(P, f) = \sum_{r=1}^{n} \frac{r^3 a^3}{n^3} \Delta x_r$

$= \sum_{r=1}^{n} \frac{r^3 a^3}{n^3} \left[\frac{a}{n}\right] \quad = \frac{a^4}{n^4} \sum_{r=1}^{n} r^3 \quad = \frac{a^4}{n^4}\left[\frac{n^2(n+1)^2}{4}\right]$

So, $\int_0^{\bar{a}} f(x)dx = \lim_{n\to\infty} U(P, f) = \frac{a^4}{4}$

And $L(P, f) = \sum_{r=1}^{n} \frac{(r-1)^3 a^3}{n^3} \left[\frac{a}{n}\right] \quad = \frac{a^4}{n^4}[1 + 2^3 + 3^3 + \cdots (n-1)^3]$

$= \frac{a^4}{n^4}\left[\frac{(n-1)^2 n^2}{4}\right]$

So, $\int_{\underline{0}}^{a} f(x)dx = \lim_{n\to\infty} L(P, f) = \frac{a^4}{4}$

Results:

(1) If f is a bounded function on $[a,b]$ and P is a partition of $[a,b]$, then $m(b-a)\le L(P,f)\le U(P,f)\le M(b-a)$, where

$M_i=\sup\limits_{a\le x\le b} f(x), m_i=\inf\limits_{a\le x\le b} f(x), \overline{\int_a^b} f(x)dx, \underline{\int_a^b} f(x)dx$ exist in $\mathbb{R}$ and

$m(b-a)\le \underline{\int_a^b} f(x)dx \le \overline{\int_a^b} f(x)dx \le M(b-a)$

(2) Every constant function on [a, b] is Riemann integrable on [a, b].

(3) If f is a bounded real valued function defined on $[a,b]$ and P_1, P_2 are partitions of $[a,b]$, then $L(P_1,f)\le U(P_2,f)$, hence deduce that $\underline{\int_a^b} f(x)dx \le \overline{\int_a^b} f(x)dx$.

(4) A continuous function on $[a,b]$ is Riemann integrable on $[a,b]$

(5) A monotonic function on $[a,b]$ is Riemann integrable on $[a,b]$.

(6) If a bounded function f defined on $[a,b]$ has only finitely many points of discontinuity on $[a,b]$, then f is Riemann integrable on $[a,b]$.

(7) If the set of points of discontinuities of a bounded function f on $[a,b]$ is countable, then f is Riemann integrable on $[a,b]$.

(8) If the set of points of discontinuity of a bounded function f on $[a,b]$ has finite number of limit points then f is Riemann integrable on $[a,b]$.

Example : Let f be a function defined on [0, 1] by $f(x)=\begin{cases}1, & x\ne \frac{1}{2}\\ 0, & x=\frac{1}{2}\end{cases}$

Show that f is Riemann integrable on [0, 1] and evaluate $\int_0^1 f(x)dx$

Solution: We know that $f(x)$ is discontinuous at ½

It is discontinuous at countable point. $\Rightarrow f(x)$ is Riemann integrable

and $\int_0^1 f(x)dx=\int_0^{1/2} f(x)dx+\int_{1/2}^1 f(x)dx=\frac{1}{2}+\frac{1}{2}=1$

REAL ANALYSIS

2. The Integral as a Limit of the Sum:

Definition Let f be a function defined on $[a,b]$ and let P = $\{x_0, x_1, ..., x_n : a = x_0 < x_1 < ... < x_n = b\}$ be a partition of $[a,b]$, then

(i) $S(P,f) = \sum_{i=1}^{n} f(t_i)\Delta x_i$, where $t_i \in [x_{i-1}, x_i]$, is called a Riemann sum of f on $[a,b]$ relative to P.

(ii) $\lim_{\|P\|\to 0} S(P,f) = l$ if $\forall \in > 0, \exists \delta > 0$ such that $|S(P,f) - l| < \in$, for every Riemann sum of f on $[a,b]$ relative to partition P of $[a,b]$ with $\|P\| < \delta$.

Note:

(1) If f is Riemann integrable on $[a,b]$, then $\int_a^b f(x)dx = \lim_{n\to\infty} \sum_{i=1}^{n} hf(a+ih)$, where $nh = b-a$.

(2) If $f, g \in R[a,b]$ and α, β∈ [a,b]then that $\alpha f + \beta g \in R[a,b]$ and

(3) If $f, g \in R[a,b]$, then $f - g \in R[a,b]$ and $\int_a^b (f-g)(x)dx = \int_a^b f(x)dx - \int_a^b g(x)dx$

For example Find the sum of the series $\frac{n}{n^2} + \frac{n}{n^2+1^2} + \frac{n}{n^2+2^2} + ... + \frac{n}{n^2+(n-1)^2} + ...$ using integral.

Solution: Here general term = $\frac{n}{n^2+r^2}, r = 0, 1, 2, ...$

$$= \frac{1}{n}\left[\frac{1}{1+\left(\frac{r}{n}\right)^2}\right] \quad ...(1) \quad = \frac{dx}{1+x^2} \text{ [Replace } \frac{1}{n} \text{ by } dx \text{ and } \frac{r}{n} \text{ by } x \text{]}$$

For limit of integration, lower limit $= \lim_{n\to\infty} \frac{n}{n^2} = 0$ [Put r = 0 in general term]

Upper limit $= \lim_{n\to\infty} \frac{n}{n^2+(n-1)^2} = 1$ [Put r = n – 1 in general term]

$\therefore$ Required sum $= \int_0^1 \frac{dx}{1+x^2} = \tan^{-1} x\Big|_0^1 = \frac{\pi}{4}$

3. Function of Bounded Variation

Total variation: Let P = $\{x_0, x_1, ..., x_n : a = x_0 < x_1 < ... < x_n = b\}$ be a partition of $[a,b]$ and sub-intervals can be taken as $[x_0, x_1], [x_1, x_2], ..., [x_{i-1}, x_i], ..., [x_{n-1}, x_n]$

If $f(x)$ is a function defined on $[a,b]$, then corresponding to each partition P of $[a,b]$, $\exists$ a sumi.e. $\sum_{i=1}^{n} |f(x_i) - f(x_{i-1})|$ Supremum of such sums i.e. $\sup_P \sum_{i=1}^{n} |f(x_i) - f(x_{i-1})|$ (where supremum is taken over P) is known as variation or total variation in $[a,b]$

Notation: $v(f, a, b) = \sup_P \sum_{i=1}^{n} |f(x_i) - f(x_{i-1})|$

Example : Total variation of f(x) = sin x on [0, 2π]

We know that sinx is monotonic on [0 , π/2] , [π/2 , 3π/2] and [3π/2 , 2π]

Then total variation is $\left|\sin\left(\frac{\pi}{2}\right) - \sin 0\right| + \left|\sin\left(\frac{3\pi}{2}\right) - \sin\left(\frac{\pi}{2}\right)\right| + \left|\sin(2\pi) - \sin\left(\frac{3\pi}{2}\right)\right|$

$= 1 + 2 + 1 = 4$

Function of bounded variation: Let f be a function and if variation of the function is bounded in a given interval, then function is known as function of bounded variation in that interval.

Results on bounded variation:

(1) If $f(x)$ is bounded monotonic function in $[a,b]$, then $f(x)$ is a function of bounded variation.

(2) If a function is differentiable in $[a,b]$ and its derivative is bounded in $[a,b]$, then $f(x)$ is a function of bounded variation and also the variation is given by $v(f, a, b) \le M(b-a)$, where M is upper limit of derivative of a function.

(3) If $f(x)$ is a function of bounded variation in an interval, then it is bounded in that interval, i.e., function of bounded variation ⇒ function is bounded.

(4) If $f(x)$ is function of bounded variation in $[a,b]$ and if $c \in (a,b)$ be any point, then $f(x)$ is a function of bounded variation in $[a,c]$ and also in $[c,b]$. Moreover $v(f,a,b) = v(f,a,c) + v(f,c,b)$

(5) If $f(x)$ is of bounded variation on $[a,b]$, then it can be expressed as difference of two monotonic non-decreasing functions and conversely [This is necessary and sufficient condition]

For example: $f(x)=\{x\};x\in[0,5]$

$f(x)=x-[x]$, is non-decreasing function

$\Rightarrow f(x)$ is difference of two non-decreasing functions $\Rightarrow f(x)$ is bounded variation

(6) If f is of bounded variation, then it is R-integrable, but converse is not true.

(7) A function of bounded variation need not be continuous in that interval.

Example : $f(x)=[x]: \quad x\in[0,3]$

Where $[x]$ is a greatest integer function of xis of bounded variation but not continuous on [0, 3]

4. Improper Integral

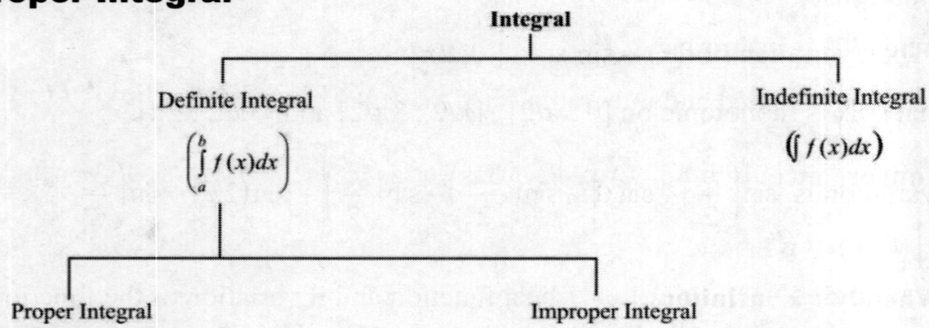

4.1 Definitions:

(1) **Improper integrals:** A function f on $[a,b]$ is Riemann Integrable if

(i) Domain of the function f is the closed interval $[a,b]$, a,b $\in$ R, a < b

(ii) f is bounded on $[a,b]$.

If either a or b (or both) become infinite or f is not bounded on $[a,b]$, then $\int_a^b f(x)\,dx$ is called an improper integral.

(2) **Proper integral:** If f is bounded on $[a,b]$ such that f is Riemann integrable on $[a,b]$, then $\int_a^b f(x)\,dx$ is called a proper integral.

(3) **Point of infinite discontinuity:** Let f be a real function with domain E and c $\in$ R, c is called a point of infinite discontinuity of the function f if $s>0, f$ is not bounded on $(c-s,c+s)\cap E$.

For example, c is a point of infinite discontinuity of the function $f(x)=\dfrac{1}{x-c}$

Note: It is assumed once for all that if a function f has no point of infinite discontinuity in [a, b], a,b∈R, a<b then f is bounded and integrable on $[a,b]$ and $\int_a^b f(x)\,dx$ is a proper integral.

4.2 Types of Improper Integral:

(1) **Improper integral of first kind:** A definite integral $\int_a^b f(x)\,dx$ is said to be improper if either a or b or both are infinite so that interval is unbounded but the function is bounded.

Examples $\int_0^\infty \sin x^2\,dx, \int_0^\infty e^{-x}\,dx$

(i) **When 'b' is infinite:**

Let f be bounded and integrable for $x \geq a$, then $\int_a^\infty f(x)\,dx = \lim_{t\to\infty}\int_a^t f(x)\,dx$

Convergence: If $\lim_{t\to\infty}\int_a^t f(x)\,dx$ exists and is finite, then the improper integral $\int_a^\infty f(x)\,dx$ is said to converge, otherwise it is called divergent.

(ii) **When 'a' is infinite:**

Let f be bounded and integrable for $x \leq b$, then $\int_{-\infty}^b f(x)\,dx = \lim_{t\to\infty}\int_t^b f(x)\,dx$

Convergence: If $\lim_{t\to\infty}\int_t^b f(x)\,dx$ exists and is finite, then the improper integral $\int_{-\infty}^b f(x)\,dx$ is said to converge, otherwise it is called divergent.

(iii) **When both 'a' and 'b' are infinite:**

Let f be bounded and integrable on R, then $\int_{-\infty}^\infty f(x)\,dx = \int_{-\infty}^c f(x)\,dx + \int_c^\infty f(x)\,dx$, where 'c' is any real number

$= \lim_{t_1\to\infty}\int_{t_1}^c f(x)\,dx + \lim_{t_2\to\infty}\int_c^{t_2} f(x)\,dx$.

Convergence: If $\lim_{t_1\to\infty}\int_{t_1}^c f(x)\,dx$ and $\lim_{t_2\to\infty}\int_c^{t_2} f(x)\,dx$ both exists and are finite, then $\int_{-\infty}^\infty f(x)\,dx$ is said to converge, otherwise it is called divergent.

(2) **Improper integral of second kind:** The definite integral $\int_a^b f(x)\,dx$ is said to be improper integral of second kind if 'a' and 'b' are finite but f has one or more point of infinite discontinuity, i.e., f is not bounded on $[a,b]$

Examples $\int_0^1 \frac{dx}{x^2}, \int_1^2 \frac{dx}{2-x}$

(i) Let 'a' be the only point of infinite discontinuity of a function f in $[a,b]$. The improper integral $\int_a^b f(x)dx$ is defined as $\int_a^b f(x)dx = \lim_{\epsilon\to 0^+}\int_{a+\epsilon}^b f(x)dx$

Convergence: If $\lim_{\epsilon\to 0^+}\int_{a+\epsilon}^b f(x)dx$ exists and is finite, then the improper integral $\int_a^b f(x)dx$ is said to converge (exist) at a, otherwise it is called divergent.

(ii) Let 'b' be the only point of infinite discontinuity of a function f in $[a,b]$. The improper integral $\int_a^b f(x)dx$ is defined as $\int_a^b f(x)dx = \lim_{\epsilon\to 0^-}\int_a^{b-\epsilon} f(x)dx$

Convergence: If $\lim_{\epsilon\to 0^-}\int_a^{b-\epsilon} f(x)dx$ exists and is finite, then the improper integral $\int_a^b f(x)dx$ is said to converge (exist) at b, otherwise it is called divergent.

(iii) Let 'a' and 'b' are the only point of infinite discontinuity of a function f in $[a,b]$. The improper integral $\int_a^b f(x)dx$ is defined as

$\int_a^b f(x)dx = \lim_{\epsilon\to 0^+}\int_{a+\epsilon}^c f(x)dx + \lim_{\epsilon'\to 0^+}\int_a^{b-\epsilon'} f(x)dx$, where $a<c<b$.

Convergence: The improper integral $\int_a^b f(x)dx$ is said to converge if both $\lim_{\epsilon\to 0^+}\int_{a+\epsilon}^c f(x)dx$ and $\lim_{\epsilon'\to 0^+}\int_a^{b-\epsilon'} f(x)dx$ exists and are finite, otherwise it is called divergent.

We may write $\int_a^b f(x)dx = \int_a^c f(x)dx + \int_c^b f(x)dx,\ a<c<b$.

(3) Improper integral of third kind:

The definite integral $\int_a^b f(x)dx$ is said to be an improper integral of third kind if the integrand of 'f' unbounded, i.e., 'a' or 'b' or both are infinite and f is also unbounded on $[a,b]$

For Example

$\int_0^\infty \frac{e^{-x}}{\sqrt{x}}dx, \int_1^\infty \frac{dx}{(x-1)^2}$ are improper integral of third kind

(4) **Comparison test:** Let $\int_a^b f(x)dx$ and $\int_a^b g(x)dx$ be two improper integrals, where

(i) f and g has only point of discontinuity at $x = a$

(ii) f and g be two non negative functions

(iii) $f(x) \le g(x)\ \forall x \in [a,b]$, then

(i) $\int_a^b f(x)dx$ converges if $\int_a^b g(x)dx$ converges.

(ii) $\int_a^b g(x)dx$ diverges if $\int_a^b f(x)dx$ diverges.

(5) **Limit comparison Test:** Let $I_1 = \int_a^b f(x)dx$ are $I_2 = \int_a^b g(x)dx$ two improper integrals, f and g defined on $[a,b]$ and $x = a$ is the only point of infinite discontinuity for both f and g, where f and g are positive functions on $[a,b]$.

$\lim_{x \to a} \frac{f(x)}{g(x)} = l$ [l is a non-zero and finite number], then both integrals converge or diverge together.

(6) **Cauchy's Integral Test:** If for $x \ge 1$, $f(x)$ is a non-negative monotonically decreasing integrable function of x such that $f(n) = u_n$ for all positive integral value of n, then the series $\sum_{n=1}^{\infty} u_n$ and the improper integral $\int_1^{\infty} f(x)dx$ converge or diverge together.

(7) Every absolutely convergent integral is convergent, i.e., $\int_a^b f(x)dx$ converges if $\int_a^b |f(x)|dx$ converges.

(8) **p-test :**

(i) $\int_0^1 \frac{1}{x^p} dx$ is convergent if p< 1 and is divergent if p≥ 1.

(ii) $\int_a^{\infty} \frac{1}{x^p} dx$ is convergent if p> 1 and is divergent if p≤ 1
$a>0$

(iii) $\int_a^b \frac{1}{(x-a)^p} dx$ is convergent if p< 1 and is divergent if p≥ 1

(iv) $\int_a^b \frac{1}{(b-x)^p} dx$ is convergent if p< 1 and is divergent if p≥ 1

REAL ANALYSIS

(9) Test for Convergence of the Integrals of a Product :

1. **Let a function** f be integrable on [a, x]; ∀ x > a & the integral $\int_a^\infty f(x)dx$ is absolutely convergent & a function ϕ be bounded on [a, ∞] & integrable on [a, x]; ∀ x>a.

   Then the integral $\int_a^\infty f(x).\phi(x)dx$ is absolutely convergent.

2. **Abel's Test :**

   Let

   (i) a function ϕ be monotonic and bounded on [a, ∞] &

   (ii) The integral $\int_a^\infty f(x)dx$ be convergent.

   Then the integral $\int_a^\infty f(x).\phi(x)dx$ is convergent.

3. **Dirichlet's test :**

   Let

   (i) a function ϕ be monotonic & bounded on [a, ∞] & $\lim_{x\to\infty} \phi(x)dx = 0$.

   (ii) The integral $\int_a^x f(x)dx$ be on [a, x]; ∀ x>a

   Then the integral $\int_a^\infty f(x)\phi(x)dx$ is convergent.

Example Examine the convergence of $\int_1^2 \frac{dx}{(2-x)^3}$

Solution: By p-test it is a divergent.

SOLVED QUESTIONS

1. Consider the identity function f(x) = x on I : = [0, 1].
   Let P_n be the partition that divides I into n equal parts. If U(f, P_n) and L(f, P_n) are the upper and lower Riemann sums, respectively, and A_n = U(f, P_n) – L(f, P_n) then

   (a) $\lim_{n\to\infty} nA_n = 0$ (b) $\sum_{n=1}^{\infty} A_n$ is convergent

   (c) A_n is strictly monotonically decreasing. (d) $\sum_{n=1}^{\infty} A_n A_{n+1} = 1$

   Ans. **(c), (d)** **CSIR NET NOV 2020**

Explanation :

f(x) = x

$$P_n = \left[0, \frac{1}{n}, \frac{2}{n}, \ldots \frac{r-1}{n}, \frac{r}{n} \ldots\ldots \frac{n}{n} = 1\right]$$

Let r^{th} subinterval $\left[\frac{r-1}{n}, \frac{r}{n}\right]$, where $\Delta x_r = 1/n$

$$\Rightarrow \quad U(f, P_n) = \sum_{r=1}^{n} M_r \Delta x_r = \sum_{r=1}^{n} \frac{r}{n} \cdot \frac{1}{n} = \frac{1}{n}\left\{\frac{1}{n} + \frac{2}{n} + \ldots + \frac{n-1}{n} + 1\right\}$$

$$L(f, P_n) = \sum_{r=1}^{n} m_r \Delta x_r = \sum_{r=1}^{n} \frac{r-1}{n} \cdot \frac{1}{n} = \frac{1}{n}\left[0 + \frac{1}{n} + \ldots + \left(\frac{n-2}{n}\right) + \frac{n-1}{n}\right]$$

$$A_n = \frac{1}{n}\left\{\frac{1}{n} + \frac{2}{n} + \ldots + \frac{n-1}{n} + 1\right\} - \frac{1}{n}\left[0 + \frac{1}{n} + \ldots + \left(\frac{n-2}{n}\right) + \frac{n-1}{n}\right]$$

$$A_n = \frac{1}{n}[1] = \frac{1}{n}$$

$\Rightarrow \quad \sum \frac{1}{n}$ is divergent and 1/n is strictly monotonically decreasing.

Now, $\sum_{n=1}^{\infty} \frac{1}{n(n+1)} = \sum_{n=1}^{\infty} \frac{(n+1)-n}{n(n+1)}$

$$= \sum_{n=1}^{\infty}\left(\frac{1}{n} - \frac{1}{n+1}\right) \qquad = \frac{1}{1} - \frac{1}{2} + \frac{1}{2} - \frac{1}{3} + \frac{1}{3} - \frac{1}{4} \ldots$$

$= 1$

$\Rightarrow$ Option (c) and (d) are true.

2. For any two non-negative n, k define $f_{n,k}(x)$ on [0, 1] by

$$f(x) = \begin{cases} x^n \sin\left(\frac{\pi}{2x}\right) - x^k, & x \neq 0 \\ 0, & x = 0 \end{cases}$$

In which of the following cases is the function $f_{n,k}$, a function of bounded variation?

(a) For all $n \geq 1$ and for all $k \geq 0$
(b) For all $n \geq 0$ and $k = 0$
(c) For all $n \geq 1$ and for all $k \geq 2$
(d) For all $n \geq 2$ and for all $k \geq 0$

Ans. (d) **CSIR NET NOV 2020**

Explanation :

$$g(x) = \begin{cases} x^n \sin\left(\frac{\pi}{2x^m}\right) & x \neq 0 \\ 0 & x = 0 \end{cases}$$ is of bounded variation iff $n \geq m + 1$.

REAL ANALYSIS

3. Define f : [0,1] → [0,1] by

$$f(x)=\begin{cases}1 & \text{if} \quad x=0\\ \dfrac{1}{n} & \text{if} \quad x=\dfrac{m}{n} \text{ for some m,n} \in \text{N with m} \le \text{n and gcd(m,n)=1}\\ 0 & \text{if} \quad x\in[0,1] \text{ is irrational}\end{cases}$$

and define g : [0,1] → [0,1] by $g(x)=\begin{cases}0 & if\ x=0\\ 1 & if\ x\in(0,1]\end{cases}$

Then which of the following is true **IIT JAM 2022**

(a) f is Riemann integrable in [0,1]
(b) g is Riemann integrable in [0,1]
(c) fog is Riemann integrable in [0,1]
(d) gof is Riemann integrable in [0,1]

Ans . (a),(b),(c)

Explanation :

(a) We know that f(x) is discontinuous at countable point So , f(x) is Riemann integrable in [0,1]

(b) g(x) is discontinuous at x = 0
So , g(x) is Riemann integrable in [0,1]

(c) Now $fog(x)=f[g(x)]=\begin{cases}f[g(0)]=f[0]=1 & if\ x=0\\ f[g(x)]=f[1]=1 & if\ x\, is\, irrational\\ f[g(x)]=f[1]=1 & if\ x\, is\, rational\end{cases}$

⇒ fog(x) = 1 isRiemann integrable in [0,1]

(d) $gof(x)=g[f(x)]=\begin{cases}1 & if\ x\in Q\cup\{0\}\\ 0 & if\ x\notin Q\end{cases}$

gof(x) is notRiemann integrable in [0,1]

> **Trick :** Let $f(x)=\begin{cases}x^{\alpha}\sin\left(\dfrac{1}{x^{\beta}}\right) & x\neq 0\\ 0 & x=0\end{cases}$ is a function of bounded variation
>
> (i) If α>βand (ii) If α≤β then f(x) is not of bounded variation.

4. Let $f(x)=\begin{cases}x\sin\left(\dfrac{1}{x}\right) & x\neq 0\\ 0 & x=0\end{cases}$ & x ∈ [0, 1] and g(x) = xf(x) for x ∈ [0, 1]. Then which of the following are TRUE? **CSIR NET JUNE 2019**

(a) f is of B.V.
(b) f is not of B.V.
(c) g is of B.V.
(d) g is not of B.V.

Ans. (b), (c)

Explanation :

In f(x) $\Rightarrow \alpha = \beta = 1 \Rightarrow$ f(x) is not of B.V.

$$g(x) = \begin{cases} x^2 \sin\left(\frac{1}{x}\right) & x \neq 0 \\ 0 & x = 0 \end{cases}$$

$\alpha > \beta \Rightarrow$ g(x) is of bounded variation.

Trick : (i) Unbounded function is never of bounded variation
(ii) Every polynomial is of bounded variation for every interval.

5. Which of the following are of bounded variation? **CSIR NET JUNE 2018**

(a) $x^2 + x + 1 \quad x \in (-1, 1)$ (b) $\tan\left(\frac{\pi x}{2}\right) \quad x \in (-1,1)$

(c) $\sin\left(\frac{x}{2}\right) \quad x \in (-\pi, \pi)$ (d) $\sqrt{1-x^2} \quad x \in (-1,1)$

Ans. **(a), (c), (d)**

Explanation :

Option (a) is correct.

Every polynomial is continuous & bounded in bounded domain .

$\Rightarrow$ so it is of bounded variation.

Option (c) is correct.

$\sin\left(\frac{x}{2}\right) \; x \in (-\pi, \pi)$ is continuous + bounded in given domain. So it is of bounded variation.

Option (d) is correct.

$\sqrt{1-x^2}$ is continuous + bounded in (-1, 1)

So, it is of bounded variation.

Option (b) is incorrect.

Unbounded function is never of bounded variation.

$\tan\left(\frac{\pi x}{2}\right) \; x \in (-1,1)$ is unbounded

6. $f(x) = \begin{cases} x^2 \cos\frac{1}{x} & x \neq 0 \\ 0 & x = 0 \end{cases}$ f : [-1, 1] $\to$ IR. Then **CSIR NET DEC 2017**

(a) f is bounded variation on [-1, 1] (b) f' is of bounded variation on [-1, 1]

(c) $|f'(x)| \leq 1 \;\; \forall x \in [-,1\ 1]$ (d) $|f'(x)| \leq 3 \; \forall x \in [-1, 1]$

Ans. (a)

Explanation :

Option (a) is correct.

We know that f(x) is of bounded variation $[\therefore \alpha>\beta]$

Option (b) is incorrect.

$$f'(x)=\begin{cases}2x\cos\frac{1}{x}+\frac{x^2}{x^2}\sin\frac{1}{x} & x\neq 0\\ 0 & x=0\end{cases} \qquad f'(x)=\begin{cases}2x\cos\frac{1}{x}+\sin\frac{1}{x} & x\neq 0\\ 0 & x=0\end{cases}$$

Theorem : Let f(x) is continuous on [a, b] and f' exist in [a, b] and f' is bounded in (a, b) then f(x) is of bounded variation on [a, b].

Option (a) is correct.

Option (b) : f'' (x) is not bounded function ⇒ f'(x) is not of bounded variation.

Let $\quad |f'(x)|=\left|2x\cos\frac{1}{x}+\sin\frac{1}{x}\right|\le\left|2x\cos\frac{1}{x}\right|+\left|\sin\frac{1}{x}\right|\le 2|x|\left|\cos\frac{1}{x}\right|+\left|\sin\frac{1}{x}\right|$

$$\left[\begin{array}{l}\left|\sin\frac{1}{x}\right|\le 1\\ \left|\cos\frac{1}{x}\right|\le 1 \qquad \le 2+1\le 3\\ |x|\le 1\end{array}\right.$$

7. Given $f:\left[\frac{1}{2},2\right]\to R$ a strictly increasing function we put g(x) = $f(x)+f\left(\frac{1}{x}\right)$ x∈[1,2]. Consider a partition of [1, 2] & let U(P, g) & L(P, g) denote the upper Reimann sum & lower Riemann sum of g then **CSIR NET DEC 2018**

(a) for a suitable f we can have U(P, g) = L(P, g) (b) for a suitable f we can have U(P, g) ≠ L(P, g)

(c) U(P, g) ≥ L(P, g) ∀ f (d) U(P, g) < L(P, g) ∀ f

Ans. (a), (b)

Explanation :

Let g be any function then ∀P(partion)

U(P, g) ≥ L(P, g)

But ∄ any g s.t. U(P, g) < L(P, g)

Option (d) is incorrect.

Option (a) is correct.

$$\text{Let } f(x)=\begin{cases}x & x\in\left[\frac{1}{2},1\right]\\ 2-\frac{1}{x} & x\in[1,2]\end{cases}$$

Now $g(x) = f(x) + f\left(\frac{1}{x}\right)$ $x \in [1, 2]$

$$g(x) = 2 - \frac{1}{x} + \frac{1}{x} = 2 \qquad \left[\because x \in [1,2], \frac{1}{x} \in \left[\frac{1}{2},1\right], f\left(\frac{1}{x}\right) = \frac{1}{x}, x \in \left[\frac{1}{2},1\right]\right]$$

$\Rightarrow$ g(x) = 2 (constant) function for this f(x)
& U(P, g) = L(P, g) for every constant function.
Option (b) is correct.

Let $f(x) = x$ $\qquad x \in \left[\frac{1}{2},2\right]$

Then $g(x) = x + \frac{1}{x}$ $\qquad x \in [1, 2]$

$$P = \left\{1, \frac{3}{2}, 2\right\}$$

$I_1 = \left[1, \frac{3}{2}\right]$ $\qquad \Delta x_1 = \frac{1}{2}$

$I_2 = \left[\frac{3}{2}, 2\right]$ $\qquad \Delta x_2 = \frac{1}{2}$

Now $m_1 = 1$, $\qquad M_1 = 3/2$
$m_2 = 3/2$ $\qquad M_2 = 2$

$U(P, g) = \Sigma M_i \Delta x_i = M_1\Delta x_1 + M_2\Delta x_2 = \frac{3}{2}\left(\frac{1}{2}\right) + 2\left(\frac{1}{2}\right) = \frac{3}{4} + 1 = \frac{7}{4}$

$L(P, g) = \Sigma m_i \Delta x_i = m_1\Delta x_1 + m_2\Delta x_2 = 1\left(\frac{1}{2}\right) + \frac{3}{2}\left(\frac{1}{2}\right) = \frac{1}{2} + \frac{3}{4} = \frac{5}{4}$

$U(P, g) \neq L(P, g)$

p-Test

(i) $\int_0^1 \frac{1}{x^p} dx$ is convergent $p < 1$ & is divergent if $p \geq 1$

(ii) $\int_a^\infty \frac{1}{x^p} dx$ is convergent $p > 1$ & divergent if $p \leq 1$

8. Let a be positive real number which of the following integrals are convergent.

(a) $\int_0^a \frac{1}{x^4} dx$ (b) $\int_0^a \frac{1}{\sqrt{x}} dx$ (c) $\int_4^\infty \frac{1}{x \log x} dx$ (d) $\int_5^\infty \frac{1}{x(\log x)^2} dx$

Ans. **(b), (d)** **CSIR NET JUNE 2015**

REAL ANALYSIS

Explanation :

Option (a) and (b) are second type improper integral

By using of p-test. $\int_0^a \frac{dx}{x^p}$ is convergent if p < 1 & is divergent if p ≥ 1

Option (a) is incorrect. $\int_0^a \frac{dx}{x^4}$ is divergent.

Option (b) is correct. $\int_0^a \frac{dx}{\sqrt{x}}$ is convergent..

Option (c) is incorrect. $\int_4^\infty \frac{dx}{x(\log x)} = \lim_{u\to\infty} \int_4^u \frac{dx}{x(\log x)}$

let log x = t $\frac{dx}{x} = dt \qquad = \lim_{u\to\infty} \int_4^u \frac{dt}{t} = \lim_{u\to\infty} [\log t]_4^u$

Put value of t $\lim_{u\to 0} [\log(\log x)]_4^u = \lim_{u\to 0} [\log(\log u) - \log\log(4)] \to \infty$

Option (d) is correct. $\int_5^\infty \frac{dx}{x(\log x)^2} \Rightarrow let \ \log x = t$

$\frac{dx}{x} = dt \quad = \lim_{u\to\infty} \int_5^u \frac{dt}{t^2} = \lim_{u\to\infty} \left[-\frac{1}{t}\right]_5^u \quad = -\lim_{u\to\infty} \left[\frac{1}{\log x}\right]_5^u \quad = -\lim_{u\to\infty} \left[\frac{1}{\log u} - \frac{1}{\log 5}\right]$ = finite

= convergent.

9. Define $f:[0,1]\to[0,1]$ by $f(x) = \frac{2^k - 1}{2^k}$ for $x \in \left[\frac{2^{k-1}-1}{2^{k-1}}, \frac{2^k-1}{2^k}\right], k \geq 1$. Then f is a Riemann –integrable function such that **CSIR NET DEC 2011**

(a) $\int_0^1 f(x)\,dx = \frac{2}{3}$ (b) $\frac{1}{2} < \int_0^1 f(x)\,dx < \frac{2}{3}$ (c) $\int_0^1 f(x)\,dx = 1$ (d) $\frac{2}{3} < \int_0^1 f(x)\,dx < 1$

Ans. (a)

Explanation :

Given $f:[0,1]\to[0,1]$

$f(x) = \frac{2^k - 1}{2^k}$

$$\int_0^1 f(x)\,dx = \sum_{k=1}^{\infty} \int_{\frac{2^{k-1}-1}{2^{k-1}}}^{\frac{2^k-1}{2^k}} f(x)\,dx$$

$$= \sum_{k=1}^{\infty} \frac{2^k-1}{2^k}\left[\frac{2^k-1}{2^k} - \frac{2^{k-1}-1}{2^{k-1}}\right] = \sum_{k=1}^{\infty} \frac{2^k-1}{2^k} \times \frac{1}{2^{k-1}}\left[\frac{2^k-1-2^k+2}{2}\right]$$

$$= \sum_{k=1}^{\infty} \frac{2^k-1}{2^{2k-1+1}} \times 1 = \sum_{k=1}^{\infty} \frac{2^k-1}{2^{2k}} = \sum_{k=1}^{\infty} \frac{1}{2^k} - \sum_{k=1}^{\infty} \frac{1}{2^{2k}}$$

$$= \frac{\frac{1}{2}}{1-\frac{1}{2}} - \frac{\frac{1}{4}}{1-\frac{1}{4}} = 1 - \frac{1}{3} = \frac{2}{3}$$

10. Which of the conditions below imply that a function $f : [0, 1] \to R$ is necessarily of bounded variation? **CSIR NET DEC 2012**

(a) f is a monotone function on $[0,1]$

(b) f is a continuous and monotone function on $[0,1]$

(c) f has a derivative at each $x \in (0,1)$

(d) f has a bounded derivative on the interval $(0,1)$

Ans. (a), (b), (d)

Explanation :

We know that if f is a monotonic function on $[a,b]$. Then f is of bounded variation on $[a,b]$

So, option (a) and (b) are correct.

Also, if derivative of $f(x)$ is bounded on (a,b) then f is of bounded variation.

So, option (d) is also correct

11. Let $L = \int_0^1 \frac{dx}{1+x^8}$. Then **CSIR NET DEC 2013**

(a) $L < 1$ (b) $L > 1$ (c) $L < \frac{\pi}{4}$ (d) $L > \frac{\pi}{4}$

Ans. (a), (d)

Explanation :

Given, $L = \int_0^1 \frac{dx}{1+x^8}$

As $0 < x < 1, x^8 < x^2 \Rightarrow 1 + x^8 < 1 + x^2$

$$\Rightarrow \frac{1}{1+x^8} > \frac{1}{1+x^2}$$

$$L=\int_0^1 \frac{dx}{1+x^8} > \int_0^1 \frac{dx}{1+x^2} = \tan^{-1} x\Big|_0^1 = \tan^{-1}1-\tan^{-1}0=\frac{\pi}{4} \Rightarrow L > \frac{\pi}{4}$$

So, option (d) is correct

Also, $\frac{1}{1+x^8}$ is maximum when $x=0$

$$\Rightarrow \frac{1}{1+x^8} < \frac{1}{1+0} \Rightarrow \int_0^1 \frac{dx}{1+x^8} < \int_0^1 dx \Rightarrow L<1$$

So, option (a) is correct.

REAL ANALYSIS

12. Let α, p be real numbers and $\alpha > 1$. **CSIR NET JUNE 2014**

(a) If $p>1$, then $\int\limits_{-\infty}^{\infty} \frac{1}{|x|^{p\alpha}} dx < \infty$

(b) If $p>\frac{1}{\alpha}$, then $\int\limits_{-\infty}^{\infty} \frac{1}{|x|^{p\alpha}} dx < \infty$

(c) If $p<\frac{1}{\alpha}$, then $\int\limits_{-\infty}^{\infty} \frac{1}{|x|^{p\alpha}} dx < \infty$

(d) For any $p \in \mathbb{R}$, we have $\int\limits_{-\infty}^{\infty} \frac{1}{|x|^{p\alpha}} dx = \infty$

Ans. (d)

Explanation :

We will check by options

For option (a)

Given $\alpha > 1$ and $p>1$

Take $\alpha=2, p=2$ $\quad \therefore I=\int_{-\infty}^{\infty} \frac{1}{x^4} dx = \int_{-\infty}^{0} \frac{1}{x^4} dx + \int_0^{\infty} \frac{1}{x^4} dx = I_1 + I_2$

Clearly I_1 and I_2 both are divergent] $\quad \therefore \int\limits_{-\infty}^{\infty} \frac{1}{|x|^{p\alpha}} dx = \infty$

Thus, option (a) is incorrect

With above example option (b) is also incorrect.

For option (c)

Take $\alpha=2, p=-2$ $\quad \therefore I=\int_{-\infty}^{\infty} \frac{1}{|x|^{-4}} dx = \int_{-\infty}^{\infty} x^4 dx$

Clearly $I=\infty$

Thus, option (c) is incorrect.

As all other options are incorrect.

$\therefore$ option (d) is correct.

13. Consider the improper Riemann integral $\int_0^x y^{-\frac{1}{2}}dy$. This integral is :

(a) continuous in $[0,\infty)$ (b) continuous only in $(0,\infty)$

(c) discontinuous in $(0,\infty)$ (d) discontinuous only in $\left(\frac{1}{2},\infty\right)$

Ans. (a) **CSIR NET JUNE 2016**

Explanation :

The given integral is $\int_0^x y^{-\frac{1}{2}}dy$

Let $f(x)=\int_0^x y^{-\frac{1}{2}}dy \Rightarrow f(x)=\left[2\sqrt{y}\right]_0^x \Rightarrow f(x)=2\sqrt{x}$, which is continuous in $[0,\infty)$

$\therefore$ Option (a) is correct

14. If $\int_{\underline{a}}^{b} f(x)dx$ and $\int_a^{\bar{b}} f(x)dx$ are lower and upper Riemann integral on $[a,b]$ then

(a) $\int_{\underline{a}}^{b} f(x)dx \geq \int_a^{\bar{b}} f(x)dx$ (b) $\int_{\underline{a}}^{b} f(x)dx = \int_a^{\bar{b}} f(x)dx$

(c) $\int_{\underline{a}}^{b} f(x)dx \leq \int_a^{\bar{b}} f(x)dx$ (d) none of these

Ans. (c)

Explanation :

Let f be a function on $[a,b]$ and p be partition of $[a,b]$

$$L[p,f]\leq U[p,f]$$

Then $\sup L[p,f]\leq Inf\, U[p,f]$

$$\int_{\underline{a}}^{b} f(x)dx \leq \int_a^{\bar{b}} f(x)dx$$

15. Let f : [a, b] → R be a monotonic function, then which of the following is true?

(a) f is of bounded variation (b) f is unbounded

(c) the set of discontinuities of f are uncountable (d) none of these

Ans. (a)

Explanation :

Result : If a function is monotonic on closed interval then this function is of bounded variation

16. The value of integral $\int_1^4 [\log x]dx$ where $[\ ]$ denotes greatest integer is

(a) $\log 4$ (b) $\frac{1}{2}$ (c) $4-e$ (d) $4+e$

Ans. (c)

Explanation :

We know that

$$\therefore [x]=\begin{cases}\vdots & \vdots\\ 0 & 0\le x<1\\ 1 & 1\le x<2\\ 2 & 2\le x<3\\ 3 & 3\le x<4\\ \vdots & \vdots\end{cases}$$

Then $[\log x]=\begin{cases}\vdots & \vdots\\ 0 & 1\le x<e\\ 1 & e\le x<e^2\\ \vdots & \vdots\end{cases}$

$$\int_1^4[\log x]\,dx$$

$$=\int_1^e 0dx+\int_e^4 1dx$$

$$=0+[x]_e^4$$

$$=4-e$$

UNSOLVED QUESTIONS

1. Define f on [0, 1] by $f(x)=\begin{cases}x^2 & x\in Q\\ x^3 & x\notin Q\end{cases}$. Then **CSIR NET JUNE 2016**

(a) $f\notin R[0,1]$

(b) $f\in R[0,1]\ \&\ \int_0^1 f(x)dx=\frac{1}{3}$

(c) $f\in R[0,1]\ \&\ \int_0^1 f(x)dx=\frac{1}{4}$

(d) $\frac{1}{4}=\underline{\int_0^1} f(x)dx<\overline{\int_0^1} f(x)dx=\frac{1}{3}$

2. The limit $\lim_{n\to\infty}\sum_{k=2}^{n}\frac{1}{k^3-k}$ is equal to **IIT-JAM 2015**

(a) 0.20 (b) 0.25 (c) 0.30 (d) – 0.25

3. $f:[0,1]\to R$ such that $f(x)=\begin{cases}0, & x \text{ is rational}\\ 1, & x \text{ is irrational}\end{cases}$. Then

(a) the upper and lower integrals of f doesn't exist

(b) f is not Riemann integrable (c) f is Riemann integrable (d) none of these

4. f is of bounded variation on $[a,b]$ iff

(a) f is the difference of two monotonic real valued function on $[a,b]$

(b) f is the product of two monotonic real valued function on $[a,b]$

(c) f is the quotient of two monotonic real valued function on $[a,b]$

(d) none of these

5. If f is an absolutely continuous function on $[a,b]$, then

(a) f is of a bounded variation (b) f is not of bounded variation

(c) f is not continuous (d) none of these

6. If $f(x)=\begin{cases} x^2\cos\frac{1}{x}, & if\ x>0 \\ 0, & if\ x\le 0 \end{cases}$, then in [0, 1]

(a) $f(x)$ is of bounded variation and $f'(x)$ exists $\forall x\in[0,1]$

(b) $f(x)$ is of bounded variation and $f'(0)$ doesn't exist.

(c) $f(x)$ is not of bounded variation and $f(x)$ is not continuous

(d) $f(x)$ is discontinuous at $x=0$

7. Let $f(x)=\begin{cases} 1, & \text{x is rational in [0, 1]} \\ -1, & \text{x is irrational in [0,1]} \end{cases}$ then in [0, 1]

(a) $f(x)$ is continuous everywhere (b) $f(x)$ is Riemann integrable

(c) $f(x)$ is not Riemann integrable (d) $f(x)$ is continuous only at the irrationals.

8. The integral $\int_{-1}^{1}\frac{dx}{x^3}$

(a) doesn't exist (b) exist (c) divergent (d) oscillating finitely

9. Let f : [0, 2] → R be defined by $f(x)=\begin{cases} 0, & x\ is\ irrational \\ 1/2, & x\ is\ rational \end{cases}$

Let A and B be the upper Riemann integral and the lower Riemann integral respectively of f on [0, 2] then

(a) A = B = 0 (b) A = 1 and B = 0 (c) A = 1/2 and B = 0 (d) A = B = 1/2

10. The total variation of the function $\sin x$ on the interval $[0,2\pi]$ is

(a) 1 (b) 2 (c) 2π (d) 4

11. $f(x)=\begin{cases} x\sin\frac{\pi}{x}, & when\ 0<x\le 1 \\ 0, & when\ x=0 \end{cases}$, then in [0, 1]

(a) $f(x)$ is continuous on [0, 1]and f is of bounded variation on[0, 1]

(b) $f(x)$ is continuous on [0, 1]

(c) f is not of bounded variation on[0, 1] (d) $f(x)$ is not continuous on [0, 1]

12. $\int_0^1 \frac{dx}{x^3}$ is not

(a) divergent (b) convergent (c) oscillating infinitely (d) oscillating finitely

13. If the derivative f' exist and is bounded on $[a,b]$ then

(a) the function f is of bounded variation on $[a,b]$

(b) the function f is not of bounded variation on $[a,b]$

(c) the function f is not necessary continuous

(d) the function f is necessary continuous

14. If c is fixed point in [0, 1]. The function $f:[0,1]\to R$ defined by $f(x)=1$ if $x<c$ and $f(x)=2$ if $x\geq c$ & it is given that the Riemann integral $\int_0^1 f(x)dx$ is 3/2. Then the value of c is not equal

(a) 1/2 (b) 1/3 (c) 1/4 (d) 1

15. For a positive real number p, let $(f_n : n=1,2,...)$ be sequence of functions defined on [0, 1] by $f_n(x)=\begin{cases} x^{p+1}, & if\ 0\leq x\leq \frac{1}{n} \\ \frac{1}{x^p}, & if\ \frac{1}{n}<x\leq 1 \end{cases}$, let $f(x)=\lim_{n\to\infty} f_n(x), x\in[0,1]$. Then on [0, 1]

(a) f is Riemann integrable **GATE-2007**

(b) the improper integral $\int_0^1 f(x)dx$ converge for $p\geq 1$

(c) the improper integral $\int_0^1 f(x)dx$ converge for $p<1$ (d) f_n converges uniformly

Answer Key

1	2	3	4	5	6	7	8	9	10
a,d	b	b	a	a	c	c	a	b	d
11	12	13	14	15					
b,c	b,c,d	a,d	b,c,d	c					

Chapter 6

Uniformly Convergence

1. Sequence of functions:

Let $E \subset R$ If for every positive integer n, there is a function $f_n : E \to R or C$, then $\{f_n\}$ is called a sequence of functions defined on E.

Example : $f_n(x) = \sin \frac{x}{n}$

$< f_n(x) >= \sin x, \sin \frac{x}{2}, \sin \frac{x}{3}, \ldots\ldots$ is a sequence of function

Point wise Limit : A sequence of function $\{f_n\}$ is said to have a limit. If limit of $\{f_n\}$ exist for all $x \in E$.

Point wise convergence: A sequence $\{f_n\}$ of functions defined on E is said to if for every $x \in E$, the real sequence $\{f_n(x)\}$ converges to $f(x)$, we write $\{f_n\} \to f$ (pointwise) and $\lim_{n\to\infty} f_n(x) = f(x)\ \forall x \in E$.

Example (1) $f_n(x) = x^n \qquad x \in [0, 1]$

At $x = 0 \qquad f_n(0) = 0$

At $x = 1 \qquad f_n(1) = 1$

At $\alpha \in (0, 1)$

$$f_n(\alpha) = \alpha^n$$

then $f_n(x) = x^n$, $x \in [0, 1]$ is pointwise convergent.

Example (2) $f_n(x) = x + \frac{\sin x}{n} \qquad x \in [0,1]$

let $\alpha \in [0, 1]$ then $f_n(\alpha) = \alpha + \sin\left(\frac{\alpha}{n}\right)$

$\lim_{n\to\infty} f_n(x) = \alpha \qquad \alpha \in [0, 1]$

$f(x) = x$

where $f(x) = \lim_{n\to\infty} f_n(x)$

Uniformly convergent :

A sequence $\{f_n\}$ of functions defined on E is said to converge uniformly on E to a real function f defined on E if $\forall \in > 0, \exists$ a positive integer n_0 (depending on $\in$ only) such that $|f_n(x)-f(x)| < \in \ \forall n \geq n_0$ and $\forall x \in E$. We write $\{f_n\} \to f$ uniformly on E.

M_n test for uniform convergence: If $\{f_n\}$ is a sequence of functions defined on E which converges pointwise to $f(x)$ on E. Let $M_n = \sup_{x\in E}|f_n(x)-f(x)|$

Then, $\{f_n\} \to f$ uniformly on E if and only if $M_n \to 0$ as $n \to \infty$.

Example : $f_n(x) = \dfrac{n}{n+x}; x \in [0,1]$;

Here f(x) = 1 We define $M_n = \sup_{x\in S}\left|\dfrac{n}{n+x} - 1\right| = \sup_{x\in S}\left|\dfrac{x}{n+x}\right|$

Here $g(x) = \dfrac{x}{n+x}$ Then $g`(x) = \dfrac{n+x-x}{(n+x)^2} = \dfrac{n}{(n+x)^2} > 0$

$\Rightarrow$ g(x) is increasing.

So, Supremum of g(x) attain at x = 1. $\Rightarrow$ $M_n = \dfrac{1}{1+n}$

Hence $M_n \to 0$ as $n \to \infty$

By M_n - test this sequene of function is uniformly convergent on [0, 1]

1.3 Results on uniform convergence of sequence:

(1) Let $\{f_n\}$ be a sequence of functions defined on E and f be a function on E. Then, $\{f_n\} \to f$ uniformly on E $\Rightarrow \{f_n\} \to f$ (pointwise)

(2) If $\{f_n\} \to f$ uniformly on E and c is a limit point of E such that $\lim_{x\to c} f_n(x) = A_n$, for every n, then $\{A_n\}$ converges and $\lim_{n\to\infty} A_n = \lim_{x\to c} f(x)$. In other words, $\lim_{n\to\infty}\lim_{x\to c} f_n(x) = \lim_{x\to c}\lim_{n\to\infty} f_n(x)$

(3) If $\{f_n\}$ is a sequence of continuous functions on E and $\{f_n\} \to f$ uniformly on E, then f is continuous on E.

(4) If $\{f_n\}$ is a sequence of continuous functions converging pointwise to a function 'f'on E, but 'f' is not continuous on E. Then $\{f_n\}$ is not uniformly convergent.

Let $f_n(x) = \dfrac{1}{1+nx}, 0 \leq x \leq 1 \ \therefore f(x) = \lim_{n\to\infty} f_n(x) = \begin{cases} 0, & when\, 0 < x \leq 1 \\ 1, & when\ x = 0 \end{cases}$

Here, '0' is point of discontinuity. Hence, sequence $\{f_n(x)\}$ doesn't converge uniformly in [0, 1].

(5) Let $\{f_n\}$ be a sequence of differentiable functions on $[a,b]$ and $\{f_n(x)\}$ converges for some $x_0 \in [a,b]$. If $\{f_n'\}$ converges uniformly on $[a,b]$, then there function f on $[a,b]$ such that $\{f_n\} \to f$ uniformly on $[a,b]$ and $f'(x) = \lim_{n\to\infty} f_n'(x)\ \forall x_0 \in [a,b]$.

(6) Let $\{f_n\} \to f$ uniformly on [a,b]

If $f_n(x)$ is Reimann integrable function $\Rightarrow$ f(x) is Riemann integrable function $\Rightarrow$

$$\lim_{n\to\infty} \int_a^b f_n(x)dx = \int_a^b \lim_{n\to\infty} f_n(x)dx = \int_a^b f(x)dx$$

1.4 **Dinni's Theorem:** If $\{f_n(x)\}$ is continuous and monotonic increasing (decreasing) sequence of functions converging to continuous function $f(x)$ on $[a,b]$ (or on any compact set)

Then convergence of $\{f_n(x)\}$ in $[a,b]$ (or on any compact set) is uniform.

1.5 **Weistrass approximation theorem** $\Rightarrow$ If f(x) is continuous on [a, b] then $\exists$ a sequence of function $<f_n(x)>$ which converges to f uniformly

2. Series of functions:

Let $\{f_n\}$ be a sequence of functions defined on E.

(i) $\sum_{n=1}^{\infty} f_n$ or $\sum_{n=1}^{\infty} f_n(x)$ is called a series of functions.

(ii) $S_n = \sum_{i=1}^{n} f_i$ or $S_n(x) = \sum_{i=1}^{n} f_i(x)$ is called partial sum of the series $\sum_{n=1}^{\infty} f_n$ or $\sum_{n=1}^{\infty} f_n(x)$

Example : $\sum_{n=1}^{\infty} \frac{\sin nx}{n^p} \quad p>1,\ x\in R$

Example : $\sum_{n=1}^{\infty} \frac{\cos nx}{n^p} \quad p>1,\ x\in R$

Pointwise and Uniform Convergence of series: A series $\sum_{n=1}^{\infty} f_n$ of functions is said to converge pointwise or converge uniformly on E according as partial sum sequence $\{S_n\}$ or $\sum_{n=1}^{\infty} f_n$ converges pointwise or converges uniformly on E.

Example : $u_n(x) = \frac{nx}{1+n^2x^2} - \frac{(n-1)x}{1+(n-1)^2x^2}$ then $\sum u_n(x) = ?$

Solution : Let $\sum u_n(x)$ be a series of function then Sop's

$S_n(x) = u_1(x) + u_2(x) + \ldots\ldots + u_n(x)$

$$S_n(x) = \frac{x}{1+x^2} + \left[\frac{2x}{1+4x^2} - \frac{x}{1+x^2}\right] + \left[\frac{3x}{1+9x^2} - \frac{2x}{1+4x^2}\right] + \left[\frac{nx}{1+n^2x^2} - \frac{(n-1)x}{1+(n-1)^2x^2}\right]$$

$S_n(x) = \frac{nx}{1+n^2x^2}$ be a sequence of function is not uniformly convergent because $M_n = 1/2$ so $\sum u_n(x)$ is not uniformly convergent.

2.2 Results on uniform convergence of series:

(1) Weierstrass's M test: Let $\sum_{n=1}^{\infty} f_n$ be a series of functions defined on E and for every n, $|f_n(x)| \le M_n \ \forall x \in E$ If the series $\sum_{n=1}^{\infty} M_n$ converges, then $\sum_{n=1}^{\infty} f_n$ converges uniformly on E.

(2) If $\{f_n\}$ is sequence of continuous functions on E such that $\sum_{n=1}^{\infty} f_n = f$, i.e., $\sum_{n=1}^{\infty} f_n(x) = f(x) \ \forall x \in E$ and $\sum_{n=1}^{\infty} f_n$ converges uniformly on E, then f is continuous on E.

Example : $\sum_{n=1}^{\infty} \frac{\cos nx}{n^4}$; x ∈ R

We know that $\left|\frac{\cos nx}{n^4}\right| \le \frac{1}{n^4}$

We also know that

$\sum \frac{1}{n^4}$ is convergent series by p-test.

$\Rightarrow \sum \frac{\cos nx}{n^4}$ is uniformly convergent

(3) Abel's test: If

i. $\sum_{n=1}^{\infty} u_n(x)$ is uniformly convergent on $[a,b]$

ii. $\{v_n(x)\}$ is monotonically decreasing for every $\forall x \in [a,b]$

iii. $\{v_n(x)\}$ is uniformly bounded on $[a,b]$, i.e., $\exists$ a positive real number K such that $|v_n(x)| < K \ \forall x \in [a,b]$, then $\sum_{n=1}^{\infty} u_n(x) v_n(x)$ converges uniformly on $[a,b]$.

(4) Dirichlet Test→ Let $\sum u_n(x)\, V_n(x)$ be a series of function.

& (a) $<V_n(x)> \xrightarrow{\text{uniformly convergenc e}} 0$

(b) $<V_n(x)>$ is monotonically decreasing.

(c) The Sops of $\sum u_n(x)$ is uniformly bounded then $\sum u_n(x)\, V_n(x)$ is uniformly convergence.

(5)Dinni's theorem for series: If $\sum f_n(x)$ is series of non-negative terms converging to sum function $f(x)$ in $[a,b]$, then if $f_n(x)$ and $f(x)$ is continuous, then convergence is uniform

SOLVED QUESTIONS

1. Let $f_n : (0, 1) \to (0, 1)$ be a sequence of function. Then which of the following are true?

CSIR NET JUNE 2023

(a) $f_n(x) = x^n$ is uniformly convergent

(b) $f_n(x) = \dfrac{x^n}{\log(n+1)}$ is uniformly convergent

(c) $f_n(x) = \dfrac{x^n}{1+x^n}$ is uniformly convergent

(d) $\dfrac{x^n}{1+x^n}$ is not uniformly convergent

Ans. **(b)**

Explanation :

(a) $\exists < a_n > = \left\langle 1-\frac{1}{n} \right\rangle$ s.t. $f_n(a_n) = \left(1-\frac{1}{n}\right)^n = \frac{1}{e}$

Hence given $f_n(x) = x^n$ is not uniformly convergent.

(c) For $f_n(a_n) = \dfrac{\left(1-\frac{1}{n}\right)^n}{1+\left(1-\frac{1}{n}\right)^n} = \dfrac{1/e}{1+1/e} = \dfrac{1/e}{\frac{1+e}{e}}$

Option (c) is not true.

REAL ANALYSIS

(b) PWL $f(x) = 0$

Using Mn test $M_n = \sup_{x\in(0,1)} \left|\frac{x^n}{log(n+1)}\right| \to 0$

Hence $\frac{x^n}{log(1+n)}$ is uniformly convergent.

(d) PWL $f(x) = 0$

Using Mn-test $M_n = \sup_{x\in(0,1)} \left|\frac{x^n}{1+nx^n}\right| \to 0$.

Hence $\frac{x^n}{1+nx^n}$ is uniformly convergent.

2. The series $\sum_{n=1}^{\infty} \frac{(-1)^n \sin nx}{n^{\log_e n}}$, $x \in R$ converges

(a) Only for $x = 0$
(b) Uniformly only for $x \in [-\pi, \pi]$
(c) Uniformly only for $x \in R / \{n\pi : n \in Z\}$
(d) Uniformly for all $x \in R$

Ans. (d) **CSIR NET NOV 2020**

Explanation :

Note that $\left|\frac{(-1)^n \sin(nx)}{n^{\log_e n}}\right| \le \frac{1}{n^{\log_e n}}$ for $n \ge m$, for some $m \in N$

We know that $\log_e n \ge 2$, for $n \ge e^2$

Then $n^{\log_e n} \ge n^2 \Rightarrow \frac{1}{n^{\log_e n}} \le \frac{1}{n^2}$ Hence $\left|\frac{(-1)^n \sin(nx)}{n^{\log_e n}}\right| \le \frac{1}{n^{\log_e n}} \le \frac{1}{n^2}$

Since, $\sum \frac{1}{n^2}$ is convergent by p-test.

$\Rightarrow$ By Weierstrass M-test $\sum_{n=1}^{\infty} \frac{(-1)^n \sin(nx)}{n^{\log_e n}}$ is converges uniformly for all $x \in R$

$\Rightarrow$ Option (d) is true.

3. Consider two series $A(x) = \sum_{n=0}^{\infty} x^n (1-x)$ and $B(x) = \sum_{n=0}^{\infty} (-1)^n x^n (1-x)$ where $x \in [0, 1]$.

Which of the following statements are true?

(a) Both A(x) and B(x) converge pointwise.
(b) Both A(x) and B(x) converge uniformly.
(c) A(x) converges uniformly but B(x) does not.
(d) B(x) converges uniformly but A(x) does not.

Ans. (a), (d) **CSIR NET NOV 2020**

Explanation :

A(0) = 1 and A(1) = 0

$$A(x)=\sum_{n=0}^{\infty}x^n(1-x)$$

$$=(1-x)\frac{1}{1-x} \qquad \left[\because \sum x^n=\frac{1}{1-x}\right]$$

$$=1$$

$\Rightarrow$ A(x) converges to $\begin{cases}0 & x=0 \text{ or } x=1\\ 1 & \text{otherwise}\end{cases}$

$\Rightarrow$ A(x) converges to discontinuous function.

$\Rightarrow$ Sequence of continuous function converges to discontinuous function.

$\Rightarrow$ A(x) converges pointwise but not uniformly.

B(0) = B(1) = 0 and $B(x)=\frac{1-x}{1+x}\ \forall x\neq 1,0$.

Now note that $\sum(-1)^n\le 2$ and $x^n(1-x)$ is decreasing and tending to 0.

$\Rightarrow$ By Dirichlet test for series of function $\sum_{n=0}^{\infty}(-1)^n x^n(1-x)$ converges uniformly to $\frac{1-x}{1+x}$.

$\Rightarrow$ Option (a) and (d) are true.

4. For non-negative integers k ≥ 1. Define $f_k(x)=\frac{x^k}{(1+x)^2}\ \forall x\ge 0$.

Which of the following statements are true?

(a) For each k, f_k is a function of bounded variation on compact intervals.

(b) For every k, $\int_0^{\infty}f_k(x)dx<\infty$.

(c) $\lim_{k\to\infty}\int_0^1 f_k(x)dx$ exists

(d) The sequence of functions f_k converge uniformly on [0, 1] as k →∞.

Ans. (a), (c) **CSIR NET FEB 2022**

Explanation :

Given that $f_k(x)=\frac{x^k}{(1+x)^2}\ \forall x\ge 0, k\ge 1$

(a) We know that $f_k(x)=\dfrac{x^k}{(1+x)^2}\ \forall x\ge 0$ is continuous on compact interval for every k ≥ 1. Hence For each k, f_k is a function of bounded variation on compact intervals

(b) Take k = 1

$$\Rightarrow \int_0^\infty \frac{x}{(1+x)^2}dx=\int_0^\infty \frac{1}{(1+x)}dx-\int_0^\infty \frac{1}{(1+x)^2}dx =[\log(1+x)]_0^\infty+\left[\frac{1}{1+x}\right]_0^\infty \to \infty$$

⇒ Option (b) is false.

(c) Note that $f_k(x) \ge f_{k+1}(x)$ for all $x \in [0, 1]$

⇒ f_k is monotonically decreasing and $f=\begin{cases}0, & x\in(0,1)\\ \frac{1}{4}, & x=1\end{cases}$.

Using monotone convergence theorem

$$\lim_{k\to\infty}\int_0^1 f_k(x)dx=\int_0^1 \lim_{k\to\infty} f_k(x)dx=\int_0^1 f(x)dx=0. \Rightarrow \text{ Option (c) is true.}$$

(a) Since f_k is sequence of continuous function and converges to $f=\begin{cases}0 & x\in(0,1)\\ 1/4 & x=1\end{cases}$.

Which is discontinuous function.

⇒ Convergence of f_k cannot be uniform

⇒ Option (d) is false.

5. Suppose that $\{f_n\}$ is a sequence of real-valued functions on R. Suppose it converges to a continuous function f uniformly on each closed and bounded subset of R. Which of the following statements are true?

(a) The sequence $\{f_n\}$ converges to f uniformly on R.

(b) The sequence $\{f_n\}$ converges to f pointwise on R.

(c) For all sufficiently large n, the function f_n is bounded.

(d) For all sufficiently large n, the function f_n is continuous.

Ans. (b) **CSIR NET NOV 2020**

Explanation :

Consider $f_n(x)=\dfrac{x}{n}, x\in R$

Clearly $\dfrac{x}{n}\to 0$ for any compact set D, Now $M_n=\dfrac{x}{n}$

Take a_n = n then $M_n=\dfrac{n}{n}=1\nrightarrow 0$

$\Rightarrow$ f_n doesn't converges uniformly on R.

$\Rightarrow$ Option (a) is not true.

Since $f_n(x) = \frac{x}{n}, x \in R$ is not bounded for any n.

$\Rightarrow$ Option (c) is false.

Now for option (d)

Consider this example $f_n(x) = \begin{cases} x/n & x \in \text{irrational} \\ 0 & x \in \text{rational} \end{cases}$

$\Rightarrow$ Only option (b) is true.

6. Let $g_n(x) = \frac{nx}{1+n^2x^2}$ x ∈ [0, ∞). Which of the following is TRUE as n →∞?

CSIR NET DEC 2019

(a) g_n converge pointwise to 0 but not uniformly (b) g_n converge uniformly to 0.

(c) $g_n(x)$ pointwise convergent to x , ∀ x∈[0, ∞)

(d) $g_n(x)$ pointwise convergent to $\frac{x}{1+x^2}$ ∀ x∈ [0, ∞)

Ans. **(a)**

Explanation :

PWL $g(x) = \{0 \quad x \in [0, \infty)$

Mn – Test

$$M_n = \lim_{n\to\infty} \sup_{x\in[0,\infty)} \left| \frac{nx}{1+n^2x^2} - 0 \right|$$

where $g(x) = \frac{nx}{1+n^2x^2}$

we have to check $\sup\limits_{x\in[0,\infty)}$ for maximum value of $\frac{nx}{1+n^2x^2}$.

$$g'(x) = \frac{n(1+n^2x^2) - nx(2n^2x)}{(1+n^2x^2)^2} = \frac{n+n^3x^2 - 2n^3x^2}{(1+n^2x^2)^2} = \frac{n-n^3x^2}{(1+n^2x^2)^2}$$

$$g'(x) = \frac{n(1-n^2x^2)}{(1+n^2x^2)^2} = 0 \text{ at } x = \pm\frac{1}{n}$$

$$\text{so } M_n = \lim_{n\to\infty} \left| \frac{n\left(\frac{1}{n}\right)}{1+n^2\left(\frac{1}{n^2}\right)} \right| = \frac{1}{2} \Rightarrow M_n \to \frac{1}{2}$$

not uniformly convergent

Option (c) is incorrect.

$g_n(x) \to 0$

Option (d) is incorrect.

7. The series $\sum_{n=1}^{\infty} 50\frac{\sin nx}{n(n+1)}$ is

(a) uniformly convergent (b)not uniformly convergent

(c) uniformly convergent only in closed interval (d)None of these

Ans. (a)

Explanation :

As $\left|50\frac{\sin nx}{n(n+1)}\right| \le \frac{50}{n(n+1)}$

Let $M_n = \frac{50}{n(n+1)}$ then $\sum M_n$ is convergent in any interval.

$\therefore$ By Weierstrass-M-test, it is uniformly convergent in any interval.

8. Let $f_n, g_n : [0,1] \to R$ be defined by $f_n(x) = x^2(1-x^2)^{n-1}$ and $g_n(x) = \frac{1}{n^2(1+x^2)}$

forn∈NThen, on [0, 1].

(a) $\sum_{n=1}^{\infty} f_n(x)$ converges uniformly but not $\sum_{n=1}^{\infty} g_n(x)$

(b) $\sum_{n=1}^{\infty} g_n(x)$ converges uniformly but not $\sum_{n=1}^{\infty} f_n(x)$

(c) both $\sum_{n=1}^{\infty} f_n(x)$ and $\sum_{n=1}^{\infty} g_n(x)$ converge uniformly

(d)neither $\sum_{n=1}^{\infty} f_n(x)$ nor $\sum_{n=1}^{\infty} g_n(x)$ converges uniformly

Ans. (b)

Explanation :

$\sum_{n=1}^{\infty} f_n(x)$ is not uniformly convergent on [0,1], $\sum_{n=1}^{\infty} g_n(x)$ is uniformly convergent

$\because f_n(x) = x^2(1-x^2)^{n-1}$. Let $F_n(x)$ be sequence of partial sums of $\sum_{n=1}^{\infty} f_n(x)$

$$\Rightarrow F_n(x)=x^2\left[1+(1-x^2)+(1-x^2)^2+...+(1-x^2)^n\right]$$

$$=x^2\left[\frac{1-(1-x^2)^{n+1}}{1-(1-x^2)}\right]=1-(1-x^2)^{n+1}$$

$$f(x)=\lim_{n\to\infty}\left(1-(1-x^2)^{n+1}\right)=\begin{cases}0 & for\ x=0\\ 1 & for\ 0<x\le 1\end{cases}$$

Since sequence function f_n is continuous throughout [0, 1] $\forall n$, but limit function f is discontinuous at $x=0 \Rightarrow \sum f_n(x)$ is not uniformly convergent on [0,1]

Now, $|g_n(x)|=\left|\frac{1}{n^2(1+x^2)}\right|\le\left|\frac{1}{n^2}\right|\forall n\in\text{N}$ and $\sum_{n=1}^{\infty}\frac{1}{n^2}$ converges

$\Rightarrow$ By Weierstrass M-Test, $\sum_{n=1}^{\infty} g_n(x)$ converges uniformly to g.

9. Let $S=\left\{x\in R: x\ge 0, \sum_{n=1}^{\infty} x^{\sqrt{n}}<\infty\right\}$. Then the supremum of S is

(a) 1 (b) 1/e (c) 0 (d) ∞

Ans. (a)

Explanation :

Given $S=\left\{x\in R: x\ge 0, \sum_{n=1}^{\infty} x^{\sqrt{n}}<\infty\right\}$. Let $u_n=x^{\sqrt{n}}$

$$\lim_{n\to\infty} u_n=\lim_{n\to\infty} x^{\sqrt{n}}=\begin{cases}0 & if\ x<1\\ 1 & if\ x=1\\ \infty & if\ x>1\end{cases}$$

So, S = [0, 1). Hence supremum of S is 1

10. If $f_n(x)=x^n$ for all in [0, 1), then the sequence $\{f_n\}$

(a) convergepointwise to the zero function on [0, 1)

(b) converge uniformly on [0, 1) (c) does not converge pointwise

(d) converge pointwise to a discontinuous function

Ans. (a, b)

Explanation :

Given that $f_n(x)=x^n$ $x\in[0,1)$ Then point wise limit $f(x)=0$ $\forall x\in[0,1)$

M_n test

$$\lim_{n\to\infty} M_n=\lim_{n\to\infty}\sup_{x\in[0,1)}\{x^n\}=0$$

$\Rightarrow f_n(x)$ is uniformly convergent on [0, 1)

REAL ANALYSIS

UNSOLVED QUESTIONS

1. For $n \geq 1$ let $f_n(x) = xe^{-nx^2}$ $x \in R$

Then the sequence <f_n> is **CSIR NET JUNE 2017**

(a) uniformly convergent on R
(b) uniformly convergent only on compact subset of R
(c) bounded and not uniformly convergent on R
(d) a sequence of unbounded functions.

2. Let $f_n(x) = \dfrac{1}{1+n^2x^2}$ for $n \in N$, $x \in R$. Which of the following are TRUE?

CSIR NET JUNE 2017

(a) f_n converge pointwise on [0, 1] to a continuous function.
(b) f_n converge uniformly on [0, 1]
(c) f_n converge uniformly on $\left[\frac{1}{2},1\right]$ (d) $\lim\limits_{n\to\infty}\int_0^1 f_n(x)dx = \int_0^1\left(\lim\limits_{n\to\infty} f_n(x)\right)dx$

3. Which one of the following(s) statement is/ are

(a) The series $\sum_{n=0}^{\infty} x^n$ converges for each $x \in [-1,1]$

(b) The series $\sum_{n=0}^{\infty} x^n$ uniformly converges for each $x \in [-1,1]$

(c) The series $\sum_{n=1}^{\infty} \frac{x^n}{n}$ converges for each $x \in [-1,1]$

(d) The series $\sum_{n=1}^{\infty} \frac{x^n}{n^2}$ uniformly converges for each $x \in [-1,1]$

4. For $n \geq 1$, consider the sequence of function $f_n(x) = \dfrac{1}{2nx+1}, g_n(x) = \dfrac{x}{2nx+1}$ on the open interval (0, 1). Consider the statement. **CSIR NET DEC 2017**

(I) The sequence <f_n> converges uniformly on [0, 1]
(II) The sequence <g_n> converges uniformly on (0, 1)

Then

(a) (I) is true (b) (I) is false (c) (I) is false & (II) is true (d) Both are true

5. Define $f_1, f_2 : [0, 1] \to R$ by $f_1(x) = \sum_{n=1}^{\infty} \dfrac{x\sin(n^2x)}{n^2}$ and $f_2(x) = \sum_{n=1}^{\infty} x^2(1-x^2)^{n-1}$. Then

GATE-2015

(a) f_1 is continuous but f_2 is NOT continuous
(b) f_2 is continuous but f1 is NOT continuous
(c) both f_1 and f_2 are continuous (d) neither f_1 nor f_2 is continuous

6. Find out which of the following series converges uniformly for $x \in (-\pi, \pi)$

CSIR NET JUNE 2016

(a) $\sum_{n=1}^{\infty} \frac{e^{-n|x|}}{n^3}$ (b) $\sum_{n=1}^{\infty} \frac{\sin x}{n^5}$ (c) $\sum_{n=1}^{\infty} \left(\frac{x}{n}\right)^n$ (d) $\sum_{n=1}^{\infty} \frac{1}{[(x+\pi)n]^2}$

7. Let $\{f_n(x)\}$, where $f_n(x) = e^{-nx}$, be a sequence of real valued function. Then

(a) $\{f_n(x)\}$ ispointwise convergent in [0, k], k > 0

(b) $\{f_n(x)\}$ is uniformly convergent in [0, k], k > 0

(c) $\{f_n(x)\}$ is not uniformly convergent in [a, b], a > 0, b > 0 (d) None of these

8. Which of the following sequence doesn't converge uniformly?

(a) $\{f_n(x)\}$ where $f_n(x) = e^{-nx}$, $x \geq 0$

(b) $\{f_n(x)\}$ where $f_n(x) = nx.e^{-nx^2}$, $x \in [0,1]$

(c) $\{f_n(x)\}$ where $f_n(x) = 1-(1-x^2)^n$, $x \in [0, \sqrt{2}]$

(d) $\{f_n(x)\}$ where $f_n(x) = \tan^{-1} nx$, $x \geq 0$

9. Let $f_n(x) = \frac{x^n}{1+x}$ and $g_n(x) = \frac{x^n}{1+nx}$ for $x \in [0, 1]$ and $n \in N$. Then on the interval [0, 1]

GATE - 2010

(a) both $\{f_n\}$ and $\{g_n\}$ converges uniformly

(b) neither $\{f_n\}$ nor $\{g_n\}$ converges uniformly.

(c) $\{f_n\}$ converges uniformly but $\{g_n\}$ does not converge uniformly.

(d) $\{g_n\}$ converges uniformly but $\{f_n\}$ does not converge uniformly.

10. Consider the power series $\sum_{n=1}^{\infty} \frac{x^n}{\sqrt{n}}$ and $\sum_{n=1}^{\infty} \frac{x^n}{n}$. Then

GATE – 2010

(a) Both converge on (-1, 1] (b) Both converge on [-1, 1)

(c) Exactly one of them converges on (-1, 1] (d) None of them converges on [-1, 1)

11. Consider all sequences $\{f_n\}$ of real valued continuous functions on $[0, \infty)$. Identify which of the following statements are correct.

CSIR NET DEC 2015

(a) If $\{f_n\}$ converges to f pointwise on $[0, \infty)$, then $\lim_{n\to\infty} \int_0^{\infty} f_n(x)dx = \int_0^{\infty} f(x)dx$.

(b) If $\{f_n\}$ converges to f uniformly on $[0, \infty)$, then $\lim_{n\to\infty} \int_0^{\infty} f_n(x)dx = \int_0^{\infty} f(x)dx$

(c) If $\{f_n\}$ converges to f uniformly on $[0,\infty)$,then f is continuous on $[0,\infty)$

(d) There exist a sequence of continuous functions $\{f_n\}$ which converge to f uniformly on $[0,\infty)$ but $\lim_{n\to\infty}\int_0^\infty f_n(x)dx \neq \int_0^\infty f(x)dx$

12. Let $\{f_n\}$ be a sequence of continuous real valued function define on $[0,\infty)$. Suppose $f_n(x)\to f(x)$ for all $x\in[0,\infty)$ and that f is integrable. Then

(a) $\int_0^\infty f_n(x)dx \to \int_0^\infty f(x)dx$ as $n\to\infty$

(b) If $f_n\to f$ uniformly on $[0,\infty)$ then $\int_0^1 f_n(x)dx \to \int_0^1 f(x)dx$

(c) If $f_n\to f$ uniformly on $[0,\infty)$ then $\int_0^\infty f_n(x)dx \to \int_0^\infty f(x)dx$

(d)If $\int_0^1 |f_n(x)-f(x)|\,dx \to 0$ and if $f_n\to f$ uniformly on $[0,\infty)$ then $f_n \to f$ uniformly on $[0,1]$

13. Let $\{b_n\}$ and $\{c_n\}$ be sequences of real numbers. Then a necessary and sufficient condition for the sequence of polynomials $f_n(x) = b_nx + c_nx^2$ to converge uniformly to 0 on the real line is **CSIR NET DEC 2014**

(a) $\lim_{n\to\infty} b_n = 0$ and $\lim_{n\to\infty} c_n = 0$

(b) $\sum_{n=1}^{\infty}|b_n| < \infty$ and $\sum_{n=1}^{\infty}|c_n| < \infty$

(c) There exists a positive integer N such that $b_n = 0$ and $c_n = 0$ for all $n > N$

(d) $\lim_{n\to\infty} c_n = 0$

14. Let $f_n(x) = (-x)^n$, $x \in [0, 1]$. Then decide which of the following are true.

(a) There exist a pointwise convergent subsequence of f_n.

(b) f_n has no pointwise convergent subsequence.

(c) f_n converges pointwise everywhere

(d) f_n has exactly one pointwise convergent subsequence.

15. Let $\{f_n\}$ be a sequence of continuous functions on R

(a) If $\{f_n\}$ converges to f pointwise on R then $\lim_{n\to\infty}\int_{-\infty}^{\infty} f_n(x)dx = \int_{-\infty}^{\infty} f(x)dx$

(b) If $\{f_n\}$ converges to f uniformly on R then $\lim_{n\to\infty}\int_{-\infty}^{\infty} f_n(x)dx = \int_{-\infty}^{\infty} f(x)dx$

(c) If $\{f_n\}$ converges to f uniformly on R then f is continuous on R

(d) There exists a sequence of continuous functions $\{f_n\}$ on R, such that $\{f_n\}$ converges to f uniformly on R, but $\lim_{n\to\infty}\int_{-\infty}^{\infty} f_n(x)dx \neq \int_{-\infty}^{\infty} f(x)dx$

16. Let $f_n(x) = x^{1/n}$ for $x \in [0, 1]$. Then

(a) $\lim_{n\to\infty} f_n(x)$ exists for all $x \in [0, 1]$

(b) $\lim_{n\to\infty} f_n(x)$ defines a continuous function on [0, 1]

(c) $\{f_n\}$ converges uniformly on [0, 1]

(d) $\lim_{n\to\infty} f_n(x) = 0$ for all $n \in [0, 1]$

17. Let $f_n(x) = \begin{cases} 1 - nx; & for \quad x \in \left[0, \frac{1}{n}\right] \\ 0; & for \quad x \in \left[\frac{1}{n}, 1\right] \end{cases}$. Then

(a) $\lim_{n\to\infty} f_n(x)$ defines a continuous function on [0, 1]

(b) $\{f_n\}$ converges uniformly on [0, 1]

(c) $\lim_{n\to\infty} f_n(x) = 0$ for all $x \in [0, 1]$

(d) $\lim_{n\to\infty} f_n(x)$ exists for all $x \in [0, 1]$

18. Let $f_n : [1, 2] \to [0, 1]$ be given by $f_n(x) = (2 - x)^n$ for all non-negative integers n. Let $f(x) = \lim_{n\to\infty} f_n(x)$ for $1 \le x \le 2$. Then which of the following is true?

(a) f is a continuous function on [1, 2]

(b) f_n converges uniformly to f on [1, 2] as $n \to \infty$

(c) $\lim_{n\to\infty} \int_1^2 f_n(x)dx = \int_1^2 f(x)dx$

(d) For any $a \in (1, 2)$ we have $\lim_{n\to\infty} f_n'(a) \neq f'(a)$.

Answer Key

1	2	3	4	5	6	7	8
a	c,d	d	b,c	a	a,b,c	a,b	a,b,c,d
9	10	11	12	13	14	15	16
d	b	c,d	b	c	a	c,d	a
17	18						
d	c						

REAL ANALYSIS

Chapter 7

Function of Several Variables

REAL ANALYSIS

Let $S \subseteq R^n$, then a map $f : S \to R$ is called real valued function on n variables.
Example : Let $f : R^2 \to R$ s.t. $f(x, y) = x + y$

Example : Let $f: R^2 \to R$ s.t. $f(x,y)=\begin{cases}\dfrac{xy}{x^2+y^2}; & (x,y)\neq(0,0)\\ 0 & (x,y)=(0,0)\end{cases}$

1. Definition :

Neighbourhood of a point: Let $(a, b) \in R^2$ be a point.
The set $N(a, b) = \{(x, y) : x \in R, y \in R, a - h < a < a + h, b - k < b < b + k\}$ is called a rectangular neighbourhood of the point (a,b), where h and k are arbitrary small positive real numbers.

1.2 Limit Point: A point (a, b) is called a limit point of a set of points $S \subseteq R^2$, if every neighbourhood (a, b) contains an infinite number of points of S. The limit point is also called condensation point.

For example: The point (0, 0) is a limit point of the set $\left\{\frac{1}{m},\frac{1}{n}: m,n\in R\right\}$

1.3 Limit of a function of two variables:
Let $S \subseteq R^2$and $f : S \to R$be a real valued function and $(a, b) \in R^2$ be a limit point of S.
Then $f(x,y)\to l$, as$(x,y)\to(a,b)$,or $\lim\limits_{(x,y)\to(a,b)} f(x,y)=l$ if there is a point $l\in$ Rwith the following property:
For every $\in>0$, there exists a $\delta>0$ such that $|f(x,y)-l|<\varepsilon$, for all $(x,y)\in S$ for which $|x-a|<\delta,|y-b|<\delta$

Example If $f(x,y)=\begin{cases}3x+4y, & (x,y)\neq\left(e,\frac{1}{e}\right)\\ 6, & elsewhere\end{cases}$

Then, $\lim\limits_{(x,y)\to\left(e,\frac{1}{e}\right)} f(x,y)$ is

(a) $e(3e^2+4)$ (b) $e^{-1}(3e^2+4)$ (c) $e(3+4e^2)$ (d) $e^{-1}(3+4e^2)$

Solution: (b)

We have, $\lim_{(x,y)\to\left(e,\frac{1}{e}\right)} f(x,y) = 3e + 4\frac{1}{e} = e^{-1}\left(3e^2+4\right)$

Non-existence of Limit:

If we can find two functions (commonly known as paths) $\phi_1(x)$ and $\phi_2(x)$ such that the limits of $f(x,\phi_1(x))$ and $f(x,\phi_2(x))$ are different, then the simultaneous limit of $f(x,y)$ does not exist.

Example Show that $\lim_{(x,y)\to(0,0)} \frac{2xy^2}{x^2+y^2}$ does not exist.

Solution: Suppose that the point (x,y) approaches (0,0) along $x = my^2$ and let y→0, then

$\lim_{y\to 0} f\left(my^2, y\right) = \lim_{y\to 0} \frac{2my^4}{\left(m^2+1\right)y^2} = \frac{2m}{1+m^2}$, which is different for different values of m.

Convert function from cartisian form to polar form :

Let f(x, y) is a function of two variable in cartisian form, then put x = r cos θ& y = r sin θ
Then f(x, y) convert into f(r, θ) which is called in polar form.

Example : Let $f(x,y) = \frac{xy^2}{x^2+y^2}$

Put x = r cos θ& y = r sin θ

Then $f(r,\theta) = \frac{r\sin^2\theta\cos\theta}{\sin^2\theta + \cos^2\theta}$

$= \frac{r\cos\theta\sin^2\theta}{1}$

Bounded function :

A function of two variables f(x, y) is said to be bounded iff ∃ M > 0 s.t. |f(x, y)| ≤ M; for all (x, y)

2. Continuity of a function at a point:

Let S ⊆ R² and $(a,b) \in S$ and f : S → Rbe a real valued function of two variables, then f is said to be continuous at point $(a,b) \in S$, if $\lim_{(x,y)\to(a,b)} f(x,y) = f(a,b)$

Example Prove that the function f : R²→ R defined by $f(x,y) = \frac{\left(y^2-x^2\right)xy}{x^2+y^2}$ for $f(x,y) \neq (0,0)$ and f (0, 0) = 0 is continuous at the point (0, 0).

Solution: Firstly, we show that $\lim_{(x,y)\to(0,0)} f(x,y) = 0$. For given $\in > 0$ and $(x,y) \neq (0,0)$, we have

$$|f(x,y)-0| = \left|\frac{(y^2-x^2)xy}{x^2+y^2}\right| = |y^2-x^2|.\frac{|xy|}{x^2+y^2} \quad ...(i)$$

Now, for all x and y, we have $(|x|-|y|)^2 \geq 0$ implies $|x|^2+|y|^2-2|x||y| \geq 0 \Rightarrow x^2+y^2 \geq 2|xy|$

implies $\frac{|xy|}{x^2+y^2} \leq \frac{1}{2}$, if $(x,y) \neq (0,0)$ ...(ii)

Using equations (i) and (ii) for all $(x,y) \neq (0,0)$, we have

$|f(x,y)-0| \leq \frac{1}{2}|y^2-x^2| \leq \frac{1}{2}|y^2+x^2| = \frac{1}{2}(x^2+y^2) < \in$

if $x^2 < \in, y^2 < \in$, i.e., $|x| < \sqrt{\in}, |y| < \sqrt{\in}$

Thus if we take $\delta = \sqrt{\in}$, then for any given $\in > 0, \exists \delta > 0$ such that $|f(x,y)-0| < \in$, whenever

$|x| < \delta$ and $|y| < \delta$

Since $\lim\limits_{(x,y)\to(0,0)} f(x,y) = 0$ and it is given that $f(0, 0) = 0$

$\Rightarrow \lim\limits_{(x,y)\to(0,0)} f(x,y) = f(0,0)$, i.e., $f(x,y)$ is continuous at the origin.

3. Directional Derivative:

Let f be a function of n-variables defined on D and let u = $(u_1, u_2, ..., u_n)$ be a unit vector then the directional derivative of f at $\alpha = (x_1, x_2, ..., x_n)$ in the direction of u denoted by $D_u f(x_1, x_2, ..., x_n)$ is defined by $D_u f(\alpha) = \lim\limits_{h\to 0} \frac{f(\alpha+hu)-f(\alpha)}{h}$ if limit exist,

Where $\alpha = (x_1, x_2, ..., x_n)$, $u = (u_1, u_2, ..., u_n)$

$$\alpha + hu = (x_1 + hu_1, x_2 + hu_2 ..... x_n + hu_n)$$

Example : Let $f(x) = \begin{cases} \frac{x^2 y}{x^2+y^2}; & (x,y) \neq (0,0) \\ 0; & (x,y) = (0,0) \end{cases}$

Then direction derivative of f(x, y) at (0, 0) in the direction (1, 1)

i.e. $\quad D_{(0,0)} f(1,1) = \lim\limits_{h\to 0} \frac{f(\alpha+hu)-f(\alpha)}{h}$

Given that $\quad \alpha = (0,0)$ & $u = (1, 1)$

So, unit vector of $\vec{u} = \left(\frac{1}{\sqrt{2}}, \frac{1}{\sqrt{2}}\right)$.

Now $\quad \alpha + hu = \left(\frac{h}{\sqrt{2}}, \frac{h}{\sqrt{2}}\right) \qquad \lim_{h\to 0} \dfrac{f\left(\frac{h}{\sqrt{2}}, \frac{h}{\sqrt{2}}\right) - f(0,0)}{h}$

$$\Rightarrow \quad \lim_{h\to 0} \frac{\frac{h^2.h}{2\sqrt{2}}}{\frac{h^2}{2}+\frac{h^2}{2}} \cdot \frac{1}{h} \quad \Rightarrow \quad \lim_{h\to 0} \frac{h^3}{2\sqrt{2}.h^3} = \frac{1}{2\sqrt{2}}$$

4. Partial derivative :

(i) **Partial derivative w.r.t. x :** Directional derivative of f(x, y) at point p(a, b) in the direction of u = (1, 0). It is denoted by $f_x(a, b)$. i.e.,

$$f_x(a,b) = \lim_{h\to 0} \frac{f(a+h,b) - f(a,b)}{h}$$

(ii) **Partial derivative w.r.t. y :** Directional derivative of f(x, y) at point p(a, b) in the direction of u = (0, 1). It is denoted by $f_y(a, b)$. i.e.,

$$f_y(a,b) = \lim_{h\to 0} \frac{f(a,b+h) - f(a,b)}{h}$$

Example: If $f(x,y) = \begin{cases} \dfrac{x^3+y^3}{x-y}, & x \neq 0 \\ 0, & x = y \end{cases}$, then show that f is discontinuous at the origin but the partial derivative exists at the origin.

Solution: Suppose, $(x,y)\to(0,0)$ through the curve y $= x - mx^3$, we get

$$\lim_{(x,y)\to(0,0)} f(x,y) = \lim_{(x,y)\to(0,0)} f\left(x, x-mx^3\right) = \lim_{x\to 0} \frac{x^3 + \left(x-mx^3\right)^3}{x - \left(x - mx^3\right)}$$

$$= \lim_{x\to 0} \frac{x^3\left\{1+\left(1-mx^2\right)^3\right\}}{mx^3} = \lim_{x\to 0} \frac{1}{m}\left\{2 - m^3x^6 - 3mx^2 + 3m^2x^4\right\}$$

$$= \lim_{x\to 0} \left\{\frac{2}{m} - m^2x^6 - 3x^2 + 3mx^4\right\} = \frac{2}{m}$$

which depends on m, therefore the simultaneous limit does not exist. Hence, the function $f(x,y)$ is discontinuous at the origin.

Now, we have $f_x(0,0)=\lim_{h\to 0}\dfrac{f(0+h,0)-f(0,0)}{h}=\lim_{h\to 0}\dfrac{\frac{h^3}{h}-0}{h}=\lim_{h\to 0}h=0$ and

$$f_y(0,0)=\lim_{k\to 0}\frac{f(0,0+k)-f(0,0)}{k}=\lim_{k\to 0}\frac{\frac{k^3}{-k}-0}{k}=0$$

Clearly, partial derivatives exist at the origin.

5. Differentiability of function of several variables:

Let f : D → R be a real valued function defined on an open set $D \subset R^2$, then f is said to be differentiability at a point (a, b) ∈ D. If $f(a + h, b + k) - f(a, b) = Ah + Bk + \sqrt{h^2+k^2}\,\phi(h, k)$

Where A & B are partial derivative i.e. $A = f_x(a, b)$ & $B = f_y(a, b)$ s.t. $\lim_{(h,k)\to(0,0)}\phi(h,k)=0$

Equivalently, f is differentiable at (a, b) if

$f(a + h, b + k) - f(a, b) = Ah + Bk + h\phi_1(h, k) + k\phi_2(h,k)$
where $A = f_x(a, b)$ & $B = f_y(a, b)$
s.t. $\lim_{(h,k)\to(0,0)}\phi_1(h,k)=\lim_{(h,k)\to(0,0)}\phi_2(h,k)=0$

Example Let f(x, y) = |xy|; (x, y) ∈ R^2

We know that

$$f_x(0,0)=\lim_{h\to 0}\frac{f(h,0)-f(0,0)}{h}=0 \text{ and } f_y(0,0)=\lim_{h\to 0}\frac{f(0,h)-f(0,0)}{h}=0$$

Now $f(h,k)-f(0,0)=\sqrt{h^2+x^2}\,\phi(h,k)$

$|hk|=\sqrt{h^2+k^2}\,\phi(h,k)$

Taking path k = mh

So, $\lim_{h\to 0}\phi(h,k)=\lim_{h\to 0}\dfrac{|mh^2|}{h\sqrt{1+m^2}}=0$

Note:

(1) A necessary condition for a function $f(x,y)$ to be differentiable at any point (a,b) partial derivatives of $f(x,y)$ exist at (a,b).

(2) If $f(x,y)$ is differentiable at (a,b), then it is also continuous at (a,b) but vice-versa need not be true. Also $f(x,y)$ is not continuous $\Rightarrow f(x,y)$ is not differentiable

6. Functions of several variables as a transformation:

If a function of several variables is transformed into a set of functions, then the transformation can also be expressed as $f(x_1,x_2,....,x_n)=(\phi_1(x_1,x_2,...,x_n),....,\phi_m(x_1,x_2,....,x_n))$

Jacobian:

Let f : D $\subseteq$ $R^n \to R^m$ defined by f(x_1, x_2, ..., x_n) = (ϕ_1, ϕ_2, ..., ϕ_m) where $\phi_i : R^n \to R$ is a real valued function, where $1 \le i \le m$. Then matrix of derivative at x = c is denoted by D(f(c)) and it is defined as

$$D(f(c)) = \begin{vmatrix} \frac{\partial \phi_1}{\partial x_1} & \frac{\partial \phi_1}{\partial x_2} & \cdots & \frac{\partial \phi_1}{\partial x_n} \\ \vdots & \vdots & \vdots & \vdots \\ \frac{\partial \phi_m}{\partial x_1} & \frac{\partial \phi_m}{\partial x_2} & \cdots & \frac{\partial \phi_m}{\partial x_n} \end{vmatrix}_{m\times n} = (a_{ij}) \qquad a_{ij} = \frac{\partial \phi_i}{\partial x_j}\begin{pmatrix} 1 \le i \le m \\ 1 \le j \le n \end{pmatrix}$$

Example : f : $R^2 \to R^3$

$$f(x, y) = (\underset{\phi_1}{x+y}, \underset{\phi_2}{x-y}, \underset{\phi_3}{x.y}) \quad \text{at } c = (1, 1)$$

$$D(f(c)) = \begin{vmatrix} 1 & 1 \\ 1 & -1 \\ y & x \end{vmatrix}_{(1,1)} = \begin{vmatrix} 1 & 1 \\ 1 & -1 \\ 1 & 1 \end{vmatrix}$$

Inverse function theorem:

If $f(x)$ is continuous differentiable function at any point c and derivative map of $f'(x)$ at c, i.e., [Df(c)] is invertible, then function is invertible in some neighbourhood of c

Example Consider a function f : $R^2 \to R^2$ given by $f(x, y) = (x^2 - y^2, 2xy)$

Solution:

$|Df| = 4(x^2 + y^2) \neq 0$ if (x, y) $\neq$ (0, 0)and is equal to zero if (x, y) = (0, 0). If we take any point in R^2except (0, 0), then $\exists$ some neighbourhood of that point in which $f(x, y)$ is invertible.

7. Extreme Values:

Definition: A point $(a_1, a_2, ..., a_n)$is said to be an extreme point and $f(a_1, a_2, ..., a_n)$an extreme value of a function f : $R^n \to R$ if for every point$(x_1, x_2, ..., x_n)$, other than $(a_1, a_2, ..., a_n)$of some neighbourhood of $(a_1, a_2, ..., a_n)$, the difference $f(x_1, x_2, ..., x_n) - f(a_1, a_2, ..., a_n)$keeps the same sign. The extreme value is minimum or maximum according as the sign is positive or negative.

Necessary Condition for Extreme Values:

The necessary condition for $f(a_1, a_2, ..., a_n)$to be an extreme value of the function f is that the partial derivatives $f_{x_1}, f_{x_2}, ..., f_{x_n}$ exists in any case. If the partial derivatives of vanish at

$(a_1, a_2, ..., a_n)$ then the point $(a_1, a_2, ..., a_n)$ is called a stationary point. Thus, the stationary points are determined by solving the following n equations simultaneously.

$$f_{x_1}(x_1, x_2, ..., x_n) = 0$$
$$f_{x_2}(x_1, x_2, ..., x_n) = 0$$
$$\vdots$$
$$f_{x_n}(x_1, x_2, ..., x_n) = 0$$

For a function f of nindependent variables $(x_1, x_2, ..., x_n)$ the condition can be given in a more compact form. Hence, if $(a_1, a_2, ..., a_n)$ is a stationary point, then, $df(a_1, a_2, ..., a_n) = 0$

$\Rightarrow f_{x_1}(a_1, a_2, ..., a_n)dx_1 + f_{x_2}(a_1, a_2, ..., a_n)dx_2 + ... + f_{x_n}(a_1, a_2, ..., a_n)dx_n = 0$

$\Rightarrow$ The coefficients of the differential $dx_1, dx_2, ..., dx_n$ of independent variables are separately equal to zero

Sufficient condition for extreme values:

For function $f(x, y)$ of two independent variables sufficient conditions for (a, b) to be an extreme point are that

(i) $df(a,b) = f_x dx + f_y dy = 0 \Rightarrow f_x = f_y = 0$ and

(ii) $d^2 f(a,b) = f_{xx}(dx)^2 + f_{yy}(dy)^2 + 2f_{xy}dxdy$, keep the same sign for arbitrary values of dx, dy; the extreme point being a maxima or minima according as the sign of $d^2 f$ is negative or positive. The point at which $d^2 f$ does not keep the same sign will be neither maxima nor minima.

Remark: The conditions for $d^2 f$ keeps the same sign may be stated in terms of matrices as follows

(i) $d^2 f$ will be always negative if and only if $f_{xx} > 0$ and $\begin{vmatrix} f_{xx} & f_{xy} \\ f_{yx} & f_{yy} \end{vmatrix} > 0$ and

(ii) $d^2 f$ will be always positive if and only if $f_{xx} < 0$ and $\begin{vmatrix} f_{xx} & f_{xy} \\ f_{yx} & f_{yy} \end{vmatrix} < 0$

Note: The points at which $f(x, y)$ has neither maxima nor minima are called saddle points.

Remark :

Let f : $R^2 \to R$ be a real valued function in two variables which posses continuous second order partial drivatives and $f_{xx}(a, b) = A$, $f_{xy}(a, b) = B$, $f_{yy}(a, b) = C$ then

(i) f has minimum at (a, b) if $AC - B^2 > 0$, $A > 0$

(ii) f has maximum at (a, b) if $AC - B^2 > 0$, $A < 0$.

(iii) f has neither maximum value nor minimum value at (a, b) if $AC - B^2 < 0$

(iv) f may or may not have an extremum at (a, b) if $AC - B^2 = 0$.
(This case is often called the doubtful case, which needs further investigation)

Additional Points:

(1) Let a function $f : S \to R$ (where $S \subseteq R$) is differentiable at $(a,b) \in S$, then f has partial derivatives f_x and f_y at (a,b).

(2) If either of f_x, f_y does not exist at (a,b), then f is not differentiable at (a,b)

(3) If a function $f : S \to R$ (where $S \subseteq R$) is differentiable at $(a,b) \in S$, then f is continuous at (a,b).

(4) A continuous function need not be differentiable.

(5) If a function $f : S \to R$ (where $S \subseteq R$) has continuous partial derivatives f_x and f_y at $(a,b) \in S$, then f is differentiable at (a,b).

(6) If a function f is not differentiable at (a,b), then f_x and f_y cannot be continuous at (a,b).

SOLVED QUESTIONS

1. Let $f : R^2 \to R$ be given by **GATE 2021**

$$f(x,y) = \begin{cases} \sqrt{x^2+y^2}\,\sin(y^2/x) & if x \neq 0 \\ 0 & if x = 0 \end{cases}$$

Consider the following statements:

P: f is continuous at (0,0) but f is NOT differentiable at (0, 0).

Q: The directional derivative $D_u f(0, 0)$ of f at (0, 0) exists in the direction of every unit vector $u \in R^2$.

Then

(a) Both P and Q are TRUE
(b) P is TRUE and Q is FALSE
(c) P is FALSE and Q is TRUE
(d) Both P and Q are FALSE

Ans. (a)

Explanation :

$$\lim_{(x,y)\to(0,0)} \sqrt{x^2+y^2}\,\sin\left(\frac{y^2}{x}\right)$$

Put $x = r\cos\theta$ and $y = r\sin\theta$

$$\Rightarrow \quad \lim_{r\to 0} r.\sin\left(r\frac{\sin^2\theta}{\cos\theta}\right) = 0 = f(0,0)$$

So f is continuous at (0, 0)

Note that $f_x(0, 0) = \lim_{h\to 0}\frac{f(h,0)-f(0,0)}{h} = \lim_{h\to 0}(0) = 0$

Similarly $f_y(0, 0) = \lim_{k\to 0}\frac{f(0,k)-f(0,0)}{k} = \lim_{k\to 0}(0) = 0$

For Differentiability

$f(0+h, 0+k) - f(0, 0) - hf_x - kf_y = \sqrt{h^2+k^2}\,\phi(h, k)$

$$\sqrt{h^2+k^2}\sin\left(\frac{k^2}{h}\right) = \sqrt{h^2+k^2}\,\phi(h,k)$$

$$\Rightarrow \quad \phi(h,k) = \frac{\sqrt{h^2+k^2}}{\sqrt{h^2+k^2}}\sin\left(\frac{k^2}{h}\right) = \sin\left(\frac{k^2}{h}\right)$$

Note that $\lim_{(h,k)\to(0,0)} \sin\left(\frac{k^2}{h}\right)$ doesn't exist choose path $h = mk^2$

So, f is not differentiable

Now let us calculate directional derivative

$$D_u f(0,0) = \lim_{h\to 0}\frac{f(0+hu_1, 0+hu_2 - f(0,0))}{h}$$

$$= \lim_{h\to 0}\frac{f(hu_1, hu_2)}{h} \qquad = \lim_{h\to 0}\frac{\sqrt{h^2u_1^2+h^2u_2^2}.\sin\frac{(hu_2)^2}{hu_1}}{h}$$

$$= \lim_{h\to 0}\frac{h\sqrt{u_1^2+u_2^2}}{h}\sin\left(\frac{hu_2^2}{u_1}\right) \qquad = \lim_{h\to 0}\sqrt{u_1^2+u_2^2}\sin\left(\frac{hu_2^2}{u_1}\right) = 0$$

So, directional derivative exist in every direction.

Hence, option (a) is true.

2. Define $f(x,y) = \begin{cases} \dfrac{x^3}{x^2+y^2} & for\ (x,y) \neq (0,0) \\ 0 & for\ (x,y) = (0,0) \end{cases}$.

Which of the following statements are true? **CSIR NET NOV 2020**

(a) f is discontinuous at (0, 0)

(b) f is continuous at (0, 0)

(c) All directional derivatives of f at (0, 0) exists.

(d) f is not differentiable at (0, 0)

Ans. (b), (c)

Explanation :

$$f(x,y)=\begin{cases}\dfrac{x^3}{x^2+y^2} & for\ (x,y)\neq(0,0)\\ 0 & for\ (x,y)=(0,0)\end{cases}$$

Let x = r cos θ, y = r sin θ

$\Rightarrow \quad f(r\cos\theta, r\sin\theta)=\dfrac{r^3\cos^3\theta}{r^2}=r\cos^3\theta$

as r → 0, r $\cos^3\theta$ → 0

Now $f_x(0,0)=\lim\limits_{h\to 0}\dfrac{\left[\dfrac{h^3}{h^2}-0\right]}{h}=1$ and $f_y(0, 0) = 0$

Let us check differentiability

$$\lim_{(h,k)\to(0,0)}\frac{f(0+h,0+h)-hf_x(0,0)-kf_y(0,0)}{\sqrt{h^2+k^2}}$$

$$=\lim_{(h,k)\to(0,0)}\frac{\dfrac{h^3}{h^2+k^2}-h}{\sqrt{h^2+k^2}}=\lim_{(h,k)\to(0,0)}\frac{(h^3-h^3-k^3h)}{\sqrt{h^2+k^2}}=\lim_{(h,k)\to(0,0)}\frac{-k^2h}{\sqrt{h^2+k^2}}=0$$

$\Rightarrow$ f(x, y) is differentiable at (0,0)

$\Rightarrow$ All directional derivative of f at (0, 0) exist.

3. Define $f(x,y)=\begin{cases}\dfrac{x^2-y^2}{x^2+y^2} & for\ (x,y)\neq(0,0)\\ 0 & for\ (x,y)=(0,0)\end{cases}$.

Which of the following statements are true? **CSIR NET NOV 2020**

(a) f is continuous at (0, 0) (b) f is bounded in a neighbourhood of (0, 0)

(c) f is not bounded in any neighbourhood of (0, 0) (d) f has all directional derivatives at (0, 0)

Ans. (b)

Explanation :

$$f(x,y)=\begin{cases}\dfrac{x^2-y^2}{x^2+y^2} & for\ (x,y)\neq(0,0)\\ 0 & for\ (x,y)=(0,0)\end{cases}$$

Along y = mx

$$f(x,mx)=\frac{x^2-m^2x^2}{x^2+m^2x^2}=\frac{1-m^2}{1+m^2}$$

REAL ANALYSIS

$\Rightarrow \quad \lim_{x\to 0} f(x,mx) = \frac{1-m^2}{1+m^2}$ [depends upon m]

$\Rightarrow$ Limit is not unique. $\Rightarrow$ f is not continuous at (0, 0)

$\Rightarrow$ Option (a) is not true.

Let $x = r\cos\theta,\ y = r\sin\theta$

$\Rightarrow$ $|f(r\cos\theta, r\sin\theta)|$

$$= \left|\frac{r^2[\cos^2\theta - \sin^2\theta]}{r^2}\right| = |\cos^2\theta - \sin^2\theta| \le |\cos^2\theta| + |\sin^2\theta| \le 1 + 1 = 2$$

$\Rightarrow$ f(x, y) is bounded. $\Rightarrow$ Option (c) is false and (b) is true.

(d) Let us find directional derivative in direction at (u_1, u_2)

$$\lim_{h\to 0}\frac{f(0+hu_1, 0+hu_2) - f(0,0)}{h}$$

$$= \lim_{h\to 0}\frac{\frac{h^2[u_1^2 - u_2^2]}{h^2[u_1^2 + u_2^2]}}{h} = \lim_{h\to 0}\frac{\frac{(u_1^2 - u_2^2)}{(u_1^2 + u_2^2)}}{h}$$ [Doesn't exist finitely]

$\Rightarrow$ Directional derivative doesn't exist.

$\Rightarrow$ Only option (b) is true.

4. Let $f : R^2 \to R$ be a function s.t. $f(x,y) = \begin{cases} \frac{xy}{x^2+y^2} & (x,y) \ne (0,0) \\ 0 & (x,y) = (0,0) \end{cases}$ Which of the following is/are TRUE? **CSIR NET DEC 2019**

(a) f(x, y) is continuous at (0, 0)

(b) $\int_0^1\int_0^1 f(x,y)\,dx\,dy$ is finite

(c) f is bounded in neighbourhood of (0,0)

(d) f is continuous at (0, 1)

Ans. (b), (c), (d)

Explanation :

Option (a) is incorrect.

We have to find $\lim_{(x,y)\to(0,0)} f$ along y = mx

$$\Rightarrow \lim_{x\to 0}\frac{mx^2}{x^2 + m^2x^2} = \frac{m}{1+m^2}$$

$\Rightarrow$ limit doesn't exist at (0, 0) $\Rightarrow$ function is not continuous at (0, 0)

Option (d) is correct.

$f(x,y) = \frac{xy}{x^2+y^2}$ at (0, 1)

f(x, y) is continuous at (0, 1)

Option (b) is correct.

$$\int_0^1\int_0^1 \frac{xy}{x^2 - y^2}dxdy$$

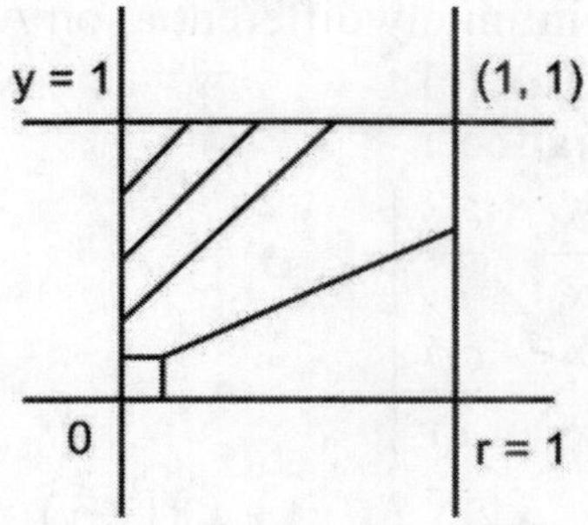

Let $\begin{matrix} x = r\cos\theta \\ y = r\sin\theta \end{matrix} \Rightarrow x^2 + y^2 = r^2$

Where dxdy = rdrd θ

$$\Rightarrow \int_r\int_\theta \frac{r^2 \sin\theta.\cos\theta}{r^2} rdrd\theta$$

$$\Rightarrow \int_r\int_\theta r\frac{\sin 2\theta}{2} drd\theta$$

θ = 0 to π/2 r = 0 to 1

$$\frac{1}{2}\int_0^1\int_0^{\pi/2} r\sin 2\theta d\theta dr \qquad \frac{1}{2}\int_0^1\left[\frac{-r\cos 2\theta}{2}\right]_0^{\pi/2} dr$$

$$\frac{1}{2}\int_0^1 -\frac{r}{2}\{\cos\pi - \cos 0\}dr = \frac{1}{2}\int_0^1\frac{r}{2}(2)dr \qquad \frac{1}{2}\left[\frac{r^2}{2}\right]_0^1 = \frac{1}{4}$$

Option (c) is correct.

$f(r,\theta) = \frac{\sin 2\theta}{2}$ is bounded nbd of (0, 0)

5. Consider the function $f(x,y) = \frac{x^2}{y^2}$ then the derivative of the function at (1, 1) along the direction (1, 1) is **CSIR NET JUNE 2016**

(a) 1 (b) 2 (c) -2 (d) 0

Ans. (d)

Explanation :

$$Du\, f(\alpha) = \lim_{h\to 0}\frac{f(\alpha + hu) - f(\alpha)}{h}$$

Here α = (1, 1) & u (1, 1)

α + hu = (1, 1) + h(1, 1) = (1 + h, 1 + h)

$$\text{Duf(x)} = \lim_{h\to 0}\frac{f(1+h,1+h) - f(1,1)}{h} = \lim_{h\to 0}\frac{\frac{(1+h)^2}{(1+h)^2} - \frac{1}{1}}{h} = 0$$

6. Let A = {(x, y) ∈ R^2 : x + y ≠ -1}. Define f : A → R^2

$f(x,y) = \left(\frac{y}{1+x+y}, \frac{x}{1+x+y}\right)$. Then **CSIR NET DEC 2015**

(a) The determinant of Jacobian of f does not vanish on A.
(b) f is infinitely differential on A (c) f is 1 – 1 (d) f(A) = R^2

Ans. (b), (c)

Explanation :

$$J=\begin{vmatrix}\frac{\partial\phi_1}{\partial x} & \frac{\partial\phi_1}{\partial y}\\ \frac{\partial\phi_2}{\partial x} & \frac{\partial\phi_2}{\partial y}\end{vmatrix} \qquad J=\begin{vmatrix}\frac{-y}{(1+x+y)^2} & \frac{1+x+y-y}{(1+x+y)^2}\\ \frac{1+x+y-x}{(1+x+y)^2} & \frac{-x}{(1+x+y)^2}\end{vmatrix}$$

$$J=\frac{xy}{(1+x+y)^4}-\frac{(1+x)(1+y)}{(1+x+y)^2} \qquad J=\frac{xy-x-y-xy-1}{(1+x+y)^4}=\frac{-1}{(1+x+y)^3}\Rightarrow J\neq 0 \text{ on A}$$

Option (b) is correct.

$J\neq 0$ on A $\Rightarrow$ f is infinitely differential on A.

Option (c) is correct. $J\neq 0$ on A $\Rightarrow$ f is 1 – 1

Option (d) is incorrect.

Let $f(x,y)=\left(\frac{x}{1+x+y},\frac{y}{1+x+y}\right)=(a,b)$

There does not exits preimage of $\left(\frac{1}{2},\frac{1}{2}\right)$.

7. Let f(x, y) = $\log(\cos^2(e^{x^2}))+\sin(x+y)$. Then $\frac{\partial}{\partial y}\frac{\partial}{\partial x}f(x,y)$ is

(a) $\frac{\cos(e^{x^2})-1}{1+\sin^2(e^{x^2})}-\cos(x+y)$ (b) 0 (c) – sin (x + y) (d) cos (x + y)

Ans. (c) **CSIR NET JUNE 2016**

Explanation :

$$\frac{\partial}{\partial y}.\frac{\partial}{\partial x}f(x,y)=\frac{-2\cos(e^{x^2})\sin(e^{x^2}).x2xe^{x^2}}{\cos^2(e^{x^2})}+\cos(x+y)$$

$$=\frac{\partial}{\partial y}.\frac{\partial}{\partial x}f(x,y)=0-\sin(x+y) \qquad = -\sin(x+y)$$

8. A function f(x, y) on R^2 has the following partial derivatives

$\frac{\partial f}{\partial x}(x,y)=x^2,\frac{\partial f}{\partial y}(x,y)=y^2.$

Then **CSIR NET DEC 2016**

(a) f has directional derivatives in all directions everywhere
(b) f has a derivative at all points
(c) f has directional derivative only along the direction(1, 1) everywhere

(d) f does not have directional derivatives in any direction everywhere

Ans. (a), (b)

Explanation :

As both partial derivatives of f(x, y) exists and are continuous, hence function is differentiable everywhere on R^2. $\Rightarrow$f has directional derivatives in all directions everywhere.

9. Let A be a matrix and $|A| \neq 0$. Define a function F(x, y) = <Ax, y> where <x, y> denoted the inner product. Let DF(x, y) denote the derivative of F at (x, y) which is a s.t. from $R^n \times R^n \to R$. Then **CSIR NET DEC 2018**

(a) If $x \neq 0$ then $DF(x, 0) \neq 0$ (b) If $y \neq 0$ then $DF(0, y) \neq 0$

(c) If $(x, y) \neq (0, 0)$ then $DF(x, y) \neq 0$ (d) If x = 0 or y = 0 then DF(x, y) = 0

Ans. (a), (b), (c)

Explanation :

Given that $|A| \neq 0$

We choose $A = \begin{bmatrix} a & b \\ c & d \end{bmatrix}$ then $F(x, y) = \left\langle \begin{bmatrix} a & b \\ c & d \end{bmatrix}\begin{bmatrix} x_1 \\ x_2 \end{bmatrix}, \begin{bmatrix} y_1 \\ y_2 \end{bmatrix} \right\rangle$

$= F(x, y) = (ax_1 + bx_2)y_1 + (cx_1 + dx_2)y_2$

DF(x, y) evaluated on (u, v), where $(u, v) = ((u_1, u_2), (v_1, v_2))$ is

$$\left(\frac{\partial F}{\partial x_1}\right)u_1 + \left(\frac{\partial F}{\partial x_2}\right)u_2 = \left(\frac{\partial F}{\partial y_1}\right)v_1 + \left(\frac{\partial F}{\partial y_1}\right)v_2$$

$= (ay_1 + cy_2)u_1 + (by_1 + dy_2)v_2 + (ax_1 + bx_2)v_1 + (cx_1 + dx_2)v_2.$

(a) $DF(x, 0) = (ax_1 + bx_2)v_1 + (cx_1 + dx_2)v_2$, which is zero iff $ax_1 + bx_2 = 0$ and $cx_1 + dx_2$. (v_1 and v_2 are arbitrary).

Have $\det\begin{bmatrix} a & b \\ c & d \end{bmatrix} \neq 0 \Leftrightarrow x_1 = 0, x_2 = 0$ i.e. x = 0.

Hence if $x \neq 0 \Rightarrow DF(x, 0) \neq 0$. Option (a) is correct.

(b) $DF(0, y) = (ay_1 + cy_2)u_1 + (by_1 + dy_2)v_2$ which is zero iff $ay_1 + cy_2 = 0$, $by_1 + dy_2 = 0$. Where $\det\begin{bmatrix} a & b \\ c & d \end{bmatrix} = \det\begin{bmatrix} a & b \\ c & d \end{bmatrix} \neq 0$.

$\Leftrightarrow y_1 = 0, y_2 = 0$ i.e. y = 0. Hence if $y \neq 0$, then $DF(0, y) \neq 0$.

Option (b) is correct.

(c) $DF(x, y) = (ay_1 + cy_2)u_1 + (by_1 + dy_2)v_2 + (ax_1 + bx_2)v_1 + (cx_1 + dx_2)v_2$ which is zero iff $ay_1 + cy_2 = 0$, $by_1 + dy_2 = 0$, $ax_1 + bx_2 = 0$ and $cx_1 + dx_2 = 0$.

Have $\det\begin{bmatrix} a & c \\ b & d \end{bmatrix} = \det\begin{bmatrix} a & b \\ c & d \end{bmatrix} \neq 0$

$\Leftrightarrow$ $y_1 = 0, y_2 = 0, x_1 = 0, x_2 = 0$

$\Leftrightarrow$ $x = 0, y = 0 \Leftrightarrow (x, y) = (0, 0)$

Hence if (x, y) ≠ (0, 0), then DF(x, y) ≠ 0.
Option (c) is correct.

(d) From above DF(x, y) = 0 ⇔ (x, y) = (0, 0). Hence if both x = 0, y = 0, then DF(x, y) = 0. So, option (d) is false.
Correct options are (a), (b) and (c).

10. Let f : $R^2 \to R$ be a function given by f(x, y) = ($x^3 + 3xy^2 - 15x - 12y$, x + y).
Let S = {(x, y) ∈ R^2; f is locally invertible at (x, y)}.Then
(a) S = R^2 / {0, 0} (b) S is open in R^2 (c) S is dense in R^2
(d) R^2 / S is countable.
Ans. (b), (c)
Explanation :

$$\text{Jacobian } J = \begin{vmatrix} 3x^2+3y^2-15 & 6xy-12 \\ 1 & 1 \end{vmatrix} \neq 0$$

Given that f is locally invertible.
J = $3x^2 + 3y^2 - 15 - 6xy + 12 \neq 0$ $= x^2 + y^2 - 2xy - 1 \neq 0$
$(x - y)^2 - 1 \neq 0$ $\Rightarrow x - y \neq 1$
$y \neq x - 1$
S = R^2 / A
Where A = {(x, y) | y = x – 1}
Option (b) and (c) are correct.

11. Let f : $R^2 \to R$ be define by $f(x,y) = \begin{cases} (x^2+y^2)\sin\left(\dfrac{1}{x^2+y^2}\right) & (x,y) \neq 0 \\ 0 & (x,y) = (0,0) \end{cases}$

Consider the following statement.
I. The partial derivative $\dfrac{\partial f}{\partial x}, \dfrac{\partial f}{\partial y}$ exist at (0, 0) but are unbounded in any nbd of (0, 0)
II. f is continuous but not differential at (0, 0)
III. f is continuous at (0, 0)
IV. f is differential at (0, 0)
Which of the following is/are TRUE?
(a) I & II only (b) I & IV only (c) IV only (d) III only
Ans. (a)
Explanation :

I. $\dfrac{\partial f}{\partial x}$ = partial derivative along direction (1, 0) at (0, 0)

$$= \lim_{h \to 0} \frac{f(h,0) - f(0,0)}{h}$$

$$= \lim_{h\to 0} \frac{h^2 \sin \frac{1}{h^2} - 0}{h} = \lim_{h\to 0} h \sin \frac{1}{h^2} = 0$$

$$\frac{\partial f}{\partial y} = \lim_{h\to 0} \frac{f(0,h) - f(0,0)}{h} \qquad = \lim_{h\to 0} \frac{h^2 \sin \frac{1}{h^2} - 0}{h} = 0$$

$\frac{\partial f}{\partial x}, \frac{\partial f}{\partial y}$ exist at (0, 0)

We have to check it is bounded or unbounded in the neighbourhood of (0, 0)

$$\lim_{(x,y)\to(0,0)} (x^2 + y^2) \sin \frac{1}{x^2 + y^2}$$

Along y = mx

$$\lim_{x\to 0} x^2(1+m^2) \sin \frac{1}{x^2(1+m^2)} = 0$$

But $x^2(1+m^2)\left[\frac{1}{x^2(1+m^2)} - \frac{1}{3![x^2(1+m^2)]^3} + ...\right]_{(0,0)}$

$= \infty$

$\Rightarrow$ This is unbounded of (0, 0)

II. $\lim_{(x,y)\to(0,0)} f(x,y) = 0 \quad f(0)$

f(x, y) is continuous.

& f(h, k) – f(0, 0) - hα - kβ

$$= (h^2 + k^2) \sin \frac{1}{h^2 + k^2} = \sqrt{h^2 + k^2}\, \phi(h,k)$$

Where $\phi(h,k) = \sqrt{h^2 + k^2} \sin \frac{1}{h^2 + k^2}$

$$\lim_{(h,k)\to(0,0)} = \lim_{(h,k)\to(0,0)} \sqrt{h^2 + k^2} \sin \frac{1}{h^2 + k^2}$$

Along k = mh

$$\sqrt{h^2 + m^2h^2} \sin \frac{1}{h^2 + m^2h^2} = h\sqrt{1+m^2} \sin \frac{1}{h^2 + m^2h^2} = 0$$

$\Rightarrow$ is differential at (0, 0)

12. Let $f(x,y)=kxy-x^3y-xy^3$ for (x, y) $\in R^2$, , where k is a real constant. The directional derivative of f at the point (1, 2) in the direction of the unit vector $u=\left(\frac{-1}{\sqrt{2}},\frac{-1}{\sqrt{2}}\right)$ is $\frac{15}{\sqrt{2}}$. The value of k is

(a) 2 (b) 4 (c) 1 (d) -2

Ans. (b)

Explanation :

$f(x,y)=kxy-x^3y-xy^3$

Directional derivative of f at the point (1, 2) in the direction of $u=\left(\frac{-1}{\sqrt{2}},\frac{-1}{\sqrt{2}}\right)$ is given by

$$\lim_{t\to 0}\frac{f(x+tu)-f(x)}{t-0}\Rightarrow \lim_{t\to 0}\frac{f\left((1,2)+t\left(-\frac{1}{\sqrt{2}},-\frac{1}{\sqrt{2}}\right)\right)-f(1,2)}{t}=\frac{15}{\sqrt{2}}$$

$$\Rightarrow \lim_{t\to 0}\frac{f\left(1-\frac{t}{\sqrt{2}},2-\frac{t}{\sqrt{2}}\right)-f(1,2)}{t}=\frac{15}{\sqrt{2}},$$

where

$$f\left(1-\frac{t}{\sqrt{2}},2-\frac{t}{\sqrt{2}}\right)=k\left(1-\frac{t}{\sqrt{2}}\right)\left(2-\frac{t}{\sqrt{2}}\right)-\left(1-\frac{t}{\sqrt{2}}\right)^3\left(2-\frac{t}{\sqrt{2}}\right)-\left(1-\frac{t}{\sqrt{2}}\right)\left(2-\frac{t}{\sqrt{2}}\right)^3$$

and $f(1,2)=(2k-2-2^3)$

$$\Rightarrow \lim_{t\to 0}\frac{\left(1-\frac{t}{\sqrt{2}}\right)\left(2-\frac{t}{\sqrt{2}}\right)\left(k-\left(1-\frac{t}{\sqrt{2}}\right)^2-\left(2-\frac{t}{\sqrt{2}}\right)^2\right)-(2k-10)}{t}=\frac{15}{\sqrt{2}}$$

$$\Rightarrow \lim_{t\to 0}\frac{\left(2-\frac{3t}{\sqrt{2}}+\frac{t^2}{2}\right)\left(k-\left(1-\frac{t}{\sqrt{2}}\right)^2-\left(2-\frac{t}{\sqrt{2}}\right)^2\right)-(2k-10)}{t}=\frac{15}{\sqrt{2}}$$

Using L' Hospital rule and solving, we get

$$2\left(0-2\left(-\frac{1}{\sqrt{2}}\right)-2.2\left(-\frac{1}{\sqrt{2}}\right)\right)+\frac{-3}{\sqrt{2}}(k-1-4)=\frac{15}{\sqrt{2}}$$

$$\Rightarrow 2\left(\frac{6}{\sqrt{2}}\right)-\frac{3}{\sqrt{2}}(k-5)=\frac{15}{\sqrt{2}}$$

$$\Rightarrow\left(\frac{12}{\sqrt{2}}\right)-\frac{3}{\sqrt{2}}(k)+\frac{15}{\sqrt{2}}=\frac{15}{\sqrt{2}}$$

$\Rightarrow$ k = 4

13. If $f(x,y)=4xy-x^3y-xy^3$ for (x, y) $\in$ R^2. The value of f at local minimum in the rectangular region $R=\left\{(x,y)\in R^2 : |x|<\frac{3}{2}, |y|<\frac{3}{2}\right\}$ is

(a) – 2 (b) – 3 (c) – 7/8 (d) 0

Ans. (a)

Explanation :

Using result, $f(x,y)$ attains a local minima at (a,b) if $f_x(a,b)=f_y(a,b)=0$ and $f_{xx}(a,b)f_{yy}(a,b)-f_{xy}(a,b)^2>0$ and $f_{xx}(a,b)$ or $f_{yy}(a,b)>0$

As $f(x,y)=4xy-x^3y-xy^3$

$\therefore f_x=4y-3x^2y-y^3\Rightarrow y=0$ or $3x^2+y^2=4$ and

$\therefore f_y=4x-x^3-3xy^2=0\Rightarrow x=0 \; or \; x^2+3y^2=4$

$\Rightarrow f$ may attains its local minima at $(0,0),(\pm1,\pm1)$

$f_{xx}=-6xy,\; f_{yy}=-6xy,\; f_{xy}=4-3x^2-3y^2$

$\therefore f_{xx}\;f_{yy}-f_{xy}^2=36x^2y^2-(4-3x^2-3y^2)^2$

$=12-(4-3-3)^2 \; at \; (\pm1,\pm1)$

$=12-4=8>0 \; and \; f_{xx}\;f_{yy}-f_{xy}^2=-4<0 \; at \; (0,0)$

$f_{xx}(-1,1)=6>0 \; and \; f_{xx}(1,-1)=6>0$

$\therefore f$ attains its local minima at (– 1,1) and (1, – 1)

$\therefore f(1,-1)=-4+1+1=-2 \; and \; f(-1,1)=-4+1+1=-2$

14. Define f(x, y) = $\begin{cases}\frac{x^3}{x^2+y^2},(x,y)\neq(0,0)\\ 0,\;(x,y)=(0,0)\end{cases}$,

which of the following statements are true?

(a) f is discontinuous at (0, 0) (b) f is continuous at (0, 0)

(c) all directional derivative of f at (0, 0) exist (d) f is not differentiable at (0, 0)

Ans. **(b), (c), (d)** **CSIR NET JUNE 2020**

Explanation :

$$f(x, y) = \begin{cases} \dfrac{x^3}{x^2+y^2}, (x,y) \neq (0,0) \\ 0, (x,y) = (0,0) \end{cases},$$ Put y = mx, $f(x, mx) = \dfrac{x^3}{x^2+(mx)^2} = \dfrac{x}{1+m^2}$.

$\lim\limits_{(x,y)\to(0,0)} f(x,mx) = \lim\limits_{(x,y)\to(0,0)} \dfrac{x}{1+m^2} = 0$, so, f is continuous at (0,0). Option (b) is correct and Option (a) is false.

Directional derivative of f at (0, 0)

$$= \lim_{h\to 0} \frac{f(c+hu)-f(c)}{h} = \lim_{h\to 0} \frac{f(hu)-f(0)}{h}$$

$$= \lim_{h\to 0} \frac{f(hu_1, hu_2)}{h} = \lim_{h\to 0} \frac{(hu_1)^3}{h[(hu_1)^2+(hu_2)^2]}$$

$$= \lim_{h\to 0} \frac{h^3u_1^3}{h^3[u_1^2+u_2^2]} = \lim_{h\to 0} \frac{u_1^3}{[u_1^2+u_2^2]} = \frac{u_1^3}{[u_1^2+u_2^2]},$$ all directional derivative of f at (0, 0) exist, Option (c) is correct.

(d) All directional derivative of f at (0, 0) exist, so, f is differentiable at (0,0), Option (d) is incorrect. Option (b), (c) and (d) are correct.

UNSOLVED QUESTIONS

1. Let f : $R^2 \to R$ be the function defined as follows : **IIT JAM 2023**

$$f(x,y) = \begin{cases} (x^2-1)^2 \cos^2\left(\dfrac{y^2}{(x^2-1)^2}\right) & if \quad x \neq \pm 1 \\ 0 & if \quad x = \pm 1 \end{cases}$$

Then number of points of discontinuity of f(x, y) is

(a) 0 (b) 1 (c) 10 (d) infinite

2. Let f : $R^2 \to R$ be defined as follows : **IIT JAM 2023**

$$f(x,y) = \begin{cases} \dfrac{x^4y^3}{x^6+y^6} & if \quad (x,y) \neq (0,0) \\ 0 & if \quad (x,y) = (0,0) \end{cases}$$ Then

(a) $\lim\limits_{t\to 0} \dfrac{f(t,2t)-f(0,0)}{t} = \dfrac{1}{3}$ (b) $\left(\dfrac{\partial f}{\partial x}\right)_{(0,0)} = 0$

(c) $\left(\frac{\partial f}{\partial y}\right)_{(0,0)} = 0$ (d) $\lim_{t\to 0} \frac{f(t,t)-f(0,0)}{t} = \frac{1}{2}$

3. Let $f(x,y) = \begin{cases} \frac{xy}{(x^2+y^2)^\alpha} & (x,y) \neq (0,0) \\ 0, & (x,y) = (0,0) \end{cases}$. Then which one of the following is TRUE for f at the point (0, 0)? **IIT – JAM 2018**

(a) For $\alpha = 1$, f is continuous but not differentiable.

(b) For $\alpha = 1/2$, f is continuous and differentiable.

(c) For $\alpha = 1/4$, f is continuous and differentiable.

(d) For $\alpha = 3/4$, f is neither continuous nor differentiable.

4. The function $f(x,y) = 3x^2y + 4y^3 - 3x^2 - 12y^2 + 1$ has a saddle point

(a) (0,0) (b) (0, 2) (c) (1.1) (d) (– 2, 1)

5. Let $f(x,y) = \begin{cases} x\sin\left(\frac{1}{y}\right) + y\sin\left(\frac{1}{x}\right), & xy \neq 0 \\ 0, & xy = 0 \end{cases}$. Then

(a) $\lim_{x\to 0}\lim_{y\to 0} f(x,y)$ does not exists (b) $\lim_{y\to 0}\lim_{x\to 0} f(x,y)$ does not exists

(c) $\lim_{(x,y)\to(0,0)} f(x,y)$ exists and equals to 0

(d) $\lim_{x\to 0} f(x,y)$ and $\lim_{y\to 0} f(x,y)$ both do not exist

6. The function $f(x,y) = x^3 + y^3 - 3x - 12y + 20$ has

(a) minima at (1,2) (b) maxima at (-1,-2)

(c) neither maxima nor minima at (1,-2) and (-1,2)

(d) the saddle points are (-1,2) and (1,-2)

7. For f(x, y) $\in R^2$, let $f(x,y) = \begin{cases} \frac{2xy}{x^2+y^2}, & if\,(x,y) \neq 0 \\ 0, & if\,(x,y) = 0 \end{cases}$. Then

(a) f_x and f_y exist at (0, 0) and f is continuous at (0, 0)

(b) f_x and f_y do not exist at (0, 0) but f is discontinuous at (0, 0)

(c) f_x and f_y do not exist at (0, 0) and f is continuous at (0, 0)

(d) f_x and f_y exist at (0, 0) but f is discontinuous at (0, 0)

8. Let if possible $\alpha = \lim_{(x,y)\to(0,0)} \frac{\sin(x^2+y^2)}{x^2+y^2}$, $\beta = \lim_{(x,y)\to(0,0)} \frac{x^2-y^2}{x^2+y^2}$. Then

(a) α exists but β does not exist (b) α does not exist but β exists
(c) α, β do not exist (d) both α, β exist

9. Let $f(x,y) = \begin{cases} \dfrac{2xy}{x^2+y}, & (x,y) \neq (0,0) \\ 0, & (x,y) = (0,0) \end{cases}$ Then

(a) $\lim_{x \to 0} f(x,0) = f(0,0)$ (b) $\lim_{y \to 0} f(0,y) = f(0,0)$
(c) $\lim_{(x,y)\to(0,0)} f(x,y) = f(0,0)$ (d) $\lim_{(x,y)\to(0,0)} f(x,y) \neq (0,0)$

10. Let $f(x,y) = \begin{cases} \dfrac{1}{x^2+y^2}, & (x,y) \neq (0,0) \\ 0, & x = y = 0 \end{cases}$ then,

(a) f is differentiable at (0, 0) (b) f is not differentiable at (0, 0)
(c) f is continuous at (0, 0) (d) none of these

11. Let $f(x, y) = x^y$, then
(a) $f_x(a, 0) = 0$, where a is constant (b) $f_y(e, 0) = 0$
(b) $f_{xy}(1, 0) = 0$ (d) $f_{yz}(1, 1) = 0$

12. Let $f(x,y) = \begin{cases} |x|+|y|, & (x,y) \neq (0,0) \\ 0, & (x,y) = (0,0) \end{cases}$. Then
(a) f(x, y) is continuous at (0, 0) (b) f(x, y) is discontinuous at (0, 0)
(c) f(x, y) is differentiable at (0, 0) (d) none of these

13. If $f(x, y) = xy(x^2 - y^2)/(x^2 + y^2)$, when $(x, y) \neq (0, 0)$ and $f(0, 0) = 0$.
(a) $f_{xy}(0, 0) = 1$ (b) $f_{xy}(0, 0) = -1$ (c) $f_{xy}(0, 0) = f_{yx}(0, 0)$
(d) $f_{xy}(0, 0) \neq f_{yx}(0, 0)$

14. Let $f(x,y) = \begin{cases} \dfrac{x^2 y}{x^3+y^3}, & x^3+y^3 \neq 0 \\ 0, & x = y = 0 \end{cases}$. Then,

(a) $f(x,y)$ is continuous at (0,0) (b) $f(x,y)$ is discontinuous at (0,0)
(c) $f(0,y)$ is continuous at origin (d) $f(x,0)$ is continuous at origin

Answer Key

1	2	3	4	5	6	7	8	9	10
a	b,c,d	c	d	a,b,c,d	a,b,c,d	d	a	a,b,d	b
11	**12**	**13**	**14**						
a	a	a,b,d	b,c,d						

Chapter 8

Metric Space

1. Metric Space

Consider a non-empty set X. A function d from X × X into the set of Real numbers Ris called a metric or a distance function if for all x, y, $z \in X$ satisfies the following conditions:

(i) $d(x, y) \geq 0$ (non-negativity)

(ii) $d(x, y) = 0$ iff x=y

(iii) $d(x, y) = d(y, x)$ (symmetry)

(iv) $d(x, z) \leq d(x, y) + d(y, z)$ (Triangle Inequality)

The pair (X, d) is called metric-space and d(x, y) is called distance between points x and y Also d(x, y) need not be the shortest distance between x and y $\in X$.

1.1 SOME DEFINITIONS

Let X Be a metric space and E be any subset of X

(1) **Neighbourhood of a point:** Neighbourhood of an element p is a set $N_r(p) = \{x \in X : d(p,x) < r, r > 0\}$. The numbers r is called radius of $N_r(p)$.

(2) **Limit Point:** An element P is a limit point of the set E if every neighbourhood of p contains an element $q \neq p$ such that $q \in E$. The set of all limit point of E is called derived set and is denoted by E'.

(3) **Isolated Point:**If $p \in E$ and p is not a limit point of E, then p is called an isolated point of E.

(4) **Closed Set:** E is closed if every limit point ofEis a point of E.

(5) **Interior Point:** An element p is an interior point of E if there is a neighbourhood N of p such that $p \in N \subset E$. The set of all interior points of E is called interior of E and is denoted by E^0.

(6) **Open et:** E is open if every element of E is an interior point of E.

(7) **Perfect Set:** E is perfect if E is closed and if every point of E is a limit point of E. Thus every closed interval of R is perfect.

Note: If P be a non-empty perfect set in R^n. Then P is uncountable.

(8) **Closure of a set:** Closure of E is the set $\bar{E} = E \cup E'$.

(9) **Bounded Set:**E is bounded if there is a real number M and an element $q \in X$ such that $d(p, q) < M$ for all $p \in E$

(10) **Dense Set:** E is dense in X if every element of X either Limit Point of E or a point of E (or both)

(11) **Non-dense set:**E is called non-dense set or nowhere dense set if $\left(\bar{E}\right)^0 = \phi$

Examples:

(1) Let (X, d) be a metric space. Then ϕ, X are both open and closed sets.

(2) In R (usual metric) the set Q of rational numbers is neither closed nor open.

(12) Let X be a non-empty set and A ⊂ X then diameter of A is diam A = Sup {d(x, y) x, y∈ A}

(13) The distance between two subsets A, B of X is d(A,B) = inf {d(x, y) : x∈ A, y∈ B}.

(14) Let A is a set and A ⊂ X is called convex if $\lambda x+(1-\lambda)y \in A, x \in A, y \in A, 0<\lambda<1$.

(15) Let a ∈ R^n and let r > 0. An open or closed ball with centre at a and radius r is the set of all $x \in R^n$ such that $|x-a|<$ r or $|x-a| \leq$ r andis denoted by B(a , r) or B[a , r]

Examples:

(i) The most important examples of metric spaces are Euclidean Space R^n. The distance in R^n is defined by d(x,y) = |x – y| (x, y ∈ R^n)

(ii) $d(x,y)=\max\{|x_1-y_1|,|x_2-y_2|,...,|x_n-y_n|\}$ where x = ($x_1, x_2, ..., x_n$), y = ($y_1, y_2, ..., y_n$) ∈ R^n. It can be easily verified that 'd'defined above is a metric on R^n (maximum metric)

(iii) Let X be a non-empty set. Define $d(x,y)=\begin{cases}0 & \text{if } x=y \\ 1 & \text{if } x\neq y\end{cases}$

It can be easily verified that this 'd' defined is a metric and is called discrete metric.

(iv) Define d: $R^2 \times R^2 \rightarrow$ Ras $d(a,b)=\sum_{i=1}^{2}|a_i-b_i|$ where a = (a_1, a_2), b = (b_1, b_2) ∈ R^2..

(v) Let X = C [a, b] is the set of all continuous real functions on [a,b]. Define ρ : X × X → R as $\rho(f,g)=\int_a^b|f(x)-g(x)|dx$. Then ρ is a metric on X

2. Metric Subspace

Consider a metric space (X,ρ). Let Y be a non empty subset of X. Then the restriction of ρ to Y×Y defines a metric on Y and we call Y to be metric subspace of X.

3. Convergence in metric space

Let (X, d) be a metric space.

A sequence {p_n} in a metric space X is said to converge if there is a point p∈ X such that for every ∈ >0 there exist an integer N_0 such that for n ≥ N_0 implies that d(p_n, p) < ∈.

We say that {p_n} converges to p, or that p is the limit of {p_n} and write $p_n \rightarrow p$, or

$\lim_{n\to\infty} p_n \rightarrow p$

If {p_n} does not converge, it is said to be divergent

Note: Convergent sequences are bounded, but bounded sequence in R^k need not be convergent. However, there is one important case in which convergence is equivalent to boundedness, this happens for monotonic sequence in R.

4. Cauchy sequence in metric space

A sequence $\{p_n\}$ in a metric space X is said to be a Cauchy sequence if for every $\in > 0$, there is a positive integer N_0 such that $d(p_m, p_n) < \in$ for $n \geq N_0$ and $m \geq N_0$.

Results on Cauchy sequence:

(1) In any metric space X, every convergent sequence is a Cauchy sequence.

(2) In R^k, every Cauchy sequence converges.

5. Complete metric space

A metric space in which every Cauchy sequence converges is said to be complete.

Examples

(1) All Euclidean spaces are complete.

(2) The space Q of rational numbers, with the standard metric given by $d(x,y)=|x-y|$ (i.e., the absolutevalue of the difference) is not complete.

Result On complete metric:

Every closed subset E of a complete metric space X is complete. (Every Cauchy sequence in E is a Cauchy sequence in X, hence it converges to some $p \in X$ and actually $p \in E$, since E is closed).

6. Limit of a function

Definition 1: Let X and Y be two metric spaces. Suppose $E \subset X$, $f: E \rightarrow Y$, and p is a limit point of E. We write $f(x) \rightarrow q$ as $x \rightarrow p$, or $\lim_{x \to p} f(x) = q$ iff for every $\in > 0$ there exists a $\delta > 0$ such that $d_Y(f(x), q) < \in$ for all point $x \in E$ for which $0 < d_X(x, p) < \delta$. The symbols d_X and d_Y refer to the metrics in X and Y respectively.

Definition 2: Let X and Y be metric spaces. Suppose $E \subset X$, $f: E \rightarrow Y$, and p is a limit point of E Then $\lim_{x \to p} f(x) = q$ and only if $\lim_{n \to \infty} f(p_n) = q$ for every sequence $\{p_n\}$ in E such that $p_n \neq p$, $\lim_{n \to \infty} p_n = p$

7. Continuity in metric space

Definition 1: Suppose X and Y are metric spaces, $E \subset X$, $p \in E$ and $f: E \rightarrow Y$. Then f is said to be continuous at p iff $\lim_{x \to p} f(x) = f(p)$ i.e., for every $\in > 0$ there exists a $\delta > 0$ such that $d_Y(f(x), f(p)) < \in$ forall point $x \in E$ for which $d_X(x, p) < \delta$.

Definition 2: A mapping f of a metric space X into a metric space Y is continuous on X if and only if $f^{-1}(V)$ is open in X for every open set V in Y.

Definition 3: A mapping f from a metric space X to a metric space Y is said to be continuous at a point $x \in X$ provided for any sequence $\{x_n\}$ in X if $\{x_n\} \rightarrow x$ then $\{f(x_n)\} \rightarrow f(x)$

Note:

(1) If f is continuous at every point of E, then f is said to be continuous on E. It should be noted that f has to be defined at the point p in order to be continuous at p.

(2) If p is an isolated point of E, then our definition implies that every function f which has E as its domainof definition is continuous at p. For, no matter which $\in > 0$ we can

choose $\delta > 0$ so that the only point $x \in E$ for which $d_X(x, p) < \delta$ is $x = p$ then d_Y $(f(x), f(p)) = 0 < \in$

(3) f: $X \rightarrow Y$ be continuous iff $f^{-1}(U)$ is closed in X for every closed set in U in Y.

8. Uniform continuity

Let f be a mapping of a metric space X into a metric space Y we say that f is uniformly continuous on X if for every $\in > 0$, there exists $\delta > 0$ such that $d_Y(f(p), f(q)) < \in$ for all p and q in X for which $d_X(p, q) < \delta$

Difference between continuity and uniform continuity:

(1) Uniform continuity is a property of a function on a set, whereas continuity can be defined at a single point.

(2) Evidently, every uniformly continuous function is continuous but a continuous function need not be uniformly continuous.

For example: Let (R, d) be a metric space where d is usual metric. Define $g : R \rightarrow R$ by metric space $g(x) = x^2$. Then g is continuous but not uniformly continuous

9. Compactness

Open cover: Let X be a metric space and $E \subseteq X$. Then open cover of a set E is the collection $\{G_\alpha\}$ of open subsets of X such that $E \subseteq \bigcup_\alpha G_\alpha$.(i.e. E is covered by the union of these open sets).

Compact set: A metric space is said to be compact if every open cover of the metric space has a finite sub-cover (sub-cover is also a cover of the set and is a subset of the cover).

Relatively compact: A subset Y of a metric space X is said to be relatively compact if $\bar{Y}$ is compact.

Example : Consider metric space (R , d) and subset A of R, where A = $[0, \infty)$. Then A is not compact.

Solution: As $A = [0, \infty) \subseteq \bigcup_{n=1}^{\infty} (-1, n)$, is infinite open cover. But there does not exist any finite sub-cover. Thus A is not compact

Note:

(1) If any set A is unbounded, then A cannot be compact.

(2) If we have metric space (R, d) and A be any subset which is closed as well as bounded, then that subset need not be compact.

9.1. Heine-Borel property of Real line:

Every closed and bounded subset of R is compact.

9.2 Results on Compact metric space:

(1) Suppose f is a continuous mapping of a compact metric space X into a metric space Y. Then f(X) is compact i.e. continuous image of a compact set is compact.

(2) If f is a continuous mapping of a compact metric space X into R^k, then f(x) is closed and bounded. Thus, f is bounded

(3) Suppose f is a continuous 1-1mapping of a compact metric space X onto a metric space Y. Then the Inverse mapping f^{-1} defined on Y by $f^{-1} f(x)) = x; x \in X$, is a continuous mapping of Yonto X

(4) Every compact metric space is complete.

(5) If X is a compact metric space and if $\{p_n\}$ is a Cauchy sequence in X, Then $\{p_n\}$ converges to some point of X

10. Connectedness

Separated Sets: Two subsets A and B of a metric space X are said to be separated if both $A \cap \bar{B}$ and $\bar{A} \cap B$ are empty i.e., if no point of A lies in the closure of B and no point of B lies in the closure of A.

For example Consider metric space (R, d) and its two subsets A = (– ∞, 1), B = [1, ∞) $\bar{A} = (-\infty, 1]$, $\bar{A} \cap B = \{1\}$. But $A \cap B = \phi$

Path: Let A be any non-empty subset of metric space (X, d) and x, y$\in$A, then a continuous function f: [0, 1] →A, where f(0) = x and f(1) =y is called a path in A from x toy.

Path-Connected: A subset A of X is said to be path-connected if and only if for all x, y ∈ A, there ispath in A from x toy

Connected Set: If (X, d) is a metric space and A$\subseteq$X, then A is said to be connected if it cannot be written as disjoint union of two non empty separated sets.

Disconnected Set: If a subset A of X is not connected, then it is said to be disconnected.

Results on connected sets:

(1) If f is continuous mapping of a metric space X into a metric space Yand if E is a connected space X, then f(E) is connected.

(2) A subset E of the real line R is connectedif and only if it has the following property if x$\in$ E, y ∈ E, and x < z < y, then z ∈ E.

(3) Any finite subset of Ris connected if it is singleton and disconnected if it has two or more element

(4) Every connected subset of Ris either an interval or a singleton set.

(5) Ris a connected set.

(6) Subsets of two separated sets are themselves separated.

(7) Two closed (open) sets are separated iff they are disjoint.

(8) The union of two connected sets, having non-empty intersection, is connected.

(9) Let X and Y are connected metric space then X×Y is connected.

(10) The closure of connected set is connected.

(11) $R^2$is connected

(12) Q × Q is disconnected.

(13) Any discrete space with more than one point is disconnected.

(14) Let A be countable subset of R^n; n>1 then R^n -A, is connected.

REAL ANALYSIS

11. Homomorphism

Let (X,d_1) and (Y, d_2) be two metric spaces. A function $f:X\to Y$ is said to be homeomorphism if

(i) f is both one-one and onto

(ii) f and f^{-1} are both continuous.

Homeomorphic Spaces: A metric space (X,d_1) is said to homeomorphic to another metric space (Y, d_2), if there exists a homeomorphism of X onto Y

Results on Homeomorphisms:

(1) Let (X, d_1) and (Y, d_2) be metric spaces and f be a mapping from X onto Y. Then f is homeomorphism from X to Y iff the following conditions hold:
   (i) f is one-one and onto
   (ii) fis continuous
   (iii) for each open set G in X, f(G) is open in Y.

(2) The homeomorphism is an equivalence relation in the collection of all metric spaces.

Examples

(1) The open interval (-1,1) is homeomorphic to (a, b).

(2) The intervals (a,b) and (c,d) in (R, d) are homeomorphic.

12. Isometry

Let (X, d_1) and (Y, d_2) be metric spaces. A function $f:X\to Y$ is said to beisometry if $d_1(x, y) = d_2(f(x), f(y))\ \forall x,y \in X$

For example:

(i) Let(X, d) be a metric space. Then identity mapping of X to itself is an isometry.

(iii) The mapping $f: R \to R$defined by $f(x)=3x+4$ is not an isometry.

13. Normed linear space

A normed linear space is a linear space on which there is defined a norm that is a function which assigns to each element x in the space a real number $\|x\|$ in such a way that:

(i) $\|x\|\geq 0$ and $\|x\|=0 \Leftrightarrow x=0$ (ii) $\|x+y\|\leq\|x\|+\|y\|$ (iii) $\|\alpha x\|=|\alpha|.\|x\|$

A normed linear space is a metric space with respect to the induced metric defined by $d(x,y)=\|x-y\|$

SOLVED QUESTIONS

1. The set X = R with the metric $d(x,y)=\dfrac{|x-y|}{1+|x-y|}$ is

GATE-2010

(a) bounded but not compact (b) bounded but not complete
(c) complete but not bounded (d) compact but not complete

Ans. **(a)**

Explanation :

$$d(x,y)=\frac{|x-y|}{1+|x-y|}$$

We know that $|x-y|<1+|x-y|$

$\Rightarrow \dfrac{|x-y|}{1+|x-y|}<1$ $\Rightarrow d(x,y)<1$ $\forall$ x, y $\in$ R

$\Rightarrow d(x,y)$ is bounded.

We know that (R, e) is a metric space, where e(x, y) = |x – y| $\forall$ x,y $\in$ R

Consider $d(x,y)=\dfrac{|x-y|}{1+|x-y|}=\dfrac{e(x,y)}{1+e(x,y)}\ \forall x,y\in R$

Then d and e are equivalent matrices.If possible,let (R, d)is compact metric space, then every d-open cover of Rhas finite Subcover.

Consequently, every e-open cover of Rhas a finitesubcover.

$\Rightarrow$ (R, e) is compact, which is contradiction.

Hence (R, d) is not compact

$\Rightarrow$ (A) is correct

2. The set $\left\{\dfrac{x^2}{1+x^2};x\in R\right\}$ is **IIT-JAM 2014**

(a) connected but not compact in R (b) compact but not connected in R
(c) compact and connected in R (d) neither compact nor connected in R

Ans. **(a)**

Explanation :

$$S=\left\{\frac{x^2}{1+x^2};x\in R\right\}$$

Which is connected, bounded but not closed as it's derived set is [0, 1]

So it is not compact.

3. Let X be a connected subset of real numbers. If every element of X is irrational, then the cardinality of X is **CSIR NET DEC 2011**

(a) infinite (b) countably infinite (c)2 (d)1

Ans. (d)

Explanation :

We know that in R, connected subsets are either intervals or singletons.If X is an interval, then it will contain rationals also, but X contains only irrationals. So X is singleton. Thus cardinality of X is '1'

4. Let (X, d) be a metric space and let $A \subseteq X$. For $x \in X$, define $d(x, A) = \inf\{d(x, a) : a \in A\}$ if for all $x \in X$, d(x, A) = 0, then which of the following assertions must be true? **CSIR NET DEC 2011**

(a)A is compact (b) A is closed (c) A is dense in X (d) A = X

Ans. (c)

Explanation :

Given that(X, d) is a metric space and $A \subseteq x$. Now, for all $x \in X, d(x, \text{A}) = 0 \forall x \in X$

$\Rightarrow$ Either $x \in A$ or x is a limit point of A $\Rightarrow \bar{A} = X \Rightarrow$ Ais dense in X.

So option (c) is correct.

Further let (R, d)be the usual metric space, i.e., X = R and A = Q $\subseteq$ R

Clearly, d(X, Q) = 0 $\forall$ x$\in$R

But Q is neither closed not compact and Q $\neq$ Ri.e., A $\neq$ X

So option (a), (b) and (d) are incorrect

5. Let $D_{(a,b)}(r) = \{(x, y) : (x-a)^2 + (y-b)^2 < r\}$. Which of the following subsets of R are connected? **CSIR NET JUNE 2012**

(a) $D_{(0,0)}(1) \cup \{(1,0)\} \cup D_{(2,0)}(1)$ (b) $D_{(0,0)}(1) \cup D_{(2,0)}(1)$

(c) $D_{(0,0)}(1) \cup \{(1,0)\} \cup D_{(0,2)}(1)$ (d) $D_{(0,0)}(1) \cup D_{(0,2)}(1)$

Ans. (a)

Explanation :

Given $D_{(a,b)}(r) = \{(x, y) : (x-a)^2 + (y-b)^2 < r\}$

Here $D_{(0,0)}(1) \cup \{(1,0)\} \cup D_{(2,0)}(1) = \{(x, y) : x^2 + y^2 < 1\} \cup \{(1,0)\} \cup \{(x, \text{y}) : (x-2)^2 + y^2 < 1\}$

For option (a)

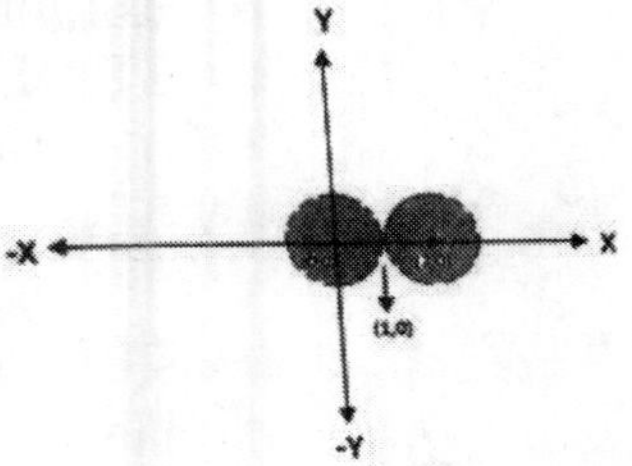

From diagram it is clear that above set is a connected set

So option (a) is correct

For option (b)

$$D_{(0,0)}(1) \cup D_{(2,0)}(1) = \{(x,y): x^2 + y^2 < 1\} \cup \{(x,y): (x-2)^2 + y^2 < 1\}$$

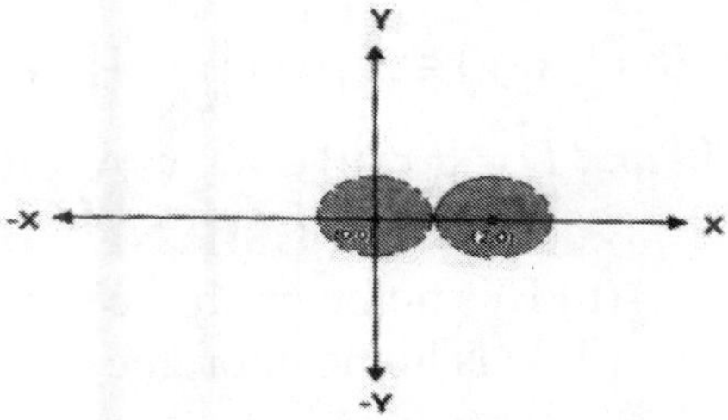

From diagram it is clear that, aboveset is disconnected as (1, 0) is not in the set

For option (c)

$$D_{(0,0)}(1) \cup \{(1,0)\} \cup D_{(0,2)}(1) = \{(x,y): x^2 + y^2 < 1\} \cup \{(1,0)\} \cup \{(x,y): x^2 + (y-2)^2 < 1\}$$

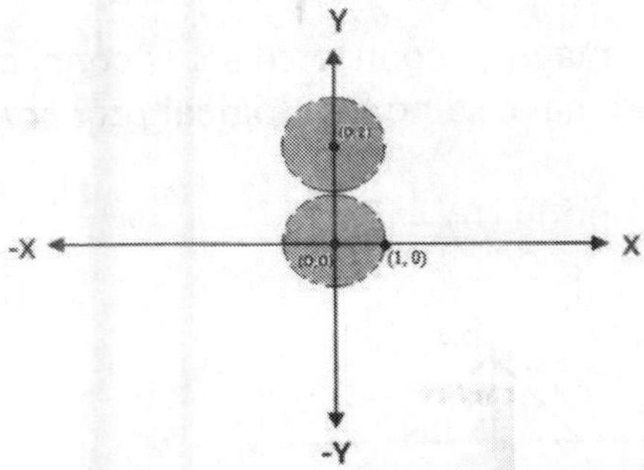

Clearly, above set is disconnected as (0, 1) is not in set.

For option (d) $D_{(0,0)}(1) \cup D_{(0,2)}(1) = \{(x,y): x^2 + y^2 < 1\} \cup \{(x,y): x^2 + (y-2)^2 < 1\}$

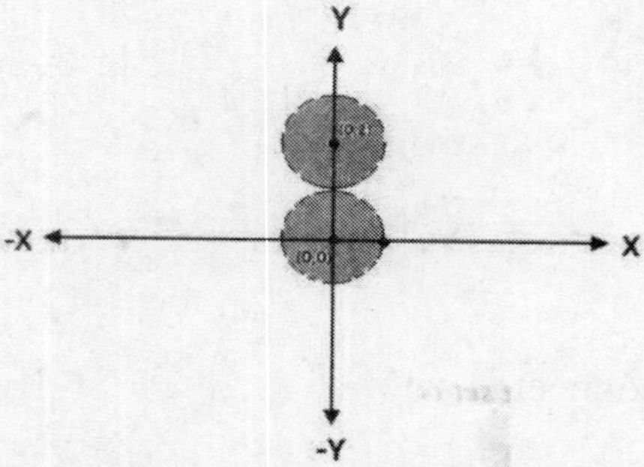

Clearly,above set is disconnected as (0, 1) is not in set.

6. Let $X=[-1,1]\times[-1,1]$. Which of the following subsets of R are connected?

$A=\{(x,y)\in X : x^2+y^2=1\}$ $\quad B=\{(x,y)\in X : |x|+|y|=1\}$

$C=\{(x,y)\in X : xy=0\}$ *and* $D=\{(x,y)\in X : x=\pm y\}$. Then

CSIR NET DEC 2012

(a) A is homeomorphic to B
(b) B is homeomorphic to C
(c) C is homeomorphic to D
(d) D is homeomorphic to A

Ans. **(a), (c)**

Explanation :

We know that a set Y is homeomorphic to Z, if $\exists$ a function $f:Y\to Z$ such that

(i) f is continuous, (ii) f is one - one and onto, or

(iii) f^{-1} is continuous or for each closed set H in Y, $f(H)$ is closed in Z.

Also we know that in a continuous function, image of connected set is connected or we can say two sets are homeomorphic if both have same topological properties.

Graphs of set A, B, C, D are given below

For option (a)

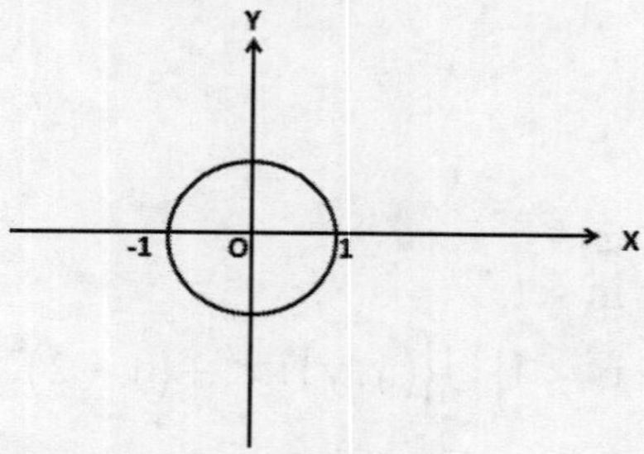

For option (b)

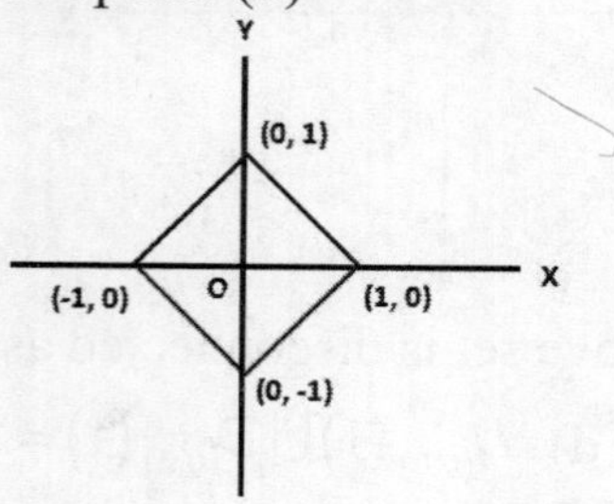

For option (c)

For option (d)

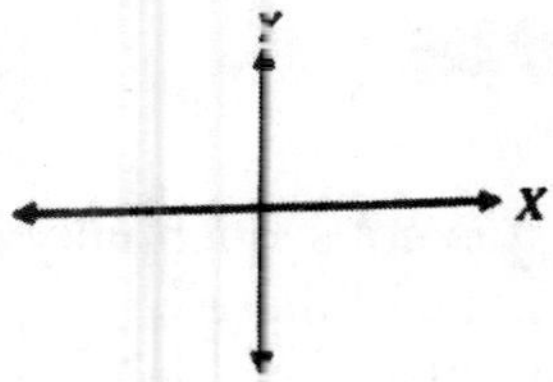

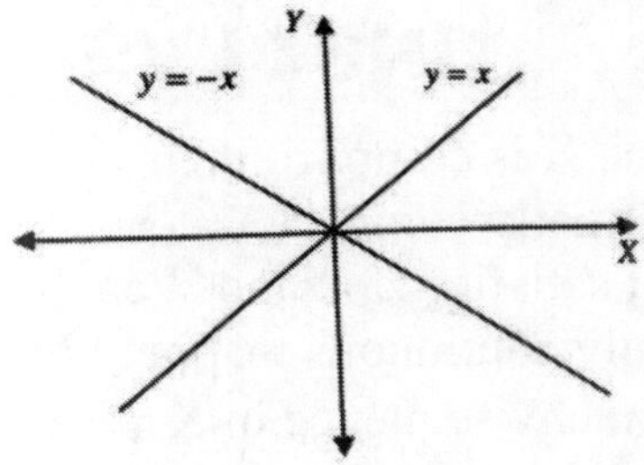

From diagram (A) and (B) it is clear that sets A and B are homeomorphic.
Thus,option(a) is correct
Also set D can be obtained just by rotating set C by angle 45°
As we know, every rotation isometry and every isometryis homeomorphic.
So set C and D are also homeomorphic.
Thus,cption(d) is also correct
B is not homeomorphic to C, as if(0, 0) (one point) is removed from C, then it is divided into four components, whereas B have only one connected component after removal of any one point. Similarly D cannot be homeomorphic to A.
Thus options(b), (d) are incorrect

7. For any two metric spaces (X, d_X), (Y, d_Y) a map $f: X \to Y$ is said to be a closed map if whenever F is closed in X, then f(F) is closed in Y. For any subset B of a metric space, B is given the induced metric. The metric on $X \times Y$ is given by $d((x, y), (x', y')) = \max\{d_X(x, x'), d_Y(y, y')\}$.
Which of the following are true?
(a) For any subset $A \subseteq X$ the inclusion map $i : A \to X$ is closed
(b) The projection map $p_1: X \times Y \to X$ given by $p_1(x, y) = x$ is closed
(c) Suppose that $f: X \to Y$, $g: Y \to Z$ are continuous maps. If $g o f: X \to Z$ is closed map then $g|_{f(x)} : f(x) \to Z$ is closed. Here $g|_{f(x)}$ means the map g restricted to f(X)
(d) If $f : X \to Y$ takes closed balls into closed sets then f is closed
Ans. (c) **CSIR NET JUNE 2020**
Explanation :
For any two metric spaces (X, d_X), (Y,d_Y) a map $f : X \to Y$ is said to be a closed map if whenever F is closed in X, then f(F) is closed in Y. For any subset B of a metric space, B is given the induced metric. The matric on $X \times Y$ is given by $D((x,y), (x',y')) = \max\{d_X(x,x'),d_Y(y,y')\}$. Suppose that $f: X \to Y$, $g: Y \to Z$ are continuous maps. If $g o f: X \to Z$ is a closed map then $g|_{f(X)} : f(X) \to Z$ is closed. Here $g|_{f(x)}$ means the map g restricted to f(X), Option (c) is correct.

UNSOLVED QUESTIONS

1. A metric space X is compact, then
(a) it is sequentially compact
(b) it is not sequentially compact
(c) it does not satisfies Bolzano Weierstrass
(d) none of these

2. If f is uniformly continuous mapping of metric space X into metric space Y, then
(a) $\langle x_n \rangle$ is Cauchy sequence in X $\Rightarrow \langle f(x_n) \rangle$ is Cauchy sequence in Y
(b) $\langle x_n \rangle$ is a convergent sequence in X $\Rightarrow \langle f(x_n) \rangle$ is a convergent sequence in Y
(c) $\langle x_n \rangle$ is Cauchy sequence in X $\Rightarrow \langle f(x_n) \rangle$ is not Cauchy sequence in Y
(d) none of these

3. Let $\langle x_n \rangle$ be a sequence from metric space (X, ρ)
A : x is a limit of x_n, then x is a cluster point.
B : x is a cluster point of x_n, then x is a limit of x_n.
(a) only A is true (b) only B is true (c) A and B both true (d) A and B both false

4. That collection C of open intervals of the form $\left(\frac{1}{n}, \frac{2}{n}\right)$, n = 2, 3,..... is an open covering of the open interval (0, 1), then
(a) finite sub-collection covers (0, 1)
(b) finite sub-collection does not cover (0, 1)
(c) the union of intervals is (0, 1)
(d) the union of intervals is (0, 0)

5. Which of the following (s) is /are true?
(a) A metric space X has the Bolzano-Weierstrass property, if X is sequentially compact.
(b) A metric space X is sequentially compact, if X has Bolzano-Weierstrass property.
(c) A metric space X has the Bolzano-Weierstrass property iff X is sequentially compact.
(d) A metric space X is sequentially compact, if X does not satisfies Bolzano-Weierstrass property.

6. If S is a subset of R^n and if S is closed and bounded, then
(a) every infinite subset of S has an accumulation point in S.
(b) every infinite subset of S has no accumulation point in S.
(c) If S is bounded, then its subset is bounded.
(d) If S is bounded, then its subset is not bounded.

Answer Key

1	2	3	4	5	6
a	a	a	b,c	a,b,c	a,c

Chapter 1

Basic Concepts Of Complex Numbers Topology

1. Complex Number

Definition : A number of the form a + ib, where a, b are real numbers and $i=\sqrt{-1}$, is called a complex number.

For example, 5 + 3i, -1 + i, 0 + 4i, 4 + 0i etc., are complex numbers. A complex number, z = a + ib is said to be purely real if Im(z) = 0 and purely imaginary if Re(z) = 0. The complex number 0 = 0 + i0 is both purely real and purely imaginary.

Integral powers of iota (i) :

Since $i=\sqrt{-1}$, hence, we have $i^2 = -1$, $i^3 = -i$ and $i^4 = 1$.

To find the value of i^n (n > 4), first divide n by 4.

Let q be the quotient and r be the remainder, i.e., n = 4q + r, where $0 \le r \le 3$.

$i^n = i^{4q+r} = (i^4)^q.(i)^r = (1)^q.(i)^r = i^r$

In general, we have $i^{4n} = 1$, $i^{4n+1} = i$, $i^{4n+2} = -1$, $i^{4n+3} = -i$, where n is any integer.

Algebra of Complex Numbers

The sum, difference, product and quotient of any two complex numbers is a complex number.

i.e. for any two given complex numbers z_1 and z_2, $z_1 + z_2$, $z_1 - z_2$, $z_1.z_2$, $\frac{z_1}{z_2}(z_2 \ne 0)$ are complex numbers.

Algebraic operations with complex numbers :

(i) **Addition** $(z_1 + z_2)$: (a + ib) + (c + id) = (a + c) + i(b + d)

(ii) **Subtraction** $(z_1 - z_2)$: (a + ib) – (c + id) = (a – c) + i(b – d)

(iii) **Multiplication** $(z_1 . z_2)$: (a + ib)(c + id) = (ac – bd) + i(ad + bc)

(iv) **Division** $\left(\frac{z_1}{z_2}\right) : \frac{a+ib}{c+id}$ (where at least one of c and d is non-zero)

$$\frac{a+ib}{c+id} = \frac{(a+ib)}{(c+id)} \cdot \frac{(c-id)}{(c-id)} = \frac{(ac+bd)}{c^2+d^2} + \frac{i(bc-ad)}{c^2+d^2}$$

Complex Plane :

The complex number z = a + ib = (a, b) is represented by a point P whose coordinates are referred to rectangular axes XOX' and YOY' which are called real and imaginary axis respectively. This plane is called Argand plane or Argand diagram or complex plane of Gaussian plane.

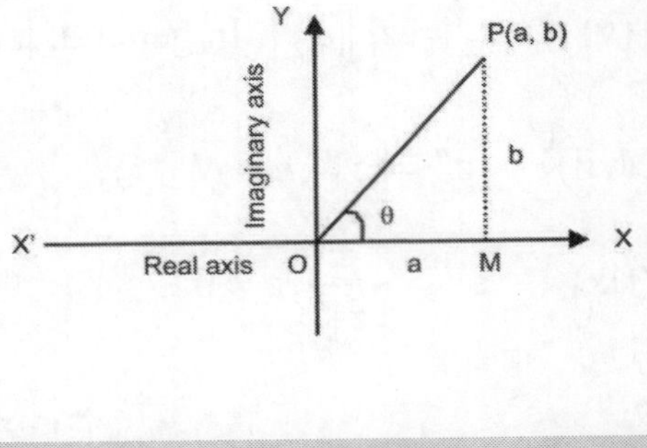

COMPLEX ANALYSIS

Conjugate of a Complex Number

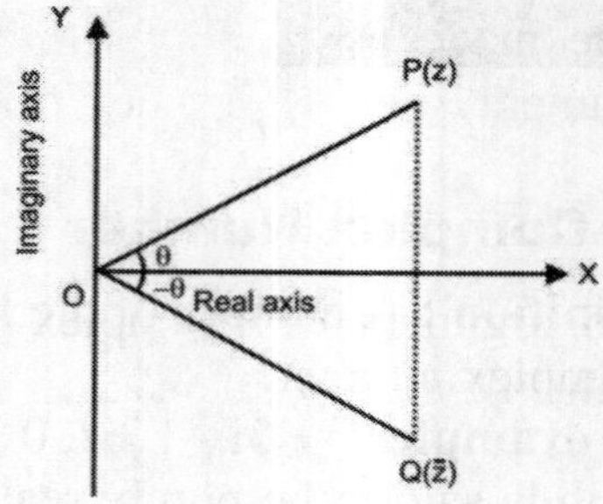

Definition : If there exists a complex number z = a + ib; a, b ∈ R, then its conjugate is defined as $\bar{z} = a - ib$.

Hence, we have $\text{Re}(z) = \dfrac{z+\bar{z}}{2}$ and $\text{Im}(z) = \dfrac{z-\bar{z}}{2i}$.

Geometrically, the conjugate of z is the reflection or point image of z in the real axis.

Properties of conjugate : If z, z_1 and z_2 are complex numbers, then we have the following results :

(i) $\overline{(\bar{z})} = z$ (ii) $\overline{z_1 + z_2} = \bar{z}_1 + \bar{z}_2$

(iii) $\overline{z_1 - z_2} = \bar{z}_1 - \bar{z}_2$ (iv) $\overline{z_1 z_2} = \bar{z}_1 \bar{z}_2$. In general $\overline{z_1.z_2.z_3.....z_n} = \bar{z}_1\bar{z}_2\bar{z}_3....\bar{z}_n$

(v) $\overline{\left(\dfrac{z_1}{z_2}\right)} = \dfrac{\bar{z}_1}{\bar{z}_2}, z_2 \neq 0$ (vi) $(\bar{z})^n = \overline{(z^n)}$

(vii) $z + \bar{z} = 2\text{Re}(z) = 2\text{Re}(\bar{z})$ purely real (viii) $z - \bar{z} = 2i\ \text{Im}(z)$ = purely imaginary

(ix) $z\bar{z} = |z|^2 = |\bar{z}|^2$ = purely rea (x) $z_1\bar{z}_2 + \bar{z}_1 z_2 = 2\text{Re}(z_1\bar{z}_2) = 2\text{Re}(\bar{z}_1 z_2)$

Modulus of a Complex Number

Modulus of a complex number z = a + ib is defined as a positive number given by $|z| = \sqrt{a^2 + b^2}$, where a, b are real numbers. Geometrically, |z| represents the distance of point P from the origin, i.e. |z| = OP.

If |z| = 1, the corresponding complex number is known as **unimodular complex number.** Clearly, z lies on a circle of unit radius having centre (0, 0).

Properties of modulus :

If z, z_1, z_2 are complex numbers, then we have the following results :

(i) $|z| \geq 0 \Rightarrow |z| = 0$ if $z = 0$ and $|z| > 0$ if $z \neq 0$ (ii) $-|z| \leq \text{Re}(z) \leq |z|$ and $-|z| \leq \text{Im}(z) \leq |z|$

(iii) $|z| = |\bar{z}| = |-z| = |-\bar{z}| = |zi|$ (iv) $z\bar{z} = |z|^2 = |\bar{z}|^2$

(v) $|z_1 z_2| = |z_1||z_2|$. In general, $|z_1 z_2 z_3....z_n| = |z_1||z_2||z_3|...|z_n|$ (vi) $\left|\dfrac{z_1}{z_2}\right| = \dfrac{|z_1|}{|z_2|}, (z_2 \neq 0)$

(vii) $|z^n| = |z|^n, n \in N$ (viii) $|z| = 1 \Leftrightarrow \bar{z} = \dfrac{1}{z}$

(ix) $z^{-1} = \dfrac{\bar{z}}{|z|^2}, z \neq 0$

2. Argument or Amplitude of a Complex Number

Let z = a + ib be any complex number. If this complex number is represented geometrically by a point P, then the angle made by the line OP with real axis is known as argument or amplitude of z and is expressed as $\arg(z) = \theta = \tan^{-1}\left(\dfrac{b}{a}\right), \theta = \angle POM$.

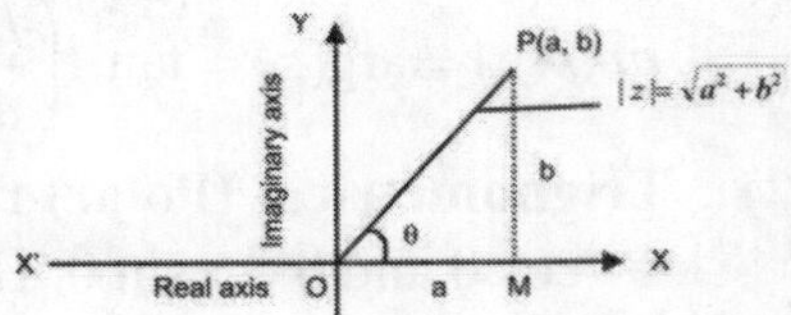

Also, argument of a complex number is not unique, since if θ is a value of the argument, then so is 2nπ + θ, where n ∈ Z. i.e. arg z = {θ + 2nπ; n ∈ Z}

Principal value of Arg z :

The value θ of the argument, which satisfies the inequality -π < θ ≤ π is called the principal value of argument, where $\theta = \tan^{-1}\left|\dfrac{b}{a}\right|$ (acute angle) and principal values of argument of z will be θ, π - θ, -π + θ and -θ as the point z lies in the 1st, 2nd, 3rd and 4th quadrants respectively.

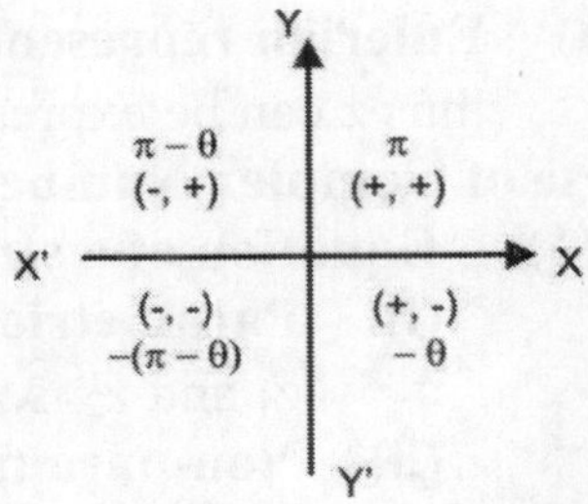

Properties of arguments (amplitudes) :

(i) arg (any real positive number) = 0 + 2nπ; n ∈Z

(ii) arg (any real negative number) = π + 2nπ; n ∈ Z

(iii) $\arg(iy) = \begin{cases} \dfrac{\pi}{2} + 2n\pi, n \in Z, & if\ y > 0 \\ \dfrac{-\pi}{2} + 2n\pi, n \in Z, & if\ y < 0 \end{cases}$

(iv) $\arg(\bar{z}) = -\arg(z) = \arg\left(\dfrac{1}{z}\right)$

(v) $\arg(-z) = \arg(z) \pm \pi$

(vi) $\arg(z^n) = n \arg(z)$

Some Representations of a Complex Number

A complex number can be represented in the following forms :

(1) Geometrical representation (Cartesian representation) : The complex number z = a + ib is represented by a point P whose coordinates are referred to rectangular axes XOX' and YOY' which are called real and imaginary axis respectively. We can also say that a complex number z represents a vector because it has magnitude as well as direction.

Distance of any complex number from the origin is called the modulus of complex number and is denoted by $|z|, i.e. |z| = \sqrt{a^2 + b^2}$. (Angle of any

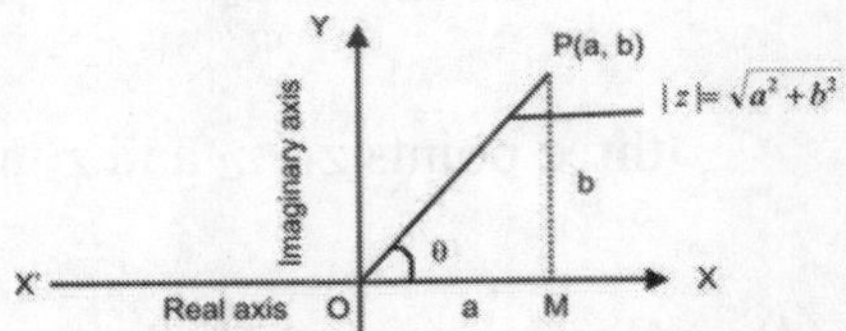

complex number with positive direction of x-axis is called amplitude of z, i.e. $amp(z) = \arg(z) = \tan^{-1}\left(\frac{b}{a}\right))$

(2) Trignometrical (Polar) representation : In figure, consider ΔPOM, let OP = r, then a = r cos θ and b = r sin θ. Hence, z can be expressed as z = r(cos θ + i sin θ), where r = |z| and θ = principal value of argument of z. For general values of the argument z = r[cos(2nπ + θ) + i sin (2nπ + θ)]. Sometimes, (cos θ + i sin θ) is written as short as cis θ.

(3) Eulerian representation (Exponential form) : Since we have $e^{i\theta} = \cos\theta + i\sin\theta$ and thus z can be expressed as $z = re^{i\theta}$, where |z| = r and θ = arg(z).

Use of Complex Numbers in Co-ordinate Geometry :

(1) Equation of a straight line :

(i) Parametric form : Equation of a straight line joining the points having affixes z_1 and z_2 is $z = t z_1 + (1 - t)z_2$, where t ∈ R.

(ii) Non-parametric form : Equation of a straight line joining the points having affixes z_1 and z_2 is $\begin{vmatrix} z & \bar{z} & 1 \\ z_1 & \bar{z}_1 & 1 \\ z_2 & \bar{z}_2 & 1 \end{vmatrix} = 0 \Rightarrow z(\bar{z}_1 - \bar{z}_2) - \bar{z}(z_1 - z_2) + z_1\bar{z}_2 - z_2\bar{z}_1 = 0$.

Or $\frac{z - z_1}{z_2 - z_1} = \frac{\bar{z} - \bar{z}_1}{\bar{z}_2 - \bar{z}_1}$.

(iii) General equation of a straight line : The general equation of a straight line is of the form $\bar{a}z + a\bar{z} + b = 0$, where 'a' is a complex number and 'b' is a real number.

(2) Area of a triangle : Area of a triangle ABC with vertices A(z_1), B(z_2) and C(z_3) is given by $\Delta = \frac{1}{4}\begin{vmatrix} z_1 & \bar{z}_1 & 1 \\ z_2 & \bar{z}_2 & 1 \\ z_3 & \bar{z}_3 & 1 \end{vmatrix}$.

(3) Condition for collinearly : Three points z_1, z_2 and z_3 will be collinear if there exists a relation $az_1 + bz_2 + cz_3 = 0$ (a, b and c are real), such that a + b + c = 0. In other words, three points z_1, z_2 and z_3 are collinear if $\begin{vmatrix} z_1 & \bar{z}_1 & 1 \\ z_2 & \bar{z}_2 & 1 \\ z_3 & \bar{z}_3 & 1 \end{vmatrix} = 0$.

(4) Equation of a circle :

(i) The equation of a circle whose centre is at point having affix z_0 and radius r is $|z - z_0| = r$.

(ii) If the centre of the circle is at origin and radius r, then its equation is |z| = r.

Triangle inequalities :

In any triangle, sum of any two sides is greater than the third side and difference of any two sides is less than the third side. By applying this basic concept to the set of complex numbers we have the following results :

(1) $|z_1 + z_2| \le |z_1| + |z_2|$ (2) $|z_1 - z_2| \le |z_1| + |z_2|$

(3) $|z_1 + z_2| \ge | |z_1| - |z_2| |$ (4) $|z_1 - z_2| \ge | |z_1| - |z_2| |$

Geometry of complex numbers :

(i) **Distance formula :** Let $z_1 = x_1 + iy_1$ and $z_2 = x_2 + iy_2$ be two complex numbers represented by points P and Q respectively in Arg and plane, then

$$PQ = \sqrt{(x_2 - x_1)^2 + (y_2 - y_1)^2} = |(x_2 - x_1) + i(y_2 - y_1)| = |z_2 - z_1|$$

De Moivre's Theorem :

(1) If n is any rational number, then $(\cos\theta + i\sin\theta)^n = \cos n\theta + i\sin n\theta$.

(2) If $z = (\cos\theta_1 + i\sin\theta_1)(\cos\theta_2 + i\sin\theta_2)(\cos\theta_3 + i\sin\theta_3) \ldots (\cos\theta_n + i\sin\theta_n)$, then $z = \cos(\theta_1 + \theta_2 + \theta_3 + \ldots + \theta_n) + i\sin(\theta_1 + \theta_2 + \theta_3 + \ldots + \theta_n)$, where $\theta_1, \theta_2, \theta_3, \ldots, \theta_n \in R$.

Roots of a Complex Number :

1. **n^{th} roots of a complex number z :**

Let $z = r(\cos\theta + i\sin\theta)$ be a complex number. By using De Moivre's theorem, n^{th} roots having n distinct values of such a complex number are given by

$$z^{1/n} = r^{1/n}\left[\cos\left(\frac{2k\pi + \theta}{n}\right) + i\sin\left(\frac{2k\pi + \theta}{n}\right)\right], \text{ where } k = 0, 1, 2, \ldots, (n-1).$$

2. **n^{th} roots of unity :**

n^{th} roots of unity are given by the solution set of the equation $z^n = 1 = \cos 0 + i\sin 0 = \cos 2k\pi + i\sin 2k\pi$

$z = [\cos 2k\pi + i\sin 2k\pi]^{1/n}$.

$$z = \cos\frac{2k\pi}{n} + i\sin\frac{2k\pi}{n}, \text{ where } k = 0, 1, 2, \ldots, (n-1).$$

Properties of nth roots of unity :

(i) Let $\alpha = \cos\frac{2\pi}{n} + i\sin\frac{2\pi}{n} = e^{i(2\pi/n)}$, n^{th} roots of unity can be expressed in the form of a sequence, i.e., $1, \alpha, \alpha^2, \ldots, \alpha^{n-1}$. Clearly, the sequence is G.P. with common ratio α, i.e., $e^{i(2\pi/n)}$.

(ii) The sum of all n roots of unity is zero, i.e., $1 + \alpha + \alpha^2 + \ldots + \alpha^{n-1} = 0$.

(iii) Product of all n roots of unity is $(-1)^{n-1}$.

(iv) Sum of pth power of n roots of unity

$$1 + \alpha^p + \alpha^{2p} + \ldots + \alpha^{(n-1)p} = \begin{cases} 0, & \text{when p is not multiple of n} \\ n, & \text{when p is a multiple of n} \end{cases}$$

COMPLEX ANALYSIS

The n, n^{th} roots of unity if represented on a complex plane, locate their positions at the vertices of a regular polygon of n sides inscribed in a unit circle having centre at origin, one vertex on positive real axis.

Euler's Formula :

$e^{i\theta} = \cos\theta + i\sin\theta$ ....(1) and $e^{-i\theta} = \cos\theta - i\sin\theta$ .... (2)

From (1) and (2), $\cos\theta = \dfrac{e^{i\theta}+e^{-i\theta}}{2}$ and $\sin\theta = \dfrac{e^{i\theta}-e^{-i\theta}}{2i}$

Thus, for an integer n, $(e^{i\theta})^n = e^{i(n\theta)} = \cos n\theta + i\sin n\theta$ and $(e^{i\theta})^{-n} = e^{i(-n\theta)} = \cos n\theta - i\sin n\theta$.

SOLVED QUESTIONS

COMPLEX ANALYSIS

1. If $\left|e^{e^z}\right| = 1$ for a complex number z = x + iy. Then which of the following is true

CSIR NET SEP. 2022

(a) x = nπ for some integer n (b) x = $(2n+1)\frac{\pi}{2}$ for some integer n

(c) y = nπ for some integer n (d) y = $(2n+1)\frac{\pi}{2}$ for some integer n

Ans. (d)

Explanation:

Given that $\left|e^{e^z}\right| = 1$ $\Rightarrow \left|e^{e^{x+iy}}\right| = 1 \Rightarrow \left|e^{e^x\cos y + ie^x\sin y}\right| = 1$

$\Rightarrow \left|e^{e^x\cos y}e^{ie^x\sin y}\right| = 1 \Rightarrow \left|e^{e^x\cos y}\right|\left|e^{ie^x\sin y}\right| = 1$ $\Rightarrow \left|e^{e^x\cos y}\right| = 1$

We know that $e^x \neq 0$ but cosy = 0 for y = $(2n+1)\frac{\pi}{2}$ Hence option (d) is true

2. The principal value of $(-1)^{(-2i/\pi)}$ is **GATE-2018**

(a) e^2 (b) e^{2i} (c) e^{-2i} (d) e^{-2}

Ans. (a)

Explanation :

$(-1)^{(-2i/\pi)} = \left(e^{i\pi}\right)^{-2i/\pi} = e^{(-2i\times i)} = e^2$

So, option (a) is correct.

3. Let n be a positive integer. For a real number R > 1, let $z(\theta) = Re^{i\theta}$, $0 \le \theta \le 2\pi$. The set $\{\theta\in[0,2\pi]: |z(\theta)^n+1| = |z(\theta)|^n - 1\}$ contains which of the following sets?

(a) $\{\theta \in [0,2\pi]: \cos n\theta = 1\}$ (b) $\{\theta \in [0,2\pi]: \sin n\theta = 1\}$

(c) $\{\theta \in [0,2\pi]: \cos n\theta = -1\}$ (d) $\{\theta \in [0,2\pi]: \sin n\theta = -1\}$

Ans. (c) **CSIR NET JUNE 2020**

Explanation :

For a real number R > 1, let $z(\theta) = Re^{i\theta}$, $0 \le \theta \le 2\pi$, the set $\{\theta \in [0,2\pi) : |z(\theta)^n + 1| = z(\theta)|^n - 1\}$ contains which of the following sets where n be a positive integer.

$S = \{\theta \in [0, 2\pi) : |z(\theta)^n + 1| = |z(\theta)|^n - 1\}$

Let $\theta \in S \Rightarrow |z(\theta)^n + 1| = |z(\theta)|^n - 1$

$\Rightarrow |(Re^{i\theta})^n + 1| = |Re^{i\theta}|^n - 1$ $\Rightarrow |(Re^{i\theta})^n + 1|^2 = (|Re^{i\theta}|^n - 1)^2$

$\Rightarrow |R^n e^{in\theta} + 1|^2 = (R^n - 1)^2$ $\Rightarrow (R^n e^{in\theta} + 1)(R^n e^{-in\theta} + 1) = R^{2n} - 2R^n + 1$

$\Rightarrow R^{2n} + R^n e^{in\theta} + R^n e^{-in\theta} + 1 = R^{2n} - 2R^n + 1 \Rightarrow R^{2n} + R^n e^{in\theta} + R^n e^{-in\theta} + 1 = R^{2n} - 2R^n + 1$

$\Rightarrow R^n(e^{in\theta} + e^{-in\theta}) = -2R^n$ $\Rightarrow (e^{in\theta} + e^{-in\theta}) = -2$

$\Rightarrow 2\dfrac{(e^{in\theta} + e^{-in\theta})}{2} = -2$ $\Rightarrow 2\cos n\theta = -2$

$\Rightarrow \cos n\theta = -1$, Option (c) is correct and Option (a), (b) and (d) are incorrect.

4. If z_1 and z_2 are distinct complex numbers such that $|z_1| = |z_2| = 1$ and $z_1 + z_2 = 1$, then the triangle in the complex plane with z_1, z_2 and -1 as vertices

CSIR NET JUNE 2013

(a) must be equilateral (b) must be right angled

(c) must be isosceles, but not necessarily equilateral

(d) must be obtuse angled.

Ans. (a)

Explanation :

Take $z_1 = -\omega = \frac{1}{2} - \frac{\sqrt{3}i}{2}, z_2 = -\omega^2 = \frac{1}{2} + \frac{\sqrt{3}i}{2}, z_3 = -1.$

Here $|z_1| = |z_2| = 1$ and $z_1 + z_2 = 1$

Evaluate $|z_1 - z_2|$, $|z_2 - z_3|$ and $|z_3 - z_1|$

$|z_1 - z_2| = \left|\frac{1}{2} - \frac{\sqrt{3}i}{2} - \frac{1}{2} - \frac{\sqrt{3}i}{2}\right| = \sqrt{3}$ $|z_2 - z_3| = \left|\frac{1}{2} + \frac{\sqrt{3}i}{2} + 1\right| = \left|\frac{3}{2} + \frac{\sqrt{3}i}{2}\right| = \sqrt{3}$

$|z_3 - z_1| = \left|-1 - \frac{1}{2} + \frac{\sqrt{3}i}{2}\right| = \left|\frac{-3}{2} + \frac{\sqrt{3}i}{2}\right| = \sqrt{3}$ $\Rightarrow |z_1 - z_2| = |z_2 - z_3| = |z_3 - z_1|$

$\Rightarrow$ length of three sides of a triangle are equal.

$\Rightarrow$ the triangle with z_1, z_2 and -1 as vertices is equilateral.

So, options (b), (c), (d) are incorrect.

We are left with option (a), which is correct one.

5. The value of Re(f(z)) for $f(z) = 2iz + 6\bar{z}$ is, where z = x + iy

(a) 6x + 2y (b) 6x – 2y (c) 6x / 2y (d) none of these

Ans. (b)

Explanation :

Given that $f(z) = 2iz + 6\bar{z}$. = 2i (x + iy) + 6(x – iy) = 2ix – 2y + 6x – 6iy

Re[f(z)] = 6x – 2y

Option (b) is correct.

6. Let C be the set of complex numbers. Let for any z = (x, y) in C, $z.\bar{z} = z$, then $\bar{z}$ is equal to

(a) (0, 1) (b) (1, 0) (c) (-1, 1) (d) (1, 1)

Ans. (b)

Explanation :

$z.\bar{z} = z \quad \Rightarrow \bar{z} = 1 = x - iy = 1 + i.0 \Rightarrow (1,0)$

7. $(\sin\theta + i\cos\theta)^6 =$

(a) sin 6θ + i cos 6θ (b) cos 6θ - i sin 6θ

(c) –cos 6θ + i sin 6θ (d) sin 6θ - i cos 6θ

Ans. (c), (d)

Explanation :

$(\sin\theta + i\cos\theta)^6 = -1[\cos\theta - i\sin\theta]^6 = -1(e^{-i\theta})^6 \quad = -1e^{-6i\theta} = -[\cos 6\theta - i\sin 6\theta]$

Option (c) and (d) are correct.

8. The number $\sqrt{2}e^{ix}$ is

(a) A rational number (b) A transcendental number

(c) An irrational number (d) An imaginary number

Ans. (c) **CSIR NET JUNE 2012**

Explanation :

Let $x = \sqrt{2}.e^{ix}$

$= \sqrt{2}(\cos\pi + i\sin\pi) = \sqrt{2}(-1 + i.0) = -\sqrt{2}$ Which is an irrational number.

9. Which of the following is the imaginary part of a possible value of $\ell n(\sqrt{i})$?

(a) π (b) π/2 (c) π/4 (d) π8

GATE-2011

Ans. (c)

Explanation :

$$log\sqrt{i} = log\left(e^{i\frac{\pi}{2}}\right)^{1/2} = log\, e^{i\frac{\pi}{4}} = i\frac{\pi}{4} \qquad \therefore Im(log\sqrt{i}) = \frac{\pi}{4}$$

UNSOLVED QUESTIONS

1. The minimum possible value of $|z|^2 + |z-3|^2 + |z-6i|^2$, where z is a complex number and $i = \sqrt{-1}$ is **CSIR NET JUNE 2013**

(a) 15 (b) 45 (c) 30 (d) 20

2. $Arg(z) + Arg(\bar{z})$, when z ≠ 0 is

(a) 0 (b) -π/2 (c) π/2 (d) πi

3. Value of Arg π is
(a) real (b) 0 (c) -1 (d) 1

4. If $\frac{4+3i}{3-4i} = x+iy$, then $\frac{x}{y}$ is equal to
(a) 0 (b) 1 (c) 4/3 (d) 4/5

5. Complex form of $\sqrt{3+4i}$ is
(a) π/3 (b) π/6 (c) 2π/3 (d) 5π/6

6. Principal value of argument of (cos 1200° + i sin 1200°) is
(a) 300^c (b) 120° (c) -150° (d) 180°

7. The polar form of the complex number 1 – i is
(a) $\sqrt{2}e^{\frac{\pi}{4}i}$ (b) $e^{\frac{\pi}{4}i}$ (c) $\sqrt{2}e^{-\frac{\pi}{4}i}$ (d) $e^{\frac{\pi}{4}i}$

8. Locus of the point z satisfying the equation |iz – 1| + |z – i| = 2 is
(a) straight line segment (b) a circle
(c) an ellipse (d) a pair of straight lines

9. Let w, w^2 be complex cube roots of unity. Then the determined

$$D = \begin{vmatrix} x+1 & \omega & \omega^2 \\ \omega & x+\omega^2 & 1 \\ \omega^2 & 1 & x+\omega \end{vmatrix}$$ is divisible by

(a) x (b) x^5 (c) x^7 (d) x^6

10. The points $z_1 = 1 + i$, $z_2 = 1 - i$ and $z_3 = 2 + I$ represents a triangle whose area is
(a) 1 (b) 2 (c) 3 (d) 4

ANSWER KEY

1	2	3	4	5	6	7	8	9	10
c	a	b	a	c	b	c	a	a	a

Chapter 2

Elementary Functions, Analytic Function and C-R Equations

1. Some Definitions

(1) δ-neighbourhood : A δ-neighbourhood of a point z_0 in the complex plane is defined as the set of points given by $S = \{z \in C : |z - z_0| < \delta, \delta > 0\}$, where δ is called the radius of neighbourhood of z_0 and the set S is denoted by $N_\delta(z_0)$. A deleted δ-neighbourhood of z_0 is a δ-neighbourhood of z_0 in which the point z_0 is excluded. We denote it by $N_\delta(z_0)/z_0 = \{z \in C : 0 < |z - z_0| < \delta\}$.

(2) Limit Point: A point z_0 is called limit point of a set S if every δ-neighbourhood of z_0 contains a point of S other than z_0. It follows that S contains infinite number of points.

{For example : $z_0 = 0$ is a limit point of $S = \left\{1, \frac{1}{2^2}, \frac{1}{3^2}, ..., \frac{1}{n^2}, ...\right\}$ *but* $0 = z_0 \notin S$.

The limit point of set S is also called cluster point or point of accumulation of S.

(3) Derived set: The set of all limit points of a set is called derived set of S and is denoted by S'.

(4) Isolated points: Points in a set S which are not limit points are called isolated point of the set S.

(5) Closed set: A set S is said to be closed if S contains all its limit points, i.e., if each limit point of S belongs to S. For example, the set of al points z such that $|z| \le r$ is a closed set.

(6) Bounded set: A set in complex plane is bounded if there exists a finite positive number M such that $|z| \le M$ for every point in S. A set which is not bounded in called an unbounded set.

(7) Compact set: A set $S \subset C$ is said to be compact if it is bounded and closed.

(8) Interior points:

A point z_0 is called an interior point of a set S, if there exists a δ-neighbourhood of z_0, all of whose points belong to S, i.e., there exists a $\delta > 0$ such that $\{z \in C : |z - z_0| < \delta\} \subset S$.

The set of all interior points of S is denoted by Int S.

(9) Open set: A set S in C is said to be open if every point of S is interior point of S. For example, the set of point z such that $|z| < r$ is an open set.

Note: The complement of an open set is a closed set and vice-versa.

(10) Convex set: A set S in R^2 is said to be convex set if for any two point X_1, X_2, $\in S$, the line segment joining X_1 and X_2 is contained in the set S, i.e., $X = \lambda X_1 + (1-\lambda)X_2 \subseteq S$ for $0 \le \lambda \le 1$.

(11) Connected set : A set S is said to be connected if it can be expressed as a union of any two disjoint non-empty sets S_1, S_2 such that either S_1 contains a limiting point of S_2 or S_2 contains a limiting point of S_1. Hence, an open set S is said to be connected if any two points of the set can be joined by a path consisting of straight line segment (i.e. a polygonal path), all points of which are in S. for example, every interval of real axis is connected.

(12) Domain or open region : An open connected set is called an open region or domain.

(13) Region: A domain together with some, none or all of its limit points is referred to as a region .

(14) Closed region: The closure of an open region or domain is called closed region.

2. Elementary Function :

Complex function: f(z) is a function of a complex variable z and is denoted by w, i.e., w = f(z) = u + iv, where u and v are real functions of x and y and are called real and imaginary parts respectively of f(z).

Polynomial functions: Polynomial functions are defined by $w = a_n z^n + a_{n-1} z^{n-1} + ... + a_1 z + a_0 = P(z)$, where $a_n \neq 0$. $a_{n-1}, a_{n-2}, ...a_0$ are complex constants and n is a positive integer called the degree of the polynomial P(z).

Properties of polynomials:

(1) The order of a zero of a polynomial equals the order of its first non-vanishing derivative. Suppose, z = a is a zero of order m of polynomial P(z), then $P(z) = (z-a)^m Q(z)$, $Q(a) \neq 0$. Differentiating both sides successively in times and putting z = a, we get $P'(a) = p''(a) = ... = P^{(m-1)}(a) = 0$ and $P^{(m)}(a) \neq 0$.

(2) If all the zeros of a polynomial lie in a half plane, then all the zeros of its derivative also lie in the same half plane (Luca's Theorem).

(3) If S is the set of all zeros of a polynomial, then all the zeros of its derivative lie on convex hull of S (Gauss Theorem).

(4) Let $P(z) = a_0 + a_1 z + a_2 z^2 + ... + a_n z^n$, where $n \geq 1$ and $a_n \neq 0$ so that P(z) is polynomial of degree one or greater, then the equation P(z) = 0 has atleast one root (Fundamental theorem of algebra).

Rational algebraic functions: Rational algebraic functions are defined by $w = \dfrac{P(z)}{Q(z)}$, where P(z) and Q(z) are polynomials without common zeros. w is sometimes called a rational transformation. The real and imaginary parts of w are rational functions of x and y. If we take P(z) = az + b and Q(z) = 1, then w = az + b is called linear transformation. The special case $w = \dfrac{az+b}{cz+d}$, where $ad - bc \neq 0$ is often called a bilinear or fractional linear transformation.

Exponential function: Exponential function is defined by $w = e^z = e^{x+iy} = e^x (\cos y + i \sin y)$, where e = 2.71828... is the natural base of logarithms. Series expansion of exponential function is given by exp (z) or $e^z = 1 + \frac{z}{1!} + \frac{z^2}{2!} + ... + \frac{z^n}{n!} + ...\infty = \sum_{n=0}^{\infty} \frac{z^n}{n!}$ ....(1)

Properties of exponential function:

(i) $e^z = e^{x+iy} = e^x . e^{iy} = e^x \ (\cos y + i \sin y) \Rightarrow | e^z | = | e^x | = e^x \left(\because e^x > 0\right)$ *and* $\arg\left(e^z\right) = y$

(ii) $e^{\bar{z}} = \left(\overline{e^z}\right)$ (iii) $e^{z_1} e^{z_2} = e^{z_1+z_2}, \frac{e^{z_1}}{e^{z_2}} = e^{z_1-z_2}$

(iv) e^z is periodic function having imaginary period $2\pi i$, $\because e^{z+2i\pi} = e^x e^{i(y+2\pi)} = e^x e^{iy} = e^z$

(v) Exponential form of $z = re^{i\theta}$ (vi) $e^z \neq 0 \ \forall \ z \in C \ \left(\because | e^z | e^x > 0 \forall x \in R\right)$

Trigonometric function: The trigonometric or circular functions sinz, cosz, etc., in terms of exponential functions are defined as follows:

$$\sin z = \frac{e^{iz} - e^{-iz}}{2i} \qquad \cos z = \frac{e^{iz} + e^{-iz}}{2} \qquad \sec z = \frac{1}{\cos z} = \frac{2}{e^{iz} + e^{-iz}}$$

$$\text{cosec} z = \frac{1}{\sin z} = \frac{2i}{e^{iz} + e^{-iz}} \qquad \tan z = \frac{\sin z}{\cos z} = \frac{e^{iz} - e^{-iz}}{i\left(e^{iz} + e^{-iz}\right)} \qquad \cot z = \frac{\cos z}{\sin z} = \frac{i\left(e^{iz} + e^{-iz}\right)}{e^{iz} - e^{-iz}}$$

Hyperbolic functions: The hyperbolic functions in terms of exponential functions are defined as follows:

$$\sinh z = \frac{e^z - e^{-z}}{2} \qquad \cosh z = \frac{e^z + e^{-z}}{2} \qquad \text{sech}\, z = \frac{1}{\cosh z} = \frac{2}{e^z + e^{-z}}$$

$$\text{cosech} z = \frac{1}{\sinh z} = \frac{2}{e^z - e^{-z}} \qquad \tanh z = \frac{\sinh z}{\cosh z} = \frac{e^z - e^{-z}}{e^z + e^{-z}} \qquad \coth z = \frac{\cosh z}{\sinh z} = \frac{e^z + e^{-z}}{e^z - e^{-z}}$$

Logarithmic function: If $z = e^w$, then we write w = log z, called the logarithm of complex number z. Thus the logarithmic function is the inverse of the exponential function and can be defined by $w = \log z = \{\ln |z| + i(\theta + 2k\pi), k = 0, \pm 1, \pm 2, ...,\}$ where $z = re^{i\theta} = re^{i(\theta + 2k\pi)}$, $k \in$ **Z**.

The function z^α: where α may be complex, is defined as $z^\alpha = e^{\alpha \ln z}$, sinceh $\ln z = \ln |z| + i(\phi + 2n\pi)$ where φ denotes the principal value of arg(z).

$$\therefore z^\alpha = e^{\alpha\left(\ln|z| + i(\phi + 2n\pi)\right)} = e^{\left(\alpha \ln|z| + i\alpha\phi + \alpha i 2n\pi\right)} = e^{\alpha \ln|z|} \ e^{i\alpha\phi} \ e^{2in\pi\alpha}$$

Algebraic: If w is a solution of the polynomial equation

$P_n(z)w^n + P_{n-1}(z)w^{n-1} + ... + P_1(z)w + P_0(z) = 0$, where $P_n \neq 0$, $P_i(z)$ are polynomials in z for all i and n is a positive integer, then w = f(z) is called an algebraic function of z such type of function involves only algebraic operations like addition, subtraction, multiplication and division as well as fractional or rational exponents.

For example: $w = z^{1/3}$ is a solution of equation $w^3 - z = 0$ and so is an algebraic function of z.

Transcendental functions: Any function which cannot be expressed as a solution of an equation is called a transcendental function. The logarithmic, trigonometric and hyperbolic functions and their corresponding inverse are examples of transcendental functions.
The function considered above, together with functions derived from them by a finite number of algebraic operations involving addition, subtraction, multiplication, division are called elementary functions.

3. Analyticity and C-R equations

Limit of a function:

Let $f : D \to \mathbf{C}$ be any function in a domain D given by $w = f(z) = u(x, y) + iv(x, y)$
Let $z_0 \in \mathbf{C}$ be a limit point of D, then l is the limit of f at $z=z_0$ if for any given $\in > 0$, there exists a $\in > 0$ such that $| f(z) - l | < \in$, whenever $0 < | z - z_0| < \in$ in D and we write $\lim\limits_{z\to z_0} f(z)=l$.

Note : 1. $\lim\limits_{z\to z_0} f(z)$ *exists iff* $\lim\limits_{(x,y)\to(x_0,y_0)} u(x,y)$ *and* $\lim\limits_{(x,y)\to(x_0,y_0)} v(x,y)$ *exist.*

2. The limit of a function f(z), if exists, must be unique.

Example : $\lim\limits_{z\to\infty} \dfrac{iz^2}{(z-1)^2}$

Solution : $\lim\limits_{z\to\infty} \dfrac{iz^2}{z^2\left(1-\dfrac{1}{z}\right)^2} \Rightarrow \text{I}$ i.e. $\lim\limits_{z\to\infty} \dfrac{iz^2}{(z-1)^2} = i$

L'Hospital Rule

Let f(z) and g(z) be analytic in a region containing the point z_0 and suppose that $f(z_0) = g(z_0) = 0$ but $g'(z_0) = 0$. Then L' Hospital rule states that $\lim\limits_{z\to z_0} \dfrac{f(z)}{g(z)} = \dfrac{f'(z_0)}{g'(z_0)}$. In case $f'(z_0) = g'(z_0) = 0$, the rule may be extended.

Example: The $\lim\limits_{z\to 0} \dfrac{\bar{z}^2}{z}$ is

(a) 0 (b) 1 (c) 1/2 (d) does not exist

Solution : (d)

We have $f(z)=\dfrac{\bar{z}^2}{z}=\dfrac{(x-iy)^2}{x+iy}=\dfrac{x^2-y^2-2ixy}{x^2+y^2}$

$\therefore u(x,y)=\dfrac{x^2-y^2}{x^2+y^2}$ *and* $v(x,y)=\dfrac{-2xy}{x^2+y^2}$

Approaching (0, 0) along y = mx, we get

$\lim\limits_{x\to 0} u(x,mx)=\lim\limits_{x\to 0}\dfrac{x^2-m^2x^2}{x^2+m^2x^2}=\dfrac{1-m^2}{1+m^2}$ which depends on m.

Hence, $\lim_{(x,y)\to(0,0)} u(x,y)$ does not exist and similarly $\lim_{(x,y)\to(0,0)} v(x,y)$ does not exist.

Hence, $\lim_{z\to 0} \frac{\bar{z}^2}{z}$ does not exist.

Continuity of a point:

A function f(z) is said to be continuous at z_0 if $\lim_{z\to z_0} f(z) = f(z_0)$

A function f(z) is said to be continuous at any interior point z_0 of region R, if for all $\in > 0$, there exists $\in > 0$ such that $|f(z) - f(z_0)| < \in$, whenever $|z - z_0| < \in$.

Example : $f(z) = \begin{cases} \frac{z \operatorname{Re} z}{|z|} & z \neq 0 \\ 0 & z = 0 \end{cases}$ then show that f(z) is continuous at z = 0.

Solution : We know that if f(z) is continuous at z_0 then $\lim_{z\to z_0} f(z) = f(z_0)$.

So, $\lim_{z\to 0} f(z) = f(0) = 0$ [$\because$ f(0) = 0]

Find $\lim_{z\to 0} \frac{z \operatorname{Re} z}{|z|} =$.

$f(z) = \frac{z \operatorname{Re} z}{|z|}$ [$\because$ z = x + iy ReZ = x $|z| = \sqrt{x^2+y^2}$]

Then $\lim_{z\to 0} \frac{x(x+iy)}{\sqrt{x^2+y^2}}$

We can use concept of function of two variable

So y = mx

$\lim_{z\to 0} \frac{x(x+imx)}{\sqrt{x^2+m^2x^2}}$ $\lim_{x\to 0} \frac{x^2(1+mi)}{x\sqrt{(1+m^2)}} \Rightarrow 0$ here we have to find

Limit at z = 0 then y = mx is set of all line which passes through z = 0 and limit of f(z) through each line are same so f(z) has limit at z = 0.

i.e. $\lim_{z\to 0} f(z) = 0 = f(0)$ f(z) is continuous at z = 0

Continuity in a region:

A function f(z) is said to be continuous in a region if it is continuous at all point of the region.

Result:

(1) f(z) is a continuous function in a domain D, it means f(z) is continuous at every $z \in D$.

(2) f(z) = u(x, y) + iv(x,y) is continuous function of z iff u and v are continuous functions of x and y.

(3) Composition mapping of continuous function is continuous.

Differentiability

If f(z) is single-valued in some region R of the z-plane, the derivative of f(z) is defined as

$$f'(z) = \lim_{z \to z_0} \frac{f(z) - f(z_0)}{z - z_0} \quad \text{........(1)}$$

Provided that the limit exists finitely.

Example : Let f(z) = $|z|^2$ at $z_0 = 0$

$\Rightarrow \quad \lim_{z \to 0} \frac{|z|^2}{z} = \lim_{z \to 0} \bar{z} = 0 \quad \Rightarrow$ It is differentiable at $z_0 = 0$

Example : Let f(z) = $\bar{z}$ at $z_0 = 0$ $\quad \Rightarrow \quad \lim_{z \to 0} \frac{\bar{z} - 0}{z - 0} = \lim_{z \to 0} \frac{\bar{z}}{z} = \lim_{(x,y) \to (0,0)} \frac{x - iy}{x + iy}$

Limit does not exist $\quad \Rightarrow$ It is not differentiable at $z_0 = 0$

4. Cauchy-Riemann Equations

A necessary condition that w = f(z) = u (x, y) + iv (x, y) is analytic in a region R is that, in R, u and v satisfy the Cauchy-Riemann equations, i.e.

$$\frac{\partial u}{\partial x} = \frac{\partial v}{\partial y}, \frac{\partial u}{\partial y} = -\frac{\partial v}{\partial x}$$

If the partial derivatives as given in (2) are continuous in R, then the Cauchy-Reiemann (c-R equations are sufficient conditions for f(z) to be analytic in R.

Example : $f(z) = x^3 + i(1 - y)^3$

Here $\quad u = x^3$ & $v = (1 - y)^3$

$u_x = 3x^2$, $u_y = 0$, $v_x = 0$ & $v_y = -3(1 - y)^2$

Here $\quad u_y = -v_x$

But $\quad u_x = 3x^2 = v_y = -3(1 - y)^3$ only at the curve $x^2 = (y - 1)^3$.

i.e. $\quad$ C-R equation satisfy at (0, 1), (1, 2) etc.

Polar form C-R equations:

If f(z) = u + iv is an analytic function and $z = re^{i\theta}$, where u, v, r, θ are all real, then the Cauchy Riemann equations are $\frac{\partial u}{\partial r} = \frac{1}{r}\frac{\partial v}{\partial \theta}, \frac{\partial v}{\partial r} = -\frac{1}{r}\frac{\partial u}{\partial \theta}$

Complex form of C-R equation :

Let f(z) = u + iv be a differentiable function, and $\frac{\partial f}{\partial \bar{z}} = 0$ at z_0 then the function satisfied C-R equation at z_0.

Example : Let $f(z) = \bar{z}$, then $\frac{\partial f}{\partial z} = 1 \neq 0$. ⇒ C-R equation not satisfy in complex plane.

Results:

(i) f(z) is analytic ⇒ f(z) is satisfy C-R equation

(ii) f(z) is analytic ⇒ f(z) is differential at uncountable point

(iii) If f(z) is differential at uncountable point then f(z) need not be analytic.

Example:- $f(z) = x^2 + iy^2$

$u = x^2$ & $v = y^2$

$u_x = 2x$ $\quad v_x = 0$

$u_y = 0$ $\quad v_y = 2y$

$u_x = v_y \Rightarrow y = x$

$v_x = u_y$

i.e. f(z) satisfy C-R equation at the all points of y = x line.

Let f(z) is differential at z_0 but f(z) is not differential at each points of nbd of z_0. So f(z) is not analytic.

(iv) If f(z) is differential at countable point then f(z) never analytic.

(v) If f(z) non-constant then $f(\bar{z})$ & $\overline{f(z)}$ never analytic.

(vi) If f(z) is analytic & $f(\bar{z})$ is also analytic then f(z) is constant function.

5. Harmonic Functions

If the second order partial derivatives of u and v with respect to x and y exist and are continuous in a region R, and $\frac{\partial^2 u}{\partial x^2}+\frac{\partial^2 u}{\partial y^2}=0, \frac{\partial^2 v}{\partial x^2}+\frac{\partial^2 v}{\partial y^2}=0$ then u and v are called harmonic functions.

Harmonic Conjugate: The function g is said to be a harmonic conjugate of h if g and h are harmonic and the first order partial derivatives of g and h satisfy the C-R equations.

Example : Let $f(x) = x^2 - y^2$ & $g(x) = 2xy$

Here $\frac{\partial^2 f}{\partial x^2}+\frac{\partial^2 f}{\partial y^2}=0$ and $\frac{\partial^2 g}{\partial x^2}+\frac{\partial^2 g}{\partial y^2}=0$.

$\Rightarrow$ Both are harmonic function and $u_x = v_y$ & $u_y = -v_x$.

So, they satisfy C-R equation. $\Rightarrow$ g is harmonic conjugate of f(x, y)

Results:(a) If f(z) = u + iv is analytic in a domain D, then v is a harmonic conjugate of u.

(b) Two real valued functions u(x, y) and v(x, y) are harmonic conjugates of each other if and only if they are constants.

Analytic Function

A function f(z) is said to be analytic at a point z_0 if there exists a neighbourhood $|z - z_0| < \square$ such that f' (z) exists at all points of the neighbourhood.

Entire function: A complex function f is said to be entire if it is analytic in the whole complex plane. Thus, the sum and the product of two or more entire functions are also entire functions.

Example : z, sinz, cosz all are entire function.

Construction of an analytic function

Method 1 : Milne-Thomson's method.

(i) If u(x, y) is known, we know that

$f(z) = u + iv$ $\quad f'(z) = u_x + v_x i$ .... (i)

By C-R equation $\quad f'(z) = u_x - u_y i$

Integration w.r.t. z $\quad \Rightarrow \quad f(z)=\int u_x(z,0)dz - i\int u_y(z,0)dz$

(ii) If v(x, y) is given from equation (i)

$$f'(z) = v_y + iv_x$$

Integrating w.r.t. z $f(z) = \int v_y(z,0)dz + i\int v_x(z,0)dz$

Example : The analytic function if real part of this function is 2xy.

i.e. $u = 2xy$

$u_x = 2y$ & $u_y = 2x$

We know that $f(z) = \int u_x(z,0)dz - i\int u_y(z,0)dz \quad = \int 0dz - i\int 2zdz$

$f(z) = -iz^2$

We can check

$f(z) = -i(x + iy)^2 \quad = 2xy - i(x^2 - y^2)$

Method 2 : (Using total derivative concept) in this method if u is given we will try to find its harmonic conjugate v by using total derivative concept and C-R equations as follows

Let u is given then by C-R equation $u_x = v_y$ and $u_y = -v_x$

$\therefore$ v = v(x, y) so by using total derivative concept

$$dv = \left(\frac{\partial v}{\partial x}\right)dx + \left(\frac{\partial v}{\partial y}\right)dy = \left(-\frac{\partial u}{\partial y}\right)dx + \left(\frac{\partial u}{\partial x}\right)dy$$

Now by integrating, we can find v.

So analytic function can be give as f(z) = u + iv

COMPLEX ANALYSIS

SOLVED QUESTIONS

1. Let f(z) be an entire function then which of the following is/are entire?

CSIR NET JUNE 2023

(a) $f(\bar{z})$ (b) $\overline{f(z)}$ (c) $\overline{f(\bar{z})}$ (d) $f(\bar{z}) + \overline{f(z)}$

Ans. (c)

Explanation :

We know that if f(z) be an entire function then $f(\bar{z})$ and $\overline{f(z)}$ never entire

but $\overline{f(\bar{z})}$ is an entire function

2. Let f be an entire function that satisfies $|f(z)| \le e^y$ for all $z \in C$, z = x + iy, x, y $\in$ R. Which of the following is true?

CSIR NET JUNE 2023

(a) $f(z) = c\, e^{-iz}$ for some $c \in \mathbb{C}$ with $|c| \le 1$

(b) $f(z) = c\, e^{iz}$ for some $c \in \mathbb{C}$ with $|c| \le 1$

(c) $f(z) = e^{-ciz}$ for some $c \in \mathbb{C}$ with $|c| \le 1$

(d) $f(z) = e^{ciz}$ for some $c \in \mathbb{C}$ with $|c| \le 1$

Ans. (a)

Explanation :

(a) $f(z) = c\,e^{-i(x+iy)} = c\,e^{-ix}e^{y}$

$|f(z)| = |ce^{y}| = |c|\,|e^{y}| = |c|e^{y}$

Given that $|f(z)| \le e^{y}$ $\Rightarrow |c|e^{y} \le e^{y}$ $\Rightarrow |c| \le 1$

Option (a) is true.

(b) $|f(z)| \le e^{y}$ $\Rightarrow |ce^{ix}e^{-y}| \le e^{y} \Rightarrow |c|e^{-y} \le e^{y}$ $\Rightarrow |c| \le e^{2y}$

We can't say that $|c| \le 1$

Similar option (c) & (d) are incorrect.

3. For any complex valued function f let D_f denote the set on which the function f satisfies Cauchy-Riemann equations. Identity the function for which D_f is equal to C.

CSIR NET FEB 2022

(a) $f(z) = \dfrac{z}{1+|z|}$ (b) $f(z) = (\cos \alpha x - \sin \alpha y) + i(\sin \alpha x + \cos \alpha y)$, where $z = x + iy$

(c) $f(z) = \begin{cases} e^{-1/z^4} & if \quad z \ne 0 \\ 0 & if \quad z = 0 \end{cases}$ (d) $f(z) = x^2 + iy^2$ where $z = x + iy$

Ans. (c)

Explanation :

(a) $f(z) = \dfrac{z}{1+|z|} = \dfrac{x}{1+\sqrt{x^2+y^2}} + i\dfrac{y}{1+\sqrt{x^2+y^2}}$

$$u_x = \frac{(1+\sqrt{x^2+y^2}) - x\dfrac{2x}{2\sqrt{x^2+y^2}}}{(1+\sqrt{x^2+y^2})^2} = \frac{\sqrt{x^2+y^2}+y^2}{(1+\sqrt{x^2+y^2})^2\sqrt{x^2+y^2}}$$

$$u_y = \frac{-x.\dfrac{2y}{2\sqrt{x^2+y^2}}}{(1+\sqrt{x^2+y^2})^2} = \frac{-xy}{(1+\sqrt{x^2+y^2})^2\sqrt{x^2+y^2}}$$

$$v_x = \frac{-xy}{(1+\sqrt{x^2+y^2})^2\sqrt{x^2+y^2}}$$

$$v_y = \frac{\sqrt{x^2+y^2}+x^2}{(1+\sqrt{x^2+y^2})^2\sqrt{x^2+y^2}}$$

$D_f = \{(0, 0)\}$

(b) $u = \cos \alpha x - \sin \alpha y$ and $v = \sin \alpha x + \cos \alpha y$

$u_x = -\alpha \sin \alpha x$, $v_x = \alpha \cos \alpha x$, $u_y = -\alpha \cos \alpha y$, $v_y = -\alpha \sin \alpha y$

$D_f = \{z = \alpha x + i\alpha y \in C;\ \alpha x = 2n\pi + \alpha y\}$

(c) $f_{\bar{z}} = 0$ for all $z \in C$

So, $D_f = C$

(d) $f(z) = x^2 + iy^2$

$u_x = 2x,\ u_y = 0$

$v_x = 0,\ v_y = 2y$

$D_f = \{z = x + iy \in C : x = y\}$

So, option (c) is correct.

4. Define $f : C \to C$ by $f(z) = | |z|^2 - 1|^2$. Which of the following statements is true?

CSIR NET NOV 2020

(a) f is complex differentiable at $z \in \{0, 1, i\}$ (b) f is complex differentiable only at z = 0

(c) f is complex differentiable only at z = 1

(d) f is complex differentiable only for $z \in \{0, 1\}$

Ans. **(a)**

Explanation :

$$f(z) = ||z|^2 - 1|^2 = (|z|^2 - 1)\overline{(|z|^2 - 1)}$$

$$f(z) = (z\bar{z} - 1)\overline{(z\bar{z} - 1)}$$

$$f(z) = (z\bar{z} - 1)^2$$

$$\frac{\partial f}{\partial \bar{z}} = 2(z\bar{z} - 1)z$$

C-R equation is satisfies at $z\bar{z} = 1 (|z| = 1)$ and z = 0

So, f is not differentiable other than z = 0 and |z| = 1

(i) Check at z = 0 $\quad \lim_{z \to 0} \frac{(z\bar{z} - 1)^2 - 1}{z - 0} = 0$

$\Rightarrow$ f(z) is differentiable at z = 0

(ii) Check at z = 1 $\quad \lim_{z \to 1} \frac{(z\bar{z} - 1)^2 - 0}{z - 1} = 0$

$\Rightarrow$ f(z) is differentiable at z = 1

(iii) Check at z = I $\quad \lim_{z \to i} \frac{(z\bar{z} - 1)^2 - 0}{z - i} = 0$

$\Rightarrow$ f(z) is differentiable at z = i

Hence All Option are incorrect

5. Let f(z) be a non-constant entire function and z = x + iy. Let u(x, y), v(x, y) denote its real and imaginary parts respectively. Which of the following statements is false?

(a) $u_x = v_y$ and $u_y = -v_x$ (b) $u_y = v_x$ and $u_x = -v_y$

(c) $|f'(x + iy)|^2 = u_x(x, y)^2 + v_x(x, y)^2$ (d) $|f'(x + iy)|^2 = u_y(x, y)^2 + v_y(x, y)^2$

Ans. **(b)** **CSIR NET FEB 2022**

COMPLEX ANALYSIS

Explanation :

f(z) be non-constant entire function

$f(z) = u(x, y) + iv(x, y)$

By C-R equations $u_x = v_y$ and $u_y = -v_x$

Option (a) is correct and (b) is incorrect and $f'(z) = u_x + iv_x$.

$\Rightarrow \quad |f'(z)|^2 = u_x^2 + v_x^2$

Using C – R equation $|f'(z)|^2 = v_x^2 + u_y^2$

Option (c) and (d) are correct statements.

Only option (b) is false.

6. Consider the real function of two real variables given by $u(x, y) = e^{2x}$ [sin 3x cos 2y cosh 3y – cos 3x sin 2y sinh 3y]. Let v(x, y) be the harmonic conjugate of u(x, y) such that v(0, 0) = 2. Let z = x + iy and f(z) = u (x, y) + iv(x, y), then the value of 4 + 2if (iπ) is **GATE - 2021**

(a) $e^{3\pi} + e^{-3\pi}$ (b) $e^{3\pi} - e^{-3\pi}$ (c) $-e^{3\pi} + e^{-3\pi}$ (d) $-e^{3\pi} - e^{-3\pi}$

Ans. (c)

Explanation:

Here u (0, 0) = 0,

and $f(z) = u(x, y) + iv(x, y)$

So $f(0) = 0 + i2 = 2i$ ... (1)

We are given with

$u(x, y) = e^{2x}$ (sin3xcos2ycosh3y – cos 3x sin2y sinh 3y)

and v(x, y) be harmonic conjugate of u(x,y) such that v(0,0) = 2.

Then $f(z) = u(x,y) + iv(x,y) \Rightarrow f(0) = u(0,0) + iv(0,0) = 2i$

We know that If u(x ,y) is given than

$$f(z) = \int u_x(z,0)dz - i\int u_y(z,0)dz \quad \text{...(1)}$$

$u(x, y) = e^{2x}$ [sin 3x cos 2y cosh 3y – cos 3x sin 2y sinh 3y]

$\Rightarrow u_x(x, y) = e^{2x}$ [3cos 3x cos 2y cosh 3y +3 sin 3x sin 2y sinh 3y] $+ 2e^{2x}$ [sin 3x cos 2y cosh 3y – cos 3x sin 2y sinh 3y]

$\Rightarrow u_x(z, 0) = e^{2z}[3\cos 3z + 0] + 2e^{2z}[\sin 3z - 0]$

$\Rightarrow u_x(z, 0) = e^{2z}[3\cos 3z + 2\sin 3z]$ ...(2)

$u_y(x, y) = e^{2x}$ [-2 sin 3x sin 2y cosh 3y + 3sin 3x cos 2y sinh 3y – 2cos 3x cos 2y sinh 3y –3cos 3x sin 2y cosh 3y] $\Rightarrow u_y(z, 0) = 0$

...(3)

From (1)

$$f(z) = \int e^{2z}(3\cos 3z + 2\sin 3z)dz - i\int 0.dz$$

$$f(z) = e^{2z}\sin 3z + c$$

Using condition f(0) = 2i Then $f(z) = e^{2z}\sin 3z + 2i$

$4 + 2if(i\pi) = 4 + 2i[e^{2\pi i}\sin(3\pi i) + 2i]$

$= 4 + 2i(i\sinh 3\pi + 2i)$ $\qquad = 4 - 2\sinh 3\pi - 4$

$= -2\sinh 3\pi$ $\qquad = -2\left(\frac{e^{3\pi} - e^{-3\pi}}{2}\right) = e^{-3\pi} - e^{3\pi}$

7. Let f be an analytic function on C such that $f(z)^2 + f'(z)^2 = 1$. Consider the following set X = { z : f'(z) = 0 } and Y = { z : f''(z) + f(z) = 0 }. Which of the following are true **CSIR NET SEP 2022**

(a) Either X or Y has limit point (b) If Y has a limit point then f must be constant

(c) If X has a limit point then f must be constant (d) f(z) ∈ { -1,1} for all z

Ans. **(a), (c)**

Explanation:

Given that $f(z)^2 + f'(z)^2 = 1$

Then f(z) is either constant or f(z) is sinz

Case – 1

If f(z) = sinz

Then $X = \left\{(2n+1)\frac{\pi}{2}; n \in Z\right\}$ and Y = C

(a) is true and (b) is false

If X has a limit point then X = C ⇒ f(z) = constant

8. For z ∈ C, let $f(z) = \begin{cases} \bar{z}^2 / z, & if\ z \neq 0 \\ 0, & otherwise \end{cases}$.

Then which of the following statements is false?

(a) f(z) is continuous everywhere

(b) f(z) is not analytic in any open neighbourhood of zero.

(c) zf(z) satisfies the Cauchy-Riemann equations at zero.

(d) f(z) is analytic in some open subset of C. **CSIR NET DEC 2019**

Ans. **(d)**

Explanation :

(a) $\lim_{z \to 0} \frac{\bar{z}^2}{z} = 0 = f(0)$ Hence, f(z) is continuous.

(b) $\frac{\partial f}{\partial \bar{z}} = \frac{2\bar{z}}{z} = 0$ only at z = 0, so f(z) is not differentiable at other than 0.

f is nowhere analytic.

(c) zf(z) = $\bar{z}^2 = g(z)$

$\frac{\partial g}{\partial \bar{z}} = 2\bar{z} = 0$ only at z = 0,

⇒ zf(z) satisfied C-R equation at 0.

(d) We know that f is nowhere analytic

Option (d) is false statement.

So, (d) is true.

9. Consider the polynomial $f(z) = z^2 + az + p^{11}$, where $a \in Z \setminus \{0\}$ and $p \geq 13$ is a prime. Suppose that $a^2 \leq 4p^{11}$. Which of the following is true? **CSIR NET DEC 2019**

(a) f has a zero on the imaginary axis

(b) f has a zero for which the real and imaginary parts are equal

(c) f has distinct roots

(d) f has exactly one real root

Ans. (c)

Explanation :

Given that $f(z) = z^2 + az + p^{11}$ is a quadratic equation

Here B = a and $C = p^{11}$ compare with $Az^2 + Bz + C = 0$

$B^2 - 4AC = a^2 - 4p^{11} \leq 0$ (Given)

But for $p \geq 13$,

$a^2 - 4p^{11} \neq 0$, hence $a^2 < 4p^{11}$

⇒ f has distinct roots

10. Let P(z), Q(z) be two complex non-constant polynomials of degree m, n respectively. The number of roots of P(z) = P(z) Q(z) counted with multiplicity is equal to:

(a) min { m, n}

(b) max {m, n}

(c) m + n

(d) m - n

CSIR NET JUNE -2016

Ans. (c)

Explanation :

Consider $P(z) = (z-1)^2$ and $Q(z) = z^3$.

Thus, P(z) and Q(z) are two complex non-constant polynomials of degree m = 2 and n = 3 respectively

⇒ P(z) = P(z) Q(z) ....(1)

$\Rightarrow (z-1)^2 = (z-1)^2 z^3$ $\Rightarrow (z-1)^2 = (1-z)^3 = 0$

⇒ The number of roots of equation (1) counted with multiplicity is equal to 5

⇒ Options (a), (b), (d) are incorrect. Hence, option (c) is correct.

11. Consider the polynomial $P(z) = \left(\sum_{n=0}^{5} a_n z^n\right)\left(\sum_{n=0}^{9} b_n z^n\right)$, where $a_n, b_n \in \mathbf{R}\ \forall$ n, $a_5 \neq 0$, $b_9 \neq 0$. Then counting roots with multiplicity we can conclude that P(z) has

(a) at least two real roots

(b) 14 complex roots

(c) no real roots

(d) 12 complex roots.

CSIR NET DEC – 2016

Ans. (a), (b)

COMPLEX ANALYSIS

Explanation :

Given $P(z) = \left(\sum_{n=0}^{5} a_n z^n\right)\left(\sum_{n=0}^{9} b_n z^n\right)$ $a_n, b_n \in \mathbf{R}\ \forall$ n, $a_5 \neq 0$, $b_9 \neq 0$, is a polynomial of degree 14 in **C**. Thus by fundamental theorem of algebra all roots exist in **C**.

$\Rightarrow$ option (b) is correct.

Now $\sum_{n=0}^{5} a_n z^n$, has at least one real root and similarly $\sum_{n=0}^{9} b_n z^n$ has at least one real root.

Otherwise a_n and b_n will no longer remain real.

$\Rightarrow P(z)$ has atleast 2 real roots

Thus, option (a), (b) are correct and (c), (d) are incorrect.

12. Let f be a non-constant holomorphic function in the unit disc $\{z \in C : |z| < 1\}$ such that f(0) = 1. Then it is necessary that

(a) there are infinitely many point z in the unit disc such that |f (z) | = 1

(b) f is bounded

(c) there are at most finitely many points z in the unit disc such that |f (z) | = 1

(d) f is a rational function. **CSIR NET DEC-2013**

Ans. (a)

Explanation :

Take f(z) = z + 1

Here f(0) =1

Clearly, f(z) is unbounded

$\Rightarrow$ option (b) is incorrect.

Also |f(z) | = |z + 1|

|f(z)| = 1 $\Rightarrow$ |z +1| = 1, which represents a unit circle centered at (-1, 0)

$\Rightarrow$ There are infinitely many z which satisfies |f(z) | = 1

Thus, option (a) is correct and (c) is incorrect.

Further, take f(z) = e^z,

Here f(0) = 1 and f(z) is analytic in the unit disc. But f(z) is not a rational function

$\Rightarrow$ option (d) is incorrect.

13. Let f : C $\to$ C be a holomorphic function and let u be the real part of f and v be the imaginary part of f. Then $|f'(x + iy)|^2$ is equal to **CSIR NET DEC. 2017**

(a) $u_x^2 + u_y^2$ (b) $u_x^2 + v_x^2$ (c) $v_y^2 + u_y^2$ (d) $v_y^2 + v_x^2$

Ans. (All)

Explanation :

f(z) = u + iv then f'(z) = ux + ivx

u(x) = vy and uy = -vx [$\therefore$ Given that f(z) is analytic, so C-R equation satisfy]

$$f'(z) = ux - iuy \qquad f'(z) = vy + ivx \qquad f'(z) = vy - iuy$$

So $|f'(z)| = \sqrt{4x^2 + vy^2}$ $\quad |f'(z) = \sqrt{vy^2 + vx^2}$ $\quad |f'(z)| = \sqrt{vy^2 + vy^2}$

$|f'(z)|^2 = 4x^2 + vy^2$ $\quad |f'(z)|^2 = vx^2 + vy^2$ $\quad |f'(z)|^2 = uy^2 + vy^2$

and $f'(z) = ux + ivx$

$|f'(z)|^2 = ux^2 + vx^2$

14. Let $p(z) = z^n + a_{n-1}z^{n-1} + \ldots.. + a_0$ where $a_0, a_1, \ldots.. a_{n-1}$ are complex number and let $q(z) = 1 + a^{n-1}z + a_0z^n$ if $|p(z)| \leq 1$ for all z with $|z| \leq 1$. **CSIR NET DEC. 2017**

(a) $|q(z)| \leq 1 \quad \forall$ z with $|z| \leq 1$

(b) q(z) is a constant $|z| \leq 1$

(c) $p(z) = z^n\ \forall$ z

(d) p(z) is constant polynomial.

Ans. (a), (b), (c)

Explanation :

Given that $p(z) \leq 1 \quad \forall z$ with $|z| \leq 1$

$p(z) = z^n + a_{n-1}z^{n-1} \ldots\ldots a_0$

Suppose $a_0, a_1, \ldots. a_{n-1}$ all are zero then above cond^n satisfy

$p(z) = z^n$ and $q(z) = 1$

(a) $q(z) = 1\ \forall z$ with $|z|<1$

(b) q(z) is a constant polynomial

(c) $p(z) = z^n$, $\forall n \in C$

(d) $p(z) = z^n$

15. Let $u(x, y) = x^3 - 3xy^2 + 2x$ then which of the following function v is u + iv a holomorphic function **CSIR NET DEC 2014**

(a) $v = y^3 - 3x^2y + 2y$ (b) $v = 3x^2y - y^3 + 2y$ (c) $v = x^3 - 3xy^2 + 2x$ (d) $v = 0$

Ans. (b)

Explanation :

Given that f(z) is holomorphic so C-R equation satisfy

$u(x) = v_y$ & $u_y = -v_x$

Given $u = x^3 - 3xy^2 + 2x$, $u_x = 3x^2 - 3y^2 + 2 = vy$, $v_y = 3x^2 - 3y^2 + 2$

Integral w.r.t. y $\quad v = 3x^2y - y^3 + 2y + k(x)$

Option (a), (c), (d) discard. Now $vx = 6xy + k'(x)$ and $uy = -6xy$

We know that $vx = -uy$

$k' = 0$

$k = 0$

So, $v = 3x^2y - y^3 + 2y$

16. Let p(z), q(z) be two non-zero complex polynomials. Then $p(z)\overline{q(z)}$ is analytic if and only if **CSIR NET JUNE 2013**

(a) p(z) is constant

(b) p(z)q(z) is constant

(c) q(z) is a constant

(d) $\overline{p(z)}q(z)$ is a constant.

Ans. (c)

Explanation :

Given that p(z) and q(z) are complex polynomials and $p(z)\overline{q(z)}$ is analytic function.

$\Rightarrow \overline{q(z)}$ is also analytic. ....(1)

Given that q(z) is analytic function. ....(2)

We know that if q(z) and $\overline{q(z)}$ both are analytic function then q(z) must be constant.

From (1) & (2) q(z) is constant.

17. Let F be a real valued harmonic function on C, that is, f satisfies the equation $\frac{\partial^2 f}{\partial x^2}+\frac{\partial^2 f}{\partial y^2}=0$. Define the functions $g=\frac{\partial f}{\partial x}+i\frac{\partial f}{\partial y}, h=\frac{\partial f}{\partial x}+i\frac{\partial f}{\partial y}$. Then

(a) g and h are both holomorphic functions.

(b) g is holomorphic, but h is need not be holomorphic.

(c) h is holomorphic, but g need not be holomorphic.

(d) both g and h are identically equal to the zero function.

Ans. (b) **CSIR NET JUNE 2015**

Explanation :

Given f : C → R is harmonic function i.e. $\frac{\partial^2 f}{\partial x^2}, \frac{\partial^2 f}{\partial y^2}, \frac{\partial^2 f}{\partial x \partial y}$ exist and are continuous

Also, $\frac{\partial^2 f}{\partial x^2}+\frac{\partial^2 f}{\partial y^2}=0$

Now, define $g=\frac{\partial f}{\partial x}+i\frac{\partial f}{\partial y}$ and $h=\frac{\partial f}{\partial x}+i\frac{\partial f}{\partial y}$

Consider example

f(z) = xy = i0

$f_x = y, f_{xx} = 0, f_y = x, f_{yy} = 0 \Rightarrow \frac{\partial^2 f}{\partial x^2}+\frac{\partial^2 f}{\partial y^2}=0 \Rightarrow$ f is harmonic.

Now, $h=\frac{\partial f}{\partial x}+i\frac{\partial f}{\partial y}=y+ix=i(x-iy)=i\bar{z}$ $g=\frac{\partial f}{\partial x}-i\frac{\partial f}{\partial y}=y-ix$

Both are not identically zero.

(b) To show g is holomorphic (analytic)

$g=\frac{\partial f}{\partial x}+i\frac{\partial f}{\partial y}=u+iv$ (let) where $u=\frac{\partial f}{\partial x}, v=\frac{\partial f}{\partial y}$

$u_x=\frac{\partial^2 f}{\partial x^2}, v_y=-\frac{\partial^2 f}{\partial y^2}$

$\therefore \quad \frac{\partial^2 f}{\partial x^2}+\frac{\partial^2 f}{\partial y^2}=0 \quad \Rightarrow \quad u_x = v_y$ Now, $u_y = \frac{\partial^2 f}{\partial x \partial y}, v_x = \frac{-\partial^2 f}{\partial y \partial x}$

$\Rightarrow$ uy = -vx $\Rightarrow$ g satisfies C – R equation

and u_x, u_y, v_x, v_y exists and are continuous

$\Rightarrow$ g is differentiable on C $\Rightarrow$ g is analytic on C

Similarly, we can show that h is not analytic by showing that it does not satisfies C.R. equation.

18. Let f : C → C be a non-constant entire function and let image (f) = {w ∈ C : ∃ Z ∈ C s.t. f(z) = w}. Then **CSIR NET DEC 2018**

(a) The interior of image (f) is empty.

(b) image (f) intersected every line passing through origin.

(c) ∃ a disc in C which is disjoint from image (f)

(d) image (f) contain all its limit point.

Ans. **(a)**

Explanation :

By picard method ⇒ Image (f) = C or image (f) = C/{α}

Interior point : Let S ⊆ C be a set then α ∈ S is called interior point if ∃ nbd of α s.t. it contain in S.

The set of all interior point is called interior.

19. Consider the set F = {f : C → C | f is entire function, |f'(z)| ≤ |f(z)| ∀ z∈C}. Then which of the following is TRUE? **CSIR NET JUNE 2019**

(a) F is finite set

(b) F is infinite set

(c) F = {$\beta e^{\alpha x}$, β∈C, α∈C}

(d) F = {$\beta e^{\alpha x}$, β∈C , |α|≤1}

Ans. **(b), (d)**

Explanation :

Given that |f'(z)| ≤ |f(z)| Choose $f(z) = e^{2x}$

Then $f'(z) = 2e^{2x}$ Here |f'(z)| ≤ |f(z)|

So we choose always $f(z) = \beta e^{\alpha x}$, |α| ≤ 1

So, (b) and (d) are true.

20. Let f = u + iv be an entire function. If the Jacobian matrix $Ja = \begin{bmatrix} u_x(a) & u_y(a) \\ v_x(a) & v_y(a) \end{bmatrix}$ is symmetric ∀ a∈C. Then **CSIR NET JUNE 2017**

(a) f is polynomial

(b) f is polynomial of degree ≤ 1

(c) f is necessary constant function

(d) f is polynomial of degree > 1

Ans. **(a), (b)**

Explanation :

Option (a) : Let f(z) = z = x + iy

$Ja = \begin{bmatrix} 1 & 0 \\ 0 & 1 \end{bmatrix}$ is symmetric $\forall$ a∈C

Option (b) : Degree of f(z) ≤ 1

Option (c) : f is not necessary constant function.

Option (d) : Let $f(z) = z^2 = x^2 - y^2 + i2xy$ $\quad Ja = \begin{bmatrix} 2x & -2y \\ 2y & 2x \end{bmatrix}$

is not symmetric $\forall$ a∈C

it is symmetric only a = 0

21. For a positive integer p, consider the holomorphic function $f(z) = \dfrac{\sin z}{z^p}$ for z ∈ C\ {0}. For which of the values of p does there exist a holomorphic function g: C\{0} → C such that f(z) = g'(z) for all z ∈ C\ {0}?

(a) All even integers (b) All odd integers

(c) All multiples of 3 (d) All multiples of 4

Ans. (b) **CSIR NET JUNE 2020**

Explanation :

$$f(z) = \frac{\sin z}{z^p} \text{ for } z \in C\backslash\{0\}.$$

$$\frac{\sin z}{z^p} = \frac{z - \frac{z^3}{3!} + \frac{z^5}{5!} - \frac{z^7}{7!} + ...}{z^p} = \frac{z}{z^p} - \frac{z^3}{z^p \times 3!} + \frac{z^5}{z^p \times 5!} - \frac{z^7}{z^p \times 7!} + ...$$

$$= \frac{1}{z^p - 1} - \frac{1}{z^{p-3} \times 3!} + \frac{1}{z^{p-5} \times 5!} - \frac{1}{z^{p-7} \times 7!} + ... = z^{1-p} - \frac{z^{p-3}}{3!} + \frac{z^{p-5}}{5!} - \frac{z^{p-7}}{7!} + ...$$

For p = 1, $\dfrac{\sin z}{z^p} = \dfrac{\sin z}{z} = z^{1-1} - \dfrac{z^{3-1}}{3!} + \dfrac{z^{5-1}}{5!} - \dfrac{z^{7-1}}{7!} + ... = 1 - \dfrac{z^2}{3!} + \dfrac{z^4}{5!} - \dfrac{z^6}{7!} + ...$, a polynomial in z, is entire and exist primitive (anti derivative), eliminate Option (a), (c) and (d). So, option (b) is correct.

UNSOLVED QUESTIONS

1. Let f(z) be an entire function such that for some constant α, |f(z)| ≤ a |z|³ for |z| ≥ 1 f(z) = f(iz) for all z ∈ **C**. Then **GATE-2006**

(a) $f(z) = \alpha z^3$ for all z ∈ **C** (b) f(z) is a constant

(c) f(z) is a quadratic polynomial (d) no such f(z) exists

2. Which of the following is not the real part of an analytic function? **GATE-2006**

(a) $x^2 - y^2$ (b) $\dfrac{1}{1 + x^2 + y^2}$ (c) cos x cosh y (d) $x + \dfrac{x}{x^2 + y^2}$

COMPLEX ANALYSIS

3. Define f: **C** → **C** by $f(z)=\begin{cases}0, & \text{if } Re(z)=0 \text{ or } Im(z)=0 \\ z, & \text{otherwise}\end{cases}$. Then the set of points where f is analytic is

(a) {z : Re(z) ≠ 0 and Im (z) ≠ 0}
(b) {z : Re(z) ≠ 0}
(c) {z : Re(z) ≠ 0 or Im (z) ≠ 0}
(d) {z : Im(z) ≠ 0} **GATE-2007**

4. Let S be the open unit disk and f : S → **C** be a real valued analytic function with f(0) = 1. Then the set {z∈S : f(z) ≠ 1} is

(a) empty
(b) non-empty finite
(c) countably infinite
(d) uncountable **GATE-2008**

5. Let u(x, y) = 2x (1-y) for all real x and y. Then a function v(x, y), so that f(z) = u(x, y) + iv (x, y) is analytic, is **GATE-2010**

(a) $x^2 - (y-1)^2$ (b) $(x-1)^2 - y^2$ (c) $(x-1)^2 + y^2$ (d) $x^2 + (y-1)^2$

6. Let f(z) be analytic on D = {z ∈ **C** : |z – 1| < 1} such that f(1) = 1. If $f(z) = f(z^2)$ for all z ∈ D, then which one of the following statement is not correct?

(a) $f(z)=[f(z)]^2$ for all $z \in D$
(b) $f\left(\frac{z}{2}\right)=\frac{1}{2}f(z)$ for all $z \in D$
(c) $f(z^3)=[f(z)]^3$ for all $z \in D$
(d) $f'(1)=0$ **GATE-2010**

7. Let f(z) be entire function such that |f(z)| < K |z|, ∀ z ∈ **C**, for some K > 0. If f(1) = i, the value of f(i) is: **GATE-2011**

(a) 1 (b) -1 (c) i (d) -i

8. Let f be an entire function on **C** such that |f(z)| ≤ 100 log |z| for each z with |z| ≥ 2. If f(i) = 2i, then f(1). **GATE-2013**

(a) must be 2
(b) must be 2i
(c) must be i
(d) cannot be determined from the given data

9. Let f : **C** → **C** be a complex valued function given by f(z) = u (x, y) + iv (x, y). Suppose that v (x, y) = $3xy^2$. Then

(a) f cannot be holomorphic on **C** for any choice of u.
(b) f is holomorphic on **C** for a suitable choice of u.
(c) f is holomorphic on **C** for all choices of u.
(d) v is not differentiable as a function of x and y. **CSIR NET JUNE-2011**

10. Let f : **C** → **C** be an analytic function. For z = x + iy, let u, v : $\mathbf{R}^2$ → **R** be such that u(x, y) = Re f(z) and v(x, y) = Im f(z). Which of the following are correct?

(a) $\frac{\partial^2 u}{\partial x^2}+\frac{\partial^2 u}{\partial y^2}=0$ (b) $\frac{\partial^2 v}{\partial x^2}+\frac{\partial^2 v}{\partial y^2}=0$ (c) $\frac{\partial^2 u}{\partial x\partial y}-\frac{\partial^2 u}{\partial y\partial x}=0$ (d) $\frac{\partial^2 v}{\partial x\partial y}+\frac{\partial^2 v}{\partial y\partial x}=0$

CSIR NET JUNE-2013

11. The function $f(z)=|z|^2+i\bar{z}+1$ is differentiable at **GATE 2014**

(a) i (b) 1 (c) –i (d) no point in C

12. Let $u(x,y)=x^3+ax^2y+bxy^2+2y^3$ be a harmonic function and $v(x,y)$ its harmonic conjugate. If $v(0,0)=1$ then $|a+b+v(1,1)|$ is equal to. **GATE 2016**

(a) 2 (b) 5 (c) 10 (d) 20

13. The points of unit circle |z| = 1 forms a/an ?

(a) open set (b) closed set (c) semi-open set (d) none of these

14. The complement of the unit circle |z| = 1 is

(a) an open set (b) closed set

(c) neither open nor closed (d) none of these

15. The principal value of $(i)^{-2i}$ is

(a) $e^{-\pi}$ (b) e^{π} (c) $e^{-\pi/2}$ (d) $e^{-2\pi}$

16. Principal value of log i is

(a) $\frac{i\pi}{2}$ (b) $-\frac{i\pi}{2}$ (c) $\frac{\pi}{2}$ (d) $\frac{i}{2}$

17. The value of sin (log i^i) is

(a) 1 (b) -1 (c) π (d) -π

18. Real part of the principal value of $i^{\log(1+i)}$ is

(a) $e^{-\frac{\pi^2}{8}} cos\left(\frac{\pi}{4} log\,2\right)$ (b) $e^{\frac{\pi^2}{8}} log\,2$ (c) $e^{\frac{\pi^2}{8}} cos\left(\frac{\pi}{4} log\,2\right)$ (d) $e^{\frac{\pi^2}{8}}$

19. Let z_1 = -z, z_2 = iz, z_3 = z – iz. The area of the triangle in the Gaussian plane is

(a) $\frac{3}{2}|z|^2 i$ (b) $2|z|^2 i$ (c) $\frac{1}{2}|z|^2 i$ (d) none of these

20. The maximum value of |z|, when z satisfies the condition $\left|z-\frac{2}{z}\right|=2$ is

(a) $\sqrt{3}-1$ (b) $\sqrt{3}$ (c) $\sqrt{3}+1$ (d) $\sqrt{2}+\sqrt{3}$

21. $|z-4i|^2+|z+4i|^2=10$ will represent

(a) circle (b) ellipse (c) hyperbola (d) none of these

22. The complex numbers z_1 = 1 + 2i, z_2 = 4 – 2i and z_3 = 1 – 6i does not form the vertices of a /an

(a) right angled triangle (b) isosceles triangle (c) equilateral triangle (d) scalene triangle

23. The principal value of $\left(1+\sqrt{3}i\right)^i$ is

(a) $i\,log\left(1+\sqrt{3}i\right)$ (b) $e^{i\ln 2-\pi/3}$ (c) $e^{i\ln 2-i\pi/6}$ (d) none of these

24. If the set S is open, then which of the following is not always true?
(a) S does not contain its boundary points (b) S contain its boundary points
(c) S have finite boundary points (d) S contains all its limits points

25. $f(z) = |z|^2$ is continuous everywhere but nowhere differentiable
(a) except at origin (b) except at $|z| = 1$ (c) in **C** (d) except at $z = 1$

26. If a function f(z) is analytic at a point $z = z_0$, then which of the following statements is/are true?
(a) f(z) is differentiable at z_0 (b) f(z) is not defined at z_0
(c) f(z) is differentiable at infinitely many points (d) f(z) is continuous at z_0

27. The function $|\bar{z}|^2$ is
(a) analytic (b) differentiable at z = 0 (c) not analytic (d) continuous

28. If $f(z) = (x^2 + axy + by^2) + i(cx^2 + dxy + y^2)$ is analytic, then values of a, b, c, d are
(a) c = 1, d = a, b = 1 (b) d = 2, a = 2, a = -2c, b = -d/2 (c) a = 2, b = -1, c = -1, d = 2
(d) because u, v are harmonic, so a, b, c, d, are not possible to find

29. $e^{\alpha x} \cos \beta y$ is a harmonic function
(a) for all α, β (b) if $\alpha^2 + \beta^2 = 0$ (c) if $\alpha^2 - \beta^2 = 0$ (d) none of these

30. Let $f : \mathbf{C} \to \mathbf{C}$ be given by $f(z) = \begin{cases} (\bar{z})^2 / z & when\ z \neq 0 \\ 0 & when\ z = 0 \end{cases}$. Then f
(a) is not continuous at z = 0 (b) is differentiable but not analytic at z = 0
(c) is analytic at z = 0 (d) Satisfies the Cauchy-Riemann equations at z = 0

31. The function $f(z) = a_0 + a_1\bar{z} + a_2\bar{z}^2 : a_0, a_1, a_2 \in C$ (C is the field of complex numbers) $a_1, a_2 \neq 0$ is
(a) Differentiable for all $z \in \mathbf{C}$ (b) Continuous for all $z \in \mathbf{C}$
(c) Analytic for all $z \in \mathbf{C}$ (d) Not analytic for all $z \in \mathbf{C}$

ANSWER KEY

1	2	3	4	5	6	7	8	9	10	11	12
b	b	a	a	a	b	b	b	a	a,b,c	c	c
13	**14**	**15**	**16**	**17**	**18**	**19**	**20**	**21**	**22**	**23**	**24**
b	a	b	a	b	a	a	c	d	a,c,d	b	b,c,d
25	**26**	**27**	**28**	**29**	**30**	**31**					
a	a,c,d	b	a,b	c	d	b,d					

Chapter 3

Complex Integration

1. CURVE/CONTINUOUS CURVE/ARC :

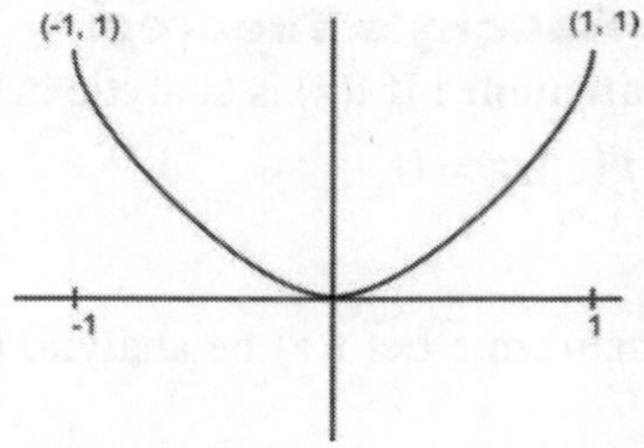

Let $[a, b] \subseteq R$ and $x = x(t)$, $y = y(t)$ are continuous function on [a, b], then $\gamma(t) = x(t) + iy(t)$, $t \in [a, b]$ is defined as a curve in complex plane. Hence the interval [a, b] is called a parametric interval of γ.

Example : Let $x = t$ and $y = t^2$; $t \in [-1, 1]$

Then $\gamma(t) = t + it^2$; $t \in [-1, 1]$

Here $y = x^2$

2. Line Integral

Let $\gamma : [a, b] \to C$ be a smooth curve and suppose f is continuous at all point of $\gamma(t)$.

Then the integral of f along γ is $\int_\gamma f(z)dz = \int_a^b f(\gamma(t))\gamma`(t)dt$.

Note : If γ is piecewise smooth curve on [a, b], then γ is union of finite number of smooth curves $\gamma_1, \gamma_2, \gamma_3, \ldots, \gamma_n$.

Then, $\int_\gamma f(z)dz = \sum_{i=1}^{n} f(z)dz$ The integrals in the sums are called Riemann integrals.

Line integral (Cartesian Form) :

If $f(z) = u(x, y) + iv(x, y)$, $z = x + iy$ is a continuous function on a smooth curve γ, where x, y, u, v are real valued.

Then using a change of variable in the definition of the Riemann integral

$$\int_\gamma f(z)dz = \int_\gamma [u(x,y)dx - v(x,y)dy] + i\int_\gamma [u(x,y)dy + v(x,y)dx].$$

Example : The integral of $f(z) = x + (y + 2)i$ between the points $z = 0$ and $z = 1 + i$.

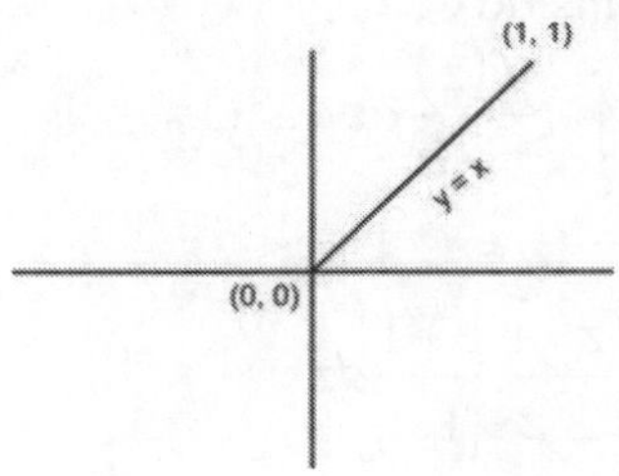

Let $x = t$ and $y = t$; $t \in [0, 1]$

So, $\gamma(t) = t(1 + i)$

$\gamma`(t) = (1 + i)dt$

Then $\int_0^1 [t + (t+2)i][1+i]dt$

$$= (1+i)\int_0^1 [t + i(t+2)]dt \quad \Rightarrow \quad (1+i)\left[\frac{t^2}{2} + i\frac{t^2}{2} + 2it\right]_0^1$$

$$\Rightarrow \quad (1+i)\left[\frac{1}{2}+\frac{i}{2}+2i\right] \qquad \Rightarrow \quad (1+i)\left[\frac{1}{2}+i\left(\frac{5}{2}\right)\right]$$

$$=\frac{1}{2}-\frac{5}{2}+\left(\frac{1}{2}+\frac{5}{2}\right)i \qquad \Rightarrow \quad -2+3i$$

3. Cauchy's Theorem

Statement : If f(z) is analytic in a simply connected domain D and C is any closed curve inside D, then

$$\oint_C f(z)dz = 0$$

Theorem : Let f(z) be analytic in a simply-connected region R, then $\int_a^z f(z)dz$ is independent of the path in R joining a and z.

Theorem: Cauchy's theorem for multiple connected regions: Let f(z) be analytic in a region bounded by two simple closed curves C an C_1 [where C_1 lies inside C as in fig. Below] and on these curves.

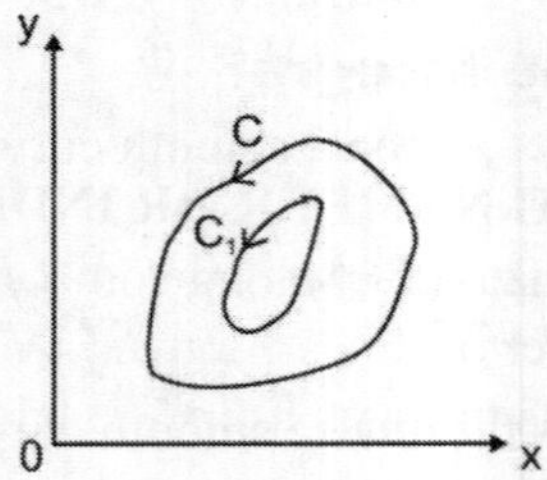

Then $\oint_C f(z)dz = \oint_{C_1} f(z)dz$, where C and C_1 are both traversed in the positive sense relative to their interiors.

Extension: If C is a closed curve and C_1, C_2, C_3...., are the other closed curves which the inside C, and if a function f(z) is analytic in the region between these curves and is continuous on C, then

$$\oint_C f(z)dz = \oint_{C_1} f(z)dz + \oint_{C_2} f(z)dz + \ldots + \oint_{C_n} f(z)dz,$$

where integral along each curve is taken in the anti-clockwise direction.

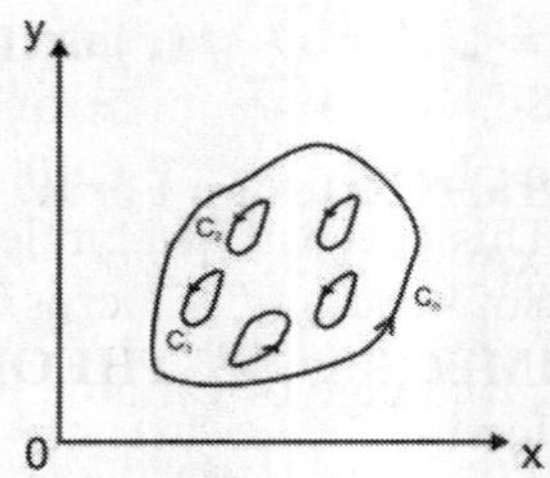

Cauchy's Integral Formula

If f(z) is analytic inside and on a simple closed curve C, taken in the positive (anti-clockwise) sense, and a is any point inside C,

Then $f(a)=\frac{1}{2\pi i}\oint_C \frac{f(z)}{z-a}dz$ .....(1)

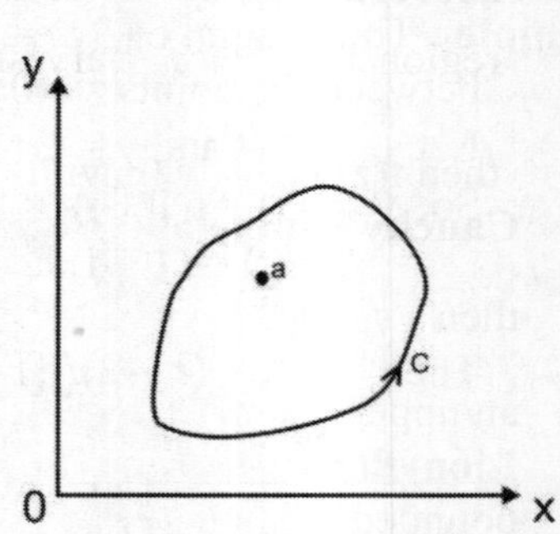

Example : Let D = {z ∈ C | |z| ≤ 2}

Then $\int_D \frac{z^2+z+1}{z-1}dz$

We know that f(z) = $z^2 + z + 1$ is analytic in D and 1 ∈ D.

So, $\int \frac{z^2+z+1}{z-1}dz = 2\pi i[3] = 6\pi i$

Cauchy's integral formula for higher order derivative :
Let f(z) is analytic within and on a closed contour C and z_0 be any interior point of C. Then all higher order derivative are analytic and given by $f^{(n)}(a) = \frac{n!}{2\pi i}\int \frac{f(z)}{(z-a)^{n+1}}dz$

i.e. $\int \frac{f(z)}{(z-a)^{n+1}} = \frac{2\pi i f^{n}(a)}{n!}$.

Example : $\int\limits_{|z|=2} \frac{z^2}{(z-1)^2}$

Solution : Here n = 1 and f(z) = z^2.

So, $\int\limits_{|z|=2} \frac{z^2}{(z-1)^2}dz = \frac{2\pi i}{1} f'(1) = 2\pi i\,[2] = 4\pi i$

WINDING NUMBER OR INDEX OF A CURVE :
Suppose that γ is a closed contour in C and α be a given point in $C \backslash \gamma$.
The number of times γ moves around α while moving in positive direction as t moves from a to b is defined as winding number of γ w.r.t. α.

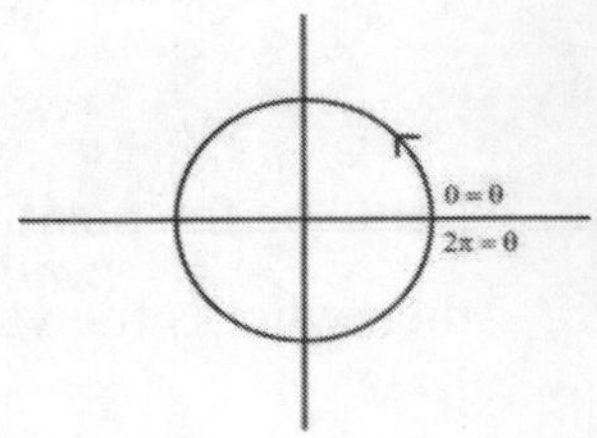

Let $z = e^{i\theta}$; $\theta \in [0, 2\pi]$
Here $|z| = 1$
$\Rightarrow$ This curve is circle with radius 1 and centred at (0, 0)
So, winding number of this function curve is 1.

Example : Let $z = e^{2i\theta}$; $\theta \in [0, 2\pi]$
This curve is also circle with radius 1 and centred at (0, 0)
So, winding number is 2.

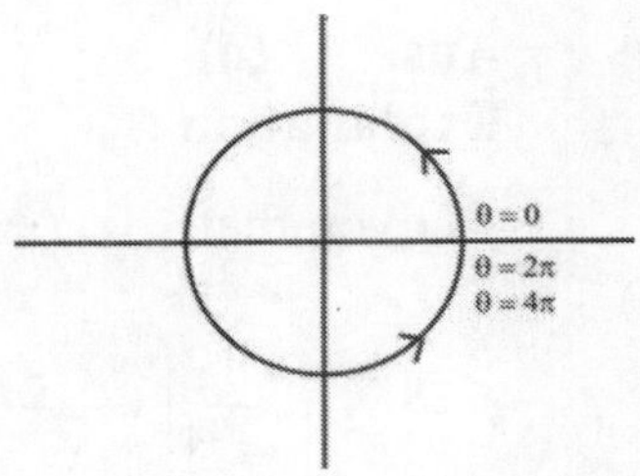

SOME IMPORTANT THEOREMS
The following is a list of some important theorems which are consequences of Cauchy's integral formulae.

1. **Morera's theorem:** If f(z) is continuous in a simply-connected region R and for every simple closed curve in R, $\oint_C f(z)dz = 0$, then f(z) is analytic within R. It is a converse of Cauchy theorem.
2. **Cauchy's inequality:** If f(z) is analytic inside an on a circle C of radius r and centered at z = a, then $|f^{n}(a)| \le \frac{M.n!}{r^n}, n = 0,\ 1,\ 2,\ \ldots,$ where M is a constant such that |f(z)| < M on C, i.e., M is an upper bound of |f(z)| on C.
3. **Liouville's theorem:** Suppose that for all z in the entire complex plane, if f(z) analytic and bounded, i.e., |f(z)| < M for some constant M. Then f(z) must be a constant.
4. If $g(z) = \overline{f(z)}$, where f(z) is an analytic function, then g(z) will be analytic iff f (z) = constant.

5. **Maximum modulus theorem:** If f(z) is analytic inside and on a simple closed curve C and is not identically equal to a constant, then |f(z)| attains its maximum value only on C. In other words. If f is analytic in a domain D such that |f(z)| ≤ |f(a)| holds for all z□D, then f is constant function.
6. **Minimum modulus theorem:** If f(z) is non-constant analytic function inside and on a simple closed curve C and f(z) ∈ 0 inside C, then |f(z)| attains its minimum value only on C. In other words, if f is analytic in a domain D, f(z) ∈ 0 on D and a is any point in D, such that |f(z)|≤f(a)| holds for all z∈D, then f is a constant function.

Note :

Pole : Let f(z) be a complex valued function define on domain D, then z = a (a ∈ D) is called a pole of order n if $\lim\limits_{z\to a}(z-a)^n f(z) = \text{finite}$.

If n = 1 then z = a is called a simple pole.

In general if $\lim\limits_{z\to a} f(z) = \infty$ then z = a is a pole of f(z).

Remark : We will discuss about singularity in detail in chapter – 5.

COMPLEX ANALYSIS

SOLVED QUESTIONS

1. Let C be the positive oriented circle in the complex plane of radius 3, centered at origin then what is the value of integral $\int_C \frac{dz}{z^2(e^z - e^{-z})}$. **CSIR NET JUNE 2023**

(a) $i\frac{\pi}{2}$ (b) $-i\frac{\pi}{2}$ (c) $i\frac{\pi}{6}$ (d) $-i\frac{\pi}{6}$

Ans. **(d)**

Explanation :

Given that $f(z) = \frac{1}{z^2}(e^z - e^{-z})^{-1}$

$$\Rightarrow \frac{1}{z^2}\left[\left(1+z+\frac{z^2}{2!}+......\right)-\left(1-z+\frac{z^2}{2!}-......\right)\right]^{-1} \Rightarrow \frac{1}{z^2}\left[2z+\frac{2z^3}{3!}+\frac{2z^5}{5!}+.....\right]^{-1}$$

$$\Rightarrow \frac{1}{2z^3}\left[1+\left(\frac{z^2}{3!}+\frac{z^4}{5!}+.......\right)\right]^{-1} \Rightarrow \frac{1}{2z^3}\left[1-\left(\frac{z^2}{3!}+\frac{z^4}{5!}+......\right)+\frac{1}{2!}\left(\frac{z^2}{3!}+\frac{z^4}{5!}+.....\right)^2+.....\right]$$

Hence Residue of $f(z) = -\frac{1}{12}$ Now $\int_C \frac{dz}{z^2(e^z - e^{-z})} = 2\pi i\left(-\frac{1}{12}\right) = -\frac{i\pi}{6}$

2. Let γ be the positively oriented circle in the complex plane given by {z ∈ C : |z – 1| = 1}. Then $\frac{1}{2\pi i}\int_\gamma \frac{dz}{z^3 - 1}$ equals **CSIR NET NOV 2020**

(a) 3 (b) 1/3 (c) 2 (d) 1/2

Ans. (b)

Explanation : $\frac{1}{2\pi i}\int_{\gamma}\frac{dz}{z^3-1}$ $z^3 - 1 = (z-1)(z-\omega)(z-\omega^2)$ or $(z-1)(z^2+z+1)$

Where ω is primitive third root of unity.
To given curve contain only one singularity i.e. at $z_0 = 1$.
ω and ω^2 are outside the curve.

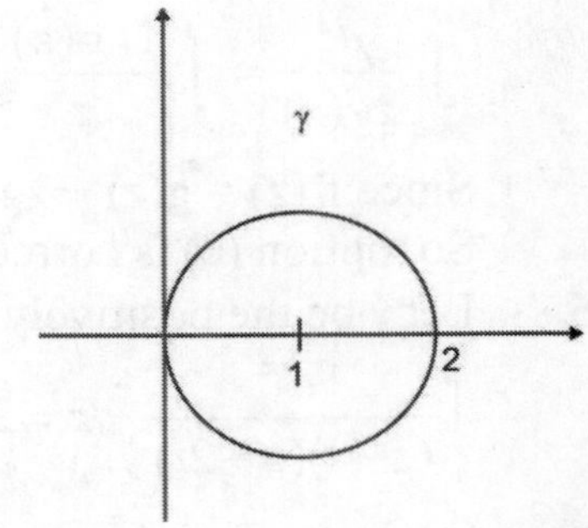

$f(z) = \frac{1}{(z^2+z+1)}$ is analytic inside and on the curve γ.

$$\frac{1}{2\pi i}\int \frac{1}{(z-1)(z^2+z+1)}dz$$

$$= f(1) = \frac{1}{(1+1+1)} = \frac{1}{3}$$

3. Let γ be the positively oriented circle in the complex plane given by $\{z \in C : |z-1| = 1/2\}$. The line integral $\int_{\gamma}\frac{ze^{1/z}}{z^2-1}dz$ equals **CSIR NET NOV 2020**

(a) $i\pi e$ (b) $-i\pi e$ (c) πe (d) $-\pi e$

Ans. (a)

Explanation :

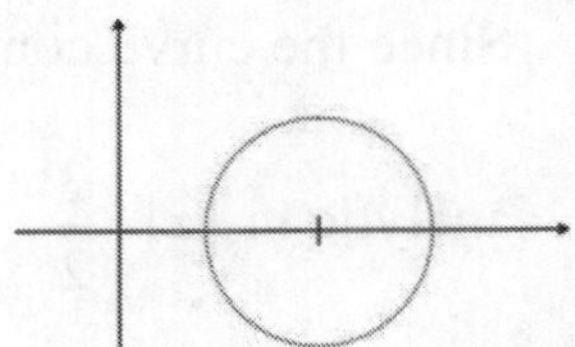

$$\int_{\gamma}\frac{ze^{1/z}}{z^2-1}dz \quad \int \frac{ze^{1/z}}{(z-1)(z+1)}dz$$, γ is the curve.

$$\gamma = \left\{z \in C : |z-1| = \frac{1}{2}\right\}$$

Only $z_0 = 1$ is the singularity that lies inside the curve.

So, the remaining part of integral i.e., $f(z) = \frac{ze^{1/z}}{(z+1)}$ is analytic inside and on the γ.

$$\therefore \quad \int \frac{ze^{1/z}}{(z+1)(z+1)}dz = 2\pi i f(1) \quad = 2\pi i\frac{1.e^1}{(1+1)} = \pi i e$$

So, option (a) is correct answer.

4. Let f be a holomorphic function on the disc $\{z \in C : |z| < 2\}$. Assume that the only zero of f in the closed unit disc $\{z \in C : |z| \le 1\}$ is a simple zero at the origin. Let γ be the positively oriented circle $\{z \in C : |z| = 1\}$. The integral $\int_{\gamma}\frac{dz}{f(z)}$ equal

CSIR NET NOV 2020

(a) $2\pi i f'(0)$ (b) $2\pi i f''(0)$ (c) $2\pi i / f'(0)$ (d) $2\pi i / f''(0)$

Ans. (c)

Explanation :

Given that f(z) has only simple zero at origin in the closed unit disc $\{z \in C : |z| \le 1\}$

So , f(z) = zg(z), where $g(z) \neq 0$ for all closed unit disk.

$$\int_{|z|=1} \frac{dz}{zg(z)} = \int_{|z|=1} \frac{1/g(z)}{z} = \frac{2\pi i}{g(0)} = \frac{2\pi i}{f`(0)}$$

Since f`(z) = g(z) + zg`(z) f`(0) = g(0) + 0 f`(0) = g(0)

So, option (c) is correct.

5. Let γ be the positively oriented circle $\{z \in C : |z| = 3/2\}$. Suppose that $\int_{\gamma} \frac{e^{i\pi z}}{(z-1)(z-2i)^2} dz = 2\pi i c$. Then |c| equals **CSIR NET FEB 2022**

(a) 2 (b) 5 (c) 1/2 (d) 1/5

Ans. (d)

Explanation :

$\gamma = \{z \in C : |z| = 3/2\}$

$$\int_{\gamma} \frac{e^{i\pi z}}{(z-1)(z-2i)^2} dz = 2\pi i c$$

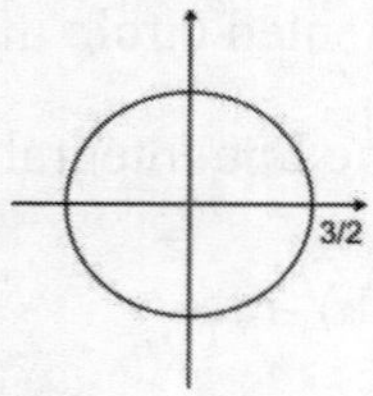

Since the curve contains only one singularity z = 1 inside, we see $f(z) = \frac{e^{i\pi z}}{(z-2i)^2}$ is analytic in $|z| = \frac{3}{2}$.

So, by Cauchy Integral formula $\int \frac{e^{i\pi z}}{(z-1)(z-2i)^2} dz = 2\pi i\ f(1)$

$$\Rightarrow \quad c = f(1) = \frac{e^{i\pi}}{(1-2i)^2} = \frac{-1}{(1-4-4i)} \quad c = \frac{1}{(3+4i)} = 1 \quad |c| = \frac{1}{|3+4i|} = \frac{1}{5}$$

Option (d) is correct.

General Complex Integration :

Let C : z(t) = x(t) + iy(t) Then $\int_C f(z)dz = \int_a^b f[z(t)]z'(t)dt$

6. $I = \int_C \text{Re}(z^2)dz$ C is the contour from 0 to 2 + 4i along parabola $y = x^2$.

(a) $\frac{-56}{15} - \frac{160i}{12}$ (b) 1 + 2i (c) $\frac{-55}{14} - \frac{159i}{11}$ (d) 2 + i

Ans. (a)

Explanation :

Given that $Re(z^2) = x^2 - y^2$ and curve parabola $y = x^2$

So C : z(t) = x(t) + iy(t)

$x(t) = t$, $y(t) = t^2$ $t \in [0, 2]$ then $z(t) = t + it^2$

$z`(t) = (1 + 2it)$

We know that $\int_C f(z)dz = \int_C f(z(t))z'(t)dt$ $\int_0^2 f(z)dz = \int_0^2 (t^2 - t^4)(1+2it)dt$

$$\int_0^2 (t^2 - t^4)dt + 2i\int_0^2 (t^3 - t^5)dt \Rightarrow \left[\frac{t^3}{3} - \frac{t^5}{5}\right]_0^2 + 2i\left[\frac{t^4}{4} - \frac{t^6}{6}\right]_0^2$$

$$\Rightarrow \left[\frac{8}{3} - \frac{32}{5}\right] + 2i\left[\frac{16}{4} - \frac{64}{6}\right] \Rightarrow \left[\frac{40-96}{15}\right] + 2i\left[\frac{48-128}{12}\right]$$

$$\Rightarrow \left[\frac{-56}{15}\right] - \frac{160i}{12}$$

Note : $\oint_C f(z) = 0$ if f(z) is analytic function over C (curve).

7. Find the value $I = \int_C zdz$ where C : |z| = 1

(a) 0 (b) 1 (c) 2 (d) 3

Ans. (a)

Explanation :

We know that |z| = 1 ⇒ $Z = e^{i\theta}$ $\theta \in [0, 2\pi]$

⇒ $Z(t) = e^{it}$ $t \in [0, 2\pi]$ $Z'(t) = ie^{it}$

$$\text{Now } \int_C f(z)dz = i\int_0^{2\pi} e^{it}e^{it}dt = i\int_0^{2\pi} e^{2it}dt = \frac{1}{2}\left[e^{4\pi i} - e^0\right] = \frac{1}{2}[1-1] = 0$$

8. Let C denote the unit circle centred at the origin in C.

Then $\frac{1}{2\pi i}\oint_C |1 + z + z^2|^2 \, dz$, where integral is taken anticlockwise along C.

CSIR NET JUNE 2017

(a) 0 (b) 1 (c) 2 (d) 4

Ans. (c)

Explanation :

Given that |z| = 1 (circle is centre at origin) $\Rightarrow |z|^2 = 1 \Rightarrow z\bar{z} = 1 \Rightarrow \bar{z} = \frac{1}{z}$.

$$I = \int_C |1 + z + z^2|^2 \, dz = \int_C (1 + z + z^2)\overline{(1 + z + z^2)}dz$$

COMPLEX ANALYSIS

$\int_C (1+z+z^2)(1+\bar{z}+\bar{z}^2)dz \quad \int_C (1+\bar{z}+\bar{z}^2+z+z\bar{z}+z\bar{z}^2+\bar{z}^2+z\bar{z}^2+z^2\bar{z}^2)dz$

Here $\bar{z}=\frac{1}{z}, \qquad z\bar{z}=|z|^2=1$

So, $\int_C \left(1+\frac{1}{z}+\frac{1}{z^2}+z+1+z+\frac{1}{z^2}+\frac{1}{z}+1\right)dz$

$\int_C (1+z+1+z+1)dz+\int_C \left(\frac{1}{z}+\frac{1}{z}+\frac{2}{z^2}\right)dz$

Here $\int_C (1+z+1+z+1)dz=0 \qquad [\therefore \int f(z)=0$, if f(z) is analytic]

$0+\int_C \frac{2}{z}dz+\int_C \frac{2}{z^2}dz=2 \qquad \frac{1}{2\pi i}\oint_C |1+z+z^2|^2\, dz=\frac{1}{2\pi i}2\pi i\times 2=2$

(By using Cauchy integral formula)

9. Let C be the circle of radius 2 with centre at then origin oriented in anti-clockwise. Then the $\oint_C \frac{dz}{(z-1)^2}$ is equal to **CSIR NET DEC 2017**

(a) $\frac{1}{2\pi i}$ (b) $2\pi i$ (c) 1 (d) 0

Ans. (d)

Explanation :

We know that $\oint \frac{f(z)}{(z-z_0)^{n+1}}=2\pi i\ f^n(z_0)$

and given that $\oint \frac{1}{(z-1)^2}dz$ here n = 1 and f(z) = 1 then $= 2\pi i\, f'(1) = 2\pi i \times 0 = 0$

10. Consider the polynomial p(z), q(z) in C and let $I_{p,q}=\oint_\gamma p(z)\overline{q(z)}\,dz$ Where γ denotes the closed contour $\gamma(t)=e^{it}$, $t \in [0, 2\pi]$. Then **CSIR NET DEC 2018**

(a) $I_{z^m,z^n}=0$ ∀ positive integer with m ≠ n (b) $I_{z^n,z^n}=2\pi i$ ∀ positive integer n.

(c) $I_{p,1}=0$ ∀ polynomial p. (d) $I_{p,q}=p(0).\overline{q(0)}$ ∀ polynomial p & q

Ans. (c)

Explanation :

Option (a) :

$$I_{z^m,z^n} = \oint_C z^m \bar{z}^{-n} = \oint_C \frac{z^m}{z^n} dz$$

$\therefore$ Given that $|z| = 1 \Rightarrow |z|^2 = 1$

$z\bar{z} = 1 \Rightarrow \bar{z} = \frac{1}{z}$

If n > m then $\oint_C \frac{1}{z^{n-m}} dz$

Take n = 1, m = 0

$$\oint \frac{1}{z} dz = 2\pi i$$

Option (a) is wrong for n = 1, m = 0

Option (b) :

$$I_{z^n,z^n} = \oint_C z^n \bar{z}^n = \oint_C \frac{z^n}{z^n} dz \Rightarrow \oint_C dz = 0$$

$[\therefore \oint f(z)dz = 0$ if f(z) is analytic$]$

Option (c) : $I_{p,1} = \oint p(z)dz = 0$

p(z) is polynomial and analytic.

Option (d) :

$$I_{p,q} = \oint_C p(z)\overline{q(z)}\, dz$$

Take q(z) is constant and p(z) = z + 1

Then $\oint (z+1).1\, dz = 0$

(z + 1) is analytic

But p(0) = 1 = q(0)

$$I_{p,q} \neq p(0).\overline{q(0)}$$

11. Let $\gamma(t) = 3e^{it}$, $0 \le t \le 2\pi$ be the positive oriented circle of radius 3 centred at origin. The vaue of λ for which $\oint_\gamma \frac{\lambda}{z-2} dz = \oint_\gamma \frac{1}{z^2 - 5z + 4} dz$ is **CSIR NET DEC 2018**

(a) $\lambda = -1/3$ (b) $\lambda = 0$ (c) $\lambda = 1/3$ (d) $\lambda = 1$

Ans. (a)

Explanation :

$\oint_\gamma \frac{1}{z^2 - 5z + 4} dz$ $\quad z^2 - 5z + 4 = z^2 - z - 4z + 4$ $\quad z(z - 1) - 4(z - 1)$

$\Rightarrow$ (z – 4) (z – 1)

$\oint_\gamma \frac{dz}{(z-4)(z-1)}$ Given that $|\gamma| = 3$

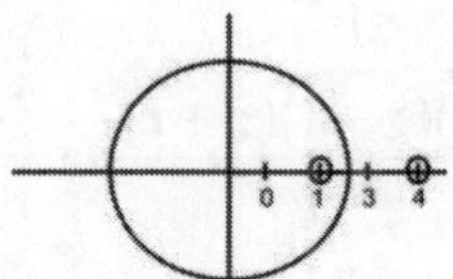

1 is in circle and 4 outside the circle. $\oint \frac{dz/z-4}{z-1}$

here $f(z) = \frac{1}{z-4}$ $\quad = 2\pi i\, f(1) = \frac{2\pi i}{-3}$

$\oint \frac{dz}{z^2 - 5z + 4} = \frac{-2\pi i}{3}$ ..... (1) and $\oint \frac{\lambda}{z-2} dz$

2 is inside circle = $2\pi i$ f(z) = $\lambda 2\pi i$ ....(2)

Given that (1) = (2) $\lambda = -\frac{1}{3}$

12. The value of integral $\oint\limits_{|1-z|=1} \frac{e^z}{z^2-1} dz$ is **CSIR NET JUNE 2018**

(a) 0 (b) $(\pi i)e$ (c) $(\pi i)e - (\pi i)e^{-1}$ (d) $(e + e^{-1})$

Ans. (b)

Explanation :

Given that a circle $|z - 1| = 1$

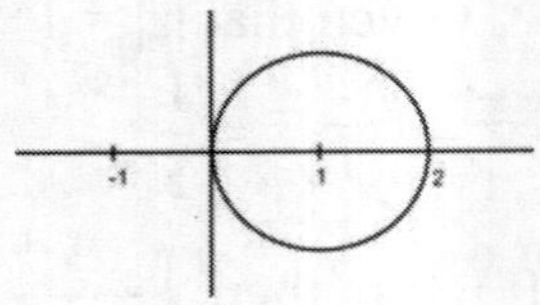

$$\oint \frac{e^z}{z^2-1} = \int \frac{e^z}{(z-1)(z+1)}$$

1 inside in the circle. -1 outside the circle.

$$\oint \frac{\frac{e^z}{z+1}}{(z-1)} dz \text{ Here } f(z) = \frac{e^z}{z+1}$$

We know that $\oint \frac{f(z)}{(z-a)} dz = 2\pi i f(a)$ $\Rightarrow 2\pi i\ f(1) = (\pi i)e$

13. Let C be the counter clockwise circle of radius 1/2 centred at i.Then $\oint\limits_C \frac{dz}{z^4-1}$ is

(a) $-\pi/2$ (b) $\pi/2$ (c) $-\pi$ (d) π

Ans. (a) **CSIR NET JUNE 2019**

Explanation :

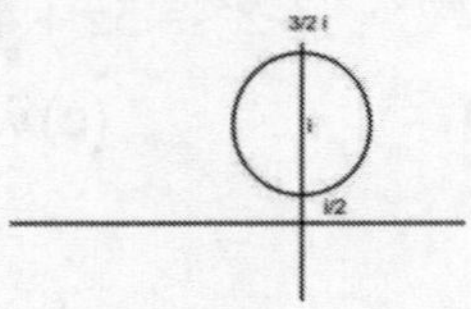

Circle $|z - i| = 1/2$

and $\oint \frac{dz}{z^4-1}$

$$\oint \frac{dz}{(z-1)(z+1)(z+i)(z-i)}$$

Here 1, -1, -i all outside the circle.

$$\oint \frac{\frac{1}{(z+1)(z-1)(z+i)}}{z-i} dz = 2\pi i\ f(i) = \frac{2\pi i}{(i+1)(i-1)(2i)} = \frac{\pi}{(i)^2-1} = -\frac{\pi}{2}$$

14. Given a real number a > 0, consider the triangle with vertices 0, a, a + ia if triangle is given counter clockwise then $\oint\limits_{\Delta} \text{Re}(z) dz$ is equal to **CSIR NET JUNE 2019**

(a) 0 (b) $i\frac{a^2}{2}$ (c) ia^2 (d) $i\frac{3a^2}{2}$

Ans. (b)

Explanation :

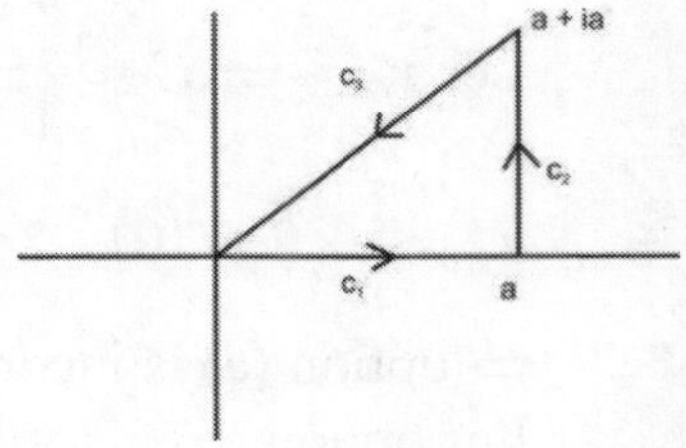

Given triangle

For c_1

a curve z(t) = x(t) + iy(t)

z(t) = t t ∈ [0, a] ∴ x = t, y = 0

$$I = \oint x dz = \int_0^a t\,dt = \frac{a^2}{2}$$

For c_2 z(t) = a + it ∴ x(t) = a, y(t) = t

t ∈ [0, a] z' = i dt $I = \int_0^a ai\,dt = ai[t]_0^a = a^2 i$

For c_3

Curve y = x line z(t) = t + it t ∈ [0, a] z' = (1 + i) dt

$$I = \int_0^a t(1+i)dt = (1+i)\int_0^a tdt = \frac{a^2}{2}(1+i)$$

We know that

$$\oint_C f(z) = dz = \oint_{C_1} f(z)dz + \oint_{C_2} f(z)dz + \oint_{C_3} f(z)dz$$

$$\oint_C f(z)dz = a^2 i + \frac{a^2}{2} - \frac{a^2}{2}(1+i) \quad = \frac{a^2}{2}i$$

15. Let $I_r = \int_C \dfrac{dz}{z(z-1)(z-2)}$, where $C_r = \{z \in C : |z| = r\}, r > 0$. Then

CSIR NET DEC-2011

(a) $I_r = 2\pi i \ \ if \ \ r \in (2,3)$ (b) $I_r = \frac{1}{2} \ if \ \ r \in (0,1)$

(c) $I_r = -2\pi i \ \ if \ \ r \in (1,2)$ (d) $I_r = 0 \ if \ \ r > 3$

Ans. (d)

Explanation :

$$I_r = \int_{C_r} \frac{dz}{z(z-1)(z-2)}, \text{ where } C_r \{z \in C : |z| = r\},\ r > 0$$

Using partial fraction, $\dfrac{1}{z(z-1)(z-2)} = \dfrac{1}{2z} + \dfrac{-1}{(z-1)} + \dfrac{1}{2(z-2)}$

$$\therefore I_r = \int_{C_r} \frac{dz}{(z-1)(z-2)} = \int_{C_r} \left(\frac{1}{2z} - \frac{1}{z-1} + \frac{1}{2(z-2)}\right)dz$$

For option (a), r ∈ (2, 3)

Let $r=\frac{5}{2}\Rightarrow C_r=\left\{z\in C:|z|=\frac{5}{2}\right\}$

$$I_r=\frac{1}{2}\times 2\pi i f(0)-2\pi i\ f(1)+\frac{1}{2}\times 2\pi i\ f(2)$$

$\Rightarrow$ option (a) is incorrect

For option (b), r $\in$ (0,1)

Let $r=\frac{1}{2}\Rightarrow C_r=\left\{z\in C:|z|=\frac{1}{2}\right\}$

$$I_r=\int_{C_r}\frac{\frac{1}{(z-1)(z-2)}}{z}dz=2\imath\ f(0)=2\pi i\left(\frac{1}{(0-1)(0-2)}\right)=\frac{2\pi i}{2}=\pi i$$

$\Rightarrow$ option (b) is incorrect

For option (c), r $\in$(1, 2)

Let $r=\frac{3}{2}\Rightarrow C_r=\left\{z\in C:|z|=\frac{3}{2}\right\}$

$$I_r=\int_{C_r}\frac{1}{(z-1)(z-2)}dz=2\pi i\left[\int_{C_r}\frac{\frac{1}{(z-1)(z-2)}}{z}dz+\int_{C_r}\frac{\frac{1}{(z-2)(z)}}{z-2}dz\right]$$

$$=2\pi i\left[\lim_{z\to 0}\frac{1}{(z-1)(z-2)}+\lim_{z\to 1}\frac{1}{(z-2)(z)}\right]=2\pi i\left[\frac{1}{2}-1\right]=2\pi i\left(-\frac{1}{2}\right)=-\pi i$$

$\Rightarrow$ option (c) is also incorrect.

As all other options are incorrect

$\Rightarrow$ option (d) is correct.

16. Let $\gamma_k=\{ke^{ik\theta}:0\le\theta\le 2\pi\}$ for k = 1, 2, 3. Which of the following are necessarily correct? **CSIR NET DEC-2012**

(a) $\frac{1}{2\pi i}\int_{\gamma_k}\frac{1}{z}dz=0\ \ for\ \ k=1,2,3$ (b) $\frac{1}{2\pi i}\int_{\gamma_k}\frac{1}{z}dz=1$ (c) $\frac{1}{2\pi i}\int_{\gamma_2}\frac{1}{z}dz=4$ (d) $\frac{1}{2\pi i}\int_{\gamma_3}\frac{1}{z}dz=3$

Ans. (b), (d)

Explanation :

$\gamma_k=\{ke^{-i\theta}:0\le\theta\le 2\pi\}\ for\ k=1,2,3$ for options (a) and (b),

$\gamma_1=\{1e^{i\theta}:0\le\theta\le 2\pi\}\Rightarrow z=0$ is inside θ1 So, $\frac{1}{2\pi i}\int_{\gamma_1}\frac{1}{z}dz=\frac{1}{2\pi i}[2\pi i.1]=1$

$\Rightarrow$ option (b) is correct and (a) is incorrect

For option (c).

Take $\frac{1}{2\pi i}\int_{\gamma_2}\frac{1}{z}dz, \gamma_2 = \{2e^{i2^{\theta}} : 0 \le \theta \le 2\pi\} \Rightarrow \frac{1}{2\pi i}\int_{\gamma_2}\frac{1}{z}dz = \frac{1}{2\pi i}[2.2.\pi i(1)] = 2$

$\Rightarrow$ option (c) is incorrect.

For option (d). $\gamma_3 = \{3e^{i3\theta} :, \le \theta \le 2\pi\}$ $\frac{1}{2\pi i}\int_{\gamma_3}\frac{1}{z}dz = \frac{1}{2\pi i}[3.2\pi i(1)] = 3$

$\Rightarrow$ option (d) is correct.

17. $\int_{|z+1|=2}\frac{z^2}{4-z^2}dz =$ **CSIR NET DEC-2015**

(a) 0 (b) -2πi (c) 2πi (d) 1

Ans. (c)

Explanation :

Let $f(z) = \frac{z^2}{4-z^2} = \frac{z^2}{(2-z)(2+z)}$

z = -2 and z = 2 are singularities of f(z), but z = -2 lies inside the circle |z + 1| = 2

$$\therefore \int_{|z+1|=2}\frac{z^2}{4-z^2}dz = \int_{|z+1|=2}\frac{\frac{z^2}{2-z}}{z+2}dz = 2\pi i\frac{(-2)^2}{2-(-2)} = 2\pi i$$

$\Rightarrow$ option (c) is correct

18. Let C be the circle |z| = 3/2 in the complex plane that is oriented in the counter clockwise direction. The value of a for which $\int_C\left(\frac{z+1}{z^2-3z+2}+\frac{a}{z-1}\right)dz = 0$, is

(a) 1 (b) -1 (c) 2 (d) -2

Ans. (c) **CSIR NET DEC-2016**

Explanation :

Given $\int_{|z|=\frac{3}{2}}\left(\frac{z+1}{z^2-3z+2}+\frac{a}{z-1}\right)dz = 0$ $\Rightarrow \int_{|z|=\frac{3}{2}}\left(\frac{z+1}{(z-1)(z-2)}+\frac{a}{z-1}\right)dz = 0$

Singularity inside $|z| = \frac{3}{2}$ *is at* $z = 1$

$$\Rightarrow \int_{|z|=\frac{3}{2}}\left[\left(\frac{\frac{(z+1)}{(z-2)}}{z-1}\right)\right]+\left(\frac{a}{z-1}\right)dz = 0 \quad ......(1)$$

COMPLEX ANALYSIS

By Cauchy's integral formula, $\oint_C \frac{f(z)}{z-a}dz = f(a)2\pi i$

$\therefore$ (1), $2\pi i\left[\frac{1+1}{1-2}\right] + a\,2\pi i = 0 \Rightarrow 2\pi i[(-2)+a] = 0 \Rightarrow a = 2$

Hence, option (c) is correct.

Question of maximum and minimum modulus theorem :

19. Let f(z) = $2z^2$ -1. Then the maximum value of |f(z)| on the unit disc D = {z ∈ C : |z| ≤ 1} is **GATE-2007**

(a) 1 (b) 2 (c) 3 (d) 4

Ans. (c)

Explanation :

$|f(z)| = |2z^2 - 1| \Rightarrow 2|z|^2 + 1 = 3$ if |z| = 1, i.e., z is on the boundary

By maximum modulus principle, maximum value occurs on boundary.

For z = i, $|2z^2 - 1| = |-3| = 3$.

20. Let T be the closed unit disk and ∂T be the unit circle. Then which one of the following holds for every analytic function f : T ∈ C. **GATE-2008**

(a) |f| attains its minimum and its maximum on ∂T

(b) |f| attains its minimum on ∂T but need not attain its maximum on ∂T

(c) |f| attains its maximum on ∂T but need not attain its minimum on ∂T

(d) |f| need not attains its maximum on ∂T and also it need not attain its minimum on ∂T

Ans. (d)

Explanation :

By maximum modulus principle and minimum modulus principle |f| will attain its maximum and minimum value on ∂T, if f is non-zero and non constant on T.

21. The maximum modulus of e^{z^2} pm the set $S = \{zg \in C : 0 \le Re(z) \le 1, 0 \le 1m(z) \le 1\}$ is

(a) 2/e (b) e (c) e+1 (d) e^2

Ans. (b) **GATE-2014**

Explanation :

As e^{z^2} is analytic on S, z by the maximum modulus theorem the maximum modulus of e^{z^2} is attained on the boundary.

Let z = a + ib $\therefore e^{z^2} = e^{a^2-b^2+2abi} \Rightarrow |e^{z^2}| = e^{a^2-b^2}$

Clearly, $|e^{z^2}|$ is maximum, when $a^2 - b^2$ is maximum and maximum value of $a^2 - b^2$ is 1. [0 ≤ a ≤ 1, 0 ≤ b ≤ 1]

$\therefore$ Maximum value of e^{z^2} is e, Hence, option (b) is correct.

UNSOLVED QUESTIONS

1. Let u(x, y) be the real part of an entire function f(z) = u(x, y) + iv (x,y) for z = x = iy∈ C. If C is the positively oriented boundary of a rectangular region R in R^2, then $\oint_C \left[\frac{\partial u}{\partial y}dx - \frac{\partial u}{\partial x}dy\right] =$ **GATE-2009**

(a) 1 (b) 0 (c) 2π (d) π

2. Let f be an analytic function on $\overline{D} = \{z \in C : |z| \le 1\}$. Assume that |f(z)|≤1 for each $z \in \overline{D}$. Then, which of the following is NOT a possible value of $(e^f)''(0)$?

(a) 2 (b) 6 (c) $\frac{7}{9}e^{\frac{1}{9}}$ (d) $\sqrt{2} + i\sqrt{2}$ **GATE-2013**

3. Let u be real valued harmonic function on **C**. Le g : $R^2 \to R$ be defined by $g(x,y) = \int_0^{2\pi} u\left(e^{i\theta}(x+iy)\right)\sin\theta\, d\theta$. Which of the following statements is TRUE?

(a) g is a harmonic polynomial (b) g is a polynomial but not harmonic
(c) g is harmonic but not a polynomial
(d) g is neither harmonic nor a polynomial **GATE-2013**

4. Let $\Omega = \{z \in C : Im(z) > 0\}$ and let C be a smooth curve lying in Ω with initial point -1 + 2i and final point 1 + 2i. The value of $\int_C \frac{1+2z}{1+z}dz$ is **GATE-2014**

(a) $4 - \frac{1}{2}\ln 2 + i\frac{\pi}{4}$ (b) $-4 + \frac{1}{2}\ln 2 + i\frac{\pi}{4}$

(c) $4 + \frac{1}{2}\ln 2 - i\frac{\pi}{4}$ (d) $4 - \frac{1}{2}\ln 2 + i\frac{\pi}{2}$

5. Let $C = \{z \in C : |z - i| = 2\}$. Then $\frac{1}{2\pi}\oint_C \frac{z^2 - 4}{z^2 + 4}dz$ is equal to **GATE-2015**

(a) -2 (b) 3 (c) 4 (d) 5

6. Let γ be the triangular path connecting the points (0,0), (2, 2) and (0, 2) in the counter-clockwise direction in R^2. Then $I = \oint_\gamma \sin(x^3)dx + 6xy\, dy$ is equal to **GATE-2016**

(a) 13 (b) 14 (c) 15 (d) 16

7. If f is analytic in a ball B)a; R) and |f(z)| ≤ M ∀ z ∈ B (a; R), then

(a) $|f^{(n)}(a)| \le \frac{n!\,M}{R^n}$ (b) $|f^{(n)}(a)| \ge \frac{n!\,M}{R^n}$

(c) $|f^{(n)}(a)|\, n!\,M$ (d) None of these

8. If γ and σ are closed rectifiable curves having same initial points. If n(□, a) denotes winding number of a, then for every a ∈ {γ} ∪ {σ}

(a) $n(\gamma+\sigma;a)=n(\gamma;a)-n(\sigma;a)$ (b) $n(\gamma+\sigma;a)=n(\sigma;a)$

(c) $n(\gamma+\sigma;a)=n(\gamma;a)+n(\sigma;a)$ (d) None of these

9. If f : G→ C is analytic and $\overline{B}(a;r)\subset G$, then

(a) $f^{(n)}(a)=\frac{n!}{2\pi i}\int_{\gamma}\frac{f(w)}{(w-a)^{n+1}}dw$, where $\gamma(t)=a+re^{it}$ and $0\le t\le 2\pi$

(b) $f(a)=\frac{1}{2\pi i}\int_{\gamma}\frac{f(w)}{(w-a)}dw$, where $\gamma(t)=a+re^{it}$ and $t\in[0,2\pi]$

(c) $f^{(n)}(a)$ does not exists (d) $f^{(n)}(a)$ exists

10. If $f(a)=\int_{\gamma}\frac{3z^2+7z+1}{z-a}dz$, where γ : |z| = 2, then

(a) f(0) = 2πi (b) f'(0) = 14π i (c) f" (-1) = 16πi (d) f"(-2) = 12 πi

11. If $\int_{\gamma}\frac{z+4}{z^2+2z+5}dz=0$, then

(a) γ : |z| = 1 (b) γ : |z + 1| = 1 (c) γ : |z| = 3 (d) $\gamma:|z+1|=\frac{1}{2}$

12. If C is the boundary of the square $\{-1\le x\le 1,-1\le y\le 1\}$ described in anticlockwise direction. Then the value of the integral $\int_C\frac{z^2}{[z-(1+i)/4]^4}dz$ cannot be

(a) zero (b) iπ/2 (c) –iπ/4 (d) 2πi

13. Let $A=\int_C\frac{e^{2z}dz}{(z-1)(z-2)}$ where C is the circle |z| = 3 and $B=\int\frac{\cos\pi z}{z^2-1}dz$ around a rectangle with vertices 2±i, -2±i. Then

(a) $A=2\pi i(e^4-e^2)$ (b) B = 0 (c) A = 0 (d) B = 2□i (e⁴ – e²)

14. Let $f:D\to\overline{D}$ be an analytic function, where $D=\{z\in C:|z|<1\}$ be the open unit disc in the complex plane. Let f(z) has zero of order 3 at z = 0. Then

(a) $|f(z)|\le|z^3|\ \forall z\in D$ (b) $|f''(0|\le 2$ (c) $\left|f\left(\frac{1}{3}\right)\right|\le\frac{1}{3}$ (d) $|f'''(0)|\le 3$

15. Let Γ denote the boundary of the square whose sides lies along x = ±1 & y = ± 1 in anticlockwise direction then $\oint_{\Gamma}\frac{z^2}{2z+3}dz$ is equal to **GATE 1996**

(a) πi (b) 0 (c) 2πi (d) -2πi

16. Let $\oint_C \left[\frac{1}{(z-2)^4} - \frac{(a-2)^2}{z} + 4\right] dz = 4\pi$.

Where c is triangle whose vertices at $i, \frac{-1-i}{\sqrt{2}}, \frac{1-i}{\sqrt{2}}$ in anticlock wise direction.

Then value of a is **GATE - 2012**

(a) 1 + i (b) 3 + i (c) 2 + i (d) 4 + i

17. If a ∈ C s.t. |a| < 1 then the value of $\frac{1-|a|^2}{\pi}\int_\Gamma \frac{|dz|}{|z+a|^2}$ where Γ : |z| = 1 taken positive oriented is **GATE 2014**

(a) 1 (b) 2 (c) 3 (d) 4

18. Let C be the boundary of the triangle formed by the points (1, 0, 0), (0, 1, 0), (0, 0, 1). Then the value of the linear integral $\oint -2y\,dx + (3x - 4y^2)dy + (z^2 + 3y)dz$ is

GATE-2007

(a) 0 (b) 1 (c) 2 (d) 4

19. Let f(z) be an analytic function. Then the value of $\int_0^{2\pi} f(e^{it})cos(t)dt$ equals **GATE-2007**

(a) 0 (b) 2πf (0) (c) 2pf' (0) (d) πf' (0)

20. Let $f(z) = \sum_{n=0}^{15} z^n$ for z ∈ **C**. If C: |z – i| = 2, then $\int_C \frac{f(z)dz}{(z-i)^{15}} =$ **GATE-2009**

(a) 2πi (1 + 15i) (b) 2πi(1- 15i) (c) 4πi (1 + 15i) (d) 2πi

Answer Key

1	2	3	4	5	6	7	8
b	b	a	a	a	d	a	c
9	**10**	**11**	**12**	**13**	**14**	**15**	**16**
a,b,d	a,b	a,b,d	b,c,d	a,b	a,b,c	b	B
17	**18**	**19**	**20**				
b	d	d	a				

Chapter 4

Sequence and Series

1. Sequence

Sequence of complex numbers: A sequence is a function of positive integral variable, denoted by f(n) or $\{u_n\}$, where n = 1, 2, 3,… formed according to a definite rule. Each number in the sequence is called a term and u_n is called the nth term. The sequence is called finite or infinite according as there are finite number of terms or not. Unless otherwise specified, we shall consider infinite sequence only.

For example:

(i) The set of numbers i, i^2, i^3, …, i^{100} is a finite sequence; the nth term is given by $u_n = i^n$, = 1, 2 …, 100.

(ii) The set of numbers 1 + i, $\frac{(1+i)^2}{2!}, \frac{(1+i)^3}{3!}$,.... is an infinite sequence: the nth term is given by $u_n = \frac{(1+i)^n}{n!}, n = 1,2,3....$

Convergence of a sequence: A number l is called the limit of an infinite sequence $u_1, u_2, u_3,\ldots$ if for any positive number $\in$, we can find a positive integer N depending upon $\in$ such that $|u_n - l| < \in$ for all n $\in$ N, i.e., $u_n \to l$, if $|u_n - l| \to 0$ as a real sequence. In such case we write $\lim\limits_{n\to\infty} u_n = 1$. If the limit of a sequence exists finitely, then the sequence is called convergent, otherwise it will be divergent.

Important theorems on sequences:

Theorem : If a sequence has a limit, then it is unique.

Theorem: Let $u_n = a_n + ib_n$, n = 1, 2, 3,…, where a_n and b_n are real. Then a necessary and sufficient condition that $\{u_n\}$ converges is that $\{a_n\}$ and $\{b_n\}$ converges.

Results on limit of sequences:

If $\lim\limits_{n\to\infty} a_n = A \, and \, \lim\limits_{n\to\infty} b_n = B, then$

(i) $\lim\limits_{n\to\infty}(a_n + b_n) = \lim\limits_{n\to\infty} a_n + \lim\limits_{n\to\infty} b_n = A + B$

(ii) $\lim\limits_{n\to\infty}(a_n - b_n) = \lim\limits_{n\to\infty} a_n - \lim\limits_{n\to\infty} b_n = A - B$

(iii) $\lim\limits_{n\to\infty}(a_n b_n) = \left(\lim\limits_{n\to\infty} a_n\right)\left(\lim\limits_{n\to\infty} b_n\right) = AB$

(iv) $\lim\limits_{n\to\infty}\frac{a_n}{b_n} = \frac{\lim\limits_{n\to\infty} a_n}{\lim\limits_{n\to\infty} b_n} = \frac{A}{B}, if \ B \neq 0$

2. Infinite Series

Series of complex numbers:

Let $u_1, u_2, u_3, \ldots$ Be a given sequence. Form a new sequence $S_1, S_2, S_3\ldots$ defined by

$S_1 = u_1,\ S_2 = u_1 + u_2,\ S_3 = u_1 + u_2 + u_3,\ \vdots\ S_n = u_1 + u_2 + \ldots + u_n,$

Here, S_n is called the nth partial sum of the first n terms of the sequence $\{u_n\}$

$\sum_{n=1}^{\infty} u_n = u_1 + u_2 + \ldots + u_n + \ldots$ is called an infinite series. If $\lim_{n\to 0} u_n = u$ exists finitely, the series is called convergent and u is its sum otherwise the series is called divergent. Thus, a series is said to converge iff the corresponding sequence of partial sums is convergent.

A necessary condition for series : If a series $\sum_{n=1}^{\infty} u_n$ is convergent then $\lim_{n\to 0} u_n = 0$

Absolutely Convergence and Conditionally Convergence :

A series $\sum_{n=1}^{\infty} u_n(z)$ is called absolutely convergent if $\sum_{n=1}^{\infty} |u_n(z)|$ converges. If $\sum_{n=1}^{\infty} u_n(z)$ converges but $\sum_{n=1}^{\infty} |u_n(z)|$ does not converge, then $\sum_{n=1}^{\infty} u_n(z)$ is said to be conditionally convergent. Every absolutely convergent series is convergent.

Results : (i) A necessary and sufficient condition that $\sum_{n=1}^{\infty} (a_n + ib_n)$ converges (where a_n and b_n are real), is that $\sum_{n=1}^{\infty} a_n$ *and* $\sum_{n=1}^{\infty} b_n$ converges.

(ii) If $\sum_{n=1}^{\infty} |u_n|$ converges, then $\sum_{n=1}^{\infty} u_n$ converges, but the converse may not be true.

Some Important tests :

(i) Comparison tests:

(a) If $\sum_{n=1}^{\infty} |v_n|$ converges and $|u_n| \le |v_n|$, then $\sum_{n=1}^{\infty} u_n$ converges absolutely.

(b) If $\sum_{n=1}^{\infty} |v_n|$ diverges and $|u_n| \ge |v_n|$, then $\sum_{n=1}^{\infty} |u_n|$ diverges but $\sum_{n=1}^{\infty} u_n$ may or may not converge.

(ii) Ratio test:

$\sum_{n=1}^{\infty} u_n$ is absolutely convergent if $\lim_{n\to\infty} \left|\frac{u_{n+1}}{u_n}\right| = l < 1$ and diverges if $l > 1$. If $l = 1$, the test fails.

(iii) Dirichlet's test

Let $\sum_{n=1}^{\infty} a_n \text{ and } \sum_{n=1}^{\infty} b_n$ be two series of complex numbers such that

(i) the partial sums of $\sum_{n=1}^{\infty} a_n$ are bounded.

(ii) $\{b_n\}$ is a monotonically decreasing sequence

(iii) $b_n \to 0$ as $n \to \infty$ Then $\sum_{n=1}^{\infty} a_n b_n$ converges

(iv) Root test $\sum_{n=1}^{\infty} u_n$ is absolutely convergent if $\lim_{n\to\infty} \sqrt[n]{|u_n|} = l < 1$ and diverges if $l > 1$.

If $l = 1$, the test falls.

(v) Integral test :

If $f(x) \geq 0$ for $x \geq a$, then $\sum f(n)$ converges or diverges according as $\lim_{M\to\infty} \int_a^M f(x)dx$ converges or diverges.

Alternating series :

Let $\sum_{n=1}^{\infty} a_n$ be a series of real number such that

1. $a_n \geq 0 \ \forall \ n \in N$ 2. $a_{n+1} \leq a_n \ \ \forall \ n \in N$ 3. $\lim_{n\to\infty} a_n = 0$

Then the series $\sum_{n=1}^{\infty} (-1)^{n-1} a_n = a_1 - a_2 + a_3.....$ is convergent.

3. Power Series

A series of the form $a_0 + a_1(z - z_0) + a_2(z - z_0)^2 + ... = \sum_{n=0}^{\infty} a_n (z - z_0)^n$ .... (1) is called a power series about point z_0, where z_0, a_0, a_1, a_2, ..., a_n are complex constants and z_0 is the centre of the series; a_0, a_1, a_2, ...a_n are known as the coefficient of the series. In particular, the series $\sum_{n=0}^{\infty} a_n z^n$ is a power series about the point z_0.

Cauchy-Hadamard Theorem:

For a power series $\sum_{n=0}^{\infty} a_n (z - a)^n$ ...(1), there exists a unique real number $R(0 \leq R \leq \infty)$ such that it

(1) Converges absolutely for $|z - a| < R$

(2) Converges uniformly for $|z - a| \leq \rho$ $(0 \leq \rho < R)$

(3) Diverges for $|z - a| > R$

R is called the radius of convergence of the power series (1) and the circle |z| = R which also includes in its interior |z| < R all the values of z for which the power series (1) converges is called the circle of convergence of the series.

The radius of convergence is given by the relation : $\frac{1}{R} = \lim_{n\to\infty} |a_n|^{1/n} = \lim_{n\to\infty} \left|\frac{a_{n+1}}{a_n}\right|$

Region of Convergence:

The power series $\sum a_n z^n$, either

(i) Converges for all values of z (ii) Converges only of $z = z_0$

(ii) Converges for z in some region in the complex plane.

Theorems on Power Series:

A power series converges uniformly and absolutely in any region which lies entirely inside its circle of convergence.

Abel's Theorem

Let $\sum a_n z^n$ have radius of convergence R and suppose that z_0 is a point on the circle of convergence such that $\sum a_n z_0^n$ converges. Then $\lim_{z\to z_0} \sum a_n z^n = \sum a_n z_0^n$.

Radius of Convergence

Let $\sum a_n (z - z_0)^n$ be a power series then R.O.C. (Radius of Convegence R)

$$\frac{1}{R} = \sup \lim_{n\to\infty} \left|\frac{a_n + 1}{a_n}\right| = \sup \lim_{n\to\infty} |a_n|^{1/n}$$

Here R is R.O.C. of given series & circle of convergence is $|z - z_0| < R$

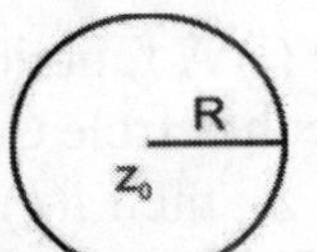

Note : The power series is convergent in the circle $|z - z_0| < R$ but at circumference of then circle, series may or may not be convergent.

4. Taylor's Series

Let f(z) be analytic inside an on a simple closed curve C. Let a and a + h be two points inside C.

$$\text{Then } f(a+h) = f(a) + hf'(a) + \frac{h^2}{2!} f''(a) + .. + \frac{h^n}{n!} f^{(n)}(a) + ... \qquad .....(1)$$

Or writing z = a + h, h = z – a, f (z) = fa) + f' (a) (z-a)

$$+ \frac{f''(a)}{2!}(z-a)^2 + .. + \frac{f^{(n)}(a)}{n!}(z-a)^n + ... \qquad ...(2)$$

$$\text{Thus, } f(z) = \sum_{n=0}^{\infty} a_n (z-a)^n, \text{ where } a_n = \frac{f^{(n)}(a)}{n!}$$

This is called Taylor's theorem and the series (1) or (2) is called Taylor's series expansion of f(a + h) or f(z). The region of convergence of the series (2) is given by $|z - a| < R$, where the radius of convergence R is the distance from a to the nearest singularity of the function f(z). On $|z - a| = R$, the series may or may not converge. For $|z - a| > R$, the series diverges.
If the nearest singularity of f(z) is at infinity, then radius of convergence is infinite, i.e., the series converges for all z. If a = 0 in (1) or (2) , the resulting series is often called Maclaurin series.

Some Special Series:

1. $e^z = 1 + z + \frac{z^2}{2!} + \frac{z^3}{3!} + ... + \frac{z^n}{n!} + ..., |z| < \infty$

2. $\sin z = z - \frac{z^3}{3!} + \frac{z^5}{5!} - ..(-1)^{n-1}\frac{z^{2n-1}}{(2n-1)!} + ...|z| < \infty$

3. $\cos z = 1 - \frac{z^2}{2!} + \frac{z^4}{4!} - ....(-1)^{n-1}\frac{z^{2n-2}}{(2n-2)!} + ...,|z| < \infty$

4. $\tan z = z + \frac{z^3}{3} + \frac{2z^5}{15} + ...,|z| < \frac{\pi}{2}$

5. $log(1+z) = z - \frac{z^2}{2} + \frac{z^3}{3} - ...(-1)^{n-1}\frac{z^n}{n} + ...|z| < 1$

6. $tan^{-1} z = z - \frac{z^3}{3} + \frac{z^5}{5} - ...(-1)^{n-1}\frac{z^{2n-1}}{2n-1} + ...,|z| < 1$

7. $(1+z)^k = 1 + kz + \frac{k(k-1)}{2!}z^2 + ... + \frac{k(k-1)..(k-n+1)}{n!}z^n + ...,|z| < 1$

Note: (i) A function f(z) can be expanded as a Taylor's series only if f(z) is analytic at all points inside the circle C. This assures the convergence of Taylor's series for f(z). If f(z) is not analytic at $z = z_0$, then f(z) is expanded about $z = z_0$ by Laurent's method.

1. Taylor series is unique, so we can find it by any method.
2. Radius of convergence of Taylor's series about z_0 of f(z) is the shortest distance of z_0 and singularity of f(z).
3. If $z_0 = 0$. Maclaurin's series

5. $(1+z)^n = 1 + nz + \frac{n(n-1)}{2!}z^2 + ...\quad R = \begin{cases} \infty & n \geq 0 \\ 1 & n < 0 \end{cases}$

6. $\frac{1}{1-z} = \sum z^n \qquad R = 1$ 7. $\frac{1}{(1-z)^2} = \sum (n+1)z^n \qquad R = 1$

Example : Find the Taylor's series of function and Radius of Convergence.

$f(z) = \frac{1}{z-1}$ about z = 0

Solution : $-\dfrac{1}{1-z}=-\sum z^n \qquad R=1$

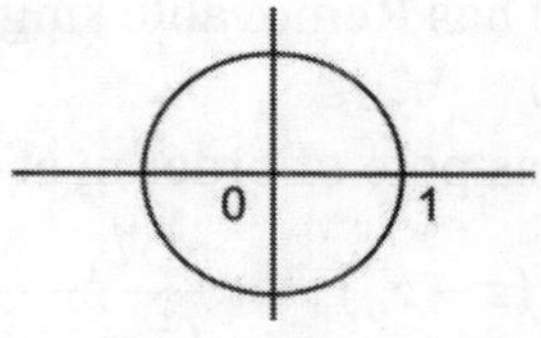

Example: Find the radius of convergence of $f(z)=\dfrac{1}{z}$ about z = 2

Solution : $\dfrac{1}{(z-2)+2}=\dfrac{1}{2\left[1+\left(\dfrac{z-2}{2}\right)\right]}=\dfrac{1}{2}\sum(-1)^n\dfrac{(z-2)^n}{2^n}$

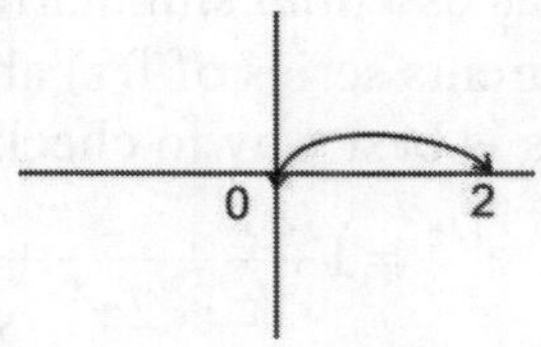

R.O.C. = 2

5. Laurent's Series

Let C_1 and C_2 be concentric circles of radii R_1 and R_2 respectively and centered at a. Suppose that f(z) is single-valued and analytic on C_1 and C_2 and in the annular region R between C_1= and C_2, shown shaded in the figure. Let a + h be any point in R. Then we have

$$f(a+h)=a_0+a_1h+a_2h^2+\ldots+\frac{a_{-1}}{h}+\frac{a_{-2}}{h^2}+\frac{a_{-3}}{h^3}+\ldots, \quad \ldots..(3)$$

$$\text{where } a_n=\frac{1}{2\pi i}\oint_{C_1}\frac{f(z)}{(z-a)^{n+1}}dz, n=0,1,2,\ldots \text{ and } a_{-n}=\frac{1}{2\pi i}\oint_{C_2}(z-a)^{-n+1}f(z)dz, n=1,2,3,\ldots \quad \ldots(4)$$

C_1 and C_2 being traversed in the positive direction with respect to their interiors.
In the above integrations we can replace C_1 and C_2 by any concentric circle C between C_1 and C_2. Then the coefficients (4) can be written in a single formula.

$$a_n=\frac{1}{2\pi i}\oint_C\frac{f(z)}{(z-a)^{n+1}}dz, n=0,\pm 1,\pm 2,\ldots. \quad \ldots.(5)$$

By taking a z = a + h, (3) can also be written as

$$f(z)=a_0+a_1(z-a)+a_2(z-a)^2+\ldots+\frac{a_{-1}}{z-a}+\frac{a_{-2}}{(z-a)^2}+\ldots, \quad \ldots..(6)$$

$$\text{where } a_n=\frac{1}{2\pi i}\oint_C\frac{f(s)}{(s-a)^{n+1}}ds,\ n=0,\ \pm 1,\ \pm 2, \quad \ldots.\ (7)$$

(3) or (6) with coefficients (4), (5) or (7) is called Laurent series expansion.
The part $a_0 + a_1(z-a) + a_2(z-a)^2 + \ldots$ is called the analytic part of the Laurent series, while the remainder of the series which consists of inverse powers of z – a is called the principal part.
Note: If the principal part is zero, then the Laurent series reduces to the Taylor series.

Results :

1. If f(z) is analytic at $z = z_0$ and $b_n = 0\ \forall\ n\in N$ i.e. Laurent's series expansion is just Taylor's series.
2. If $z = z_0$ is an isolated singularity then f(z) has Laurent's series expansion about z_0.

3. If f(z) has Removable singularity at z_0 iff principal part of series at z_0 is equal to '0' i.e.⇔ $b_n = 0 \quad \forall\, n \in N$

4. f(z) has pole of order m at $z = z_0$.⇔

$$\sum_{n=1}^{\infty} b_n (z-z_0)^{-n} = \frac{b_1}{(z-z_0)} + \frac{b_2}{(z-z_0)^2} + \ldots + \frac{b_m}{(z-z_0)^m}$$

5. f(z) has essential singularity at $z = z_0$.

⇔ In Laurents series of f(z) about z_0. Principal part has infinite many terms.

This is best way to check E.S.

e.g. $e^{1/z} = 1 + \frac{1}{z} + \frac{1}{2z^2} + \frac{1}{3!z^3} + \ldots\ldots$ about z = 0

Principal part has infinite many terms.

Example: Find Laurentz series expansion of the function $f(z) = \frac{1}{z(z-1)}$ about z = 0

Solution : Type – 1 $\quad 0 < |z| < 1$

Given $f(z) = \frac{1}{z(z-1)} = -\frac{1}{z}\left[\frac{1}{(1-z)}\right] = -\frac{1}{z}\sum z^n = -\sum_{n=0}^{\infty} z^{n-1}$

Type – 2 $\quad 1 < |z| < \infty$

$$f(z) = \frac{1}{z^2\left[1 - \frac{1}{z}\right]} = \frac{1}{z^2}\sum_{n=0}^{\infty}\frac{1}{z^n} = \sum_{n=0}^{\infty}\frac{1}{z^{n+2}}$$

SOLVED QUESTIONS

1. Let $f(z) = \frac{1}{1-z-z^2}$ be a complex function whose denominator is not equal to zero the

CSIR NET JUNE 2023

(a) f(z) is entire

(b) f(z) have simple pole

(c) $f(z) = \sum a_n z^n$ where $a_0 = 1$, $a_1 = 0$ & $a_{n+2} = a_n + a_{n+1}$

(d) $f(z) = \sum a_n z^n$ where $a_0 = 1$, $a_1 = 1$, $a_{n+2} = a_n + a_{n+1}$.

Ans. (b),(d)

Explanation:

Given that $f(z) = \frac{1}{1-z-z^2}$ So, f(z) has a pole at simple pole

Option (b) is true

If $f(z)=\sum a_n z^n$ is taylor's series .Then $a_1=\frac{f'(0)}{1!}$ Now $f'(z)=\frac{1+2z}{(1-z-z^2)^2}$

Then $f'(0)=1=a_1$ Hence option (d) is true

2. Let f be a complex function given by $f(z)=\frac{z^2+2z-4}{z}$. then radius of the convergence of f(z) at z = 1 is **CSIR NET SEPT. 2022**

(a) 0 (b) 1 (c) 2 (d) ∞

Ans . (b)

Explanation :

f(z) has a singularity at z = 0

So , radius of the convergence of f(z) at z = 1 is 1

3. Let f(z) = (1 –z) $e^{\left(z+\frac{z^2}{2}\right)}=1+\sum_{n=1}^{\infty}a_n z^n$ .Then which of the following is false

CSIR NET SEPT. 2022

(a) $f'(z)=-z^2 e^{\left(z+\frac{z^2}{2}\right)}$ (b) $a_1 = a_2$ (c) $a_n \in (-\infty, 0]$ (d) $\sum_{n=1}^{\infty}|a_n|<1$

Ans. (d)

Explanation:

(a) f(z) = (1 –z) $e^{\left(z+\frac{z^2}{2}\right)}$

$$f'(z)=-e^{\left(z+\frac{z^2}{2}\right)}+(1-z)e^{\left(z+\frac{z^2}{2}\right)}(1+z)$$

$$\Rightarrow f'(z)=-e^{\left(z+\frac{z^2}{2}\right)}+(1-z^2)e^{\left(z+\frac{z^2}{2}\right)}=-z^2 e^{\left(z+\frac{z^2}{2}\right)}$$

Option (a) is true

(b) Given that f(z) = (1 –z) $e^{\left(z+\frac{z^2}{2}\right)}=1+\sum_{n=1}^{\infty}a_n z^n=\sum_{n=0}^{\infty}a_n z^n$

We know that f(z) = (1 –z) $e^{\left(z+\frac{z^2}{2}\right)}$ is analytic function, so we can construct a tailor series

Where $a_n=\frac{f^{(n)}(0)}{n!}$ $a_1=\frac{f'(0)}{1}=0$ ...(1)

$$f''(z)=-z^2 e^{\left(z+\frac{z^2}{2}\right)}(1+z)-2ze^{\left(z+\frac{z^2}{2}\right)}$$

COMPLEX ANALYSIS

And $a_2 = \frac{f''(0)}{2} = 0$ ..(2) $\Rightarrow a_1 = a_2$

Option (b) is true

(c) We know that $a_1 = a_2 = 0$

$$f'''(z) = -(2z+3z^2)e^{\left(z+\frac{z^2}{2}\right)} - (z^2+z^3)e^{\left(z+\frac{z^2}{2}\right)}(1+z) - 2e^{\left(z+\frac{z^2}{2}\right)} - 2ze^{\left(z+\frac{z^2}{2}\right)}(1+z)$$

$f'''(0) = -2$ Then $a_3 = \frac{f'''(0)}{3!} = -\frac{2}{3!}$

We can check a_4, a_5, a_6 …. etc.

$\Rightarrow a_n \in (-\infty, 0]$

Option (c) is true

Hence Option (d) is false

COMPLEX ANALYSIS

4. Let $\frac{z}{1-z-z^2} = \sum_{n=0}^{\infty} a_n z^n$, $a_n \in R$ **GATE 2023**

for all z in some neighbourhood of 0 in C. Then the value of $a_6 + a_5$ is equal to

(a) 10 (b) 7 (c) 13 (d) 15

Ans. **(c)**

Explanation:

Given that $\frac{z}{1-z-z^2} = \sum_{n=0}^{\infty} a_n z^n$, $a_n \in R$

$$\Rightarrow \frac{z}{1-z-z^2} = \frac{z}{1-(z+z^2)} = z\left[\frac{1}{1-(z+z^2)}\right]$$

We know that $\frac{1}{1-z} = 1 + z + z^2 + z^3 + ...$

Hence $\left[\frac{1}{1-(z+z^2)}\right] = 1 + (z+z^2) + (z+z^2)^2 + (z+z^2)^3 ...$

And $z\left[\frac{1}{1-(z+z^2)}\right] = z\left[1 + (z+z^2) + (z+z^2)^2 + (z+z^2)^3 + (z+z^2)^4 + (z+z^2)^5 ...\right]$

So coefficient of z^4 and z^5 in $\left[1 + (z+z^2) + (z+z^2)^2 ...\right]$

In $(z+z^2)^2$ coefficient of z^4 is 1 and coefficient of z^5 is 0

In $(z+z^2)^3$ coefficient of z^4 is 3 and coefficient of z^5 is 3

In $(z+z^2)^4 = z^4(1+z)^4 = z^4\left(1 + 4z + \frac{4.3}{2}z^2 + ...\right)$

coefficient of z^4 is 1 and coefficient of z^5 is 4

In $(z+z^2)^5 = z^5(1+z)^5 = z^5\left(1+5z+\frac{5.4}{2}z^2+...\right)$

coefficient of z^4 is 0 and coefficient of z^5 is 1

Then $a_5 = 5$ and $a_6 = 8$ $\Rightarrow a_6 + a_5 = 13$

5. The coefficient of $\left(x-\frac{\pi}{2}\right)$ in the Taylor series expansion of the function

$$f(x)=\begin{cases}\frac{4(1-\sin x)}{2x-\pi} & x\neq\frac{\pi}{2}\\ 0 & x=\frac{\pi}{2}\end{cases} \text{ about } x=\frac{\pi}{2}.$$

CSIR NET JUNE 2019

(a) 1 (b) 2 (c) 3 (d) 4

Ans. **(a)**

Explanation :

Given that

$$f(x)=\frac{4(1-\sin x)}{2\left(x-\frac{\pi}{2}\right)}=\frac{2\left[1-\sin\left(x-\frac{\pi}{2}+\frac{\pi}{2}\right)\right]}{x-\frac{\pi}{2}}$$

$$=\frac{2}{x-\frac{\pi}{2}}\left[1-\left\{\sin\left(x-\frac{\pi}{2}\right)\cos\frac{\pi}{2}+\cos\left(x-\frac{\pi}{2}\right)\sin\frac{\pi}{2}\right\}\right]=\frac{2}{x-\frac{\pi}{2}}\left[1-\cos\left(x-\frac{\pi}{2}\right)\right]$$

$$=\frac{2}{x-\frac{\pi}{2}}\left[1-\left\{1-\frac{\left(x-\frac{\pi}{2}\right)^2}{2!}\right\}+\frac{\left(x-\frac{\pi}{2}\right)^4}{4!}+......\right]=\frac{2}{x-\frac{\pi}{2}}\left[1-1+\frac{\left(x-\frac{\pi}{2}\right)^2}{2!}+.....\right]$$

$$=\text{Coefficient of }\left(x-\frac{\pi}{2}\right)=1.$$

6. Consider the power series $f(x)=\sum\log(n)x^n$. The R.O.C. of f(x) is

(a) 0 (b) 1 (c) 3 (d) ∞

Ans. **(b)**

CSIR NET JUNE 2017

Explanation :

$a_n = \log(n)$

We know that $\frac{1}{R} = \sup \lim_{n\to\infty} \left|\frac{a_n + 1}{a_n}\right|$

Find $\lim_{n\to\infty} \left|\frac{a_{n+1}}{a_n}\right| = \lim_{n\to\infty} \left|\frac{\log(n+1)}{\log(n)}\right| \quad \left[\because \frac{\infty}{\infty}\right]$

So $\frac{\frac{d}{dn}\log(n+1)}{\frac{d}{dn}(\log(n))} = \frac{\frac{1}{n+1}}{\frac{1}{n}} = \lim_{n\to\infty} \frac{n}{1+n} = 1$

R = 1 R.O.C. of f(x) = 1

& domain of cgt. is |x| < 1

7. Let R denote the radius of convergence of power series $\sum kx^k$. Then

(a) R > 0 & the series is convergent on [-R, R]

(b) R > 0 & the series is convergent at x = -R but doesn't convergent at x = R

(c) R > 0 & the series doesn't convergent outside (-R, R)

(d) R = 0 **CSIR NET DEC 2018**

Ans. (c)

Explanation :

Here $a_k = k$

Then $\frac{1}{R} = \sup \lim_{k\to\infty} \left|\frac{a_{k+1}}{a_k}\right| = \sup \lim_{k\to\infty} \left|\frac{k+1}{k}\right| = 1$ R = 1

Option (d) incorrect.

= R = 1 then series is convergent in (-1, 1) / |x| < 1

Now we have to check at x = 1

Then series become $\sum_{k=1}^{\infty} k$ is not convergent.

At x = -1

$\sum k(-1)^k$ is also not convergent.

Option (b) is incorrect.

8. Consider the power series $f(z) = \sum_{n=1}^{\infty} (-1)^n \frac{z^{2n+1}}{(2n)!}$. Which of the following are TRUE?

(a) R.O.C. of f(z) is infinite

(b) The set {f(x) : x ∈ R} is b.d.

(c) The set {f(x) : -1<x<1} is b.d.

(d) f(z) has infinitely many zeros.

Ans. (a), (c), (d) **CSIR NET DEC 2019**

Explanation :

Given that

$$f(z)=\sum_{n=0}^{\infty}(-1)^n\frac{z^{2n+1}}{(2n)!}=z-\frac{z^3}{2!}+\frac{z^5}{4!}-\frac{z^7}{6!}......$$

$$=z\left[1-\frac{z^2}{2!}+\frac{z^4}{4!}+.......\right]=z\cos z$$

(a) z cos z is entire function so R.O.C. is infinite.

(b) f(x) = x cos x

Take x = 2nπ Then f(2nπ) = 2nπ

$\lim_{n\to\infty} 2n\pi \to \infty$ i.e. not b.d.

(c) |f(x)| = |x cosx| = |x| |cosx| ≤ |x| < 1 ⇒ |f(x)| < 1[Given that |x| < 1 x ∈ (-1, 1)

(d) f(z) = z cos z z = (2n + 1) π/2

infinite zeroes

9. Let f be a holomorphic function in the open unit disc s.t. $\lim_{z\to 1} f(z)$ does not exist. Let $\sum a_n z^n$ be a Taylor series of f(z) about z = 0 & let R be its R.O.C. Then

(a) R = 0 (b) 0 < R < 1 (c) R = 1 (d) R > 1

Ans. (c) **CSIR NET DEC 2017**

Explanation :

Given that $\lim_{Z\to 1} f(z)$ doesn't exist. $\Rightarrow f(z)=\frac{1}{1-z}$

& $\sum a_n z^n$ be a Taylor series of f(z). $=f(z)=\sum z^n$

Where $a_n = 1$ Then $\frac{1}{R}=\sup\lim_{n\to\infty}\left|\frac{a_n+1}{a_n}\right| = \frac{1}{R}=1$ R = 1

10. Let D be the open unit disc centred at 0 in C and f : D → C be an analytic function. If $f(z)=\sum a_n z^n$ is a power series of f the f is constant if

CSIR NET JUNE 2018

(a) $\bar{f}$ is analytic

(b) $u\left(\frac{1}{2}\right)\geq u(z)\ \ \forall z\in D$

(c) The set {n ∈ N | a_n = 0} is infinite

(d) For any closed curve γ in D $\int_\gamma \frac{f(z)dz}{(z-a)^2}=0\quad \forall a\in D$ With $|a|\geq\frac{1}{2}$.

Ans. (a), (b), (d)

COMPLEX ANALYSIS

Explanation :

Theorem : If f is non-constant analytic function then $\bar{f}$ is non-analytic function.

Option (a) is correct. Option (b) is correct.

Maximum Modulus Theorem : Let f(z) be non-constant analytic function then maximal value of |f(z)| attain on the boundary of D.

$u\left(\frac{1}{2}\right) \geq u(z)$ $\quad \forall z \in D \quad \frac{1}{2} \in D \quad$ D = |z| < 1

$\Rightarrow$ u is constant $\Rightarrow$ f is constant Option (c) is incorrect. $\sin z = z - \frac{z^3}{3!} + \frac{z^5}{5!}$

The set $\{a_{2n} = 0\}$ is infinite but f(z) is not constant.

Option (d) is correct.

Given that |z| < 1 $\quad \int_\gamma \frac{f(z)dz}{(z-a)^2} = 2\pi i\, f'(a) = 0 \quad \forall\, |a| \geq \frac{1}{2}$

$$= f'(a) = 0 \quad \forall\, |z| \geq \frac{1}{2}$$

By Identity theorem f'(z) = 0 $\quad \forall$ z $\in$ $\quad$ f(z) = constant

11. Let D = {z $\in$ C | |z| < 2n} and f : D $\to$ C be the function defined by

GATE-2021

$f(z) = \begin{cases} \dfrac{3z^2}{1-\cos z} & if \quad z \neq 0 \\ 6 & if \quad z = 0 \end{cases}$. If $f(z) = \sum_{n=0}^{\infty} a_n z^n$ for z $\in$ D, then $6a_2$ is

(a) -3 (b) 3 (c) 6 (d) none of these

Ans. (b)

Explanation :

$$f(z) = \frac{3z^2}{1-\cos z} = \frac{3z^2}{1-\left[1-\frac{z^2}{2!}+\frac{z^4}{4!}-\frac{z^6}{6!}-....\right]}$$

$$\Rightarrow \frac{3z^2}{\frac{z^2}{2!}-\frac{z^4}{4!}+\frac{z^6}{6!}-....} \Rightarrow \frac{3}{\frac{1}{2!}\left(1-\frac{z^2}{12}+\frac{z^4}{360}.....\right)}$$

$$\Rightarrow 6\left[1-\left(\frac{z^2}{12}-\frac{z^4}{360}+.....\right)\right]^{-1} \Rightarrow 6\left[1+\frac{z^2}{12}-\frac{z^4}{360}+.....\right]$$

a_2 = coefficient of $z^2 = \frac{1}{2}$. So, $6a_2 = 6\left(\frac{1}{2}\right) = 3$ Option (b) is correct.

UNSOLVED QUESTIONS

1. The coefficient $(z - \pi)^2$ in the Taylor series of $f(z) = \begin{cases} \dfrac{\sin z}{(z-\pi)} & z \neq \pi \\ -1 & z = \pi \end{cases}$.

GATE-2015

(a) 1/2 (b) 1/4 (c) 1/6 (d) -1/6

2. Let $\sum_{-\infty}^{\infty} a_n (z+1)^n$ be the Laurentz series expansion of $f(z) = \sin\left(\frac{z}{z+1}\right)$ then a_{-2} =

(a) 1 (b) 0 (c) cos 1 (d) $-\frac{1}{2}\sin 1$

3. The radius of convergence of $\sum_{n=0}^{\infty} \frac{\left(1+\frac{1}{n}\right)^{n^2}}{n^3} z^n$ GATE-2006

(a) e (b) 1/e (c) 1 (d) ∞

4. It is given that $\sum_{n=0}^{\infty} a_n z^n$ Converges at z = 3 + 4i. Then the radius of convergence of the power series $\sum_{n=0}^{\infty} a_n z^n$ is GATE-2000

(a) ≤5 (b) ≥5 (c) < 5 (d) > 5

5. Let f : C → C be an arbitrary analytic function satisfying f(0) = 0 and f(1) = 2. Then

GATE-2007

(a) there exists a sequence $\{z_n\}$ such that $|z_n| > n$ and $|f(z_n)| > n$
(b) there exists a sequence $\{z_n\}$ such that $|z_n| > n$ and $|f(z_n)| < n$
(c) there exists a bounded sequence $\{z_n\}$ such that $|f(z_n)| > n$
(d) there exists a sequence $\{z_n\}$ such that $z_n \to 0$ and $f(z_n) \to 2$

6. The radius of convergence of the power series $\sum_{n=0}^{\infty} 4^{(-1)^n n} z^{2n}$ is GATE-2014

(a) 0.1 (b) 0.2 (c) 0.4 (d) 0.5

7. If the power series $\sum_{n=0}^{\infty} a_n (z+3-i)^n$ converges at 5i and diverges at – 3i, the power series.

GATE-2015

(a) Converges at – 2 + 5i and diverges at 2 – 3i
(b) converges at 2 – 3i and diverges at – 2 + 5i
(c) converges at both 2 – 3i and -2 + 5i (d) diverges at both 2 – 3i and -2 + 5i

8. The power series $\sum_{0}^{\infty} 2^{-n} z^{2n}$ converges if **CSIR UGC NET JUNE-2011**

(a) $|z| \le 2$ (b) $|z| < 2$ (c) $|z| \le \sqrt{2}$ (d) $|z| < \sqrt{2}$

9. If series $\sum a_n$ converges absolutely, then

(a) $\sum a_n$ converges (b) $\sum a_n$ does not converges

(c) $\sum a_n$ diverges (d) $\sum a_n$ converges

10. The radius of convergence of the power series $f(z) = \sum \frac{n+1}{(n+2)(n+3)} z^n$ is not

(a) 1 (b) 2 (c) 3 (d) 4

11. Consider the power series $\sum_{n \ge 1} a_n z^n$, where a_n = number of divisors of n^{50}. Then the radius of convergence of $\sum_{n \ge 1} a_n z^n$ is **CSIR NET DEC-2011**

(a) 1 (b) 50 (c) 1/50 (d) 0

12. Consider the power series $\sum_{n=1}^{\infty} z^{n!}$. The radius of convergence of this series is

(a) 0 (b) ∞ (c) 1

(d) a real number greater than 1 **CSIR NET DEC-2012**

13. Let p(x) be a polynomial of the real variable x of degree k ³ 1. Consider the power series $f(z) = \sum_{n=0}^{\infty} p(n) z^n$, where z is a complex variable. Then the radius of convergence of f(z) is **CSIR NET JUNE-2014**

(a) 0 (b) 1 (c) k (d) ∞

14. Consider the following power series in the complex variable z:

$f(z) \sum_{n=1}^{\infty} n \log n \, z^n, g(z) = \sum_{n=1}^{\infty} \frac{e^{n^2}}{n} z^n$. If r, R are the radii of convergence of f and g respectively, then

(a) r = 0, R = 1 (b) r = 1, R = 0 (c) r = 1, R = ∞ (d) r = ∞, R = 1

15. The radius of convergence of the series $\sum_{n=1}^{\infty} z^{n^2}$ is **CSIR NET DEC-2016**

(a) 0 (b) ∞ (c) 1 (d) 2

Answer Key

1	2	3	4	5	6	7	8	9	10
c	d	b	b	a	d	a	d	a,d	b,c,d
11	**12**	**13**	**14**	**15**					
a	c	b	b	c					

Chapter 5

Classification of Singularities and Residue

1. Zeros of an analytic function:

The value of z for which the analytic function f(z) becomes zero is said to be zero of f(z). If f(z) is analytic in a domain D and z_0 is any point of D, then we can expand f(z) as Taylor series about $z = z_0$ *given by* $f(z) = \sum_{n=0}^{\infty} a_n (z - z_0)^n$

If $a_0 = a_1 = a_2 = ... = a_{m-1} = 0$ and $a_m \neq 0$, then f(z) is said to have a zero of order m at $z = z_0$.
A zero of order one (m = 1) is said to be a simple zero

Definition: An analytic function defined in $|z - a| < R$ is said to have zero of order $m \in N$ at point a if $f(a) = f'(a) = ... = f^{(m-1)}(a) = 0$ *and* $f(m)(a) \neq 0$.

Note : Zeros of a non-constant function are always isolated.

Example : The function $f(z) = \frac{z-2}{z^2} sin\left(\frac{1}{z-1}\right)$ has zero at z = 2, because at z = 2, f(z) = 0

Also, zeros are obtained when $sin\left(\frac{1}{z-1}\right) = 0 \Rightarrow \frac{1}{z-1} = n\pi \Rightarrow z - 1 = \frac{1}{n\pi} \Rightarrow, n = \pm 1, \pm 2, ...$

2. Branch points and branch lines:

Definition: The point z_0 is called a branch point for the complex (multiple) valued function f(z) if the value of f(z) does not return to its initial value as a close curve starting from some arbitrary point on the curve, around is traced in such a way that f varies continuously as the path is traced.

Example :. $f(z) = (z - 4)^{1/3}$ has a branch point at z = 4.

Example : $f(z) = ln\,(z^2 - z - 2)$ has branch points, where $z^2 - z - 2 = 0$,
i.e., at z = -1 and z = 2.

3. Singular point or singularity:

Let f(z) be a complex-valued function. A point $z = z_0$ is called a singular point or singularity of f(z), if f(z) is not analytic at $z = z_0$, but every neighbourhood of z_0 contains at least one point at which f is analytic. Singularities are of two types: isolated and non-isolated singularities.

Isolated and non-isolated singularities: The point $z = z_0$ is called an isolated singularity or isolated singular point of f(z) if there exists $\delta > 0$ such that the circle $|z - z_0| = \delta$ encloses no singular point other than z_0 (i.e., there exists a deleted δ neighbourhood of z_0 containing no singularity). If no such δ can be found, then z_0 is called a non-isolated singularity. If z_0 is not a singular point and there exists $\delta > 0$ such that $|z - z_0| = \delta$ encloses no singular point, then z_0 is called an ordinary point of f(z).

Example : (1) $f(z)=\frac{1}{z-1}, z=1$ is an isolated singularity of f(z) in |z-1| < 5

(2) f(z) = log z, z = 0 is non-isolated singularity of f(z).

Results : (i) Branch points of a multiple-valued functions are singular points.

(ii) If a function has finite number of singular points, then singularity at those points is always isolated.

Type of singularities:

(i) Removable singularity: If all the coefficients b_n in the Laurent series expansion of f(z) about z_0 are zero, i.e., if the principal part of Laurent's series expansion does not contain any terms, then z_0 is called removable singularity of f(z). If f(z) has a removable singularity at $z = z_0$, then there exists an analytic function g(z) at z_0 such that f(z) = g(z) for all z in some deleted neighbourhood of z_0.

For example : If $f(z)=\frac{e^z-1}{z}$, then z = 0 is a removable singularity since f(0) is not defined

but $\lim_{z\to 0}\frac{e^z-1}{z}=1$.

We define, $f(0)=\lim_{z\to 0}\frac{e^z-1}{z}=1$.

$$\frac{e^z-1}{z}=\frac{1}{z}\left[\left(1+z+\frac{z^2}{2!}+\frac{z^3}{3!}+.....\right)-1\right]=1+\frac{z}{2!}+\frac{z^2}{3!}+\frac{z^3}{4!}+......$$

Since, the Laurent series expansion of $f(z)=\frac{e^z-1}{z}$ about z = 0 has no principal part, so z = 0 is a removable singularity of f(z).

Results : If f has an isolated singularity at z_0, then $z = z_0$ is a removable singularity iff one of the following conditions hold :

(i) f is bounded in a deleted neighbourhood of z_0.

(ii) $\lim_{z\to z_0} f(z)$ exists. (iii) $\lim_{z\to z_0}(z-z_0)f(z)=0$.

(ii) Pole: If in the Laurent series expansion $\sum_{n=0}^{\infty}a_n(z-a)^n$ of f(z) in some deleted neighbourhood of a, the principal part has only a finite number of terms given by $\frac{a_{-1}}{z-a}+\frac{a_{-2}}{(z-a)^2}+...+\frac{a_{-n}}{(z-a)^n}$, where a_{-n} □ 0. then z = a is called a pole of order n. For n = 1, it is called a simple pole. If f(z) has a pole at z = a, then $\lim_{z\to a} f(z)=\infty$

Example : $f(z)=\frac{1}{(z-1)^4}$ has a pole of order 4 at z = 1.

Result : If $g(z) = (z - z_0)^n f(z) \neq 0$ and n is a positive integer, then $z = z_0$ is called a zero of order n of g(z). For n = 1, z_0 is called a simple zero. In such case, z_0 is a pole of order n of the function $\frac{1}{g(z)}$.

(iii) Essential Singularity: A singularity which is not a pole, branch point or removable singularity is called as essential singularity. If $z = z_0$ is an essential singularity of f(z), then the principal part of the Laurent series expansion has infinitely many terms. Equivalently, $z = z_0$ is an essential singularity if there does not exist any positive integer n such that $\lim_{z \to z_0} (z - z_0)^n f(z) = A \neq 0$.

Example : $f(z) = e^{\frac{1}{z-2}}$ has an essential singularity at z = 2.

Results : Let $h(z) = \frac{f(z)}{g(z)}$ where f(z) has zero of order m at z_0 and g(z) has zero of order n at z_0.

(i) If m = n then h(z) has a removable singularity at z_0.
(ii) If m > n then h(z) has a zero of order m – n at z_0.
(iii) If m < n then h(z) has a pole of order n – m at z_0.

Singularity at infinity:

Let f(z) be a function then behaviour of f(z) at ∞ is similar to behaviour of f(1/z) at 0.

(i) Pole : Let f(z) be a polynomial of order m, then f(z) has a pole of order m at ∞.

Example : Let $f(z) = z^2$.

Then $f\left(\frac{1}{z}\right) = \frac{1}{z^2}$ has a pole of order 2 at origin.

So, $f(z) = z^2$ has a pole of order 2 at ∞.

Example : $f(z) = 1 + 2z + 3z^3$ has a pole of order 3 at ∞.

(ii) Essential singularity : If f(1/z) has essential singularity at '0' then f(z) has essential singularity at ∞.

Example : $f(z) = e^z$.

$f(1/z) = e^{1/z}$ has essential singularity at '0'. So, e^z has essential singularity at ∞.

Results : (i) Limit point of the poles of a function is a non-isolated essential singularity of that function..

(ii) An analytic function f(z) is a polynomial of order n iff the only singularity of f(z) in extended complex plane is a pole of order n at ∞.

(iii) If f(z) is an entire function which has no singularity at infinity, then f(z) is constant.

4. Entire function :

A function which is analytic everywhere in the entire complex plane [i.e. everywhere except at ∞] is called an entire function or integral function.

Result :

(i) An entire function can be represented by a Taylor series which has an infinite radius of convergence.

(ii) A function which is analytic everywhere including ∞ must be a constant.

(iii) Range of entire function is densed in C.

Example : e^z , z, sinz, cosz are entire function.

Meromorphic function:

A function which is analytic everywhere in the finite plane except at a finite number of poles is called a meromorphic function.

Example:

1. $f(z)=\dfrac{z+1}{(z+2)(z-1)^2}$, is analytic everywhere in the finite plane except at the poles, z =-

2. (Simple pole) and z = 1 (pole of order two), so f(z) is a meromorphic function.

2. $f(z)=\dfrac{1}{e^z-1}$, is analytic everywhere except at poles given by $e^z = 1$, i.e. $z = 2n\pi i$, $n \in z$.

Here, singularities of f(z) in **C** are at most poles.

$\therefore$ f(z) is meromorphic

Residue

Let f(z) has a Laurent series about z = a given by $f(z)=\sum_{n=-\infty}^{\infty} a_n(z-a)^n$

$$= a_0 + a_1(z-a) + a_2(z-a)^2 + \ldots + \frac{a_{-1}}{z-a} + \frac{a_{-2}}{(z-a)^2} + \ldots$$

Then the coefficient of $\dfrac{1}{z-a} = a_{-1}$ is called a Residue of f(z) at z = a.

Let f(z) be single-valued and analytic inside and on a circle C except at the point z = a chosen as the centre of C. Then $a_{-1} = \dfrac{1}{2\pi i}\oint_C f(z)dz \Rightarrow \int_C f(z)dz = 2\pi i . a_{-1}$

Calculation of residue at finite singularities:

(i) Let z = a be a pole of f(z) of order m, then the residue of f(z) at z = a is given by

$$\operatorname{Res}_{z=a} f(z) = \lim_{z\to a} \frac{1}{(m-1)!} \frac{d^{m-1}}{dz^{m-1}}\{(z-a)^m f(z)\}$$

If m = 1, i.e., if z = a is simple pole of f(z), then $\operatorname{Res}_{z=a} f(z) = \lim_{z\to a}(z-a)f(z)$ $[\because 0! = 1]$

Example. Consider, $f(z)=\dfrac{z+1}{(z-2)^2(z-1)}$

(ii) Then, f(z) has poles at z = 2 of order 2 and at z = 1 of order 1.

Residue at $z=2=\lim_{z\to 2}\frac{1}{1!}\frac{d}{dz}\left\{(z-2)^2\left(\frac{z+1}{(z-2)^2(z-1)}\right)\right\}=\lim_{z\to 2}\frac{d}{dz}\left(\frac{z+1}{(z-1)}\right)=-2$

Residue at $z=1=\lim_{z\to 1}(z-1)\left(\frac{z+1}{(z-2)^2(z-1)}\right)=2$

Important result : If z = a is an essential singularity of f(z), then the residue of f(z) at z = a can be found by using any known series expansion.

Example. If $f(z)=\sin\frac{1}{z}$, then z = 0 is an essential singularity, as

$\sin\frac{1}{z}=\frac{1}{z}-\frac{1}{3!\,z^3}+\frac{1}{5!\,z^5}$.... ....(1) Then, $\mathop{Res}_{z=0} f(z)=\text{coefficient of } \frac{1}{z}=1$

Residue at infinity:

Let z = ∞ be an isolated singularity of f(z), then residue of f(z) at z = ∞ is given by

Res [f(z), ∞] = $-\text{Res}\left[\frac{1}{z^2}f\left(\frac{1}{z}\right);0\right]$.

Example : Let f(z) = z, then find residue of f(z) at z = ∞.

$f\left(\frac{1}{z}\right)=\frac{1}{z}\Rightarrow -\frac{1}{z^2}f\left(\frac{1}{z}\right)=-\frac{1}{z^3}$ So, $\text{Res}\left[-\frac{1}{z^3};0\right]\ -\lim_{z\to 0}\frac{1}{2!}\frac{d^2}{dz^2}(1)=0$

Working rules for computing the residue:

1. $Res(z=a)=\lim_{z\to 0}(z-a)f(z)$ for simple pole.
2. $Res(z=a)=\frac{\phi^{(m-1)}(a)}{(m-1)!}$ for pole of order m, if $f(z)=\frac{\phi(z)}{(z-a)^m}$
3. $Res(z=a)=\frac{1}{2\pi i}\int_C f(z)dz$ for pole of any order. (If 'a' is the only singularity inside C)
4. $Res(z=\infty)=-\frac{1}{2\pi i}\int_C f(z)dz=-\mathop{Res}_{z=0}\frac{1}{z^2}f\left(\frac{1}{z}\right)$
5. $Res(z=\infty)=$ negative of the coefficient of $\left(\frac{1}{z}\right)$ in the expansion of f(z) in the neighbourhood of z = ∞.
6. If $f(z)=\frac{\phi(z)}{\psi(z)}$ has a simple pole at z = a, then $Res(z=a)=\frac{\phi(a)}{\psi'(a)}$
7. Let P(z) and Q(z) be complex polynomials. If $deg(Q(z))\ge deg(P(z))+2$, *then* $\oint_C\frac{P(z)}{Q(z)}dz=0$, where C encloses all singular points of $\frac{P(z)}{Q(z)}$.

Some Theorems:

Theorem: If f(z) has a simple pole at z_0 and g(z) is analytic at z_0 such that $g(z_0) \# 0$, then $\underset{z=z_0}{Res}\, f(z)g(z) = g(z_0)\underset{z=z_0}{Res}\, f(z)$.

Theorem: If $f(z) = (z-a)^{-n}(z-b)^{-m}$, where m and n are positive integers, then $\underset{z=a}{Res}\, f(z) = -\underset{z=b}{Res}\, f(z)$.

Theorem: If function f(z) has an isolated singularity at z_0 and f(z) is even in $z-z_0$ i.e., $f(z-z_0) = f(-(z-z_0))$, then $\underset{z=z_0}{Res}\, f(z) = 0$

Theorem: Let f(z) be analytic at z_0 such that $f(z_0) \neq 0$ and g(z) has a zero or order two at z_0, then $\underset{z=z_0}{Res}\frac{f(z)}{g(z)} = \frac{6f'(z_0)g''(z_0) - 2f(z_0)g'''(z_0)}{3[g''(z_0)]^2}$

Theorem: Suppose f(z) and g(z) are both analytic at z_0 and $f(z_0) \neq 0$ and z_0 is a first order zero for $\underset{z=z_0}{Res}\frac{f(z)}{[g(z)]^2} = \frac{f'(z_0)g'(z_0) - f(z_0)g''(z_0)}{[g'(z_0)]^3}$

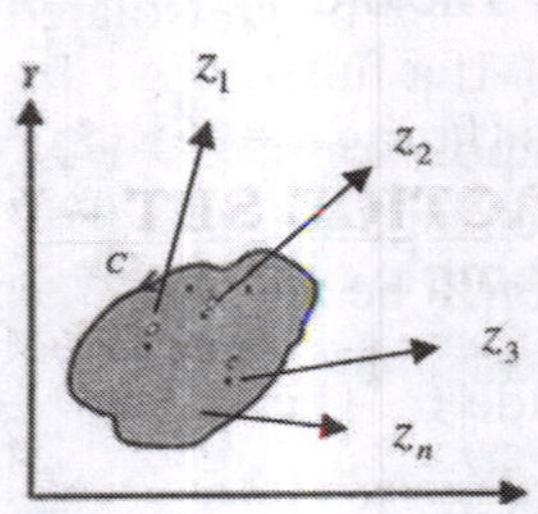

The Residue Theorem

Let f(z) be single-valued and analytic inside and on a simple closed curve C except at the singularities $z_1, z_2, z_3, \ldots z_n$ inside C which have residue given by $a_{-1}, a_{-2}, a_{-3}, \ldots$. Then the residue theorem states that $\oint_C f(z)dz = 2\pi i(a_{-1} + a_{-2} + a_{-3}, + \ldots)$, the integral of f(z) around C is $2\pi i$ times the sum of the residues of f(z) at the singularities enclosed in C. Thus $\oint_C f(z)dz = 2\pi i\sum_{m=1}^{n}\underset{z=z_m}{Res} f(z)$.

Exercise : The residue at $z = \infty$ of $f(z) = \frac{z^3}{(z-2)(z-3)(z-5)}$ is

(a) 0 (b) -10 (c) 10 (d) none of these

Solution : (b)

Reside at $(z = \infty) = -\{Res(z=2) + Res(z=3) + Res(z=5)\}$

$$= -\left\{(z-2)\frac{z^3}{(z-2)(z-3)(z-5)}\bigg|_{z=2} + (z-3)\frac{z^3}{(z-2)(z-3)(z-5)}\bigg|_{z=3} + (z-5)\frac{z^3}{(z-2)(z-3)(z-5)}\bigg|_{z=5}\right\}$$

$$= -\left\{\frac{z^3}{(z-3)(z-5)}\bigg|_{z=2} + \frac{z^3}{(z-2)(z-5)}\bigg|_{z=3} + \frac{z^3}{(z-2)(z-3)}\bigg|_{z=5}\right\}$$

$$= -\left\{\frac{8}{(-1)(-3)} + \frac{27}{-2} + \frac{125}{3\times 2}\right\} = -\left\{\frac{8}{3} - \frac{27}{2} + \frac{125}{6}\right\} = -\left\{\frac{16-81+125}{6}\right\} = -\left\{\frac{141-81}{6}\right\} = -10$$

Characterization of Rational Functions
If a single-valued function f(z) has no singularities other than pole in the extended complex plane is a rational function.

Note :

(1) A function f(z) is rational iff it is meromorphic in the extended complex plane.

(2) Let f(z) be meromorphic in C and there exist at natural number N, M > 0 and R> 0 such that $|f(z)| \le M|z|^n \; for |z| > R$. Then f is a rational function.

Characterization of Polynomials
A function which has no singularity in the extended complex plane other than a pole of order n at infinity is a polynomial of degree n.

Results :

(i) An entire function f(z) whose singularity at infinity is at the most, a pole is necessarily a polynomial.

(ii) The order of a zero of a polynomial equals the order of its first non-vanishing derivative.

(iii) If a function f(z) is analytic for all finite values of z and as $|z| \to \infty$, $|f(z)| = A|z|^k$, then f(z) is a polynomial of degree ≤ K.

The Argument Theorem
Let f(z) be analytic inside and on a simple closed curve C except for a finite number of poles inside C. Then $\frac{1}{2\pi i}\oint_C \frac{f'(z)}{f(z)}dz = N - P$, where N= $\sum_{i=1}^{k} n_i$; N is the sum of multiplicities of zeros of f(z) lying inside the region. n_i is the multiplicity of ith zero of f(z), $P = \sum_{j=1}^{l} m_j$; P is the sum of the multiplicities of poles of f(z) lying within the region and m_j is the order of jth pole inside the region.

Application:

(1) Let $f(z) = a_0 + a_1 z + a_2 z^2 + ... + a_n z^n$, where $a_n \neq 0$ and $a_0, a_1, ..., a_n$ are complex constants.
Then

(a) $\frac{1}{2\pi i}\int_\gamma \frac{zf'(z)}{f(z)}dz = -\frac{a_{n-1}}{a_n}$;

(b) $\frac{1}{2\pi i}\int_\gamma \frac{zf'(z)}{f(z)}dz = -\frac{a_{n-1}^2 - 2a_n a_{n-2}}{a_n^2}$; where γ is simple closed contour enclosing all zeros of P(z).

(2) If f(z) and g(z) be analytic inside and on a simple closed curve C such that f(z) has zeros at $n_1, n_2, ..., n_p$ and poles at $m_1, m_2, ...m\ m_q$ of order (multiplicities) $r_1, r_2, ...r_p$ and $s_1, s_2, ...$ s_q respectively, then $\frac{1}{2\pi i}\oint_C g(z)\frac{f'(z)}{f(z)}dz = \sum_{t=1}^{p} r_t g(n_t) - \sum_{t=1}^{q} s_t g(m_t)$

Important Theorems
Picard's Little theorem for meromorphic functions: Every meromorphic function □ in C that omits three distinct values a, b, c, ∈C is necessarily a constant function.

Picard's theorem: Every non-constant entire function omits atmost one complex number as its value. In other words, if an entire function omits two values, then it is constant.

For example: If f(z) is a non-constant entire function such that $|f(z)|\geq 10 \forall z \in C$., ie., f(z) omits all those values for which |f(z)| < 10.

⇒ by Picard's theorem, f(z) is constant.

Gauss theorem: If zeros of f(z) lies in a set S contained in a domain D, then zeros of f'(z) will lie on convex hull of S. Open disc and closed disc are convex sets.

Luca's theorem: If zeros of f(z) lies on upper half complex plane, then zeros of f'(z) also lies on that same upper half plane.

Riemann's theorem: Let $|f(z)|\leq M \ \forall z$ such that $0<|z-z_0|<\delta$ and f is analytic in $0<|z-z_0|<\delta$.

Define $g(z)=\begin{cases}(z-z_0)^2 f(z), & z\neq z_0\\ 0 & z=z_0\end{cases}$

Then g(z) is analytic in the disc.

Riemann's theorem on removable singularities: Let $D\subset C$ be an open subset of the complex plane. For a point a ∈ D and f, an analytic function defined on D|{a}. Then following are equivalent:

(i) f is analytically extendable to D.
(ii) f is continuously extendable to D
(iii) there exists a neighbourhood of a on which f is bounded.
(iv) $\lim\limits_{z\to a}(z-a)f(z)=0$

Argument principle: Let a function f(z) be analytic within and on a closed contour C, having N zeros inside C but no zero on C. Then $N=\frac{1}{2\pi}\Delta_c$ arg f(z), where Δ_c arg f(z) denotes the change in arg f(z), as z moves along C in the positive sense.

Example. Let $f(z)=z^3e^{z^2}$ for z ∈ C and let Γ be the circle $z=e^{i\theta}$, where θ varies from 0 to 4π. Then $\frac{1}{2\pi i}\oint_\Gamma \frac{f'(z)}{f(z)}dz=$ **GATE-2018**

(a) 5 (b) 6 (c) 7 (d) 8

Solution : (b)

$f(z)=z^3e^{z^2}$ $\qquad \Gamma: be\ circle\ z=e^{i\theta}, 0<\theta\leq 4\pi$

$$\frac{1}{2\pi i}\int_\Gamma \frac{f'(z)dz}{f(z)}=\frac{1}{2\pi i}\int_{\substack{|z|=1\\ 0\leq\theta\leq 4\pi}} \frac{f'(z)dz}{f(z)}=2\left(\frac{1}{2\pi i}\int_{\substack{|z|=1\\ 0\leq\theta\leq 2\pi}} \frac{f'(z)}{f(z)}dz\right)$$

=2 (Number of zeroes of f(z) – Numbers of poles of f(z))

$f(z)=z^3e^{z^2}$ has no poles but it has 3 zeroes. $\Rightarrow \frac{1}{2\pi i}\int_\Gamma \frac{f'(z)}{f(z)}dz=2(3-0)=6$

Mittag Leffler's expansion theorem: Suppose that the only singularities of f(z) in the finite z plane are the simple poles a_1, a_2, a_3, ... arranged in order of increasing absolute value, i.e., $0<|a_1|<|a_2|<...<|a_n|$ with residues b_1, b_2, b_3, ... respectively. Let C_N be circle of radius R_N which do not passes through any pole and $|f(z)|<M$, on C_N, where M is independent of N and $R_N \to\infty$ as $N \to\infty$ Then, Mittage Leffler's expansion theorem states that $f(z)=f(0)+\sum_{n=1}^{\infty}b_n\left\{\frac{1}{z-a_n}+\frac{1}{a_n}\right\}$

SOLVED QUESTIONS

1. For $f(z) = e^{z+1/z}$, $z \in C^*$, then Res[f(z), 0] **CSIR NET JUNE 2023**

(a) $\sum_{l=1}^{\infty}\frac{1}{l+1}$ (b) $\sum_{l=0}^{\infty}\frac{1}{l(l+1)}$ (c) $\sum_{l=1}^{\infty}\frac{1}{l!(l+1)!}$ (d) $\sum_{l=0}^{\infty}\frac{1}{l^2+1}$

Ans. **(c)**

Explanation :

Given that $f(z) = e^z e^{1/z}$

$$\Rightarrow \left(1+z+\frac{z^2}{2!}+\frac{z^3}{3!}+\frac{z^4}{4!}+......\right)\left(1+\frac{1}{z}+\frac{1}{2!z^2}+\frac{1}{3!z^3}+.....\right)$$

$$\Rightarrow \operatorname{Re}s(f(z);0)=1+\frac{1}{2!}+\frac{1}{2!3!}+\frac{1}{3!4!}+\frac{1}{4!5!}+......=\sum_{l=0}^{\infty}\frac{1}{l!(l+1)!}$$

2. Let γ be the positively oriented circle in the complex plane given by

$\{z \in C : |z-1| = ½\}$. The line integral $\int_\gamma \frac{ze^{3/z}}{z^2-1}dz$ equals

(a) iπe (b) -iπe (c) πe (d) -πe

Ans. **(a)** **CSIR NET NOV. 2020**

Explanation :

$\gamma : |z-1| = 1$, is a circle centred at (1, 0) with radius 1, $\int_\gamma \frac{ze^{1/z}}{(z^2-1)}dz=?$

$z^2-1=0 \Rightarrow z=\pm1$, z = 1 inside γ and -1 outside.

Let $f(z)=\frac{ze^{1/z}}{(z^2-1)}$ By Cauchy integral formula,

$$\int_\gamma \frac{ze^{1/z}}{(z^2-1)}dz = 2\pi i\times \text{ Sum of resudues of f(z).} \quad \int_\gamma \frac{ze^{1/z}}{(z^2-1)}dz = 2\pi i\times \lim_{z\to 1}(z-1)\frac{ze^{1/z}}{(z^2-1)}$$

$$= 2\pi i\times \lim_{z\to 1}(z-1)\frac{ze^{1/z}}{(z-1)(z+1)} \quad = 2\pi i\times \lim_{z\to 1}\frac{ze^{1/z}}{(z+1)} = \frac{2\pi ie}{2} = \pi ie$$

i.e., $\int_\gamma \frac{ze^{1/z}}{(z^2-1)}dz = i\pi e$ Option (a) is correct.

3. Consider the function f : C → C given by f(z) = e^z. Which of the following is false?

(a) f({z ∈ C : |z| < 1}) is not an open set. (b) f({z ∈ C : |z| ≤ 1}) is not an open set.

(c) f({z ∈ C : |z| = 1}) is a closed set.

(d) f({z ∈ C : |z| > 1}) is an unbounded open set. **CSIR NET – JUNE 2019**

Ans. (a)

Explanation :

Suppose f is analytic and non-constant in an open set G, then f(G) is open.

(a) {z ∈ C : |z| < 1} is open subset of C. So, its image f({z ∈ C : |z| < 1}) is open by open mapping theorem as f(z) = e^z is an entire function. So, the statement, f({z ∈ C : |z| < 1}) is not open set is a false statement, the correct answer. Option (a) is correct.

(b) {z ∈ C : |z| ≤ 1} is not open subset of C. So, its image f({z ∈ C : |z| ≤ 1}) is not open. Option (b) is incorrect.

(c) {z ∈ C : |z| = 1} is closed subset of C. So, its image f({z ∈ C : |z| = 1}) is closed. Option (c) is incorrect.

(d) {z ∈ C : |z| > 1} is open subset of C. So, its image f({z ∈ C : |z| > 1}) is open by open mapping theorem. Option (d) is incorrect.

4. The value of the integral $\int_C \frac{z^{100}}{z^{101}+1}dz$ where C is the circle of radius 2 centred at origin taken in the anti – clockwise direction is **GATE 2022**

(a) -2πi (b) 2π (c) 0 (d) 2πi

Ans. (d)

Explanation:

We can written the equation in the form $\frac{1}{101}\int_C \frac{101.z^{100}}{z^{101}+1}dz$

which is in the form of $\frac{1}{101}\int_C \frac{f'(z)}{f(z)}dz$ where f(z) = $z^{101}+1$

f(z) is polynomial of order 101 so , P = 0 and N = 101

Using Argument principle $\frac{1}{101}[101(2\pi i)] = 2\pi i$

5. Let R = {z = x + iy ∈ C ; 0 < x < 1 and -11π < y < 11π }and C be the positively oriented boundary of R. then the value of the integral $\frac{1}{2\pi i}\oint_C \frac{e^z dz}{e^z - 2}$ is **GATE - 2021**

(a) 0 (b) 11 (c) 21 (d) 13

Ans. (b)

Explanation:

We can solution of this question by Argument principle and Residue both

(1) By Argument principle

Given that $\frac{1}{2\pi i}\oint_C \frac{e^z dz}{e^z - 2}$ where $f(z) = e^z - 2$

We know that f(z) has no pole So P = 0

Now $f(z) = e^z - 2 = 0 \Rightarrow z = \log 2 + 2n\pi i$

⇒ 11 zeros lies in R $\frac{1}{2\pi i}\oint_C \frac{e^z dz}{e^z - 2} = N - P = 11$

(2) By Residue

Given that $\frac{1}{2\pi i}\oint_C \frac{e^z dz}{e^z - 2}$ where $f(z) = \frac{e^z}{e^z - 2}$

It has 11 simple poles lies in R So Residue at each pole is 1

$\Rightarrow \frac{1}{2\pi i}\oint_C \frac{e^z dz}{e^z - 2} = 11$

6. Let γ be teh positively oriented circle in the complex plane given by {z ∈ C : |z –1| = 1}. Then $\frac{1}{2\pi i}\oint_\gamma \frac{dz}{z^3 - 1}$ equals

(a) 3 (b) 1/3 (c) 2 (d) 1/2

Ans. (b) **CSIR NET NOV. 2020**

Explanation :

We need to find $\frac{1}{2\pi i}\oint_\gamma \frac{1}{z^3 - 1} dz, \gamma : |z - 1| = 1$

$z^3 - 1 = 0 \Rightarrow z = 1, w, w^2$, clearly z = 1 only lies inside |z – 1|=1. circle cantered at (1,0) with radius 1. Let $f(z) = \frac{1}{z^3 - 1}$, we have, $\frac{1}{2\pi i}\oint_\gamma \frac{1}{z^3 - 1} dz =$ Sum of residues of

COMPLEX ANALYSIS

f(z). So, $\frac{1}{2\pi i}\int_{\gamma}\frac{1}{z^3-1}$ dz= $\operatorname{Res}_{z=1} f(z) = \lim_{z\to 1}(z-1)\frac{1}{z^3-1} = \lim_{z\to 1}(z-1)\frac{1}{(z-1)(z^2+z+1)}$

$= \lim_{z\to 1}\frac{1}{(z^2+z+1)}$ Option (b) is correct.

7. Let p be a positive integer. Consider the closed curve $r(t) = e^{it}$, $0 \le t < 2\pi$. Let f be a function holomorphic in {z: |z| < R} where R > 1. If f has a zero only at z_0, $0 < |z_0| < R$, and it is of multiplicity q, then $\frac{1}{2\pi i}\int_{\gamma}\frac{f'(z)}{f(z)}z^p$ dz equals

(a) qz_0^p (b) z_0q^p (c) pz_0^q (d) z_0p^q

Ans. **(a)** **CSIR NET NOV. 2020**

Explanation :

Generalized Argument principle: In addition to Argument principle, if g is analytic in the region C, then $\frac{1}{2\pi i}\int_c\frac{f'(z)}{f(z)}g(z) = \sum_a g(a)n(C,a) - \sum_b g(b)n(C,b)$, where the first summation is again over all zeros a of f counted with their multiplicities, and the second summation is again over the poles b of f counted with their orders.

Here, $g(z) = z^p$ and f be a function holomorphic in {z: |z| < R} where R > 1, so n(C,b) = 0 ⇒ g(b)n(C,b) = 0, ie, $\frac{1}{2\pi i}\int_C g(z)\frac{f'(z)}{f(z)}dz = \sum_a g(a)n(C,a)$.

$\frac{1}{2\pi i}\int_C z^p\frac{f'(z)}{f(z)}dz = g(z_0)q = z_0^p q$, Option (a) is correct.

8. Which of the following are TRUE? **CSIR NET DEC 2018**

(a) tan z is an entire function
(b) tan z is a meromorphic on C
(c) tan z as isolated singularity at ∞
(d) tan z has non-isolated singularity at

Ans. **(b), (d)**

Explanation :

Option (a) is incorrect.

$f(z) = \tan z = \frac{\sin z}{\cos z}$ is not entire function because it has poles at $z = (2n+1)\frac{\pi}{2}$.

Option (b) is correct.

tan z has only poles, so it is meromorphic

Option (c) is incorrect.

$\tan z = \frac{\sin z}{\cos z}$ has poles at $z = (2n+1)\frac{\pi}{2}$ z

We know that ∞ is limit point of poles, so tan z has non-isolated singularity at ∞.

Note : Let f(z) be a function and it has pole at z = z_0.

$$\lim_{z\to z_0} (z-z_0)^m f(z) = \text{finite}$$

Then f(z) has a pole of order M at $z = z_0$ if f(z) has a pole of order 1 then simple pole.

9. At z = 0, the function $f(z) = exp\left(\frac{z}{1-\cos z}\right)$ has **CSIR NET JUNE-2012**

(a) a removable singularly (b) a pole (c) an essential singularity
(d) the Laurent expansion of f(z) around z = 0 has infinitely many positive and negative powers of z.

Solution: (c), (d)

$$\lim_{z\to 0} f(z) = \lim_{z\to 0} exp\left(\frac{z}{1-\cos z}\right) = exp\left(\lim_{z\to 0}\frac{z}{1-\cos z}\right) = exp\left(\lim_{z\to 0}\frac{1}{\sin z}\right) = e^{\infty}$$

which does not exist

Hence, $f(z) = exp\left(\frac{z}{1-\cos z}\right)$ has an essential singularity.

$$exp\left(\frac{z}{1-\cos z}\right) = 1 + \left(\frac{z}{1-\cos z}\right) + \left(\frac{z}{1-\cos z}\right)^2 \frac{1}{2!} + \left(\frac{z}{1-\cos z}\right)^3 \frac{1}{3!} + \ldots$$

$$= 1 + \frac{z}{1-\left(1-\frac{z^2}{2!}+\frac{z^4}{4!}-\ldots\right)} + \left(\frac{z}{1-\left(1-\frac{z^2}{2!}+\frac{z^4}{4!}-\ldots\right)}\right)^2 \frac{1}{2!} + \ldots$$

Clearly, above expansion contains infinitely many positive and negative powers of z.
Hence, option (c) and (d) are correct.

10. Consider the functions f, g : C →C defined by $f(z) = e^z$ and $g(z) = e^{iz}$.
Let S = {z ∈ C : Re z ∈ [-π, π]}. Then **CSIR NET JUNE-2012**

(a) f is an onto entire function (b) g is a bounded function on C
(c) f is bounded on S (d) g is bounded on S.

Ans. (c)

Explanation :

$$f, g : C \to C \text{ and } f(z) = e^z, g(z) = e^{iz} \text{ and } S = \{z \in C : Re\ z \in [-\pi, \pi]\}$$

$$f(z) = e^{x+iy} \Rightarrow |f(z)| = |e^{x+iy}| = |e^x||e^{iy}| = e^x \quad [\because |e^{iy}| = 1]$$

$$|f(z)| = e^x, as -\pi \le x \le \pi, e^{-\pi} \le |f(z)| \le e^{\pi} \Rightarrow f(z) \text{ is bounded on } S.$$

∴ option (c) is correct.

$$g(z) = e^{i(x+iy)} = e^{-y}.e^{ix} \Rightarrow |g(z)| = |e^{-y}.e^{ix}| = e^{-y}$$

Clearly, g(z) is not bounded on S. ⇒ option (d) is incorrect.

Also f(z) ∈ 0 for any z ∈ C ∴ f is not onto entire function.

COMPLEX ANALYSIS

⇒ option (a) is also incorrect.

As, we know non-constant entire functions are unbounded.

⇒ option (b) is also incorrect, as $g(z) = e^{iz}$, a non-constant entire function.

11. Let f be a non – constant entire function. Which of the following properties is possible for f for each $z \in C$? **CSIR NET DEC-2013**

(a) $\text{Re } f(z) = \text{Im } f(z)$ (b) $|f(z)| < 1$ (c) $\text{Im } f(z) < 0$ (d) $f(z) \neq 0$

Ans. (d)

Explanation :

Result: Picard's theorem states that a non-constant entire function cannot omit more than one value or we can say if an entire function omits more than one value then it is constant.

Using above result options (a), (b), (c) are incorrect.

As all other options are incorrect.

∴ option (d) is correct.

12. Let f, g be meromorphic functions on **C**. If f has a zero of order k at z = a and g has a pole of order m at z = 0, then g(f(z)) has **CSIR NET JUNE-2014**

(a) a zero of order km at z = a (b) a pole of order km at z = a

(c) a zero of order |k-m| at z = a (d) a pole of order |k-m| at z = a

Ans. (b)

Explanation :

Here, f has a zero of order k at z = a and g has a pole of order m at z = 0

Let $f(z) = (z - a)^k$ and $g(z) = \frac{1}{(z-0)^m} = \frac{1}{z^m}$

Then $g(f(z)) = g\left[(z-a)^k\right] = \frac{1}{\left[(z-a)^k\right]^m} = \frac{1}{(z-a)^{km}}$

So, g (f (z)) has a pole of order km at z = a

∴ option (b) is correct

13. For $z \in C$, define $f(z) = \frac{e^z}{e^z - 1}$. Then **CSIR NET JUNE-2014**

(a) f is entire (b) the only singularities of f are poles

(c) f has infinitely many poles on the imaginary axis.

(d) each pole of f is simple

Ans. **(b), (c), (d)**

Explanation :

We have $f(z) = \frac{e^z}{e^z - 1}$

For singularities of $f(z)$, *take* $e^z - 1 = 0 \Rightarrow e^z = 1 = e^{2n\pi_i} = z = 2n\pi i, n \in Z.$

Since, $\lim\limits_{z \to 2n\pi i} f(z) = \infty$ and $\lim\limits_{z \to 2n\pi i} (z - 2n\pi i) f(z) = 1$

So, f(z) has simple poles a $z = 2n\pi i, n \in Z$.

$\Rightarrow$ f has infinitely many poles on the imaginary axis and each pole of f is simple.
$\Rightarrow$ Options (b), (c), (d) are correct
f(z) has infinitely many singularities $\Rightarrow$ f is not entire.
$\Rightarrow$ Option (a) is incorrect.

14. Let $p(z)=a_0+a_1z+...+a_nz^n$ and $q(z)=b_1z+b_2z^2+...+b_nz^n$ be complex polynomials. If a_0, b_1 are non-zero complex numbers, then the residue of p(z) / q(z) at 0 is equal to

CSIR NET DEC-2014

(a) $\frac{a_0}{b_1}$ (b) $\frac{b_1}{a_0}$ (c) $\frac{a_1}{b_1}$ (d) $\frac{a_0}{a_1}$

Ans. (a)
Explanation :

We know that , if $f(z)=\frac{\phi(z)}{\psi(z)}, \phi(a)\neq 0$ and at the point z = a, □ (z) has a zero of order one, i.e. $\psi(a)=0, \psi'(a)\neq 0$, then $\underset{z=a}{res} f(z)=\frac{\phi(a)}{\psi'(a)}$

Here given that $p(z)=a_0+a_1z+...+a_nz^n$ and $q(z)=b_1z+b_2z^2+...+b_nz^n$

Therefore, $\underset{z=0}{res}\left(\frac{p(z)}{q(z)}\right)=\frac{p(0)}{q'(0)}=\frac{a_0}{b_1}$

Hence, option (a) is correct.

15. Let f be an entire function. Which of the following statements are correct?
(a) f is constant if the range of f is contained in a straight line
(b) f is constant if f has uncountably many zeros.
(c) f is constant if f is bounded on $\{z \in C : Res(z) \in 0\}$
(d) f is constant if the real part of f is bounded.

Ans. (a), (b), (d) **CSIR NET JUNE-2015**
Explanation :

According to Picard's theorem, every non-constant entire function omit at most one value in complex plane.
For option (a),
range of f is contained in a straight line implies range of f is bounded.
So, by Picard's theorem, f is constant.
$\Rightarrow$ option (a) is correct.
For option (b)
Non-constant entire function has always countably many zeros.
Only entire function that has uncountably many zeros is f(z) = 0
Thus, option (b) is correct.
For option (c).
If f is bounded on $\{z \in C : Re(z) \leq 0\}$, then we can't apply Picard's theorem

COMPLEX ANALYSIS

$\therefore$ for Picard's theorem function should be bounded for whole complex plane.
Take $f(z) = e^z$

$\therefore |f(z)| = |e^x \cdot e^{iy}| = e^x$

Clearly, f(z) is bounded on $\{z \in C : Re(z) \le 0\}$, but f(z) is non-constant.
Thus, option (c) is incorrect.
For option (d)
If the real part of f is bounded, then f is bounded for whole complex plane.
So, by Picard's theorem, f is a constant function.
Thus, option (d) is correct.

16. The residue of the function $f(z) = e^{-e^{1/z}}$ at z = 0 is **CSIR NET JUNE-2016**

(a) $1 + e^{-1}$ (b) e^{-1} (c) $-e^{-1}$ (d) $1-e^{-1}$

Ans. (c)

Explanation :

Given $f(z)=e^{-e^{\frac{1}{z}}}$, to find residue at z = 0, we have to find coefficient of $\frac{1}{z}$

Now $e^{-e^{1/z}} = e^{-\left[1+\left(\frac{1}{z}\right)+\frac{1}{2!}\left(\frac{1}{z}\right)^2+\frac{1}{3!}\left(\frac{1}{z}\right)^3+\frac{1}{4!}\left(\frac{1}{z}\right)^4+....\right]}$

$$=1-\left[1+\frac{1}{z}+\frac{1}{2!}\left(\frac{1}{z}\right)^2+...\right]+\frac{1}{2!}\left[1+\frac{1}{z}+\frac{1}{2!}\left(\frac{1}{z}\right)^2+...\right]^2-\frac{1}{3!}\left[1+\frac{1}{z}+\frac{1}{2!}\left(\frac{1}{z}\right)^2+...\right]^3$$

$$+....\left[\because e^{-z}=1-z+\frac{z^2}{2!}-\frac{z^3}{3!}+....\right]$$

$$=\frac{1}{z}\left(-1+\frac{2}{2!}-\frac{3}{3!}+\frac{4}{4!}...\right)+...=\frac{1}{z}\left(-1+\frac{1}{1!}-\frac{1}{2!}+\frac{1}{3!}-\frac{1}{4!}...\right)+....$$

$$=\frac{1}{z}\left[\left(1-\frac{1}{1!}+\frac{1}{2!}-\frac{1}{3!}\right)....\right]+.....=\frac{1}{z}\left(-e^{-1}\right)+....$$

$\therefore$ coefficient of $\frac{1}{z}$ is $-e^{-1}$ Hence residue of f(z) at z = 0 is $-e^{-1}$. $\Rightarrow$ Option (c) is correct.

17. Consider the function $F(z)=\int_1^2\frac{1}{(x-z)^2}dz\ Im(z)>0$. Then there is a meromorphic function G(z) on C that agrees with F(z) when Im (z) > 0, such that

CSIR NET JUNE -2016

(a) 1, ∞ are poles of G (z) (b) 0, 1, ∞ are poles of G(z)
(c) 1, 2 are poles of G(z) (d) 1,2 are simple poles of G(z)

Ans. (c), (d)

COMPLEX ANALYSIS

Explanation :

Given $F(z)=\int_1^2 \frac{1}{(x-z)^2}dx,\ Im\ z>0\ = \left.\frac{-1}{(x-z)}\right|_1^2 = \frac{-1}{2-z}+\frac{1}{(1-z)}=\frac{1}{(z-2)}-\frac{1}{(z-2)}$

Given F(z) = G(z) on Imz > 0

We know, if two functions agree on a connected domain, then they agree for all values of z.

$\therefore$ G(z) = F(z) $\forall$ z $\in$ C

Clearly, 1, 2 are simple poles of F(z) and hence of G(z)

$\therefore$ options (c) and (d) are correct.

18. The value of the integral $\oint_{|1-z|=1} \frac{e^z}{z^2-1}dz$ is

(a) 0 (b) $(\pi i)e$ (c) $(\pi i)e-(\pi i)e^{-1}$ (d) $(e+e^{-1})$

Ans. (b) **CSIR NET JUNE-2018**

Explanation :

By Cauchy Residue Theorem, $\oint_C f(z)dz = 2\pi i\sum_{z=a} Res\, f(z)$ where a lies inside C

$f(z)=\frac{e^z}{z^2-1}=\frac{e^z}{(z+1)(z-1)}$ has simple poles at z = ± 1 and only z = 1 lies inside C

$$\oint_{|1-z|=1} f(z)dz = 2\pi i = \underset{z=1}{Res}\, f(z) = 2\pi i \lim_{z\to 1}(z-1)f(z) = 2\pi i \lim_{z\to 1}\frac{e^z}{z+1} = 2\pi i.\frac{e}{z} = \pi i e$$

UNSOLVED QUESTIONS

1. Let S be the positive oriented circle given by |z – 3i| =2. Then the value of $\int_S \frac{dz}{z^2+4}$ is

(a) $\frac{-\pi}{2}$ (b) $\frac{\pi}{2}$ (c) $\frac{-i\pi}{2}$ (d) $\frac{i\pi}{2}$

GATE-2008

2. Consider the function $f(z)=\frac{e^{iz}}{z(z^2+1)}$. The residue of f at the isolated singular points in the upper half plane $\{z=x+iy\in C;>0\}$ is **GATE-2009**

(a) $\frac{-1}{2e}$ (b) $\frac{-1}{e}$ (c) $\frac{e}{2}$ (d) 1

3. Let C be the contour |z| = 2 oriented in the anti-clockwise direction. The value of the integral $\oint_C ze^{3/z}dz$ is **GATE-2013**

(a) 3 πi (b) 5πi (c) 7πi (d) 9πi

4. The value of $\frac{i}{4-\pi}\int_{|z|=4}\frac{dz}{z\cos(z)}$ is equal **GATE-2015**

(a) 2 (b) 3 (c) 4 (d) 5

5. At z = 0 the function $f(z)=\frac{e^z+1}{e^z-1}$ has **CSIR NET JUNE-2011**

(a) a removable singularity (b) a pole
(c) an essential singularity (d) the residue of f(z) at z = 0 is 2

6. Let f be an entire function. If Ref is bounded, then **CSIR NET JUNE-2011**

(a) Im f is constant (b) f is constant
(c) f = 0 (d) f' is a non zero constant

7. Let f be entire function such that $\lim_{|z|\to\infty}|f(z)|=\infty$. Then **CSIR NET DEC-2011**

(a) $f\left(\frac{1}{z}\right)$ has an essential singularity at 0 (b) f cannot be a polynomial
(c) f has finitely many zeros (d) $f\left(\frac{1}{z}\right)$ has a pole at 0

8. $f(z)=\frac{z-\sin z}{z}$; z = 0 is **GATE**

(a) a pole of order 3 (b) a pole of order 2
(c) an essential singularity (d) a removable singularity

9. $f(z)=\sin\left(\frac{1}{\cos\left(\frac{1}{z}\right)}\right)$ at z = 0 is **GATE**

(a) removable singularity (b) a pole
(c) a essential singularity (d) a non-isolated essential singularity

10. Consider the f(z) = $z^3(1-\cos z)$. Which of the following are TRUE? **CSIR NET-2013**

(a) f(z) has a zero of order 2 at z = 0 (b) f(z) has a zero of order 1 at 2nπ
(c) f(z) has a zero order 4 at z = 0 (d) f(z) has a zero order 2 at z = 2nπ

11. $f(z)=\frac{z-\sin z}{z}$; z = 0 is **GATE**

(a) a pole of order 3
(b) a pole of order 2
(c) an essential singularity
(d) a removable singularity

12. The residue of the function $ze^{1/z}$ at $z=\infty$ is **GATE**

(a) 1/2 (b) -1/2 (c) 1 (d) -1

13. The residue of cot z at any of its pole is **GATE**

(a) 0 (b) 1 (c) -1 (d) none

14. Consider the $f(z) = z^3(1 - \cos z)$. Which of the following are TRUE? **CSIR NET-2013**

(a) f(z) has a zero of order 2 at z = 0
(b) f(z) has a zero of order 1 at 2nπ
(c) f(z) has a zero order 4 at z = 0
(d) f(z) has a zero order 2 at z = 2nπ

15. The value of $\int_0^{2\pi} exp\left(e^{i\theta}-i\theta\right)d\theta$ equals **GATE-2006**

(a) 2πi (b) (2π) (c) π (d) iπ

16. The sum of the residues at all the poles of $f(z)=\frac{\cot \pi z}{(z+a)^2}$, where a is constant, $(a \neq 0, \pm 1, \pm 2, ...)$ is **GATE-2006**

(a) $\frac{1}{\pi}\sum_{r=-\infty}^{\infty}\frac{1}{(n+a)^2}-\pi \operatorname{cosec}^2 \pi a$
(b) $-\frac{1}{\pi}\sum_{n=-\infty}^{\infty}\frac{1}{(n+a)^2}+\pi \operatorname{cosec}^2 \pi a$
(c) $-\frac{1}{\pi}\sum_{n=-\infty}^{\infty}\frac{1}{(n+a)^2}-\pi \operatorname{cosec}^2 \pi a$
(d) $\frac{1}{\pi}\sum_{n=-\infty}^{\infty}\frac{1}{(n+a)^2}+\pi \operatorname{cosec}^2 \pi a$

17. Let $f(z)=\cos z-\frac{\sin z}{z}$ for non-zero $z \in C$ and f(0) = 0. Then f(z) has a zero at z = 0 of order **GATE-2008**

(a) 0 (b) 1 (c) 2 (d) greater than 2

18. Let $f(z)=\cos z-\frac{\sin z}{z}$ for non-zero $z \in C$. Also, let g(z) = sinhz for $z \in C$. Then $\frac{g(z)}{zf(z)}$ has a pole at z = 0 of order **GATE-2008**

(a) 1 (b) 2 (c) 3 (d) greater than 3

Answer Key

1	2	3	4	5	6	7	8	9	10
b	a	d	a	b,d	a,b	c,d	d	c	d
11	12	13	14	15	16	17	18		
d	b	b	d	b	a	c	b		

Chapter 6

Identity Theorem, Open-Mapping Theorem, Rouche's Theorem & Schwartz Lemma

1. Identity theorem:

Let f(z) be anaytic in a region D and S be any set s.t. f(a) = 0; for all a ∈ S. Suppose l is limit point of S and $l \in$ D, then f(z) ≡ 0

Example : Let f(z) is analytic in open unit disc 'D' s.t. $f\left(\frac{1}{n}\right)=0$.

Solution : Here $S=\left\{\frac{1}{n}\right\}$ and S → 0 Where 0 ∈ D, So f(z) ≡ 0.

Example : Let f(z) is analytic in open disc unit disc 'D' s.t. $f\left(2-\frac{1}{n}\right)=0$.

Solution : Here $S=\left\{2-\frac{1}{n}\right\}$ and S → 2 Where 2 ∉ D

So, we can't say that f(z) is identically zero.

Second Form of Identity Theorem :

If f(z) and g(z) are analytic in D with $f(z_n) = g(z_n)$ for each n and if limit point of z_n lies in D, Then f(z) = g(z).

Example : If f(z) is analytic in open unit disc s.t. $f\left(\frac{1}{n}\right)=\frac{1}{n^2}$.

Solution : We know that $S_n=\frac{1}{n}$ and $S_n \to 0$, 0 ∈ D So, take $\frac{1}{n}=z$ Then we get $f(z) = z^2$.

Example : If f(z) is analytic in open unit disc 'D' s.t. $f\left(2-\frac{1}{n}\right)=\left(2-\frac{1}{n}\right)^2$.

Solution : Here $S_n=2-\frac{1}{n}$ and $S_n \to 2$ but 2 ∉ D So, identity theorem is not valid.

2. Open Mapping Theorem:

In complex analysis, the open mapping theorem, states that if U is a connected open subset of the complex plane C and f: U → C is a non-constant holomorphic function, then f is an open map (i.e., maps open subsets of U to open subsets of C).

2. Consider the polynomial $P(z)=\sum_{n=1}^{N} a_n z^n, 1 \le N < \infty, a_n \in R|\{0\}$. Then with $D=\{w \in C : |w| < I\}$

(a) $P(D) \subseteq R$ (b) P(D) is open (c) P(D) is closed (d) P(D) is bounded

Solution : (b), (d)

$P(z)=\sum_{n=1}^{N} a_n z^n, 1 \le N < \infty, a_n \in R|(0)$ and $D=\{w \in C : |w| < 1\}$

$\Rightarrow$ P(z) is a polynomial function and D is open. By Open Mapping theorem, P(D) is open

$\Rightarrow$ Option (b) is correct and (c) is incorrect.

Let P(z) = z ; (a_1 = 1), N = 1 For z = i/2, P(z) = i/2 $\Rightarrow$ P (D) $\not\subset$ R

$\therefore$ Option (a) is incorrect Also, as domain is bounded and P(z) is entire.

$\Rightarrow$ P(D) is bounded. $\therefore$ Option (d) is correct.

Hence, options (b) and (d) are correct.

3. Schwartz Lemma:

If f(z) is analytic in a domain D defined by |z|<R and satisfies the condition |f(z) $\in$ M for all z $\in$ D and f(0) = 0, then $|f(z)| \le \frac{M}{R}|z|$.

Also, if equality occurs for any z, then $|f(z)| \le \frac{M}{R}|z|$.

Also, if equality occurs for any z, then $|f(z)| = \frac{M}{R} z e^{i\alpha}$, where α is a real constant.

or if D = {z : | z| < 1} and if f(z) is analytic on D with

(a) |f(z)| ≤ 1 for all z $\in$ D

(b) f(0) = 0, then |f' (0)| ≤ 1 and |f(z)| ≤ |z| for some z $\in$ 0, then f(w) = cw for all w $\in$ D.

Extension of Schwartz Lemma : If a function f(z) is such that

(i) f(z) is analytic in |z| < 1 and (ii) |f(z)| ≤ 1 for all z, |z| < 1

(iii) f(z) has zero of order m at z = 0. Then

(a) $|f(z)| \le z|^m \ \forall |z| < 1$ (b) $|f^m(0)| \le m!$

Corollary: Schwarz Pick Theorem: If f(z) is an analytic function in |z| < 1, then

(i) $|f(z)| \le \frac{|f(0)|+|z|}{1+|f(0)||z|}$ (ii) $|f'(z)| \le \frac{1-|f(z)|^2}{1-|z|^2}$

4. Rouche' theorem:

If f(z) and g(z) are analytic inside and on a simple closed curve C and if |g(z)|<|f(z)| on C, then f(z) + g(z) and f(z) have the same number of zeros inside C.

Example : Using Rouche's theorem, find the number of roots of $z^7 - 4z^3 + z - 1$, which lie interior to the unit circle |z| = 1.

Solution : Here, take f(z) = - $4z^3$ and g (Z) = $Z^7 + z - 1$.

Now, $$\left|\frac{g(z)}{f(z)}\right| = \frac{|z^7 + z - 1|}{|-4z^3|} \le \frac{|z|^7 + |z| + 1}{4|z|^3} = \frac{3}{4} < 1 \; on \; |z| = 1$$

Thus conditions of Rouche's theorem are satisfied. So f(z) = $-4z^3$ and f(z) + g(z) = $z^7 - 4z^3 + z - 1$ have the same number of zeros inside |z| =1. But f(z) has a zero of order 3 at the origin. Hence, we conclude that $z^7 - 4z^3 + z - 1$ has 3 roots inside |z| = 1.

SOLVED QUESTIONS

COMPLEX ANALYSIS

1. Let f : C → C be an entire function with $f\left(\frac{1}{n}\right) = \frac{1}{n^2}$ for all n ∈ N.

Then which of the following statements is true? **CSIR NET DEC 2019**

(a) No such f exists (b) Such as f is not unique

(c) f(z) = z^2 for all z ∈ C (d) f need not be a polynomial function

Ans. **(c)**

Explanation :

f : C → C be entire function.

Consider g(z) = z^2 entire function f(z_n) = g(z_n) when which is a convergent sequence in C.

Hence, by identity theorem f(z) = g(z) ⇒ f(z) = z^2

So, option (c) is correct.

2. Let D ⊂ C be the open unit disc {z ∈ C : |z| < 1} and H(D) be the space of all holomorphic functions on D. Consider the sets **CSIR NET FEB 2022**

$$A = \left\{ f \in H(D) : f\left(\frac{1}{n}\right) = \begin{cases} e^{-n} & \text{if n is even} \\ 0 & \text{if n is odd} \end{cases} ; \text{for } n \ge 2 \right\}$$

$$B = \left\{ f \in H(D) : f\left(\frac{1}{n}\right) = (n-2)/(n-1), n \ge 2 \right\}$$

Which of the following statement is true?

(a) Both A and B are non-empty (b) A is empty and B has exactly one element

(c) A has exactly one element and B is empty (d) Both A, B are empty

Ans. **(b)**

Explanation :

Let D be the open unit disc {z ∈ C | |z| < 1}

Let $$A = \left\{ f \in H(D) |: f\left(\frac{1}{n}\right) = \begin{cases} e^{-n} & \text{if n is even} \\ 0 & \text{if n is odd} \end{cases} ; n \ge 2 \right\}$$

We see for any f ∈ A $f\left(\frac{1}{2n+1}\right)=0 \ \forall n\geq 2$ and $\frac{1}{2n+1}\to 0\in D$.

So, f = 0 for any f ∈ A, by identity theorem.

Hence any f(z) will not satisfies $f\left(\frac{1}{2n}\right)=\frac{1}{e^{2n}}$, so A is empty.

$$B=\left\{f\in H(D)\,|\,f\left(\frac{1}{n}\right)=\frac{n-2}{n-1}, n\geq 2\right\}$$

Let $z=\frac{1}{n}$ then $n=\frac{1}{z}$ $f(z)=\dfrac{\frac{1}{z}-2}{\frac{1}{z}-1}=\dfrac{1-2z}{1-z}$

Which is analytic in D, |z| < 1.So, B contains one element.

So, we choose option (b).

3. For z ∈ C, let Re(z) denotes its real part. Let f(z) be an entire function satisfying |f(z)| ≤ |z| |Re(z)| on C. Which of the following statements are true? **CSIR NET NOV 2020**

(a) f(0) = 0 (b) f′(0) = 0

(c) The only entire function satisfying the given property is f(z) = 0

(d) There exists a non-constant entire function satisfying the given property.

Ans. **(a), (b), (c)**

Explanation :

|f(z)| ≤ |z| |Re(z)| Let z_0 be any point on imaginary axis.

i.e. $z_0 = iy$, y ∈ R So, $|f(z_0)| \leq |z_0|\,|0|$

⇒ $|f(z_0)| \leq 0$ ⇒ f(z) = 0 for all imaginary (purely)

Now by identity theorem f(z) ≡ 0 Therefore option (a), (b), (c) are correct.

Option (d) is incorrect.

4. Let f be a holomorphic function on the open unit disc D = {z ∈ C:|z|<1}. Suppose that |f| ≥ 1 on D and f(0) = i. Which of the following are possible values of $f\left(\frac{1}{2}\right)$?

(a) –i (b) i (c) 1 (d) –1

Ans. (b) **CSIR NET NOV 2020**

Explanation :

f be a holomorphic function on the open unit disc D = {z ∈ C:|z| < 1}. And |f| ≥ 1 on D and f(0) = i.

Let $g(z)=\frac{1}{f(z)}$, then since |f| ≥ 1 on D and is holomorphic in D, so is g(z) and moreover, we have |g(z)| ≤ 1 for all z in D. Hence by Liouville's theorem, g is constant, so is f.

Given $f(0) = i \Rightarrow f(z) = i$ for all $z \in D$ and $f\left(\frac{1}{2}\right) = i$, Option (b) is correct and Option (a), (c) and (d) are incorrect.

5. Let $D = \{z \in C : |z| < 1\}$ be the open unit disc and let $f: D \to D$ be a holomorphic function. Suppose that $f(0) = 0$ and $f'(0) = 0$. Which of the following are possible values of $f\left(\frac{1}{2}\right)$?

(a) 1/4 (b) –1/4 (c) 1/3 (d) –1/3

Ans. **(a), (b)** **CSIR NET NOV 2020**

Explanation :

Let $D = \{z \in C : |z| < 1\}$ be the open unit disc and let $f: D \to D$ be a holomorphic function and $f(0) = f'(0) = 0$. We need to find the possible values of $f\left(\frac{1}{2}\right)$.

We have a theorem, "$z = z_0$ is zero of order m of f(z) if $f(z_0) = f'(z_0) = ... = f^{m-1}(z_0) = 0$ and $f^m(z_0) \neq 0$".

Extension of Schwartz lemma: If f(z) be a function such that

(i) f(z) is analytic in $|z| < 1$. (ii) $|f(z)| < 1$ for all $|z| < 1$.

(iii) f(z) has zero of order m at $z = 0$.

Then $|f(z)| \leq |z|^m$ for all $|z| < 1$ and $|f^m(0)| \leq m!$.

From this we can say that f(z) has zero of order at least $m \geq 2$.

Let m = 2, then $|f(z)| \leq |z|^2$.

$$\left|f\left(\frac{1}{2}\right)\right| \leq \left|\frac{1}{2}\right|^2 = \frac{1}{4} \Rightarrow -\frac{1}{4} \leq f\left(\frac{1}{2}\right) \leq \frac{1}{4}.$$

So, Option (a) and (b) are correct and Option (c) and (d) are incorrect.

6. The number of zeros of polynomial $2z^7 - 7z^5 + 2z^3 - z + 1$ in the unit disc $\{z \in C : |z| < 1\}$ is **GATE - 2022**

(a) 2 (b) 4 (c) 5 (d) 7

Ans . **(c)**

Explanation:

Let $f(z) = -7z^5$ and $g(z) = 2z^7 + 2z^3 - z + 1$

Now $|f(z)| = 7$ for $|z| = 1$ and $|g(z)| = 4$ for $|z| = 1$

Here $|f(z)| > |g(z)|$ for $|z| = 1$

So , f(z) and f(z) + g(z) have same zeros in $|z| < 1$ f(z) has 5 zeros in $|z| < 1$

7. Let $D = \{z \in C : |z| < 1\}$ and $f : D \to C$ be defined by

$f(z) = z - 25z^3 + \frac{z^5}{5!} - \frac{z^7}{7!} + \frac{z^9}{9!} - \frac{z^{11}}{11!}$. Consider the following statements :

P : f has three zeros (counting multiplicity) in D.

COMPLEX ANALYSIS

Q: f has one zero in $U=\left\{z\in C:\frac{1}{2}<|z|<1\right\}$. **GATE 2023**

Then

(a) P is true but Q is false (b) P is false but Q is true

(c) both P and Q are true (d) both P and Q are false

Ans. (a)

Explanation :

Given that $f(z)=z-25z^3+\frac{z^5}{5!}-\frac{z^7}{7!}+\frac{z^9}{9!}-\frac{z^{11}}{11!}$

For $|z|<1$

Let $g(z) = -25z^3$ and $h(z)=z+\frac{z^5}{5!}-\frac{z^7}{7!}+\frac{z^9}{9!}-\frac{z^{11}}{11!}$.

Here $|g(z)| > |h(z)|$ on $|z| = 1$

$\Rightarrow$ g(z) and g(z) + h(z) = f(z) both have same zero.

$\Rightarrow$ f(z) has three zeros in D

Now for $\frac{1}{2}<|z|<1$

Here $|g(z)| > |h(z)|$ on $|z| = 1/2$. $\Rightarrow$ f(z) have three zero in $|z| < 1/2$.

Hence f(z) has no zero in $1/2 < |z| < 1$.

Option (a) is correct.

8. Let f be a holomorphic function on the open unit disc $D = \{z \in C : |z| < 1\}$. Suppose that $|f| \geq 1$ on D and f(0) = i. Which of the following are possible values of $f\left(\frac{1}{2}\right)$? **CSIR NET NOV 2020**

(a) –i (b) i (c) 1 (d) -1

Ans. (b)

Explanation :

Let f be holomorphic function on the open unit disc $D = \{z \in C : |z| < 1\}$. Suppose that $|f| \geq 1$ on D and f(0) = i. So as minimum of |f| is attained in the interior of domain D so by minimum – modulus principle, the function will be constant so $f\left(\frac{1}{2}\right)=i$, as f(z) = i, for all $z \in D$. So, option (b) is true.

9. Let $D \subset C$ be the open unit disc. Consider the family F of all holomorphic maps f : D $\rightarrow$ D such that f(0) = 1/2. For $f \in F$, the possible values of $|f'(0)|$ are **CSIR NET NOV 2020**

(a) 7/8 (b) 5/6 (c) 3/4 (d) 1

Ans. (c)

COMPLEX ANALYSIS

Explanation :

$f : D \to D$ holomorphic f(0) = ½ Now |f`(0)|

By Schwartz pick theorem $|f`(z)| \le \frac{1-|f(z)|^2}{1-|z|^2} = \frac{1-\frac{1}{4}}{1} = \frac{3}{4} \Rightarrow |f`(0)| \le \frac{3}{4}$

∴ Only option (c) is correct.

10. Let D be the open unit disc in C. Let $g : D \to D$ be holomorphic. g(0) = 0 & let $h(z) = \begin{cases} \frac{g(z)}{z} & , z \in D \quad z \neq 0 \\ g'(0) & z = 0 \end{cases}$. Which of the following are TRUE?

CSIR NET DEC 2016

(a) h is holomorphic on D (b) $h(D) \subseteq \overline{D}$ (c) |g'(0)| > 1 (d) $\left|g\left(\frac{1}{2}\right)\right| \le \frac{1}{2}$.

Ans. **(a), (b), (d)**

Explanation :

Option (a) Given that g(z) is analytic in D then $\frac{g(z)}{z}$ is also analytic in D/{0}

We have to check at {0}

$\lim_{z\to 0} \frac{g(z)}{z}$ $\left(\frac{0}{0} form\right)$ then $\lim_{z\to 0} \frac{g(z)}{z} = g'(0)$ (finite)

⇒ h(z) is holomorphic / analytic in D.

Option (b): Schwartz – Lemma

g(z) is analytic in D & g(0) = 0

i) $|g(z)| \le |z|^n$ ii) $|g(0)| \le n!$

Now $|h(z)| = \left|\frac{g(z)}{z}\right| \le |z|^{n-1}$ [∴ $|g(z)| \le |z|^n$ & $z \in D$ $|z| < 1$]

Then $|h(z)| \le 1$ [n = 1] $\boldsymbol{h(D) \subseteq \overline{D}}$

Option (d): $|g(z)| \le |z|^n = \left|g\left(\frac{1}{2}\right)\right| \le \left|\frac{1}{2}\right|^n = \left|g\left(\frac{1}{2}\right)\right| \le \frac{1}{2}$

Option (c) : $|g'(0)| \le \frac{1-|g(0)|^2}{1-|0|^2}$ $|g'(0)| \le 1$

11. Let $f : C \to C$ be an entire function. Suppose that f = u + iv, where u & v are real and imagery part then f is constant if **CSIR NET JUNE 2016**

(a) {u(x, y)} is bounded (b) {v(x, y)} is bounded

(c) {u(x,y) + v(x, y)} is bounded (d) $\{u^2 + v^2\}$ is bounded

Ans. **(All)**

Explanation :

Given that f(z) = u + iv We know that if f(z) is bounded then u & v are bounded

Option (a), (b) are correct.

Option (c) is correct. u & v are bounded ⇒ u + v is bounded

Option (d) is correct.

u is bounded ⇒ u^2 is bounded

v is bounded ⇒ v^2 is bounded ⇒ $u^2 + v^2$ is bounded

12. Let D be open disc in C and H(D) be the collection of all holomorphic function and let $S = \left\{ f \in H(D) \mid f\left(\frac{1}{2}\right) = \frac{1}{2}, f\left(\frac{1}{4}\right) = \frac{1}{4} .... f\left(\frac{1}{2n}\right) = \frac{1}{2n} \right\}$ and

$$T = \left\{ f \in H(D) : f\left(\frac{1}{2}\right) = f\left(\frac{1}{3}\right) = \frac{1}{2} ..... f\left(\frac{1}{2n}\right) = f\left(\frac{1}{2n+\_1}\right) = \frac{1}{2n} \right\}.$$ Then

(a) Both S and T are singleton (b) S is singleton & T = ϕ

(c) S = ϕ & T is singleton (d) S = ϕ & T = ϕ

Ans. (b) **CSIR NET JUNE 2016**

Explanation :

By using of identity theorem

For S $f\left(\frac{1}{2n}\right) = \frac{1}{2n}$ Let $\frac{1}{2n} = z \therefore$ g(z) = z (analytic) $f\left(\frac{1}{2n}\right) = g\left(\frac{1}{2n}\right)$

$\frac{1}{2n} \to 0$ and $0 \in D \Rightarrow$ f(z) = g(z) ⇒ f(z) = z (analytic) and S is singleton.

For T $f\left(\frac{1}{2n}\right) = \frac{1}{2n} = f\left(\frac{1}{2n+1}\right)$

Let $\frac{1}{2n} = z$ Then g(z) = z $f\left(\frac{1}{2n}\right) = g\left(\frac{1}{2n}\right)$ $\frac{1}{2n} \to 0 \in D$

f(z) = g(z)

f(z) = z and $f\left(\frac{1}{2n+1}\right) = \frac{1}{2n}$ here $\frac{1}{2n+1} = z$

$$2n+1 = \frac{1}{z} \qquad 2n = \frac{1}{z} - 1 = \frac{1-z}{z} \qquad \frac{1}{2n} = \frac{z}{1-z}$$

then $f(z) = \frac{z}{1-z}$ Now $f\left(\frac{1}{2n}\right) \cap f\left(\frac{1}{2n+1}\right) = \phi$

So T is empty.

13. Let f be an analytic function defined an open unit disc on C. Then f is constant if

(a) $f\left(\frac{1}{n}\right)=0 \ \forall n\geq 1$ (b) $f(z)=0 \ \forall |z|=\frac{1}{2}$

(c) $f\left(\frac{1}{n^2}\right)=0 \ \forall n\geq 1$ (d) $f(z)=0 \ z\in(-1,1)$

Ans. (All) **CSIR NET JUNE 2015**

Explanation :

Option (a) Set of zeroes of $f = S = \left\{\frac{1}{n}\right\}$.

S' = 0 ∈ D (open unit disc) By identity theorem f ≡ 0

Option (b) : $|z|=\frac{1}{2}\in|z|=1$

By identity theorem (b) is correct.

Option (c) is correct. Set of zeros of $f\left(\frac{1}{n^2}\right)$ $\frac{1}{n^2}\to 0\in D$

By identity theorem f ≡ 0 Option (d) is correct. f(z) = 0 ∀ z ∈ (-1, 1)

By identity theorem.

UNSOLVED QUESTIONS

1. Let f(z) be an analytic function on the unit disc $D\{z\in C \,||\, z|<1\}$. If $f\left(\frac{1}{n}\right)=\frac{1}{2n-1} \forall n\in Z^*$ (the set of non-zero integers), then which of the following is not correct?

(a) f (0) = 0 (b) f(z) has no singularity

(c) f(z) has simple pole at z = 2 (d) f(z) is constant function

2. Let S be the disk |z| < 3 in the complex plane and le f : S → C be an analytic function such that $f\left(1+\frac{\sqrt{2}}{n}i\right)=-\frac{2}{n^2}$ for each natural number n. Then $f\left(\sqrt{2}\right)$ is equal to

GATE-2008

(a) $3-2\sqrt{2}$ (b) $3+2\sqrt{2}$ (c) $2-3\sqrt{2}$ (d) $2+3\sqrt{2}$

3. Let f : D → D be holomorphic with f (0) = 1/2 and f(1/2) = 0, where D = {z : |z| ≤ 1}. Which of the following is correct? **CSIR NET JUNE-2011**

(a) $|f'(0)| \leq 3/4$ (b) $|f'(1/2)| \leq 4/3$

(c) $|f'(0)| \leq 3/4 \text{ and } |f'(1/2)| \leq 4/3$ (d) $f(z) = z, z \in D$

4. Let f : D → D be holomorphic function with f(0) = 0 where D is open unit disc {z∈C : |z| < 1} then **CSIR NET DEC 2012**

(a) |f'(0)| = 1 (b) $\left|f\left(\frac{1}{2}\right)\right| \leq \frac{1}{2}$ (c) $\left|f\left(\frac{1}{2}\right)\right| \leq \frac{1}{4}$ (d) $|f'(0)| \leq \frac{1}{2}$

5. Let D = {z∈C : |z| < 1}. Which of the following are correct? **CSIR NET JUNE 2013**

(a) ∃ a holomorphic f : D → D s.t. f(0) = 0 & f'(0) = 2

(b) ∃ a holomorphic f : D → D s.t. $f\left(\frac{3}{4}\right) = \frac{3}{4}$ & $f'\left(\frac{2}{3}\right) = \frac{3}{4}$.

(c) ∃ a holomorphic function f : D → D with $f\left(\frac{3}{4}\right) = \left(\frac{3}{4}\right)$ & $f'\left(\frac{2}{3}\right) = \left(-\frac{3}{4}\right)$.

(d) ∃ a holomorphic f : D → D with $f\left(\frac{1}{2}\right) = -\frac{1}{2}$ & $f'\left(\frac{1}{4}\right) = 1$.

6. Suppose f is holomorphic in an open neighbourhood of $z_0 \in C$. Given that the series $\sum_{n=0}^{\infty} f^{(n)}(z_0)$ converges absolutely, we can conclude that

(a) f is constant (b) f is a polynomial

(c) f can be extended to an entire function (d) $f(x) \in R \text{ for all } x \in R$

7. Let D = {z ∈ C : |z| < 1}. Then ∃ a non-constant analytic function on D s.t. **GATE 2015**

(a) $f\left(\sqrt{-\frac{1}{n}}\right) = 0$ (b) $f\left(1 - \frac{1}{n}\right) = 0$ (c) $f\left(\frac{1}{n}\right) = 0$ (d) $f\left(\frac{1}{2} - \frac{1}{n}\right) = 0$

8. Let f be a holomorphic function on the unit disc {z ∈ C : |z| < 1} in the complex plane. Which of the following is/are necessarily true? **CSIR NET DEC-2013**

(a) If for each positive integer n we have $f\left(\frac{1}{n}\right) = \frac{1}{n^2}$, then $f(z) = z^2$ on the unit disc

(b) If for each positive integer n we have $f\left(1 - \frac{1}{n}\right) = \left(1 - \frac{1}{n}\right)^2$, then f(z) = z^2 on the unit disc

(c) f cannot satisfy $f\left(\frac{1}{n}\right) = \frac{(-1)^n}{n}$ for each positive integer n.

(d) f cannot satisfy $f\left(\frac{1}{n}\right)=\frac{1}{n+1}$ for each positive integer n.

9. Let $f : C \to C$ be a meromorphic function analytic at z = 0 and $f\left(\frac{1}{n}\right)=\frac{n}{2n+1}$

CSIR NET JUNE 2012

(a) $f(0)=\frac{1}{2}$ (b) f has a simple pole at z = -2

(c) $f(2)=\frac{1}{4}$ (d) No such mero. Function exist.

10. Let $f : D \to D$ holomorphic with $f(0)=\frac{1}{2}$ & $f\left(\frac{1}{2}\right)=0$ where D = {z | |z| < 1}. Which of the following are correct? **CSIR NET JUNE 2012**

(a) $\left|f'\left(\frac{1}{2}\right)\right| \le \frac{4}{3}$ (b) $|f'(0)| \le 1$

(c) $\left|f'\left(\frac{1}{2}\right)\right| \le \frac{4}{3}$ & $|f'(0)| \le 1$ (d) f(z) = z

11. Let $S=\{0\}\cup\left\{\frac{1}{4n+7}:n=1,2,...\right\}$, Then the number of analytic functions which vanish only on S is **GATE-2007**

(a) infinite (b) 0 (c) 1 (d) 2

12. Let D = {z ∈ C : |z| < 1} then ∃ a holomorphic function $f : D \to \overline{D}$ with f(0) = 0 with the properties. **CSIR NET JUNE 2014**

(a) $f'(0)=\frac{1}{2}$ (b) $\left|f\left(\frac{1}{3}\right)\right|=\frac{1}{4}$ (c) $f\left(\frac{1}{3}\right)=\frac{1}{2}$ (d) $|f'(0)|=\sec\left(\frac{\pi}{6}\right)$

13. Let f be an analytic function in C. Then f is constant if the zero set of f contains the sequence **CSIR NET DEC-2015**

(a) $a_n = 1/n$ (b) $a_n=(-1)^{n-1}\frac{1}{n}$ (c) $a_n=\frac{1}{2n}$

(d) $a_n = n$ if 4 does not divide n and $a_n=\frac{1}{n}$ if 4 divides n

Answer Key

1	2	3	4	5	6	7
a,c	a	a,b,c	b	b,c	c	b
8	**9**	**10**	**11**	**12**	**13**	
a,c	a,b,c	a,b,c	b	a,b	a,b,c,d	

Chapter 7

Bilinear Transformation

1. Conformal Mapping

Let w = f(z) = u + iv be a complex valued function of a complex variable z, where z = x + iy and u = u(x, y), v = v(x, y), then some correspondence between point (u, v) in the w-plane with the point (x, y) in the z-plane is obtained which is called a mapping or transformation of points in z-plane into point of the w-plane.

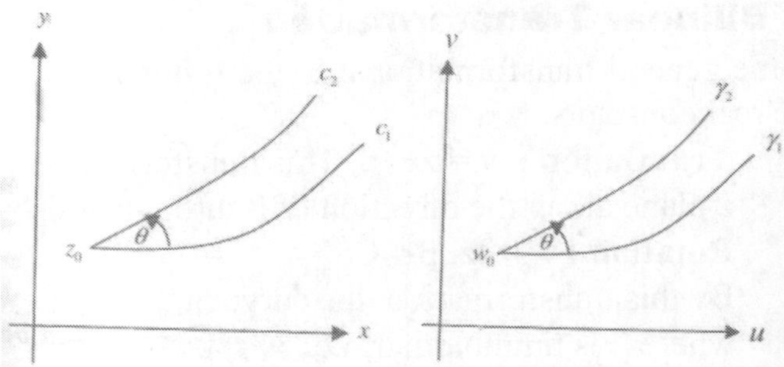

Conformal Transformation

Consider the transformation w = u(x, y) + iv(x,y) that maps a point $z_0 + x_0 + iy_0$ of the z plane to a point $w_0 = u_0 + iv_0$ of the w plane. Further let the two curves C_1 and C_2 intersecting at z_0 be mapped on two curve λ_1 and λ_2 intersecting at w_0. If the angle between the intersecting curves C_1 and C_2 at z_0 is equal to the angle between the intersecting curves λ_1 and λ_2 at both in magnitude and direction, then the transformation is called conformal transformation.

Necessary and sufficient conditions for w = f(z) to represent a conformal mapping:

(a) Let D be a domain and $w = f(z) = u(x, y) + iv(x, y)$ be a conformal transformation of D in the z-plane into a domain D' of the w-plane, the f(z) must be an analytic function of z. Thus u(x, y) and v(x, y) are differentiable functions and hence must satisfy Cauchy-Riemann equations, i.e., $\frac{\partial u}{\partial x} = \frac{\partial v}{\partial y}$ and $\frac{\partial u}{\partial y} = -\frac{\partial v}{\partial x}$

(b) Let f(z) be an analytic function of z in a region D of the z-plane and let f'(z) ≠ 0 inside D. Then, the mapping w = f(z) is conformal at the points of D.

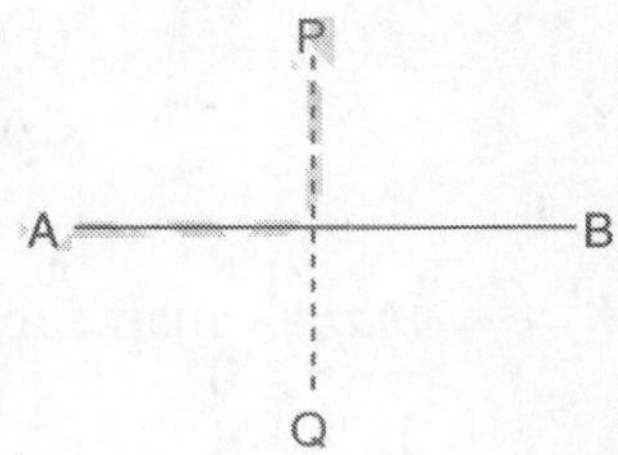

Inverse Points

Inverse points with respect to a line: Two points P and Q are said to be inverse points with respect to a line AB if Q is the image of P in AB, i.e., if the line AB is the right angle bisector of PQ.

Inverse points with respect to a circle:

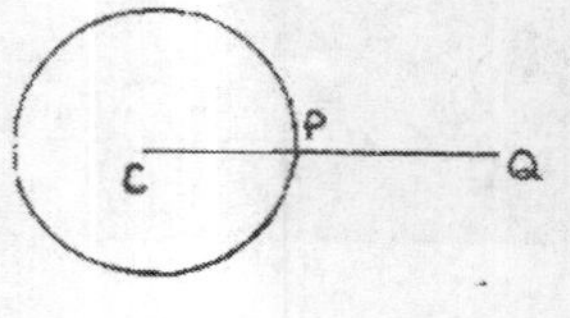

Two points P and Q are said to be the inverse points with respect to a circle S if they are collinear with the centre C and on the same side of it and if the product of their distances from the centre is equal to r^2, where r is the radius of the circle. Thus, when P and Q are the inverse points of the circle, then the three points C, P, Q are collinear, and also CP. CQ = r^2, where P and Q lie on the same side of C.

2. Bilinear Transformation

Some general transformations: In the following α, β are given complex constants while a, θ_0 are real constants.

1. **Translation :** $w = z + \beta$. This transformation will displace or translate every point in the z plane along the direction of β through a distance equal to $|\beta|$.
2. **Rotation :** $w = \beta z$, $\beta \in C$
   By this transformation, the curve in the plane are rotated through an angle equal to arg β, where β is unimoldular, i.e., $w = \beta z$ is rotation where $|\beta| = 1$.
3. **Stretching or Magnification** : $w = az$, $a \in R$
   By this transformation, curves in the z-plane are stretched (or contracted) in the direction of z in ratio |a| if $a > 1$ (or $0 < a < 1$). Contraction as a special case of stretching. This transformation is also known as homothetic transformation. $w = az$, $a \in C$ represents magnification if arg a = zero, i.e., if a is real and positive.
4. **Inversion:** $w = 1/z$. Clearly $w = (1/z)$ is the resultant of $z = 1/z$, $w = 1/z$, where the formal is reflection in the real axis and the latter is reflection in unit circle.
   Linear Transformation : The transformation $w = \alpha z + \beta$ .... (1), where α and β are given complex constants, is called a linear transformation. Thus a general linear transformation is a combination of the translation, rotation and magnification.
   Inversion / Inverse map :

$$f(z) = \frac{1}{z} \qquad f(z) = \begin{cases} \frac{1}{z} & if\ z \neq 0 \\ \infty & z = 0 \\ 0 & z = \infty \end{cases}$$

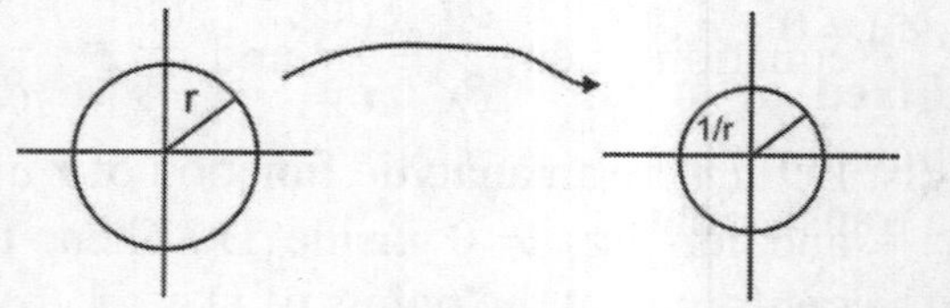

- $map\ \frac{1}{z}$ map circle $|z| = r$ onto circle $|w| = \frac{1}{r}$.
- $\frac{1}{z}$ map exterior $|z| > r$ onto interior $|w| < \frac{1}{r}$.
- $w = \frac{1}{z} \Rightarrow z = \frac{1}{w} \Rightarrow z = \frac{1}{u+iv} \quad x + iy = \frac{1}{u+iv} \times \frac{u-iv}{u-iv} \quad x = \frac{u}{u^2+v^2},\ y = \frac{-v}{u^2+v^2}$
- $a(x^2 + y^2) + bx + cy + d = 0$
  $a \neq 0$ circle

a = 0 s.t. line

$$a\left(\frac{u^2}{(u^2+v^2)^2}+\frac{v^2}{(u^2+v^2)^2}\right)+\frac{bu}{u^2+v^2}+\frac{c(-v)}{u^2+v^2}+d=0 = d(u^2+v^2)+bu-cv+a=0$$

Note :

1. $\frac{1}{z}$ maps circle (a ≠ 0) not passing through origin (d ≠ 0) to circle not passing through origin.
2. $\frac{1}{z}$ map circle (a ≠ 0) passing through origin (d = 0) to straight line not passing through origin.
3. $\frac{1}{z}$ map straight line (a = 0) not passing through origin (d ≠ 0) to circle passing through origin.
4. $\frac{1}{z}$ map strainght line (a = 0) passing through origin (d = 0) to straight line passing through origin.

Bilinear or Fractional or Mobius Transformation:

The transformation $w=\frac{az+b}{cz+d}, a,b,c,d \in C$, $ad-bc \neq 0$, where a, b, c, d, are complex numbers, is called a bilinear or mobius or fractional transformation.

Result (1) : Product of two bilinear transformations is bilinear transformation:

(2) : Inverse of a bilinear transformation is again a bilinear transformation.

Critical points: Let f(z) be a non-constant analytic function in a domain D. If f' (z_0) = 0 for some z_0 in D, then z_0 is called the critical point of the transformation. Also, the points where f' (z_0) ≠ 0 are called ordinary points.

Fixed or invariant points of a transformation: The points which coincide with their transformation are called invariant or fixed points of the transformation. Thus, fixed points of a transformation w = f(z) are obtained by putting w = z.

The invariant points of the transformation $w=\frac{az+b}{cz+d}, a,b,c,d \in C$ .....(1)

are given by $z=\frac{az+b}{cz+d}$ or $cz^2-(a-d)z-b=0$ or $z=\frac{(a-d)\pm\sqrt{M}}{2C}$ ....(2)

where M = $(a-d)^2 + 4bc$. The number of finite fixed points is one or two according as M = 0 or M ≠ 0.

Case I : When c ≠ 0

(1) If M ≠ 0, then (1) have two fixed points given by $z=\dfrac{(a-d)\pm\sqrt{M}}{2c}$

(2) If M = 0, then (1) has only one finite fixed point given by $z=\dfrac{(a-d)}{2c}$

Case II : When c = 0

(1) If d ≠ 0, then (1) becomes $w=\dfrac{az+b}{0+d}=\dfrac{a}{d}z+\dfrac{b}{d}$

The fixed point is given by $z=\dfrac{a}{d}z+\dfrac{b}{d}$ *or* $z=\dfrac{b}{d-a}$ ....(3)

(2) If a – d ≠ 0, then (2) declares that one fixed point is ∞ and (3) declares that the other fixed point is finite.

(3) If a – d = 0, then the transformation has one fixed point, i.e., ∞, according to (3).
Thus we have the following results:
(i) If c ≠ 0 and M ≠ 0, two finite fixed points.
(ii) If c ≠ 0 and M = 0, one finite fixed points.
(iii) If c = 0 and d ≠ 0, only one finite fixed points.
(iv) If c = 0 and a - d = 0, only one infinite fixed point , i.e., ∞. In this case w = az + (b/d)
(v) If c = 0 and a – d ≠ 0, one finite and the other is ∞

Normal form of a bilinear transformation:

(i) Every bilinear transformation with two finite fixed point α, β can be put in the form $\dfrac{w-\alpha}{w-\beta}=k\dfrac{z-\alpha}{z-\beta}$ where k ∈ C and if β = ∞, then it becomes w – α = k (z – α)
(a) If |k| = 1, then transformation is called elliptic.
(b) If k > 0 (≠ 1), then transformation is called hyperbolic.
(c) If $k = a.e^{i\alpha}$, where a and α are real numbers such that a ≠ 1, a > 0 and a ≠ 0, then transformation is called loxodromic.

(ii) Every bilinear transformation which has only one finite fixed point α can be put in the form $\dfrac{1}{w-\alpha}=\dfrac{1}{z-\alpha}+k, k\in C$ and if α = ∞, then it becomes w = z + k. In this case, transformation is called parabolic.

Cross ratio: If z_1, z_2, z_3, z_4 are distinct points, then the ratio $\dfrac{(z_4-z_1)(z_2-z_3)}{(z_2-z_1)(z_4-z_3)}$ is called the cross ratio of z_1, z_2, z_3, z_4 and is denoted by (z_1, z_2, z_3, z_4). A bilinear transformation preserves cross ratio. i.e. if z_1, z_2, z_3, z_4 are transformation to w_1, w_2, w_3, w_4 respectively, then $(z_1, z_2, z_3, z_4) = (w_1, w_2, w_3, w_4)$.

Equation of a circle through three given points:

Equation of a circle passing through three given points z_1, z_2 and z_3 is given by

$$\frac{(z-z_1)(z_2-z_3)}{(z_1-z_2)(z_3-z)}=\frac{(\bar{z}-\bar{z}_1)(\bar{z}_2-\bar{z}_3)}{(\bar{z}_1-\bar{z}_2)(\bar{z}_3-\bar{z})}$$ where z is any complex variable lying on circle

Some important theorems:

Theorem: Every bilinear transformation is the resultant of magnification, rotation, inversion and translation.

Theorem: The four points z_1, z_2, z_3 and z_4 are concyclic, if $\frac{(z_1-z_2)(z_3-z_4)}{(z_2-z_3)(z_4-z_1)}$ is purely real.

Theorem : The set of all bilinear transformation forms a non-abelian group under the product of transformations.

Theorem : Bilinear transformation maps inverse points w.r.t. a circle to the inverse points of transformed circle.

Theorem : Every bilinear transformation is the resultant of an even number of inversions.

Theorem : Every bilinear transformation maps the family of circles $\arg\left(\frac{z-z_1}{z-z_2}\right)=\mu(\mu\neq 0)$ through the fixed point z_1 and z_2 onto a similar family in the w-plane through two fixed points w_1 and w_2 which are the transformation of the points z_1 and z_2 respectively.

Theorem : Bilinear Transformation Theorem: Let f(z) be a function analytic at z_0 such that $f'(z_0) = 0$. If f'(z) has a zero of order k-1 (where k = 1, w, ...) at z_0. Then mapping w = f(z) magnifies the angle at the vertex z_0 by the factor k.

Theorem : Riemann Mapping Theorem: Let Ω be a simple connected domain in the complex plane (except the plane itself). Then | a conformal map which maps Ω in a one to one manner onto the unit disc $|w|<1$.

Remark: To find a bilinear transformation which maps z_1, z_2, z_3 to w_1, w_2, w_3. Let $w=\frac{az+b}{cz+d}$ be the required bilinear transformation. Then simplify $(w, w_1, w_2, w_3) = (z, z_1, z_2, z_3)$.

Example. Find the bilinear transformation mapping z = 2, 1, 0 to w = 1, 0, i respectively.

Solution. Let z be mapped to w.

Then (w, 1, 0, i) = (z, 2, 1, 0) or $\frac{(w-1)(0-i)}{(1-0)(i-w)}=\frac{(z-2)(1-0)}{(2-1)(0-z)}$

$$\Rightarrow (w-1)(-1)(-z)=(z-2)(i-w)\Rightarrow iz(w-1)=(i-w)(z-2)$$

$$\Rightarrow w[z(1+i)-2]=2iz-2i=2i(z-1)\Rightarrow w=wi\frac{(z-1)}{z(1+i)-2}$$

Some Important Formula :

1. Bilinear transformations that maps upper (lower) half to itself is given by

$$f(z)=\frac{az+b}{cz+d};\ ad-bc>0$$

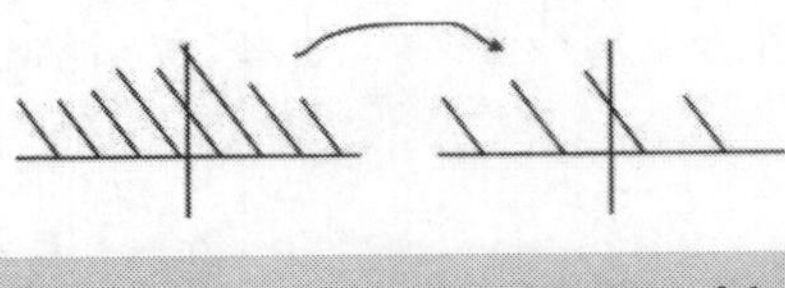

2. $f(z)=\dfrac{az+b}{cz+d}$ $\quad ad-bc<0$

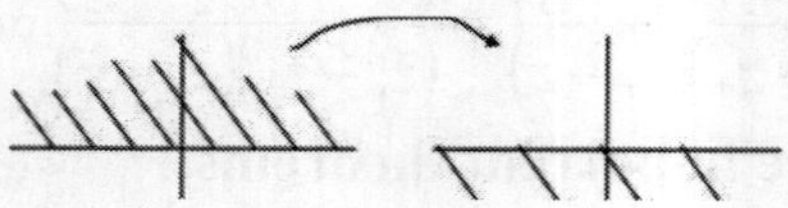

3. $f(z)=e^{\frac{-i\pi}{2}}\dfrac{az+b}{cz+d}$ $\quad$ ad – bc > 0

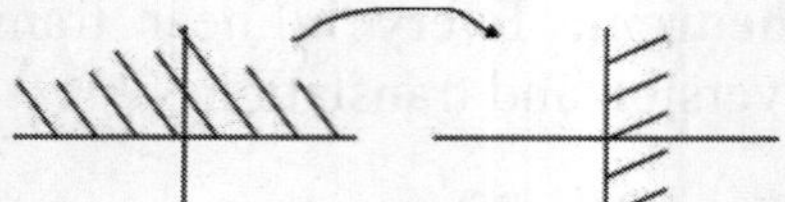

4. $f(z)=e^{\frac{i\pi}{2}}\dfrac{az+b}{cz+d}$ $\quad ad-bc>0$

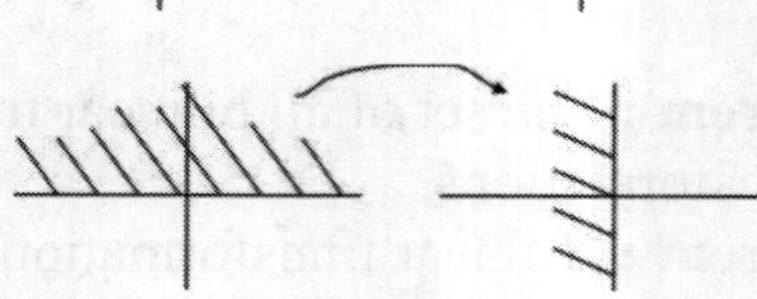

5. The most general Bilinear transformations that maps upper half onto unit disc is

$f(z)=e^{i\theta}\dfrac{z-\beta}{z-\overline{\beta}}$ where Im $\beta>0$

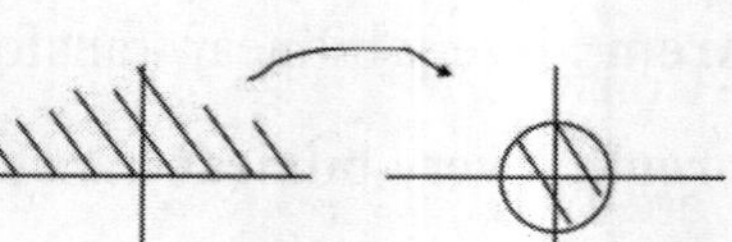

6. Bilinear transformations that maps lower half to unit disc is

$f(z)=e^{i\theta}\dfrac{z-\beta}{z-\overline{\beta}}$ $\quad$ Im $\beta<0$

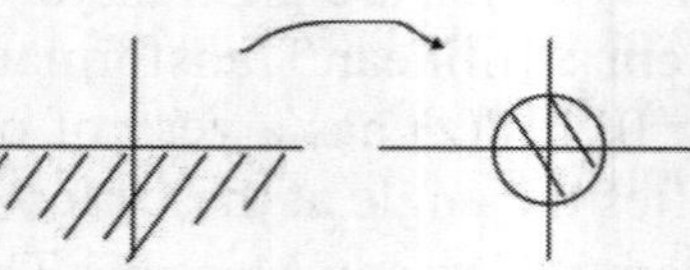

7. Bilinear transformations that maps Right half to unit disc is

$f(z)=e^{i\theta}\dfrac{z-\beta}{z+\overline{\beta}}$ $\quad$ Re(β) > 0

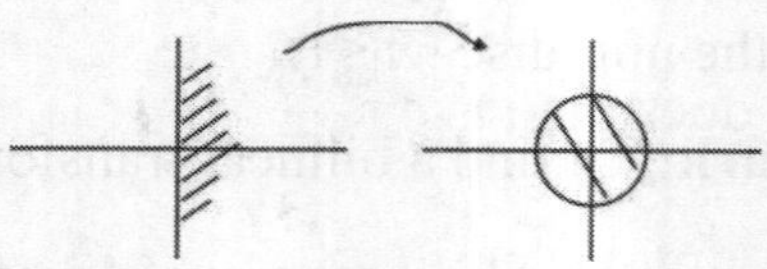

8. Bilinear transformations that maps Left half to unit disc is

$f(z)=e^{i\theta}\dfrac{z-\beta}{z+\overline{\beta}}$ $\quad$ Re(β) < 0

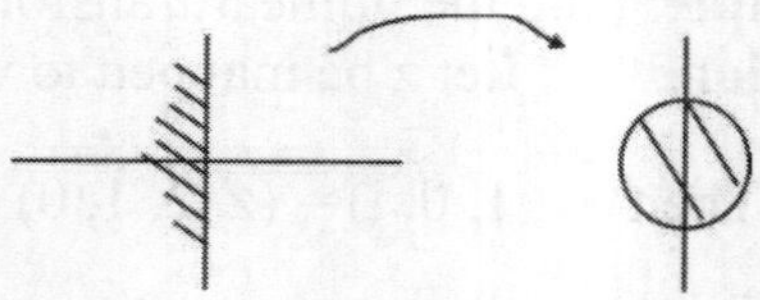

9. B.T. that maps unit disc to itself $f(z)=e^{i\theta}\dfrac{z-z_0}{1-\overline{z}_0 z}$ $\quad |z_0|<1$ and if $|z_0|>1$

SOLVED QUESTIONS

1. Let D = {z ∈ C | |z| < 1} and w ∈ D. Define F_w : D → D by $F_w(z) = \frac{w - z}{1 - \overline{w}z}$. Then which of the following are TRUE? **CSIR NET JUNE 2019**

(a) F is 1 – 1 (b) F is not 1 – 1 (c) F is onto (d) F is not onto

Ans. (a), (c)

Explanation :

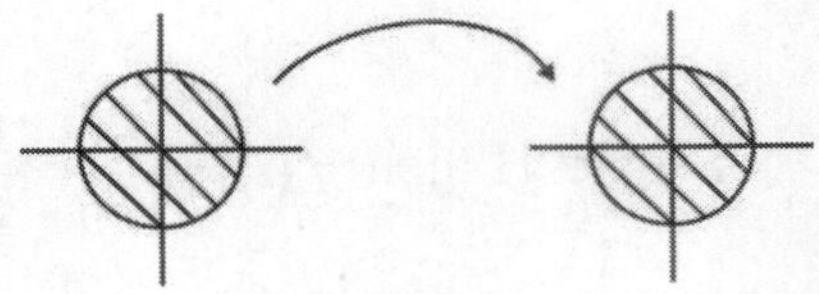

Given $F_w(z) = \frac{w - z}{1 - \overline{w}z}$ is b.T. and w ∈ D

s.t. |w| < 1

i.e. is one-one & onto both.

2. Consider the map ϕ : C/{1} → C given by $\phi(z) = \frac{1+z}{1-z}$. Then which of the following are TRUE? **CSIR NET JUNE 2018**

(a) ϕ{(z ∈ C |z| < 1} ⊆ {z ∈ C | |z| < 1} (b) ϕ {z ∈ C : Rez < 0} ⊆ {z ∈ C : Rez < 0}

(c) ϕ is onto (d) ϕ{C/{1}) = C/{-1}

Ans. (d)

Explanation :

Option (c) is incorrect.

Let ∃ z ∈ C/{1} s.t. f(z) = -1 then $\frac{1+z}{1-z} = -1 \Rightarrow$ 1 + z = -1 + z =1 = -1 contradiction.

i.e. there does not exists any element z ∈ C\{1} s.t. f(z) = -1 pre image of -1 doesn't exist.

⇒ ϕ(z) is not onto. Option (d) is correct. ϕ(C/{1}) = C/{-1}

Option (a) is incorrect.

Take z = 0 ∈ D ϕ(0) = 1 ∉ {|z| < 1}

Option (b) is incorrect. If it is true then ∃ z ∈ {Rez < 0} s.t. f(z) ∈ {Rez < 0}

$f(2) = \frac{3}{1} = -3 \notin \{z \in C \mid \text{Re}\, z \mid$ f(-1) = 0 ∉ {z∈ C : Rez < 0}

3. Let H = {z ∈ C : y > 0} for z ∈ H then which of the following are TRUE? **CSIR NET DEC 2018**

(a) $\frac{1}{z} \in H$ (b) $\frac{1}{z^2} \in H$ (c) $\frac{-z}{z+1} \in H$ (d) $\frac{z}{2z+1} \in H$

Ans. (d)

Explanation :

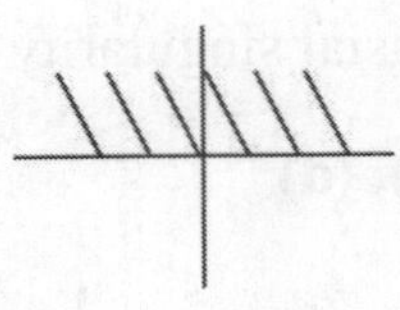

& $f(z) = \frac{1}{z}$.

Take z = i ∈ H then $f(i)=\frac{1}{i}=-i\notin H$. Option (b) is incorrect. $f(z)=\frac{1}{z^2}$

z = i ∈ H then $f(i)=\frac{1}{(i)^2}=-1$ is boundary point of H.

Option (c) is incorrect.

$f(z)=\frac{-z}{z+1}$

z = i ∈ H then $f(i)=\frac{-i}{1+i}\times\frac{1-i}{1-i}$. $=\frac{-i+(i)^2}{1+1}=\frac{-1(1+i)}{2}=-\frac{1}{2}-\frac{i}{2}\notin H$

Option (d) is correct. $f(z)=\frac{z}{2z+1}$

$z=0 \qquad f(0)=0\in H \qquad z=1\notin H \qquad f(1)=\frac{1}{3}\notin H$

$z=i\in H \qquad f(i)=\frac{i}{2i+1}\times\frac{1-2i}{1-2i}=\frac{i+2}{1+4}=\frac{2}{5}+\frac{i}{5}\in H \quad f(z)=\frac{z}{2z+1}\in H$

4. Consider the $f(z)=\frac{1}{z}, z\in C, z\neq 0$, if C denote a circle with +ve radius passing through origin then f maps C/{0} to **CSIR NET JUNE 2017**

(a) a circle
(b) a line passing through origin
(c) a line
(d) a line not passing through origin

Ans. (c), (d)

Explanation :

We know that $f(z)=\frac{1}{z}$ maps

$a(x^2+y^2)+by+cx+d=0$ to $d(u^2+v^2)+bu+cv+a=0$ and C denote the circle passing through origin i.e. d = 0.

So f maps to line which not passing through origin.

5. Let H = {z ∈ C : Im(z) > 0} be the upper half plan and D = {z ∈ C : |z| < 1} be the open unit disc. Suppose f is a mobius transformation which maps H onto D and suppose f(2i) = 0. Pick the correct statement. **CSIR NET JUNE 2016**

(a) f has a simple pole at z = -2i
(b) f satisfy the condition $f(i).\overline{f(-i)}=1$.

(c) f has Essensial singularity at z = -2i
(d) $|f(2+2i)=\frac{1}{\sqrt{5}}$.

Ans. (a), (b), (d)

Explanation :

By (5) $f(z)=e^{i\theta}\frac{z-\beta}{z-\overline{\beta}}$ $\text{Im}(\beta)>0$

Given that f(2i) = 0 So $f(z)=e^{i\theta}\frac{z-2i}{z+2i}$

Option (a) is correct.
Option (b) is correct.

$$f(i)=\frac{e^{i\theta}(-i)}{3i}=-\frac{1}{3}e^{i\theta} \quad \overline{f(-i)}=\frac{\overline{e^{i\theta}(-3i)}}{i}=-3e^{-i\theta} \quad f(i).\overline{f(-i)}=-\frac{1}{3}e^{i\theta}\times -3e^{-i\theta}=1$$

Option (c) is incorrect.
Because f has simple pole at z = -2i

Option (d) is correct.

$$|f(2+2i)|=\left|e^{i\theta}\frac{(2+2i)-2i}{2+2i+2i}\right|=\left|e^{i\theta}\frac{2}{2+4i}\right|=\left|e^{i\theta}\frac{1}{1+2i}\right|=\frac{|e^{i\theta}|}{|1+2i|}=\frac{1}{\sqrt{5}}$$

6. The image of the half plane Re(z) + Im(z) > 0 under the map $w=\frac{z-1}{z+i}$ is given by

GATE 2018

(a) Re (w) > 0 (b) Im (w) >0 (c) |w| > 1 (d) |w| < 1

Ans. (d)

Explanation:

Given that Re(z) + Im(z) > 0 $w=\frac{z-1}{z+i}$

When z = 1, we get w = 0
So, option (a), (b) and (c) are discard.
So, option (d) is correct.

7. Let f: C\ {3i} → C be defined by $f(z)=\frac{z-i}{iz+3}$. Which of the following statements about f is FALSE?

GATE-2013

(a) f is conformal on C\{3i} (b) f maps circle in C\{3i} onto circle in C
(c) All the fixed point of f are in the region {z ∈ C : Im(z) > 0}
(d) There is no straight in C\{3i} which is mapped onto a straight line in C by f

Ans. (c)

Explanation :

$f:C\setminus\{3i\}\to C \text{ be } f(z)=\frac{z-i}{iz+3}$

f(z) is conformal when f'(z) $\neq 0$ and f(z) is analytic. So f(z) is conformal in C\{3i}, then f(z) maps circles in C\ {3i} onto circles in C. Fixed points are given by $\frac{z-i}{iz+3}=z$

$$\Rightarrow iz^2+3z=z-i \Rightarrow iz^2+2z+i=0 \Rightarrow z^2-2iz+1=0$$

$$\Rightarrow z=\frac{2i\pm\sqrt{-4-4}}{2} \Rightarrow z=i\left(1\pm\sqrt{2}\right)$$

8. The image of the region {z ∈ C : Re (z) > Im(z) > 0} under the mapping $z \to e^{z^2}$ is

GATE-2013

(a) $\{w\in C: Re(w)>0, Im(w)>0\}$

(b) $\{w\in C: Re(w)>0, Im(w)>|w|>1\}$

(c) $\{w\in C:|w|>1\}$

(d) $\{w\in C: Im(w)>0,|w|>1\}$

Ans. (c)

Explanation :

The given region is $D=\{z\in C: Re(z)>Im(z)>0\}$ and the mapping is $f(z)=e^{z^2}$

Take, z = 4 + πi

$$f(z)=e^{(4+\pi i)^2}=e^{16-\pi^2+8\pi i}=e^{16-\pi^2}\left[\because e^{8\pi i}=1\right]$$

Here, Im (W) = 0

So, option (c) is correct.

9. Let Re(z) and Im(z), respectively, denote the real part and the imaginary par tof a complex number z. Let T : C ∪ {∞} → C ∪ {∞} be the bilinear transformation such that T(6) = 0, T(3 - 3i) i and T (0) = ∞. Then, the image of $D=\{z\in C:|z-3|<3\}$ under the mapping w = T(z) is

GATE-2017

(a) $\{w\in C: Im(w)<0\}$

(b) $\{w\in C: Re(w)<0\}$

(c) $\{w\in C: Im(w)>0\}$

(d) $\{w\in C: Re(w)>0\}$

Ans. (d)

Explanation :

Here, $z_1 = 6$, $w_1 = 0 = T(z_1)$ $z_2 = 3 - 3i$, $w_2 = i = T(z_2)$ $z_3 = 0$, $w_3 = \infty = T(z_3)$

Since the bilinear transformation preserves cross ratio

$$\therefore \frac{(w-w_1)(w_2-w_3)}{(w_1-w_2)(w_3-w)} = \frac{(z-z_1)(z_2-z_3)}{(z_1-z_2)(z_3-z)}$$

On substituting values and solving for w, we get $w = T(z) = \frac{6-z}{z}$

Given region is $D = \{z \in C : |z-3| < 3\}$. $Take, z = 2 \in D$ but
$T(2) = 2 \notin \{w \in C : Re(w) < 0\}$

$\Rightarrow$ option (b) is incorrect

and $2 \notin \{w \in C : Im(w) < 0\}$ ($\therefore$ T(2) $\in$ boundary of regions given in options (a) and (c))

But z = 2 $\notin$ Boundary of D

option (a) and (c) are correct. $\Rightarrow$ option (d) is correct.

10. Let p be a polynomial in 1-complex variable. Suppose all zeros of p are in the upper half plane $H = \{z \in C \mid Im(z) > 0\}$. Then **CSIR NET JUNE-2015**

(a) $Im \frac{p'(z)}{p(z)} > 0 \; for \; z \in R$ (b) $Re\, i \frac{p'(z)}{p(z)} < 0 \; for \; z \in R$

(c) $Im \frac{p'(z)}{p(z)} > 0 \; for \; z \in C, with \; Im\, z < 0$ (d) $Im \frac{p'(z)}{p(z)} > 0 \; for \; z \in C, with \; Im\, z > 0$

Ans. (a), (b), (c)

Explanation :

Take p(z) – a (z – z_1) (z – z_2) ..(z - z_k), with $z_1, z_2, \ldots z_k$ lie in H = {z $\in$ C | Im (z) > 0}

$$\frac{P'(z)}{p(z)} = \sum_{k=1}^{n} \frac{1}{z-z_k} \text{ and for all } 1 \le k \le n$$

$$\frac{1}{z-z_k} = \frac{\bar{z}-\bar{z}_k}{|z-z_k|^2} = \frac{(x-x_k)-(y-y_k)i}{|z-z_k|^2} \text{ with } z = x+iy,\, z_k = x_k + iy_k$$

$$Re \frac{P'(z)}{p(z)} = \sum_{k=1}^{n} \frac{x-x_k}{|z-z_k|^2}, Im \frac{p'(z)}{p(z)} = -\sum_{k=1}^{n} \frac{y-y_k}{|z-z_k|^2}$$

For option (a), for z $\in$ R, put y = 0, z = x

$$\text{Im } \frac{P'(z)}{p(z)} = \sum_{k=1}^{n} \frac{-y_k}{|z-z_k|^2} = \sum_{k=1}^{n} \frac{y_k}{|z-z_k|^2} > 0 \text{ as } y_k > 0 \text{ for } 1 \le k \le n$$

$\therefore$ option (a) is correct.

$$i\frac{P'(z)}{p(z)} = \sum_{k=1}^{n}\frac{(x-x_k)i}{|z-z_k|^2} + \sum_{k=1}^{n}\frac{y-y_k}{|z-z_k|^2}$$

$$\text{Re } i\frac{P'(z)}{p(z)} = \sum_{k=1}^{n}\frac{y-y_k}{|z-z_k|^2} = \sum_{k=1}^{n}\frac{y_k}{|z-z_k|^2} < 0, \textit{ for } z \in R \textit{ as } y_k > 0$$

$\therefore$ option (b) is correct. For option (c),

$$Im\frac{P'(z)}{p(z)} = -\sum_{k=1}^{n}\frac{y-y_k}{|z-z_k|^2} > 0, \text{as } y_k > 0 \textit{ for } 1 \le k \le n \textit{ and } Im\,z < 0$$

$\therefore$ option (c) is correct.

For option (d),

$$Im\frac{P'(z)}{p(z)} = -\sum_{k=1}^{n}\frac{y-y_k}{|z-z_k|^2}$$ is -ve or +ve, depends on whether Im(z) > Im(z_k) or Im(z) < Im(z_k)

$\therefore$ option (d) is correct.

UNSOLVED QUESTIONS

1. Define $H^+ = \{z \in C, y > 0\}$

$H^- = \{z \in C : y < 0\}$

$L^+ = \{z \in C : x > 0\}$

$L^- = \{z \in C : x < 0\}$

The function $f(z) = \dfrac{z}{3z+1}$. **CSIR NET JUNE 2011**

(a) maps H^+ onto H^+ and L^- onto L^- (b) maps H^+ onto H^+ and H^- onto H^+

(c) maps H^+ onto L^+ and H^- onto L^- (d) maps H^+ onto L^- and H^- onto L^+

2. Let D be the open unit disc $\{z : |z| < 1\}$ and let $f : D \to C$ be t he map define by $f(z) = e^{i\psi}\dfrac{z-a}{1-\bar{a}z}$ for $a \in D$, $\psi \in [0, 2\pi]$. Then which of the following is / are TRUE? **CSIR NET DEC 2011**

(a) $|f(e^{i\theta})| = 1$ for $\theta \in [0, 2\pi]$ (b) f maps $\{z : |z| < 1\}$ onto itself

(c) f maps $\{z : |z| < 1\}$ onto itself (d) f is one-one

3. The s.t. $w = \dfrac{2z}{z-2}$ *maps* $\{z : |z-1| < 1\}$ onto

(a) $\{w : Re(w) < 0\}$ (b) $\{w : Im(w) > 0\}$ (c) $\{w : Re(w) > 0\}$ (d) $\{w : Im(w) < 0\}$

4. Let $f(z)=\dfrac{1+z}{1-z}$. Which of the following is/are true? **CSIR NET DEC 2013**

(a) f maps $\{|z|<1\}$ onto $\{\text{Re } z>0\}$
(b) f maps $\{|z|<1;\ \text{Im } z>0\}$ onto $\{\text{Re } z>0,\ \text{Im } z>0\}$
(c) f maps $\{|z|<1 : \text{Im } z<0\}$ onto $\{\text{Re } z<0,\ \text{Im } z<0\}$
(d) f maps $\{|z|>1\}$ onto $\{\text{Im } z>0\}$

5. Let f be B.T. that maps -1 to 1, i to ∞ and $-i$ to 0 then f(1) = ? **GATE**

(a) -2 (b) $-i$ (c) i (d) -1

6. Let G_1 and G_2 be the images of the disc $\{z\in C:|z+1|<1\}$ under the transformations $W=\dfrac{(1-i)z+2}{(1+i)z+2}$ *and* $w=\dfrac{(1+i)z+2}{(1-i)z+2}$ respectively. Then **GATE-2007**

(a) $G_1=\{w\in C: Im(w)<0\}$ *and* $G_2=\{w\in C: Im(w)<0\}$
(b) $G_1=\{w\in C: Im(w)>0\}$ *and* $G_2=\{w\in C: Im(w)<0\}$
(c) $G_1=\{w\in C: |w|>2\}$ *and* $G_2=\{w\in C: |w|<2\}$
(d) $G_1=\{w\in C: |w|<2\}$ *and* $G_2=\{w\in C: |w|>2\}$

7. Transformation $f(z)=\dfrac{1}{z}$

(a) always maps a circle into a circle
(b) always maps a circle into straight line.
(c) maps circle into straight line if circle in z-plane passes through origin
(d) maps straight line into circle

8. Consider any bilinear transformation $w=\dfrac{az+b}{cz+d}(ad+bc\neq 0)$, where $\Delta=(d-a)^2+4bc$, then w has

(a) only one finite fixed point, if $c\neq 0$ and $\Delta\neq 0$
(b) two finite fixed point, if $c\neq 0$ and $\Delta\neq 0$
(c) one finite other infinite fixed point, if $c\neq 0$ and $a\neq d$
(d) only one infinite fixed point, if $c=0$ and $a\neq d$

9. For the transformation $w=\dfrac{z}{2-z}$

(a) fixed points are 0 and 2
(b) fixed points are 0 and 1
(c) transformation is parabolic
(d) transformation is hyperbolic

10. Let a, b, c, d, $\in$ R be such that ad – bc > 0. Consider the Mobius transformation $T_{a,b,c,d}(z)=\dfrac{az+b}{cz+d}$. Define $H_+=\{z\in C: Im(z)>0\}, H.=\{z\in C: Im(z)<0\}$ $R_+=\{z\in C: Re(z)>0\}, R=\{z\in C: Re(z)<0\}$. Then $T_{a,b,c,d}$ maps. **CSIR NET DEC-2008**

(a) H_+ to H_+ (b) H_+ to H_- (c) R_+ to R_+ (d) R_+ to R_-

11. Define
$H^+ = \{z \in C : y > 0\}, H^- = \{z \in C : y < 0\}, L^+ = \{z \in C : x > 0\}, L^- = \{z \in C : x < 0\}$

The function $f(z) = \dfrac{z}{3z+1}$ **CSIR NET JUNE-2011**

(a) maps H^+ onto H^+ and H^- onto H^-
(b) maps H^+ onto H^- and H^- onto H^+
(c) maps H^+ onto L^+ and H^- onto L^-
(d) maps H^+ onto L^- and H^- onto L^+

12. Let $f(z) = z + \dfrac{1}{z}$ *for* $z \in C$ *with* $z \neq 0$. Which of the following are always true?

CSIR NET DEC-2012

(a) f is an analytic function on C \ {0}
(b) f is a conformal map on C \ {0}
(c) f maps the unit circle to a subset of the real axis
(d) The image of any circle in C\ {0} is again a circle.

13. Under the transformation $w = \sqrt{\dfrac{1-iz}{z-i}}$, the region $D = \{z \in C : |z| < 1\}$ is transformed to

GATE-2010

(a) $\{z \in C : 0 < \arg z < \pi\}$
(b) $\{z \in C : -\pi < \arg z < 0\}$
(c) $\left\{z \in C : 0 < \arg < \dfrac{\pi}{2} \text{ or } \pi < \arg z < \dfrac{3\pi}{2}\right\}$
(d) $z \in C\left\{\dfrac{\pi}{2} < \arg z < \pi \text{ or } \dfrac{3\pi}{2} < \arg z < 2\pi\right\}$

Answer Key

1	2	3	4	5	6	7
a	All	a	a,b	b	b	a,c
8	**9**	**10**	**11**	**12**	**13**	
b	b,d	a	a	a,c	c	

COMPLEX ANALYSIS

Chapter 1

Groups and Subgroups

1. Binary Composition or Binary Relation :

Let S be a non-empty set. A binary relation * on S is a mapping * : $S \times S \to S$ defined as *(a, b) = a * b. i.e., if $a \in S$ and $b \in S$, then $a * b \in S$.

Whether a composition is a binary composition on a set or not, it depends on the set.

For example : usual addition '+' is binary composition for the set of rationals but not a binary composition for the set of irrational numbers as the sum of two irrational numbers may not be irrational.

2. Definition of Semigroup :

An algebraic structure (G, *), where G is a non-empty set and '*' is a binary operation on G is said to be a semigroup if it satisfies the following associativity axiom :

$a * (b * c) = (a * b) * c \ \forall\ a, b, c \in G$.

3. Definition of Monoid :

An algebraic structure (G, *), where G is a non-empty set and '*' is a binary operation on G is said to be a monoid if it satisfies the following two axioms :

(i) $a * (b * c) = (a * b) * c \ \forall\ a, b, c \in G$

(ii) $\exists e \in G$ such that $a * e = e * a = a \forall a \in G$. Here 'e' is called identity element of G.

4. Definition of Group :

An algebraic structure (G, *), where G is a non-empty set and '*' a binary operation on G is said to be a group if it is satisfies the following axioms :

(i) $a * (b * c) = (a * b) * c \ \forall\ a, b, c \in G$

(ii) $\exists e \in G$ such that $a * e = e * a = a \forall a \in G$. Here 'e' is called identity element of G.

(iii) $\forall a \in G, \exists\ b \in G$ such that $a * b = b * a = e$. Here b is called inverse of the element a and denoted by a^{-1}.

Examples of Groups :

(1) The set of integers Z is a group under addition, but not a group under multiplication.

(2) The set of rational numbers Q is a group under addition.

(3) $Q^* = Q/\{0\}$ is a group under multiplication, whereas Q itself is not a group under multiplication as '0' does not have multiplicative inverse.

(4) The set R of real numbers is a group under addition and $R^* = R/\{0\}$ is a group under multiplication.

(5) For some positive integer n, the set $Z_n = \{0, 1, 2, 3, \ldots\ldots, n - 1\}$, of all remainders when integers are divided by n, forms group under addition modulo n. The binary operation is denoted by $+_n$.

(6) The set of all matrices of order $m \times n$ is a group under addition.

(7) The set $\{1, \omega, \omega^2\}$ of cube roots of unity is a group under multiplication.
(8) $Q_8 = \{1, -1, i, -i, j, -j, k, -k\}$ is called a quaternion group.
(9) $K_4 = \{e, a, b, ab\}$ is a Klein-4 group where $a^2 = e, b^2 = e, (ab)^2 = e$

5. Some Definitions :

(1) Order of a group : The number of elements in a group is called the order of the group. If number of elements in G is finite, then G is called a finite group and if it contains infinite elements then it is called an infinite group. We write O(G) to denote order of G.
Example : $O(Z_n) = n$ and $O(Q)$ = infinite.

(2) Order of an element : Order of an element $a \in G$, is the least positive integer r such that $a^r = e$.
Note : If no such least positive integer exists for an element $a \in G$, then we say that a is of infinite order and hence G has infinite order.

Important Results on Order of an Element :

(1) If G is a finite group, then order of any element of G divides the order of G.
(2) For any two elements $a, x \in G$, $O(a) = O(x^{-1}ax)$.
(3) If for any $a \in G$, O(a) is finite, then for any integer m, $a^m = e \Rightarrow O(a)$ divides m. We write $O(a) \mid m$ to mean O(a) divides m.
(4) For any two elements $a, b \in G$, $O(ab) = O(ba)$.
(5) Let a, b be two elements of finite order in a group G such that O(a) and O(b) are co-prime. If ab = ba, then $O(ab) = O(a).O(b)$.
(6) $O(a^k) = \dfrac{O(a)}{(O(a), k)}$, where $k \in N$ and (O(a), k) is the gcd of O(a) and k.
(7) If $O(G) = 2n$, where n is odd, then $\exists\, x \in G$ such that $O(x) = 2$
(8) If p is a prime and $p \mid O(G)$, then $\exists\, x \in G$ such that $O(x) = p$.

6. Cyclic Group

A group (G, .) is said to be a cyclic group if there exists and element $a \in G$ such that every element of G is a power of a. Element a is called a generator of G and we denote G by $\langle a \rangle$, i.e. $G = \langle a \rangle = \{a^n ; n \in Z\}$. If (G, +) is a cyclic group then

$G = \{na ; n \in Z\}$.

The order of a cyclic group is same as that of the order of the generator of the group, i.e., if $G = \langle a \rangle$, then $O(G) = O(a)$.

Examples of Cyclic Groups :

(1) The multiplication group $\{1, \omega, \omega^2\}$ formed by the cube roots of unity is a cyclic group generated by ω.
(2) The multiplicative group $\{1, -1, i, -i\}$ formed by the fourth roots of unity is a cyclic group generated by i and -i.
(3) The group Z of integers under addition is an infinite cyclic group generated by either 1 or -1.
(4) $(Z_n, +_n)$ is a cyclic group of order n.

MODERN ALGEBRA

(5) $2Z = \{0, \pm2, \pm4, \pm6, \ldots\}$ is a cyclic group generated by either 2 or -2 under addition. Also, mZ for all $m \in Z^+$ is a cyclic group generated by either m or –m under addition.

Examples of Groups which are not cyclic :

(1) (Q, +), the group of rational numbers under addition.

(2) (R, +), the group of real numbers under addition.

(3) (C*, .), the group of non-zero complex numbers under multiplication.

Important Results on Cyclic Groups :

(1) Number of generators of a cyclic group G of order n = $\phi(n)$, where ϕ is Euler's ϕ -function that counts the number of positive integers less than n and relatively prime to n.

(2) Subgroup of cyclic group is cyclic but the converse need not be true.
Example : H = {+1, -1} is a cyclic subgroup of quaternion group which is a itself non-cyclic group.

(3) Every cyclic group is abelian but the converse is not true.
Example : K_4 is a abelian but not cyclic group.

(4) An infinite cyclic group has precisely two generators a and a^{-1}.

(5) Every group of prime order is cyclic.

(6) If O(G) = p.q (p and q are distinct prime numbers with p < q) and $p \nmid q - 1$, then G is cyclic.

(7) If O(G) = p.q (p and q are distict primes with p < q) and $p \mid q - 1$, then either G is cyclic or G is non-abelian group.

(8) For some integer n > 1, let G = <a> be a cyclic group of order n. Since $e = a^0, a^1, \ldots, a^{n-1}$ are all distinct and for positive integer m < n, and m is generator of G if and only if (m, n) = 1. Thus, the number of generators of G is $\phi(n)$.

7. Abelian group

A group (G, *) is said to be abelian if it satisfies commutative property, i.e. , $a*b = b*a \ \forall \ a, b \in G$.

Note : A group which is not abelian is called non-abelian.

Examples of Abelian Groups :

(1) (Z, +), (Q, +), (R, +), (C, +) are all abelian groups.

(2) $(Q^*, \bullet), (R^*, \bullet), (C^*, \bullet)$ are all abelian groups.

(3) For a fixed integer $m \geq 1$, (mZ, +) is an abelian group. Also {0} is an abelian subgroup of (Z, +).

(4) For any positive integer n, $\{Z_n, +_n\}$ is a finite abelian group of order n.

Examples of Non-Abelian Groups :

(1) $G = \{A = [a_{ij}]_{n \times n} : a_{ij} \in F, |A| \neq 0\}$, i.e. the set of all n × n matrices having non-zero determinant over C, under the operation of matrix multiplication forms a non-abelian group. This group is known as **General Linear Group** of degree n and is denoted by GL (n, C) or $GL_n(C)$.

(2) $G = \{[A]_{n\times n} : a_{ij} \in F, |A| = 1\}$, i.e. the set of all n × n matrices having determinant equal to 1 over C, under the operation of matrix multiplication forms a non-abelian group. This group is known as **Special Linear Group** of degree n and is denoted by SL(n , C) or $SL_n(C)$.

Important Results on Abelian Groups :

(1) Every group of order p^2 is abelian, where p is prime.

(2) If $O(x) = 2\ \forall\ x \in G$, then G is abelian.

8. Center of group :

The center Z(G) of a group G is defined by $Z(G) = \{z \in G / zx = xz\ \forall\ x \in G\}$.

Centralizer of an element or normalize of an element

Centralizer of an element $a \in G$, is the set of all those elements in G which commute with a. It is denoted by C(a). In symbols, $C(a) = \{g \in G / ga = ag\}$.

Important Results on Center of a Group and Centralizer of an Element

(1) If $O(G) = p^n$, where p is a prime, then $Z(G) \neq \phi$.

(2) $Z(Z(G)) = Z(G)$.

(3) Let G be a group and $a \in G$, then $C(a) = C(a^{-1})$.

(4) If $a \in Z(G)$, then $C(a) = G$.

(5) $Z(G) \subseteq C(a)$.

(6) $Z(G) = \bigcap_{a\in G} C(a)$

(7) If n | m, then mZ is a subgroup of nZ.

(8) If G is abelian, then Z(G) = G.

9. Subgroup

A non-empty subset H of G is said to be a subgroup of G if H itself is a group under same binary composition as that of G.

Important Results on Subgroups :

(1) {e} and G are subgroups of G.

(2) A non-empty subset H of a group (G, .) is a subgroup of G if and only if $a, b \in H \Rightarrow ab^{-1} \in H$ and foradditive operation $a - b \in H$.

(3) Let G be a group and H be a finite non-empty subset of G. Then H is a subgroup of G if and only if $ab \in H$ for all $a, b \in H$.

Example : (Z, +) is subgroup of (Q, +).

(4) Z(G) is a subgroup of G.

(5) C(a) is a subgroup of G, $\forall\ a \in G$.

(6) Subgroup of an Abelian group is abelian.

(7) If G is a finite abelian group and k is a divisor of O(G), then G must contain a subgroup of order k.

(8) If every proper subgroup of G is abelian, then G may not be abelian.

Example : Every subgroup of quaternion group is abelian.

(9) Z(G) is an abelian subgroup of G.

(10) If H and K are subgroups of a group G, then HK is a subgroup of G if and only if HK = KH.

(11) For any two subgroups H and K of a group G, $H \cap K$ is a subgroup of G but $H \cup K$ may not be subgroup.

Example : G = (Z, +), H = (2Z, +) and K = (3Z, +) then $H \cup K$ is not subgroup of G.

(12) **Lagrange's Theorem :** The order of any subgroup of a finite group divides the order of the group. Also, Lagrange's theorem yields that $O(x) \mid O(G)\ \forall\ x \in G$. But the converse need not be true.

Converge of Lagrange's theorem is true for finite cyclic group

(13) If H and K are finite subgroups of a group G, then $O(HK) = \dfrac{O(H)O(K)}{O(H \cap K)}$.

(14) Let G be a finite group and d | O(G), then number of cyclic subgroups of order d of G

$$= \frac{\text{number of elements of order 'd' in G}}{\phi(d)}$$

(15) Number of subgroups in Z_n is $\tau(n)$ denotes the number of positive divisors of n.

(16) Every group of order less than or equal to 5 are abelian.

SOLVED QUESTIONS

1. Let G be a group of order of 2023. Then which of the following is true.

CSIR NET JUNE 2023

(a) G is cyclic group (b) G is abelian group

(c) G is not simple group (d) g is simple group

Ans.: (b), (c)

Explanation:

$O(G) = 2023 = 7 \times 17^2$

and $7 \pm 1 \pmod{17}$

Let $g = Z_{119} \times Z_{17}$ which is not cyclic

Here $17^2 \pm 1 \pmod 7$

Hence G is abelian.

By sylow third theorem

$1 + 17k \mid 7$; k = 0

Hence G contain unique subgroup of order 17^2

Hence G is not simple

2. Let G be a group with |G| = 96. Suppose H and K are subgroups of G with |H| = 12 and |K| = 16. Then

CSIR NET JUNE-2018

(a) $H \cap K = \{e\}$ (b) $H \cap K \neq \{e\}$ (c) $H \cap K$ is Abelian

(d) $H \cap K$ is not Abelian

Ans. (b), (c)

Explanation :

Given G is a group such that |G| = 96

Let H and K be subgroups of G with |H| = 12, |K| = 16

$\Rightarrow H \cap K$ is subgroup of H as well as $K \Rightarrow O(H \cap K) \mid O(H)\ \&\ O(H \cap K) \mid O(K)$

$\Rightarrow O(H\cap K) \mid 12$ & $O(H\cap K)|16 \Rightarrow O(H\cap K) = 1,2,4$ are only possibilities which satisfy above relation

$\Rightarrow H\cap K$ is abelian group ($\because$ Group of order p^2 where p is prime abelian)

So, option (c) is correct but option (d) is incorrect.

We know $O(HK) = \dfrac{O(H)O(K)}{O(H\cap K)} = \dfrac{12\times 16}{O(H\cap K)}$

If $O(H\cap K) = 1 \Rightarrow O(HK) = 12 \times 16 > 96$, which is absurd

So, $O(H\cap K) \neq 1 \Rightarrow H\cap K$ can never be identity group.

So, option (a) is incorrect but option (b) is correct.

3. Let G be a abelian group of order 34. Let $S = \{g \in G; g^{-1} = g\}$. Then number of non-identity elements in S is –

(a) 5 (b) 2 (c) 1 (d) 0

Ans. **(c)**

Explanation :

Given that G is abelian group and $O(G) = 34 = 2.17$ then $G \cong Z_{34}$

$S = \{g\in G; g^{-1} = g\}$ = Number of self inverse elements of G

We know that if $O(g) = 2$ then g is self inverse

Number of element of order 2 in Z_{34}.

= 1

4. Let $G = \{n \in N : n \leq 55, \gcd(n, 55) = 1\}$ be the group under multiplication modulo 55. Let $x \in G$ be such that $x^2 = 26$ and $x > 30$. Then x is equal to **IIT JAM – 2019**

(a) 31 (b) 30 (c) 45 (d) 46

Ans. **(a), (d)**

Explanation :

Here $G = U(55)$

i.e. Multiplication of modulo 55.

Take $x = 31$ then $x^2 = 961$

$961 \equiv 26 \pmod{55}$

So $x^2 = 26$ for $x = 31$ and same for $x = 46$.

5. Consider the set of matrices $G = \left\{ \begin{pmatrix} s & b \\ 0 & 1 \end{pmatrix} : b \in Z, s \in \{-1,+1\} \right\}$.

Then which of the following is true? **CSIR NET JUNE 2019**

(a) G forms a group under addition

(b) G forms an abelian group under multiplication

(c) Every element in G is diagonalizable over C.

(d) G is a finitely generated group under multiplication.

Ans. **(d)**

MODERN ALGEBRA

Explanation :

(a) $I=\begin{bmatrix}0 & 0\\0 & 0\end{bmatrix}\notin G$

Option (a) is incorrect.

(b) $A=\begin{bmatrix}1 & b\\0 & 1\end{bmatrix}\in G$ and $B=\begin{bmatrix}-1 & c\\0 & 1\end{bmatrix}\in G$

$A=\begin{bmatrix}1 & b\\0 & 1\end{bmatrix}\begin{bmatrix}-1 & c\\0 & 1\end{bmatrix}=\begin{bmatrix}-1 & b+c\\0 & 1\end{bmatrix}\neq\begin{bmatrix}-1 & c\\0 & 1\end{bmatrix}\begin{bmatrix}1 & b\\0 & 1\end{bmatrix}$

Option (b) is incorrect.

(c) $A=\begin{bmatrix}1 & 1\\0 & 1\end{bmatrix}\in G$ is not diagonalizable.

Option (d) is correct.

6. Let S_5 be the symmetric group on five symbols. Then which of the following statements is false?

(a) S_5 contains a cyclic subgroup of order 6 **CSIR NET DEC. 2019**

(b) S_5 contains a non-abelian subgroup of order 8

(c) S_5 does not contain a subgroup isomorphic to $Z/2Z \times Z/2Z$

(d) S_5 does not contain a subgroup of order 7

Ans. (c)

Explanation :

(a) We know that S_5 has an element of order 6.
So, it contain a cyclic subgroup of order 6
Option (a) is incorrect.

(b) Non-abelian subgroup of order 8 is either D_4 or Q_8
In S_5 number of element of order 2 are 25 and number of element of order 4 are 30.
So, option (b) is incorrect.

(c) $|Z/2Z \times Z/2Z| = 4$ and if $|G| = 4$ then $G \approx Z_4, K_4$.
Here $Z/2Z \times Z/2Z$ is not cyclic.
So, it is not isomorphic to Z_4.
$\therefore\ Z/2Z \times Z/2Z \approx K_4$ and $K_4 = \{e, a, b, ab\}$, $|a| = |b| = |ab| = 2$.
If we take $\sigma_1 = (13)$, $\sigma_2 = (24)$ then $\sigma_1\sigma_2 = (13)(24)$.
So, $K = \{I, (13), (24), (13)(24)\} \approx K_4 \approx Z_2\times Z_2$.
$\therefore \exists$a subgroup of S_5 isomorphic to $Z_2\times Z_2$.
So, option (c) is true.

(d) S_5 does not contain an element of order 7
Option (d) is incorrect.

MODERN ALGEBRA

7. Let X be a non-empty set and P(X) be the set of all subsets of X. On P(X), define two operations * and Δ as follows: for A, B ∈ P(X), A *B = A ∩ B; AΔB = (A∪B)\(A∩B). Which of the following statements is true? **CSIR NET JUNE 2020**

(a) P(X) is a group under * as well as under Δ
(b) P(X) is a group under *, but not under Δ
(c) P(X) is a group under Δ, but not under *
(d) P(X) is neither a group under * nor under Δ

Ans. (c)

Explanation :

A,B ∈ P (X), A*B = A∩B; and AΔB = (A∪B)\(A∩B).

We know Δ is the symmetric difference. We need to verify the group axioms.

AΔB = (A∪B)\(A∩B) ⊂X ∈ P(X), closure property satisfied.

(AΔB)ΔC = AΔ(BΔC), associative.

AΔE = A ⇒ (A∪E)\(A∩E)=A, we get E = ϕ∈ P(X). Identity element exist.

AΔB = ϕ⇒ (A∪E)\(A∩E) = ϕ⇒ (A∪E) = (A∩E) ⇒ A = B, ie, all elements are its own inverse, inverse element exist.

So, P(X) is a group under Δ.

A*B = A ∩ B, closure property satisfied and is associative.

A ∩ E = A ⇒ E = X ∈ P (X), Identity element exist.

A*B = E ⇒ A ∩ B = X, which is not possible, so inverse is not exists, so, P(X) is not a group under *, Option (c) is correct.

8. Let G be finite abelian group and a ,b∈ G with order (a) = m , order (b) = n . which of the following are necessarily true **CSIR NET DEC. 2017**

(a) Order (ab) = mn
(b) Order (ab) = lcm{m,n}
(c) There is an elements of G whose order is lcm{a , b}
(d) Order(ab) = gcd(m ,n)

Ans. (c)

Explanation :

$Let\, G = (Z_9, +), a = 2, b = 4$

$Then\, m = O(a) = 9\, and\, n = O(b) = 9$

$Here \quad ab = 2 + 4 = 6 \Rightarrow O(ab) = 3$

$And \quad m.n = 81 = 0$

$\gcd(m, n) = 9 = 0$

$lcm(m, n) = 9 = 0$

So, Option (a) ,(b) and (d) are discarded

(c) Let G = K_4

O(a) = 2 , (b) = 2 ⇒O(ab) = 2 = lcm(2 , 2)

MODERN ALGEBRA

UNSOLVED QUESTIONS

1. Let G be a group of order n. Which of the following conditions imply that G is abelian?
(a) n = 15 (b) n = 21 (c) n = 36 (d) n = 125

NBHM – 2007

2. If Z(G) denotes the center of a group G, then the order of the quotient group G/Z(G) cannot be **GATE-2009**
(a) 4 (b) 6 (c) 15 (d) 25

3. Let G be a group and let H and K be two subgroups of G. If both H and K have 12 elements, which of the following numbers cannot be the cardinality of the set HK = {hk : h∈H, k∈K)? **TIFR-2014**
(a) 72 (b) 60 (c) 48 (d) 36

4. How many proper subgroups does the group Z ⊕ Z have? **TIFR-2014**
(a) 1 (b) 2 (c) 3 (d) infinitely many

5. Let H_1, H_2 be two distinct subgroups of a finite group G, each of order 2. Let H be the smallest subgroup containing H_1 and H_2. Then the order of H is **TIFR-2014**
(a) always 2 (b) always 4 (c) always 8 (d) none of the above

6. Let C be the field of complex numbers and C^* be the group of non-zero complex numbers under multiplication. Then which of the following are true? **CSIR NET DEC-2015**
(a) C^* is cyclic
(b) Every finite subgroup of C^* is cyclic
(c) C^* has finitely many finite subgroups
(d) Every proper subgroup of C^* is cyclic

7. Let G = Z/100Z and let $S = \{h \in G : O(h) = 50\}$. Then O(S) equals **TIFR – 2016**
(a) 20 (b) 25 (c) 30 (d) 50

8. If p is a prime number and G is non-abelian group of order p^3, then the incorrect statement(s) is/are
(a) Z(G) has exactly p – 1 elements
(b) Z(G) has exactly p elements
(c) Z(G) has exactly p^3 elements
(d) all of these

9. Which statements are true?
(a) Every cyclic group is abelian
(b) Every group of prime order is cyclic
(c) Every subgroup of a cyclic group is cyclic
(d) Every group of prime order is finite

10. Consider the group G = {1, -1, i, -i}, then
(a) the order of G is 4
(b) i is the generator of G
(c) 1 is the identity element of G
(d) there is no identity element in G

11. Which of the following is/are true for G to be an abelian group?
(a) $a^2 = e, \forall a \in G$
(b) $o(a) = 2, \forall a \in G$
(c) $ab = ba, \forall a, b \in G$
(d) $(ab)^{-1} = (b^{-1}a^{-1}), \forall a, b \in G$

12. Let be a group such that $(ab)^m = a^m b^m$, for three consecutive integers m and ∀ a, b ∈ G. Then
(a) G is an abelian group
(b) $a^n b = ba^n$
(c) ab = ba
(d) ab ≠ ba

MODERN ALGEBRA

13. Let G be a cyclic group of order 10. For $a \in G$, let <a> denote the subgroup generated by a. How many elements are there in the set $\{a \in G \mid \langle a \rangle = G\}$? **NBHM-2005**

(a) 3 (b) 4 (c) 5 (d) 6

14. Which of the following are subgroups of $GL_3(C)$? **NBHM - 2012**

(a) $H = \{A \in M_3(C) \mid \det(A) = 2^l, l \in Z\}$ (b) $H = \left\{\begin{bmatrix} 1 & \alpha & \beta \\ 0 & 1 & \gamma \\ 0 & 0 & 1 \end{bmatrix} \mid \alpha, \beta, \gamma \in C\right\}$

(c) $H = \left\{\begin{bmatrix} 1 & 0 & \alpha \\ 0 & 1 & 0 \\ 0 & 0 & 1 \end{bmatrix} \mid \alpha \in C\right\}$ (d) None of these

15. Let G be a non-abelian group of order 125. Then the total number of elements in $Z(G) = \{x \in G : gx = xg \text{ for all } g \in G\}$ equals **GATE-2017**

(a) 4 (b) 5 (c) 6 (d) 8

16. Consider the group $Z^2 = \{(a, b) \mid a, b \in Z\}$ under component-wise addition. Then which of the following is a subgroup of Z^2? **IIT-JAM 2018**

(a) $\{(a, b) \in Z^2 \mid ab = 0\}$ (b) $\{(a, b) \in Z^2 \mid 3a + 2b = 15\}$

(c) $\{(a, b) \in Z^2 \mid 7 \text{ divides } ab\}$ (d) $\{(a, b) \in Z^2 \mid 2 \text{ divides } a \text{ and } 3 \text{ divides } b\}$

17. Let $C^* = C / \{0\}$ denote the group of non-zero complex numbers under multiplication. Suppose $Y_n = \{z \in C \mid z^n = 1\}$, $n \in N$. Which of the following is(are) subgroup(s) of C^*?

IIT-JAM 2018

(a) $\bigcup_{n=1}^{100} Y_n$ (b) $\bigcup_{n=1}^{\infty} Y_{2^n}$ (c) $\bigcup_{n=100}^{\infty} Y_n$ (d) $\bigcup_{n=1}^{\infty} Y_n$

18. Which of the following is TRUE?

(a) Z_n is cyclic iff n is prime. (b) Every proper subgroup of Z_n is cyclic.

(c) Every proper subgroups of S_4 is cyclic. (d) None of these

Answer Key

1	2	3	4	5	6	7	8	9	10
a	c	b	d	d	b	a	a,c,d	a,b,c,d	a,b,c
11	**12**	**13**	**14**	**15**	**16**	**17**	**18**		
a,c	a,b,c	b	a,b,c	b	d	b,c,d	b		

Chapter 2

Cosets, Normal Subgroups & Class Equation

Let H be a subgroup of a group G. Then for any $a \in G$, the set $Ha = \{ha / h \in H\}$ is called right coset of H in G and the set $aH = \{ah / h \in H\}$ is called left coset of H in G.
If G is an abelian group, then $Ha = aH \ \forall \ a \in G$.

Properties of Cosets :

Let G be a group, H be a subgroup of G and a, b $\in$ G. Then

(1) $Ha = H \Leftrightarrow a \in H$.

(2) $Ha = Hb \Leftrightarrow ab^{-1} \in H$ (under multiplication) and (under addition, $a - b \in H$).

(3) $aH = bH \Leftrightarrow a^{-1}b \in H$.

(4) **Index of a Subgroup :** Let G be a group and H be a subgroup of G. Then the number of distinct (left) cosets of H in G is called the index of H in G.

The index of H in G is denoted by [G : H] i.e. $[G : H] = \dfrac{O(G)}{O(H)}$

1. Normal Subgroup

A subgroup H of a group (G, .) is said to be a normal subgroup of G is $Ha = aH \ \forall \ a \in G$ and denoted by $H \triangleleft G$.

Important Results on Normal Subgroups :

(1) A subgroup H of a group G is normal if and only if $g^{-1}hg \in H$ for every $h \in H$, $g \in G$.

(2) If H and K are two subgroups of G such that either H or K is normal in G, then HK is a subgroup of G.

(3) If H is normal in G and K is a subgroup of G, then $H \cap K$ is normal in H.

(4) If H and K are both normal in G, then $H \cap K$ is normal in G.

(5) $H \subseteq K \subseteq G$ and H is normal in G, then H is normal in K.

(6) Z(G) is a normal subgroup of G.

(7) Every subgroup of an abelian group is abelian and hence normal but converse need not be true.

Example : Every subgroup quaternion group is normal but quaternion group is non-abelian.

(8) A subgroup of index 2 in a group G is a normal subgroup of G, i.e., $H \leq G$ and $[G : H] = 2$, then H is normal subgroup of G.

Simple Group : A group $G \neq \{e\}$, which does not have any proper normal subgroup is called a simple group.

Example : Z_p is a simple group if and only if p is prime.

2. Quotient Group :

Let (G, .) be a group and N be a normal subgroup of G then we define

$\frac{G}{N}=\{Na : a \in G\}$ and in '+' operation, G/N = {N+a : a ∈ G}

G/N is quotient group of G under the operation defined as (Na)(Nb) = Nab for all Na, Nb ∈ G/N and in '+' operation (N+a) + (N+b) = N+(a+b).

Example - 1 : Q/Z is a quotient group.

Example – 2 : Let G = R, Z be normal subgroup then $\frac{R}{Z}\cong S'$, where S` is the set of all complex numbers z with |z| = 1.

Important Results on Quotient Groups :

(1) For a finite group G and for each normal subgroup N of G, $O(G/N)=\frac{O(G)}{O(N)}$

(2) If G is abelian (cyclic) and N is normal in G, then G/N is also abelian (cyclic).

(3) If G/Z(G) is cyclic, then G is abelian.

3. Normalizer of a subgroup

Let S be any non-empty subset of a group G and K be any subgroup of G. Then normalize of S in K is the set $N_K(S) = \{x \in K \mid x^{-1}Sx = S\}$. We simply write N(S) for $N_G(S)$ and call N(S), the normalize of S.

Remark : N(S) is a subgroup of G.

4. Conjugacy classes and conjugacy equivalence relation

1. **Conjugate elements :** Let G be a group and a, b ∈ G.a is said to be a conjugate of b in G if there eixts c∈G such that $a = c^{-1}bc$. We say a is related to b iff a is a conjugate of b.
2. **Conjugacy classes :** Let a ∈ G, then collection of all conjugate elements of a is congugacy class of a and denoted by C*l*(a).

Results on Conjugacy Classes :

(1) If G is a finite group and a ∈ G then $O(G)=\sum_{i=1}^{t}O[Cl(a_i)]$, where $O[Cl(a_i)]$ is the order of conjugate class to which 'a' belongs.

(2) For any finite group G, $O(G)=\sum_{a}\frac{O(G)}{O(N(a))}$, where the sum runs over elements a, taken one from each conjugate class.

5. Class equation

Let G be a finite group. The equation $O(G)=O(Z(G))+\sum_{a\notin Z(G)}\frac{O(G)}{O(N(a))}$ where the sum runs over elements a, taken one from each of those distinct conjugate classes which contain more than one element, is called the **class equation** of the group G.

Important Results :

(1) Any two cycles of length r in S_n are conjugates where S_n is a permutation group on 'n' symbol under the composition of mapping.

(2) Number of conjugacy classes in $S_n = P(n)$ = partition of an element n.

(3) Every group of order p^2 is abelian. This holds in view of the above fact that

$$O(G)=O(Z(G))+\sum_{x\notin Z(G)}\frac{O(G)}{O(N(x))}\equiv \text{Class Equation}$$

where the sum runs over the elements x, taken one from each of those disjoint-conjugate classes which contains more than one element. We now classify groups of order p^2, p-prime.

So, if G is a group and $O(G) = p^2$. Then $G \cong Z_{p^2}$ or $Z_p \oplus Z_p$

In the first case, if $\exists\ x \in G$ such that $O(x) = p^2$, then $G \cong Z_{p^2}$.

In the other case, one immediately get that $G \cong Z_p \oplus Z_p$

(4) Number of conjugacy classes in $D_n = \begin{cases} \dfrac{n+6}{2}, & \text{if n is even} \\ \dfrac{n+3}{2}, & \text{if n is odd} \end{cases}$.

(5) Let G be a non-abelian group of order p^3, then the number of conjugate classes in $G = p^2 + p - 1$, where p is a prime number.

(6) If G is abelian, then the class equation is $\underbrace{1+1+1+....+1}_{O(G)\,times}$.

(7) Let G be a non-abelian group of order 2p, where p is odd prime, then the class equation is $1+p+\underbrace{2+2+....+2}_{\left(\frac{p-1}{2}\right)times}$.

(8) If $O(G) = p^3$ where G is non abelian, p is any prime, then the class equation $\underbrace{1+1+1+....+1}_{p-times}+\underbrace{p+....+p}_{\left(p^2-1\right)times}$.

Some Class equations :

$S_3 = 1 + 2 + 3$ $\quad$ $S_4 = 1 + 6 + 8 + 6 + 3$

$S_5 = 1 + 10 + 20 + 24 + 30 + 15 + 20$

$A_4 = 1 + 4 + 4 + 3$

$A_5 = 1 + 20 + 15 + 12 + 12$

$D_3 = 1 + 2 + 3$

$D_4 = 1 + 1 + 2 + 2 + 2$

$D_5 = 1 + 2 + 2 + 5$

$D_6 = 1 + 1 + 3 + 3 + 2 + 2$

$Q_8 = 1 + 1 + 2 + 2 + 2$

SOLVED QUESTIONS

1. Which of the following are true **CSIR NET JUNE 2023**

(a) O(G) = 244 Then G has unique subgroup of order 27

(b) O(G) = 1694 Then G has unique subgroup of order 121

(c) O(G) = 154 Then G has unique subgroup of order 7

(d) $\exists$ a group of order 121 which have 2 subgroup of order 11

Ans. (b),(c)

Explanation:

(a) O (G) = 244 = $2^2 \times 61$ and 27 + 244

Hence G does not have a subgroup of order 27

(b) O(G) = 1694 = $2 \times 7 \times (11)^2$

Using sylow 3d theorem $1 + 11k \mid 14$; k =0

Hence, G has unique subgroup of order 121

(c) O(G) = 154 = $2 \times 7 \times 11$

usingsylow 3rd theorem $1 + 7k \mid 22$; k = 0, 3

Hence G have unique or 22 subgroup of order 7.

But G have 22 subgroup of order 7. Then subgroup contain elements more than its order

which is not possible

(d) We know that group of order 121 is always abelian.

(i) if $G \approx Z_{121}$ then G have unique subgroup of order 11.

(ii) If $G \approx Z_{11} \times Z_{11}$

Number of subgroup of order 11 $= \frac{120}{10} = 12$

2. Which of the following is class equation of a group of order 10

CSIR NET JUNE 2023

(a) 1 + 1 + ….. +1 (10 times)

(b) 1 + 1 + 1 + 2 + 5

(c) 1 + 1 + 3 + 5

(d) 1 + 1 + 2 + 2 + 2 + 2

Ans. (a)

Explanation:

If (G) = 10 then it have two possibility of class equation

1 + 1 + ….. +1 (10 times) or 1 + 5 + 2 + 2

3. Let $\sigma \in S_8$, where S_8 is the permutation group on 8 elements. Suppose σ is the product of σ_1 and σ_2, where σ_1 is a 4-cycle and σ_2 is a 3-cycle in S_8. If σ_1 and σ_2 are disjoint cycles, then the number of elements in S_8 which are conjugate to σ is

GATE 2023

(a) 10 (b) 11 (c) 12 (d) 13

Ans. (c)

Explanation :

$\sigma = (abcd)(efg)$

We know that $O(Cl(\sigma)) = \dfrac{O(G)}{O[N(\sigma)]}$ $\quad \dfrac{8!}{3\times 4} = \dfrac{8!}{O[N(\sigma)]}$ $\quad \Rightarrow \quad O[N(\sigma)] = 12$

4. Let G be the group of all invertible 2 × 2 upper triangular matrices (under matrix multiplication). Pick out the normal subgroups of G from the following :

(a) $H = \{A \in G : a_{12} = 0\}$

(b) $H = \{A \in G : a_{11} = 1\}$

(c) $H = \{A \in G : a_{11} = a_{22}\}$

(d) None of these

Ans. (b), (c)

Explanation :

Given that G = Group of all invertible 2 × 2 upper triangular matrices.

For option (a),

$H = \{A \in G : a_{12} = 0\}$ $\quad$ Take $B = \begin{bmatrix} 1 & 0 \\ 0 & 2 \end{bmatrix}, A = \begin{bmatrix} 2 & 1 \\ 0 & 3 \end{bmatrix}$

Clearly $B \in H, A \in G$ $\quad$ $A^{-1} = \begin{bmatrix} \frac{1}{2} & \frac{-1}{6} \\ 0 & \frac{1}{3} \end{bmatrix}$

$$\therefore ABA^{-1} = \begin{bmatrix} 2 & 1 \\ 0 & 3 \end{bmatrix}\begin{bmatrix} 1 & 0 \\ 0 & 2 \end{bmatrix}\begin{bmatrix} 1/2 & -1/6 \\ 0 & 1/3 \end{bmatrix} = \begin{bmatrix} 2 & 2 \\ 0 & 6 \end{bmatrix}\begin{bmatrix} 1/2 & -1/6 \\ 0 & 1/3 \end{bmatrix}$$

$$ABA^{-1} = \begin{bmatrix} 1 & 1/3 \\ 0 & 2 \end{bmatrix}$$

$ABA^{-1} \notin H$ $\{\because a_{12}$ of ABA^{-1} is not zero$\}$ $\Rightarrow$ H is not a normal subgroup of G

($\because$ H is a normal subgroup of G iff $\forall\, h \in H, \forall\, g \in G, ghg^{-1} \in H$)

$\Rightarrow$ (a) is incorrect.

For option (b),

$H = \{A \in G : a_{11} = 1\}$

Let $P = \begin{bmatrix} 1 & a \\ 0 & b \end{bmatrix}, Q = \begin{bmatrix} c & d \\ 0 & e \end{bmatrix}$, where $P \in H, Q \in G$ $\quad$ $Q^{-1} = \dfrac{1}{ce}\begin{bmatrix} e & -d \\ 0 & c \end{bmatrix}$

Now,

$$QPQ^{-1} = \begin{bmatrix} c & d \\ 0 & e \end{bmatrix}\begin{bmatrix} 1 & a \\ 0 & b \end{bmatrix}\begin{bmatrix} 1/c & -d/ce \\ 0 & 1/e \end{bmatrix} = \begin{bmatrix} c & ca+bd \\ 0 & be \end{bmatrix}\begin{bmatrix} 1/c & -d/ce \\ 0 & 1/e \end{bmatrix} = \begin{bmatrix} 1 & \frac{ac+bd-d}{e} \\ 0 & b \end{bmatrix}$$

As a_{11} of $QPQ^{-1} = 1 \Rightarrow QPQ^{-1} \in H$

Thus, H is normal subgroup of G $\quad \Rightarrow$ (b) is correct.

MODERN ALGEBRA

For option (c),

H = {A ∈ G : $a_{11} = a_{22}$}

Let $M = \begin{bmatrix} a & b \\ 0 & a \end{bmatrix}, N = \begin{bmatrix} c & d \\ 0 & e \end{bmatrix}$, M∈H, N ∈ G

$$NMN^{-1} = \begin{bmatrix} c & d \\ 0 & e \end{bmatrix}\begin{bmatrix} a & b \\ 0 & a \end{bmatrix}\begin{bmatrix} 1/c & -d/ce \\ 0 & 1/e \end{bmatrix} = \begin{bmatrix} ca & cb+ad \\ 0 & ea \end{bmatrix}\begin{bmatrix} 1/c & -d/ce \\ 0 & 1/e \end{bmatrix} = \begin{bmatrix} a & cd/e \\ 0 & a \end{bmatrix}$$

$a_{11} = a_{22}$ in $NMN^{-1} \Rightarrow NHN^{-1} \in H$

Thus, H is a normal subgroup of G ⇒ (c) is also correct. So (b), (c) are correct.

5. Consider the quotient group Q/Z of the additive group of rational numbers. The order of the element $\frac{2}{3} + Z$ in Q/Z is **IIT-JAM 2012**

(a) 2 (b) 3 (c) 5 (d) None of these

Ans. (b)

Explanation :

Given Q/Z is the Quotient group of the additive group of rational numbers.

Thus, identity of Q/Z is Z.

Also as $\left(\frac{2}{3}+Z\right)+\left(\frac{2}{3}+Z\right)+\left(\frac{2}{3}+Z\right)=\left(\frac{2}{3}+\frac{2}{3}+\frac{2}{3}\right)+Z=2+Z=Z$

⇒ Order of $\frac{2}{3}+Z$ is 3.

6. H = {e, (1 2) (3 4)} and K = {e, (1 2) (3 4), (1 3) (2 4), (1 4) (2 3)} are subgroups of S_4, where e denotes the identity element of S_4. Then **CSIR NET JUNE-2011**

(a) H and K are normal subgroups of S_4 (b) H is normal in K and K is normal in A_4

(c) H is normal in A_4 but not normal in S_4 (d) K is normal in S_4 but H is not

Ans. (b), (d)

Explanation :

Let H = {e, (12) (34)}

K = {e, (12) (34), (13) (24), (14) (23)} are subgroups of S_4.

Here, K is Klein-4-group, i.e.K = V_4 and V_4 is normal subgroup of A_4 as well as S_4.

Since, H ⊆ K = V_4 imply that H is subgroup of K = V_4 such that [K : H] = 2

⇒ H is normal subgroup of K.

But H is not normal subgroup of S_4 as well as A_4.

∴ Ha ≠ aH ∀ a ∈ S_4 and Ha ≠ aH ∀ a ∈ A_4

Let a = (123) ∈ A_4 ⊆ S_4

H(123) = {(123), (12) (34) (123)} = {(123), (243)}

(123)H = {(123), (123) (12) (34)} = {(123), (134)}

$\exists$ an element $a = (123) \in A_4 \subseteq S_4$ such that $Ha \neq aH \Rightarrow H$ is not normal subgroup of A_4 as well as S_4.

So, options (a) and (c) are incorrect and options (b) and (d) are correct.

7. Let $\omega = \cos\frac{2\pi}{3} + i\sin\frac{2\pi}{3}$, $M = \begin{pmatrix} 0 & i \\ i & 0 \end{pmatrix}$, $N = \begin{pmatrix} \omega & 0 \\ 0 & \omega^2 \end{pmatrix}$ and G = <M, N> be the group generated by the matrices M and N under matrix multiplication. Then **GATE-2012**

(a) $G/Z(G) \cong Z_6$ (b) $G/Z(G) \cong S_3$ (c) $G/Z(G) \cong Z_2$ (d) $G/Z(G) \cong Z_4$

Ans. (b)

Explanation :

$M = \begin{bmatrix} 0 & i \\ i & 0 \end{bmatrix}$, $N = \begin{bmatrix} \omega & 0 \\ 0 & \omega^2 \end{bmatrix}$ and $G = \langle M, N \rangle$ is the group generated by M and N under matrix multiplication. As, M, N $\in$ G

$MN = \begin{bmatrix} 0 & i \\ i & 0 \end{bmatrix}\begin{bmatrix} \omega & 0 \\ 0 & \omega^2 \end{bmatrix} = \begin{bmatrix} 0 & i\omega^2 \\ i\omega & 0 \end{bmatrix}$ $\qquad NM = \begin{bmatrix} \omega & 0 \\ 0 & \omega^2 \end{bmatrix}\begin{bmatrix} 0 & i \\ i & 0 \end{bmatrix} = \begin{bmatrix} 0 & i\omega \\ i\omega^2 & 0 \end{bmatrix}$

Here, MN $\neq$ NM. Hence G is non-abelian group.

Therefore, $\frac{G}{Z(G)}$ is not a cyclic group.

Hence, $\frac{G}{Z(G)}$ must be isomorphic to non-abelian group of order '6'

$\Rightarrow$ Options (a), (c) and (d) are incorrect. $\Rightarrow$ Option (b) is correct.

Trick :

(1) If O(G) = 2.n where n is odd.
Then class equation 2n = 1 + n + 2 + 2 + ….. + 2

(2) If O(G) = p^2 : p is prime
Then G is abelian group.
Class equation $p^2 = 1 + 1 + \ldots + 1$

(3) O(G) = 8
Then class equation either 8 = 1 + 1 + 1 ….. + 1
Or 8 = 1 + 1 + 2 + 2 + 2

8. Determine which of the following cannot be the class equation of a group.

(a) 10 = 1 + 1 + 1 + 2 + 5 (b) 4 = 1 + 1 + 2
(c) 8 = 1 + 1 + 3 + 3 (d) 6 = 1 + 2 + 3

Ans. (a), (b), (c) **CSIR NET DEC-2013**

Explanation :

For option (a) O(G) = 10 $\Rightarrow$
If G is abelian then 10 = 1 + 1 …. + 1
G is non-abelian then 1 + 5 + 2 + 2

MODERN ALGEBRA

O(G) = 4 then G is abelian. $\Rightarrow 4 = 1 + 1 + 1 + 1$

$8 = 1 + 1 + 2 + 2 + 2$

9. Let G denote the group of all 2×2 invertible matrices with entries from R.

Let $H_1 = \{A \in G : \det(A) = 1\}$ and $H_2 = \{A \in G : A \text{ is upper triangular}\}$

Consider the following statements :

P : H_1 is a normal subgroup of G. Q : H_2 is a normal subgroup of G.

Then **CSIR NET DEC 2011**

(a) Both P and Q are true (b) P is true and Q is false

(c) P is false and Q is true (d) Both P and Q are false

Ans. (b)

Explanation :

H_1 is precisely the group SL(2, Z), which we know, is normal subgroup of G.

H_2 is not normal subgroup of given group.

As $h = \begin{pmatrix} 1 & 3 \\ 0 & 2 \end{pmatrix} \in H_2, a = \begin{pmatrix} 2 & 1 \\ 1 & 1 \end{pmatrix} \in G$ But $a^{-1}ha = \begin{pmatrix} 3 & 2 \\ -1 & 0 \end{pmatrix} \notin H_2$

$\therefore$ H_2 is not normal.

10. Which of the following equations can occur as the class equation of a group of order 10? **NBHM-2015**

(a) $10 = 1 + 1 + 1 + 2 + 5$ (b) $10 = 1 + 2 + 3 + 4$

(c) $10 = 1 + 1 + \ldots.. + 1$ (10 times) (d) None of these

Ans. (c)

Explanation :

Given that O(G) = 10 = 2.5 where 5 is a prime. Then class equation of G is either

$10 = 1 + 1 + 1 + \ldots\ldots+1$ (10 times) Or $10 = 1 + 5 + 2 + 2$

11. Let $SL_2(z) = \left\{\begin{pmatrix} a & b \\ c & d \end{pmatrix} \in M_2(Z) : ad - bc = 1\right\}$ and for any prime p, let

$$r(p) = \left\{\begin{pmatrix} a & b \\ c & d \end{pmatrix} \in SL_2(Z) : \begin{matrix} a \equiv 1(\text{mod } p), d \equiv 1(\text{mod } p) \\ c \equiv 0(\text{mod } p), b \equiv 0(\text{mod } p) \end{matrix}\right\}.$$

Which of the following are true?

(a) Γ (p) is a subgroup of $SL_2(Z)$ (b) $\Gamma(p)$ is not a normal subgroup of $SL_2(Z)$

(c) Γ (p) has at least two elements (d) $\Gamma(p)$ is uncountable

Ans. (a), (c) **CSIR NET JUNE 2020**

Explanation :

$SL_2(z) = \left\{\begin{pmatrix} a & b \\ c & d \end{pmatrix} \in M_2(z): ad - bc = 1\right\}$ and for any prime p, let $\Gamma(p) =$

$$\left\{\begin{pmatrix} a & b \\ c & d \end{pmatrix} \in SL_2(Z) : \begin{matrix} a \equiv 1(\text{mod } p), d \equiv 1(\text{mod } p) \\ c \equiv 0(\text{mod } p), b \equiv 0(\text{mod } p) \end{matrix}\right\}.$$

(a) Consider $\begin{pmatrix} a_1 & b_1 \\ c_1 & d_1 \end{pmatrix}, \begin{pmatrix} a_2 & b_2 \\ c_2 & d_2 \end{pmatrix} \in \Gamma(p)$, then $\begin{pmatrix} a_1 & b_1 \\ c_1 & d_1 \end{pmatrix}, \begin{pmatrix} a_2 & b_2 \\ c_2 & d_2 \end{pmatrix}^{-1}$

$= \frac{1}{a_2d_2 - c_2b_2}\begin{pmatrix} a_1 & b_1 \\ c_1 & d_1 \end{pmatrix}\begin{pmatrix} d_2 & -b_2 \\ -c_2 & a_2 \end{pmatrix} = \begin{pmatrix} a_1 & b_1 \\ c_1 & d_1 \end{pmatrix}\begin{pmatrix} d_2 & -b_2 \\ -c_2 & a_2 \end{pmatrix}$

$= \begin{pmatrix} a_1d_2 - b_1c_2 & -a_1b_2 + b_1a_2 \\ c_1d_2 - d_1c_2 & -c_1b_2 + d_1a_2 \end{pmatrix}$. Since,

$a_1 \equiv 1 \pmod p$, $d_1 \equiv 1 \pmod p$,
$c_1 \equiv 0 \pmod p$, $b_1 \equiv 0 \pmod p$, and
$a_2 \equiv 1 \pmod p$, $d_2 \equiv 1 \pmod p$,
$c_2 \equiv 0 \pmod p$, $b_2 \equiv 0 \pmod p$, then $= \begin{pmatrix} a_1 & b_1 \\ c_1 & d_1 \end{pmatrix}\begin{pmatrix} a_2 & b_2 \\ c_2 & d_2 \end{pmatrix}^{-1} = \begin{pmatrix} a_1d_2 & 0 \\ 0 & d_1a_2 \end{pmatrix}$.

Since, $a_1d_2 \equiv 1 \pmod p$, $d_1a_2 \equiv 1 \pmod p$ and $a_1d_2 - d_1a_2 = 1$, so
$\begin{pmatrix} a_1 & b_1 \\ c_1 & d_1 \end{pmatrix}\begin{pmatrix} a_2 & b_2 \\ c_2 & d_2 \end{pmatrix}^{-1} \in \Gamma(p)$. Ie, $\Gamma(p)$ is a subgroup of $SL_2(Z)$,
Option (a) is correct.

(b) $x \in G = SL_2(Z)$ & $h \in H = \Gamma(p) \Rightarrow xhx^{-1} \in H = \Gamma(p)$ (?)

$x = \begin{pmatrix} x & y \\ z & w \end{pmatrix}$, $h = \begin{pmatrix} a & b \\ c & d \end{pmatrix}$, $x^{-1} = \begin{pmatrix} w & -y \\ -z & x \end{pmatrix}$

$xhx^{-1} = \begin{pmatrix} x & y \\ z & w \end{pmatrix}\begin{pmatrix} a & b \\ c & d \end{pmatrix}\begin{pmatrix} w & -y \\ -z & x \end{pmatrix} = \begin{pmatrix} x & y \\ z & w \end{pmatrix}\begin{pmatrix} aw-bx & -ay+bx \\ cw-dz & -cy+dx \end{pmatrix} = \begin{pmatrix} x & y \\ z & w \end{pmatrix}\begin{pmatrix} w & -y \\ -z & x \end{pmatrix}$

$= \begin{pmatrix} 1 & 0 \\ 0 & 1 \end{pmatrix} \in H = \Gamma(p)$, $\Gamma(p)$ is a normal subgroup of $SL_2(Z)$, Option (b) is incorrect.

(c) $I \equiv \Gamma(p)$ and $\begin{pmatrix} 1 & p \\ p & 1 \end{pmatrix} \in \Gamma(p)$, so $\Gamma(p)$ has at least two elements, Option (c) is correct.

(d) $\Gamma(p) = \left\{\begin{pmatrix} a & b \\ c & d \end{pmatrix} \in SL_2(Z)\right\}$, where a, b, c, d $\in$ Z, so each a, b, c, d have only countable number of choices, so does $\Gamma(p)$, Option (d) is incorrect.

Option (a) and (c) are correct.

UNSOLVED QUESTIONS

1. Which of the following groups is simple? **GATE-2010**

(a) S_3 (b) GL(2, R) (c) $Z_2 \times Z_2$ (d) A_5

MODERN ALGEBRA

2. Let G be a finite group and H be a normal subgroup of G of order 2, then the order of the centre of G is
(a) 0 (b) 1 (c) An even integer ≥ 2 (d) An odd integer ≥ 3

3. If H and K are normal subgroups of a group G, then which of the following is not true?
(a) $H \cap K$ and HK are normal in G (b) HK and $H \cup K$ are normal in G
(c) $H \cap K$ is normal but HK is not normal (d) $H \cup K$ is normal but HK is not normal

4. For any group G of order 36 and any subgroup H of G of order 4, then **CSIR NET JUNE-2012**
(a) $H \subset Z(G)$ (b) $H = Z(G)$ (c) H is normal in G (d) H is an abelian group

5. Which of the following are true?
(a) Every group of order 12 has a non-trivial proper normal subgroup.
(b) Some group of order 12 does not have a non-trivial proper normal subgroup.
(c) Every group of order 12 has a subgroup of order 6.
(d) Every group of order 12 has an element of order 12.

6. Suppose H is a normal subgroup of G then which of the following are true?
(a) If G is infinite group $\Rightarrow \frac{G}{H}$ is infinite.
(b) If G is non-abelian group $\Rightarrow \frac{G}{H}$ is non-abelian.
(c) If G is cyclic $\Rightarrow \frac{G}{H}$ is abelian. (d) G is abelian $\Rightarrow \frac{G}{H}$ is cyclic.

7. Let G be a group and $e \neq a \in G$ s.t. $O(a) = n$ (finite) and G has exactly two classes. Then which of the following are correct.
(a) Centre of G has exactly two element (b) Every subgroup of G is normal in G
(c) Number of self inverse element in G is two (d) None of these

8. Which of the following are the possible class equation of G?
(a) $14 = 1 + 7 + 3 + 3$ (b) $22 = 1 + 11 + 2 + 2 + 2 + 2 + 2$
(c) $35 = 1 + 7 + 7 + 5 + 5 + 5$ (d) $21 = 1 + 7 + 7 + 3 + 3$

9. Let G be a group. Which of the following statements are true? **NBHM-2018**
(a) The normalizer of a subgroup of G is normal subgroup of G.
(b) The centre of G is a normal subgroup of G.
(c) If H is a normal subgroup of G and is of order 2, then H is contained in the centre of G.
(d) None of these

Answer Key

1	2	3	4	5	6	7	8	9
d	c	b,c,d	d	a	c	a,b,c	b,d	b,c

Chapter 3

Homomorphism and Isomorphism

1. Homomorphism:

Let G and G' be any two groups and let 'o' and '*'denote their respective binary operations. Then a mappingf : G →G' is called a homomorphism, if f(a ° b)=f(a) *f(b) ∀ a, b∈G.

2. Isomorphism:

A homomorphism f from a group G to a group G' is called an isomorphism if f is a one-one and onto mapping.

3. Automorphism:

An isomorphism of a group G onto itself is called an automorphism of G.

4. Inner automorphism: Let a be an element of a group G. The automorphism f_a : G →G given by $f_a(x) = axa^{-1}$, x∈G is called an inner automorphism of G determined by a. Inn(G) denotes the set of all inner automorphisms of G.

Note : Set of all automorphisms from G to Gform a group under composition and is denoted by Aut (G).

5. Kernel of a Homomorphism:

Let*f* be a homomorphism of a group G into a group G' then the kernel(f)={x∈G | f(x) = e'}.

where e' is the identity of G' We denote the kernel(f) by ker(f).

Note : Let f : G → G' then ker f is a normal subgroup of G.

Important Results on Isomorphisms :

(1) Any infinite cyclic group is isomorphic to the additive group of integers.

(2) Any finite cyclic group of order n is isomorphic to Z_n.

(3) Any two finite cyclic groups of same order are isomorphic.

(4) Z/nZ is isomorphic to Z_n.

(5) If n | m, then $\frac{nZ}{mZ} \cong Z_{\frac{m}{n}}$.

Important Results on Automorphisms :

(1) For any group G, the set of the automorphisms of G, Aut(G) is a group under the composition.

(2) For any group G, Inn(G) is a normal subgroup of Aut(G). Further Inn(G) ≅G/Z(G)

(3) Inn(G) consists of the identity function if and only if G is abelian.

(4) If f : G → G and G is any infinite cyclic group, then Aut(G) ≅Z_2.

(5) Aut (Z_n) = U(n)

MODERN ALGEBRA

(6) Aut $(K_4) = S_3$

(7) Aut $(S_n) = S_n$; $n \neq 6$

Important Theorems on Homomorphism and Isomorphism :

Theorem. Let G be a finite abelian group of order n. A mapping $f : G \to G$ defined by $f(x) = x^m$ is an isomorphism if gcd (m, n) = 1.

Thereom. Homomorphic image of an abelian group is abelian but the converse of the above theorem is not true.

Example : $f : S_4 \to Z_4$ be a homomorphism and $\ker(f) = A_4$, then $f(S_4) \cong Z_2$.

Theorem. The homomorphic image of cyclic group is cyclic, but converse of the above therem is not true.

Theorem. Fundamental Theorem of Homomorphism : Let G be a group. If N is a normal subgroup of G, then G/N is a homomorphic image of G. Conversely, if any group G' is a homomorphic image of G, then G' is a isomorphic to some quoteient group of G. In fact, if f is a homomorphism of G onto G', then $G' \cong G/\ker f$.

Theorem. First Theorem of Isomorphism : Let f be a homomorphism of a group G onto a group G' and H = ker f, K' be any normal subgroup of G' and $K = \{x \in G | f(x) \in K'\}$. Then K is a normal subgroup of G containing H and $G/K \cong G'/K'$.

$f^{-1}(k^1) = K$ then $\frac{G}{K} \cong \frac{G'}{K'}$

Result of Automorphism :

Aut (Z_n) is always abelian group as $\text{Aut}(Z_n) \approx U(n)$

(i) Number of automorphism in $S_n = n!$ $n \geq 3$

(ii) Number of automorphism in $D_n = n.\phi(n)$ $n \geq 3$

(iii) Aut $(Z_p \times Z_p \ldots\ldots Z_p) \approx GL(n, Z_p)$
(n-copies)

(iv) $Inn(G) \cong \frac{G}{Z(G)}$.

SOLVED QUESTIONS

1. Suppose ϕ, G G' be a surjective group homomorphism. Which of the following is/are true.

CSIR NET JUNE 2023

(a) if G is abelian then G' is abelian

(b) if G is cyclic then G' is cyclic

(c) if $H \subseteq G$ then $\phi(H) \subseteq G'$

(d) if H is normal subgroup of G then $\phi(H)$ is normal subgroup of G'

Ans. **(a), (b)**

Explanation:

Given that $f : G \to G'$ is on to then by FTH $\frac{G}{\ker \phi} \approx G'$

we know that Every quotient group of abelian and cyclic group is abelian and cyclic.

(c) If $H \subseteq G$ then $ab^{-1} \in H$; $\forall a,b \in H$

Now suppose $\phi(a), \phi(b) \in \phi(H)$ then $\phi(a)[\phi(b)]^{-1} = \phi(a)\,\phi(b^{-1})$

$= \phi(ab^{-1}) \in \phi(H)$

$\Rightarrow \phi(H)$ is subgroup of G'

(d) Given that H is normal subgroup of G then $xhx^{-1} \in H$ for all $x \in G$ & $h \in x$

Now $\phi(x)\phi(h)[\phi(\lambda)]^{-1} = \phi(xhx^{-1}) \in \phi(H)$ where $\phi(x) \in G'$

Hence $\phi(H)$ is normal subgroup of G'

2. For any group G, Aut(G) denote the group of automorphism of G. Which of the following are true?

(a) If G is finite, then Aut(G) is finite (b) If G is cyclic, then Aut(G) is cyclic

(c) If G is infinite, then Aut(G) is infinite

(d) If Aut (G) is isomorphic to Aut(H) then G is isomorphic to H.

CSIR NET DEC-2018

Ans. **(a)**

Explanation :

(b) Let G = Z_8, then Aut(Z_8) $\cong$ U(8) $\approx Z_2 \times Z_2$

We know that $Z_2 \times Z_2$ is not cyclic.

Option (b) is incorrect.

(c) Let G = Z, then Aut(Z) $\cong Z_2$

Option (c) is incorrect.

(d) Let $G_1 = Z_8$ and $G_2 = Z_{12}$

Then Aut(Z_8) $\cong$ U(6) $\approx Z_2 \times Z_2$

andAut(Z_{12}) $\cong$ U(12) $\approx Z_2 \times Z_2$

Here Aut(Z_8) and Aut(Z_{12}) are isomorphic but $Z_8 \not\approx Z_{12}$.

Option (d) is incorrect.

(a) Option (a) is correct.

3. Let G be the cyclic group of order 8 and H = S_3 be the permutation group of 3 elements.Which of the following statements are necessarily true?

CSIR NET NOV 2020

(a) There exists no non-trivial group homomorphism from G to H

(b) There exists no injective group homomorphism from G to H

(c) There exists no surjective group homomorphism from G to H

(d) There are more than 20 different group homomorphism from G to H

Ans. **(b),(c)**

Explanation :

Number of Normal subgroup of Z_8 are {0} , Z_2 , Z_4 , Z_8

(i) $\frac{Z_8}{\{0\}} = Z_8$ which is not subgroup of S_3

MODERN ALGEBRA

(ii) $\frac{Z_8}{Z_2} = Z_4$ which is not subgroup of S_3

(iii) $\frac{Z_8}{Z_4} = Z_2$ which is subgroup of S_3

Here n_1 = number of subgroup Z_4 in Z_8 is 1 , n_2 = number of subgroup Z_2 in S_3 are 3 , $n_3 = O(Aut(Z_2)) = 1$

Then $n_1n_2n_3 = 3$

(iv) $\frac{Z_8}{Z_8} = \{0\}$ which is subgroup of S_3

(a) There is a non – trivial homomorphism
(a) is incorrect

(b) There is no injective homomorphism
(b) is correct

(c) There is no surjective homomorphism
(b) is correct

(d) There are toral 4 homomorphism

4. Let G = S_3 be the permutation group . Then **CSIR NET JUNE 2018**

(a) G is isomorphic to a subgroup of a cyclic group
(b) There exist a cyclic group H such that G maps homomorphically onto H
(c) G is product of cyclic groups
(d) There exists a non – trivial group homomorphism from G to the additive group (Q ,+)

Ans. (b)

Explanation:

(a) G is non cyclic group and every subgroup of cyclic group is cyclic
(b) Suppose H = Z_2
Then a mapping f : $S_3 \rightarrow Z_2$

By FTH $\frac{S_3}{A_3} \approx Z_2$

There exist a cyclic group H such that G maps homomorphically onto H
(c) Product of cyclic group is either cyclic or abelian
(d) $f : S_3 \rightarrow (Q,+)$
We know that finite group of quotient group is always finite

By FTH $\frac{S_3}{Kerf} \approx \{0\}$ [{0} is only finite subgroup of Q]

Hence Ker f = S_3
Then f is trivial homomorphism

Trick : Number of group homomorphism $f : Z_m \to Z_n$ are gcd(m, n)

5. How many homomorphism $f : Z_4 \to Z_8$

(a) 1 (b) 2 (c) 3 (d) 4

Ans. (d)

Explanation :

gcd(4, 8) = 4

Total number of homomorphism = 4

Another method :

Possible order of element in Z_8 are 1, 2, 4, 8 and possible order of element in Z_4 are 1, 2, 4.

Common order of element in Z_4 & Z_8 are 1, 2, 4

Number of element of order 1 in $Z_8 = \phi(1) = 1$

Number of element of order 2 in $Z_8 = \phi(2) = 1$

Number of element of order 4 in $Z_8 = \phi(4) = 2$

Total number of homomorphism = 1 + 1 + 2 = 4

Trick : (i) Number of homomorphism from $f : Z_m \to Z_n \times Z_k$ = gcd(m, n) × gcd(m, k)

(ii) Number of homomorphism from $f : Z_m \times Z_n \to Z_k$ = gcd(m, k) × gcd(n, k)

6. How many homomorphism from $f : Z_2 \times Z_4 \to Z_8$

(a) 1 (b) 2 (c) 3 (d) 8

Ans. (d)

Explanation :

Possible order of element in Z_2 = 1, 2

Possible order of element in Z_4 = 1, 2, 4

Possible order of element in Z_8 = 1, 2, 4, 8

Common order of element in $Z_2 \times Z_8$ are 1, 2

Number of element of order 1 in Z_8 = 1

Number of element of order 2 in $Z_8 = \phi(2) = 1$

Total number of homomorphism from Z_2 to Z_8 = 1 + 1 = 2

Common order of element in Z_4 & Z_8 are 1, 2, 4

Number of element of order 1 in Z_8 = 1

Number of element of order 2 in Z_8 = 1

Number of element of order 4 in $Z_8 = \phi(4) = 2$

Total number of homomorphism from Z_4 to Z_8 are 1 + 1 + 2 = 4

∴ Total number of homomorphism from $Z_2 \times Z_4 \to Z_8$ are 2 × 4 = 8

Trick 1 : $f : Z_m \times Z_n \to Z_k \times Z_l$

Number of homomorphism = gcd(m, k) × gcd(m, l) × gcd(n, k) × gcd(n, l)

Trick 2 : $f : Z_m \to Z_n$ and n | m then number of onto homomophism = $\phi(n)$

7. $f : Z_{20} \to Z_{10}$. How many onto homomorphism.

(a) 1 (b) 2 (c) 3 (d) 4

Ans. (d)

Explanation :

Number of onto homomorphism = $\phi(10) = 4$

8. $f : Z \to Z$. How many onto homomorphism.

(a) 1 (b) 2 (c) 3 (d) 4

Ans. (b)

Explanation :

$$\left.\begin{array}{l} f(x) = x \\ f(x) = -x \end{array}\right\} = \text{Im } f \approx Z$$

Exactly two onto homomorphism.

> **Trick :** Number of homomorphism from $f : Z_n \to Z$ is exactly 1.

9. $f : Z_4 \to Z$. How many homomorphism?

(a) 1 (b) 2 (c) 3 (d) 4

Ans. (a)

Explanation :

Possible order of ker f = 1, 2, 4

Case – I If ker f $\approx \{0\}$

by FTH

$\frac{Z_4}{\{0\}} \cong Z_4$ and Z_4 is not subgroup of Z.

Case – II ker f $\approx Z_2$

$\frac{Z_4}{\ker f} \approx Z_2$ is not subgroup of Z.

Case – III ker f $\approx Z_4$

$\frac{Z_4}{\ker f} \approx \{0\}$ is subgroup of Z.

Total number of homomorphism = 1

> **Trick :** $f : S_3 \to Z_n$
>
> Number of homomorphism = $\begin{cases} 1 & n \text{ is odd} \\ 2 & n \text{ is even} \end{cases}$.

10. How many homomorphism from $f : S_3 \to Z_6$

(a) 1 (b) 2 (c) 3 (d) 4

Ans. (b)

MODERN ALGEBRA

Explanation :

Case – I Ker f ≈ I $\frac{S_3}{I} \approx S_3$ is not subgroup of Z_6.

Case – II Ker f ≈ A_3 $\frac{S_3}{A_3} \approx Z_2$ is subgroup of Z_6.

Case – III ker f = S_3 $\frac{S_3}{S_3} \approx I_{Z_1} = \{0\}$ is subgroup of Z_6.

Number of element of order is 1.
Total number of homomorphism = 1 + 1 = 2

11. How many homomorhism in f : $GL(2,F_2) \to U(7)$
(a) 1 (b) 2 (c) 3 (d) 6
Ans. (b)
Explanation :
$O[GL(2, F_2)] = (p^n - 1)(p^n - p^{n-1})\ldots = (2^2 - 1)(2^2 - 2) = 3 \times 2 = 6$
i.e., $GL(2,F_2) \approx S_3$
f : $S_3 \to U(7)$
i.e. f : $S_3 \to Z_6$
Number of homomorphism = 2

12. How many inner automorphism in Z_{10}.
(a) 1 (b) 2 (c) 4 (d) 8
Ans. (a)
Explanation :
$G = Z_{10}$

$$Inn(G) \approx \frac{G}{Z(G)} \text{ and } O[Inn(G)] = \frac{O(G)}{O(Z(G))} = \frac{O(Z_{10})}{O(Z(Z_{10}))} = \frac{10}{10} = 1$$

13. Let G be a group. Which of the following statements are true? **NBHM-2017**
(a) Let H and K be subgroups of G of order 3 and 5 respectively. Then $H \cap K = \{e\}$, where e is the identity element of G.
(b) If G is an abelian group of odd order, then $\varphi(x) = x^2$ is an automorphism of G.
(c) If G has exactly one element of order 2, then this element belongs to the center of G.
(d) None of these
Ans. (a), (b), (c)
Explanation :
For statement (a),
Given O(H) = 3, O(K) = 5
As H and K are subgroups of G ∴ $H \cap K$ is also a subgroup of G
$H \cap K \subseteq H$, $H \cap K \subseteq K$ ∴ $O(H \cap K) | O(H)$ and $O(H \cap K) | O(K)$

MODERN ALGEBRA

$\Rightarrow O(H\cap K) \mid (O(H), O(K)) \Rightarrow O(H\cap K)|(3, 5) \Rightarrow O(H\cap K)|1 \Rightarrow O(H\cap K) = 1 \Rightarrow H\cap K = \{e\}$

$\therefore$ Statement (a) is true.

For statement (b), Let G be an abelian group of odd order (say 2m + 1) under multiplication.

Clearly, $\phi(x) = x^2 \forall\ x\in G$ is a homomorphism.

For one-one, let $\phi(x) = \phi(y)$, i.e., $x^2 = y^2$

$\Rightarrow x = xe = xx^{2m+1} = x^{2m+2} = (x^2)^{m+1} = (y^2)^{m+1} = y^{2m+2} = yy^{2m+1} = ye = y$

$\therefore$ 'ϕ' is one-one.

For onto,

For any $g \in G$, $g = ge = gg^{2m+1} = g^{2m+2}$, $\exists\ g^{m+1}$ such that $\phi(g^{m+1}) = (g^{m+1})^2 = g$

$\therefore \phi$ is onto.

$\therefore$ Given mapping $\phi(x) = x^2$ is an automorphism.

Thus, statement (b) is true.

For statement (c),

Let 'a' be the only element in G of order $2 \Rightarrow a^2 = e$

Also $O(a) = O(xax^{-1})$ for any $x\in G \Rightarrow xax^{-1} = a \forall x \in G$ [$\because$ there is only one element of order 2)

$\Rightarrow xa = ax\ \forall\ x \in G \Rightarrow$ a belongs to center of G

$\therefore$ Statement (c) is true.

14. Let G_1 and G_2 be two group s.t. $O(G_1)$ and $O(G_2)$ are relatively prime. If $f : G_1 \to G_2$ is homomorphism then

(a) f is trivial (b) f is onto (c) f is one-one (d) f is an isomorphism

Ans. (a)

Explanation :

Let $G_1 \approx Z_6$ and $G_2 \cong Z_7$ then $f : Z_6 \to Z_7$

Number of homomorphism is gcd (6, 7) = 1 (unique $\Rightarrow$ trivial)

$\Rightarrow$ only trivial homomorphism.

Trick : Aut $(Z_n) \cong U(n)$

15. Let G be a group of order 65 then the order of Aut (G)

(a) 1 (b) 64 (c) 48 (d) 24

Ans. (c)

Explanation :

Given that $O(G) = 65 = 13 \times 5$

$5 \nmid 13 - 1 \Rightarrow 5 \nmid 12 \Rightarrow$ G is cyclic

$G \cong Z_{65}$

Then $\text{Aut}(G) \approx \text{Aut}(Z_{65}) \approx U(65)$

Order of U(65) = Order of U(13 × 5) = Order of U(13) × U(5) = 12 × 4 = 48

16. Let $T(n : R) \subset M(n, R)$ be the set of all matrices whose trace is zero then the quotient space $\frac{M(n;R)}{T(n,R)}$ is isomorphic to

(a) R (b) ϕ (c) Q (d) M(n : R)

Ans. (a)

Explanation :

Let we define a map $f : M(n:R) \xrightarrow{onto} R$

s.t. f(A) = Tr(A) where $A \in M(n; R)$

then $\ker f = \{A \in M(n:R) \mid Tr(A) = 0\}$

Given that T(n : R) = ker f

By FTH $\frac{M(n:R)}{\ker f} \approx R \Rightarrow \frac{M(n:R)}{T(n:R)} \approx R$

UNSOLVED QUESTIONS

1. Upto isomorphism the number of abelian groups of order 108 is

IIT-JAM-2012

(a) 12 (b) 9 (c) 6 (d) 5

2. Let (Q, +) be the group of all rational number under addition and $(Q^+, .)$ be the group of positive non-zero rational under multiplication. Then $f\left(\frac{2016}{2015}\right) = ?$. $f : (Q, +) \to (Q^+, .)$ be homomorphism.

(a) $(2016)^{2015}$ (b) $(2015)^{2016}$ (c) $\frac{2015}{2016}$ (d) None of these

3. The value of n for which there exist is one-one homomorphism from Z_{12} to S_n.

(a) 6 (b) 7 (c) 9 (d) 11

4. Let G be a non-abelian group, $y \in G$, and let the maps f, g, h from G to itself be defined by $f(x) = yxy^{-1}$, $g(x) = x^{-1}$ and h = g o g. Then **IIT-JAM 2019**

(a) g and h are homomorphisms and f is not a homomorphism

(b) h is a homomorphism and g is not a homomorphism

(c) f is a homomorphism and g is not a homomorphism

(d) f, g and h are homomorphisms

5. Let f : (Q, +) → (Q, +) be a non-zero homomorphism. Pick out the true statements :

NBHM – 2006

(a) f is always one – one (b) f is always onto

(c) f is always a bijection (d) f need not be neither one-one nor onto.

6. Which one of the following conditions on a group G implies that G is abelian?

IIT JAM-2014

MODERN ALGEBRA

(a) The order of G is p^3 for some prime p (b) Every proper subgroup of G is cyclic
(c) Every subgroup of G is normal in G
(d) The function $f : G \to G$, defined by $f(x) = x^{-1}$ for all $x \in G$, is a homomorphism.

7. The number of non-isomorphic abelian group of order 24 is **GATE-2013**
(a) 3 (b) 4 (c) 5 (d) 6

8. Which of the following statements are true? **NBHM – 2019**
(a) If G is a finite group, then there exists $n \in N$ such that 'G is isomorphic' to a subgroup of $GL_n(R)$.
(b) There exists an infinite group G such that every element, other than the identity element, is of order 2.
(c) The group $GL_2(R)$ contains a cyclic subgroup of order 5. (d) None of these

9. Let G be a group whose presentation is $G = \{x, y \mid x^5 = y^2 = e, x^2y = yx\}$. Then G is isomorphic to **GATE - 2016**
(a) Z_5 (b) Z_{10} (c) Z_2 (d) Z_{30}

10. Let S_n be the symmetric group of n letters there doesn't exist an onto homomorphism.
(a)from S_5 to S_4 (b)from S_4 to Z_2 (c)from S_5 to Z_5 (d)from S_4 to Z_4

11. Let G be a group then the possible order of Inn(G)
(a)35 (b)121 (c)56 (d)77

12. Consider the following statement :
I : If (Q, +) be a group and $f : Q \to Q$ is a non-trivial homomorphism then f isan isomorphism.
II. : Any quotient group of cyclic group is cyclic.
III : If every subgroup of a group G is a normal subgroup then G is abelian.
IV. : Every group of order 33 is cyclic.
Which of the following are true?
(a) II and IV only (b) II & III only (c) I, II & IV only (d) I, III & IV only

13. Consider the group homomorphism $\phi : M_2(R) \to R$ given by $\phi(A) = Tr(A)$. Then the kernel of ϕ is not isomorphic to
(a)$M_2(R) / \{A \in M_2(R) : \phi(A) = 0\}$ (b)R^2 (c)R^3 (d)$GL_2(R)$

14. Let G be a non-cyclic group of order 4. Consider the statement I and II.
I. There is no 1-1 homomorphism from G to Z_8.
II. There is not onto homomorphism from $Z_8 \to G$. Then
(a) I is true (b)I is false (c)II is true (d)II is false
(b)

Answer Key

1	2	3	4	5	6	7	8	9	10
c	d	b,c,d	b,c	a,b,c	d	a	a,b,c	c	a,c,d
11	**12**	**13**	**14**						
b,c	c	a	a,c						

MODERN ALGEBRA

Chapter 4

Some Important Groups&Direct Products

1. External Direct Product :

Let $(G_1, \circ_1), (G_2, \circ_2), \ldots(G_n, \circ_n)$ be n groups. Then external direct product $G_1 \times G_2 \times \ldots \times G_n$ or $G_1 \oplus G_2 \oplus \ldots \oplus G_n$ is a group with respect to the binary operation '$\circ$' defined by $(a_1, a_2, \ldots, a_n) \circ (b_1, b_2, \ldots, b_n) = (a_1 \circ_1 b_1, a_2 \circ_2 b_2, \ldots, a_n \circ_n b_n)$

2. Internal Direct Product :

A group G is an internal direct product of its subgroups H and K if and only if

(i) elements of H and K commute.

(ii) every element of G can be expressed uniquely in the form hk, where $h \in H$ and $k \in K$.

Note : If a group G is the interval direct product of its subgroups H and K, then G is isomorphic $H \times K$.

Important Results :

(1) Order of any elements $(g_1, g_2, \ldots, g_n)$ in $G_1 \times G_2 \times \ldots \times G_n = \text{lcm}(O(g_1), O(g_2), \ldots, O(g_n))$ where $g_i \in G_i$, $1 \le i \le n$.

(2) If G is the internal direct product of its subgroups $H_1, H_2, \ldots, H_n$, then

(i) G is abelian if and only if each H_i is abelian, for all $i = 1, 2, \ldots, n$.

(ii) G is finite if and only if each H_i is finite, for all $i = 1, 2, \ldots, n$.

(iii) If $H_1, H_2, \ldots, H_n$ are cyclic, then G need not be cyclic.

(3) Let $G_1, G_2, \ldots, G_n$ be n groups. Then

(i) $G_1 \times G_2 \times \ldots \times G_n$ is finite if and only if $G_1, G_2, \ldots, G_n$ are finite. Moreover, $O(G_1 \times G_2 \times \ldots \times G_n) = O(G_1).O(G_2) \ldots .O(G_n)$

(ii) $G_1 \times G_2 \times \ldots \times G_n$ is abelian if and only if $G_1, G_2, \ldots, G_n$ are abelian.

(iii) If G_1, G_2 are cyclic, then $G_1 \times G_2$ need not be cyclic.

Example : $Z_2 \times Z_2$ is a non-cyclic abelian group.

(4) $Z(G_1 \times G_2 \times \ldots \times G_n)$ is isomorphic to $Z(G_1) \times Z(G_2) \times \ldots \times Z(G_n)$.

(5) $O(Z(G_1 \times \ldots \times G_n) = O[Z(G_1)] \times \ldots \times O[Z(G_n)]$

(6) $Z_m \times Z_n$ is a cyclic group of order mn iff $\gcd(m, n) = 1$

3. Fundamental theorem of finite abelian groups

Every finite abelian group is a direct product of cyclic groups of prime-power order. Moreover, the factorization is unique except for rearrangement of the factors.

Some examples are given in the table below, where G is a group of prime power order.

MODERN ALGEBRA

Example.

Order of G.	Partitions of O(G)	Possible direct products for G.
P	1	Z_p
p^2	2 1+1	Z_{p^2} $Z_p \times Z_p$
p^3	3 2+1 1+1+1	Z_{p^3} $Z_{p^2} \times Z_p$ $Z_p \times Z_p \times Z_p$
p^4	4 3+1 2+2 2+1+1 1+1+1+1	Z_{p^4} $Z_{p^3} \times Z_p$ $Z_{p^2} \times Z_{p^2}$ $Z_{p^2} \times Z_p \times Z_p$ $Z_p \times Z_p \times Z_p \times Z_p$

Example. If O(G) = 1176 = $2^3 \times 3 \times 7^2$, then the possible direct products for G are as follows:

$Z_8 \times Z_3 \times Z_{49}$

$Z_4 \times Z_2 \times Z_3 \times Z_{49}$

$Z_2 \times Z_2 \times Z_2 \times Z_3 \times Z_{49}$

$Z_8 \times Z_3 \times Z_7 \times Z_7$

$Z_4 \times Z_2 \times Z_3 \times Z_7 \times Z_7$

$Z_2 \times Z_2 \times Z_2 \times Z_3 \times Z_7 \times Z_7$

Note : Number of non-isomorphic abelian group of order of G = $p^m q^n r^s$ are $p(m) \times p(n) \times p(s)$ Where, p(k) is partition of k.

4. Some important groups

(1) Klein – four group

K_4 = {e, a, b, ab} is a non-cyclic abelian group of order 4.

Results : (i) Every subgoup of K_4 is cyclic and normal.

(ii) Each non-identity elements of order 2.

(2) Quaternion Group :

Q_8 = {+1, -1, +i, -i, +j, -j, +k, -k} is a quaternion group of order 8.

Results:

(i) Q_8 is non-abelian.

(ii) Q_8 is a non cyclic group whose all proper subgroups are cyclic.

(iii) Q_8 is non abelian group whose all subgroups are normal in Q_8.

(iv) Class equation

$8 = 1 + 1 + 2 + 2 + 2$

MODERN ALGEBRA

(3) Dihedral Group

$D_n = \{x, y \mid x_n = 1, y^2 = 1, yxy = x^{-1}\}$ is called dihedral group of order 2n.

Important Results on Dihedral Group :

(1) Number of elements of order 2 in $D_n = \begin{cases} n+1 & \text{if } n \text{ is even} \\ n & \text{if } n \text{ is odd} \end{cases}$.

(2) If $d \mid n$ and $d \neq 2$, then the number of elements of order d in $D_n = \phi(d)$.

(3) Number of subgroups in $D_n = \tau(n) + \sigma(n)$, where $\sigma(n)$ denotes the sum of positive divisors of n.

(4) If p is an odd prime, then any group of order 2p is either cyclic or dihedral.

(5) Let p and q be primes such that $p > q$ and G be a group of order pq. Then either G is cyclic or $G = \{x, y \mid x^p = 1, y^q = 1, yxy^{-1} = x^m\}$ where $m^q \equiv 1 \pmod p$ but $m \equiv 1 \pmod p$.

(6) Number of conjugacy classes in $D_n = \begin{cases} \dfrac{n+6}{2}, & \text{if } n \text{ is even} \\ \dfrac{n+3}{2}, & \text{if } n \text{ is odd} \end{cases}$

(7) $O[Z(D_n)] = \begin{cases} 2 & \text{if } n \text{ is even} \\ 1 & \text{if } n \text{ is odd} \end{cases}$

(8) Class equation of D_4 $\quad 8 = 1 + 1 + 2 + 2 + 2$

(9) Class equation of D_5 $\quad 10 = 1 + 5 + 2 + 2$

(4) Group of units

Let $n > 1$ be a positive integer. Then $U(n) = \{x \mid 1 \leq x \leq n, \gcd(x, n) = 1\}$ is called the group of units of n. $O(U(n)) = \phi(n)$.

Note : If $k \mid n$, then $U_k(n)$ is subgroup of U(n) where $U_k(n) = \{x \in U(n) \mid x \equiv 1 \pmod k\}$

Results on Group of Units of n :

(1) U(n) is abelian for every n.

(2) Let $n = n_1 n_2 \ldots n_k$, where $\gcd(n_i, n_j) = 1$ for $i \neq j$. Then $U(n) \cong U(n_1) \oplus U(n_2) \oplus \ldots \oplus U(n_k)$.

(3) If U(n) is cyclic, then the number of generators of $U(n) = \phi(\phi(n))$.

(4) $Aut(Z_n) \cong U(n)$. Also, if G is any finite cyclic group of order n, then $Aut(G) \cong U(n)$.

(5) The set of all automorphisms from a cyclic group need not be cyclic.

For example : If $Z_8 \to Z_8$, then $Aut(Z_8) \cong U(8) \cong Z_2 \times Z_2$, but $Z_2 \times Z_2$ is not cyclic.

(6) Let G_1 and G_2 be two groups. Then $Aut(G_1) \cong Aut(G_2) \not\Rightarrow G_1 \cong G_2$.

For example : Let $G_1 = Z_{10}$, $G_2 = Z_5$. Then $Aut(Z_{10}) \cong Z_4$ and $Aut(Z_5) \cong Z_4$.

So $Aut(Z_{10}) \cong Aut(Z_5)$ but $Z_{10} \ncong Z_5$.

MODERN ALGEBRA

Counting results			
1.	$U(2) \cong \langle 0 \rangle$	5.	U(2p) is cyclic for all odd prime p
2.	$U(2^2) \cong Z_2$	6.	U(p) is cyclic for all prime p
3.	$U(2^n) \cong Z_2 \oplus Z_{2^{n-2}} \forall\ n \geq 3$	7.	$U(2^n)$ is not cyclic for $n \geq 3$
4.	$U(p^n) \cong Z_{p^n - p^{n-1}}$ for p an odd prime		

Note : Centres of Some Important Groups :

(a) $Z(Z_n) = Z_n$

$Z(Z) = Z$

Since all the above groups under addition are abelian, their center is the group itself.

(b) If $G = GL_n(R)$ for $n > 1$, then $\left\{ A = \begin{bmatrix} a & 0 & \cdots & 0 \\ 0 & a & & \vdots \\ \vdots & & \ddots & 0 \\ 0 & \cdots & 0 & a \end{bmatrix}_{n\times n} \middle|\ 0 \neq a \in R \right\}$

and $O(Z(GL_n(R))) = \infty$

(c) If $G = GL_n(Z_p)$, p-prime for $n > 1$, then $\left\{ A = \begin{bmatrix} a & 0 & \cdots & 0 \\ 0 & a & & \vdots \\ \vdots & & \ddots & 0 \\ 0 & \cdots & 0 & a \end{bmatrix}_{n\times n} \middle|\ 0 \neq a \in Z_p \right\}$,

Where $Z_p = \{0, 1, 2, \ldots, p-1\}$ and $O[Z(GL_n(Z_p))] = p - 1$

(d) If $G = SL_n(R)$, for $n > 1$, then $Z(G) = \left\{ A = \begin{bmatrix} a & 0 & \cdots & 0 \\ 0 & a & & \vdots \\ \vdots & & \ddots & 0 \\ 0 & \cdots & 0 & a \end{bmatrix}_{n\times n} \middle|\ 0 \neq a \in R,\ \det A = 1 \right\}$

and $O[Z(SL_n(R))] = \begin{cases} 1 & \text{if } n \text{ is odd} \\ 2 & \text{if } n \text{ is even} \end{cases}$.

(e) If $G = SL_n(Z_p)$, p-prime, for $n > 1$, then

$$Z(G) = \left\{ A = \begin{bmatrix} a & 0 & \cdots & 0 \\ 0 & a & & \vdots \\ \vdots & & \ddots & 0 \\ 0 & \cdots & 0 & a \end{bmatrix}_{n\times n} \middle|\ 0 \neq a \in Z_p, \det A = 1 \right\}$$

and $O[Z(SL_n(Z_p))] = \gcd(n, p-1)$.

(f) If $G = SL_n(C)$ for $n > 1$, then $Z(G) = \left\{ A = \begin{bmatrix} a & 0 & \cdots & 0 \\ 0 & a & & \vdots \\ \vdots & & \ddots & 0 \\ 0 & \cdots & 0 & a \end{bmatrix}_{n\times n} \middle| \; 0 \neq a \in C, \det A = 1 \right\}$

and $O[Z(SL_n(C))] = n$.

(g) If $G = Q_8 = \{\pm 1, \pm i, \pm j, \pm k\}$, the Quaternion group of order 8, then $O(Z(Q_8)] = 2$.

Note : If $G = Z_m \times Z_n$, gcd (m, n) ≠ 1, then to find the number of elements of any order (say k), we take pair (a, b) such that a|m and b|n and lcm (a, b) = k, then we find the number of elements in Z_m corresponding to first tuple entry, (i.e., 'a') and number of elements in Zn corresponding to second tuple entry, i.e. 'b', in the pair, and we multiply them. In the end, we add the multiplies to get the required number of elements of order k.

S.No.	Examples of Groups	Number of elements
1.	Number of elements of order 2 in $G = Z_2 \times Z_2$.	Firstly, we take pairs (a, b) such that a|2 and b|2 and lcm(a, b) = 2 In $Z_2 \times Z_2$ Possibilities are (1, 2), (2, 1) and (2, 2) Number of elements Corresponding to (1, 2) = $\phi(1) \times \phi(2) = 1 \times 1 = 1$ Corresponding to (2, 1) = $\phi(2) \times \phi(1) = 1 \times 1 = 1$ Corresponding to (2, 2) = $\phi(2) \times \phi(2) = 1 \times 1 = 1$ Thus, number of elements of order 2 in $Z_2 \times Z_2$ is 3.
2.	Number of elements of order 5 in $G = Z_5 \times Z_{10}$	Firstly we take pairs (a, b) such that a | 5 and b | 10 and lcm(a, b) = 5 In $Z_5 \times Z_{10}$, possibilities are (5, 1), (5, 5) , (1, 5) Number of elements Corresponding to (5, 1) = $\phi(5) \times \phi(10 = 4 \times 1 = 4$ Corresponding to (1, 5) = $\phi(1) \times \phi(5) = 1 \times 4 = 4$ Corresponding to (5, 5) = $\phi(5) \times \phi(5) = 4 \times 4 = 16$ Thus, number of elements of order 5 in G is 24.
3.	Number of elements of order 2 in $G = Z_2 \times Z_2 \times Z_2$	We take triplet's (a, b, c) such that a|2, b|2 and c|2 and lcm (a, b, c) = 2 In $Z_2 \times Z_2 \times Z_2$, possibilities are (1, 2, 1), (1, 1, 2), (2, 1, 1), (2, 2, 1), (2, 1, 2), (1, 2, 2), (2, 2, 2) Number of elements Corresponding to (1, 2, 1) = $\phi(1) \times \phi(2) \times \phi(1) = 1$ Corresponding to (1, 1, 2) = $\phi(1) \times \phi(1) \times \phi(2) = 1$ Corresponding to (2, 1, 1) = $\phi(2) \times \phi(1) \times \phi(1) = 1$

MODERN ALGEBRA

		Corresponding to (2, 2, 1) = $\phi(2) \times\phi(2) \times\phi(1) = 1$ Corresponding to (1, 2, 2) = $\phi(1) \times\phi(2) \times\phi(2) = 1$ Corresponding to (2, 1, 2) = $\phi(2) \times\phi(1) \times\phi(2) = 1$ Corresponding to (2, 2, 2) = $\phi(2) \times\phi(2) \times\phi(2) = 1$ Thus, the number of elements of order 2 in $G = Z_2\times Z_2\times Z_2$ is 7.
4.	Number of elements of order 2 in $G = D_4\times Z_2$, where D_n is a dihedral group of order 2n.	Firstly, we take pairs (a, b) such that a|8 and b|2 and lcm(a, b) = 2 In $D_4\times Z_2$ the possibilities are (1, 2), (2, 1) and (2, 2) Number of elements Corresponding to (1, 2) = $1 \times\phi(2) = 1$ Corresponding to (2, 1) = $(4 + 1) \times\phi(1) = 5$ Corresponding to (2, 2) = $(4 + 1) \times\phi(2) = 5$ [$\because$ number of elements of order 2 in D_4 is 4 + 1 = 5 and number of elements of order 2 in Z_2 is $\phi(2) = 1$] Thus, the number of elements of order 2 in $D_4\times Z_2$ are 11.
5.	Number of elements of order 2 and 4 in $G = D_3\times Q_8$, where D_n is dihedral group of order 2n.	Firstly, we take pairs (a, b) such that a|6 and b|8 and lcm(a, b) = 2. In $D_3\times Q_8$, possibilities are (1, 2), (2, 1) and (2, 2) Number of elements Corresponding to (2, 1) = $3 \times 1 = 3$ Corresponding to (1, 2) = $1 \times 1 = 1$ Corresponding to (2, 2) = $3 \times 1 = 3$ Thus in G, there are 7 elements of order 2. Now we find number of elements of order 4. In $D_3\times Q_8$, possibilities are (1, 4), (2, 4) Corresponding to (1, 4) = $1 \times 6 = 6$ Corresponding to (2, 4) = $3 \times 6 = 18$ Thus, number of elements of order 4 in $G = D_3\times Q_8$ is 24.

5. Symmetric Group S_n

(i) Permutations : A permutation on a set S is a one-one mapping from S into itself.

(ii) Cyclic permutation : Let S be a finite set. A permutation f of S is said to be a cyclic permutation or a cycle if there exist elements $i_1, i_2, \dots i_t$ in such that $f(i_1) = i_2$, $f(i_2) = i_3$, …., $f(i_{t-1}) = i_t$, $f(i_t) = i_1$ and for any $j \in S$ different $i_1, i_2, \dots, i_e$, $f(j) = j$. We denote f by the symbol $(i_1 i_2 \dots . i_t)$. This notation of f is called a one-row notation. Further t is called the length of the cycle f. A cycle of length t is also called t-cycle.

MODERN ALGEBRA

Example : Let s = (123) then f (1) = 2, f(2) = 3, f(3) = 1

⇒ (123) is a cyclic permutation

Length of cycle : number of elements in any cycle is called length of cycle.

Example : (123) is a 3 length cycle.

(iii) **Transposition :** A cycle of length 2 is called a transposition.

Example : (12) is a transposition.

(iv) **Disjoint permutations :** Two permutations f and g of a set S are said to be disjoint if they satisfy the following :

(i) For any $j \in S$, $f(j) \neq j \Rightarrow g(j) = j$

(ii) For any $j \in S$, $g(j) \neq j \Rightarrow f(j) = j$ i.e., if any element of S is moved by f then it is left fixed by g and if any element of S is moved by g then it is left fixed by f.

Example : f = (12) and g = (34)

(v) **Cycles of permutation :** Let S be any finite non-empty set and let f be any permutation of S, then for a given $s \in S$, there exists a positive integer k such that s, f(s), $f^2(s)$, ….., f^{k-1} (s) are all distinct and {s, f(s), …. $f^{k-1}(s)$} is an f-orbit of s. The cyclic permutation (s, f(s), …., $f^{k-1}(s)$) is called a cycle of f.

Important Results on permutations :

(1) Any non-identity permutation of a finite set can be expressed uniquely as a product of disjoint cycles, each of length> 1, except for the order in which the cycles are written.

(2) Every permutation of a finite set S having more than one element is a product of transpositions.

(vi) **Even and Odd Permutation:** A permutation f of a finite non-empty set S is said to be even or oddaccording as *f* is expressible as a product of even or odd number of transpositions.

Symmetric Group S_n

Definition: Let S=(1,2,3,…, n) be a finite set, then S = {f|f : S → S, f is one— one and onto) forms a group, under the composition of mappings. In other words, S_n= Set of all permutations on set of n-symbols under the composition of mappings forms a group called the symmetric group. Order of symmetric group S_n = n!.

Important Results on Symmetric Group S_n :

1. S_n is non-abelian group for $n \geq 3$.
2. The symmetric group S_n may be generated by (n – 1) transpositions (12), (13), …., (1n).
3. If $f \in S_n$ such that $f = f_1 f_2 ... f_1$, where $f_1, f_2, …., f_t$ are disjoint cycles of length $r_1, r_2,…., r_t$ respectively.

   Then $O(f) = \text{lcm}(r_1, r_2, ..., r_t)$

Note : Any permutation can be written as a product of transpositions. If σ is expressed as a product of r transpositions. define sign(σ)=$(-1)^r$.

4. Let $\alpha \equiv S_n$

(i) α is an even permutation iff sgn(α) = 1.

(ii) α is an odd permutation if and only if sgn(α) = -1.

MODERN ALGEBRA

5. The number of distinct cycles of length k in S_n is $\frac{n!}{k(n-k)!}$

6. The number of distinct permutations which are product of n_i cycles of length l_i is $\frac{n!}{\prod_i n_i! \prod_i l_i^{n_i}}$ $i = 1,2,3,....$

7. **Cayley Theorem:** Every finite group is isomorphic to a permutation group.

Example. The maximum order of a permutation a in the symmetric group S_{10} is

Solution: Maximum order of a permutation in S_{10} is 30 as it contains permutation σ= (12) (345) (678910), which is clearly of order 30.

6. Alternating Group A_n

Definition: Let S_n be a symmetric group, then $A_n = \{\sigma \in S_n \mid \sigma \text{ is even permutation}\}$ forms a group of order $\frac{n!}{2}, A_n$ is called the Alternating group. Moreover A_n is subgroup of S_n.

Important Results on A_n

1. Since $[S_n: A_n] = 2$, so A_n is normal subgroup of S_n.
2. A_4 has no subgroup of order six.
3. Every element of A_n is a product of 3 cycles.
4. If a normal subgroup H of A_n contains a cycle of length 3, then $H = A_n$.

SOLVED QUESTIONS

1. Suppose s_n denote the symmetric group of n symbol. Let $\Delta = \{(\sigma, \sigma); \sigma \in s_n\} \subseteq s_n \times s_n$. The which of the following is/are true? **CSIR NET JUNE 2023**

(a) $f : s_n \times s_n \to s_n$ defined by $f(a,b) = a.b$ is a group homon orphism.

(b) Δ is a subgroup of $s_n \times s_n$ (c) Δ is a normal subgroup of $s_n \times s_n$

(d) Δ is a normal subgroup of $s_n \times s_n$ if n is prime no.

Ans. (b), (c)

Explanation:

(a) Let a, b, c, d, $\in s_n$

and (a, b), (c, d) $\in s_n \times s_n$ $f[(a,b),(c,d)] = f[(ac,bd)] = acbd$

& $f(a,b)\ f(c,d) = (a,b).(c,d) = abcd$

which is not true.

(b) Let $(\sigma, \sigma) \in \Delta$ $(\tau, \tau) \in \Delta$

Now $(\sigma, \sigma)(\tau^{-1}, \tau^{-1}) = (\sigma\tau^{-1}, \sigma\tau^{-1}) \in \Delta$

Hence Δ is subgroup

(c) Let $(\rho, \tau) \in s_n \times s_n$ then $(\rho^{-1}, \tau^{-1}) \in s_n \times s_n$

and $(s, s) \angle \Delta$

Now $(\rho,\tau)(\sigma,\sigma)(\rho^{-1},\tau^{-1}) = (\rho\sigma, \tau\sigma)(\rho^{-1},\tau^{-1}) = (\rho\sigma\rho^{-1}, \tau\sigma\tau^{-1}) \notin \Delta$

Hence Δ is not normal subgroup.

2. Let H denote the group of all 2 × 2 invertible matrices over Z_5 under usual matrix multiplication. Then the order of the matrix $\begin{bmatrix} 2 & 3 \\ 1 & 2 \end{bmatrix}$ in H is **IIT JAM-2016**

(a) 3 (b) 4 (c) 5 (d) 6

Ans. **(a)**

Explanation :

Let $A = \begin{bmatrix} 2 & 3 \\ 1 & 2 \end{bmatrix} \Rightarrow A^2 = \begin{bmatrix} 2 & 3 \\ 1 & 2 \end{bmatrix}\begin{bmatrix} 2 & 3 \\ 1 & 2 \end{bmatrix} = \begin{bmatrix} 7 & 12 \\ 4 & 7 \end{bmatrix} = \begin{bmatrix} 2 & 2 \\ 4 & 2 \end{bmatrix}$

Thus $A^3 = \begin{bmatrix} 2 & 3 \\ 1 & 2 \end{bmatrix}\begin{bmatrix} 2 & 2 \\ 4 & 2 \end{bmatrix} = \begin{bmatrix} 16 & 10 \\ 10 & 6 \end{bmatrix} = \begin{bmatrix} 1 & 0 \\ 0 & 1 \end{bmatrix}$

Hence, $A^3 = I$

∴ The order of $\begin{bmatrix} 2 & 3 \\ 1 & 2 \end{bmatrix}$ in H is 3.

3. Which of the following numbers can be orders of permutations σ of 11 symbols such that σ does not fix any symbol? **CSIR NET DEC-2011**

(a) 18 (b) 30 (c) 15 (d) 28

Ans. **(a), (b), (c), (d)**

Explanation :

Given $\sigma \in S_{11}$ such that σ does not fix any of 11 symbols.

We have,

(ii) σ can be of type (9 cycle) (2 cycle) element ⇒ O(σ) = lcm (9, 2) = 18

(iv) σ can be of type (6 cycle) (5 cycle) element ⇒ O(σ) = lcm (6, 5) = 30

(i) σ can be of type (3 cycle) (3 cycle) (5 cycle) ⇒ O(σ) = lcm (3, 3, 5) = 15

(iii) σ can be of type (7 cycle) (4 cycle) element ⇒ O(σ) = lcm (7, 4) = 28

(In all types, none of symbols is fixed)

Order of $\sigma \in S_{11}$ can be 30, 28, 18, 15.

So, all options are correct.

4. Let S_7 denote the symmetric group of all permutations of the symbols {1, 2, 3, 4, 5, 6, 7}. Pick out the true statements : **NBHM-2012**

(a) S_7 has an element of order 10 (b) S_7 has an element of order 15

(c) The order of any element of S_7 is atmost 12 (d) None of these

Ans. **(a), (c)**

Explanation :

S_7 denote the symmetric group of all permutation of symbols {1, 2, 3, 4, 5, 6, 7}

Option (a) is correct

As elements of order 10 in S_7 are (1 2 3 4 5) (6 7)

(2 3 4 5 6) (1 7)

MODERN ALGEBRA

(3 4 5 6 7) (1 2) etc.

For option (b),

In S_7, the maximum order of elements is 12.

$\Rightarrow$ Option (b) is incorrect and (c) is correct.

5. The number of elements of order 3 in symmetric group S_6 is **GATE – 2020**

(a) 81 (b) 80 (c) 100 (d) 101

Ans. (b)

Explanation :

Cycles of order 3 in S_6 are type of (abc) and (abc)(def).

Number of cycle of type (abc) are 40 and number of cycles of type (abc)(def) are 40.

So total 80 elements of order 3 in S_6.

6. Number of elements of order 8 in S_6.

(a) 0 (b) 1 (c) 1 (d) 4

Ans. (a) **GATE-2018**

Explanation :

We know that maximum order of elements in S_6 is 6.

7. Let S_n be the permutation group on n symbols and A_n be alternative group. which of the following is true?

(a) $\exists$ finite group which is not a subgroup of $S_n \forall n \geq 1$.

(b)Every finite group is a subgroup of A_n for some $n \geq 1$.

(c)Every finite group is quotient of A_n for some $n \geq 1$

(d)No finite abelian group is a quotient of S_n for $n > 3$ **CSIR NET DEC-2016**

Ans. (b)

Explanation :

Option (a)

Caley Theorem : Every group is isomorphic to subgroup of S_n.

Option (c)

Every finite group is quotient of A_n for some $n \geq 1$.

We know that

$$\text{Normal subgroupsof } A_n = \begin{cases} I, A_n & n \neq 4 \\ I, K_4, A_n & n = 4 \end{cases}.$$

So for n = 4

Only three groups are quotients of A_n other than only two groups are quotient of A_n.

$\Rightarrow \exists$ a finite group which is not a quotient of A_n.

Option (d)

For n = 4

$\exists$ K_4 subgroup of S_4 which is abelian and quotient of S_4.

Proof of option – (c)

By Caley Theorem

8. Let $\sigma = (1, 2, 3, 4, 5) \rightarrow (1, 2, 3, 4, 5)$ be a permutation s.t.

$\sigma^{-1}(j) \le \sigma(j)$ $\forall j, \quad 1 \le j \le 5$

Then which of the following are true? **CSIR NET - 2015**

(a) $\sigma \circ \sigma(j) = j$ (b) $\sigma^{-1}(j) = \sigma(j)$

(c) The set [k; $\sigma(k) \ne k$] has an even number of element

(d) The set [k; $\sigma(k) = k$] has an odd number of elements

Ans. (a), (b), (c)

Explanation :

Given that σ is element of S_5 Let $\sigma = (1\ 2\ 3) \Rightarrow \sigma^{-1} = (3\ 2\ 1)$

$\sigma^{-1}(3) = 2$ and $\sigma(3) = 1$ $\sigma^{-1}(3) \ge \sigma(3)$

i.e. $\sigma^{-1}(j) \le \sigma(j)$ condition satisfy when σ is of order 2.

Let $\sigma = (1\ 2)(3)(4)(5)$ or $\sigma = (1\ 2)(3\ 4)(5)$

Option – (4)

Set {k; $\sigma(k) = k$} = Number of self image element always odd number.

Option – (3) Set {k; $\sigma(k) \ne k$}

Always even number.

9. Let G be a non-abelian group. Let $\alpha \in G$ have order 4 and let $\beta \in G$ have order 3. Then the order of the element $\alpha\beta$ in G is

(a) is 6 (b) is 12 (c) is of the form 12k for $k \ge 2$ (d) need not be finite

Ans. (d)

Explanation :

Let G = S_7 and let $\alpha = (1234)$ and $\beta = (567)$

$\Rightarrow O(\alpha) = 4$ and $O(\beta) = 3$, then $\alpha\beta = (1234)(567)$

$\Rightarrow O(\alpha\beta) = \text{lcm}(O(\alpha), O(\beta)) = \text{lcm}(4, 3) = 12$

$\Rightarrow$ Options (a) and (c) are incorrect.

Now let G = S_5,

Take $\alpha = (2345)$ and $\beta = (12)(23) = (123)$, so that $O(\alpha) = 4$ and $O(\beta) = 3$

Now $\alpha\beta = (2345)(123) = (13)(245)$

$\therefore \alpha\beta$ is a product of two distinct cycles.

$\therefore O(\alpha,\beta) = \text{lcm}(O(\alpha), O(\beta)) = \text{lcm}(2, 3) = 6$ $\Rightarrow$ option (b) is incorrect.

As all other options are incorrect, we see that

Option (d) is correct.

In order to demonstrate the order of $\alpha\beta$ to be infinite, take $\alpha = \begin{bmatrix} 0 & 1 \\ -1 & 0 \end{bmatrix}$ and

$\beta = \begin{bmatrix} 0 & -1 \\ 1 & -1 \end{bmatrix}$ so that

$O(\alpha) = 4$, $O(\beta) = 3$ and $\alpha\beta = \begin{bmatrix} 1 & -1 \\ 0 & 1 \end{bmatrix}$ Here $O(\alpha\beta)$ = infinite $\left[\because (\alpha\beta)^n = \begin{bmatrix} 1 & -n \\ 0 & 1 \end{bmatrix}\right]$

Hence, option (d) is correct.

10. Let $G = S_3$ be the permutation group of 3 symbols. Then **CSIR NET JUNE 2018**

(a) G is isomorphic to a subgroup of a cyclic group.

(b) there exists a cyclic group H such that G maps homomorphically onto H

(c) G is a product of cyclic groups

(d) there exists a nontrivial group homomorphism from G to additive group (Q, +) of rational number.

Ans. **(b)**

Explanation :

For (a)

G is non-abelian group.

Option (a) is incorrect.

For (b) :

Let $f : S_3 \to Z_2$ and $\ker(f) = A_3$ then by FTH

$$\frac{S_3}{A_3} \cong Z_2 = f(S_3)$$

Option (b) is correct.

For (c) :

Product of cyclic group is cyclic.

Option (c) is incorrect.

Option (d) is also incorrect.

UNSOLVED QUESTIONS

1. How many proper subgroups does the group $Z \oplus Z$ have? **TIFR-2014**

(a) 1 (b) 2 (c) 3 (d) infinitely many

2. Let S_7 denote the group permutations of the set {1, 2, 3, 4, 5, 6, 7}. Which of the following is true? **CSIR NET JUNE-2018**

(a) There are no elements of order 6 in S_7 (b) There are no elements of order 7 in S_7

(c) There are no elements of order 8 in S_7 (d) There are no elements of order 10 in S_7

3. Let S_7 denote the symmetric group of all permutations of the symbols {1 2, 3, 4, 5, 6, 7}. Pick out the true statements : **NBHM - 2012**

(a) S_7 has an element of order 10 (b) S_7 has an element of order 15

(c) The order of any element of S_7 is at most 12 (d) None of these

4. In the permutation group S_6, the number of elements of order 8 is **GATE – 2018**

(a) 0 (b) 1 (c) 2 (d) 4

5. Which one of the following statements is true? **GATE – 2018**

(a) Every group of order 12 has a non-trivial proper normal subgroup.

(b) Some group of order 12 does not have a non-trivial proper normal subgroup.

(c) Every group of order 12 has a subgroup of order 6.

(d) Every group of order 12 has an element of order 12.

6. Find the number of elements of order two in the symmetric group S_4 of all permutations of the four symbols {1, 2, 3, 4}. **NBHM-2013**

(a) 8 (b) 9 (c) 10 (d) 11

MODERN ALGEBRA

7. Let S_n denote the symmetric group on n symbols. The group $S_3 \oplus (Z/2Z)$ is isomorphic to which of the following groups? **CSIR NET JUNE-2013**
(a) $Z/12Z$ (b) $(Z/6Z) \oplus (Z/2Z)$
(c) A_4, the alternating group of order 12 (d) D_6, the dihedral group of order 12

8. Given the permutation $\sigma = \begin{pmatrix} 1 & 2 & 3 & 4 & 5 \\ 3 & 1 & 2 & 5 & 4 \end{pmatrix}$, the matrix A is defined to be the one whose i^{th} column is the $\sigma(i)^{th}$ column of the identity matrix I. Which of the following is correct? **CSIR NET JUNE-2014**
(a) $A = A^{-2}$ (b) $A = A^{-4}$ (c) $A = A^{-5}$ (d) $A = A^{-1}$

9. Let a_n denote the number of those permutations σ on {1, 2, …., n} such that σ is a product of exactly two disjoint cycles. Then **CSIRNET DEC-2015**
(a) $a_5 = 50$ (b) $a_4 = 14$ (c) $a_5 = 40$ (d) $a_4 = 11$

10. Let S_9 be the group of all permutations of the set {1, 2, 3, 4, 5, 6, 7, 8, 9}. Then the total number of elements in S_9 that commute with τ = (1 2 3)(4 5 6 7) in S_9 equals **GATE – 2017**
(a) 21 (b) 22 (c) 24 (d) 25

11. Let S_n be the symmetric group on n symbols and A_n be the alternating subgroup of S_n, then
(a) A_n is normal subgroup of S_n (b) $[S_n : A_n] = 2$
(c) order of A_n , is n ! (d) A_n is the set of all odd permutations of S_n

12. Given the symmetric group S_n on four symbols. Then
(a) The cyclic group of order 2 is {(1 2), (1) (2) (3) (4) = e}
(b) The cyclic group of order 3 is {(2 3 4), (2, 4, 3), (1), (2), (3), (4) = e}
(c) The cyclic group of order 4 is {(1 3 2 4), (1 2), (3 4), (1 4 2 3), e}
(d) No cyclic group exist.

13. Consider the element $\alpha = \begin{pmatrix} 1 & 2 & 3 & 4 & 5 \\ 2 & 1 & 4 & 5 & 3 \end{pmatrix}$ of the symmetric group S_5 on five elements.
Pick out the true statements :
(a) The order of α is 5 (b) α is conjugate to $\begin{pmatrix} 4 & 5 & 2 & 3 & 1 \\ 5 & 4 & 3 & 1 & 2 \end{pmatrix}$
(c) α is the product of two disjoint cycles (d) α commutes with all elements of S_3.

Answer Key

1	2	3	4	5	6	7	8	9	10
d	c	a,c	a	a	b	d	c	a,d	c
11	**12**	**13**							
a,b	a,b,c	b,c							

MODERN ALGEBRA

Chapter 5

Sylow's Theorem

1. Simple Group :

A group which has no proper normal subgroup is called a simple group.

Example : (i) Z_p is a simple group iff p is prime.

(ii) A_5 is smallest non-abelian simple group.

2. p-subgroup :

Let p be a prime number. Then a subgroup H of a group G is called a p-subgroup if the order of each element of H is a power of p.

In particular, if each element of G is of order of a power of some fixed prime number p, then G is called a p-group.

Let G be a finite group. If $O(G) = p^k$, then as by Lagrange's theorem, for each $a \in G$. $O(a) \mid p^k$, thus O(a) is a power of p. Consequently G is a p-group. Conversely, if G is a p-group, then no prime $q \neq p$ can divide O(G), since otherwise G will contain an element of order q, some prime number different from p (Cauchy's theorem, refer point 3 of results given below). Consequently, if G is a p-group, then $O(G) = p^k$ is a power of p. Thus G is a p-group if and only if O(G) is a power of p.

Sylow p-subgroup of a finite group : Let G be a finite group and p, a prime number, such that $p^k \mid O(G)$ and $p^{k+1} \nmid O(G)$. Then any subgroup of G of order p^k is called a Sylow p-subgroup of G.

The above definition implies that all Sylow p-subgroup of G are of same order, viz, p^k.

Some Important Results :

(1) **Cauchy's Theorem for Abelian groups :** Suppose G is a finite Abelian group and $p|O(G)$, where p is a prime number. Then there exists $a(\neq e) \in G$ such that $a^p = e$. If G is finite abelian group and a positive integer k divides O(G), then G contains a subgroup of order k.

(2) **Sylow's First Theorem :** Let G be infinite group of order $n = p^k q (k \geq 1)$, where p& q are prime number such that gcd(p, q) = 1. Then for each i, $(1 \leq i \leq k)$, G has at least one subgroup of order p^i.

(3) **Cauchy's Theorem for finite groups :** If a prime number p divides the order of a finite group G, then G contains at least one element of order p.

(4) Let G be a finite group and p be a prime number that divide order of G. Then every p-subgroup of G is contained in some Sylow p-subgroup of G.

(5) **Sylow's Second Theorem :** Let G be a finite group and p be a prime number such that $p \mid O(G)$. Then all Sylow p-subgroup of G are conjugates of one another.

(6) **Sylow's Third Theorem :** Let G be a finite group and p be a prime number such that $p \mid O(G)$. Then the number of Sylow p-subgroups is of the form 1+mp, where m is some non-negative integer such that $(1 + mp) \mid O(G)$.

(7) If p is smallest prime divisor of the order of a group G and H is a subgroup of G of index p, then H is normal in G.

(8) **Frobenius Theorem:** Let G be a group of order $p^n m$, $m \geq 1$, p be a prime and gcd(p,m)=1. If p^r divides O(G), then G has a subgroup of order p^r. The number of such subgroups is congruent to 1 modulo p.

(9) A Sylow p-subgroup of a finite group G is a normal subgroup of G iff it is the only Sylow p-subgroup of G.

(10) Let O(G) = pq, where p,q are distinct primes, p<q, p does not divide q– 1. Then G is cyclic.

3. Counting of Subgroups:

(1) If O(G) = pq , p <q and G is non-abelian, then
- (i) number of subgroups of order q in G = 1
- (ii) number of elements of order q in G = $\phi(q)$
- (iii) number of subgroups of order p in G = q
- (iv) number of elements of order p in G = $q.\phi(p)$

(2) If O(G) = pq, p <q, p|q-1 and G is abelian (then cyclic), then G has a unique subgroup of order 1, p, q andpq.

(3) If O(G) =pq, p <q, p|q-1 and G is non-abelian, then
- (i) number of subgroups of order p = q
- (ii) number of subgroups of order q = 1
- (iii) number of subgroups of order 1 = 1
- (iv) number of subgroups of order pq = 1

Example 1. If O(G) = 21 and G is non-abelian, then calculate the number of subgroups of order 3 and 7 in G

Solution : Since O(G)= 21 = 3 × 7, we observe that G has 7 subgroups of order 3 and G has 1 subgroup order 7.

Example 2. Show that a group of order 35 is cyclic.

Solution : O(G) = 35 = 5 × 7

Let P be Sylow 7 – subgroup. Number of7Sylow subgroups = 1 + 7ksuch that 1+ 7k|5 so k =0.

$\therefore$ Number of Sylow 7 – subgroup is one so P is normal in G.

Let Q be Sylow 5-subgroup of G.

Number of Sylow 5-subgroup of G = $1 + 5\ell$ such that $1 + 5\ell \mid 7$

$\Rightarrow$ Number of Sylow 5-subgroups is 1. So 'Q' is normal in G.

Let P = <x> and Q = <y> and O(x) = 7 and O(y) = 5 [$\therefore$ Every group of prime order is cyclic].

Now $x^{-1}(y^{-1}xy) \in P$ and $(x^{-1}y^{-1}x)\, y \in Q \Rightarrow x^{-1}y^{-1}xy = e \Rightarrow xy = yx$

Since O(x) = 7 and O(y) = 5 and gcd(7, 5) = 1 and xy = yx, so O(xy) = 35 = O(G).

So, G is generated by $xy \in G \Rightarrow$ G is cyclic.

4. Results on simple groups

(1) Let G be a group of order p^2q, where p and q are distinct primes, then G has at least one normal Sylow subgroup. Hence, G is not Simple.

(2) If G is a simple group of odd order, then $G \cong Z_p$ for some prime p.

(3) Ifp and q are distinct primes and G is a group of order p^2q^2, then G is not simple.

(4) **Sylow Test for Non-simplicity:** Let n be a positive integer that is not prime and let p be a prime divisor of n. If 1 is the only divisor of n that is congruent to 1 modulo p, then there does not exist a simple group of order n.
All simple groups of order less than 1000 are 60 (= A_5), 168, 360(=A_6), 504, 660.

(5) **Odd Test:** An integer of the form 2n, where n is an odd number greater than 1, is not the order of a simple group.

(6) **Index Theorem:** If G is a finite group and H is a proper subgroup of G such that O(G) does not divide [G : H]!, then H contains a non-trivial normal subgroup of G. In particular, G is not simple.

(7) If G is a finite non-abelian simple group and H is a subgroup of G, then $[G : H] \geq 5$.

(8) Suppose that G is a finite group and p is the smallest prime divisor of O(G), then a subgroup H of index p in G is normal in G and hence G is not simple.

(9) Let G be a group of order pqr, $p<q<r$ being primes. Then some Sylow subgroup of G is normal. Hence, G is not simple.

(10) If $n \leq 4$, then S has no non-abelian simple subgroup.

(11) **Galois Theorem:** $A_n$is simple for $n \geq 5$.

Example . Show that no group of order 30 is simple.

Solution : Let G be a group of order 30. Note that 30 = 2.3.5.
So G has at least a Sylow 2-subgroup, a Sylow 3-subgroup and a Sylow 5-subgroup.
Number of Sylow 5-subgroups is 1 + 5k such that $(1 + 5k)|6 \Rightarrow k = 0$ or 1.
If k=0, then Sylow 5-subgroup is a normal subgroup of G.
If k = 1, number of Sylow 5-subgroups is six. Let H_1, H_2, H_3, H_4, H_5, H_6 be the Sylow 5-subgroups.
Since $O(H_i \cap H_j) = 1$ for $i \neq j$, these six Sylow 5-subgroups contain 24 elements of order 5 and there is one element of order 1.
Thus only 5 remaining elements of G can be of order 2 or 3.
Now, number of Sylow 3-subgroups of G is 1 + 3m such that $(1 + 3m) \mid 10$.
Then m = 0 or 3. If m = 0 we have only one Sylow 3-subgroup which is be normal.
If m = 3, we have 10 Sylow 3-subgroups. But these will contain 20 elements of order 3.
This is absurd in view of the arguments given above.
Hence, G has either a normal subgroup of order 5 or a normal subgroup of order 3.
Thus G is not simple.

SOLVED QUESTIONS

1. For any prime p, consider the group $G = GL_2(Z/PZ)$. Then which of the following are true? **CSIR NET 2019**

(a) G has an element of order p

(b) G has exactly one element of order p

(c) G has no p-Sylow subgroups

(d) Every elements of order p is conjugate to a matrix $\begin{pmatrix} 1 & a \\ 0 & 1 \end{pmatrix}$, where $a \in \left(\frac{Z}{pZ}\right)^*$.

Ans. **(a), (d)**

Explanation :

We have $G = GL_2(Z/pZ) = GL_2(Z_p) \Rightarrow o(G) = (p^2 - 1)(p^2 - p) = p(p^2 - 1)(p - 1)$

For (a) :

By cauchy's theorem (a) is true

Let $A = \begin{pmatrix} 1 & 1 \\ 0 & 1 \end{pmatrix} \in G$ then $A^2 = \begin{pmatrix} 1 & 2 \\ 0 & 1 \end{pmatrix}$ & $A^3 = \begin{pmatrix} 1 & 3 \\ 0 & 1 \end{pmatrix}$. $\Rightarrow$ $A^p = \begin{pmatrix} 1 & p \\ 0 & 1 \end{pmatrix} = \begin{pmatrix} 1 & 0 \\ 0 & 1 \end{pmatrix}$

So, order of O(A) = p Hence option (a) is correct.

For (b) : Let $A = \begin{pmatrix} 1 & a \\ 0 & 1 \end{pmatrix}; \gcd(a,p) = 1;$

Then O(A) = p for all a such that gcd(a, p) = 1 Hence option (b) is not correct.

For (c) :

Since there exist more than 1 element of order p.

Since order of $G = p(p-1)^2(p+1)$

Note that p divides O(G) so must have a p-Sylow subgroup.

Hence option (c) is incorrect.

For (d) :

Let $P = \left\langle \begin{pmatrix} 1 & 1 \\ 0 & 1 \end{pmatrix} \right\rangle$, then P is a p-sylow subgroup of order p.

Now let $\begin{pmatrix} p & q \\ r & s \end{pmatrix}$ is any element of order p in G.

Then $H = \left\langle \begin{pmatrix} p & q \\ r & s \end{pmatrix} \right\rangle$ is p-sylow subgroup.

We know any two p-sylow subgroups are conjugate.

Thus H is conjugate to P.

So, there exists $\begin{pmatrix} p & q \\ r & s \end{pmatrix} \in H$ is conjugate to some power of $\begin{pmatrix} 1 & 1 \\ 0 & 1 \end{pmatrix}$, .

Let $\begin{pmatrix} p & q \\ r & s \end{pmatrix} \sim \begin{pmatrix} 1 & 1 \\ 0 & 1 \end{pmatrix}^a$ for some $a \in \left(\frac{Z}{pZ}\right)^*$. $\Rightarrow$ $\begin{pmatrix} p & q \\ r & s \end{pmatrix} \sim \begin{pmatrix} 1 & a \\ 0 & 1 \end{pmatrix}$ for some $a \in \left(\frac{Z}{pZ}\right)^*$.

Hence option (d) is correct.

2. How many 3-Sylow subgroups does the symmetric group S_4 have? **CSIR NET 2019**

(a) 8 (b) 4 (c) 2 (d) 1

Ans. (b)

Explanation :

$o(S_4) = 4! = 2^3 \times 3$

So, possible number of 3-Sylow subgroups = $3k + 1 \mid 8$, k = 1, 2, 3 …. = 4

Hence option (d) is correct.

3. Let p be a prime number. The order of a Sylow-p subgroup of the group $GL_{50}(Z_p)$ of invertible 50 × 50 matrices with entries from Z_p, equals **CSIR NET DEC-2011**

(a) p^{50} (b) p^{125} (c) p^{1250} (d) p^{1225}

Ans. (d)

Explanation :

Number of p-SSG in $GLn(Z_P) = P^{\frac{n(n-1)}{2}}$.

$\Rightarrow$ Number of p-SSG in $GL_{50}(Z_P) = P^{\frac{50(50-1)}{2}} = P^{1225}$.

4. Let G be a simple group of order 60. Then **CSIR NET DEC-2015**

(a) G has six Sylow-5 subgroups (b) G has four Sylow-3 subgroups

(c) G has a cyclic subgroup of order 6 (d) G has a unique element of order 2

Ans. (a)

Explanation :

Given that G be a simple group of order 60. $\Rightarrow G \cong A_5$

We know that $O(G) = 60 = 2^2 \times 3 \times 5$

$\Rightarrow A_5$ has 3-SSG of order 3 and A_5 has 5-SSG of order 5.

(a) H be a 5-SSG of A_5

Then $H \cong Z_5$. Number of cyclic cubgroup of order 5.

$$= \frac{\text{Number of elements of order 5}}{\phi(5)} = \frac{24}{4} = 6 \quad \left[\because {}^5C_5 \mid \underline{4} = 24\right]$$

Option (a) is correct.

(b) K be a 3-SSG of order A_5. $K \cong Z_3$.

Number of cyclic subgroup of order 3 = $= \dfrac{\text{Number of elements of order 3}}{\phi(3)}$.

$$= \frac{20}{2} = 10$$

Option (b) is incorrect.

(c) A5 has no element of order 6. $\Rightarrow$ (c) is incorrect. (d) is incorrect.

5. Which of the following are incorrect?

(a) Every group of order 505 is cyclic. (b) Every group of order 51 is cyclic.

(c) Every group of order 151 is cyclic.
(d) Every group of order 30 in which 2 – SSG is unique is cyclic.
Ans. (a)
Explanation :
Let O(G) = p.q (p < q, p, q are co-prime) If $p \nmid q-1$ then G is cyclic
If $p \mid q-1$ then ∃ G which is cyclic and ∃ G which is non-abelian.
Option (a) is false statement. O(G) = 505 = 5 × 101 5 | 100
⇒ G need not be always a cyclic.
Option (b) is true statement. O(G) = 51 = 3 × 17
$3 \nmid 16$ ⇒ G is cyclic
Option (c) is true statement.
O(G) = 151
G is cyclic
Option (d) is true statement. O(G) = 2 × 3 × 5 = 30
Then order of 2-SSG is 2
⇒ 2-SSG is cyclic.
⇒ only (a) is false statement.
⇒ Option (a) is correct.

6. The number of 2 – SSG in SL(2, Z_3) is
(a)1 (b)4 (c)8 (d)16
Ans. (a)
Explanation :

$$O(SL(2,Z_3)) = \frac{(3^2-1)(3^2-3)}{3-1} = \frac{8\times 6}{2} = 24 = 2^3\times 3$$

Number of 2-SSG are 1 + 2k s.t. 1 + 2k | 3 ⇒ k = 0, k = 1
Either unique or 3, 2-SSG exist.
Option (a) is correct.

7. Let G = {A = $[a_{ij}]_{3\times 3}$: $a_{ij} \in Z_p$ & |A| ≠ 0} and $H = \left\{ A\in G : A = \begin{bmatrix} 1 & a & b \\ 0 & 1 & c \\ 0 & 0 & 1 \end{bmatrix} a,b,c\in Z_p \right\}$ then
(a) H is not p-subgroup (b) H is not p – SSG (c) H is p-SSG (d) None of these
Ans. (c)
Explanation :
Given that G ≈ GL(3, Z_p)
then $O(G) = (p^3-1)(p^3-p)(p^3-p^2)$ $= p^3(p^3-1)(p^2-1)(p-1)$
Here $p^3 \mid O(G)$ ⇒ $O(H) = p^3$ ⇒ H is p-SSG
Option (c) is correct.

8. Let G be a group of order 45. Then **CSIR UGC NET DEC-2014**
(a) G has an element of order 9
(b) G has a subgroup of order 9
(c) G has a normal subgroup of order 9
(d) G has a normal subgroup of order 5.
Ans. (b), (c), (d)
Explanation :
G be a group of order 45.

i.e., $O(G) = 45 = 3^2 \times 5$ such that $3 \nmid 5-1$ and $5 \nmid 3^2 - 1 \Rightarrow$ G is abelian

[G is a group and $|G| = p^2q$ such that $q \nmid p^2 - 1$ and $p \nmid q-1$, where p and q are distinct primes, then G is abelian.]

$\Rightarrow$ All the subgroups of G are normal subgroups and also for every divisor d of order of group G, $\exists$a subgroup of that order ($\because$ Converse of Lagrange theorem is true for finite abelian groups)

Since, 9|45 and 5|45

So, G has a subgroup of order '5' and order 9 and both are normal subgroups. So, option (b), (c) and (d) are correct. Since, G is group of order 45.

So, G can be $\cong Z_3 \times Z_3 \times Z_3$, but it has no element of order 9. So, option (a) is incorrect.

UNSOLVED QUESTIONS

1. Let G be a group of order 45. Let H be a Sylow-3 subgroup of G and K be a Sylow-5 subgroup of G. Then **GATE-2008**

(a) both H and K are normal in G. (b) H is normal in G but K is not normal in G

(c) H is not normal in G but K is normal in G (d) both H and K are not normal in G.

2. Let p be a prime number. Let $n \in N$, $n > 1$. The order of a p-Sylow subgroup of $GL_n(F_p)$ is

(a) $p^{\frac{n(n-1)}{2}}$ (b) $p^{\frac{n(n-1)}{3}}$ (c) $p^{\frac{n(n+1)}{2}}$ (d) None of these

NBHM-2019

3. Which of the following is a 5-Sylow subgroup of $GL_3(F_5)$? **NBHM-2019**

(a) $H = \left\{ \begin{bmatrix} 1 & p & 1 \\ 0 & 1 & r \\ 0 & 0 & 1 \end{bmatrix} : p,q,r \in Z_5 \right\}$ (b) $H = \left\{ \begin{bmatrix} 1 & p & q \\ 0 & 1 & r \\ 0 & 0 & 1 \end{bmatrix} : p,q,r \in Z_5 \right\}$

(c) $H = \left\{ \begin{bmatrix} 1 & 1 & 1 \\ 0 & 1 & r \\ 0 & 0 & 1 \end{bmatrix} : p,q,r \in Z_5 \right\}$ (d) $H = \left\{ \begin{bmatrix} 1 & p & q \\ 0 & 1 & r \\ 0 & 0 & 1 \end{bmatrix} : p,q,r \in Z_3 \right\}$

4. Let G be a group of order 231. The number of elements of order 11 in G is

(a) 0 (b) 2 (c) 5 (d)10

GATE-2013

5. Let G be a simple group of order 168. What is the number of subgroups of G of order 7? **CSIR NET JUNE-2013**

(a) 1 (b) 7 (c) 8 (d) 28

6. Let G be a group. Suppose $O(G) = p^2q$, where p and q are distinct prime numbers satisfying $q \equiv 1 \bmod p$. Which of the following is always true? **TIFR – 2015**
(a)G has more than one p-Sylow subgroup (b)G has a normal p-Sylow subgroup
(c)The number of q-Sylow subgroups of G is divisible by p
(d)G has a unique q-Sylow subgroup.

7. Let $S_3 = \{e, (1\ 2), (1\ 3), (2\ 3), (1\ 2\ 3), (1\ 3\ 2)\}$ be a symmetric group on three numbers 1, 2 and 3, then
(a) S_3 has 2 Sylow-2 subgroups (b) S_3 has 3 Sylow-2 subgroups
(c) S_3 has 1 Sylow-3 subgroup (d) S_3 has 2 Sylow-3 subgroups

8. Consider the symmetric group S_{20} and its subgroup A_{20} consisting of all even permutations. Let H be a Sylow-7 subgroup of A_{20}. Pick each correct statement from below. **CSIR NET JUNE-2016**
(a) $O(H) = 49$ (b) H must be cyclic (c) H is a normal subgroup of A_{20}
(d) Any Sylow-7 subgroup of S_{20} is a subset of A_{20}

9. For any prime 'p' consider the group $G = GL(2, Z_p)$
Then which of the following are true?
(a) G has an element of order p. (b) G has exactly one element of order p
(c) G has no p-SSG
(d) Every element of order p is conjugate to a matrix $\begin{bmatrix} 1 & a \\ 0 & 1 \end{bmatrix}$ where $a \in Z_p^*$

10. Let G be a simple group of order 168. What is the number of subgroups of G of order 7? **CSIR NET – JUNE 2013**
(a) 1 (b) 7 (c) 8 (d) 28

11. How many normal subgroups does a non-abelian group G of order 21 have other that the identity subgroup {e} and G? **CSIR NET – DEC 2013**
(a) 0 (b) 1 (c) 3 (d) 7

12. In the group of all invertible 4×4 matrices with entries in the field of 3 elements, any 3-Sylow subgroups has cardinality **CSIR NET – DEC 2014**
(a) 3 (b) 81 (c) 243 (d) 729

13. Let G be a group of order 125. Which of the following statements are necessarily true? **CSIR NET – JUNE 2017**
(a) G has a non-trivial abelian subgroup (b) The centre of G is a proper subgroup
(c) The centre of G has order 5 (d) There is a subgroup of order 25

Answer Key

1	2	3	4	5	6	7	8	9	10
a	a	b	d	c	b	b,c	a,d	a,d	c
11	**12**	**13**							
b	d	a,d							

MODERN ALGEBRA

Ring Structure and Properties

1. Ring Structure &Properties

Definition : A non-empty set R, together with two binary composition + and . is said to form a ring if the following axioms are satisfied :

(i) $a + b \in R$ and $a.b \in R$ whenever $a,b \in R$.

(ii) $(a + b) + c = a + (b + c) \quad \forall\, a, b, c \in R$

(iii) $a + b = b + a$ for $a, b \in R$

(iv) $\exists$ some element 0(called zero) in R s.t.
$a + 0 = 0 + a = a \quad \forall\, a \in R$

(v) For each $a \in R$, $\exists$ an element $(-a) \in R$ s.t.
$a + (-a) = 0 = (-a) + a$

(vi) $a . (b . c) = (a . b) . c \quad \forall\, a, b, c \in R$

(vii) $a . (b + c) = a.b + a.c$
$(b + c) . a = b.a + c.a \quad \forall\, a, b, c \in R$

i.e. (R, +) is abelian group and (R, .) is semi group.

Table to be remembered

S.No.	*Name of Ring*	*Addition operation*	*Multiplication operation*	*Unity*	*Commutative*
1	(C, +, .)	+	.	1	Yes
2	(R, +, .)	+	.	1	Yes
3	(Q, +, .)	+	.	1	Yes
4	(Z, +, .)	+	.	1	Yes
5.	(mZ, +, .) $m \neq \pm 1$	+	.	No	Yes
6	$R = \left\{\begin{bmatrix} a & b \\ c & d \end{bmatrix}; a,b,c,d \in R\right\}$ $ad - bc \neq 0$	Matrix Addition	Matrix Multiplication	$\begin{bmatrix} 1 & 0 \\ 0 & 1 \end{bmatrix}$	No
7.	$R = \left\{\begin{bmatrix} a & b \\ c & d \end{bmatrix}; a,b,c,d \in R\right\}$	Ordinary addition	(component) Ordinary multiplication	$\begin{bmatrix} 1 & 1 \\ 1 & 1 \end{bmatrix}$	Yes
8.	$R = \left\{\begin{bmatrix} a & a \\ a & a \end{bmatrix} a \in R\right\}$	Matrix Addition	Matrix multiplication	$\begin{bmatrix} \frac{1}{2} & \frac{1}{2} \\ \frac{1}{2} & \frac{1}{2} \end{bmatrix}$	Yes
9	$Z[i] = \{a + ib : a, b \in Z\}$	Ordinary addition of	Multiplication	1	Yes

		complex number			
10	$Z_m = \{0, 1, 2, \dots m-1\}$	Addition modulo M	Multi. Mod. M	1	Yes
11.	$R = \{0, 2, 4, 6, 8\}$	Addition mod 10	Multi Mod 10	6	Yes
12	$R=Q[\sqrt{p}]=\{a+\sqrt{p}b$: p is prime	Addition	Multiplication	1	Yes
13.	$R = \{a + ib : a, b \in Q\}$	Addition	Multiplication $a.b = 0 \; \forall \; a, b$	No	Yes
14	R = P(N) Power set of N	Symmetric difference	Intersection	N	Yes
15	(Z, *, .)	$a * b = a + b$	$a.b = a + b - ab$	0	Yes
16	R = Set of all real valued continuous functions defined on closed interval [0, 1]	$(f \,\&\, g)\, x = f(x) + g(x) \; \forall \; x \in [0, 1]$	$(f.g)(x) = f(x).g(x)$	$f(x) = 1, x \in [0, 1]$	Yes
17.	$Q[i] = \{a + ib\}$ $a,b \in Q$	Ordinary addition of complex number	Ordinary multiplications of complex number	1	Yes
18.	$R^S = \{f : s \to R\}$; $0 \neq S$ & R be a ring. Then R^S is ring.	$(f+g)(x) = f(x) + g(x)$	$(f.g)(x) = f(x) \,.\, g(x)$	No	No

Definition :

1. **Unity :** Multiplication identity in any ring is called unity. If $\exists$ e $\in$ R with a $\in$ R, a.e = e.a = a then e is said to be as unity of ring and it is denoted by 1.
2. **Trival ring :** A ring R is said to be a trival ring if R = {0}.
3. **Zero ring :** Let R be a ring then it is said to be zero ring if a.b = 0 (where 0 is additive identity) $\forall$ a, b $\in$ R
4. **Commutative Ring :** A ring R is said to be a commutative ring if a.b = b.a $\forall$ a, b $\in$ R
   Example No. 1, 2, 3, 4, 5 in above table
   Note :
   (i) Any ring of prime order is commutative ring.
   (ii) A ring of order p^2 may not be commutative
   Example : $R = \left\{ \begin{bmatrix} a & b \\ 0 & 0 \end{bmatrix} ; a,b \in Z_2 \right\}$
   The smallest non-commutative ring is of order 4.
   (iii) A ring with unity & of order p^2 then it is commutative.
5. **Unit :** Let (R, +, .) be any ring and a $\in$ R then a is said to be unit element of R, if $\exists$ b $\in$ R a.b = b.a = 1
   Example No. 1, 2, 3 of table, each non-zero elements are unit.

Example No. 6 . The set of all 2 × 2 order non-singular matrix is unit set.

Example⇒R = (Z_m, $\oplus_m$, Θ_m) under addition modulo m and multiplication modulo m

(i) Number of units in $Z_m = \phi(m)$

Let R = (Z_8, $\oplus_8$, Θ_8) then number of units in $Z_8 = \phi(8) = \phi(2^3) = 8\left(\frac{1}{2}\right) = 4$

i.e. {1, 3, 5, 7} is the set of units in Z_8.

6. **Zero divisor :** Let R be a ring and $0 \neq a \in R$ then a is said to be zero divisor if $\exists\ 0 \neq b \in R$ s.t. a.b = 0 = b.a

Example : In(Z_4, $\oplus_4$, Θ_4)

2.2 = 4 = 0

So 2 is a zero divisor.

Example No. (6) (table) $\begin{bmatrix} 0 & 1 \\ 0 & 0 \end{bmatrix}\begin{bmatrix} 1 & 0 \\ 0 & 0 \end{bmatrix} = \begin{bmatrix} 0 & 0 \\ 0 & 0 \end{bmatrix}$

both $\begin{bmatrix} 0 & 1 \\ 0 & 0 \end{bmatrix}$ & $\begin{bmatrix} 1 & 0 \\ 0 & 0 \end{bmatrix}$ are zero divisors.

Note :

(i) In Z_m, $0 \neq a \in Z_m$ is a unit iff gcd(a, m) = 1

Example : 1 & 5 are units in Z_6.

(ii) Every non-zero elements in Z_m is either unit or zero divisor.

Let R = Z_8 then

R = {0, 1, 2, 3, 4, 5, 6, 7}

H = {1, 3, 5, 7} set of units then remaining elements other than '0' are all zero divisor.

So, K = {2, 4, 6} is zero divisors set.

7. **Nilpotent element :** Let $a \in R$ then a is said to be a nilpotent element if $\exists\ n \in N$ s.t. $a^n = 0$ i.e. a.a.a. ..... a = 0 n-times

The smallest such n is called index.

Example:- 6 is nilpotent elements of index 2 in Z_{12}.

Remark:-Number of nilpotent elements in Z_m

If m = $p_1^{n_1}.p_2^{n_2}.......p_r^{n_r}$.

Then number of nilpotentsare $p_1^{n_1-1}\ p_2^{n_2-1}......p_r^{n_r-1}$.

Example:- in R = Z_{100}

Number of nilpotent elements

$m = 100 = 2^2 \times 5^2 = 2^{2-1}\ 5^{2-1} = 2 \times 5 = 10$

and the set of nilpotent element = <$p_1, p_2, p_3....p_r$>

<10> = {0, 10, 20, 30, 40, 50, 60, 70, 80, 90}

Example : In Z_8

$m = 8 = 2^3$

Number of nilpotent elements are $2^{3-1} = 4$
and <2> = {0, 2, 4, 6} are nilpotent.

8. **Indempotent element :** Let $a \in R$ then a is said to idempotent element if $a^2 = a$

Example:- 5 is idempotent in Z_{10}.

$5^2 = 25(\text{mod } 10) = 5$

Remark :- Idempotent element in $Z_m = 2^d$

d is prime divisor of m

Let $R = Z_{100}$

$m = 2^2 \times 5^2$

Number of prime divisor of 100 are 2

So $2^2 = 4$ idempotent elements.

9. **Factor :** Let R be CRU (Commutative ring with unity) and $a, b \in R$ then a is said to be factor of b if $\exists\ c \in R$ s.t. $b = ac$

10. **Irreducible element :** A non-unit, non-zero element 'a' in a commutative ring with unity ring R is said to be irreducible element if it has no proper factor.
i.e. whenever a = b.c then either b is unit or c is unit. An element which is not irreducible then it is reducible.

Example : Is 3 irreducible in $Z\left[\sqrt{-5}\right]$.

Solution : We know that $a+\sqrt{-5}b \in Z[\sqrt{5}]$ Let $3=[a+b\sqrt{-5}]\,[c+d\sqrt{-5}]$

Now take its conjugate $3=[a-b\sqrt{-5}][c-d\sqrt{-5}]$

Multiply

$3.3=[a+b\sqrt{-5}][a-b\sqrt{-5}][c+d\sqrt{-5}][c-d\sqrt{-5}]$

$9 = (a^2 + 5b^2)(c^2 + 5d^2)$

Then possibility

$(a^2 + 5b^2)$	$(c^2 + 5d^2)$	
3	3	for any $a, b, c, d \in R$
9	1	a = 3, b = 0, c = 1, d = 0
1	9	a = 1, b = 0, c = 3, d = 0

i.e. factor of 3 are

3 = 3.1 unit

Or 3 = 1.3

So 3 is irreducible element.

11. **Prime element :** Let R be commutative ring with unity then non-zero, non-unit element of R is said to be prime if $p|ab \Rightarrow p|a$ or $p \mid b$

Question - 1 : Is 2 prime in Z_8.

Solution : If $2 \mid a.b$ then either $2|a$ or $2|b$.

So 2 is prime element.

Like $2 \mid 6 \Rightarrow 2 \mid 2.3$ i.e. $2 \mid 2$

Question 2 : Pick out the units in $Z[\sqrt{3}]$.

(a) $-7+4\sqrt{3}$ (b) $2-\sqrt{3}$ (c) $5+3\sqrt{3}$ (d) $-3-2\sqrt{3}$

Solution : Let $a \in R$ and a is said to be a unit in R then $\exists\ b \in R$ s.t. a.b = 1

Option (a) is unit

Let $-7+4\sqrt{3}$ is unit in $Z[\sqrt{3}]$.

So $\dfrac{1}{-7+4\sqrt{3}} \in Z[\sqrt{3}]$.

Rationalize $\dfrac{1}{-7+4\sqrt{3}} \times \dfrac{-7-4\sqrt{3}}{7-4\sqrt{3}} = \dfrac{7-4\sqrt{3}}{+49-48} = \dfrac{-7-4\sqrt{3}}{1} \in Z[\sqrt{3}]$

Option (b) is not unit $\dfrac{1}{5+3\sqrt{3}} \times \dfrac{5-3\sqrt{3}}{5-3\sqrt{3}} = \dfrac{5-3\sqrt{3}}{25-27} = \dfrac{5-3\sqrt{3}}{-2} = \dfrac{-5}{2} + \dfrac{3}{2}\sqrt{3} \notin Z[\sqrt{3}]$

but if $a+\sqrt{3}b$ $a \in Z, b \in Z$.

Option (c) is unit $\dfrac{1}{2-\sqrt{3}} \times \dfrac{2+\sqrt{3}}{2+\sqrt{3}} = \dfrac{2+\sqrt{3}}{4-3} = \dfrac{2+\sqrt{3}}{1} \in Z[\sqrt{3}]$

Option (d) is not unit

$\dfrac{1}{-3-2\sqrt{3}} \times \dfrac{-3+2\sqrt{3}}{-3+2\sqrt{3}} = \dfrac{-3+2\sqrt{3}}{9-12} = \dfrac{-3}{-3} + \dfrac{2}{-3}\sqrt{3} = 1 - \dfrac{2}{3}\sqrt{3} \notin Z[\sqrt{3}]$ $\quad \dfrac{2}{3} \notin Z$

2. Integral Domain :

A commutative ring with unity without zero divisors is called an integral domain. Example No. 1, 2, 3, 4 (from table) are Intergal Domain.

Skew Field (Division Ring) : A ring with unity (R, +, .) is said to be skew field if each non-zero element has multiplicative inverse.

Field : A commutative skew field is a field.

Characteristic of a ring : The smallest positive integer n is said to be characteristic of a ring R

If na = 0 $\forall\ a \in R$ i.e. if a + a + a ….. + a = 0 (n-times)

If no such positive integer exist then characteristic of R is 0 and it is denoted by 0.

Properties :

1. Let R be a ring with unity 1.
   Then characteristic of R is equal to the order of 1 under addition provided it is finite. If it is not finite then char R = 0.
2. The characteristic of an integral domain and field is either zero or prime.
3. The characteristic of finite integral domain is of the form p^2,where p is prime.

Some important structure

$(Z_m, \oplus_m, \Theta_m)$ is commutative ring with unit $\Rightarrow$

(i) Number of units = $\phi(m)$

(ii) Let U (Z_m) be the set of units then U(Z_m) will form the group.

(ii) $U(Z_m) = \{x \in Z_m \mid \gcd(x,m) = 1\}$

(iv) Every non-zero element is either unit or zero divisor.

(v) Number of zero divisors = $m - (\phi(m) + 1)$

(vi) Z_m is Integral Domain iff m is prime.

(vii) Number of associative of $a \in Z_m$ are $\phi(O(a))$

(viii) Number of nilpotent element in $Z_{p^n} = p^{n-1}$

(ix) If $m = p_1^{n_1} p_2^{n_2} ....p_n^{n_n}$. Then number of nilpotent elements in Z_m.

$= p_1^{n_1-1} p_2^{n_2-1} ....p_n^{n_n-1}$

(x) $a \in Z_m$ is nilpotent iff every prime divisor of m divides a

(xi) Number of idempotent elements in $Z_m = 2^k$ where k is number of distinct prime divisor of m.

Note :- V = {Set of all function from R to R}

Where R is ring.Then (V, +, .) is a also a ring.

(i) Properties :

(i) V has unity if R has unity.

(ii) V is commutative if R is commutative ring.

(iii) V is never integral domain even if R is integral domain.

(iii) V always have infinite zero divisor.

$$f(x) = \begin{cases} 0 & x \le a \\ x & x > a \end{cases} \qquad g(x) = \begin{cases} x & x \le a \\ 0 & x > a \end{cases} \qquad \text{then } f.g = 0$$

3. Matrix Ring :

$M_n(R)$ the set of all n × n matrices with entries from ring R form a ring with respect to ordinary matrix addition and ordinary matrix multiplication.

Properties :

(i) If R is non-trival ring and $n \ge 2$ then $M_n(R)$ is not commutative with unity.

(ii) $M_n(R)$ has zero-divisor for all non-trival ring R whenever $n \ge 2$

(iii) If $n \ge 2$ then if A is strictly upper triangular matrix or strictly lower triangular matrix then $A^n = 0$

Example : Let R = $M_2(R)$

and $A = \begin{bmatrix} 0 & 1 \\ 0 & 0 \end{bmatrix}$ then $A^2 = \begin{bmatrix} 0 & 0 \\ 0 & 0 \end{bmatrix} \Rightarrow A^2 = 0$

A is nilpotent element of index 2.

Example : Let $A = \begin{bmatrix} 0 & 1 \\ 0 & 0 \end{bmatrix}$ and $B = \begin{bmatrix} 1 & 0 \\ 0 & 0 \end{bmatrix}$ then $A.B = \begin{bmatrix} 0 & 0 \\ 0 & 0 \end{bmatrix}$

A & B are zero-divisors.

C[0, 1] = {Set of all continuous function from [0, 1] to R}

(i) C[0, 1] is CRU

(ii) There are infinite number of units.

(iii) There are infinite zero-divisor.

$$f(x)=\begin{cases}0 & 0<x\le a\\ x & a<x\le 1\end{cases}\qquad g(x)=\begin{cases}x & 0\le x\le a\\ 0 & a<x\le 1\end{cases}$$

So f(x) . g(x) = 0 both are zero divisor.

SOLVED QUESTIONS

1. Let p be an odd prime such that $p \equiv 2 \pmod 3$. Let F_p be the field with p elements. Consider the subset E of $F_p \times F_p$ given by $E = \{(x, y) \in F_p \times F_p : y^2 = x^3 + 1\}$

Which of the following are true? **CSIR NET NOV 2020**

(a) E has atleast two elements

(b) E has atmost 2p elements

(c) E can have p^2 elements

(d) E has atleast 2p elements

Ans. (a), (b)

Explanation :

For any p, (0, 1), (2, 3) ∈ E always, so E has atleast two elements.

So, option (a) is true.

Also (0, 0) ∉ E, for every p, hence E cannot have p^2 elements.

So, option (c) is false.

Now we need $y^2 = x^3 + 1$ and x, y $\in F_p$ as $F_p \cong Z_p$ for every p, $y^2 = x^3 + 1$ is possible only if y > x, and there are $\frac{p^2 - p}{2}$, such elements.

For p = 5, we have {(0, 1), (0, 2), (0, 3), (0, 4), (1, 2), (1, 3), (1, 4), (2, 3), (2, 4), (3, 4)}

Out of these only (0, 1), (2, 3) are working.

So for p = 5, E = {(0,1), (2, 3)}

Hence option (d) is false.

Now we can observe that if (a, b) ∈ E for some F_p then (a, c) ∉ E for any c ≠ b and a has p choices.

So, such elements are atmost p and (b) is true.

2. Let S denote the set of all the prime numbers (p) with the property that the matrix $\begin{bmatrix}91 & 31 & 0\\ 29 & 31 & 0\\ 79 & 23 & 59\end{bmatrix}$ has an inverse in the field $\frac{Z}{pZ}$. **CSIR NET DEC. 2015**

(a) S = {31} (b) S = {31, 59} (c) S = {7, 13, 59} (d) S is infinite

Ans. (d)

Explanation :

Let A be a given matrix and $|A| \neq 0$ then A has an inverse given that

$$A = \begin{bmatrix} 91 & 31 & 0 \\ 29 & 31 & 0 \\ 79 & 23 & 59 \end{bmatrix} \begin{matrix} \to R_1 \\ \to R_2 \\ \to R_3 \end{matrix}$$

Operation $R_1 \to R_1 - R_2$ $\quad A \sim \begin{bmatrix} 62 & 0 & 0 \\ 29 & 31 & 0 \\ 79 & 23 & 59 \end{bmatrix}$

$|A| = 62 \times 31 \times 59 = 2 \times (31)^2 \times 59$

Option (a) $\quad S = \{31\} \quad \Rightarrow \quad |A| = 0$

Option (b) $\quad S = \{31, 59\} \quad \Rightarrow \quad |A| = 0$

Option (c) $\quad S = \{7, 13, 59\} \Rightarrow \quad |A| = 0$

Option (d)

If we choose any p other than 2, 31 and 59 then $|A| \neq 0$

$\Rightarrow$ Infinite choice of p.

Trick (a): Let $(R, \oplus, \Theta)$ be a ring then a set of all units of R is a group under composition Θ.
(b) $(Z_n^*, .)$ is isomorphic to U(n) where $Z_n^* = Z_n \setminus \{0\}$.

3. Let $G = \left(\dfrac{Z}{25Z}\right)$ be the group of units i.e. the element that have a multiplicative inverse in ring $\left(\dfrac{Z}{25Z}\right)$. Which of the following is a generator of G.

CSIR NETJUNE-2016

(a) 3 (b) 4 (c) 5 (d) 6

Ans. (a)

Explanation :

Given that $R = \dfrac{Z}{25Z} = (Z_{25}, \oplus_{25}. \Theta_{25})$ be a ring then a set of all units of Z_{25} is again a group under composition Θ_{25}.

So here group of units U(25)

Then generator of U(25) is 3.

4. Let R be a non-zero ring with identity s.t. $a^2 = a$ for all $a \in R$. Which of the following statement are true?

CSIR NET JUNE-2017

(a) There is no such ring

(b) $2a = 0 \quad \forall\, a \in R$

(c) $3a = 0 \quad \forall\, a \in R$

(d) $\dfrac{Z}{2Z}$ is a subring of R

Ans. (b), (d)

MODERN ALGEBRA

Explanation :

Definition of Boolean ring : A ring is said to be Boolean ring if all elements of R is idempotent

i.e. $a^2 = a \ \forall a \in R$

e.g. $Z_2 \times Z_2$ and Z_2

(b) We know that characteristic of boolean ring is always 2

So $2.a = 0 \quad \forall a \in R$

(d) Z_2 is subring of every Boolean ring

5. Which of the following is integral domain?

(a) $\left\{\begin{bmatrix} 0 & x \\ y & 0 \end{bmatrix} : x, y \in Z\right\}$ (b) $\left\{\begin{bmatrix} x & 0 \\ 0 & y \end{bmatrix} : x, y \in Z\right\}$

(c) $\left\{\begin{bmatrix} x & 0 \\ 0 & x \end{bmatrix} : x \in Z\right\}$ (d) $\left\{\begin{bmatrix} x & y \\ z & w \end{bmatrix} : x, y, z, w \in Z\right\}$

Ans. **(c)**

Explanation :

Option (a) is incorrect $\begin{bmatrix} 0 & 1 \\ 0 & 0 \end{bmatrix}\begin{bmatrix} 0 & 2 \\ 0 & 0 \end{bmatrix} = \begin{bmatrix} 0 & 0 \\ 0 & 0 \end{bmatrix}$

$\Rightarrow$ It has a zero divisor. $\Rightarrow$ It is not integeral domain.

Option (b) is incorrect. $\begin{bmatrix} 1 & 0 \\ 0 & 0 \end{bmatrix}\begin{bmatrix} 0 & 0 \\ 0 & 1 \end{bmatrix} = \begin{bmatrix} 0 & 0 \\ 0 & 0 \end{bmatrix}$

Result : If the complex analytic functions defined on connected domain then it will be a ring.

6. Which of the following rings are integral domains? **NBHM-2013**

(a) R[x], the ring of all polynomials in one variable with real coefficients.

(b) $M_n(R)$

(c) The ring of complex analytic functions defined on the unit disc of the complex plane (with pointwise addition and multiplication as the ring operations).

Ans. **(a), (c)**

Explanation :

We know that

(a) R[x] is an integral domain ($\because$ Ris an integral Domain)

$\Rightarrow$(a) is correct

(b) Let $A = \begin{bmatrix} 0 & 1 \\ 0 & 0 \end{bmatrix}$ and $B = \begin{bmatrix} 1 & 0 \\ 0 & 0 \end{bmatrix}$ $A, B \in M_2R)$ (forn =2) $AB = \begin{bmatrix} 0 & 0 \\ 0 & 0 \end{bmatrix}$

A and B are non-zero matrices, but $AB = 0 \Rightarrow M_2(R)$ is not an integral Domain

$\Rightarrow$(b) is incorrect

(c) Unit disc is a connected domain. $\Rightarrow$ (c) is correct.

UNSOLVED QUESTIONS

1. Which one of the following is true? **IIT JAM 2012**
(a) The characteristic of the ring 6Z is 6. (b) The ring 6Z has a zero divisor.
(c) The characteristic of the ring (Z/6Z) ×6Z is zero.
(d) The ring 6Z × 6Z is an integral domain.

2. Let R be a non-zero ring with identity such that $a^2 = a$ for all $a \in R$. Which of the following statements are true? **CSIR NETJUNE-2017**
(a) There is no such ring . (b) 2a=0 for all $a \in R$.
(c) 3a = 0 for all $a \in R$ (d) Z/2Z is a subring of R.

3. For which of the following primes p, does the polynomial $x^4 + x + 6$ have a root of multiplicity > 1 over a field of characteristic p? **CSIR NET DEC 2011**
(a) p = 2 (b) p = 3 (c) p = 5 (d) p = 7

4. For n ≥ 1,let $(Z/nZ)^*$ be the group of units of (Z/nZ). Which of the following groups are cyclic? **CSIR NET DEC 2015**
(a) $(Z/10Z)^*$ (b) $(Z/23Z)^*$ (c) $(Z/100Z)^*$ (d) $(Z/163Z)^*$

5. Let p be a prime. Pick each correct statement from below. Up to isomorphism
(a) There are exactly two abelian groups of order p^2.
(b) There are exactly two groups of order p^2
(c) There are exactly two commutative rings of order p^2
(d) There is exactly one integral domain of order p^2. **CSIR NET JUNE 2016**

6. Pick out the units in $Z[\sqrt{3}]$ **NBHM – 2005**
(a) $-7+4\sqrt{3}$ (b) $5+3\sqrt{3}$ (c) $2-\sqrt{3}$ (d) $-3-2\sqrt{3}$

7. Pick out the integral domains from the following list of rings : **NBHM – 2005**
(a) $\{a+b\sqrt{5} \mid a,b \in Q\}$
(b) The ring of continuous functions from [0, 1] into R
(c) The ring of complex analytic functions on the disc $\{z \in C \mid |z| < 1\}$
(d) The polynomial ring Z[x]

8. Pick out the correct statements from the following list :
(a) The element $2 \in Z[\sqrt{-5}]$ is irreducible in $Z[\sqrt{-5}]$.
(b) Units of the ring $Z[\sqrt{-5}]$ are the units of Z.
(c) The element 2 is a prime element in $Z[\sqrt{-5}]$.
(d) None of these

Answer Key

1	2	3	4	5	6	7	8
c	b,d	a,c,d	a,d	a,b,d	a,c	a,c,d	b,c

Chapter 7 Subring and Ideals

1. Subring:

Let (R, +, .) be a ring. Then a non-empty subset S of R is called a subring of R if (S, +,.) is a ring itself.

SubringTest:

Let S be a non – empty subset of R. Then S is subring of R if and only if

(i) $ab \in S \; ;\forall a,b \in S$

(ii) $a - b \in S \; ;\forall a, b \in S$

Similarly we can define subfield of a field & subdivision ring of a division ring.

Example : (i)A subring of Z is type of nZ. (ii) {0} and R always a subring of R.

Properties:

- The intersection of any family of subrings of a ring R is a subring of R.
- $S = \{(r, r) \mid r \in R\}$ is subring of $R \times R$ where R is a ring.
- Center of a division ring is a field.

Sum of Ring : If S and T are two subring of a ring R, then their sum is defined as

$S + T = \{a + b : a \in S, b \in T\}$

- Sum of two subrings of a ring may not be subring.

  Example : Consider the ring, $M_2(Z)$

  Let $S = \left\{\begin{bmatrix} a & 0 \\ b & 0 \end{bmatrix} | a,b \in Z\right\}$ and $T = \left\{\begin{bmatrix} 0 & c \\ 0 & 0 \end{bmatrix} | c \in Z\right\}$

  Clearly, S and T are subrings of $M_2(Z)$.

  But S + T is not a subring as it is not closed under multiplication.

- In a subring ax = b can have more than one solution.
- If S is a subring of ring R then,If char R is finite the char S ≤ char R

2. Ideal :

- **Left Ideal :** Let R be a ring and I be a subring of R, then I is said to be left ideal of R if for each $r \in R$ and $a \in I$ we have $ra \in I$.
- **Right Ideal :** Let R be a ring and I be a subring of R, then I is said to be right ideal of R. If for each $a \in I$, $r \in R$ we have $ar \in I$.
- **Ideal :** A subring I of a ring R is said to be ideal.

  If I is both left ideal and right ideal.
- **IdealTest:** A non- empty subset A of a ring R is an ideal of R if

  (i) $a - b \in A$ whenever $a,b \in A$.

  (ii) ra & $ar \in A$ whenever, $a \in A$ & $r \in R$.

Properties :

- Let R be a ring then subring R and {0} are ideals of R and called improper ideal of R.
- nZ is ideal of Z for every $n \in N$.
- $M_n(kZ)$ is ideal of $M_n(Z)$ $\forall k \in Z$.
- In a commutative ring R, every left ideal or right ideal is ideal of R.
- Arbitrary intersection of ideals of R is an ideal of R.
- The sum oftwo ideals ofaring Ris an idealof R.
- If I and J are two ideals ofa ring R, then their product IJdefined as
  $kJ = \{a_1b_1 + ... + a_nb_n : a_i\in I, b_j\in J, 1 \leq i \leq n, 1 \leq j \leq n, n \in N\}$
  is an ideal of R.Moreover $IJ \subseteq I + J$
- Let R be a ring with unity 1 andIbean ideal of R. If $1 \in I$ then I=R.
- Let F be a field, then {0} and Fare only ideals ofF.
- Let R be a ring, then centre of ring Z(R)is subring of Rbut it need not be ideal of R.
- If R is a commutative ring then $N(R) = \{a \in R \mid a^m = 0$ for some $m \geq 1\}$
  is an ideal in R
  Here N(R) iscalled **nil-radical**of R.
- $N(Z_n)$ i.e. nil-radical of Z_n is non-zero iff n is divisible by square of a prime.
- If R_1 and $R_2$are rings with unity then every ideal of $R_1 \times R_2$ is of the form $I_1\times I_2$where I_1isan idealof $R_1$and I_2 is an ideal of R_2. '
- If R is a ring with unity then every ideal of $M_n(R)$ is of the form $M_n(I)$ where I is an ideal of R.
- If R is commutative ring with unity and $a_1,a_2....a_n\in R$ then
  $I = Ra_1 + Ra_2 + ... + Ra_n$ is an ideal in A.
- Let Rbea ring with unityand I be proper ideal of R, then no element of I can have multiplicative inverse.
- Let I be an ideal of ring Rand Sbe subringofRThen $I\cap S$isideal of S.
- **Simple Ring:** A ring Ris called a simple ring. IfR has no proper ideals i.e.. the only ideals of R are {0}and R.
  Example : (i) (Q, +, .) is a simple ring., (ii) (Z_p, +, .) is a simple ring iff p is prime.

Result:

- Every division ring is simple ring.
- Characteristic of simple ring is either zero or prime.
- If D is division ring, then $M_{n\times n}(D)$ is simple.
- **Maximal Ideal**
  An ideal M in a ring R iscalled maximal ideal ofRif $M\neq R$ and if $\exists$ideal UofRsuch that $M\subseteq U\subseteq R$. Theneither M=U or U = R.
- **Prime Ideal :** Let R be a commutative ring. An ideal P is called a prime ideal if $P\neq R$ and whenever $ab \in P$ then either $a \in P$ or $b \in P$.

Results:

- $pZ \times Z = \{(px, y) \mid x, y \in Z\}$ is maximal ideal and prime ideal of $Z \times Z$ where p isprime.
- $\{(0, 0), (0, 1)\}$ and $\{(0, 0), (1, 0)\}$ are the maximal ideals and prime ideals of $Z_2 \times Z_2$.
- $A = \left\{ f \in C[0,1] : f\left(\frac{1}{2}\right) = 0 \right\}$ is maximal ideal and prime ideal of R = C[0, 1].
- Let R be commutative ring with unity. Then R is field if and only if {0} is maximal ideal of R.
- An ideal M = nZ = {nx : x ∈ Z} is maximal ideal of Z if and only if n is prime number.
- Let R be a commutative ring with unity. Then an ideal M of R is maximal if and only if $\frac{R}{M}$ is a field.
- Let R be a commutative ring. Then the Ideal P of R is prime ideal in R if and only If $\frac{R}{P}$ is an integral domain.
- Let R be a commutative ring with unity then every maximal ideal of R is prime ideal of R. But the converse need not be true.
  Example : {0} is prime ideal in Z but not maximal ideal.
- Let R be finite commutative ring with unity. An ideal of I of R is maximal if and only if I is prime ideal of R.
- Let R be commutative ring with ideal I then if P is prime ideal of I, then P is prime ideal of R.
- Let R be Boolean ring. Then each proper prime ideal P of R is maxamal ideal.
  Boolean ring : A ring R is said to be Boolean ring if $\{a^2 = a, \forall a \in R\}$
- **Quotient Rings:** Let R be a ring and I be an ideal of R.
  Then $\frac{R}{I} = \{r + I \mid r \in R\}$.= set of all cosets of I in R forms a ring, called the Quotient Ring or Residue Class ring of R by I. Under the operation $(r_1 + I)+(r_2+I) = (r_1 + r_2) + I$, $r_1, r_2 \in R$ and $s(r + I) = sr + I$
  Properties:
  Let R be a ring and Ube an ideal of R.
- If R is a commutative ring, then $\frac{R}{U}$ is also commutative ring (not conversely)
- If Rhas unity 1, then $\frac{R}{U}$ hasunity 1+U.
- If R is Boolean ring then $\frac{R}{U}$ is also Boolean ring.

- Let R be a CRU and $P \neq R$ is an ideal of ring R. Then it is an integral domain if and only if P is prime Ideal.
- The ideals of $\frac{R}{I}$ are of the form $\frac{K}{I}$, where K is an ideal of R and $I \subset K$.
- Let R be a CRU & $M \neq R$ be an ideal in R. Then $\frac{R}{M}$ is a field if and only if M is maximal ideal.

Result of Gaussian Ring : $Z[i] = \{a + ib \mid a, b \in Z\}$

(i) Unity = 1 (ii) Units = ±1, ±i (iii) quotient ring

Let <a − ib> is an ideal of Z[i] Then $\frac{Z[i]}{< a+ib >} \approx Z_{a^2+b^2}$ if gcd (a, b) = 1

(iv) $\frac{Z[i]}{< n >} \approx Z_n[i]$ (v) $\frac{Z[i]}{< a+ib >} \approx$ not I.D. if gcd(a, b) ≠ 1

(vi) $\frac{Z[i]}{< ka+kbi >} \approx Z_k[i] \times Z_{a^2+b^2}$

SOLVED QUESTIONS

1. Let C[0, 1] be the ring of all real valued continuous function on [0, 1].
Let A = {f ∈c[0, 1] : f(1/4) = f(3/4) = 0}. **CSIR NET DEC 2019**
Then which of the following statements are true?
(a) A is an ideal in C[0, 1] but is not a prime ideal in C[0, 1]
(b) A is a prime ideal in C[0, 1] (c) A is a maximal ideal in C[0, 1]
(d) A is a prime ideal in C[0, 1], but is not a maximal ideal in C[0, 1]

Ans. (a)

Explanation :

C[0, 1] be the ring of real values continuous functions on [0, 1] and

$$A = \left\{ f \in c[0,1] : f\left(\frac{1}{4}\right) = f\left(\frac{3}{4}\right) = 0 \right\}$$

Let f, g ∈ a, then $(f-g)\left(\frac{1}{4}\right) = f\left(\frac{3}{4}\right) - g\left(\frac{1}{4}\right) = 0 - 0 = 0$

And $(f-g)\left(\frac{3}{4}\right) = f\left(\frac{3}{4}\right) - g\left(\frac{3}{4}\right) = 0 - 0 = 0$

Thus f − g ∈ A also if f ∈ A and g ∈c[0, 1]

Then $(g,f)\left(\frac{1}{4}\right) = g\left(\frac{1}{4}\right).f\left(\frac{1}{4}\right) = g\left(\frac{1}{4}\right)0 = 0$ $(f.g)\left(\frac{1}{4}\right) = f\left(\frac{1}{4}\right).g\left(\frac{1}{4}\right) = 0.g\left(\frac{1}{4}\right) = 0$

Thus f .g and g . f both are in C[0, 1].
Thus A is an ideal.

MODERN ALGEBRA

Now consider two functions $f(x)=\left(x-\frac{1}{4}\right)$ & $g(x)=\left(x-\frac{3}{4}\right)$ then

$$h(x)=f(x).g(x)=\left(x-\frac{1}{4}\right)\left(x-\frac{3}{4}\right)\in A$$

But neither f(x) nor g(x) belongs to A.
So, A is not prime ideal.
Thus (a) is correct and (b) is false.
Since, in a commutative ring with unity every maximal ideal is prime ideal so, it is not maximal ideal as well. So, (c) and (d) are also false.

2. Let R is non-zero CRU such that every Ideal of R is prime ideal then
(a) If R is finite then char(R) = prime
(b) 0 & 1 are only idempotent.
(c) R is integral domain
(d) R is field.

Ans. (a), (b), (c), (d)

Explanation :
if every proper ideal of non zero CRU, R is a prime ideal, then R is a field.
(d) is correct
Therefore R is integral Domain (c) is correct
Also Only idempotent elements in field are 0 and 1. $\therefore (b) is correct$
Also (a) is correct
Because finite field has characteristic prime

Definition : Principal ideal domain : If every ideal of a integral domain is principal ideal then this integral domain is called a principal ideal domain.

3. Let R be commutative ring with unity. Which of the following is true?
(a)If R has finitely many prime ideals, then R is a field.
(b)If R has finitely many ideals, then R is finite.
(c)If R is PID then every subring of R with unity is PID
(d)If R is I.D. which has finitely many ideals then R is field.

Ans. (d) **CSIR NET JUNE 2019**

Explanation :
Option (a) is incorrect. For $R = Z_{15}$
Option (b) is incorrect. For R = Q
Option (c) is incorrect. For $R = Z[x] \subseteq Q[x]$
Option (d) is correct.

4. Which of the following are prime ideals in the ring C[0, 1]
(a) $J=\{f\in C[0,1] \mid f(x)=0\ \forall \frac{1}{3}\le x\le\frac{2}{3}\}$
(b) $J=\{f\in C[0,1] \mid f\left(\frac{1}{3}\right)=f\left(\frac{1}{2}\right)=0\}$
(c) $J=\{f\in C[0,1] \mid f\left(\frac{1}{3}\right)=0\}$
(d)None of these

Ans. (c)

Explanation :

Let R = C[0, 1] and $J = \{f \in C[0, 1] \mid f(x_0) = 0\}$

Then J is prime as well as maximal ideal.

5. Let R be finite non-zero CRU then which of the following statements are true?

(a) Any non-zero element of R is either unit or zero divisor.

(b) There may exist a non-zero element of R which is neither unit nor zero divisor.

(c) Every prime ideal of R is maximal.

(d) If R has no zero divisor then order of any additive subgroup of R is a prime power.

Ans. (a), (c), (d) **CSIR NET 2016**

Explanation :

Result : In finite CRU every non-zero element is either zero divisor or unit.

(a) is correct. (b) is incorrect.

(c) (i) In finite CRUevery prime ideal $\Leftrightarrow$ maximal ideal.

(ii) but in CRU every maximal ideal $\Rightarrow$ prime ideal

Option (c) is correct.

For Option (d) :

Given that R has no-zero divisor and finite CRU

$\Rightarrow$ R is finite I.D. $\Rightarrow$ R is field.

And we know that order of every field is prime power.

Option (d) is correct.

Trick : Let R , CRU be a ring and I be an ideal of R

(i) $\frac{R}{I}$ is field iff I is maximal ideal.

(ii) $\frac{R}{I}$ is I.D. iff I is prime ideal.

(iii) If R is finite CRU and $\frac{R}{I}$ is I.D. $\Rightarrow$ I prime ideal

6. Let $R = Z \times Z \times Z$ and $I = Z \times Z \times \{0\}$ then which of the following statement is correct?

GATE 2012

(a) I is maximal but not prime ideal

(b) I is prime but not maximal ideal

(c) I is prime as well as maximal ideal

(d) I is neither prime nor maximal ideal

Ans. ( )

Explanation :

Given that $R = Z \times Z \times Z$ and $I = Z \times Z \times \{0\}$. Then $\frac{R}{I} = \frac{Z \times Z \times Z}{Z \times Z \times \{0\}} \approx Z$.

Here $\frac{R}{I}$ is I.D. $\Rightarrow$ I is prime ideal and R is not finite ring.

So I never maximal ideal.

7. Which one of the following ideal of the ring Z[i] is not maximal ideal?

GATE 2009

(a) <1 + i> (b) <1 – i> (c) <2 + i> (d) <3 + i>

Ans. (d)

Explanation :

We know that $\dfrac{Z[i]}{<a+ib>} \approx Z_{a^2+b^2}$ if gcd(a, b) = 1

Option (a) $\dfrac{Z[i]}{<1+i>} \approx Z_{1^2+1^2} \approx Z_2$

Z_2 is I.D. and field. So, <1 + i> is maximal ideal.

Option (b) $\dfrac{Z[i]}{<1-i>} \approx Z_2$

So <1 – i> is maximal ideal

Option (c) $\dfrac{Z[i]}{<2+i>} \approx Z_{2^2+1} \approx Z_5$ Z_5 is field.

<2 + i> is maximal field.

Option (d) $\dfrac{Z[i]}{<3+i>} \approx Z_{10}$ i.e. Z_{10} is not field.

So, <3 + i> not maximal ideal.

8. Pick out the true statement(s):

NBHM-2010

(a) The set of all 2 ×2 matrices with rational entries (with the usual operations of matrix addition and matrix multiplication) is a ring which has no non-trivial ideals.

(b) Let R=C[0, 1] be a ring with the usual operations of pointwise addition and pointwise multiplication. Let $I = \{f : [0,1] \to R \mid f(1/2) = 0\}$. Then I is a maximal ideal.

(c) Let R be a commutative ring and let P be a prime ideal of R. Then R/P is an integral domain.

(d) None of these

Ans. (a), (b), (c)

Explanation :

Set of all 2 ×2 matrices form a ring with unity which has no non-trivial ideals (Entries from rationals)

$M_2(F)$ is simple iff F is field.

⇒ $M_2(Q)$ is simple. ⇒ It has no proper ideal.

(b) R = C [0, 1] is a ring with usual operations of pointwise addition and pointwise multiplication

$$I=\left\{f:[0,1]\to R \mid f\left(\frac{1}{2}\right)=0\right\}=\left\langle x-\frac{1}{2}\right\rangle \quad \text{Define } \phi: C[0,1]\to R \text{ by } \phi(g(x))=g\left(\frac{1}{2}\right)$$

$\Rightarrow$Kernel consists all those polynomial having $x-\frac{1}{2}$ as a factor

Also, the map is onto

By fundamental theorem of homomorphism, $\dfrac{C[0,1]}{\left\langle x-\frac{1}{2}\right\rangle}\cong R$

As R is afield, therefore $\left\langle x-\frac{1}{2}\right\rangle$ is a maximal ideal.

(c) Let(a+P)(b+P)=P a,b$\in$R
ab+P = P $\Rightarrow$ab$\in$P
Now as P is prime ideal $\Rightarrow$either a $\in$P or b $\in$P $\Rightarrow$Either (a + P) = P or (b + P) = P
$\therefore$ R/P is an integral domain.

9. Pick out the true statements: **NBHM-2011**

(a) Let R be a commutative ring with unity. Let M be an ideal such that every element of R not in M is a unit. Then R/M is afield.

(b) Let R be as above and let M be an ideal such that R/M is an integral domain. Then M is a prime ideal.

(c) Let R=C[0, 1] be the ring of real-valued continuous functions on [0,1] with respect to pointwise addition and pointwise multiplication. Let M = {f $\in$R |f (0) = f (1) = 0}, then M is a maximal ideal.

Ans. (a), (b)

Explanation :

Let R be a commutative ring with unity

R - M = set of all units in R $\Rightarrow$M contain all non-units

Now, ideal in a ring is generated by non-unit or contain only non-unit

Now as all the non-unit belongs to M, i.e., no non unit left in R so that we cannot make more ideal

M is maximal ideal

If M is maximal ideal then R/M is field.

(b) R is commutative ring with unity
Given R/M is an integral domain = M is a prime ideal.
$\therefore$options (a) and (b) are true.

(c) R = C[0, 1] is ring of real-valued continuous functions on [0, 1] w. r. t. pointwise addition and poiniwise multiplication
LetM= {f$\in$R | f(0) =f(1) = 0}. ConsiderM'= {f$\in$ R | f(0) = 0}
Now, M $\subseteq$M' and we know M' is maximal ideal

MODERN ALGEBRA

And hence M is not a maximal ideal.

$\therefore$ option (c) is false.

10. For two ideals I and J in a commutative ring R, define I : J = {a ∈R:a J⊂i}. In the ring Z of all integers, if I =12Z and J=8Z, find I : J. **NBHM-2008**

(a) Z (b) 2Z (c) 3Z (d) 4Z

Ans. (c)

Explanation :

Giventhat I and Jare ideals ofRand I :J ={a ∈R :aJ⊂I)

12Z :8Z =(a ∈Z | a .8 Z ⊂ 12Z}

let a = 0 0.8Z ={0}⊂12Z

let a = 3 3.8Z ⊂ 12Z

Takea =3n where n∈Z 3n. 8Z⊂ 12Z

Take a isanyintegerbuta ≠3.n,n∈Z.

If a is not multiple of 3 then a . 8Z cannot produce all elements of 12Z.

So if a ∈3Z then a . 8Z ⊄ 12 Z.

Thus 12Z: 8Z =3Z.

11. The number of maximal ideals in Z_{27} is **GATE-2007**

(a) 0 (b) 1 (c) 2 (d) 3

Ans. (b)

Explanation :

$Z_{27} = Z_{3^3}$

Number of maximal ideals = Number of distinct prime divisors of 27 = 1

(Result : Number of maximal ideals in Z_{p^n} = Number of distinct prime divisors of p^n)

UNSOLVED QUESTIONS

1. Let R be a commutative ring with 1 (unity) which is not a field. Let I ⊂ R be a proper ideal such that every element of R not in I is invertible in R. Then the number of maximal ideals of R is

(a) 1 (b) 2 (c) 3 (d) infinite

2. Let R be a subring of Qcontaining 1. Then which of the following is necessarily true?

(a) R is a principal ideal domain (P.I.D.) **CSIR NET DEC-2017**

(b) R contains infinitely many prime ideals.

(c) R contains a prime ideal which is not a maximal ideal.

(d) for every maximal ideal M in R, the residue field R/M is finite.

3. Let a, b, c, d be real numbers with a <c <d <b. Consider the ring C [a, b], with pointwise addition and multiplication. If S = {f∈C[a,b] : f (x) = 0 for all x ∈[c, d]}, then **GATE-2014**

(a) S is NOT an ideal of C [a, b]

(b) S is an ideal of C[a, b] but NOT a prime ideal of C[a, b]
(c) S is a prime ideal of C[a, b] but NOT a maximal ideal of C[a, b]
(d) S is a maximal ideal of C[a, b]

4. Let R be a finite non-zero commutative ring with unity. Then which of the following statements are necessarily true? **CSIRNET DEC-2016**
(a) Any non-zero element of R is either a unit or a zero divisor.
(b) There may exist a non-zero element of R which is neither a unit nor a zero divisor.
(c) Every prime ideal of R is maximal.
(d) If R has no zero-divisors then order of any additive subgroup of R is a prime power.

5. Let $M_3(R)$ be the ring of all 3 × 3 real matrices. If I, J $\subseteq M_3(R)$ are defined as

$$I = \left\{ \begin{pmatrix} a & b & c \\ 0 & 0 & 0 \\ 0 & 0 & 0 \end{pmatrix} | a,b,c \in R \right\}, J = \left\{ \begin{pmatrix} a & 0 & 0 \\ b & 0 & 0 \\ c & 0 & 0 \end{pmatrix} | a,b,c \in R \right\}, \text{ then}$$ **GATE-2006**

(a) I is a right ideal and J is a left ideal
(b) I and J are both left ideals
(c) I and J are both right ideals
(d) I is a left ideal and J is a right ideal

6. P : R = {f | f : R → R} is a ring under point wise addition and pointwise multiplication.
Q : S = {f | f : R → R} is a ring under point wise addition and composition of function.
Then choose correct.
(a) P & Q both are true
(b)Q false because composition is not distributive over addition
(c)S is ideal of R
(d)None of these

7. Which of the following is not maximal ideal of (Z, +, .)
(a)(2Z, +, .)
(b)(7Z, +, .)
(c)(11Z, +, .)
(d)(27Z, +, .)

8. Consider the ring Z[x] & I = {f(x) ∈ Z[x] | f(0) is divisible by 2} then –
(a) I be maximal & prime ideal both
(b) $\left|\frac{R}{I}\right|$ = prime
(c) $\frac{R}{I}$ is field with characteristic '0'
(d) all of the above

9. Let R be a ring s.t. every element is idempotent then
(a) Every prime ideal is maximum ideal
(b) every maximal ideal is prime ideal
(c) if R| > 2 ⇒ R never I.D.
(d)If |R| > 2 ⇒ R never field.

Answer Key

1	2	3	4	5	6	7	8	9	
a	a	b	a,c,d	a	b	d	a,b	a,b,c,d	

MODERN ALGEBRA

Chapter 8

Ring Homomorphism & Ring Isomorphism

1. Ring Homomorphism :

Definition: Let (R, +, .)&(R', ⊕, Θ)two rings then a map $\phi:R \to R'$ is defined as ring homomorphism if

(i) $\phi(a+b) = \phi(a)\oplus\phi(b); \forall a,b \in R$ (ii) $\phi(a.b) = \phi(a)\Theta\phi(b); \forall a.b \in R$

Results: Let $\phi : R \to R'$ be a ring homomorphism then

- $\phi(0) = 0'$
- $\phi(-a) = -\phi(a)$; $\forall a \in R$
- If $\phi(r) = a$, then $O(a) \mid O(r)$
- For any $r \in R$ & any positive integer n,$\phi(nr) = n\phi(r)$ and $\phi(r^n) = (\phi(r))^n$
- $\phi(A) = \{\phi(a) \mid a \in A\}$ is a subring of R'. If A is subring of R.
- $\phi^{-1}(B) = \{r \in R \mid \phi(r) \in B\}$ is an ideal of R. If B is ideal of R'.
- If R is commutative, then $\phi(R)$ is commutative.
- If $a \in R$ is nilpotent element, then $\phi(a) \in R'$is nilpotent
- If $a \in R$is idempotent clement, then $\phi(a) \in R'$is idempotent.
- **Kernel of Homomorphism:**

Letϕ be a homomorphism from a ring R to a ring R'.

Then ker$\phi$is defined as: $\ker\phi = \{r \in R \mid \phi(r) = 0\}$

Properties :

- $\ker \phi = \{0\}$ if and only if f is one – one .
- $\ker \phi$ is an ideal in R.

2. Isomorphic Rings:

Two rings R & R' are said to be isomorphic if it is one-one onto homomorphism between them.

Properties :

If a homomorphism from R to R' then:

- Every isomorphic image of
  (i) Integral domain is integral domain (ii) Fieldis field
  (iii) Division ring is division ring.
  The homomorphic image of commutative ring is commutative.
- Let R be commutative ring and suppose
  $px = 0; \forall x \in R$ where p isa prime number.
  Then the mapping $\phi(x)=x^p;\forall x \in R$
  is a homomorphism &this map is known as Frobenius map.
- The relation of being isomorphic is an equivalence relation.

MODERN ALGEBRA

3. Isomorphism theorems

1. **First Theorem of Isomorphism (Fundamental Theorem of Ring Homomorphism):**

   Let ϕbe a onto ring homomorphism from R to R', then the mapping $\psi : \frac{R}{\ker \phi} \to R'$ givenby $\psi(r+\ker\phi)=\phi(r)$ is an isomorphism. In symbols $\frac{R}{\ker \phi} \cong R'$

2. **Second Theorem of Isomorphism :**

   Let A,Bbe two idealsofa ring R, then $\frac{A+B}{A} \cong \frac{B}{A \cap B}$

3. **Third Theorem of Isomorphism:**

   Let A and B be two ideals of ring R such that $B \subseteq A$. Then $\frac{R}{A} \cong \frac{R/B}{A/B}$

- **Embedding of Ring:** A ring R is said to be embedded in to a ring R' if there is an isomorphism from Rinto R'.i.e., R is isomorphic to a sub ring of R'. We also say that R' is an extension ring or over ring of R.
  Result: Every Ring can be embedded in a ring with unity.
- **Prime Field:** A field that does not hove proper subfield is known as prime field
  Observation: \\
- $Z_p$is prime field if p is prime,
- Q is prime field.

SOLVED QUESTIONS

1. Let X = (0, 1) be the open unit interval and C(X, R) be the ring of continuous functions from X to R. For any $x \in (0, 1)$, let $l(x) = \{f \in C(X, R) \mid f(x) = 0\}$. Then which of the following are true?

   (a) $l(x)$ is a prime ideal (b) $l(x)$ is a maximal ideal

   (c) Every maximal ideal of C(X, R) is equal to $l(x)$ for some $x \in X$.

   (d) C(X, R) is an interval domain. **CSIR NET JUNE 2019**

   Ans. **(a), (b)**

   Explanation :

   $X = (0, 1)$& $l(x) = \{f \in C(X, R) \mid f(x) = 0\}$

   Define $\phi : C(X, R) \to R$ $\phi(f(t)) = f(x)$ then $\ker(\phi) = lx$

   Then ϕ is ring homomorphism. Also ϕ is onto.

   $\therefore \forall\, y \in R$ define $g(t) = t - x + y$ then $\phi(g(t)x = g(x) = x - x + y = y \Rightarrow \phi$ is onto.

   Apply FTRH $\therefore \frac{C(X,R)}{l(x)} \cong R$: R is a field.

   So $l(x)$ is maximal and prime ideal of C(X,R)

2. The number of ring homomorphisms from $Z_2 \times Z_2$ to Z_4 is equal to

GATE – 2015

(a) 1 (b) 2 (c) 3 (d) 4

Ans. (a)

Explanation :

Let f be a ring homomorphism from $Z_2 \times Z_2$ to Z_4

We know that image of idempotent element is idempotent and there is only one idempotent element in $Z_2 \times Z_2$ and Z_4.

$\Rightarrow$ There is only one ring homomorphism which is possible.

> **Trick :** Total number of maximal ideals in quotient ring $Q[x] / (x^m - 1)$ is equal to number of divisors of m.

3. Let Q[x] be the ring of polynomials over Q. Then the total number of maximal ideals in the quotient ring $Q[x] / (x^4 - 1)$ equals **GATE – 2017**

(a) 0 (b) 1 (c) 2 (d) 3

Ans. **(d)**

Explanation : Number of divisor of 4 = 3

4. $S = \{f \mid f : Z_{12} \to Z_{20}$ ring homomorphism) and $T = \{f \mid f : Z_5 \to Z_{20}$ ring homomorphism$\}$ then

(a) $|S| \leq |T|$ (b) $|S| \geq |T|$ (c) $|S| < |T|$ (d) $|S| > |T|$

Ans. **(a), (b)**

Explanation :

$|S|$ = Number of ring homomorphism from $Z_{12} \to Z_{20}$

$f(1) = r \Rightarrow O(r) \mid O\, f(1)$ $\Rightarrow O(r) \mid 12$

idempotent element in Z_{20} 5×2^2

0 1 5 16 idempotent elements

↓ ↓ ↓ ↓

1 20 4 5 order of idempotent elements

1 | 12 and 4 | 12 there are two ring homomorphism $|S| = 2$

$|T|$ = Number of ringhomorphism from $Z_5 \to Z_{20}$

$f(1) = r$

$O(r) \mid O[f(1)]$

$O(r) \mid 5$

Idem. Element in Z_{20}

0 1 5 16 Idempotent elements

↓ ↓ ↓ ↓

1 20 4 5 Order of idempotent elements

1 | 5 5 | 5

$|T| = 2$

Option (a) & (b) is correct.

5. If (R, +, .) and (R', +, .) are two rings and f : (R, +, .) → (R', +, .) be a onto ring homomorphism then choose incorrect.
(a) If R' be I.D. ker f is max. ideal of R (b) If R' be finite I.D. then ker f is max. ideal of R
(c) If R' is field, ker f is prime ideal of R (d) If R' is I.D. then ker f is prime ideal
Ans. **(a)**
Explanation :
Given that f : R → R' be a onto ring homomorphism. ⇒ f(R) = R'
By fundamental theorem of ring homomorphismm.
Let f : R → R' be a ring homomorphism.

Then $\frac{R}{\ker f} \approx f(R) \approx R'$ (onto) $\Rightarrow \frac{R}{\ker f} \approx R'$

We know that if R' is I.D. then ker f is prime ideal and R' is field then ker f is maximal ideal.
Option (a) is correct option.

6. Let f : R' → R be a homomorphism of R' into R. Then
(a) ker (f) is an ideal of R
(b) ker (f) is an ideal of R'
(c) If ker (f) = {0} then f is isomorphism
(d) If ker (f) = R' ⇒ f(x) is not 0
Ans. **(b), (c)**
Explanation :
Option (b) is correct.
ker (f) is an ideal of R'
Option (c) is correct.
If ker(f) = 0
Then f is isomorphism
Option (d) is incorrect
If ker (f) = R' ⇒ f(x) =0

7. The number of ring homomorphisms from Z[x, y] to $F_2[x]/(x^3 + x^2 + x + 1)$ equals
(a) 2^6 (b) 2^8 (c) 1 (d) 2^9
Ans. **(a)** **TIFR – 2019**
Explanation :
The ring Z[x, y] has two generators x and y.
The images of x, y under homomorphism f : Z[x, y] → $F_2[x] / (x^3 + x^2 + x + 1)$ can be arbitrar elements of co-domain and they uniquely determined F.
As the ring $F_2[x]/(x^3 + x^2 + x + 1)$ has 8 elements, we have total of 8 × 8 ring homorphisms.

8. Consider the following statements :
I :For every surjective ring homomorphism $\phi : R \to S$, we have $\phi(R^X) = S^X$.
II :Every non-commutative ring has atleast 10 elements.
Which of the following is correct?
(a) I and II both are true
(b) Both are false.
(c) I is true and II is false
(d) I is false and II is true.

Ans. **(b)**

MODERN ALGEBRA

Explanation :

Take R = Z and S = Z_5 $\phi : Z \to Z_5$ defined by $\phi(a) = a + 5Z$ where $a \in Z$.

ϕ is surjective ring homomorphism. But $\phi(Z^*) = Z_5$ (not Z_5^*)

I is false. Every non-commutative ring may not have atleast 10 elements.

II is false.

UNSOLVED QUESTIONS

1. If Z[i] is the ring of Gaussian integers, the quotient Z[i]/<3 - i> is isomorphic to **GATE-2011**

(a) Z (b) Z/3Z (c) Z/4Z (d) Z/10Z

2. Letp,q be distinct primes. Then **CSIR NETJUNE-2014**

(a) $\frac{Z}{p^2qZ}$ has exactly 3 distinct ideals. (b) $\frac{Z}{p^2qZ}$ has exactly 3 distinct prime ideals.

(c) $\frac{Z}{p^2qZ}$ has exactly 2 distinct prime ideals. (d) $\frac{Z}{p^2qZ}$ has a unique maximal ideal.

3. The number of non-trivial ring homomorphisms from Z_{12} to Z_{28} is **CSIR NET JUNE 2012**

(a) 1 (b) 3 (c) 4 (d) 7

4. Let Z[i] denote the ring of Gaussian integers. For which of the following values of n is the quotient ring Z[i] / nZ[i] an integer domain? **CSIR NET DEC. 2011**

(a) 2 (b) 13 (c) 19 (d) 7

5. Let R be a non-zero commutative ring with units 1_R. Define the characteristic of R to be the order of 1_R in (R, +) if it is finite and to be zero if the order of 1_R in (R, +) is infinite. We denote the characteristic of R by char(R). In the following, let R and S be non-zero commutative rings with unity. Then

CSIR NET DEC 2013

(a) Char(R) is always a prime number

(b) If S is a quotient ring of R, then either char(S) divides char(R), or char(S) = 0

(c) If S is a subring of R containing 1_R then char(R) = char(R)

(d) If char (R) is a prime number, then R is a field.

Answer Key

1	2	3	4	5
d	c	a	c,d	c

MODERN ALGEBRA

Chapter 9

Rings of Polynomial – R[X]

1. Ring of Polynomials

- **Definition:** Let Rbe a ring, then the set $R[x]=\{a_nx^n+ a_{n-1}x^{n-1}....+a_1x+a_0: a_i \in R, n \in N \cup \{0\}\}$ forms a ring with respect to addition and multiplication of polynomials.
- **Degree of a Polynomial :**
Let $f(x) = a_0 + a_1x + ... + a_nx^n \in R[x]$, we say that f(x) isa non-zero polynomial if atleast one of the coefficients $a_0, a_1, ..., a_n$. is non-zero, thenwe say that f(x) hasdegree nif$a_n \neq 0$. We write it as deg f(x) = n or deg (f) = n.
Properties :
    - Degree of the zero polynomial is not defined.
    - Iff(x)+g(x)≠ 0, then deg(f(x) .g(x))≤max{deg f(x).deg g(x)}
    - Iff(x).g(x)≠ 0,then deg(f(x).g(x))≤degf(x) + degg(x)
    - If ring R is CRU, then R(x)is CRU. Moreover, unity of R [x] is same as that of R.
    - If R be a commutative ring with unity then units of R are also units of R[x].
    - If R is an integral domain, then units of Rand R[x] are same.
        - R[x] never be a field.
        - If R is an integral domain, then R[x] is also an integral domain.
        - If R is a non-trivial ring. Then R[x]is always infinite ring.
    - If I is ideal of ring R. then I[x] is ideal of ring R[x].
    - If R be CRU & A be an ideal of R. Then $\frac{R[x]}{A[x]} \cong \frac{R}{A}[x]$.
    - If R be a ring, then $\frac{R[x]}{\langle x \rangle} \cong R$, where ⟨x⟩ is ideal generated by x.
    - If R is CRU & Pbe any prime ideal of R. Then P[x]is prime ideal of R[x].
    - If R is a ring, then R can be embedded in R[x] i.e., every ring R is isomorphic tc a subring of R[x]
    - If R & R'are two isomorphic rings then R[x] and R'[x] are also isomorphic rings
- **Division Algorithm:**
Let R be a field and f(x).g(x)∈R[x]with g(x)≠0.Then there exist unique polynomials q(x) and r(x) in R[x]such that f(x)=g(x)q(x) + r(x) and either
r(x) = 0 or degr(x)<deg g(x).

MODERN ALGEBRA

2. Irreducible Polynomial and Reducible Polynomial:

Let R be an integral domain. A polynomial $f(x) \in R[x]$ of positive degree (i.e. deg≥ 1) is said to be an irreducible polynomial over R if it can not be expressed as product of two polynomials of positive degree.

In other words, if whenever $f(x)=g(x)h(x)$ then deg $g = 0$ or deg $h= 0$.

Example:- $f(x) = 2x^2 + 6$ is irreducible over Q.

Sol. $f(x) = 2(x^2 + 3) = g(x) . h(x)$

$g(x) = 2 \in Q[x]$ is unit in $Q[x]$ so $f(x) = 2x^2 + 6$ is irreducible over Q.

Note : If $f(x) = g(x) . h(x)$ and $g(x)$ is unit then the behavior of $f(x)$ is depending on $h(x)$ i.e. if $h(x)$ is irreducible then $f(x)$ is irreducible & if $h(x)$ is reducible then $f(x)$ is reducible.

Note:A polynomial of positive degree which is not irreducible is called reducible over R.

Properties.

- Polynomials of degree 1 are irreducible over field F.
- Every irreducible element in R[x]is an irreducible polynomial, where R is an integral domain with unity.
- If F is a field then every irreducible polynomial of F[x] is irreducible element of F[x]and conversely.
- **Irreducibility Tests**
  (i) If a polynomial $f(x)$ is ofdegree >1 and $f(a)= 0$ for some $a\in F$.
  Then f(x) isreducibleoverFwhere F is a field
  (ii) **Reducibility Test for degree 2 and 3:**
  Let F be a field if $f(x)\in F[x]$and deg$f(x)=2$ or 3. then f(x) is reducible over Fif and only if f(x) has a zero in F.
  (iii) Let $f(x)\in Z[x]$, if f(x) is reducible over Z, then it is reducible over Q.
  (iv) **Mod p irreducibility Test:**
  Let p be a prime and suppose that $f(x)\in Z[x]$with deg$f\geq 1$.
  Let f(x) be the polynomial in $Z_p[x]$ obtained from f(x) by reducing all the coefficientsof f(x) modulopiff(x)is irreducible over Z_p
  and deg$f(x)$=deg $f_p(x)$then f(x) is irreducible over Q.
  (v) **Eisenstein's Criterion:**
  Let $f(x)=a_nx^n + a_{n-1}x^{n-1} + \ldots.. + a_0\in Z[x]$.
  If $\exists$a prime p such that $p|a_0 . p|a_2, \ldots. p|a_{n-1}$and $p^2 \nmid a_0$, $p \nmid a_n$, then f(x) is irreducible over Q.

Example:- $f(x) = 3 + 6x + 12x^2 + x^3 \in Z[x]$ is irreducible over Q.

Sol. $p = 3$; Here $3 \mid 3$, $3 \mid 6$, $3 \mid 12$ but $3 \nmid 1$ and $3^2 \nmid 3$ Then f(x) is irreducible over Q.

Results :

- Maximal ideals of Z[x]are of the form $<p, x>= \{pf(x) +xg(x) \mid f(x). g(x) \in Z[x]\}$ where pis prime integer.

- If pis prime number, then the number of reducible polynomials of the form $x^2 + ax + b$ over $Z_p$are $\frac{p(p+1)}{2}$.
- If p is prime number. Then number of irreducible polynomials over Z_p of the form $x^2+ ax +b$ are $\frac{p(p-1)}{2}$
- $\langle x\rangle$is prime ideal of Z[x] and $\langle x\rangle$ is not maximal ideal of Z[x].

Some Important Theorems:

- LetFbe a field and $p(x) \in F[x]$. Then $\langle p(x)\rangle$ is maximal ideal in F[x] if and only if p(x) is irreducible over F.
- Let F be a field and p(x) is irreducible polynomial over F. Then $\frac{F[x]}{\langle p(x)\rangle}$ is a field.
- Let $f(x) = a_nx^n + a_{n-1}x^{n-1} + ...+ a_1x + a_0 \in Z[x]$ and $a_n \neq 0$. Let $q = \frac{r}{p} \in Q$ be such that $f(q) = 0$, where rand q are relatively prime, Then r divides a_0, and p divides a_n.

SOLVED QUESTIONS

1. Consider the polynomial $f(x) = x^2 + 3x - 1$. Which of the following statements are true?
CSIR NET JUNE 2020

(a) f is irreducible in $Z[\sqrt{13}]$. (b) f is irreducible over Q

(c) f is reducible over $Q[\sqrt{13}]$. (d) $Z[\sqrt{13}]$ is a unique factorization domain

Ans. (a) ,(b), (c)

Explanation :

(a) $f(x) = x^3 + 3x - 1$ Then $x = \frac{-3 \pm \sqrt{13}}{2} = -\frac{3}{2} \pm \frac{1}{2}\sqrt{13}$

So, f(x) is irreducible in $Z[\sqrt{13}]$.

(b) $f_2(x) = x^3 + x + 1$ It is irreducible over Z_2. So, it is irreducible over Q.

(c) f is reducible over $Q[\sqrt{13}]$.

(d) We can use this example to eliminate option $12 = 4 \times 3 = (-1+\sqrt{13})(1+\sqrt{13})$.

2. Let $f = a_0 + a_1X + ... + a_nX^n$ be a polynomial with $a_i \in Z$ for $0 \leq i \leq n$.
Let p be a prime such that $p \mid a_i$ for all $1 < i \leq n$ and p^2 does not divide a_n.
Which of the following statements are true? **CSIR NET FEB 2022**

(a) f is always irreducible (b) f is always reducible

(c) f can sometimes be irreducible and can sometimes be reducible.

(d) f can have degree 1

Ans. (c)

MODERN ALGEBRA

Explanation :

$f = a_0 + a_1X + \ldots\ldots + a_nX^n$ be a polynomial with $a_i \in Z$

Let p be a prime such that $p \mid a_i$, $\forall\ 1 < i \leq n$, $p^2 \nmid a_n$.

(a) Consider $p(x) = 2x^2 + 2x$, then $p(x) = x(2x + 2)$ is reducible in Z[x],
So, option (a) is incorrect.

(b) For $p(x) = 2x^2 + 1$, we see p(x) is irreducible in Z[x], so option (b) is incorrect.

(c) So, by option (a) and (b), f can be reducible and f can irreducible.

(d) Since $p \mid a_i$ and $p^2 \nmid a_n$ for all $1 < i \leq n$
So, $n \geq 2$
So, p(x) cannot be of degree 1.

3. Let $c \in Z_3$ be such that $\dfrac{Z_3[x]}{\langle x^3 + cx + 1\rangle}$ is a field. Then c is equal to

GATE 2015

(a) 1 (b) 2 (c) 0 (d) 3

Ans. (b)

Explanation :

$\dfrac{Z_3[x]}{\langle x^3 + cx + 1\rangle}$ is a field $\Leftrightarrow x^3 + cx + 1$ is irreducible polynomial in Z_3

When c = 0, then $x^3 + 1$ is reducible as x = 2 is the root of $x^3 + 1$

When c = 1, then $x^3 + x + 1$ is reducible as x = 1 is the root of $x^3 + x + 1$

When c = 2, then $x^3 + 2x + 1$ is irreducible

Hence, c = 2 is answer

4. Let $f(x) = x^3 + 2x^2 + 1$ and $g(x) = 2x^2 + x + 2$ then over Z_3.

(a) Both are irreducible (b) f(x) is irreducible but g(x) is not

(c) g(x) is irreducible but f(x) is not (d) neither f(x) nor g(x) is irreducible

Ans. (b) **CSIR NET JUNE – 2012**

Explanation :

Given that $f(x) = x^3 + 2x^2 + 1 \in Z[x]$

$Z_3 = \{0, 1, 2\}$

$f(0) = 1$, $f(1) = 4 = 1$ $f(2) = 8 + 8 + 1 = 17 = 2$

i.e. f(x) is irreducible over Z_3.

Again $g(x) = 2x^2 + x + 2 \in Z[x]$

$Z_3 = \{0, 1, 2\}$

$g(0) = 2$, $g(1) = 5 = 2$ $g(2) = 12 = 0$

i.e. g(x) is reducible over Z_3.

5. Let R be the $\frac{Z[x]}{<(x^2+x+1)(x^3+x+1)>}$ and I be the ideal generated by 2 in R. What is the cardinality of the ring R. **CSIR NET JUNE 2015**

(a) 27 (b) 32 (c) 64 (d) Infinite

Ans. (d)

Explanation :

$$R=\frac{Z[x]}{<(x^2+x+1)(x^3+x+1)>}$$

$= \{I + a_0 + a_1x + a_2x^2 + a_4x^4 \mid a_i\in Z\}$

then cardinality of R is infinite.

Result : Highest Degree of irreducible factor of $x^n - 1$ is $\phi(n)$

6. Which of the following is an irreducible factor of $x^{12} - 1$ over Q? **CSIR NET 2015**

(a) $x^8 + x^4 + 1$ (b) $x^4 + 1$ (c) $x^4 - x^2 + 1$ (d) $x^5 - x^4 + x^3 - x^2 + x - 1$

Ans. (c)

Explanation :

Here $\phi(12) = 4 \Rightarrow$ Option (a) and (d) are incorrect.

(b) If $x^4 + 1 = 0$ is a factor then, $x^4 = -1 \Rightarrow x^{12} = -1 \Rightarrow x^{12} + 1 = 0$

Hence, $x^4 + 1$ is not a factor of $x^{12} - 1$.

7. $\frac{Z_2[x]}{<(x^3+x^2+1)>}$ is **GATE-2009**

(a) afield having 8 elements (b) afield having 9 elements

(c) an infinite field (d) not afield

Ans. (a)

Explanation :

$p(x)=x^3+ x^2 +1$is apolynomial in $Z_2[x]$. It is of degree 3 and $p(0)= 0+0+1=1$

$p(1) = 1 + 1 + 1 \equiv 1(\text{mod } Z_2) \Rightarrow$0 and 1 are not roots ofp(x)=0

$\therefore$ p(x) is irreducible in $Z_2[x]$ and ideal (p(x)) is maximal ideal in $Z_2[x]$

$\therefore \frac{Z_2[x]}{<(x^3+x^2+1)>}$ is field with $2^3 = 8$ elements.

(**Fact:** If p(x) is an irreducible polynomial of degree k in $Z_p[x]$, then $Z_p[x]/ <p(x)>$ is a finite field with p^kelements).

8. Consider Z_5 as field modulo 5 and let $f(x) = x^5 + 4x^4 + 4x^3 + 4x^2 + x + 1$. Then the zeros of f(x) over Z_5 are 1 and 3 with respective multiplicity. **GATE-2007**

(a) 1 and 4 (b) 2 and 3 (c) 2 and 2 (d) 1 and 2

And. (d)

MODERN ALGEBRA

Explanation :

In $Z_5[x]$

$f(x) = x^5 + 4x^4 + 4x^3 + 4x^2 + x + 1$, $f'(x) = 0 + x^3 + 2x^2 + 3x + 1$

$f''(x) = 3x^2 + 4x + 3$, $f'''(x) = x + 4$

$f(1) = 1 + 4 + 4 + 4 + 1 + 1 = 0$

$f'(1) = 1 + 2 + 3 + 1 + 2 \neq 0$ $\quad\therefore$ multiplicity of 1 is 1.

$f(3) = 3 + 4 + 3 + 1 + 3 + 1 = 0$, $f'(3) = 2 + 3 + 4 + 1 = 0$, $f''(3) = 2 + 2 + 3 = 2 \neq 0$

$\therefore$ 3 is zero of multiplicity 2.

9. Which of the following polynomials are irreducible in Z[x]?**CSIR NET JUNE 2017**

(a) $x^4 + 10x + 5$ (b) $x^3 - 2x + 1$ (c) $x^4 + x^2 + 1$ (d) $x^3 + x + 1$

Ans. (a), (d)

Explanation :

Option (a) :

By Eisenstein's criteria,

$x^4 + 10x + 5$ is irreducible over Q. $\Rightarrow$ $x^4 + 10x + 5$ is irreducible over Z.

Option (b) :

x = 1 is a root of polynomials $x^3 - 2x + 1$

Then it is reducible over Z.

Option (c) :

$\therefore$ $(x^4 + x^2 + 1) = (x^2 + x + 1)(x^2 - x + 1)$

Hence, it is reducible over Z.

Option (d) :

By root test it is irreducible over Z.

10. Let $f(x) \in Z[x]$ be a monic polynomial. Then the roots of f

CSIR NET JUNE 2018

(a) Can belong to Z (b) Always belong to $(R/Q) \cup Z$

(c) Always belong to $(C/Q) \cup Z$ (d) Can belong to (Q/Z)

Ans. (a), (c)

Explanation :

$x + 1 = 0$ $\quad x = -1 \in Z$ so option (a) is correct.

Option (b) is incorrect because roots may belong to C.

e.g. $x^2 + 1 = 0$

Hence, option (c) is correct.

Option (d) is incorrect.

Because if $f(x) \in Z[x]$ has a rational root then it will be integer.

Trick : f(x) is monic polynomial in Z[x]

f(x) is irreducible over Z $\Leftrightarrow$ it is irreducible over Q.

11. Let $f(x) \in Z[x]$ be a monic polynomial degree n. Then, which of the following are true?

(a) If f(x) is irreducible in Z[x], then it is irreducible in Q[x]

(b) If f(x) is irreducible in Q[x], then it is irreducible in Z[x]
(c) If f(x) is reducible in Z[x], then it has a real root.
(d) If f(x) has a real root, then it is reducible in Z[x]. **CSIR NET DEC 2019**

Ans. (a), (b)

Explanation :

Given that f(x) ∈ Z[x] be a monic polynomial degree n
Option (a) & (b) are correct.
For option (c) :
$x^4 + 3x^2 + 2 \in Z[x]$
$x^4 + 3x^2 + 2 = (x^2 + 1)(x^2 + 2)$ is reducible in Z but it has no real root.
For option (d) :
$x^2 - 2$ has a real root but it is irreducible in Z[x].
Option (d) is incorrect.

UNSOLVED QUESTIONS

1. Consider the ideal $I = \langle x^2 + 1, y \rangle$ in the polynomial ring C [x, y]. Which of the following tatements is true? **CSIR NET JUNE-2017**
(a) I is a maximal ideal. (b) I is a prime ideal but not a maximal ideal.
(c) I is a maximal ideal but not a prime ideal.
(d) I is neither a prime ideal nor a maximal ideal.

2. Let R = Q[x]/I, where I is the ideal generated by $1 + x^2$. Let y be the coset of x in R. Then **CSIR NET JUNE -2012**
(a) $y^2 + 1$ is irreducible over R (b) $y^2 + y + 1$ is irreducible over R.
(c) $y^2 - y + 1$ is irreducible over R. (d) $y^3 + y^2 + y + 1$ is irreducible over R.

3. Let $f(x) = x^3 + 2x^2 + x - 1$. Determine in which of the following cases f is irreducible over the field K. **CSIR NET DEC-2013**
(a) K =Q, the field of rational numbers (b) K = R, the field of real numbers
(c) K $=F_2$, the finite field of 2 elements (d) K $=F_3$, the finite field of 3 elements

4. Let I_1 be the ideal generated by x^4+3x^2+2 and I_2 be the ideal generated by x^3+1 in Q[x]. If $F_1=Q[x]/I_1$ and $F_2=Q[x]/I_2$ then **CSIR NET JUNE -2011**
(a) F_1 and F_2 are fields (b) F_1 is afield, but F_2 is not afield
(c) F_1 is not afield while F_2 is afield (d) Neither F_1 nor F_2 is afield

5. Let $\langle p(x) \rangle$ denote the ideal generated by the polynomial p(x) in Q[x]. If $f(x) = x^3 + x^2 + x + 1$ and $g(x) = x^3 - x^2 + x - 1$, then **CSIR NET JUNE -2011**
(a) $\langle f(x)\rangle + \langle g(x)\rangle = \langle x^3 + x\rangle$ (b) $\langle f(x)\rangle + \langle g(x)\rangle = \langle f(x).g(x)\rangle$
(c) $\langle f(x)\rangle + \langle g(x)\rangle = \langle x^2 + 1\rangle$ (d) $\langle f(x)\rangle + \langle g(x)\rangle = \langle x^4 - 1\rangle$

Answer Key

1	2	3	4	5
d	c	a,c	d	c

Chapter 10

PID, UFD, ED and Field

Principal ideal : An ideal generated by a single element is said to be principal ideal.

Examples: (i) Every ideal of Z_m is principal ideal.

(ii) Every ideal of Z is principal ideal is of the form nZ, $\forall n \in Z$.

Observation : If (R, +, .) is ring and If (R, +) is cyclic group then every ideal is P.I.

Principal Ideal Ring : A commutative ring with unity is defined as principal ideal ring if every ideal is principal ideal.

Example:- Z_m& Z both are pricipal ideal ring.

1. Principal Ideal Domain :

An integral domain R is said to be principal ideal domain if each ideal I of R is P.I.

i.e. $I = \langle a \rangle = \{ar \mid r \in R\}$

(i)every field is P.I.D. but conversely need not be true.

Example : Q[x]

(ii)$Z_p[x]$ is P.I.D. if p-prime.

(iii)Z[x] is not P.I.D. ($\exists, I = \langle 2, x \rangle$ which is not PI

Norm on an I.D. : Let R be an I.D. then function $N : R \to Z^+ \cup \{0\}$ with N(0) = 0 is called norm on domain if $N(a) > 0 \ \forall\ 0 \neq a \in R$, N is called +ve norm also

Example:- in Z[i]

$N(a + ib) = a^2 + b^2 \quad \forall\ a + ib \in Z[i]$ is a norm.

2. Euclidean Domain :

Let R be an integral domain if $\exists$ a norm on R s.t.

(i) $N(a.b) \geq N(a) \quad \forall\ a, b \neq 0 \in R$

(ii) For any a, $0 \neq b \in R$ $\exists$ q, r $\in$ R s.t. a = bq + r, r = 0 or N(r) < N(b) then R is said to be Euclidean domain.

Example:- (i) Every field is E.D.

(ii) Z, Z[i]& $Z[\sqrt{2}]$ are E.D.

3. Unique Factoriation Domain (UFD) :

An I.D. is called UFD if it satisfy the properties.

(i) Each non-zero element of R is either unit or can be expressed as a product of finite number of irreducible element of R.

Example :- (i) Z[x] is UFD.

(ii) Every field is UFD.

(iii) For n > 3, $Z[\sqrt{-n}]$ is never UFD

MODERN ALGEBRA

Observation :

(i) Field $\Rightarrow$ ED $\Rightarrow$ PID $\Rightarrow$ UFD $\Rightarrow$ ID

(ii) If R is field $\Rightarrow$ R[x] is Euclidean domain.

(iii) Let R be PID and $a \in R$ then a is prime element $\Leftrightarrow$ a is irreducible element.

(ii) Let R be PID then an ideal I = <a> is maximal ideal iff a is irreducible element of R.

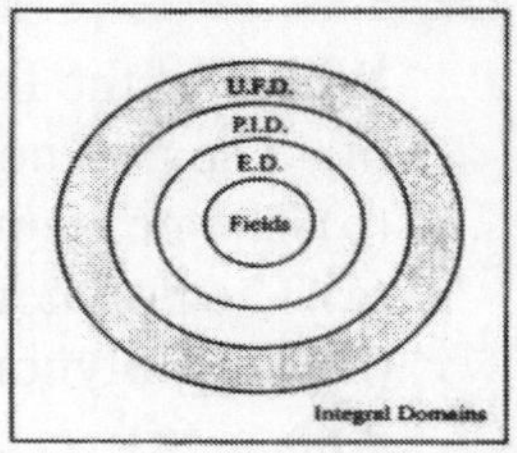

Note : Recall that all the containmentsare proper.

- **Results Over ED, PID, UFD**
    - IfRisafield. then R[x]is ED.
    - If $A = \langle a \rangle$ and $B = \langle b \rangle$, then $AB = \langle ab \rangle$
    - Let R be PID& $a \in R$, then a isprime element ifa is Irreducible element.
    - LetR beaPID, whichisnotafield,then an ideal $I = \langle a \rangle$ is maximal Ideal iff a is an irreducible element of R.
    - Let Rbe a PID and I be non-zero ideal such that $I \neq R$, then I is prime ideal if and only if I is maximal ideal.
    - If R is a PID,then every ascending chain of ideals $\langle a_1 \rangle \subseteq \langle a_2 \rangle \subseteq \ldots\ldots \langle a_n \rangle \subseteq \ldots\ldots$ is finite.
    - The quotient ring of a PID by a prime ideal is again a PID.
    - If Risan ED with norm Nand $a \in R$ be arbitrary then a is unit if and only if N(a) = N(1).
    - R is a UFD $\Leftrightarrow$ R[x] is a UFD.
    - In UFD, every pairof non-zero elements have a gcd and lcm.

4. Content of a Polynomial

Let Rbe a PID and $f(x) = a_n x^n + a_{n-1} x^{n-1} + \ldots\ldots + a_1 x + a_0$ such that $a_n \neq 0$ be a non-zero polynomial over R.

The content of f(x) is greatest common divisor of coefficients $a_0, a_1 \ldots\ldots a_n$ and is denoted by c(f).

Result: If R is UFD and $f(x), g(x) \in R[x]$, then c(fg)= c(f).c(g)

- **Primitive Polynomial:**

Let Rbea principal ideal domain, then a nonzero polynomial f(x) over R is said to be primitive ifthe content of f(x) isunity ofR.

Results:

- **Gauss Lemma:**Let R bet a UFD, then the product of two primitive polynomials in R[x]is a primitive polynomial in R[x].
- If R is UFD, then any $f(x) \in R[x]$ is an irreducible element if and only if either f(x) is an irreducible element of Ror f(x) is an irreducible primitive polynomial of R[x].

SOLVED QUESTIONS

1. Which of the following statements is NOT true? **CSIR NET NOV 2020**

(a) The polynomial ring Z[x] is a Principal Ideal domain (PID)
(b) The polynomial ring Q[x] is a Principal Ideal Domain (PID)
(c) The polynomial ring Z[x] is a Unique Factorization Domain (UFD)
(d) The polynomial ring Q[x] is a Unique Factorization Domain (UFD)

Ans. (a)

Explanation :

For any ring of polynomials R[x]; it is PID if and only if R is a field.
Now, Q is a field but Z is not a field.
So, option (a) is false and (b) is true.
For any ring of polynomials R[x]; it is UFD if and only if R is a UFD.
Now both Z and Q are UFD so option (c) and (d) are true.
But we have choose the incorrect statements.
Thus option (a) is correct answer.

2. Let F[X] be the polynomial ring in one variable over a field F. Then which of the following statements are true? **CSIR NET DEC 2019**

(a) F[X] is a UFD
(b) F[X] is a PID
(c) F[X] is a Euclidean domain
(d) F[X] is a PID but it is not an Euclidean domain

Ans. (a), (b), (c)

Explanation :

Given F is a field, So F[X] is Euclidean domain.
Thus (c) is true.
Since every Euclidean domain is PID and every PID is UFD
So, (a) and (b) is also true.

Trick :

(i) Let R[x,y] be a polyring with two variable and <x> be an ideal then $\frac{R[x,y]}{<x>} \cong R[y]$.
If R is field ⇒ R[y] is ED, PID, UFD, ID.

(ii) $R_1 \times R_2$ never ID.

(iii) Z[x] never PID because it has an ideal <4, x> which is not principal ideal.

(iv) Let R = M_n(R) then it is never ID because it has zero divisors.

3. Which of the following rings are PID. **CSIR NET JUNE 2013**

(a) $\frac{Q[x,y]}{<x>}$ (b) Z ⊕Z (c) Z[x]

(d) M_2(Z), be the ring of all 2 × 2 matrices with entries in Z.

Ans. (a)

MODERN ALGEBRA

Explanation :

Option (a) :

$\frac{Q[x,y]}{<x>} \cong Q[y]$ and Q is field

$\Rightarrow$ Q[y] is PID.

Option (b) :

$Z \oplus Z$ never I.D. $\Rightarrow$ never PID

(1, 0)(0, 1) = (0, 0)

i.e. it has zero divisors.

Option (c) :

Z[x] never PID.

Option (d) : Let $R = M_2(Z)$

$$\begin{bmatrix} 0 & 1 \\ 0 & 0 \end{bmatrix}\begin{bmatrix} 1 & 0 \\ 0 & 0 \end{bmatrix} = \begin{bmatrix} 0 & 0 \\ 0 & 0 \end{bmatrix}$$

$\Rightarrow$ It has zero divisors.

R is never ID, R never PID.

Trick : Let R be ED but not field so we can take R = Z and Z[x] is UFD but not PID.

4. Let R be a ED but not field. Then the polynomial ring R[x] is always

(a) ED (b) PID but not ED (c) UFD but not PID (d) Not a UFD

Ans. (c) **CSIR NET DEC-2011**

Explanation :

Given that R be ED but not field then we can take R = Z. So polynomial ring R[x] = Z[x] and we know that Z[x] is UFD but not PID not ED not field.

So option (c) is true.

Trick : Let F be a finite field and $|F| = p^n$ then number of subfields are $\tau(n)$

Example:- let $|F| = 3^4$then number of subfields are $\tau(2^2) = 3$and divisor of 4 are 1, 2, 4 then cardinality of subfield are 3, 3^2, 3^4

5. Let p be prime. How many distinct subring (with unity) of cardinality p does the field F_{p^2} have

CSIR NET JUNE 2016

(a) 0 (b) 1 (c) p (d) p^2

Ans. (b)

Explanation :

Given that $|F| = p^2$ then number of subfiled of F = $\tau(2) = 2$ like F_p and F_{p^2}.

i.e. Number of subring (with unity) of the field of cardinality p is 1.

Trick : R = Q(Rational)

(i) Q is field

(ii) Q has only two ideal and {0} is only maximal & prime ideal.

6. Let R be a subring of Q containing 1 (unity) the which of the following is true?

CSIR NET DEC. 2017

(a) R is PID (b) R contain infinitely many prime ideals

(c) R contain a prime ideal which is not a maximal.

(d) for every maximal ideal M in R, the residue field R/M is finite.

Ans. (a)

Explanation :

Given that R is subring of Q containing 1 then we can choose R = Q

Option (b) : Q has only one prime ideal {0}

Option (c) : Q has only one maximal ideal as {0} which is prime.

Option (d) : Maximal ideal in Q is {0} then $\frac{Q}{\{0\}} \cong Q$ (Infinite)

Option (a) : Q is field ⇒ Q is also PID.

> **Trick :** (i) ED ⇒ PID ⇒ UFD ⇒ ID i.e. not ID ⇒ not UFD ⇒ not PID ⇒ not ED
> (ii) $Z[\sqrt{-n}]$ never UFD if n ≥ 3

7. Which of the following are ED? **CSIR NET DEC. 2017**

(a) $Z[\sqrt{-3}] = \{a + b\sqrt{-3}; a, b \in Z\}$ (b) Z[x]

(c) $R[x^2, x^3] = \left\{ f = \sum_{i=0}^{n} a_i x^i \in R \mid a_i = 0 \right\}$ (d) $\left(\frac{Z[x]}{<2, x>} \right)[y]$

Ans. (d)

Explanation :

Option (a)

We know that $Z[\sqrt{-n}]$ is never UFD for n ≥ 3 ⇒ $Z[\sqrt{-3}]$ not UFD ⇒ not ED

Option : (b)

Z[x] never PID. ⇒ Z[x] is not ED

Option : (c)

R[x] is a ring and <x> is an ideal of R[x]

Given that a ring $R[x^2, x^3]$

So it has an ideal $<x^2, x^3>$

Which is not a principal ideal because it is generated by two elements.

⇒ $R[x^2, x^3]$ is not PID ⇒ not ED

Option (d) : $\left(\frac{Z[x]}{<2, x>} \right)[y] \approx Z_2[y]$

Z_2 is field ⇒ $Z_2[y]$ is ED.

8. Which of the following are PID? **CSIR NET JUNE 2017**

(a) Q[x] (b) Z[x] (c) (Z/6Z)[x] (d) (Z/7Z)[x]

Ans. (a), (d)

Explanation :

(a) If R is field ⇒ R[x] is ED then R[x] is PID (b) Z[x] never PID (c) $Z_6[x]$ never ID

It has zero divisors

⇒It is not PID

(d) $Z_7[x]$ is PID because Z_7 is field.

9. Which of the following are true? **CSIR NET JUNE 2018**

(a) Every subring of ID is ID. (b) Every subring of UFD is UFD

(c) Every subring of PID is PID (d) Every subring of ED is ED.

Ans. (a)

Explanation :

Let R = R[x] and subring of R is Z[x] R[x] is ED, PID, UFD, ID because R is field.

But Z[x] is UFD & ID But it is neither PID nor ED

So, (c), (d) discarded. Option (b) :

Let R = C(complex number) and subring = $Z = [\sqrt{-5}]$

C is UFD but $Z[\sqrt{-5}]$ is not UFD

10. Which of the following statements are true? **CSIR NET DEC-2014**

(a) Z[x] is Principal Ideal Domain

(b) Z[x, y]/<y + 1> is a unique factorization domain

(c) If R is a principal ideal domain and p is a non-zero prime ideal, then R/P has finitely many ideals.

(d) If R is principal ideal domain, then any subring of R containing I is again a principal ideal domain.

Ans. (b), (c)

Explanation :

For option (a) Z[X] has an ideal <2, x> which is not principal ideal.

Option (a) is incorrect. For option (b), $\frac{Z[X,Y]}{<Y+1>} \cong Z[X]$

Z[X] is UFD. For option (c).

R is P.I.D and P is non-zero prime ideal. P is maximal ideal

($\because$ In case of P.I.D, every non-zero prime ideal $\Leftrightarrow$maximal ideal) $\Rightarrow$R/P is field.

We know that field has only two ideals i.e., {0} and itself

$\Rightarrow$R/P has finitely many prime ideals.

So, option (c) is true. For option (d):

R = R[x] isP.I.D

Let Z[x] be subring of R[x] with unity 1 which is not P.I.D (as explained in option (1)).

So, option (d) is incorrect.

UNSOLVED QUESTIONS

1. Which ofthefollowing rings are principal ideal domains (P.I.D.s)? **CSIR NET DEC-2017**

(a) Z[X]/<x^2 + 1> (b) Z[X] (c) C[X,Y] (d) R[XY]/<x^2+ 1,Y>

2. Let R be a Euclidean domain such that R is not a field. Then the polynomial ring R[X] is

MODERN ALGEBRA

(a) a Euclidean domain (b) a P.I.D. but not E.D.
(c) a U.F.D. but not P.I.D. (d) not U.F.D.

3. Consider the following statement :

I : The ring $Z[\sqrt{-1}]$ is a UFD

II : The ring $Z[\sqrt{-1}]$ is a PID

III : In the $Z_2[x]$, the ideal generated by $x^3 + x + 1$ is maximal

IV : $<x^6 + 1>$ is a prime ideal in $Z_3[x]$.

Which of the following are true?

(a) I, II & III only (b) I and III only (c) I, II & IV only (d) II & III only

4. Let F be a field of 8 elements and $A = \{x \in F \mid x^7 = 1$ and $x^k \neq 1$ for all natural numbers $k < 7\}$. Then the number of elements in A is

CSIR NET JUNE 2012

(a) 1 (b) 2 (c) 3 (d) 6

5. Consider the ring $R = Z[\sqrt{-5}] = \{a + b\sqrt{-5} : a, b \in Z\}$ and the element $\alpha = 3 + \sqrt{-5}$ of R. Then

(a) α is prime (b) α is irreducible (c) R is not a unique factorization domain
(d) R is not an integral domain **CSIR NET DEC 2012**

6. Let R be a commutative ring with unity, such that R[X] is a UFD. Denote the ideal of R[X] by I. Pick each correct statement from below : **CSIR NET JUNE 2016**

(a) I is prime (b) If I is maximal, then R[X] is a PID
(c) If R[X] is a Euclindean domain, then I is maximal
(d) If R[X] is a PID, then it is a Euclidean domain.

7. Consider the following statements

1. Every PID is ED
2. The group of units in the ring Z_{37} is cyclic
3. There is a field with 65 elements

(a) Only 1 is true (b) only 2 is true (c) only 3 is true (d) All of these is false

8. Pick out the cases where the given ideal is a maximal ideal.

(a) The ideal 15Z in Z.
(b) The ideal $T = \{f : f(0) = 0\}$ in the ring C[0,1] of all continuous real valued functions on the interval [0, 1]
(c) The ideal generated by $x^3 + x + 1$ in the ring of polynomials $F_3[x]$, where F_3 is the field of three elements.
(d) None of these **NBHM – 2009**

Answer Key

1	2	3	4	5	6	7	8
a,d	c	a	d	b,c	a,b,c,d	b	b

Chapter 11

Field Theory

1. Field Extension &Galois

- **Definition:** If F is a subfield of a field E; then " E is called an 'extension field ofF" or simply an extension of F.

  Note: If E is an extension of F; then E is a vector space over F; and dimension of E over Fis usually written as [E : F].

  Note : If [E: F]is finite then we say that E is a finite extension of F.

  Example :$[Q(\sqrt{2}):Q]=2$. As $\{1, \sqrt{2}\}$ is a basis.

Result: If [E: K] is finite and [K : F] is finite then [E:F]=[E:K][K:F]

Definition: Let F be a field and let Ebe an extension field of F such that $\alpha \in E$ then $F(\alpha)$ denotes smallest field containing Fand α.

Minimal Polynomial:

Let Fbe a field and let E be an extension field of F such that $\alpha \in E$ then the monic irreducible polynomial of least degree $p(x) \in F[x]$ such that $p(\alpha)=0$ is called the Minimal Polynomial of α over F.

Example : Let E = C and F = R

For minimal polynomial of $\alpha = i$ is $x^2 + 1$

Notation: If E is an extension of Fand $\alpha \in E$ has minimal polynomial of degree n over Fthen $[F(\alpha):F] = n$.

Theorem: Let $\alpha \in F$ be a algebraic over a subfield K of F and let g(x) be the minimal polynomial of α over K. Then,

(i) g(x) is irreducible in K[x]

(ii) For $f(x) \in K[x]$;we have $f(\alpha) = 0$ iffg(x) divides f(x).

(iii) g(x) is the polynomial of least degree having α as a root.

Theorem: Let p(x) be an irreducible polynomial in F[x]. Then ∃ an extension E of Fin which p(x) has a root."

Theorem (Kronecker Theorem)

"Letf(x) be a non-constant polynomial & $f(x) \in F[x]$. Then ∃ an extension EofFin which f(x) hasaroot."

Algebraic Extension :

Algebraic Element :

Let E be an extension of F. An element $\alpha \in E$ is said to be 'algebraic over F if ∃ elements a_0, $a_1, \ldots, a_n \in F (n \geq 1)$

(Not all equal to 0) such that $a_0 + a_1\alpha + \ldots + a_n\alpha^n = 0$

MODERN ALGEBRA

Algebraic Extension :

IfFis a subfield of field Ethen Eis called Algebraic Extension of F if all the elements of E are algebraic over F.

Result-1: "Every finite extension is algebraic extension but converse is not true."

Splitting Field: Let f(x) be a polynomial in F[x] of degree ≥ 1. Then an extension Kof F is called a splitting field of f(x) over F If

(i) f(x) factors into linear factors in K[x]i.e., $f(x)=c(x - x_1)...(x - \alpha_n)$;$\alpha_i \in K$.

(ii) K is generated over F by the roots $\alpha_1, ..., \alpha_n$ of f (x) in K.

Theorem: Any two splitting fields off(x) over a field Fare isomorphic.

Examples : Let p be a prime; $f(x) = x^p - 1 \in Q[x]$then Q has asplittingfield $Q(\alpha)$ where $\alpha \neq 1$ and$\alpha^p = 1$ also $[Q(\alpha); Q] = p - 1$

Result :

The following are equivalent

Every irreducible polynomial in F[x] which has a root in Esplits into linear factors in E.

Let Cbe an algebraic extension of F; then Cis an algebraic closure of F if every non constant polynomial in F[x] splits over C.

Normal Extension :

An extension Eoffield Fis called normal if Esatisfies the condition that if p(x) is an irreducible polynomial over p[x] and αbe a root of p(x) and if $\alpha \in E$ then all the roots of p(x)must be in E.

i.e. p(x) must splits into linear factors in E.

Properties :

C is a normal extension ofR.

R is not a normal extension of Q because $\exists$an irreducible polynomial overQ, $p(x)=x^4 - 2$ and p(x) hasa root in $R\{2^{1/4}\}$ but p(x) does not have all the roots in R.

In general any extension E of a field F; such that [E:F]=2 is a normal extension.

Every finite extension of a finite field is normal.

Important Results :

Let $f(x) \in F[x]$ be a polynomial of degree ≥ 1; with α as a root; then α is a multiple root iff $f'(\alpha)=0$.

Any irreducible polynomial f(x) over a field of characteristic 0 has simple roots; and any irreducible polynomial f(x) over a field F of characteristic $p \neq 0$ has multiple rootsiff $\exists g(x) \in F[x]$.suchthat $f(x) = g(x^p)$,

If F is finite field with p^n elements and m is positive Integer then $\exists$an extension field E of F such that (E:F)=m.

A finite field Fof p^n elements has exactly one subfield with p^melements for each divisior m of n.

Separable Extension :

Separable polynomial : An irreducible polynomial $f(x) \in F[x]$ is called a separable polynomial if all its roots are simple.

Separable Element :Let Ebe an extension of a field F. An element $\alpha \in E$ that is algebraic over F is called separableover F if its minimal polynomial over F is separable.

Separable Extension : An algebraic extension E of a field F is called a "separable extension" if each element of E is separablc over F.

Results :

Any polynomial over a field of characteristic 0 is separable.

Therefore if F is a field of characteristic 0 then any algebraic extension of F is separable.

Irreducible polynomials over finite fields have distinct roots. Hence any algebraic extension of a finite field is separable.

2. Galois Extension :

A finite, normal and separable extension Eof a field Fis called a Golois extension ofF.

GALOIS GROUP

Definition:

Let E be an extension field of the field F.

An Automorphism of Eisa ring isomorphism from E onto E.

The Automorphism group of E fixing Fis the set of all Automorphism of E that take every element of F to itself and it is denoted by Gal(E / F) is called Galois group of Eover F.

Theorem: If F is a finite extension of a field F then $|Gal(E/F)| \leq [E : F]$.

Theorem: Let E be a finite separable extension of a field F;then following are equivalent

(i) E is a normal extension of F.

(ii) Fis the fixed field of Gal(E/F)

(iii) $[E : F] = |Gal(E / F)|$

n

n^{th}Cyclotonic polynomial:

For any positive Integer n; the nth Cyclotomic polynomial is given by

$\Phi_n(x) = (x - \omega_1)(x - \omega_2) \ldots\ldots (x - \omega_s)$

where $\omega_1, \ldots, \omega_s$,are the primitive n^{th}roots of unity.

Results:

- Splitting field $x^n - 1$over Q is $Q(\omega)$.
- For every positive integer n; $x^n - 1 = \prod_{d/n} \Phi_d(x)$

  where the product runs over all positive divisors dof n.
- For every positive integer n; $\Phi_n(x)$ has integer coefficients.
- The cyclotomic polynomials $\Phi_n(x)$ arc irreducible over Z.
- Let wbe a primitive n^{th} root of unity; then $Gal(Q(\omega)|Q) \cong U(n)$.
- If p is a prime and k is a positive integer then $\Phi_{p^k}(x) = \Phi_p(x^{p^{k-1}})$
- The finite field $F_p$contains a primitive n^{th}rootof 1 iff$p \equiv 1 (mod n)$.

SOLVED QUESTIONS

1. Let I denote the ideal generated by $x^4 + x^3 + x^2 + x + 1$ in $Z_2[x]$ and $F = Z_2[x]/I$. Then
(a) F is an infinite field. (b) F is a finite field of 4 elements.
(c) F is a finite field of 8 elements. (d) F is a finite field of 16 elements.

Ans. (d) **GATE-2006**

Explanation :

Note that $I = \langle x^4 + x^3 + x^2 + x + 1\rangle$ is maximal in $Z_2[x]$ as $x^4 + x^3 + x^2 + x + 1$ is irreducible in $Z_2[x]$ and $F = Z_2[x]/I = \{ax^3 + bx^2 + cx + d;\ a, b, c, d \in Z_2\}$ and F contains 2^4 elements.

2. Consider the algebraic extension $E = Q(\sqrt{2}, \sqrt{3}, \sqrt{5})$ of the field Qof rational numbers. Then [E: Q], the degree of E over Q is **GATE-2010**
(a)3 (b)4 (c)7 (d)8

Ans. (d)

Explanation :

$[Q(\sqrt{2},\sqrt{3},\sqrt{5}) : Q]$

$= [Q(\sqrt{2},\sqrt{3},\sqrt{5}) : Q(\sqrt{2},\sqrt{3})] \times [Q(\sqrt{2},\sqrt{3}) : Q(\sqrt{2})] \times [Q(\sqrt{2}) : Q] = 2\times 2\times 2 = 8$

3. The degree of the extension $Q(\sqrt{2}+\sqrt[3]{2})$ over the field $Q(\sqrt{2})$ is
(a)1 (b)2 (c)3 (d)6

Ans. (c) **CSIR NET JUNE – 2011**

Explanation :

Let $\alpha = \sqrt{2}+\sqrt[3]{2} \Rightarrow \alpha - \sqrt{2} = \sqrt[3]{2} = 2^{1/3}$

Take cube on both sides, we get $(\alpha - \sqrt{2})^3 = 2 \Rightarrow \alpha$ satisfies $(x-\sqrt{2})^3 - 2 = 0$ over $Q(\sqrt{2})$ and also $(x-\sqrt{2})^2 - 2$ is irreducible polynomial over $Q(\sqrt{2})$

So, $[Q(\sqrt{2}+\sqrt[3]{2}) : Q(\sqrt{2})] = 3$

So, option (c) is true.

4. Let ω be a complex number such that $\omega^3 = 1$ and $\omega \neq 1$. Suppose L is the field $Q(\sqrt[3]{2}, \omega)$ generated by $\sqrt[3]{2}$ and ωover the field Q of rational numbers. Then the number of subfields K of L such that $Q \not\subset K \not\subset L$ is
(a)2 (b)3 (c)4 (d)5

Ans. (c) **CSIR NET DEC-2012**

Explanation :

Given $L = Q(\sqrt[3]{2}, \omega)$is field, where 'w' is primitive cube root of unity.

$Q(\sqrt[3]{2}, \omega)$is afield extension and $[Q(\sqrt[3]{2}, \omega) : Q] = [Q(\sqrt[3]{2}, \omega) : Q(\omega)][Q(\omega) : Q] = 3\times 2 = 6$

$\therefore$ Degree offield extension $Q(\sqrt[3]{2}, \omega)/Q$ is '6'

So using Galois theory, Galois group of$(Q(\sqrt[3]{2}, \omega)/Q)$ is of order '6' and it is isomorphic to S_3.

$\text{Gal}\left[Q\left(\sqrt[3]{2},\omega\right)/Q\right] \cong S_3$

Using Fundamental theorem of Galois theory, which states that there is one-one correspondence between subgroups of Galois group and intermediate sub fields ofthefield extension.

The number of proper subflelds K of $L = Q\left(\sqrt[3]{2},\omega\right)$ such that Q ⊆ K ⊆L is equal to number of proper subgroups of Gal $\left[Q\left(\sqrt[3]{2},\omega\right)/Q\right] \cong S_3$ which is equal to 4

(∴ proper subgroups of S_3 are A_3, {e, (12)}, {e, (13)} and {e, (23)})

So, option (c) is correct.

5. Which of the following is true? **CSIR NET JUNE- 2012**

(a) sin (7^0) is algebraic over Q.
(b) cos(π/17) is algebraic over Q.
(c) $\sin^{-1}(1)$ is algebraic over Q.
(d) $\sqrt{2}+\sqrt{\pi}$ is algebraic over Q(π)

Ans. **(a), (b), (d)**

Explanation :

For option (a)

$\alpha = \sin 7^0 = \sin\frac{7\pi}{180}$ Consider $\xi = e^{i\frac{7\pi}{180}} = \cos\frac{7\pi}{180} + i\sin\frac{7\pi}{180}$ $\xi^{180} = -1$

⇒ξsatisfies $x^{180}+1$ ⇒ξis algebraic over Q

⇒Q(ξ) is algebraic extension of $Q \Rightarrow \frac{\xi-\xi^{-1}}{2i} = \sin\frac{7\pi}{180}$ is algebraic over Q

(as scalar multiple of algebraic sum of two algebraic number is also algebraic over Q).

$\Rightarrow \sin\frac{7\pi}{180} = \sin 7°$ is algebraic over Q.

So option (a) is correct.

For option (b) $\alpha = \frac{\cos\pi}{17}$ Consider $\xi = e^{\frac{i\pi}{17}} = \cos\frac{\pi}{17} + i\sin\frac{\pi}{17}$

$\Rightarrow \xi^{17} = -1$ ⇒ξsatisfies $x^{17}+1$ ⇒ξis algebraic over Q

⇒Q(ξ) is algebraic extension over Q

$\Rightarrow \frac{\xi+\xi^{-1}}{2} = \cos\frac{\pi}{17}$ isalgebraicoverQ

So, option (b) is correct.

For option (c)

$\alpha = \sin^{-1}(-1) = \frac{\pi}{2}$ is transcendental over Q

as there does not exist any polynomial over Qwhose root is $\frac{\pi}{2}$

So, option (c) is incorrect

MODERN ALGEBRA

For option (d)

$\alpha = \sqrt{2} + \sqrt{\pi} \qquad \Rightarrow \alpha^2 = 2 + \pi + 2\sqrt{2\pi} \Rightarrow (\alpha^2 - (2+\pi)) = 2\sqrt{2\pi}$

$\Rightarrow (\alpha^2 - (2+\pi))^2 = 4 \times 2\pi \Rightarrow (\alpha^2 - (2+\pi))^2 - 8\pi = 0$

$\Rightarrow \alpha$ satisfiesf(x) = $(x^2-(2+\pi))^2$- 8 over Q(π) $\Rightarrow \alpha = \sqrt{2} + \sqrt{\pi}$ is algebraic over Q(π)

So, option (d) is correct.

6. Let F and F' be two finite fields of order q and q' respectively. Then:

(a) F' contains a subfield isomorphic to F if and only ifq≤ q'

(b) F' contains a subfield isomorphic to F if and only ifq divides q'.

(c) If the g.c.d. of q and q' is not 1, then both are isomorphic to subfields of some finite field L.

(d) Both F and F' are quotient rings of the ring Z[x].

Ans. (c), (d) **CSIR NET DEC-2013**

Explanation :

Given F and F' are twofinitefields of order q and q' respectively.

We know that, if F is field of orderp^n then subfield ofF has order p^m where m|n.

Take F to be afield of order p^3= q and F' afield of order p^4 = q'.

Then q≤q'and q|q',

but F' does not contain any subfield which is isomorphic to F as $3 \nmid 4$.

So, options (a) and (b) are incorrect.

If gcd (q, q') is not equal to 1, then q = p^m and q' = $p^n$for some m, n ∈(any finite field has order a power ofprime).

Thus, both F and F' are isomorphic to subfield ofsomefinitefield L such that o(L) = $p^{\text{lcm}(m,n)}$

So, option (c) is correct

We know that every finite field of characteristic 'p' is isomorphic to quotient ring

$\dfrac{Z_p[x]}{< f(x) >}$ wheref(x) ∈$Z_p[x]$ is irreducible polynomial of degree 'n'.

Let $\tilde{f}(x) \equiv f(x)$ over$Z_p[x]$, then $F = \dfrac{Z[x]}{< p, \tilde{f} >}$

So, both F and F' are quotient rings of ring Z[x]

So, option (d) is correct.

7. Let F=$F_3[x]/(x^3+2x-1)$, where F_3 is the field with 3 elements. Which of the following statements are true? **CSIR NET JUNE -2013**

(a) F is a field with 27 elements. (b) F is a separable but not a normal extension of F_3

(c) The automorphism group of F is cyclic.

(d) The automorphism group of F is abelian but not cyclic.

Ans. (a), (c)

Explanation :

For option (a)

$F=\dfrac{F_3[x]}{<x^3+2x-1>}$, where F is field with '3' elements

Sincef(x) = $x^3 + 2x - 1$ is irreducible polynomial over F_3,

⇒F is afield with '27' elements.

($\dfrac{F_p(x)}{<f(x)>}$ is afield with p^n elements iff(x) is irreducible polynomial of degree n over F_p)

So, option (a) is correct.

For option (b),

Note that an algebraic field extension $F \subseteq K$ is called normal extension if K is the splitting field of a polynomial $f(x) \in F[x]$

Since (i) F is finite extension over F_3 of degree 3

(ii) F is splitting field of the polynomial $x^{27}-x$over F_3

⇒ F is normal extension of F_3.

(IfK is field extension of Z_p of degree n then K is splitting field of f(x) = x^{p^n} - xover Z_p)

So, option (b) is incorrect.

For option (c) and (d)

Since, F is a finite field with $27 = 3^3$ elements

The automorphism group of finite field is a cyclic group and is generated by the Frobenius endomorphism.

So, option (c) is correct and option (d) is incorrect.

8. Pick the correct statements **CSIR NET JUNE-2014**

(a) $Q(\sqrt{2})$and Q(i) are isomorphic as Q-vector spaces.

(b) $Q(\sqrt{2})$ and Q(i) are isomorphic as fields.

(c) $Gal_Q\, Q(\sqrt{2})/Q) \cong Gal_Q(Q(i)/Q)$

(d) $Q(\sqrt{2})$ and Q(i) are both Galois extensions of Q.

Ans. (a), (c), (d)

Explanation :

$[Q(\sqrt{2}):Q]=2$

$and\ [Q(i):Q]=2$

So $Q(\sqrt{2})$ and Q(i) are isomorphic as Q-vector spaces

option (a) is correct

For option (b)

Suppose $Q(\sqrt{2})$and Q(i) are isomorphic fields as ϕ: $Q(\sqrt{2})\to Q(i)$ be an isomorphism

MODERN ALGEBRA

$\phi(a+b\sqrt{2})=a+b\phi(\sqrt{2})$

Using the fact that ϕ_Q acts as identity map

We have $(\phi(\sqrt{2}))^2=\phi(\sqrt{2})\phi(\sqrt{2})=\phi(\sqrt{2}.\sqrt{2})=\phi(2)=2\Rightarrow\phi(\sqrt{2})=\pm\sqrt{2}$ which is wrong as $\pm\sqrt{2}\notin Q(i)$

So, $Q(\sqrt{2})$ and Q(i) cannot be isomorphic as fields

So, option (b) is incorrect.

For option (c),

Since, $Q(\sqrt{2}):Q]=$ degree of irreducible polynomial satisfied by $\sqrt{2}$ over Q = 2

Parallely, [Q(i): Q] = 2

Using Galois theory, we get $O(Gal(Q(\sqrt{2})/Q))$ and O(Gal(Q(i)/Q) = 2

$\Rightarrow Gal(Q(\sqrt{2})/Q)\cong Z_2$ *and* $Gal(Q(i)/Q)\cong Z_2$

$\Rightarrow Gal(Q(\sqrt{2})/Q)\cong Gal(Q(i)/Q)$

(As every group of order '2' is isomorphic to Z_2)

So, option (c) is correct.

For option (d)

Since $Q(\sqrt{2})/Q$ and Q(i)/Q both are algebraic extensions and also any automorphism $\sigma:Q(\sqrt{2})\to Q(\sqrt{2})$ and $\tau:Q(i)\to Q(i)$ fixes the elements of Qi.e. fixed field is Q

$\Rightarrow Q(\sqrt{2})/Q$ andQ(i) / Qboth are Galois extensions.

$\Rightarrow$So, option (d) is correct.

9. Let $w=\cos\frac{2\pi}{10}+i\sin\frac{2\pi}{10}$. LetK=Q($w^2$) and letL =Q(w). Then

(a) [L: Q] = 10 (b)[L : K]=2 (c)[K : Q] =4 (d)L=K

Ans. (c), (d)

Explanation :

$$w=\cos\frac{2\pi}{10}+i\sin\frac{2\pi}{10}=e^{\frac{i2\pi}{10}}$$

w is 10th root of unity

Now, K = Q(w^2) andL = Q(w)

Clearly $w^2\in Q(w)\Rightarrow Q(w^2)\subseteq Q(w)$

Claim: w $\in Q(w^2)$.

Nowas$w^5=-1, w^6=-w, (-w^2)^3=w\Rightarrow w\in Q(w^2)\Rightarrow Q(w)\subseteq Q(w^2)$

$\therefore$ we get Q(w) = Q(w^2) [Q(w) :Q] = $\phi(10)=\phi(5)\ \phi(2)=4$.

[By using the result that degree of the cyclotomic extension is $\phi(n)$ where n is primitive n^{th} root of unity].

So, option (a) is ruled out as L = K.

So, [K :Q]= 4 and option (b) is also ruled out.

10. Which of the following quotient rings are fields?

(a) $F_3[X]/<(X^2+X+1)>$, where F_3 is thefinitefield with 3 elements.

(b) $Z[X]/<(X-3)>$ (c) $Q[X]/<(X^2+X+1)>$

(d) $F_2[X]/<(X^2+X+1)>$ where F_2 is the finite field with 2 elements.

Ans. (a), (c)

Explanation :

We know the result if F is field then F[x] is P.I.D and $<f(x)>$ is maximal in F[x] iff f(x) is irreducible over F and $F[x]/(<f(x)>)$ is field iff f(x) is maximal.

For option (a),

$f(x) = 1 + x + x^2$ is not irreducible over F_3 as$f(1) = 3 \equiv 0$ in F_3

(and we know that to check irreducibility of polynomials of degree 2 or 3, we just need to check roots)

So $F_3[x]/(x^2+x+1)$ is not afield.

For option (b),

$Z[x]/(x-3)$ is not afield[$\therefore Z[x]$ is not P.I.D.]

For option (c),

$Q[x]/(x^2+x+1)$where$f(x)=x^2+x+1$. Wewillcheck$f(x+1)=(x+1)^2+x+1+1=x^2+2x+1+x+1+1=x^2+3x+3$

Thus, f(x +1) is irreducible by using Eisenstein irreducibility criteria

$\Rightarrow$f(x) is irreducible $Q[x]/(x^2+x+1)$ is field.

For option (d),

$f(x) = 1 + x + x^2$ is irreducible over $F_2[x]$

$\therefore f(0) = 1 \neq 0$,so$f(1)=3\neq 0$over$F_2$

$\Rightarrow F_2[x]/<x^2+x+1>$is afield.

11. Suppose (F, +, .) is the finite field with 9 elements. Let G = (F, +) and H =(F\{0}) denote the underlying additive and multiplicative groups respectively. Then

(a) $G \cong (Z/3Z) \times (Z/3Z)$ (b) $G \cong (Z/9Z)$

(c) $H \cong (Z/2Z) \times (Z/2Z) \times (Z/2Z)$ (d) $G \cong (Z/3Z) \times (Z/3Z)$ and H (Z/8Z)

Ans. (a), (d) **CSIR NET JUNE -2014**

Explanation :

Given (F, +, .) is finite field with $9 = 3^2$elements

Result: If 'F' is field of order p^n, then (F, +) as a group under addition will not be cyclic group as any element added to itself 'p' times will become identity.

So $(F, +) \cong Z_3 \times Z_3$ So, option (a) is correct and (b) is incorrect

Also, if F isfinitefield, then multiplicative group F* = F{0} is a cyclic group

$\Rightarrow H = F^* = F/\{0\} \cong Z_8$

So, option (c) is incorrect and option (d) is correct.

UNSOLVED QUESTIONS

1. Let $F \subseteq C$ be the splitting field of $x^7 - 2$ over Q and $z = e^{2\pi i/7}$, a primitive seventh root of unity. Let $[F:Q(z)] = a$ and $[F:Q(\sqrt[7]{2})] = b$ Then **CSIR NET JUNE-2013**

(a) a=b=7 (b)a=b=6 (c)a>b (d)a<b

2. Let F be a field of 8 elements and $A = \{x \in F \mid x^7 = 1$ and $x^k \neq 1$ for all natural numbers $k < 7\}$. Then the number of elements in A is

(a)1 (b)2 (c)3 (d)6

3. Let G denote the group of all the automorphisms of the field $F_{3^{100}}$ that consists of 3^{100} elements. Then the number of distinct subgroups of G is equal to

CSIR NET JUNE -2014

(a) 4 (b) 3 (c) 100 (d) 9

4. The number of subfields of afield of cardinality 2^{100} is **CSIR NET JUNE-2015**

(a) 2 (b) 4 (c) 9 (d) 100

5. Let F be afinite field and let K/F be afield extension of degree 6. Then the Galois group of K/F isisomorphic to **CSIR NET DEC-2017**

(a) the cyclic group of order 6. (b) the permutation group on (1,2,3).
(c) the permutation group on (1,2,3,4,5,6). (d) the permutation group on (1).

6. Which of the following(s) is/are true?

(a) $[Q(\sqrt{2}+\sqrt{3}):Q] = 4$ (b) $[Q(\sqrt{2}\sqrt{3}):Q] = 2$
(c) $[Q(\sqrt[3]{4}):Q] = 3$ (d) $[Q(\sqrt{2}+\sqrt{3}):Q] = 2$

7. Which of the following statement(s) is/are true for a field F?

(a) F[x] is an integeral domain (b) F[x] is Euclidean domain
(c) F[x] is principal ideal domain (d) F[x] is not a group

8. Let K be a field, L, a finite extension of K and M a finite extension of L. Then

(a) [M : K] = [M : L] + [L : K] (b) [M : K] = [M : L] [L : K]
(c) [M : L] divides [M : K] (d) [L : K] divides [M : K]

9. Let p and q be two distinct primes. Pick the correct statements from the following :

(a) $Q(\sqrt{p})$ is isomorphic to $Q(\sqrt{q})$ as fields.
(b) $Q(\sqrt{p})$ is isomorphic to $Q(\sqrt{-q})$ as vector spaces over Q
(c) $[Q(\sqrt{p},\sqrt{q}):Q] = 4$ (d) $Q(\sqrt{p},\sqrt{q}) = Q(\sqrt{p}+\sqrt{q})$ **NBHM – 2006**

Answer Key

1	2	3	4	5	6	7	8	9
c	d	d	c	a	a,b,c	a,b,c	b,c,d	b,c,d

MODERN ALGEBRA

Chapter 12

Combination &Number Theory

INTRODUCTION

In this chapter, we shall study some counting principles along with some concepts associated with integers which are also important from an abstract point of view in modern algebra.

1. Combination

Principle of inclusion and exclusion

If A and B are two sets, then

(i) $n(A\cup B)= n(A)+ n(B) - n(A\cap B)$; it is also known as principle of inclusion and exclusion

(ii) $n(A\cup B)= n(A)+ n(B) \Leftrightarrow$ A, B are disjoint sets.

(iii) $n(A-B)+ n(A\cap B)= n(A)$

(iv) $n(A\Delta B)= n(A)+ n(B) - 2n(A\cap B)$, where Δ denotes the symmetric difference of sets

(v) $n(A\cup B\cup C)=n(A)+n(B)+n(C) -n(A\cap B) -n(B\cap C) -n(A\cap C)+n(A\cap B\cap C)$.

Example. In a colony if 70% people read Hindustan Times, 80% people read Times of India and 10% people prefer to stay away from newspaper, then prove that 60% of people read both newspapers and 30% ofpeople who read only one newspaper.

Solution: $n(H)=70$, $n(T) =80$, $n(H'\cap T')= 10$

$\Rightarrow n(H\cup T)=100 - 10 =90$,

$\Rightarrow n(H\cap T) =70 + 80 - 90= 60$.

$\Rightarrow$60% ofpeople read both newspapers

$\Rightarrow n(H'\cap T)+n(H\cap T')= 10 +20=30$

$\Rightarrow$30% people read only one newspaper

Note: $\text{Max}\{n(A),n(B)\} \le n(A\cup B)\le \text{Min}\ \{n(U), n(A)+ n(B)\}$

Permutations:

The arrangements of a number of things taking some or all of them at a time are called permutations. The total number of permutations of n distinct things taking r $(1\le r\le n)$ at a time is denoted by by P(n, r) and is equal $\frac{n!}{(n-r)!}$.

Combinations :

The selections in the form of groups of a number of objects taking some or all of them at a time are called combinations. The total number of combinations of n distinct things taken r $(1 \le r \le n)$ at a timeis denoted by ^nC_ror by C(n, r) and is equal to $\frac{n!}{r!(n-r)!}$.

Results on Combinations:

(i)For 1 ≤r≤n, $^nC_r = \frac{^nP_r}{r!}$. (ii)For1≤r≤n, $^nC_r = \frac{n(n-1)....(n-r+1)}{1.2.....r}$

(iii)In particular, $^nC_0 = {}^nC_n = 1$ (iv) If 1 ≤r≤n, then $^nC_r = {}^nC_{n-r}$

(v) If 1 ≤ r <n, then $^nC_r + {}^nC_{r-1} = {}^{n+1}C_r$

Some important results :

1. If n is a positive integer and p is a prime number, then the exponent of p in n! is given by

$\left[\frac{n}{p}\right]+\left[\frac{n}{p^2}\right]+\left[\frac{n}{p^3}\right]+.....$, where [.] denotes the greatest integer function.

For example, the exponent ofprime number 5 in 234! is equal to

$$\left[\frac{234}{5}\right]+\left[\frac{234}{(5)^2}\right]+\left[\frac{234}{(5)^3}\right]+\left[\frac{234}{(5)^4}\right]+.......=[46.9]+[9.36]+[1.872]+[0.3744]+.......$$

=46+9+1+0+... =56.

2. The number of ways of dividing n identical items among r persons, each one of whom can receive 0,1,2,3,…, n items is $^{n+r-1}C_{r-1}$.

3. If n distinct items are arranged in a row, then the number of ways in which they can be arranged so that no one of them occupies the place assigned to it is

$$n!\left[1-\frac{1}{1!}+\frac{1}{2!}-\frac{1}{3!}+.....+(-1)^n\frac{1}{n!}\right]$$ = the number of derangements.

MODERN ALGEBRA

2. Number theory

Divisibility

Let a and b be any two integers with a ≠ 0. Then a is said to divide b (written as a|b), f there exists an integer c such that b = ac and also a is called a divisor or a factor of b and b is called a multiple of a. If no such integer can be found we say a does not divide b.

Note: If a|b, then∃ an integerc such that b = ac, i.e., b=(-a)(-c). i.e, - a|b. Thus for all practical purposes we can limit our attention to positive divisors integers. As an immediate consequence of the definition, we have

Result of Divisibility: For non-zero integers a, b, c, d the following hold:

(1) a|0, 1|aanda|a.

(2) a|1⇒a=±1.

(3) a|b andc|d ⇒ac|bd.

(4) a|b and b|c ⇒a|c.

(5) a|bandb|a⇒a= ±b or |a| = |b|.

(6) a|banda|c⇒a|bx +cyforanyintegersxandy.

(7) a|bandb≠0 ⇒ |a| ≤ |b|.

The division algorithm

Given integers a and b with $b \neq 0$, there exist unique integers q and r satisfying $a = bq + r$, $0 \leq r < |b|$.

Note: If b does not divide a then $0 < r < |b|$. If $b|a$, then $r = 0$.

Example. Square of an integer is of the type 4k or 4k + 1.

Solution: Any integer is either of the form 2q or 2q+ 1.

$\therefore$ its square is of the form $4q^2$ or $4q^2 + 4q + 1$ i.e., of the form 4k or 4k + 1.

Greatest Common Divisor (gcd)

Let a and b be two integers with atleast one of them different from zero. Then the greatest common divisor of a and b denoted by (a, b) is the positive integer d satisfying

(i) $d|a$ and $d|b$.

(ii) if $c|a$ and $c|b$ then $c|d$.

We denote gcd(a,b) by (a,b).

Results on gcd:

(1) For any integer $a \neq 0$, $(a, a) = a$.

(2) If $a|b$ then $(a, b) = a$.

(3) Let a and b be two integers such that atleast one of them is non-zero. Then there exist integers x and y such that $(a, b) = ax + by$.

(4) gcd is always unique.

(5) $(ma, mb) = m(a, b)$, where m is positive integer or $(ma, mb) = |m|\,(a, b)$ where m is any integer.

(6) $(a,b) = (b,a) = (a,-b)$

(7) (i) If $d|a$ and $d|b$, then $\left(\frac{a}{b}, \frac{b}{d}\right) = \frac{1}{d}(a,b)$

(ii) If d (a, b), then $\left(\frac{a}{d}, \frac{b}{d}\right) = 1$

(8) $(a, b) = 1 \Leftrightarrow \exists$ integers x and y such that $ax + by = 1$.

(9) If $(a, m) = 1$, $(b, m) = 1$ then $(ab,m) = 1$

(10) **Euclid's lemma:** For $a, b, c \in Z$. If $c|ab$ and $(c, b) = 1$ then $c|a$.

Remark: If $(c, b) \neq 1$ the result is not true $\therefore 8|72$ i.e. $8|8.9$

$8|8$ but $8\nmid 9 \therefore (8,8) \neq 1$

(11) Product of 'r' consecutive integers is divisible by r!

Example. If $a|m$, $b|m$ and $(a, b) = 1$, then $ab|m$.

Solution: $a|m \Rightarrow m = ax$ for some integer x.

$b|m \Rightarrow b|ax$.

Also $(a, b) = 1$

$\therefore$ By Euclid's lemma, $b|x$

$\therefore x = by$ for some integer y.

Thus we get that m = ax = aby.

$\therefore$ ab|m.

Coprime: If a and b are integers not both zero, then a, b are relatively prime if (a, b) = 1.

Least common multiple (lcm)

Let a and b two non-zero integers. The least common multiple of two non-zero integers a and b denoted by l.c.m (a, b) or [a, b] is the positive integer m satisfying

(i) a|m and b|m,

(ii) if a|c and b|c with c>0, then m|c.

Same Parity: Two integers are said to be of same parity if they are both odd or even.

Results on LCM:

(1) Let m be a positive integer, then [ma, mb] = m[a, b] when a, b are non-zero integers.

(2) [a, b] (a, b) = |ab|, i.e., product of two numbers = lcm × gcd (of the two numbers)

Example (a,4)=2, (b,4)=2 then prove that (a+b,4) = 4

Solution: (a, 4)=2 $\Rightarrow$2|a

$\Rightarrow a = 2a_1$ for some a_1, $2 \nmid a_1$.

(b, 4) =2$\Rightarrow$2|b

$\Rightarrow b = 2b_1$ for some integer b_1, $2\nmid b_1$

So a_1 and b_1 are both odd.

$\therefore a + b = 2(a_1 + b_1)$ is an even number and equals 4k for some integer k.

Prime numbers and composite numbers

Prime Numbers: An integer p> 1 is called a prime number f there is no divisor d of p satisfying 1<d <p i.e., p and 1 are the only divisors of p. The only even integer that is a prime is 2.

Composite Number: A positive integer n>1 which is not a prime number is called a composite number.

Results on prime numbers:

(1) Every integer n> 1 can be expressed as a product of primes.

(2) If p|ab and p is a prime number, then p|a or p|b.

(3) If $p|a_1a_2,...a_n$ then $p|a_1$ or $p|a_2$ or ...or $p|a_n$.

Congurences a is congruent to b modulo m:

We say that an integer a is congruent to b modulo m if m|a-b where m is a non-negative integer and we write it as a ≡b (mod m), m ∈Z, m ≠ 0

If $m\nmid a - b$, then $a \not\equiv b \pmod m$

Note: As m|a – b$\Rightarrow$ -m|a-b. So, we can concentrate on modulo positive integers only.

Results on congruences:

(1) If a ≡b (mod m), then a and b leave same remainder r when divided by m and a≡ r(mod m)

(2) If a, b, c, x, y are integers and m is a positive integer, then

(i) a≡a(mod m).

(ii) a≡b(mod m)$\Leftrightarrow$b≡a(mod m) $\Leftrightarrow$a – b≡ 0(mod m) $\Leftrightarrow$m|a-b.

MODERN ALGEBRA

(iii) $a \equiv b \pmod m$ and $b \equiv c \pmod m \Rightarrow a \equiv c \pmod m$.

Note: Congruency (mod m) is an equivalence relation

(iv) $a \equiv b \pmod m, c \equiv d \pmod m \Rightarrow ax+cy \equiv (bx+dy) \pmod m$ for integer x,y

(v) $a \equiv b \pmod m,\ c \equiv d \pmod m \Rightarrow ac \equiv bd \pmod m$

(vi) $a \equiv b \pmod m \Rightarrow ac \equiv bc \pmod m$ for integer c.

(vii) $a \equiv b \pmod m$ and $d|m$ then $a \equiv b \pmod d$

(viii) $a \equiv b \pmod m \Rightarrow a^k \equiv b^k \pmod m$ for any positive integer k

Note: The converse is not true

$\therefore 4^2 \equiv 8^2 \pmod 3$ as $3 \mid (16 - 64)$ but $4 \not\equiv 8 \pmod 3$.

(3) (i) $ax \equiv ay \pmod m \Leftrightarrow x \equiv y \pmod m$ iff $(a,m) = 1$.

(ii) $x \equiv y \pmod{m_i}, 1 \le i \le r$. Then $x \equiv y \pmod{[m_1.m_2...m_r]}$

(4) If $b \equiv c \pmod m$, then $(b, m) = (c, m)$.

(5) **Euler's Theorem:** Let m be a positive integer and $(a,m) = 1$, then $a^{\phi(m)} \equiv 1 \pmod m$, where $\phi(m)$ is Euler's - ϕfunction.

(6) **Fermat's Theorem.** Let p be a prime and $a \in Z$ be such that $p \nmid a$.
Then $a^{p-1} \equiv 1 \pmod p$. In general $\forall\ a \in Z$, $a^p \equiv a \pmod p$.

(7) Let p be a prime, then $x^2 \equiv 1 \pmod p \Rightarrow x \equiv \pm 1 \pmod p$.

Example. Find the remainder when $S = 1! + 2! + 3! + ... + 10000!$ is divided by 4.

Solution: As 4!, 5!, 6! ... are all divisible by 4 i.e., $k! \equiv 0 \bmod 4\ \forall k \ge 4$.

$\therefore 1! + 2! + 3! \pmod 4 \equiv 9 \pmod 4$, i.e., $S \equiv 1 \pmod 4$

So, remainder = 1.

Linear congruence

An expression of the type $ax \equiv b \pmod m$ is called linear congruence. By a solution of this linear congruence, we mean an integer x_0 such that $ax_0 \equiv b \pmod m$.

Note: By the number of solutions of the congruence $ax \equiv b \pmod m$, we mean the number of incongruent integers x such that $ax \equiv b \pmod m$

Important Results on Linear Congruences:

(1) The linear congruence $ax \equiv b \pmod n$ has a solution $\Leftrightarrow d|b$, where $d = (a, n)$.
Also if $d|n$, then it has d mutually incongruent solutions (mod n).

(2) **Chinese Remainder Theorem:**
Let $n_1, n_2, \ldots, n_r$ be pairwise coprime positive integers and $a_1, a_2, \ldots, a_r$ be any integers, then the system congruences:

$x \equiv a_1 \pmod n$

$x \equiv a_2 \pmod{n_2}$

|

$x \equiv a_r \pmod{n_r}$

has a simultaneous solution which is unique modulo $n_1 n_2 ... n_r$.

(3) If $(a, m) = 1$, then $ax \equiv b \pmod m$ has a solution x_1 and if x_2 is any other solution, then $x_2 \equiv x_1 \pmod m$.

MODERN ALGEBRA

Result: Linear congruences $ax \equiv b(\text{mod}\, m)$ has a solution iff $(a,m) = d|b$. It has exactly 'd' incongruent solution modulo m.

(4) **Wilson's theorem:** An integer p is a prime if and only if $(p-1)! \equiv -1(\text{mod}\, p)$

Example $p = 13$

$(p-1)! = 1.2.3...12$

$2.7 \equiv 1 \pmod{13}$

$3.9 \equiv 1 \pmod{13}$

$4.10 \equiv 1 \pmod{13}$

$5.8 \equiv 1 \pmod{13}$

$6.11 \equiv 1 \pmod{13} \Rightarrow (12)! \equiv -1 \pmod{13}$

(5) Let p be a prime, then $x^2 \equiv -1 \pmod{p}$ has a solution if $p = 2$ or $p \equiv 1 \pmod{4}$.

Example. Find an integer x such that

$x \equiv 1 \pmod{3}$,

$x \equiv 2 \pmod{5}$,

$x \equiv 3 \pmod{7}$.

Solution: We have $n = 3.5.7 = 105$

There $N_1 = 35$, $N_2 = 21$, $N_3 = 15$

We are to find x_1, x_2, x_3 such that $N_1x_1 \equiv 1 \pmod{3}$, $N_2x_2 \equiv 1 \pmod{5}$, $N_3x_3 \equiv 1 \pmod{7}$.

Now, $35\, x_1 \equiv 1 \pmod{3}$ i.e. $2x_1 \equiv 1 \pmod{3}$

i.e., $x_1 = 2$.

$N_2x_2 \equiv 1 \pmod{5}$

$21\, x_2 = 1 \pmod{5} \Rightarrow x_2 \equiv 1 \pmod{5}$

i.e., $x_2 = 1$

$N_3x_3 \equiv 1 \pmod{7}$

$15x_3 \equiv 1 \pmod{7} \Rightarrow x_3 \equiv 1 \pmod{7}$

i.e. $x_3 = 1$

Now, take $x = 1.35.2 + 2.21.1 + 3.15.1 = 70 + 42 + 45 = 157$.

So that $157 \equiv 52 \pmod{105}$

Thus, the required solution is 52.

Example. Show that there is no integer x such that

$x \equiv 29 \pmod{52}$,

$x \equiv 19 \pmod{72}$.

Solution: Suppose an integer x such that $x \equiv 29(\text{mod}\, 13.2^2)$ .... (1)

and $x \equiv 19(\text{mod}\, 2^2.3^2)$ .... (2)

$(1) \Rightarrow x \equiv 29 \pmod{13}$ and $x \equiv 29(\text{mod}\, 2^2)$.

$\Rightarrow x \equiv 3 \pmod{13}$ and $x \equiv 1 \pmod{4}$

$(2) \Rightarrow x \equiv 19(\text{mod}\, 2^3)$ and $x \equiv 19(\text{mod}\, 3^2)$.

$\Rightarrow x \equiv 3 \pmod{8}$ and $x \equiv 1 \pmod{9}$

But $x \equiv 1 \pmod{4}$ and $x \equiv 3 \pmod{8}$ cannot hold simultaneously

MODERN ALGEBRA

Since x ≡ 3(mod8) ⇒x ≡ 3 (mod4)

∴Given congruence has no solution.

Example. Write a single congruence equivalent to the pair of congruence x ≡1 mod 4 and x ≡2 mod 3.

Solution: x ≡1 mod 4⇒4|x –1⇒12|3x – 3

x≡2mod3⇒3|x – 2= 12|4x – 8

⇒12|(4x – 8) - (3x – 3)

⇒12|x – 5⇒ x ≡5 (mod 12) OR x≡1 (mod 4) ⇒x⇒5 (mod 4)⇒4|x – 5

x≡2(mod3) ⇒x≡5(mod3) ⇒3|x – 5 ⇒12|x – 5⇒x≡5(mod12).

(6) The number of incongruent solutions of 18x ≡30 (mod 42) is given by (18, 42) = 6 and as 6|30,

∴the given congruence has 6 solutions.

Note that 42|18x – 30 ⇒x_0= 4 is a solution.

$\therefore\ x_0, x_0+\frac{n}{d}, x_0+\frac{2n}{d}, ..., x_0+\frac{5n}{d}$ are all the mutually incongruent solutions modulo n, namely 4, 11, 18, 25, 32, 39.

Example Any integer, which is a square, must have 0,1,4,5,6 or 9 at unit place.

Solution: Any positive integer N can be written as $N = a_0 + a_1 10 + a_2 10^2 + ... + a_n 10^n$

⇒ N = 10a + b where $a = a_1 + a_2 10 + ... + a_n 10^n$ and $b = a_0$.

⇒ $N^2 = 100a^2 + 20ab + b^2 = 10A + b^2$ where $A = 10a^2 + 2ab$

Now, b^2= 0,1,4,9,16,25,36,49,64,81

∴digit at unit place of N^2= digit at unit place of b^2= 0,1,4,5,6 or 9.

Special divisibility test

Divisibility test is short way of determining whether a given integer is divisible by a fixed divisor without performing division. Let $N = a_n .... a_2 a_1 a_0$ be an integer.

$N = a_n 10^n + ... + a_2 10^2 + a_1 10 + a_0$ is the decimal representation of N.

Then the following table lists the divisibility test for integer till 15.

DIVISIBILITY TEST FOR (1 TO 15)

DIVISOR	DIVISIBILITY CONDITION	EXAMPLES
1	No condition. Every integer is divisible by 1	1|a ∀ a∈Z
2	The unit digit is even (0,2,4,6 or 8)	1498; 8 is even
3	The sum of digits is divisible by 3	1233 is divisible by 3 as 1+2+3+3=9 is divisible by 3.
4	The number formed by its unit and tens place is divisible by 4	5432 is divisible by 4 as last two digit '32' is divisible by 4.
5	The unit digit is 0 or 5	15625 is divisible by 5 as last digit is 5

MODERN ALGEBRA

6	If it is divisible by 2 and 3	23208 is divisible by 3 as 2+3+2+0+8 is divisible by 3and last digit is 8, it is divisible by 2. Hence divisible by 6
7	Subtract 2 times the last digit from remaining number and the result must be divisible by7	343 is divisible by 7 as $34 - 3\times2 = 28$ is divisible by 7
8	If the last three digit is divisible by 8	19136 is divisible by 8 as last three digit 136 is divisible by 8
	Add four times the hundreds digit to twice the tens digit to the unit plane digit and the result must be divisible by '8'.	19136; $4 \times 1 + 2 \times 3 + 1 \times 6 = 16$ is divisible by 8 $\Rightarrow$ 19136 is divisible by 8
9	The sum of its digits is divisible by 9	931122 is divisible by 9 as sum of digits 9+3+1+1+2+2=18 is divisible by 9
10	The last digit is 0	9182430 is divisible by 10
11	Subtract the last digit from the remaining leading number. If result is divisible by 11. Then so was the first number and apply this over and over again as necessary.	9845 is divisible by 11 $\therefore$984-5=979 $\Rightarrow$97-9=88 is divisible by 11. Hence 9845 is divisible by 11.
12	If it is divisible by 3 and 4	107580 is divisible by 3 and 4. Hence it is divisible by 12
13	Add four times the last digit to the remaining leading number. If result is divisible by 13, then so was the first number. Apply this rule over and over again as necessary.	59397 is divisible by 13 $5939+ 7 \times 4 = 5967$ 596+28=624 $62+4\times4=78$ is divisible by 13.
14	If it is divisible by 7 and 2	13482 is divisible by 2 and 7 $\Rightarrow$13482 is divisible by 14
15	If it is divisible by 3 and 5	6840 is divisible by 3 and 5 $\Rightarrow$it is divisible by 15

Number theoretic functions

1. Arithmetic Functions: Any function whose domain is the set of positive integers is called an arithmetic function. The value of a number theoretic function need not be an integer.

2. **Euler's ϕfunction:** The number of positive integer less than 'n' which are coprime to n denoted by $\phi(n)$ is called Euler's ϕfunction.
So, $\phi(n)$ = cardinality of $\{j : 1 \le j<n, (j,n) = 1)$

3. **Multiplicative Function:** A number theoretic function is called multiplicative if $f(mn) = f(m)f(n)$ whenever $(m,n) = 1$.

4. **Number of divisors $\tau(n)$ and sum of divisors $\sigma(n)$:** Given a positive integer n, let $\tau(n)$ (or $d(n)$) denotes the number of positive divisors of n and $\sigma(n)$ denotes the sum of positive divisors of n. i.e.,
i.e., $\tau(n)=\sum_{d|n}1,\ \sigma(n)=\sum_{d|n}d$

Note: If $p_1^{k_1}p_2^{k_2}.....p_r^{k_r}$ is the prime factorization of n> 1, then the positive divisors of n are precisely the integersdoftheform $d = p_1^{\alpha_1}p_2^{\alpha_2}....p_r^{\alpha_r}$, where $0 \le \alpha_i \le k_i$, i = 1,2,...., r.

SOLVED QUESTIONS

1. Let n be a fixed positive integer and let C_r stand for the usual binomial coefficient; i.e., the number of ways of choosing r objects from n objects. Evaluate: $C_1 + 2C_2 + \dots nC_n$.

NBHM-2009

(a) $(n^{2(n-1)})$ (b) $(n2^{(n+1)})$ (c) $n + 1$ (d) $n - 1$

Ans. (b)

Explanation :

Take n = 3

The number of ways of choosing r objects from 3 objects is C_r

Now $^3C_1+2.{}^3C_2+3.{}^3C_3=\frac{3!}{1!(3-1)!}+2.\frac{3!}{2!(3-2)!}+3.\frac{3!}{3!0!}$

$=\frac{3!}{2!}-2.\frac{3!}{2!}+3=3+6+3=12+3.2^2$

Take n = 4

$\Rightarrow$Given relation is $C_1+ 2C_2+ 3C_3+ 4C_4= {}^4C_1+ 2.{}^4C_2 + 3.{}^4C_3 + 4.{}^4C_4$
$= 4+2.6+3.4+4=4 + 12 + 12 + 4 = 4.2^3$ and so on .....

If we take n terms, then we get $C_1+ 2C_2+ \dots\dots\ {}^nC_n= n\ 2^{n-1}$

2. Pick out the true statements.
(a) The sum of r consecutive positive integers is divisible by r
(b) The product of r consecutive positive integers is divisible by r!.
(c) For each positive integer r, there exist r consecutive positive integers which are all composite.
(d) None of these

Ans. (b), (c)

MODERN ALGEBRA

Explanation :

Take r = 4

Sum of 4 consecutive positive integers =1+2+3+4=10

10 is not divisible by r=4

⇒option (a) is incorrect

We take any r(an integer) and consider the sequence

(r+1)!+2, (r+1)!+3, (r+1)!+4+... (r+1)!+(r+l)

All these numbers are composite

So, for any r, we can make a sequence of r consecutive numbers which are composite

⇒option (c) is correct.

Now consider r consecutive positive integers (n+1), (n +2) ... (n+r)

We have $(n+1)(n+2)......(n+r)=\dfrac{(n+r)!}{n!}=r!\ {}^{n+r}C_r$

Now since ${}^{n+r}C_r$is always a positive integer

⇒r! divides product of r consecutive positive integers

∴option (b) is correct.

3. The number of 4 digit numbers with no two digits common is

(a) 4536 (b) 3024 (c) 5040 (d) 4823

Ans. (a) **CSIR NET JUNE-2011**

Explanation :

The number of '4' digit numbers with no two digit common = the total number of '4' digit number without repetition.

At first place, we have 9 choices (0 will not come at first place otherwise, number will become 3 digit number).

At 2nd place, we have 9 choices (Now 0 is included and number that has come at first place will not come at 2nd place as repetition is not allowed).

Similarly at 3rd place, we have 8 choices

At 4th place, we have 7 choices.

Total numbers = 9×9×8×7 = 4536

So, option (a) is correct.

4. The unit digit of 2^{100} is **CSIR NET JUNE -2011**

(a) 2 (b)4 (c)6 (d)8

Ans. (c)

Explanation :

$2^1\equiv 2$ $2^2\equiv 4$ $2^3\equiv 8$ $2^4\equiv 6$

$2^5\equiv 2$ $2^6\equiv 4$ $2^7\equiv 8$ $2^8 = 6$ and so on ….

In general $2^{4k+1}\equiv 2$

$2^{4k+2}\equiv 4$

$2^{4k+3}\equiv 8$

$2^{4k+4}\equiv 6$

100 is of type 4k+4

So, unit digit of 2^{100} is 6

So, option (c) is correct.

5. Which of the following statements are true? **NBHM-2012**

(a) The 9th power of any positive integer is of the form 19m or 19m ± 1

(b) For any positive integer n, the number $n^{13} - n$ is divisible by 2730

(c) The number 18! + 1 is divisible by 437 (d) None of these

Ans. (a), (b), (c)

Explanation :

We will prove that the 9th power of any integer is of the form 19m or 19m±1

Any integern = 19m+a, $0 \le a < 19$

$$n^9 = (19m + a)^9 = (19m)^9 + 19a(19m)^8 + 171a^2(19m)^7 + ... + 19a^8(19m) + a^9$$

Clearly, we can factor 19 out of the first 19 terms.

Now, we will show a^9, $0 \le a \le 19$ is of the required form

$0^9 = 0 = 19(0)$

$1^9 = 1 = 19(0) + 1$

$2^9 = 512 = 19(27) - 1$

$3^9 = 19683 = 19(1036) - 1$

$4^9 = (2^2)^9 = [19(27)-1]^2 = 19[19(27)^2 - 2(27)] + 1$

Similarly, we can prove for other values of a

⇒option (a) is correct

Take a number 2730

$2730 = 2 \times 3 \times 5 \times 7 \times 13$

By Fermat 's little theorem,

$n^7 \equiv n \pmod 7$

$n^{13} \equiv n^7 . n^6 \equiv n . n^6 \equiv n \pmod 7$

$n^{13} \equiv (n^5)^2 . n^3 \equiv n^2 . n^3 \equiv n \pmod 5$

$n^{13} \equiv (n^3)^4 . n \equiv n^4 . n \equiv n^3 . n^2 \equiv n^3 \equiv n \pmod 3$

$n^{13} \equiv n \pmod{13}$

$\Rightarrow 2 | n^{13} - n,\ 3 | n^{13} - n,\ 5 | n^{13} - n,\ 7 | n^{13} - n,\ 13 | n^{13} - n$

So $2.3.5.7.13 \mid n^{13} - n \Rightarrow n^{13} \equiv n \pmod{2.3.5.7.13} \Rightarrow n^{13} \equiv n \pmod{2730}$

So option (b) is also correct.

Now, we will prove 18! + 1 is divisible by 437.

$437 = 19 \times 23$

By Wilson's theorem,for any prime p, $(p-1)! \equiv -1 \pmod p$,

19 is a prime number, so by Wilson 's theorem

$(19-1)! \equiv 1 \pmod{19} \Rightarrow 19 | 18! + 1$

$\Rightarrow 18! \equiv -1 \pmod{19}$

Also 23 is prime, so by Wilson's theorem

$\Rightarrow 22! \equiv -1 \pmod{23}$

MODERN ALGEBRA

As (18)! (19) (20) (21) (22) = 22!, we note that
$19 \equiv -4 \pmod{23}$,
$20 \equiv -3 \pmod{23}$,
$21 \equiv -2 \pmod{23}$,
$22 \equiv -1 \pmod{23}$,
$\Rightarrow (18)! \equiv (-4)(-3)(-2)(-1)(-1) \pmod{23}$ $\Rightarrow (18)! \equiv -1 \pmod{23}$
$\Rightarrow 23 \mid (18)!+1$ $\Rightarrow (19)(23) \mid 18!+1$ $\Rightarrow 437 \mid 18!+1$
Thus, options (a), (b) and (c) are correct.

6. For a positive integer m, let $\phi(m)$ denote the number of integers K such that $1 \le K \le m$ and GCD (K, m) = 1. Then which of the following statements are necessarily true?
(a) $\phi(n)$ divides n for every positive integer n.
(b) n divides $\phi(a^n - 1)$ for all positive integers a and n.
(c) n divides $\phi(a^n - 1)$ for all positive integers a and n such that GCD (a, n) = 1.
(d) a divides $\phi(a^n - 1)$ for all positive integers a and n such that GCD (a, n) = 1.

Ans. (b), (c) **CSIR NET DEC –2012**

Explanation :
For option (a),
Let n = 3, $\phi(n) = 2$
But $2 \nmid 3 \Rightarrow \phi(n)$ does not divide 'n 'for every integer n
So, option (a) is incorrect.
For option (b),
To prove: $n \mid \phi(a^n - 1)$, $\forall$positive integers a and n
$a^n - 1 \mid a^n - 1$ $\forall$a, n positive integers.
$\Rightarrow a^n \equiv 1 \pmod{a^n - 1}$ and 'n' is least positive integer.
$\Rightarrow$Order(a) = n (mod(a^n- 1)) ... (1)
Since, gcd(a, a^n-1) = 1
$\therefore$ ifgcd(a, a^n- 1) = d
$\Rightarrow d \mid a, d \mid a^n - 1 \Rightarrow d \mid a^n, d \mid a^n - 1 \Rightarrow d \mid a^n - a^n + 1 = d \mid 1 \Rightarrow d = 1$
By Euler's theorem, $a^{\phi(an-1)} \equiv 1(\text{mod } (a^n - 1))$
$\Rightarrow O(a) \mid \phi(a^n - 1) \Rightarrow n \mid \phi(a^n - 1)$for all positive integers a and n
So, option (b) is true.
For option (c),
Since, $n \mid \phi(a^n - 1)$ $\forall$a, n positive integers.
In particular, $n \mid \phi(a^n - 1)$, for all positive integers a and n such that gcd(a,n) = 1.
So, option (c) is true.
For option (d),
Let a = 3, n = 2 such that (3,2) = 1
$a^n - 1 = 3^2 - 1 = 8$
$\phi(a^n - 1) = \phi(8) = 4but3 \nmid 4$

$\Rightarrow$'a' does not divide $\phi(a^n - 1)$for all positive integers a and n such that gcd(a,n) = 1
So option (d) is incorrect.

7. Consider the congruence $x^n \equiv 2 \pmod{13}$. This congruence has a solution for x if
(a)n=5 (b)n=6 (c)n=7 (d)n=8

Ans. (a, c) **CSIR NET JUNE-2013**

Explanation :

We have a result, that says $x^n \equiv a \pmod m$ has a solution iff $a^{\phi(m)/d} \equiv 1 (\text{mod} m)$,
where d=gcd (n, $\phi(m)$)
So $x^n \equiv 2 (\text{mod } 13)$ has a solution iff $2^{\phi(13)/d} \equiv 1 \pmod{13}$, where d = (n, 12)
$\Rightarrow 2^{12} \equiv 1(\text{mod } 13)$, d=gcd(n, 12)
Ifn=5andn=7thend=1 ($\because$ gcd(5,12)=landgcd(7,12,)=1)
$\Rightarrow 2^{12} \equiv 1 \pmod{13}$ which is true always
(By using Euler's theorem)
So, option (a) and (c) are correct.
If n = 6, thend=gcd(6, 12) =6
$2^{12/6} \equiv 2^2 \equiv 1 \pmod{13}$
So, option (b) is incorrect.
If n = 8, thend = gcd(8, 12) =4
$2^{12/4} \equiv 2^3 \equiv 1 \pmod{13}$
So, option (d) is incorrect.

8. A class consists of 4 boys and 4 girls. A committee that consists of a president, a vice-president and a secretary is to be chosen among the 8 students of the class. Let a denote the number of ways of choosing the committee in such a way that there is at least one boy and at least one girl. Let b denote the number of ways of choosing the committee in such a way that the number of girls is greater than or equal to that of boys. Then

CSIR NET JUNE -2014

(a) a=288 (b)b = 168 (c)a = 144 (d)b = 192

Ans. (a), (b)

Explanation :

We are given a class consisting of 4 girls and 4 boys.
a = number of ways of choosing the committee in such a way that the committee has atleast one boy
and atleast one girl = $({}^4C_2.{}^4C_1+{}^4C_1.{}^4C_2).3!$ ....(1)
(We have 2 cases only, Case-1= 2 boys and 1 girl Case-2 = 2 girls and 1 boy)
Also when we have choosen 3 students among 8 students, then first student has 3-choices
(president/V.P/secretary)
2^{nd}student has 2 choices and then last student has one choice.
Total ways = 3!
That is why, we have to multiply by 3! in equation (1))

$\Rightarrow a = (24 + 24)6 = 48 \times 6 = 288$

Similarly $b = ({}^4C_3 . {}^4C_0 + {}^4C_2 . {}^4C_1).3! = (4 + 24)6 = 28 \times 6 = 168$

So, options (a) and (b) are correct.

9. For positive integers m and n, let $F_n = 2^{2n} + 1$ and $G_m = 2^{2m} - 1$. Which of the following statements are true? **CSIR NET JUNE-2014**

(a) F_n divides G_m whenever m >n.
(b) GCD $(F_n, G_m) = 1$, whenever $m \neq n$
(c) GCD $(F_n, F_m) = 1$, whenever $m \neq n$.
(d) G_m divides F_n whenever $m < n$

Ans. **(a), (c)**

Explanation :

For positive integer m and n

$F_n = 2^{2n} + 1$ and $G_m = 2^{2m} - 1$

For option (1),

Form >n

let $m = n + k$, $k > 0$

So, $G_m = (x)^{2k} - 1$ where $x = 2^{2n}$ $= (x+1)\underbrace{\left(x^{2^k-1} - x^{2^k-2} + x^{2^k-3} .... -1\right)}_{=\alpha}$

$\Rightarrow G_m (2^{2n} + 1).\alpha = F_n.\alpha$ $\qquad \Rightarrow F_n$ divides G_m

So, option (a) is true.

For option (b)

Take n = 1, m = 2

$F_n = F_1 = 5$ and $G_m = G_2 = 15$

$\Rightarrow Gcd(F_n, G_m) = Gcd(F_1, G_2) \neq 1$ for $m \neq n$

So, option (b) is incorrect.

For option (c)

let $(F_m, F_n) = d$ where $m \neq n$

without loss of generality, suppose m >n

Since, both F_m and F_n are odd, therefore 'd' is odd.

Now $\dfrac{F_m - 2}{F_n} = \dfrac{\left(2^{2^n} + 1\right) - 2}{2^{2^n} + 1} = \dfrac{\left(2^{2^n}\right)^{2^{m-n}} - 1}{2^{2^n} + 1} = \dfrac{x^k - 1}{x - 1}$, where $x = 2^{2n}$ and $K = 2^{m-n}$

$= x^{K-1} - x^{K-2} + x^{K-3}$ ….. -1.

= an integer quantity

$\Rightarrow F_n | F_m - 2$

Also $d | F_n$ and $d | F_m$

$\Rightarrow d | F_m - 2$ and $d | F_m \Rightarrow d | F_m - (F_m - 2) \Rightarrow d | 2 \Rightarrow d = 1$ or 2

But 'd' is odd, $\therefore$ d = 1

So, $(F_n, F_m) = 1$ for $m \neq n$

So, option (c) is true

For option (d)

Take m = 1 and n = 2

Then, $G_m = G_1 = 3$ and$F = F_2 = 17$ and $3 \nmid 17$

So, option (d) is incorrect.

10. Let φ(n) be the cardinality of the set {a|1 ≤ a ≤ n, (a, n) = 1} where (a, n) denotes the gcd of a and n. Which of the following is NOT true?

(a) There exist infinitely many n such that φ(n) >φ(n + 1)

(b) There exist infinitely many n such that φ(n) <φ(n + 1)

(c) There exist N ∈ N such that N > 2 and for all n > N, φ(N) <φ(n)

(d) The set $\left\{\frac{\varphi(n)}{n} : n \in N\right\}$ has finitely many limit points.

Ans. (d) **CSIR NET JUNE 2020**

Explanation :

φ(n) be the cardinality of the set {a|1≤ a ≤ n, (a, n) = 1}. We need to find the wrong statement. We have, for every prime p, φ(p) = p –1.

(a) There exist infinitely many n such that φ(n) >φ(n + 1) (?)
We have there are infinitely many primes, so if we fix n = p, then we get infinitely many primes satisfying φ(p) >φ(p + 1), correct statement, Option (a) is incorrect.

(b) There exist infinitely many n such that φ(n) <φ(n + 1)
if we fix n + 1 = p, then we get infinitely many primes satisfying φ(p – 1) < φ(p), ie, there exist infinitely many n such that φ(n) <φ(n +1), correct statement, Option (b) is incorrect.

(c) There exist N ∈ N such that N > 2 and for all n > N, φ (N) <φ(n)
If N = 6, then φ(6) = 2, then for all n > N, φ(N) <φ(n), correct statement, since we need the wrong statement, Option (c) is incorrect.

(d) If n = p, $\lim_{n\to\infty} \frac{\varphi(n)}{n} = \lim_{p\to\infty} \frac{p-1}{p} = 1.$

If n = 2p, $\varphi(2p) = 2p\left(1-\frac{1}{2}\right)\left(1-\frac{1}{p}\right) = 2p \times \frac{1}{2} \times \frac{p-1}{p} = p - 1$

$$\lim_{n\to\infty} \frac{\varphi(n)}{n} = \lim_{p\to\infty} \frac{\varphi(2p)}{2p} = \lim_{p\to\infty} \frac{p-1}{2p} = \frac{1}{2}.$$

If n = 3p, $\varphi(3p) = 3p\left(1-\frac{1}{3}\right)\left(1-\frac{1}{p}\right) = 3p \times \frac{2}{3} \times \frac{p-1}{p} = 2(p-1)$

$$\lim_{n\to\infty} \frac{\varphi(n)}{n} = \lim_{p\to\infty} \frac{\varphi(3p)}{3p} = \lim_{p\to\infty} \frac{2(p-1)}{3p} = \frac{2}{3}.$$

Similarly, if n = k_p, k∈ N, the we get infinitely many limit points, statement in option (4) is incorrect, so, Option (d) is correct.

MODERN ALGEBRA

UNSOLVED QUESTIONS

1. The last two digits of$(7)^{81}$ are **CSIR NET DEC-2012**
(a) 07 (b) 17 (c) 37 (d) 47

2. What is the smallest positive integer in the set $(24x + 60y + 2000z \mid x, y, z \in Z\}$?
CSIR NET JUNE-2013
(a)2 (b)4 (c) 6 (d) 14

3. An ice cream shop sells ice cream in five possible flavors : Vanilla, Chocolate, Strawberry, Mango and Pineapple. How many combinations of three scoop cones are possible? [Note: The repetition of flavors is allowed but the order in which the flavors are chosen does not matter.] **CSIR NET JUNE - 2014**
(a) 10 (b) 20 (c) 35 (d) 243

4. If n is positive integer such that the sum of all positive integers 'a' satisfying $1 \le a \le n$ and g.c.d. (a, n) = 1 is equal to 240n, then the number of summands namely, $\phi(n)$, is
CSIR NET JUNE -2014
(a) 120 (b) 124 (c) 240 (d) 480

5. Find the number of divisors of N = 2520 (excluding unity and N) **NBHM-2015**
(a) 41 (b) 43 (c) 45 (d) 46

6. Which of the following primes satisfy the congruence $(a)^{24} \equiv 6a + 2 \pmod{13}$?
CSIR NET JUNE -2015
(a) 41 (b) 47 (c) 67 (d) 83

7. Which of the following intervals contains an integer satisfying the following three congruences.
$x \equiv 2 \pmod 5$, $x \equiv 3 \pmod 7$ andx $\equiv 4 \pmod{11}$ **CSIR NET DEC-2015**
(a) [401, 600] (b) [601, 800] (c) [801, 1000] (d) [1001, 1200]

8. Given a natural number n>1 such that $(n-1)! \equiv 1 \pmod n$. We can conclude that
(a) n k, where p is prime, k> 1. **CSIR NET DEC-2016**
(b) n = pq, where p and q are distinct primes.
(c) n = pqr, where p, q and r are distinct primes.
(d) n = p, where p is aprime

9. The remainder when 98! is divided by 101 is equal to
GATE-2016
(a) 49 (b) 50 (c) 20 (d) 51

10. Let S be the set of all integers from 100 to 999 which are neither divisible by 3 nor divisible by 5.The number of elements in S is **CSIR UGC JUNE -2017**
(a) 480 (b) 420 (c) 360 (d) 240

Answer Key

1	2	3	4	5	6	7	8	9	10
a	b	c	d	d	a,c	b,d	d	b	a

MODERN ALGEBRA

Chapter 1

Introduction to differential equations and differential equations of first order

Definition : An equation involving derivatives with respect to an independent variable and involve dependent variable is called an Ordinary Differential Equation.

Order of Differential Equation : The order of highest derivatives is called the order of Differential Equation.

Example : $\frac{d^2y}{dx^2} + y = 0$ is 2nd order Differential Equation.

Degree of Differential Equation : The degree of the highest derivatives is called degree of the Differential Equation.

Example : $\left(\frac{d^2y}{dx^2}\right)^2 + \frac{dy}{dx} + y = 0$ is 2nd degree Differential Equation.

Linear differential equation : The Differential Equation is called linear if (i) every dependent variable and its derivatives occurs in the first degree only
(ii) No product of dependent variable and its derivatives.

Non-linear Differential Equation : The Differential Equation which is not linear is called non-linear Differential Equation.

Example : (i) $\frac{d^2y}{dx^2} + \frac{dy}{dx} + y = 0$ is linear Differential Equation.

(ii) $y\left(\frac{dy}{dx}\right)^2 + xy = 0$ is called non-linear Differential Equation.

Solution of Differential Equation : Any relation between the dependent and independent variables, when substituted in the Differential Equation, reduces it to an identity is called a solution of Differential Equation.

Example : $\frac{dy}{dx} = 2y$ ..... (1)

then $y = ce^{2x}$ is the solution substitute in (1) $y = ce^{2x} \Rightarrow \frac{dy}{dx} = 2ce^{2x}$

Then $2ce^{2x} = 2ce^{2x}$ Which is true.

(i) General solution : Let $f(x, y, y_1, y_2 \ldots\ldots y_n) = 0$ ..... (1)

Where $y_1 = \frac{dy}{dx}, \ldots\ldots\ldots, y_n = \frac{d^ny}{dx^n}$ be an nth order Ordinary Differential Equation.

Example. General solution of $y'' - 3y' + 2y = 0$
is $y = C_1 e^x + C_2 e^{2x}$

DIFFERENTIAL EQUATION

where C_1, C_2 are arbitrary constant.
Order of equation = 2 = arbitrary constant.

(ii) **Particular solution :** A solution obtained from general solution by giving particular value of arbitrary constant is called particular solution.

(iii) **Singular solution :** A solution which cannot be obtained from any general solution by any choice of constant is called singular solution.

Let $y = (x + c)^2$ is the general solution of $\left(\frac{dy}{dx}\right)^2 - 4y = 0$

We can also verified that y = 0 is solution of this but y = 0 cannot be obtained by any choice of c.

$\Rightarrow$ y = 0 is singular solution.

1. EQUATIONS OF FIRST ORDER AND FIRST DEGREE

A General differential equation of first order and first degree is of the form

$\frac{dy}{dx} f(x, y)$ *or* $Mdx + Ndy = 0$ where M, N are functions of x and y both.

Following are some of the standard methods to solve differential equations:

1. Variable separable form
2. Reducible to variable separable form
3. Homogeneous equations
4. Reducible to homogeneous form
5. Linear differential equations of first order
6. Reducible to linear form

Case I: Variable separable: An equation whose variables are separable and can be put in the form *g(x)dx+h(y)dy=0* is called an equation of variable separable form. Integrating, $\int g(x)dx + \int h(y)dy = c$, where c is an arbitrary constant. This is the general solution of the differential equation.

Case II: Equations reducible to variable separable: To solve the equation

$$\frac{dy}{dx} = f(ax + by + c) \qquad ..(1)$$

Method to solve:

Put $ax + by + c = z$ so that $a + b\frac{dy}{dx} = \frac{dz}{dx}$, *i.e.*, $\frac{dy}{dx} = \frac{1}{b}\left(\frac{dz}{dx} - a\right)$

Putting in (1), $\frac{1}{b}\left(\frac{dz}{dx} - a\right) = f(z) \Rightarrow \frac{dz}{dx} = a + bf(z)$, which is of the type "Variable Separable" and hence can be solved.

Case III: **Homogeneous Equations:**

Definition: Homogeneous Function: A function *f(x,y)* is said to be homogeneous of nth degree in x and y if it can be put in the form $x^n f\left(\frac{y}{x}\right)$.

Example. $\phi(x,y)=x^4+x^2y^2+xy^3+y^4$ is a homogeneous function of degree four.

Definition: Homogeneous Function: A equation in x and y is said to be homogeneous equation if it can be put in the form $\frac{dy}{dx}=\frac{f(x,y)}{\phi(x,y)}$, where *f(x,y)* and ϕ*(x, y)* are both homogenous functions of the same degree in x and y.

(i) To solve the equation $\frac{dy}{dx}=\frac{f(x,y)}{\phi(x,y)}$, where *f(x,y)* and ϕ*(x,y)* are both homogeneous functions of the same degree in x and y.

(i) put y = vx so that $\frac{dy}{dx}=v+x\frac{dv}{dx}$ (ii) Proceed as in Case I.

(ii) Sometimes the equation is not homogeneous, even then it can be solved by the same method as for homogenous equations.

Example. Solve $x^2y\,dx-(x^3+y^3)\,dy=0$

Solution : The given differential equation is

$$x^2ydx-(x^3+y^3)dy=0\Rightarrow\frac{dy}{dx}=\frac{x^2y}{x^3+y^3}\qquad\text{........(1)}$$

Since each of the functions x^2y and x^3+y^3 are homogenous functions of degree 3, so the given differential equation (1) is homogenous.

Putting $y=vx$ *and thus* $\frac{dy}{dx}=v+x\frac{dv}{dx}$ in (1), we get $v+x\frac{dv}{dx}=\frac{vx^3}{x^3+v^3x^3}$

$$\Rightarrow v+x\frac{dv}{dx}=\frac{v}{1+v^3}\quad\Rightarrow x\frac{dv}{dx}=\frac{v}{1+v^3}-v\ \Rightarrow x\frac{dv}{dx}=\frac{v-v-v^4}{1+v^3}$$

$$\Rightarrow x\frac{dv}{dx}=-\frac{v^4}{1+v^3}\Rightarrow x(1+v^3)dv=-v^4dx\Rightarrow\frac{1+v^3}{v^4}dv=-\frac{dx}{x}\Rightarrow\left(\frac{1}{v^4}+\frac{1}{v}\right)dv=-\frac{dx}{x}$$

Integrating both sides, we get $\frac{v^{-3}}{-3}+\log v=-\log x+C$

$$\Rightarrow-\frac{1}{3v^3}+\log v+\log x=C\ \Rightarrow-\frac{1}{3}\frac{x^3}{y^3}+\log\left(\frac{y}{x}.x\right)=C\ [\because v=y/x]$$

$$\Rightarrow-\frac{1}{3}.\frac{x^3}{y^3}+\log y=C,\ \text{which is the required solution.}$$

Case IV: **Equations reducible to homogeneous equations**

To solve the equation $\frac{dy}{dx}=\frac{a_1x+b_1y+c_1}{a_2x+b_2y+c_2}$

(1) For, $\frac{a_1}{a_2}\neq\frac{b_1}{b_2}$, we have

Method to solve:

(i) Put $x = X + h$, $y = Y + k$.

(ii) Equate the constant terms of numerator and denominator to zero and find the values of h and k.

(iii) Proceed as in case III.

(2) For, $\frac{a_1}{a_2} = \frac{b_1}{b_2}$, we have

Method to solve:

(i) Put $a_1 x + b_1 y = v$

(ii) The equation gets reduced to variable separable form in v and x.

Example. Solve $\frac{dy}{dx} = \frac{x+y+2}{x-y-3}$

Solution : Given $\frac{dy}{dx} = \frac{x+y+2}{x-y-3}$ here $\frac{a_1}{a_2} \neq \frac{b_1}{b_2}$

Let $x = X + h, y = Y + k \Rightarrow \frac{dy}{dx} = \frac{dY}{dX} \Rightarrow \frac{dY}{dX} = \frac{X+Y+(h+k+2)}{X-Y+(h-k-3)}$

i.e., choose h, k such that h + k + 2 = 0 ....(i)

h – k – 3 = 0 ......(ii)

Solving equation (i) and (ii), we get $h = \frac{1}{2}; k = -\frac{5}{2} \Rightarrow \frac{dY}{dX} = \frac{X+Y}{X-Y} = \frac{1+\frac{Y}{X}}{1-\frac{Y}{X}}$

Putting $Y = Xv \Rightarrow \frac{dY}{dX} = v + X\frac{dv}{dX}$

$$\Rightarrow v + X\frac{dv}{dX} = \frac{1+v}{1-v} \Rightarrow \frac{dX}{X} = \frac{1-v}{1+v^2}dv \Rightarrow \frac{dv}{1+v^2} - \frac{vdv}{1+v^2} = \frac{dX}{X}$$

$$\Rightarrow 2\log X + \log\left(1+\frac{Y^2}{X^2}\right) - \log c = 2\tan^{-1}\left(\frac{Y}{X}\right) as\ v = \frac{Y}{X}$$

$$\Rightarrow log\{(X^2+Y^2)/c\} = 2\,tan^{-1}\left(\frac{Y}{X}\right) \Rightarrow X^2+Y^2 = ce^{2\tan^{-1}\left(\frac{Y}{X}\right)}$$

$$\Rightarrow \left(x-\frac{1}{2}\right)^2 + \left(y+\frac{5}{2}\right)^2 = ce^{2\,tan^{-1}\left(y+\frac{5}{2}/x-\frac{1}{2}\right)}$$, where c is an arbitrary constant.

DIFFERENTIAL EQUATION

Case V: Linear differential equations of first order

Definition: The differential equation of the form $\frac{dy}{dx} = P(x)y + Q(x)$ ....(1) is said to be linear differential equation of first order, where y is the dependent variable and x is the independent variable. If Q(x) = 0, then (1) is called **homogeneous linear differential equation, otherwise it is called non-homogeneous linear differential equation.**

The solution of $\frac{dy}{dx} + Py = Q,$ where P and Q are functions of x only, is given by

$y.e^{\int Pdx} = \int e^{\int Pdx} Qdx + C,$ where C is a constant and integrating factor $= e^{\int Pdx}$.

Example. Solve $(1+x^2)\frac{dy}{dx} + 2xy - 4x^2 = 0$ subject to the initial condition y(0) = 0.

Solution : The given differential equation can be written as $\frac{dy}{dx} + \frac{2x}{1+x^2} y = \frac{4x^2}{1+x^2}$ .....(i)

This is a linear differential equation of the form $\frac{dy}{dx} + Py = Q,$ where

$P = \frac{2x}{1+x^2}$ and $Q = \frac{4x^2}{1+x^2}$

We have , I.F. = $e^{\int Pdx} = e^{\int 2x/(1+x^2)dx} = e^{\log(1+x^2)} = 1 + x^2$

The solution is given by $y(1+x^2) = \int 4x^2 dx + C \Rightarrow y(1+x^2) = \frac{4x^3}{3} + C$

Hence, $y = 0,$ when x = 0, So, 0 = 0 + C ⇒ C = 0.

Hence, $y(1+x^2) = \frac{4x^3}{3}$ is the required solution.

Case VI: Equations reducible to linear differential equation

To Solve the equation $\frac{dy}{dx} + Py = Qy^n$, where P and Q are functions of x.

Method to solve

(i) Dividing throughout by y^n.

(ii) Putting $y^{1-n} = z$ and get the linear equation in z.

(iii) Proceed as in Case V.

Example. The solutions of (dy/dx) + 2xy = $2xy^2$, is

(a) $y = (cx)/(1+e^{-x^2})$ (b) $y = 1/(1-ce^{x})$

(c) $y = 1/(1+ce^{x^2})$ (d) $y = (cx)/(1+e^{x^2})$

Solution : (c)

DIFFERENTIAL EQUATION

Dividing by y^2, we get $y^{-2}\frac{dy}{dx}+2xy^{-1}=2x$

Putting $y^{-1}=v$, so that $-y^{-2}(dy/dx)=dv/dx,\ or\ (dv/dx)-2xv=-2x$

here P = -2x and Q = -2x.

$\text{I.F.}=e^{\int Pdx}=e^{-2\int xdx}=e^{-x^2}$

So, solution is $ve^{-x^2}=\int(-2x)e^{-x^2}dx=\int e^t dt, putting(-x^2)=t$ .....(1)

or $y^{-1}e^{-x^2}=e^t+c=e^{-x^2}+c$, (using (1))

or $y^{-1}=1+ce^{x^2}\ or\ y=1/(1+ce^{x^2})$, which is the required solution.

2. Exact Differential Equation

Definition: The equation Mdx + Ndy = 0 is exact if $\frac{\partial M}{\partial y}=\frac{\partial N}{\partial x}$ and its solution is

$$\underset{(treating\,y\,as\,constant)}{\int Mdx}+\int(terms\ of\ N\ not\ containing\ x)dy=c,$$ where c is constant.

Integrating Factors

Differential equation is multiplied by a function μ(x,y) so that the resulting differential equation becomes exact, then μ(x,y) is called an integrating factor and is denoted by I.F. Numbers of integrating factors of Mdx + Ndy = 0 is infinite.

Five Rules for finding integrating factors

Rule I : If the differential equation Mdx + Ndy = 0 is homogeneous in x and y i.e., if M, N are homogeneous functions of the same degree in x and y, then $\frac{1}{Mx+Ny}$ is an integrating factor provided $Mx + Ny \neq 0$.

Rule II : If the differential equation Mdx + Ndy = 0 is of he form $f_1(xy)y\,dx + f_2(xy)\,x\,dy = 0$, then $\frac{1}{Mx-Ny}$ is an integrating factor provided $Mx - Ny \neq 0$.

Rule III : If the differential equation Mdx + Ndy = 0, $\frac{\frac{\partial M}{\partial y}-\frac{\partial N}{\partial x}}{N}$ is a function of x only = f(x) (say), then $e^{\int f(x)dx}$ is an integrating factor. If $\frac{\frac{\partial M}{\partial y}-\frac{\partial N}{\partial x}}{N}$ is not free from y, then there does not exist any I.F. which is function of x only.

Rule IV : If the differential equation Mdx + Ndy = 0, $\dfrac{\dfrac{\partial N}{\partial x}-\dfrac{\partial M}{\partial y}}{M}$ is a function of y only $= f(y)$ (say), then $e^{\int f(y)dy}$ is an integrating factor. If $\dfrac{\dfrac{\partial N}{\partial x}-\dfrac{\partial M}{\partial y}}{M}$ is not free from x, then there does not exist any I.F. which is function of y only.

Rule V : If the differential equation is $x^a y^b(mydx+nxdy)+x^r y^s\ (pydx+qxdy)=0$, then $x^h y^k$ is an integrating factor, where $\dfrac{a+h+1}{m}=\dfrac{b+k+1}{n},\dfrac{r+h+1}{p}=\dfrac{s+k+1}{q}$.

Fundamental Theorem of Differential Equation :

Let $p(x)\dfrac{dy}{dx}+Q(x)y=R(x)$ be a first order first degree differential equation.

Then $\exists$ x_0, y_0 s.t. $y(x_0) = y_0$ this differential equation has unique solution if $p(x) \neq 0$ $\forall\ x \in [a, b]$

SOLVED QUESTIONS

1. Let $y_0 > 0$, $z_0 > 0$ and $\alpha > 1$. Consider the following two differentiable equations:

(*) $\begin{cases}\dfrac{dy}{dt}=y^{\alpha}\ for\,t>0,\\ y(0)=y_0\end{cases}$, (**) $\begin{cases}\dfrac{dz}{dt}=-z^{\alpha}\ for\,t>0,\\ z(0)=z_0\end{cases}$ we say that the solution if it exists for all t > 0. Which of the following statements is true?

(a) Both (*) and (**) have global solutions

(b) None of (*) and (**) have global solutions

(c) There exists a global solution for (*) and there exists a $T < \infty$ such that $\lim\limits_{t\to T}|z(t)| = +\infty$

(d) There exists a global solution for (**) and there exists a $T < \infty$ such that $\lim\limits_{t\to T}|y(t)| = +\infty$

Ans. (d) **CSIR NET JUNE 2020**

(*) $\begin{cases}\dfrac{dy}{dt}=y^{\alpha}\ for\,t>0,\\ y(0)=y_0\end{cases}$, (**) $\begin{cases}\dfrac{dz}{dt}=-z^{\alpha}\ for\,t>0,\\ z(0)=z_0\end{cases}$

$y_0 > 0$, $z_0 > 0$ and $\alpha > 1$.

Set $y_0 = 1$, $z_0 = 1$ and $\alpha = 2$, then (**) $\Rightarrow \dfrac{dz}{dt} = -z^2$, $z(0) = 1$

DIFFERENTIAL EQUATION

$\Rightarrow -\frac{1}{z} = -t + c$, $z(0) = 1 \Rightarrow c = -1$, then

$-\frac{1}{z} = -t - 1 \Rightarrow z = \frac{1}{t+1}$, since t > 0, this solution exists globally. Option (b) is incorrect.

$(*) \Rightarrow \frac{dy}{dt} = y^2$, $y(0) = 1 \Rightarrow -\frac{1}{y} = t + c_1$

$y(0) = 1 \Rightarrow c_1 = -1$, then $y = \frac{1}{1-t'}$ not exist globally as at t = 1, y is not defined.

Option (a), (b) and (c) are incorrect, Option (d) is correct.

2. Consider the ODE $ty' - 3y = t^2y^{1/2}$, $y(1) = 1$. Find the value of y(2)

CSIR NET NOV. 2020

(a) 14 (b) 16 (c) 0 (d) 8

Ans. (b)

Explanation :

The given ordinary differential equation is $y^{-1/2}y' - \frac{3y^{1/2}}{t} = t, y(1) = 1$.

This is Bernoulli equation,.

putting $y^{1/2} = x$ and $\frac{y^{-1/2}}{2}dy = dx$

So, transformed equation will be $\frac{dx}{dt} + \frac{3x}{2t} = \frac{t}{2}$

This is linear in x, so equation will be

$$I.F. = e^{\int \frac{-3}{2t}dt} = e^{\frac{-3}{2}\log t} \Rightarrow \frac{1}{t^{3/2}}$$

$$x\frac{1}{t^{3/2}} = \int \frac{t}{2}\frac{1}{t^{3/2}}dt + c \quad x\frac{1}{t^{3/2}} = \int \frac{1}{2\sqrt{t}}dt + c = \frac{1}{2}(2t^{1/2}) + c$$

$$x.t^{-3/2} = \sqrt{t} + c \qquad x = t^2 + ct^{3/2}$$

Converting it into y

$$\sqrt{y} = t^2 + ct^{3/2}$$

$$1 = 1 + c \Rightarrow c = 0$$

$$\sqrt{y} = t^2 \Rightarrow y = t^4$$

So y(2) = 16

So, option (b) is correct.

3. Consider the ordinary differential equation y' = y (y – 1) (y – 2). Which of the following statement is true?

DIFFERENTIAL EQUATION

(a) If y (0) = 0.5 then y is decreasing (b) If y(0) = 1.2 then y is increasing
(c) If y(0) = 2.5 then y is unbounded (d) If y(0) < 0 then y is bounded below

Ans. (c)

Explanation :

Given differential equation is $\frac{dy}{dx} = y(y-1)(y-2)$

there are three equilibrium solutions, i.e., y = 0, y = 1 and y = 2

and $\frac{dy}{dx} < 0 \, or \, \frac{dy}{dx} > 0$ if $y \in (-\infty, 0) \cup (1,2)$ *or* $y \in (0,1) \cup (2, \infty)$ respectively

$\therefore$ if $y(0) \in (-\infty, 0) \cup (1,2), then \; y(x)$ is decreasing function and if $y(0) \in (0,1) \cup (2,\infty),$ *then* $y(x)$ is increasing function.

$\therefore$ option (a), (b) and (d) are incorrect. Hence, option (c) is correct

4. $\frac{dy}{dx} + \left(\frac{2x+1}{x}\right) y = e^{-2x}$; x > 0; y(x) be the solution of the D.E. then

(a) y $\to \infty$ as x $\to \infty$ (b) y $\to -\infty$ as x $\to \infty$ (c) y $\to$ 0 as x $\to \infty$(d) y is unbounded

Ans. (c)

Explanation :

Given D.E. is 1st order linear D.E. I.F. $e^{\int\left(\frac{2x+1}{x}\right)dx} = e^{\int\left(2+\frac{1}{x}\right)dx} = e^{2x+\log x} = xe^{2x}$

Then solution $y.I.F. = \int Q.IF \, dx$ $\qquad yxe^{2x} = \int e^{-2x} xe^{2x} dx$

$xye^{2x} = \frac{x^2}{2} + k$ $\qquad y(x) = \frac{xe^{-2x}}{2} + \frac{e^{-2x}}{x} k$

y $\to$0 as x $\to\infty$ Option (c) is correct.

5. If g(x, y)dx + (x + y)dy = 0 is an exact differential equation and if g(x, 0) = x^2 then the general solution of DE is

(a) $2x^3 + 2xy + y^2 = c$ (b) $2x^3 + 6xy + 3y^2 = c$ (c) $2x + 2xy + y^2 = c$

(d) $x^2 + 2xy + y^2 = c$

Ans. (b)

Explanation :

If Mdx + Ndy = 0 is exact DE, then $\frac{\partial M}{\partial y} = \frac{\partial N}{\partial x}$

Given that g(x, y) dx + (x + y)dy = 0 is exact.

$\frac{\partial g}{\partial y} = 1 \Rightarrow g = y + c \Rightarrow g(x, 0) = x^2 = c \Rightarrow g(x, y) = x^2 + y$

Then $(x^2 + y)$ dx + (x + y)dy = 0

The solution is given by $\int\limits_{y-\text{constant}} (x^2+y)dx + \int\limits_{\text{terms not containing x}} (x+y)dy$

$\frac{x^3}{3} + yx + \frac{y^2}{2} = c \quad \Rightarrow \quad 2x^3 + 6xy + 3y^2 = c$

6. The non-zero value of n for which the differential equation $(3xy^2 + n^2x^2y)dx + (nx^3 + 3x^2y)dy = 0$ become exact is

(a) -3 (b) -2 (c) 2 (d) 3

Ans. (d)

Explanation :

Let M(x, y)dx + N(x, y)dy = 0 be a differential equation and

It is exact iff $\frac{\partial M}{\partial y} = \frac{\partial N}{\partial x} \Rightarrow M = 3xy^2 + n^2x^2y$ and $N(x, y) = nx^3 + 3x^2y$

$\frac{\partial M}{\partial y} = 6xy + n^2x^2$ and $\frac{\partial N}{\partial x} = 3nx^2 + 6xy$

$\frac{\partial M}{\partial y} = \frac{\partial N}{\partial x} \Rightarrow n^2x^2 = 3nx^2 = n(n-3) = 0$ either n = 0 or n = 3

Option (d) is correct.

7. The equation of the curve satisfying $\sin y \frac{dy}{dx} = \cos y(1 - x\cos y)$ and passing through the origin is **IIT-JAM-2013**

(a) sec y = x + 1 (b)cosy = x+1 (c) secx = y+1 (d) cosx = y+1

Ans. (a)

Explanation :

Given, $\sin y \frac{dy}{dx} = \cos y(1 - x\cos y)$

$\Rightarrow \frac{dy}{dx} = \cot y - x\cos y.\cot y \Rightarrow \frac{dy}{dx} - \cot y = -x\cos y.\cot y$

$\Rightarrow \sec y \tan y \frac{dy}{dx} - \sec y = -x$ ...(i)

Put $v = \sec y \Rightarrow \frac{dv}{dx} = \sec y \tan y \frac{dy}{dx}$ $\quad \therefore (1) becomes \frac{dv}{dx} - v = -x$

I.F = $e^{-\int dx} = e^{-x}$ $\therefore Complete\ solution\ is\ v.e^{-x} = -\int e^{-x}.xdx + c$

$\Rightarrow v.e^{-x} = x.e^{-x} + e^{-x} + c \Rightarrow v = x + 1 + ce^{x} \Rightarrow \sec y = (x+1) + ce^{x}$

As the curve passes through origin, i.e., (0,0)

So, $\sec 0 = (0 + 1) + ce^0 \Rightarrow 1 = 1 + c \Rightarrow c = 0$

Hence, the required equation of curve satisfying the given differential equation is sec y = x + 1.

8. The integrating factor (IF) of the differential equation $\left(\frac{e^{-2\sqrt{x}}}{\sqrt{x}}-\frac{y}{\sqrt{x}}\right)\frac{dx}{dy}=1$ is

CUET 2022

(a) $e^{2\sqrt{x}}$ (b) $e^{-2\sqrt{x}}$ (c) $e^{\sqrt{x}}$ (d) $e^{-\sqrt{x}}$

Ans. (a)

Explanation :

$\frac{dy}{dx}+\frac{y}{\sqrt{x}}=\frac{e^{-2\sqrt{x}}}{\sqrt{x}}$ Then I.F. $e^{\int\frac{1}{\sqrt{x}}dx}=e^{\int x^{-1/2}dx}=e^{2\sqrt{x}}$

9. The integrating factor of the differential equation x dy – (y + xy^2(1 + logx))dx = 0 is

CUET 2022

(a) x (b) x^2 (c) logx (d) e^x

Ans. (a)

Explanation :

Given that xdy – (y + xy^2 + xy^{2}logx) dx = 0

Here $\frac{dy}{dx}-\frac{y}{x}=y^2(1+\log x)$ $\Rightarrow$ $\frac{1}{y^2}\frac{dy}{dx}-\frac{1}{xy}=(1+\log x)$

Let $\frac{1}{y}=t$ then $-\frac{1}{y^2}\frac{dy}{dx}=\frac{dt}{dx}$ $\Rightarrow$ $-\frac{dt}{dx}-\frac{t}{x}=(1+\log x)$

$\Rightarrow$ $\frac{dt}{dx}+\frac{t}{x}=-(1+\log x)$ I.F. $e^{\int\frac{1}{x}dx}\Rightarrow e^{\log x}=x$

UNSOLVED QUESTIONS

1. Integrating factor of $(x^7y^2+3y)dx+(3x^8y-x)dy=0$ *is* x^my^n, *where* **GATE-2002**

(a) m = - 7, n = 5 (b) m = -1, n = 5 (c) m = -7, n = 1 (b) m = - 7, n = -1

2. The general solution of the differential equation dy/dx + tan y tan x = cos x sec y, is

(a) $2\sin y=(x+c)\sec x$ (b) $\sin y=(x+c)\cos x$

(c) $\cos y=(x+c)\sin x$ (d) $\sec y=(x+c)\cos x$ **GATE-2001**

3. Which of the following is an integrating factor of xdy – ydx = 0?

(a) $\frac{1}{x^2}$ (b) $\frac{1}{x^2+y^2}$ (c) $\frac{1}{xy}$ (b) $\frac{x}{y}$ **GATE-2001**

4. Let y(x) be a continuous solution of the initial value problem y` + 2y = f(x), y(0) = 0, where $f(x)=\begin{cases}1, & 0\le x\le 1\\ 0 & x>1\end{cases}$. Then $y\left(\frac{3}{2}\right)$ is equal to **CSIR NET JUNE 2015**

(a) $\frac{\sinh(1)}{e^3}$ (b) $\frac{\cosh(1)}{e^3}$ (c) $\frac{\sinh(1)}{e^2}$ (d) $\frac{\cosh(1)}{e^2}$

5. The solution of differential equation ydx + xdy = 0 is

(a) xy = c (b) x = y + c (c) x – y = c (d) none of these

6. The general solution of the differential equation $\frac{dy}{dx}=\frac{x}{y}$ is

(a) $x^2+y^2=a^2$ (b) $x^2-y^2=a^2$ (c) $x=-y$ (d) $x+y=a$

7. Which of the following differential equations is not linear ?

(a) $(1+y)\frac{dy}{dx}+xy=\sin x$ (b) $y\frac{dy}{dx}x+x=0$

(c) $(1+y)\frac{dy}{dx}+\sin x=0$ (d) $\frac{d^2y}{dx^2}+x^2\frac{dy}{dx}+(1+x)^3y=\sin x+e^x$

8. The general solution of the differential equation $\frac{dx}{x}=\tan y dy$ is

(a) $\log x=-\log\cos y+C$ (b) $\log(x\cos y)=C$ (c) $\log(x\sin y)=C$

(d) $x\cos y=e^C$

9. Differential equation $x(1-x^2)dy+(2x^2y-y-ax^3)dx=0$ has an integrating factor

(a) $e^{\int\frac{2x^2-1}{1-x}dx}$ (b) $e^{\int\frac{2x^2-1}{x}dx}$ (c) $e^{\int\frac{2x^2-1}{x(1-x^2)}dx}$ (d) none of the above

10. The solution of the differential equation $\frac{dy}{dx}=\frac{1-x}{y}$ represents

(a) a family of circles centered at (1, 0) (b) a family of circles centered at (0, 0)

(c) a family of circles centered at (-1, 0) (d) a family of straight lines with slope -1

11. The solution of the differential equation $(x+y-2)dy/dx=(x+y)$ is

(a) $(y+x)=\log(y-x+1)+c$ (b) $(y-x)=\log(x+y-1)+c$

(c) $y-2x=\log(x+y-1)+c$ (d) $y+2x=\log(x+y+1)+c$

12. Which of the following is not the integrating factor for the differential equation $(x+1)\frac{dy}{dx}-y=e^{3x}(x+1)^2$?

(a) $\frac{1}{x+1}$ (b) $x+1$ (c) $\frac{1}{x^2+1}$ (d) x

13. The general solution of the differential equation $dy/dx + y\frac{d\phi}{dx} = \phi(x)\frac{d\phi}{dx}$, where ϕ is a function of x alone, is not given by

(a) $y = \phi + ce^{-\phi}$ (b) $y = \phi + 1 - ce^{-\phi}$ (c) $y = \phi - 1 + ce^{\phi}$ (d) $y = \phi - 1 + ce^{-\phi}$

14. Which of the following information(s) is/are true for the differential equation?

$E : (x^3 + y^3)dx - (xy^2)dy = 0$

(a) E is an exact differential equation (b) Integrating factor of E is x^{-4}

(c) Solution of E is $\log x - \frac{y^3}{3x^3} = C$ (d) Solution of E is $\log x - \frac{x^3}{3y^3} = C$

15. Let y be the solution of y` + y = |x|, $x \in R$ y(-1) = 0. Then y(1) is equal to

GATE – 2016

(a) $\frac{2}{e} - \frac{2}{e^2}$ (b) $\frac{2}{e} - 2e^2$ (c) $2 - \frac{2}{e}$ (d) 2 – 2e

16. If x^3y^2 is an integrating factor of $(6y^2 + axy)dx + (6xy + bx^2)dy = 0$, where $a, b \in R$, then

GATE – 2017

(a) 3a – 5b = 0 (b) 2a – b = 0 (c) 3a + 5b = 0 (d) 2a + b = 0

17. The general solution of the differential equation xy` = $y + \sqrt{x^2 + y^2}$ for x > 0 is given by (with an arbitrary positive constant k)

GATE – 2018

(a) $ky^2 = x + \sqrt{x^2 + y^2}$ (b) $kx^2 = x + \sqrt{x^2 + y^2}$

(c) $kx^2 = y + \sqrt{x^2 + y^2}$ (d) $ky^2 = y + \sqrt{x^2 + y^2}$

18. Let y(x) be the solution of the differential equation $(xy + y + e^{-x})dx + (x + e^{-x})dy = 0$ satisfying y(0) = 1. Then y(-1) is equal to

IIT JAM – 2017

(a) $\frac{e}{e-1}$ (b) $\frac{2e}{e-1}$ (c) $\frac{e}{1-e}$ (d) 0

19. An integrating factor of the differential equation

$\left(y + \frac{1}{3}y^3 + \frac{1}{2}x^2\right)dx + \frac{1}{4}(x + xy^2)dy = 0$ is

IIT JAM – 2018

(a) x^2 (b) 3 $\log_e x$ (c) x^3 (d) 2 $\log_e x$

20. The solution(s) of the differential equation $\frac{dy}{dx} = (\sin 2x)y^{1/3}$ satisfying y(0) = 0 is (are)

IIT JAM - 2018

(a) y(x) = 0 (b) $y(x) = -\sqrt{\frac{8}{27}}\sin^3 x$ (c) $y(x) = \sqrt{\frac{8}{27}}\sin^3 x$ (d) $y(x) = \sqrt{\frac{8}{27}}\cos^3 x$

21. If $x^h y^k$ is an integrating factor of the differential equation $y(1 + xy)\,dx + x(1 - xy)dy = 0$, then the ordered pair (h, k) is equal to **IIT JAM – 2019**

(a) (-2, -2) (b) (-2, -1) (c) (-1, -2) (d) (-1, -1)

22. The general solution of the differential equation $\frac{dy}{dx} + y = f(x)$, where $f(x) = \begin{cases} x, & 0 \le x \le 1 \\ 0, & x \ge 1 \end{cases}$ and y(0) = 0 is

(a) $y = \begin{cases} 2(1-e^{-x}), & 0 \le x \le 1 \\ 2e^{-x}(e-1), & x \ge 1 \end{cases}$ (b) $y = \begin{cases} 2(e^{-x}-1), & 0 \le x \le 1 \\ 2e^{-x}(e-1), & x \ge 1 \end{cases}$

(c) $y = \begin{cases} 2(1-e^{-x}), & 0 \le x \le 1 \\ 2e^{-x}(1-e), & x \ge 1 \end{cases}$ (d) none of these

23. The solution of the equation $\frac{dy}{dx} = (\sin 2x)y^{1/3}$ satisfying y(0) = 0.

(a) $y(x) \ne 0$ (b) $y(x) = -\sqrt{\frac{6}{27}}\sin^3 x$ (c) $y(x) = \sqrt{\frac{8}{27}}\sin^3 x$ (d) $y(x) = \sqrt{\frac{8}{27}}\cos^3 x$

24. The solution of differential equation $\frac{dy}{dx} - x\tan(y-x) = 1$

(a) $\log\sin(y - x) = x^2 + c$ (b) $\log\sin(y-x) = \frac{x^2}{2} + c$

(c) $\log\sin(y-x) = \frac{x^2}{2} + c$ (d) None of these

25. Let y(x) be the solution of differential equation $x^4\frac{dy}{dx} + 4x^3 y + \sin x = 0$

$y(\pi) = 1, x > 0$ then $y\left(\frac{\pi}{2}\right)$ is

(a) $\frac{10(1+\pi^4)}{\pi^4}$ (b) $\frac{12(1+\pi^4)}{\pi^4}$ (c) $\frac{14(1+\pi^4)}{\pi^4}$ (d) $\frac{16(1+\pi^4)}{\pi^4}$

Answer Key

1	2	3	4	5	6	7	8	9	10	11	12
c	b	a,b,c	c	a	a	a,b,c	a,b,d	c	a	b	b,c,d
13	**14**	**15**	**16**	**17**	**18**	**19**	**20**	**21**	**22**	**23**	**24**
a,b,c	b,c	a	a	c	b	c	a,b,c	a	d	c	b
25											
d											

Chapter 2

General Theory Of Linear Differential Equation

1. DEFINITIONS

(i) Linear Combination: If $f_1, f_2, \ldots f_n$ are n functions defined on the interval I and $c_1, c_2, \ldots. c_n$ are n arbitrary constants, then the function $c_1f_1 + c_2f_2 + \ldots + c_nf_n$ is called linear combination of $f_1, f_2, \ldots f_n$ over I.

For example : $e^{-x} + 2e^{x} - 4e^{2x}$ is a linear combination of e^{-x}, e^{x}, e^{2x} [Hence $c_1 = 1, c_2 = 2, c_3 = -4$].

(ii) Linearly Dependent: The functions $f_1, f_2 \ldots.., f_n$ of x are said to be linearly dependent over an interval I iff there exist constants $c_1, c_2, \ldots., c_n$ (not all zero) such that $c_1f_1 + c_2f_2 + \ldots.. + c_nf_n = 0$ for all x in I.

For example : The functions x and 2x are linearly dependent on [0, 1] because there exist constants ($c_1 = 2, c_2 = -1$) s.t. $c_1f_1 + c_2f_2 = 0$ i.e. $(2)(x) + (-1)(2x) = 0$.

(iii) Linearly Independent : The functions $f_1, f_2, \ldots.., f_n$ are of x are said to be linearly independent over an interval I iff there exist constants $c_1, c_2, \ldots.., c_n$ such that $c_1f_1 + c_2f_2 + \ldots.. + c_nf_n = 0$ for all x in I, then $c_i = 0 \ \forall \ i = 1, 2, \ldots., n$.

For example : The functions x and x^2, are linearly independent on [0, 1]

2. GENERAL LINEAR DIFFERENTIAL EQUATION

Definition: A general linear differential equation of order n is

$$P_0\frac{d^ny}{dx^n} + P_1\frac{d^{n-1}y}{dx^{n-1}} + P_2\frac{d^{n-2}y}{dx^{n-2}} + \ldots + P_ny = Q \quad \ldots. (1)$$

where $P_0 \neq 0, P_1, P_2, \ldots. P_n$ and Q are functions of x defined on some interval I. When $Q = 0$, then (1) is said to be homogeneous. When $Q \neq 0$, then (1) is said to be non-homogeneous.

Existence and Uniqueness Theorem

Statement: If $P_1, P_2, \ldots., P_n$ and Q are continuous functions of x over an open interval I and if $x_0 \in I$ be the real number and $y_0, y_1, \ldots, y_{n-1}$ are arbitrary real numbers, then there exists one and only one function; say $y^{(n)} + P_1y^{(n-2)} + \ldots +$
$P_ny = Q$ s.t. $\phi(x_0) = y_0, \phi'(x_0) = y_1, \ldots \phi^{n-1}(x_0) = y_{n-1'}$

3. Homogenous Linear Differential Equation

Equation (1) is said to be homogeneous if $Q = 0$.

DIFFERENTIAL EQUATION

Theorem. Let $f_1(x)$ and $f_2(x)$ be two linearly independent solution of

$$P_0\frac{d^2y}{dx^2}+P_1\frac{dy}{dx}+P_2y=0 \quad .....(1)$$

over an open interval I, where P_0, P_1, P_2 are all continuous functions of x and $P_0(x) \neq 0$ on I. Then $f = \alpha f_1 + \beta f_2$, where α, β are some constants is also its solution.

4. Non-homogenous linear equation

Equation (1) is said to be non-homogeneous if $Q \neq 0$.

Let $f_1(x)$ and $f_2(x)$ be linearly independent solution of $P_0\frac{d^2y}{dx^2}+P_1\frac{dy}{dx}+P_2y\neq 0$ then $f = \alpha f_1+\beta f_2$ is a solution iff $\alpha+\beta=1$

5. The Wronskian :

Definition : The Wronskian of n functions $y_1(x)$, $y_2(x)$ ...... $y_n(x)$ is denoted by w(x)

$$\text{or } w(y_1, y_2, ..., y_n)=\begin{vmatrix} y_1 & y_2 .... y_n \\ y'_1 & y_2' .... y'_n \\ \vdots & \\ y_1^{(n-1)} & y_2^{(n-1)} ...... y_n^{(n-1)} \end{vmatrix}$$

Let $a_0(x)\, y'' + a_1(x)\, y' + a_2(x)\, y = 0$

Where $a_0(x)$, $a_1(x)$, $a_2(x)$ are continuous and $a_0(x) \neq 0\ \forall x$

If $y_1(x)$ and $y_2(x)$ are solution

Then $w(y_1,y_2)=\begin{vmatrix} y_1 & y_2 \\ y_1' & y_2' \end{vmatrix}=y_1y_2'-y_2y_1'$

$w(y_1,y_2)=Ae^{-\int\frac{a_1(x)}{a_0(x)}dx}$ is called Able's formula.

Able's Formula:

RESULTS:

(1) If $w(y_1, y_2, ... y_n) \neq 0$ then $y_1, y_2 ...... y_n$ are L.I. solution.

(2) If $w(y_1, y_2, ... y_n) = 0$ then $y_1, y_2, ...y_n$ are LD solution.

$y = c_1y_1 + c_2y_2 + ... c_ny_n$ is also a solution.

If $c_1 + c_2 + ... + c_n = 1$(for non-homogeneous D E)

SOLVED QUESTIONS

1. Consider the 2nd order ODE $x''+p(t)x'+q(t)x=0$ and let x_1, x_2 be two solutions of this ODE in [a, b]. Which of the following statements are true for the Wronskian w of x_1, x_2? **CSIR NET FEB 2022**

(a) W = 0 in (a, b) implies that x_1, x_2 are independent

(b) W can change sign in (a, b)

(c) $W(t_0) = 0$ for some $t_0 \in (a, b)$ implies that W = 0 in (a, b)

(d) W(t₀) = 1 for some $t_0 \in (a, b)$ implies that W = 1 in (a, b)

Ans. **(c)**

Explanation :

Consider the given second order ODE

$x''+p(t)x'+q(t)x=0$

x_1 and x_2 are two solution of given ODE in [a, b]

We know that Wronskian either identically zero or never zero

Option (c) is the correct answer.

2. Consider the ODE

$u''(t)+P(t)u'(t)+Q(t)u(t)=R(t), t\in[0,1]$

There exist continuous function P, Q and R defined on $[0,1]$ and two solutions u_1 and u_2 of the ODE such that the Wronskian W of u_1 and u_2 is

(a) $W(t)=2t-1, \quad 0\le t\le 1$ (b) $W(t)=\sin 2\pi t, \quad 0\le t\le 1$

(c) $W(t)=\cos 2\pi t, \quad 0\le t\le 1$ (d) $W(t)=1, \quad 0\le t\le 1$

Ans. **(d)**

Explanation :

We know that Wronskian W is either identically zero or never zero.

Option (d) is correct.

Important Result :

Let y(x) be a non-trivial solution of y`` + q(x)y = 0, where q(x) > 0 ∀ x > 0 &

$\int_1^{\infty} q(x)dx=\infty$. Then

(a) y(x) has infinite number of zeroes on R^+(positive x-axis)

(b) y`(x) has infinite zeros.

3. Let y be a solution of $(1 + x^2)y'' + (1 + 4x^2)y = 0$, x > 0, y(0) = 0. Then y(x) has

(a) infinitely many zeros in [0, 1] (b) infinitely many zeros in [1, ∞)

(c) at least n zeros in [0, nπ] ∀ n∈N (d) at most 3n zeros in [0, nπ] ∀ n∈N

Ans. **(b), (c), (d)** **CSIR NET DEC 2019**

Explanation :

$y''+\left(\frac{1+4x^2}{1+x^2}\right)y=0$ where q(x) = $\frac{1+4x^2}{1+x^2}$, ∀x > 0 Here, $\int_1^{\infty} q(x)dx=\infty$

⇒ Option (b) is correct. For (c) and (d)

We take q(x) = 1 and y(x) = sin x

Option (c) and (d) is correct.

4. Consider the ordinary DE y`` + P(x) y` + Q(x) y = 0, Where P and Q are smooth functions. Let y_1 and y_2 be any two solution of the ODE. Let w(x) be the Wronskian. Then which of the following is always true. **CSIR NET JUNE 2018**

(a) If y_1 & y_2 are LD then $\exists\ x_1, x_2$ s.t. $w(x_1) = 0$ and $w(x_2) \neq 0$

(b) If y_1 & y_2 are LI then $w(x) = 0\ \forall\ x$

(c) If y_1 & y_2 are LD then $w(x) \neq 0\ \forall\ x$ (d) If y_1 & y_2 are LI then $w(x) \neq 0\ \forall\ x$

Ans. (d)

Explanation :

We know that If $y_1, y_2, \ldots, y_n$ are LI $\Leftrightarrow w(y_1, y_2, \ldots, y_n) \neq 0$

Option (d) is correct.

5. If $y_1(x)$ and $y_2(x)$ be two solution of DE (cos x) y`` + (sin x) y` $-(1+e^{-x^2})y = 0$, $\forall x \in \left(\frac{-\pi}{2}, \frac{\pi}{2}\right)$ With $y_1(0) = \sqrt{2}$, $y`_1(0) = 1$, $y_2(0) = -\sqrt{2}$, $y_2`(0) = 2$. Then the Wronskian of $y_1(x)$ and $y_2(x)$ at $x = \frac{\pi}{4}$ is **CSIR NET DEC 2018**

(a) $3\sqrt{2}$ (b) 6 (c) 3 (d) $-3\sqrt{2}$

Ans. (c)

Explanation :

Given that (cos x)y`` + (sin x)y` – $(1+e^{-x^2})$y = 0 Here $a_1(x)$ = sin x, $a_0(x)$ = cosx

By Able's formula

$$W(y_1, y_2) = Ae^{-\int \frac{\sin x}{\cos x}dx} = Ae^{-\int \tan x dx} = Ae^{-\log \sec x} = Ae^{\log \cos x} = A \cos x$$

$$W(y_1 y_2)_{x=0} = \begin{vmatrix} y_1(0) & y_2(0) \\ y'_1(0) & y_2'(0) \end{vmatrix} = \begin{vmatrix} \sqrt{2} & -\sqrt{2} \\ 1 & 2 \end{vmatrix} = 2\sqrt{2} + \sqrt{2} \Rightarrow A = 3\sqrt{2}$$

$W(y_1, y_2) = 3\sqrt{2}\cos x$

Then we have to find $w(y_1, y_2)$ at $x = \frac{\pi}{4}$ $W' = 3\sqrt{2}\cos\frac{\pi}{4} = 3\sqrt{2}\left[\frac{1}{\sqrt{2}}\right] = 3$

Trick : Let $a_0(x)y'' + a_1(x)y' + a_2(x)y = Q(x)$ be a non homogeneous LDE and y_1 & y_2 are two solutions then $y(x) = c_1y_1 + c_2y_2$ be also solution if $c_1+c_2=1$

6. Three solutions of a certain second order non-homogenous linear differential equation are

$$y_1(x) = 1 + xe^{x^2},\ y_2(x) = (1+x)e^{x^2} - 1,\ y_3(x) = 1 + e^{x^2}.$$

Which of the following is (are) general solution(s) of the differential equation? **CSIR NET DEC 2018**

(a) $(C_1+1)y_1 + (C_2 - C_1)y_2 - C_2y_3$, where C_1 and C_2 are arbitrary constants

DIFFERENTIAL EQUATION

(b) $C_1(y_1-y_2)+C_2(y_2-y_3)$, where C_1 and C_2 are arbitrary constants

(c) $C_1(y_1-y_2)+C_2(y_2-y_3)+C_3(y_3-y_1)$, where C_1, C_2 and C_3 are arbitrary constants

(d) $C_1(y_1-y_3)+C_2(y_3-y_2)+y_1$, where C_1 and C_2 are arbitrary constants

Ans. (a), (d)

Explanation :

Any linear combination of solutions of a non-homogeneous second order linear differential equation is again a solution if the sum of the coefficients is one.

(a) Solution is $(C_1+1)y_1+(c_2-c_1)y_2-c_2y_3$

Sum of coefficients $=c_1+1+c_2-c_1-c_2=1$

(b) Solution is $c_1y_1+(c_2-c_1)y_2-c_2y_3$ Sum of coefficients $=c_1+c_2-c_1-c_2=0$

(c) Solution is $(c_1-c_3)y_1+(c_2-c_1)y_2+(c_3-c_2)y_3$

Sum of coefficients $=c_1-c_3+c_2-c_1+c_3-c_2=0$

(d) Solution is $(c_1+1)y_1-c_2y_2+(c_2-c_1)y_3$ Sum of coefficients $C_1+1-C_2+C_2-C_1=1$

Hence, options (a) and (d) are correct.

7. Let P,Q be continuous real valued functions defined on $[-1,1]$ and $u_i:[-1,1]\to R, i=1,2$ be solutions of the ODE:

$\frac{d^2u}{dx^2}+P(x)\frac{du}{dx}+Q(x)d=0, x\in[-1,1]$ satisfying $u_1\geq 0, u_2\leq 0$ and $u_1(0)=u_2(0)=0$.

Let w denote the Wronskian of u_1 and u_2, then **CSIR NET DEC 2014**

(a) u_1 and u_2, are linearly independent (b) u_1 and u_2, are linearly dependent

(c) $w(x)=0$ for all $x\in[-1,1]$ (d) $w(x)\neq 0$ for some $x\in[-1,1]$

Ans. (b),(c)

Explanation :

Since u_1 and u_2 have a common zero.

$\Rightarrow$ They are linearly dependent $\Rightarrow W(x)=0; \forall x\in[-1,1]$

Hence, option (b) and (c) are correct.

8. Let $Y_1(x)$ and $Y_2(x)$ defined on $[0,1]$ be twice continuously differentiable functions satisfying $Y''(x)+Y'(x)+Y(x)=0$. Let $W(x)$ be the Wronskian of Y_1 and Y_2 and satisfy $W\left(\frac{1}{2}\right)=0$. Then

(a) $W(x)=0$ for $x\in[0,1]$ (b) $W(x)\neq 0$ for $x\in[0,1/2)\cup(1/2,1]$

(c) $W(x)>0$ for $x\in(1/2,1]$ (d) $W(x)<0$ for $x\in[0,1/2)$

Ans. (a)

DIFFERENTIAL EQUATION

Explanation :
All the necessary conditions are satisfied so if $W(x)$ is zero at a point, then $W(x) \equiv 0$.

Since $W\left(\frac{1}{2}\right) = 0 \Rightarrow W(x) = 0$ Hence, option (a) is correct.

9. Let $y_1(x) = x^3$ and $y_2(x) = x^2|x|$ for $x \in R$.
Consider the following statements :

(P) : $y_1(x)$ and $y_2(x)$ are linearly independent solutions of $x^2 \frac{d^2y}{dx^2} - 4x\frac{dy}{dx} + 6y = 0$ on R.

(Q) : The Wronskian $y_1(x)\frac{dy_2(x)}{dx} - y_2(x)\frac{dy_1(x)}{dx} = 0$ for all $x \in R$. Which of the above statements hold TRUE? **GATE 2017**

(a) Both P and Q (b) Only P (c) Only Q (d) Neither P nor Q

Ans. (a)

Explanation :
Given $y_1(x) = x^3$ and $y_2(x) = x^2|x|$

Now, Wronkskian of y_1 and y_2 is $W(y_1, y_2) = \begin{vmatrix} y_1 & y_2 \\ y'_1 & y'_2 \end{vmatrix} = \begin{vmatrix} x^3 & x^2|x| \\ 3x^2 & 3x|x| \end{vmatrix} = 0$

$\therefore$ $W(y_1, y_2) = 0 \Rightarrow$ Statement (Q) is true.

And y_1 and y_2 are solutions of $x^2\frac{d^2y}{dx^2} - 4x\frac{dy}{dx} + 6y = 0$

So, we cannot conclude that y_1 and y_2 is L.D. or L.I.
Now we have to check further, Consider the linear combinations $c_1y_1 + c_2y_2 = 0$, where c_1, c_2 be arbitrary constants and $x \in R$

Case I : Let $x > 0$ $c_1y_2 + c_2y_2 = 0$

$\Rightarrow c_1x^3 + c_2x^2|x| = 0 \Rightarrow c_1x^3 + c_2x^3 = 0 \Rightarrow (c_1 + c_2)x^3 = 0$ $c_1 = -c_2$....(1)

Case II : Let $x < 0$ $c_1y_1 + c_2y_2 = 0$

$\Rightarrow c_1x^3 - c_2x^3 = 0 \Rightarrow (c_1 - c_2)x^3 = 0 \Rightarrow c_1 = c_2$....(2)

(1) and (2) is possible only if $c_1 = c_2 = 0$ $\Rightarrow c_1y_2 + c_2y_2 = 0 \Rightarrow c_1 = 0 = c_2$

$\Rightarrow$ y_1 and y_2 are two L.I. solutions of differential equation $x^2\frac{d^2y}{dx^2} - 4x\frac{dy}{dx} + 6y = 0 \Rightarrow$ statement P is true. Hence, option (a) is correct.

10. Let $x : [0, 3\pi] \to R$ be a non-zero solution of the ODE $x''(t) + e^{t^2}x(t) = 0$ for $t \in [0, 3\pi]$ then the cardinality of the set $\{t \in [0, 3\pi] \mid (x(t) = 0\}$ **CSIR NET DEC 2016**

(a) equal to 1 (b) ≥ 2 (c) $= 2$ (d) ≥ 3

Ans. (b), (d)

Explanation :

We have to find that the cardinality of set of solution $\{x(t) \mid t \in [0, 3\pi]\}$

Given that $x''(t)+e^{t^2}x(t)=0$ be a 2nd order DE

Then there are two LI solution exist. $x_1(t)$ & $x_2(t)$ are both LI.

Given DE is homogeneous. $\Rightarrow$ every L.C. of x_1, x_2 are solution of DE

Then the set A = $\{x_1(t), x_2(t), c_1x_1(t) + c_2x_2(t)\}$

Option (b) is correct. Option (d) is correct.

11. Let $y_1(x)$ and $y_2(x)$ form a fundamental set of solution of

$\frac{d^2y}{dx^2}+p(x)\frac{dy}{dx}+q(x)y=0, a \le x \le b$, where $p(x)$ and $q(x)$ are real valued continuous functions on $[a,b]$. If x_0 and x_1, with $x_0 < x_1$, are consecutive zeros of $y_1(x)$ in (a,b), then

(a) $y_1(x)=(x-x_0)q_0(x)$, where $q_0(x)$ is continuous on $[a,b]$ with $q_0(x_0)\neq 0$

(b) $y_1(x)=(x-x_0)^2 p_0(x)$, where $p_0(x)$ is continuous on $[a,b]$ with $p_0(x_0)\neq 0$

(c) $y_2(x)$ has no zeros in (x_0,x_1) (d) $y_2(x_0)=0$ but $y_2'(x_0)\neq 0$

Ans. (a)

Explanation :

$\because$ y_1 and y_2 form a fundamental set of solutions. $\Rightarrow$ They are continuous.

Also, we have

(1) y_1 and y_2 are linearly independent. (2) y_1 and y_2 cannot have common zero.

(3) y_1 and y_2 cannot have repeated zeroes.

(4) Given that x_0 and x_1 are consecutive zeros of $y_1(x)$. $\Rightarrow$ $y_2(x)$ has a zero in (x_0,x_1)

Option (a): If x_0 is a zero of y_1 and Taylor series expansion exists about x_0

Then $y_1(x)=(x-x_0)q_0(x)$ Where $q(x)$ is continuous & $q_0(x_0)\neq 0$

$\therefore$ Option (a) is true. Option (b) is false by (3)

Option (c) is false, since there exists exactly one root of $y_2(x_0)$ cannot be zero (by(2)).

12. $xy'' + 2y' + xe^x y = 0$. Let y_1 and $y_2$are L.I. solution if $w(x) = y_1(x)\, y_2'(x) - y_2(x)\, y_1'(x)$ with $w(1) = 2$.Find $w(5) =$

(a) $\frac{2}{13}$ (b) $\frac{2}{19}$ (c) $\frac{2}{25}$ (d) None of these

Ans. (c)

Explanation :

Let y_1 and y_2 are two solution and w(x) be Wronskian then

$$w(y_1,y_2)=\begin{vmatrix} y_1(x) & y_2(x) \\ y'_1(x) & y'_2(x) \end{vmatrix} = y_1y_2'-y_2y_1'$$

We know that Abel's formula

$$w(x) = Ae^{-\int \frac{a_1(x)}{a_0(x)}dx} = Ae^{-\int \frac{2}{x}dx} = Ae^{-2\log x} = Ae^{\log\left(\frac{1}{x^2}\right)} \quad w(x) = \frac{A}{x^2}$$

Given that w(1) = 2 $\quad 2 = \frac{A}{1} \Rightarrow A = 2 \quad w(x) = \frac{2}{x^2} \quad w(5) = \frac{2}{25}$

13. The Wronskian of the function 1, sinx, cosx is **CUET 2022**

(a) 0 (b) 1 (c) 1/2 (d) -1

Ans. (d)

Explanation :

$$W = \begin{vmatrix} 1 & \sin x & \cos x \\ 0 & \cos x & -\sin x \\ 0 & -\sin x & -\cos x \end{vmatrix} \Rightarrow W = 1(-\cos^2 x - \sin^2 x) = -1$$

UNSOLVED QUESTIONS

1. Let $y_1(x)$ and $y_2(x)$ be two solutions of $\left(1-x^2\right)\frac{d^2y}{dx^2} - 2x\frac{dy}{dx} + \sec x.y = 0$ with Wronskian W(x). If $y_1(0) = 1, \left(\frac{dy_1}{dx}\right)_{x=0} = 0$ and $W\left(\frac{1}{2}\right) = \frac{1}{3}$, then $\left(\frac{dy_2}{dx}\right)_{x=0}$ equals **GATE 2006**

(a) 1/4 (b) 1 (c) 3/4 (d) 4/3

2 Let $y = \phi(x)$ and $y = \psi(x)$ be solutions of $y'' - 2xy' + \left(\sin x^2\right)y = 0$ such that **GATE – 2004** $\phi(0) = 1, \phi'(0) = 1$ and $\psi(0) = 1, \psi'(0) = 2$. Then the value of the Wronskian $W(\phi, \psi)$ at $x = 1$ is

(a) 0 (b) 1 (c) e (d) e^2

3. If y_1 and y_2 be solution of y`` + p(x)y` + q(x)y = 0. Where p(x) and q(x) are continuous functions of x and the Wronskian $w(y_1, y_2) = 0$ then

(a) Linear dependent (b) Linear independent (c) Proportional

(d) None of these

4. Let $y_1(x)$ and $y_2(x)$ from a complete set of solutions to the differential equation $y'' - 2xy' + \sin\left(e^{2x^2}\right)y = 0, x \in [0,1]$ with $y_1(0) = 0, y_1'(0) = 1, y_2(0) = 1, y_2'(0) = 1$. Then the Wronskian $W(x)$ of $y_1(x)$ and $y_2(x)$ at $x = 1$ is **CSIR NET JUNE 2014**

(a) e^2 (b) $-e$ (c) $-e^2$ (d) e

5. Consider the ordinary differential equation $y'' + P(x)y' + Q(x)y = 0$ where P and Q are smooth functions. Let y_1 and y_2 be two solutions of the ODE. Let W(x) be the corresponding Wronskian. Then which of the following is always true?

(a) If y_1 and y_2 are linearly dependent then $\exists x_1, x_2$ such that $W(x_1)=0$ and $W(x_2)\neq 0$
(b) If y_1 and y_2 are linearly independent then $W(x)=0 \forall x$
(c) If y_1 and y_2 are linearly dependent then $W(x)\neq 0 \forall x$
(d) If y_1 and y_2 are linearly independent then $W(x)\neq 0 \forall x$

6. If $y_1(x)$ and $y_2(x)$ are solutions of $y'' + x^2 y' + (1 - x) y = 0$ such that $y_1(0) = 0$, $y'_1(0) = -1$ and $y_2(0) = -1$, $y'_2(0) = 1$, then the Wronskian $W(y_1, y_2) = R$
(a) is never zero
(b) is identically zero
(c) is zero only at finite number of points
(d) is zero at countably infinite number of points

7. If y_1 and y_2 are linearly independent solutions of the homogeneous differential equation $L(y)=y''+p_1(x)y'+p_2(x)y=0$ and $W(x)$ represents their Wronskian. Then $p_1(x)$ and $p_2(x)$ are not given by

(a) $p_1(x)=\frac{y_1y''_2-y''_1y_2}{W(x)}, p_2(x)=\frac{y'_1-y''_2-y_1''y_2'}{W(x)}$

(b) $p_1(x)=-\left[\frac{y_1y''_2-y''_1y_2}{W(x)}\right], p_2(x)=\frac{y'_1-y''_2-y_1''y_2'}{W(x)}$

(c) $p_1(x)=\frac{y_1y''_2-y''_1y_2}{W(x)}, p_2(x)=-\left[\frac{y'_1-y''_2-y_1''y_2'}{W(x)}\right]$

(d) $p_1(x)=-\left[\frac{y_1y''_2-y''_1y_2}{W(x)}\right], p_2(x)=-\left[\frac{y'_1-y''_2-y_1''y_2'}{W(x)}\right]$

8. The value of the Wronskian of the functions x^2, $3x + 2$, $2x + 3$ is **GATE – 1999**
(a) 0
(b) $2 + x$
(c) $10 + x^2$
(d) -10

9. Let y be a non-trivial solution of the boundary value problem $y''+xy=0, x\in[a,b]$, $y(a)=y(b)=0$, then **CSIR NET DEC 2013**
(a) $b>0$
(b) y is monotone in $(\alpha,0)$ if $\alpha<0<b$
(c) $y'(\alpha)=0$
(d) y has infinitely many zeros in $[a,b]$

10. Let P b a continuous function on R and W the Wronskian of two linearly indepei lent solutions y_1 and y_2 of the ODE: **CSIR NET JUNE 2015**

$$\frac{d^2y}{dx^2}+(1+x^2)\frac{dy}{dx}+P(x)y=0, x\in R$$ Let $W(1)=a$, $W(2)=b$, $W(3)=c$, then

(a) $a<0$ and $b>0$
(b) $a<b<c$ or $a>b>x$
(c) $\frac{a}{|a|}=\frac{b}{|b|}=\frac{c}{|c|}$
(d) $0<a<b$ and $b>c>0$

Answer Key

1	2	3	4	5	6	7	8	9	10
a	c	a	b	d	b	a,c,d	d	a,b	b,c

Chapter 3

Linear Diffrential Equation

1. Linear differential equation with constant coefficients

Definition: A linear differential equation with constant coefficients is one in which the dependent variable and its differential coefficients occur only in the first degree and are not multiplied together.

$$P_0\frac{d^n y}{dx^n}+P_1\frac{d^{n-1}y}{dx^{n-1}}+P_2\frac{d^{n-2}y}{dx^{n-2}}+...+P_n y=Q$$

(Where $P_0, P_1, P_2, ..., P_n$ are constants and Q is a function of x) is the linear equation of the nth order.

- If $y = y_1, y = y_2, ..., y = y_n$ are n linearly independent particular solutions of the equation. $P_0\frac{d^n y}{dx^n}+P_1\frac{d^{n-1}y}{dx^{n-1}}+P_2\frac{d^{n-2}y}{dx^{n-2}}+...+P_n y=0$, where $P_0, P_1, P_2, .., P_n$ are constants, then the complete solution is $y = c_1y_1 + c_2y_2 + .. + c_ny_n$.

Auxiliary Equation (A.E.): Auxiliary equation is obtained by equating to zero the symbolic co-efficients of y.

Thus, $D^n+a_1D^{n-1}+a_2D^{n-2}+...+a_n=0$ is the auxiliary equation.

Method to solve the equation: $P_0\frac{d^n y}{dx^n}+P_1\frac{d^{n-1}y}{dx^{n-1}}+P_2\frac{d^{n-2}y}{dx^{n-2}}+...+P_n y=0$, where $P_0, P_1, P_2, ... P_n$ are constants.

(i) Write the equation in the symbolic form as : $(P_0D^n + P_1D^{n-1} + P_2D^{n-2} +... + P_n)y=0$.
(ii) Write down the auxiliary equation (A.E.) as: $P_0m^n + P_1m^{n-1} + ... + P_n = 0$. Solve it for m
(iii)From the roots of A.E. write down the corresponding part of the complete solution (C.S.) as follows:

Roots of A.E.	Corresponding parts of C.S.
One real root m_1.	$c_1e^{m_1x}$
Two real and distinct roots m_1, m_2	$c_1e^{m_1x}+c_2e^{m_2x}$
Two real and equal roots m_1, m_1	$e^{m_1x}(c_1+c_2x)$
Three real and equal roots m_1, m_1, m_1	$e^{m_1x}(c_1+c_2x+c_3x^2)$
One pair of imaginary and different roots $\alpha \pm i\beta$	$e^{\alpha x}(c_1 \cos\beta x+c_2 \sin\beta x)$
Two pairs of imaginary and equal roots $\alpha\pm i\beta, \alpha\pm i\beta$	$e^{\alpha x}[\cos\beta x(c_1+c_2x)+\sin\beta x(c_3+c_4x)]$

Let Diffrential Equation is non-homogeneous i.e. $y'' + ay' + by = Q(x)$ then general solution = Complementary Function + Particular Integral

Standard Cases of particular integral:

Determination of PI of $f(D)y = Q$ $\quad y = \frac{1}{f(D)}Q$

(1) General method for PI $\frac{1}{D-\alpha}Q = e^{\alpha x}\int Q\, e^{-\alpha x}dx$ $\quad$ (2) $\frac{1}{(D-\alpha)^n}e^{\alpha x} = \frac{x^n}{n!}e^{\alpha x}$

(3) $\frac{1}{(D-\alpha)^r\,\phi(D)}e^{\alpha x} = \frac{1}{\phi(a)}\frac{x^r}{r!}e^{\alpha x}$

(4) $\frac{1}{\phi(D^2)}\sin ax = \frac{1}{\phi(-a^2)}\sin ax$

if $\phi(-a^2)$ is non-zero.

If $\phi(-a^2) = 0$ then $\frac{1}{\phi(D^2)}\sin ax = x\frac{1}{\frac{d}{dD}(D^2)}\sin ax$

Same for cos ax

(5) $\frac{1}{f(D)}e^{\alpha x}V = e^{\alpha x}\frac{1}{f(D+\alpha)}V$ $\quad$ (6) $\frac{1}{f(D)}xV = x\frac{1}{f(D)}V + \left(\frac{d}{dD}\frac{1}{f(D)}\right)V$

(7) Method to evaluate $\frac{1}{f(D)}x^m$, where m is positive integer, is as follows:

(i) From f(D), take the lowest degree term outside. Then the remaining factor will be of the type $[1 \pm \phi(D)]$.

(ii) Take $[1 \pm \phi(D)]$ to the numerator and expand it by Binomial Theorem $\left(\because D^{m+1}(x^m) = 0\right)$

(iii) Operate each term on x^m.

Remember:

(i) $(1-D)^{-1} = 1 + D + D^2 + \ldots to\ \infty$ $\quad$ (ii) $(1+D)^{-1} = 1 - D + D^2 - \ldots to\ \infty$

(iii) $(1-D)^{-2} = 1 + 2D + 3D^2 + \ldots to\ \infty$ $\quad$ (iv) $(1-D)^{-3} = 1 + 3D + 6D^2 + \ldots to\ \infty$

Example. Solve $\left(D^4 + 2D^3 - 3D^2\right)y = 3e^{2x} + 4\sin x$

Solution: $A.E.\ is\ m^4 + 2m^3 - 3m^2 = 0$ $\quad \Rightarrow m = 0, 0, 1, -3$

C.F. is $y = (c_1 + c_2 x) + c_3 e^x + c_4 e^{-3x}$

On putting $D = 2$, $P.I.\ of\ 3e^{2x}\ is\ \frac{3}{20}e^{2x}$

P.I. of $4\sin x = \frac{4\sin x}{D^2\left(D^2 + 2D - 3\right)}$

$$=\frac{4\sin x}{(-1)(-1+2D-3)}=-2\frac{\sin x}{(D-2)}=-2\frac{(D+2)\sin x}{D^2-4}=\frac{-2}{-1-4}(\cos x+2\sin x)=\frac{2}{5}(2\sin x+\cos x)$$

$\therefore$ General Solution is y = C.F. + P.I.

$$\Rightarrow y=(c_1+c_2x)+c_3e^x+c_4e^{-3x}+\frac{2}{5}(2\sin x+\cos x)+\frac{3}{20}e^{2x}$$

Example: Solve $\frac{d^2y}{dx^2}-2\frac{dy}{dx}+y=x^2e^{3x}$

Solution: A.E. is $m^2-2m+1=0 \quad \Rightarrow m=1,1$

$\therefore C.F\ is\ y=(c_1+c_2x)e^x$

P.I for x^2 e^{3x} will be $\frac{x^2e^{3x}}{(D-1)^2}=e^{3x}\left\{\frac{1}{(D+3-1)^2}\right\}x^2=e^{3x}\left\{\frac{1}{(D+2)^2}\right\}x^2$

$$=\frac{e^{3x}}{4}\left\{\left(1+\frac{D}{2}\right)^{-2}\right\}x^2=\frac{e^{3x}}{4}\left\{1-2\left(\frac{D}{2}\right)+3\left(\frac{D^2}{4}\right)...\right\}x^2$$

$$=\frac{e^{3x}}{4}\left(x^2-2x+\frac{3}{4}.2\right)=\frac{e^{3x}}{8}(2x^2-4x+3)$$

2. Linear differential equations with variable coefficients

Definition: A homogeneous linear equation of the form

$P_0\frac{d^ny}{dx^n}+P_1\frac{d^{n-1}y}{dx^{n-1}}+P_2\frac{d^{n-2}y}{dx^{n-2}}+...+P_ny=Q$, where $P_0,P_1,P_2,...,P_n$ and Q are functions of x is called linear differential equation with variable coefficients.

Couchy's Linear Equation

Definition: A homogeneous linear equation is of the form

$P_0x^n\frac{d^ny}{dx^n}+P_1x^{n-1}\frac{d^{n-1}y}{dx^{n-1}}+...+P_ny=Q(x)$, where P_0, P_1, …, P_n are all real constants and Q(x) is a function of x, is called Cauchy's linear equation.

Method to solve

(i) Put $x = e^z$, i.e., z = log x, x > 0

(ii) Put $\frac{d}{dz}=D$ so that $xD=D, x^2D^2=D(D-1),..., x^nD^n=D(D-1)...(D-n+1)$

Putting these in the given equation, we get

$[P_0D(D-1)...(D-n+1)+P_1D(D-1)...(D-n+2)+...+P_n]y=Q(e^z)$, which is linear equation with constant coefficients and solve for y in terms of z.

(iii) Put z = log x, and we get the required solution.

DIFFERENTIAL EQUATION

Example. Solve $x^4\frac{d^3y}{dx^3}+2x^3\frac{d^2y}{dx^2}-x^2\frac{dy}{dx}+xy=1$

Solution: The given equation may be written as $x^3\frac{d^3y}{dx^3}+2x^2\frac{d^2y}{dx^2}-x\frac{dy}{dx}+y=\frac{1}{x}$

Putting x = e^z, the equation becomes $[D(D-1)(D-2)+2D(D-1)-D+1]y=e^{-z}$

Or $(D-1)^2(D+1)y=e^{-z}$

A.E. is $(m-1)^2(m+1)=0$ or m = 1, 1, -1

$\therefore$ C.F. $=(c_1+c_2z)e^z+c_3e^{-z}=(c_1+c_2\log x)x+c_3x^{-1}$

$$\text{P.I.}=\frac{1}{(D-1)^2(D+1)}e^{-z}=\frac{1}{(D+1)}\left(\frac{1}{(D-1)^2}e^{-z}\right)=\frac{1}{D+1}\left(\frac{1}{4}e^{-z}\right)=\frac{1}{4}e^{-z}\frac{1}{D-1+1}1$$

$$=\frac{1}{4}ze^{-z}=\frac{1}{4x}\log x.$$

Hence, the solution is $y=(c_1+c_2\log x)x+\frac{c_3}{x}+\frac{1}{4x}\log x$

3. Linear Ordinary Differential Equations of second order with variable coefficients:

An equation of the form $\frac{d^2y}{dx^2}+P\frac{dy}{dx}+Qy=R$ .....(1)

Where P, Q, R are continuous functions of x only, is called Linear Ordinary Differential Equation of second order.

The complete solution when one integral is known:

Let y = u be a known integral in the C.F. of (1) then Second part of the complementary function is $u\int\frac{e^{-\int Pdx}}{u^2}dx$ and particular integral is $=u\int\frac{e^{-\int Pdx}}{u^2}\left[\int Rue^{\int Pdx}dx\right]dx$

(i) To find one integral in C.F. by inspection, i.e. to find a solution of $D^2y+PDy+Qy=0$

1. y = x is a solution of (1) if P + Qx = 0
2. y = x^2 is a solution of (I) if 2 + 2Px + Qx^2 = 0
3. y = e^x is a solution of (1) if 1 + p + Q = 0
4. y = e^{-x} is a solution of (1) if 1 – P + Q = 0
5. y = e^{mx} is a solution of (1) if m^2 + mP + Q = 0
6. y = x^m is a part of C.F. if m(m-1) + Pmx + Qx^2 = 0

Example. $x\frac{d^2y}{dx^2}-2(x+1)\frac{dy}{dx}+(x+2)y=(x-2)e^x$

Solution $$\frac{d^2y}{dx^2}-\frac{2(x+1)}{x}\frac{dy}{dx}+\frac{x+2}{x}y=\left(\frac{x-2}{x}\right)e^x$$

Here $1+P+Q=1-\frac{2(x+1)}{x}+\frac{x+2}{x}=0 \Rightarrow y=e^x$ is a part of the C.F.

Putting $y=ve^x \Rightarrow \frac{dy}{dx}=e^x\left(\frac{dv}{dx}+v\right) \Rightarrow \frac{d^2y}{dx^2}=e^x\left[\frac{d^2v}{dx^2}+2\frac{dv}{dx}+v\right]$

Equation (1)

$$\Rightarrow e^x\left[\frac{d^2v}{dx^2}+\frac{2dv}{dx}+v\right]-\frac{2(x+1)}{x}e^x\left[\frac{dv}{dx}+v\right]+\frac{(x+2)}{x}ve^x=\frac{(x-2)}{x}e^x$$

$$\Rightarrow \frac{d^2v}{dx^2}+\frac{dv}{dx}\left(2-\frac{2(x+1)}{x}\right)+v\left(1-\frac{2(x+1)}{x}+\frac{x+2}{2}\right)=\frac{x-2}{x}$$

$$\Rightarrow \frac{dp}{dx}+\left(-\frac{2}{x}\right)p=\frac{x-2}{x}\left[\because \frac{dv}{dx}=p\right]$$

I.F. $=e^{\int -\frac{2}{x}dx}=\frac{1}{x^2} \qquad \therefore \frac{p}{x^2}=\int\frac{x-2}{x^3}dx=\int\frac{dx}{x^2}-\int\frac{2dx}{x^3}=-\frac{1}{x}+\frac{1}{x^2}+c_1$

$$\Rightarrow p=c_1x^2-x+1 \Rightarrow v=c_2+\frac{c_1x^3}{3}-\frac{x^2}{2}+x$$

Here, complete solution is $y=ve^x=e^x\left(x-\frac{x^2}{2}+\frac{c_1}{3}x^3+c_2\right)$

Removal of the first derivative (Reduction to Normal form)

Let $\frac{d^2y}{dx^2}+P\frac{dy}{dx}+Qy=R$ ....(1)

Its normal form is $\Rightarrow \frac{d^2v}{dx^2}+Xv=z$ .....(2)

Which is in normal form where $X=Q-\frac{1}{2}\frac{dP}{dx}-\frac{P^2}{4}$, $z=\text{R}e^{\frac{1}{2}\int Pdx}$

Variation of parameter

Let u & v are two solution of y`` + py` + qy = 0 then its general solution

$$y(x)=u\int\frac{v.R}{W(v,u)}dx+v\int\frac{-u.R}{W((v,u)}dx+C_1u+C_2v$$

Where $W(v,u)=\begin{vmatrix} v & u \\ v' & u' \end{vmatrix}$

Example. If y(x) be a P.I. of y`` + y = cosec x $x \in \left(0, \frac{\pi}{2}\right)$ then

(a) y(x) is convex on $\left(0, \frac{\pi}{2}\right)$ (b) concave on $\left(0, \frac{\pi}{2}\right)$ (c) Positive (d) y(x) is Negative

Solution : (a), (d)

Given that y'' + y = cosec x

Then find the solution of y'' + y = 0

$m^2 + 1 = 0$

m = ±I $\qquad$ $y(x) = c_1 \cos x + c_2 \sin x$

here u = cosx & v = sin x

$$W(v,u) = \begin{vmatrix} \sin x & \cos x \\ \cos x & -\sin x \end{vmatrix} = -1 \neq 0$$

Then P.I.

$$y_p(x) = u\int \frac{v.R}{W(v,u)}dx - v\int \frac{u.R}{W(v,u)}dx$$

$$= \cos x \int \frac{\sin x.\operatorname{cosec} x dx}{-1} - \sin x \int \frac{\cos x.\operatorname{cosec} x dx}{-1} = -\cos x \int dx + \sin x \int \cot x dx$$

$y_p(x) = -x \cos x + \sin x . \log \sin x$

Option (c), (d)

Given that $x \in \left(0, \frac{\pi}{2}\right)$ $\quad y_p(x) = \frac{(-x)}{-ve}\frac{(\cos x)}{+ve} + \frac{\sin x}{+ve}\frac{\log \sin x}{-ve}$ $\qquad y_p(x) < 0$

Option (d) is correct. $\qquad$ Option (c) is incorrect. $\qquad$ For (a) and (b)

Given that y`` + y = cosec x $\Rightarrow y'' = \frac{\operatorname{cosec} x}{+ve} - (y)$ $\Rightarrow$ y`` > 0 $\Rightarrow$ y(x) is convex.

Ordinary and singular points of second order DE

P(x) y`` + Q(x) y` + R(x) y = 0

A point $x_0 \in R$ s.t. $P(x_0) = 0$ or $Q(x_0) = R(x_0) = \infty$ then x_0 is called singular point.

(i) Regular singular point :

$$\lim_{x \to x_0} (x - x_0)\frac{Q(x)}{P(x)} = finite \quad \text{and} \quad \lim_{x \to x_0} (x - x_0)^2 \frac{R(x)}{P(x)} = finite$$

(iii) **Irregular singular point :** A point which is not a regular singular point.

SOLVED QUESTIONS

1. Consider the differential equation $x^2y'' - 2x(x + 1)y' + 2(x + 1)y = 0$.

CSIR NET NOV 2020

If a polynomial is a solution then the degree of the polynomial is equal to

(a) 1 $\qquad$ (b) 2 $\qquad$ (c) 3 $\qquad$ (d) 4

Ans. (a)

Explanation :

$x^2y'' - 2x(x+1)y' + 2(x+1)y = 0$ .... (i)

$$\Rightarrow \quad y'' - \frac{2x(x+1)}{x^2}y' + \frac{2(x+1)}{x^2}y = 0$$

$$\Rightarrow \quad y'' - \frac{2(x+1)}{x}y' + \frac{2(x+1)}{x^2}y = 0 \quad ..... (*)$$

Comparing (*) with $y'' + P(x)y' + Q(x)y = 0$

We get $P = \frac{-2(x+1)}{x}$ & $Q = \frac{2(x+1)}{x^2}$

Now $P + Qx = \frac{-2(x+1)}{x} + x.\frac{2(x+1)}{x^2}$

$= \frac{-2(x+1)}{x} + \frac{2(x+1)}{x^2} = 0$

$\Rightarrow \quad P + Qx = 0$

So, y = x is a solution of (i) and y = x is a polynomial of degree 1.

So, option (a) is correct.

2. Consider a DE $x^3y'' + x^2y' - xy = 0$ if y_1& y_2 are two LI solution of DE then choose the incorrect.

(a) $w(y_1, y_2)$ is continuous on R

(b) DE has unique solution if y(0) = 0, y'(0) = 1

(c) DE has no solution if y(0) = 0 = y'(0)

(d) DE has unique solution if y(1) = 0 = y'(1)

Ans. **(a), (c)**

Explanation :

Let $a_0y'' + a_1y' + a_2y = 0$

Then $w(y_1, y_2) = e^{-\int \frac{a_1}{a_0}dx}$

So here $w(y_1, y_2) = Ae^{-\int \frac{1}{x}dx} = Ae^{-\log x}$

$= \frac{A}{x}$ is not continuous on R.

Option (a) is incorrect.

Given that $x^2y'' + xy' - y = 0$

A.E. $m^2 - 1 = 0$

$\Rightarrow m = \pm 1$

$$y(x) = c_1x + \frac{c_2}{x} \quad ..... (1)$$

Condition $y(0) = 0 \Rightarrow c_2 = 0$ and $y'(0) = 1$

$\Rightarrow c_1 = 1$ Then $y(x) = x$ unique solution.

Option (b) is correct. Equation (1) $y(x)=c_1x+\frac{c_2}{x}$

Given condition y(0) = 0 = y`(0) is trivial condition.

$\Rightarrow$ y ≡ 0 (trivial solution always exist) Option (c) is incorrect.

Equation (1) $y(x)=c_1x+\frac{c_2}{x}$

$y(1) = 0 \Rightarrow c_1 + c_2 = 0$ $y'(1) = 0 \Rightarrow c_1 - c_2 = 0$ $c_1 = c_2 = 0$

y(x) = 0 unique solution

Option (d) is incorrect.

3. Consider the differential equation $\frac{d^2y}{dx^2}-2\tan x\frac{dy}{dx}-y=0$ defined on $\left(-\frac{\pi}{2},\frac{\pi}{2}\right)$.

Which among the following are true? **CSIR NET DEC-2017**

(a) there is exactly one solution

$$y=y(x)\, with\, y(0)=y'(0)=1\, and\, y\left(\frac{\pi}{3}\right)=2\left(1+\frac{\pi}{3}\right)$$

(b) there is exactly one solution y = y(x) with y(0) = 1,

$$y'(0)=-1\; and\; y\left(-\frac{\pi}{3}\right)=2\left(1+\frac{\pi}{3}\right)$$

(c) any solution y = y (x) satisfies y" (0) = y (0)

(d) If y_1 and y_2 are any two solutions then $(ax+b)y_1=(cx+d)y_2$ for some a, b, c, d, ∈ R

Ans. (a), (b), (c), (d)

Explanation :

Given differential equation is y" – 2 tan x y ' – y = 0 ...(1)

Comparing (1) with y" + Py' + Qy = R $\Rightarrow P=-2\tan x,\ Q=-1,\ R=0$

To remove the first derivative from (1) we choose

$u=e^{-\frac{1}{2}\int P\,dx}=e^{-\frac{1}{2}\int -2\tan x}=\sec x$ $\Rightarrow u=\sec x$

Let the General Solution is y(x) = sec x. v(x)

Substitute the values of y(x), y'(x) and y"(x) in (1)

we get v"(x) = 0 ⇒ v(x) = ax + b ∴ y(x) = (ax + b) sec x ... (2)

∴ for option (a).

y(0) = y'(0) = 1 $\Rightarrow a=1, b=1, as\ y'(x)=(ax+b)\sec x\tan x+a\sec x$

$\therefore y(x)=(x+1)\sec x$

$\Rightarrow y\left(\frac{\pi}{3}\right)=2\left(1+\frac{\pi}{3}\right)$ ⇒ option (a) is correct.

For option (b),

$y(0)=1,\ y'(0)=-1$ $\Rightarrow a=-1,\ b=1 \Rightarrow y(x)=(1-x)\sec x$

$$\Rightarrow y\left(-\frac{\pi}{3}\right)=2\left(1+\frac{\pi}{3}\right)$$ $\therefore$ option (b) is also correct.

For option (c)

$y''(0)=b$ and $y(0)=b \Rightarrow y''(0)=y(0)$ for any solution y = y(x)

$\therefore$ option (c) is also correct.

For option (d)

Let $y_1=(c_1x+c_2)\sec x$ and $y_2=(d_1x+d_2)\sec x$ $\therefore (ax+b)y_1=(cx+d)y_2$

$\Rightarrow (ax+b)(c_1x+c_2)=(cx+d)(d_1x+d_2)$

if $c_1=c, c_2=d, d_1=a, d_2=b$, then equality holds

$\therefore$ option (d) is also correct.

4. The method of variation of parameter to solve the differential equation $y'' + p(x)y' + q(x)y = r(x)$ where $x \in I$ and p(x), q(x), r(x) are non-zero continuous function on an interval I seek a particular solution of the form $y_p(x) = v_1(x)y_1(x) + v_2(x)y_2(x)$ where y_1 & y_2 are two LI solution of $y'' + p(x)y' + q(x)y = 0$ and v_1 & v_2 are function to be determined. Which of the following are TRUE? **CSIR NET DEC 2018**

(a) The Wronskian of y_1 and y_2 is never zero

(b) v_1, v_2 and $v_1y_1 + v_2y_2$ are twice differential

(c) v_1 & v_2 may not be twice differential but $v_1y_1 + v_2y_2$ be twice differential.

(d) The solution set of $y'' + py' + qy = r(x)$ of the form $ay_1 + by_2 + y_p$

Ans. **(a), (c), (d)**

Explanation :

By the variation of parameter

General solution of $y'' + py' + qy = r(x)$ is

$$y(x)=u\int\frac{v.R}{W(v,u)}dx+v\int\frac{-u.R}{W(v,u)}+c_1u+c_2v$$

Where u & v are LI solution.

and $y_p(x)=u\int\frac{v.R}{W(v,u)}-v\int\frac{u.R}{W(v,u)}$ is particular integral.

Then $y(x) = c_1u + c_2v + y_p(x)$

Option (d) is correct. Option (a) is correct.

Let y_1 & y_2 are LI solution then $W(y_1, y_2) \neq 0$

For (b), (c) $y_p(x) = v_1y_1 + v_2y_2$

where $v_1=\int\frac{y_2R}{W(y_2,y_1)}dx$ & $v_2=\int\frac{y_1R}{W(y_2,y_1)}dx$

where v_1 & v_2 may not be twice differential for some R & $W(y_2, y_1)$

but $y_p(x) = v_1y_1 + v_2y_2$ is always twice differential.

Option (c) is correct.

5. Let x, x + e^x and 1 + x + e^x be solutions of a linear second order ordinary differential equation with constant coefficients. If y(x) is the solution of the same equation satisfying y(0) = 3 and y`(0) = 4, then y(1) is equal to **CSIR NET JUNE 2019**

(a) e + 1 (b) 2e + 3 (c) 3e + 2 (d) 3e + 1

Ans. (d)

Explanation :

Let the differential equation be $\frac{d^2y}{dx^2}+a\frac{dy}{dx}+by=Q(x)$ ..... (1)

Let $y_1 = x$, $y_2 = x + e^x$ and $y_3 = 1 + x + e^x$

Since y_1, y_2, y_3 are solution of (1)

Then, a + bx = Q(x) ..... (2)

$e^x + a(1 + e^x) + b(x + e^x) = Q(x)$ ...... (3)

$e^x + a(1 + e^x) + b(1 + x + e^x) = Q(x)$ ....(4)

$\Rightarrow$ a = Q(x)

Using the value of a and b in (3), we get

$e^x + Q(x)(1 + e^x) = Q(x)$ $\Rightarrow$ Q(x) = -1

Hence, the differential equation is $\frac{d^2y}{dx^2}-\frac{dy}{dx}=-1$

The corresponding auxiliary equation is

$m^2 - m = 0$ $\Rightarrow$ $m(m - 1) = 0$ $\Rightarrow$ $m = 1, 0$

Therefore the solution of the homogeneous part of the differential equation

$\frac{d^2y}{dx^2}-\frac{dy}{dx}=-1$ is

$c_1 + c_2e^x$ and particular integral is $\frac{1}{D^2-D}.1=\frac{1}{D}(1-D)^{-1}.1=\frac{1}{D}=x$

$\Rightarrow$ The complete solution is $y(x) = c_1 + c_2e^x + x$

Since y(0) = 3 and y`(0) = 4

$\Rightarrow$ $c_1 + c_2 = 3$ and $c_2 + 1 = 4$ $\Rightarrow$ $c_1 = 0$ and $c_2 = 3$ $\Rightarrow$ $y(x) = 3e^x + x$

$\Rightarrow$ y(1) = 3e + 1

Hence, option (d) is correct.

6. Consider the differential equation $(x-1)y''+xy'+\frac{1}{x}y=0$. Then

(a) x = 1 is the only singular point (b) x = 0 is the only singular point

(c) both x = 0 and x = 1 are singular points (d) neither x = 0 nor x = 1 are singular points

Ans. (c)

Explanation :

Given equation is $(x-1)y''+xy'+\frac{1}{x}y=0 \Rightarrow y''+\frac{x}{x-1}y'+\frac{1}{x(x-1)}y=0$ ....(1)

By comparing (1) with $y''+P(x)y'+Q(x)y=0$,

we get $P(x)=\frac{x}{x-1}; Q(x)=\frac{1}{x(x-1)}$

Now for

$$x=0; \lim_{x\to0}(x-0)P(x)=\lim_{x\to0} x.\frac{x}{x-1}=0, \lim_{x\to0}(x-0)^2Q(x)=\lim_{x\to0}\frac{x^2}{x(x-1)}=0$$

and for $x=1; \lim_{x\to1}(x-1)P(x)=\lim_{x\to1}(x-1)\frac{x}{(x-1)}=1, \lim_{x\to1}(x-1)^2 Q(x)$

$$=\lim_{x\to1}(x-1)^2\frac{1}{x(x-1)}=0$$

$\therefore$ Both 0 and 1 are regular singular points.

7. For the differential equation $4x^3y''+6x^2y'+y=0$, point at infinity is

GATE-1997

(a) regular singular point (b) irregular singular point
(c) no singular point (d) none of these

Ans. **(a)**

Explanation : Given, $4x^3y''+6x^2y'+y=0$

We transform the independent variable x to t by the relation $t=\frac{1}{x} or\, x=\frac{1}{t}$

Now $y'=\frac{-1}{x^2}\frac{dy}{dt}\Rightarrow y''=\frac{1}{x^4}\frac{d^2y}{dt^2}+\frac{2}{x^3}\frac{dy}{dt}$ and given differential equation (i) transform

to $4t\frac{d^2y}{dt^2}+2\frac{dy}{dt}+y=0$

The point at ∞ is transformed to the origin

From equation (ii), we note that the origin is regular singular point.

Hence, the point at ∞ is a regular singular point of the given equation.

8. Let $y(x)$ be the solution of $x^2y''(x)-2y(x)=0, y(1)=1, y(2)=1$.

Then the value of $y(3)$ is **CSIR NET JUNE-2019**

(a) $\frac{11}{21}$ (b) 1 (c) $\frac{17}{21}$ (d) $\frac{11}{7}$

Ans. **(d)**

Explanation :

Given that 2nd order linear differential equation with variable coefficients hence, the auxiliary equation is

$$\Rightarrow m^2-m-2=0 \Rightarrow m=\frac{-(-1)\pm\sqrt{1+8}}{2}=\frac{1\pm3}{2}=-1,2 \Rightarrow y(x)=\frac{c_1}{x}+c_2x^2$$

Given that $y(1)=1$ and $y(2)=1$, we get

$c_1+c_2=1$ and $\frac{c_1}{2}+4c_2=1 \Rightarrow y(x)=\frac{6}{7x}+\frac{x^2}{7} \Rightarrow y(3)=\frac{2}{7}+\frac{9}{7}=\frac{2+9}{7}=\frac{11}{7}$

9. A particular integral of the differential equation y`` + 2y` + 2y = e^{e^x} is

(a) $e^{e^x}e^{-x}$ (b) $e^{e^x}e^{-2x}$ (c) $e^{x^2}e^{2x}$ (d) $e^{e^x}e^x$

Ans. (b) **IIT-JAM 2018**

Explanation :

y`` + 3y` + 2y = e^{e^x}

Particular Integral $(P.I.)=\frac{1}{D^2+3D+2}(e^{e^x})$

$P.I.=\frac{1}{(D+1)(D+2)}e^{e^x}$ $P.I.=\left[\frac{1}{D+1}-\frac{1}{D+2}\right]e^{e^x}$

$P.I.=\frac{1}{D+1}e^{e^x}-\frac{1}{D+2}e^{e^x}$ P.I. = I_1 - I_2

$I_1=\frac{1}{(D+1)}e^{e^x}=v$ Let $e^{e^x}=(D+1)v$

$\frac{dv}{dx}+v=e^{e^x}$

$I.F.=e^{\int dx}=e^x$ $\Rightarrow$ $v.e^x=\int e^{e^x}.e^x dx$

Let $e^x=t, e^x dx=dt$

$v.e^x=e^t$

$v.e^x=e^{e^x}$

$v=e^{e^x-x}$ $\Rightarrow$ $I_1=e^{e^x-x}$

$I_2=\frac{1}{D+2}e^{e^x}=v$

Let $e^{e^x}=(D+2)v$ $\Rightarrow$ $\frac{dv}{dx}+2v=e^{e^x}$

$\Rightarrow$ I.F. = e^{2x}

$v.e^{2x}=\int e^{2x}.e^{e^x}dx$ $v.e^{2x}=te^t-e^t$

$v.e^{2x}=e^x e^{e^x}-e^{e^x}$ $v.e^{2x}=(e^x-1)e^{e^x}$

$I_2=v=e^{-2x}(e^x-1)e^{e^x}$

P.I. = $I_1 - I_2$

$e^{e^x - x} - e^{-2x}(e^x - 1)e^{e^x}$ $e^{-x}e^{e^x} - (e^{-x} - e^{-2x})e^{e^x}$

$(e^{-x} - e^{-x} + e^{-2x})e^{e^x}$ $P.I. = e^{-2x}.e^{e^x}$

10. The solution of the differential equation $\frac{d^2y}{dx^2} - 4y = x\sinh x$ is

CUET 2022

(a) $y = c_1e^{2x} + c_2e^{-2x} - \frac{x}{3}\sinh x - \frac{2}{9}\cosh x$ (b) $y = c_1e^{-2x} + c_2e^{x} + \frac{x}{3}\sinh x - \frac{2}{9}\cosh x$

(c) $y = c_1e^{x} + c_2e^{-x} - \frac{x}{3}\sinh x - \frac{2}{9}\cosh x$ (d) $y = c_1e^{2x} + c_2e^{-2x} - \frac{x}{3}\sinh x + \frac{2}{9}\cosh x$

Ans. (a)

Explanation :

C.F. $y(x) = c_1e^{2x} + c_2e^{-2x}$ P.I. $\frac{1}{D^2 - 4}x\sinh x$

$\Rightarrow \frac{1}{D^2-4}\left(\frac{xe^x - xe^{-x}}{2}\right) \Rightarrow \frac{e^x}{2}\left[\frac{1}{(D+1)^2 - 4}x\right] - \frac{e^{-x}}{2}\left[\frac{1}{(D-1)^2 - 4}x\right]$

$\Rightarrow \frac{e^x}{2}\left[\frac{x}{D^2 + 2D - 3}\right] - \frac{e^{-x}}{2}\left[\frac{x}{D^2 - 2D - 3}\right]$

$\Rightarrow -\frac{e^x}{6}\left[\left(1 - \frac{D^2 + 2D}{3}\right)^{-1}x\right] + \frac{e^{-x}}{6}\left[\left(1 - \frac{D^2 - 2D}{3}\right)^{-1}_x\right]$

$\Rightarrow -\frac{e^x}{6}\left[x + \frac{2}{3}\right] + \frac{e^{-x}}{6}\left[x - \frac{2}{3}\right] \Rightarrow -\frac{x}{3}\left(\frac{e^x - e^{-x}}{2}\right) - \frac{2}{9}\left(\frac{e^x + e^{-x}}{2}\right)$

$= -\frac{x}{3}\sinh x - \frac{2}{9}\cosh x$

UNSOLVED QUESTIONS

1. The differential equation whose linearly independent solutions are cos 2x, sin 2x and e^x, is

GATE – 2001

(a) $(D^3 + D^2 + 4D)y = 0$ (b) $(D^3 - D^2 + 4D - 4)y = 0$

(c) $(D^3 + D^2 - 4D - 4)y = 0$ (d) $(D^3 - D^2 - 4D + 4)y = 0$

2. Suppose $y_p(x) = x\cos(2x)$ is a particular solution of $y'' + ay = -4\sin(2x)$. Then the constant α equals

GATE-2007

(a) -4 (b) -2 (c) 2 (d) 4

3. If y(t) is a solution of the differential equation y`` + 4y = 2e^t, then $\lim_{x\to\infty} e^{-t}y(t)$ is equal to

(a) $\frac{2}{3}$ (b) $\frac{2}{5}$ (c) $\frac{2}{7}$ (d) $\frac{2}{9}$

4. Let y(x) be the solution of the differential equation $\frac{d}{dx}\left(x\frac{dy}{dx}\right)=x; y(1)=0, \left.\frac{dy}{dx}\right|_{x=1}=0$.

Then y(2) is

(a) $\frac{3}{4}+\frac{1}{2}\ln 2$ (b) $\frac{3}{4}-\frac{1}{2}\ln 2$ (c) $\frac{3}{4}+\ln 2$ (d) $\frac{3}{4}-\ln 2$

5. The number of arbitrary constants in the complete primitive of differential equation $\frac{d^5y}{dx^5}+2\frac{d^4y}{dx^4}=0$ is/are not

(a) 5 (b) 4 (c) 1 (d) 6

6. For the differential equation $\frac{d^4y}{dx^4}-2\frac{d^3y}{dx^3}+\frac{d^2y}{dx^2}=0$

(a) every solution converges to as x → ∞ (b) every solution is bounded on [0, ∞]
(c) general solution has countable number of zeros in [0, ∞]
(d) there exists a solution, which is not bounded on [0, ∞]

7. The singular points of the differential equation x^3(x – 1)y`` – 2(x – 1)y` + 3xy = 0

NBHM - 2013

(a) x = 0 is irregular singular point (b) x = 1 is regular singular point
(c) x = 0 is regular singular point (d) x = 1 is irregular singular point

8. Consider the equation of an ideal planar pendulum given by $\frac{d^2x}{dt^2}=\sin x$, where x denotes the angle of displacement. For sufficiently small angles of displacement, the solution is given by (where a,b are constant) **CSIR NET JUNE 2013**

(a) $x(t)=a\cosh t+b\sinh t$ (b) $x(t)=a+bt$
(c) $x(t)=ae^t+be^{2t}$ (d) $x(t)=a\cosh t+b\sinh t$

9. Let y(x) = xV(x) be a solution of D.E. x^{2}y`` – 3xy` + 3y = 0 if V(0) = 0 and V(1) = 1, then V(-2) is

(a) 1 (b) 2 (c) 3 (d) 4

Answer Key

1	2	3	4	5	6	7	8	9
a	b,d	b	b	b,c,d	a	d	d	d

Chapter 4

Uniqueness and Existence of Fofd (Ist Order & Ist Degree Differential Equation)

Let $\frac{dy}{dx} = f(x, y)$ s.t. $y(x_0) = y_0$ ...(1)

Then the necessary and sufficient condition such that. (1) . has a solution Iff f(x, y) is integrable.

Examples. Let $f(x, y) = \lambda \sin(x + y)$ such that . $y(0) = 1$ & $\lambda \in R$. Then

(a) No – solution in neighbourhood of 0
(b) A solution in R if $|\lambda| < 1$
(c) A solution in neighbourhood of 0
(d) A solution in R only if $|\lambda| > 1$

Solution : (b), (c) **CSIR NET JUNE 2016**

Here $f(x, y) = \lambda \sin(x + y)$ is integrable for any $\lambda \in R$

Then it has solution in R

Option (d) is incorrect

A solution in R only if $|\lambda| > 1$

1. The Lipschitz Condition :

Let f(x, y) be a function defined for (x, y) in a set S then f is said to satisfy Lipschitz condition on S if $\exists$ $k \in Z^+$

s.t. $|f(x, y_1) - f(x, y_2)| \le k|y_1 - y_2|$ $\forall$ $(x, y_1), (x, y_2) \in S$.

Then constant k is called Lipschitz constant.

Theorem : If f is real valued function defined on S where S is either a rectangle.

$\{(x, y) : |x - x_0| \le a, |y - y_0| \le b; (a, b > 0)\}$ or a strip

$\{(x, y) : |x - x_0| \le a, |y| < \infty; (a > 0)\}$

Such that $\left|\frac{\partial f}{\partial y}\right|$ exist and $\left|\frac{\partial f}{\partial y}\right| \le k$ $(x, y) \in S$

Then f(x) satisfy a lipschitz condition on S with lipschitz constant k.

2. Existence and Uniqueness theorem :

If the Inial Value Problem $y` = f(x, y), y(x_0) = y_0$

Then function is continuous in a rectangle R

$R : |x - x_0| \le a, |y - y_0| \le b$

And satisfying lipschitz condition in R then $\exists$ a unique solution $y = y(x)$ on an interval $I : |x - x_0| \le h$ of the IVP

Where $h = \min\left\{a, \frac{b}{M}\right\}$ & $M = \max |f(x, y)|$ in R

SOLVED QUESTIONS

1. The initial value problem $x`(t) = 3x^{2/3}$, $x(0) = 0$ has **CSIR NET JUNE 2023**

(a) Unique solution (b) No solution
(c) Infinite many solution (d) Only two solution

Ans. (c)

Explanation :

Given that $\frac{dx}{dt} = 3x^{2/3}, x(0) = 0$

We know that $\frac{dx}{dt} = x^{\alpha}; x(b) = 0$

has infinitely solution iff $\alpha \in (0, 1)$; $b \in R$

2. Consider two DE **CSIR NET JUNE 2023**

P : $\frac{dx}{dt} = \sqrt{x}, x(0) = 0$

Q : $\frac{dy}{dt} = -\sqrt{y}, y(0) = 0$

Then

(a) P has unique solution (b) Q has unique solution
(c) P has infinite solution (d) Q has infinite solution

Ans. (c),(d)

Explanation:

For P

We know that $\frac{dx}{dt} = x^{\alpha}; x(b) = 0$

has infinitely solution iff $\alpha \in (0, 1)$; $b \in R$

For Q

$\frac{dy}{dt} = Ay^{\alpha}; y(b) = 0 \,; A < 0$

has unique solution

3. Let k be a positive integer. Consider the differential equation $\begin{cases} \frac{dy}{dt} = y^{\frac{5k}{5k+2}} \text{ for } t > 0, \\ y(0) = 0 \end{cases}$

which of the following statements is true?

(a) It has a unique solution which is continuously differentiable on $(0, \infty)$
(b) It has at most two solutions which are continuously differentiable on $(0, \infty)$
(c) It has infinitely many solutions which are continuously differentiable on $(0, \infty)$
(d) It has no continuously differentiable solution on $(0, \infty)$

Ans. (c) **CSIR NET JUNE 2020**

Explanation :

$$\frac{dy}{dt}=y^{\frac{5k}{5k+2}}=f(x,y)(let)$$

$$\frac{\partial f}{\partial y}=\frac{5k}{5k+2}y^{\frac{5k}{5k+2}-1}=\frac{5k}{5k+2}y^{\frac{5k}{5k+2}-1}=\frac{5k}{5k+2}y^{\frac{-2}{5k+2}}=\frac{5k}{5k+2}\frac{1}{y^{\frac{2}{5k+2}}}$$, y(0) = 0 implies,

not defined. So, it has no solution or infinitely many solution ie, not unique solution, Option (a) and (b) are false.

$$y^{\frac{5k}{5k+2}-1}dy=dx\Rightarrow\frac{y^{\frac{-5k}{5k+2}+1}}{\frac{-5k}{5k+2}+1}=x+c \qquad \frac{y^{\frac{2}{5k+2}}}{\frac{2}{5k+2}}=x+c, y(0)=0\Rightarrow c=0$$

So, infinitely many solutions. Option (c) is correct and option (d) is wrong.

4. Let $\frac{dy}{dx}=y^2$ s.t. y(1) = 1 then unique solution exist in an interval

(a) $\left(\frac{3}{4},\frac{5}{4}\right)$ (b) (1, 2) (c) $\left(\frac{3}{4},2\right)$ (d) $(-\infty, \infty)$

Ans. (a)

Explanation :

Here f(x, y) = y^2 $\left|\frac{\partial f}{\partial y}\right|=2y$ We have to construct a rectangle

R : $|x-1|\le a$, $|y+1|\le b$ Where $x_0 = 1$ & $y_0 = -1$ a, b $\in R^+$

In this rectangle $\left|\frac{\partial f}{\partial y}\right|$ is bounded.

$\Rightarrow$ f(x, y) satisfy lipschitz condition. Now f(x, y) = y^2

M = Max f(x, y) in S

= Max $\{y^2\}$ is S $|y+1|\le b$, $-b\le y+1\le b$, $-1-b\le y\le b-1$

M = $(-1-b)^2=(1+b)^2$

Then $h=\min\left\{a,\frac{b}{(1+b)^2}\right\}$ Take $\phi(b)=\frac{b}{(1+b)^2}$

We have to find the maximum and minimum value.

$$\phi'(b)=\frac{(1+b)^2-2b(1+b)}{(1+b)^3}=0 \qquad =\frac{1+b^2+2b-2b-2b^2}{(1+b)^3}=0 \quad = 1-b^2=0$$

$b=\pm1$ At b = 1 $\Rightarrow \phi(b)=\frac{1}{4}$

$\Rightarrow h = \min\left\{a, \frac{1}{4}\right\}$ We take a ∈ R^+ $\Rightarrow a = \frac{1}{4}$ s.t. $h = \frac{1}{4}$

Then unique solution in $|x-1| \le \frac{1}{4}$

$\frac{-1}{4} \le (x-1) \le \frac{1}{4}$ $\frac{3}{4} \le x \le \frac{5}{4}$

5. Consider the IVP $\frac{dy}{dx} = xy^{1/3}$, y(0) = 0. Then which of the following are TRUE?

(a) The function f(x, y) = $xy^{1/3}$ doesn't satisfy a lipschitz condition with respect to y in neighbourhood of y = 0

(b) There exist a unique solution.

(c) No solution

(d) More than one solution

CSIR NET JUNE 2013

Ans. (a), (d)

Explanation :

Let f(x, y) = $xy^{1/3}$ Option (a) is correct. Then $\frac{\partial f}{\partial y} = \frac{1}{3}xy^{-2/3}$

Then $\left|\frac{\partial f}{\partial y}\right|$ is unbounded in neighbourhood of '0' ⇒ not satisfy lipschitz condition.

Option (c) is incorrect.

Given condition y(0) = 0 ⇒ Trivial solution always exist. y = 0

Option (b) $\frac{dy}{dx} = xy^{1/3}$ $y^{-1/3}$ dy = xdx $-\frac{3}{2}y^{2/3} = x + k$ ⇒ y(0) = 0⇒ k = 0

-3/2 $y^{-2/3}$ = x ⇒ $y^{-2/3}$ = -2x/3 is also a solution.

Option (b) is incorrect. ⇒ not unique solution.

Option (d) is correct. More than one solution.

6. Let y : R → R be differentiable satisfy the ODE $\frac{dy}{dx} = f(y)$, y(0) = y(1) = 0 where f : R → R is lipschitz continuous function. Then

CSIR NET DEC 2014

(a) y(x) = 0 iff x ∈ {0, 1}

(b) y is bounded.

(c) y is strictly increasing

(d) $\frac{dy}{dx}$ is unbounded.

Ans. (b)

Explanation :

Let f(y) = 0 is lipschitz function then y ≡ 0 is a solution.

Then option (b) is correct.

7. For the IVP $\frac{dy}{dx} = y^2 + \cos^2 x$ y(0) = 0

CSIR NET JUNE 2015

DIFFERENTIAL EQUATION

Then the largest interval of existence of t he solution predicted by picard's theorem is

(a) [0, 1] (b) [0, 1/2] (c) [0, 1/3] (d) [0, 1/4]

Ans. (b)

Explanation :

Given that $x_0 = 0 = y_0$ then the rectangles

R : $|x| \le a, |y| \le b$ where $a, b \in R^+$

$M = \max |f(x, y)|$ where $f(x, y) = |y^2 + \cos^2 x| \le |y|^2 + |\cos^2 x| \le b^2 + 1$

$M = |b^2 + \cos^2 a| = b^2 + 1$

Then $h = \min\left\{a, \dfrac{b}{1+b^2}\right\}$ Let $\phi(b) = \dfrac{b}{1+b^2}$

$\phi'(b) = 0 = \dfrac{1+b^2-2b^2}{(1+b^2)^2} =$ $b = \pm 1$ Where $b \in R^+ \Rightarrow b \ne -1 \Rightarrow b = 1$

Then $\phi(b) = \dfrac{1}{2}$ $\Rightarrow h = \min\left\{a, \dfrac{1}{2}\right\}$

For $a \ge \dfrac{1}{2}$ $h = \dfrac{1}{2}$ $\Rightarrow |x| \le \dfrac{1}{2}$ $\Rightarrow x \in \left[\dfrac{-1}{2}, \dfrac{1}{2}\right]$

Largest interval $\left[0, \dfrac{1}{2}\right]$

8. The initial value problem $y' = 2\sqrt{y}, y(0) = a$, has **CSIR NET JUNE 2015**

(a) A unique solution if $a < 0$ (b) No solution if $a > 0$

(c) Infinitely many solutions if $a = 0$ (d) A unique solution if $a \ge 0$

Ans. (c)

Explanation :

Consider the given IVP $y' = 2\sqrt{y}, y(0) = a$

Then for $a = 0$, the given IVP has infinitely many solutions of the form

$y(x) = \begin{cases} (x-c)^2 & if\ x \ge c \\ 0 & if\ x < c \end{cases}$ provided $c \ge 0$ Hence, option (c) is correct.

9. Let $u(t)$ be a continuously differentiable function taking non-negative values for $t > 0$ satisfying $u'(t) = 3[u(t)]^{2/3}$ and $u(0) = 0$. **CSIR NET JUNE 2014**

Which of the following are possible solutions of the above equation?

(a) $u(t) = 0$ (b) $u(t) = t^3$

(c) $u(t) = \begin{cases} 0 & for\ 0 < t < 1 \\ (t-1)^3 & for\ t \ge 1 \end{cases}$ (d) $u(t) = \begin{cases} 0 & for\ 0 < t < 3 \\ (t-3)^3 & for\ t \ge 3 \end{cases}$

Ans. (a),(b),(c),(d)

Explanation :

$u'(t)=3[u(t)]^{2/3}; u(0)=0$ ....(1)

By method of separation of variables, $\Rightarrow u(t)=(t+c)^3$

c is arbitrary constant using $u(0)=0 \Rightarrow c=0$ $\Rightarrow u(t)=t^3$ is a solution of (1),

$u(t)=0$ is also solution of (1)

By shifting the curve $u(t)=t^3$ on the x-axis

we know that $u(t)=\begin{cases} 0 & ; 0<t<r \\ (t-r)^3 & ; t \geq r \end{cases}$ is also a solution of (1) $\forall r \in R^+$

Hence, options (a),(b),(c) and (d) are correct.

10. Solve $\frac{dy}{dx}=y^{\frac{1}{2}}$ with y(0) = 0

(a)unique solution (b)two solution (c) no solution (d) infinite solution

Ans. (d)

Explanation : $\frac{1}{2} \in (0,1)$ and a=0

Infinite solution

11. Let P be a polynomial of degree N, with N ≥ 2.Then the initial value problem u'(t) = P(u(t)), u(0) = 1, has always **CSIR NET JUNE 2011**

(a) A unique solution in R

(b) N number of distinct solutions in R

(c) No solution in any interval containing θ for some P

(d) A unique solution in an interval containing 0

Ans. (d)

Explanation :

Given u' (t) = P(u(t)), u(0) = 1

Consider $P(x) = x^2$ $\therefore u'(t) = u^2(t)$ $\Rightarrow \frac{du}{dt}=u^2 \Rightarrow \frac{du}{u^2}=dt$,

Integrate on both sides $\frac{-1}{u}=t+C$

Also $u(0)=1 \Rightarrow C=-1$ $\frac{-1}{u}=t-1 \Rightarrow C=-1$

Which does not exist at t = 1 but exist in an interval containing 0.

∴ Option (d) is true.

12. Consider the equation $\frac{dy}{dt}=(1+f^2(t))y(t), y(0)=1: t \geq 0$

Where f is a bounded continuous function on $[0,\infty)$. Then **CSIR NET DEC 2011**

(a) This equation admits a unique solution $y(t)$ and further $\lim_{t\to\infty} y(t)$ exists and is finite
(b) This equation admits two linearly independent solutions
(c) This equation admits a bounded solution for which $\lim_{t\to\infty} y(t)$ does not exist
(d) This equation admits a unique solution $y(t)$ and further $\lim_{t\to\infty} y(t)=\infty$

Ans. **(d)**

Explanation :

$\frac{dy}{dt}=(1+f^2(t))y(t); y(0)=1; t\geq 0$ Let $f(t)=0$ [continuous and bounded]

Differential equation becomes

$y'(t)=y; y(0)=1 \qquad \Rightarrow y(t)=e^t \qquad \lim_{t\to\infty} y(t)=\infty$

Hence, option (d) is correct.

13. Let $y: R\to R$ be differentiable and satisfy the ODE: $\left.\begin{aligned}\frac{dy}{dx}=f(y), x\in R\\ y(0)=y(1)=0\end{aligned}\right\}$

where $y: R\to R$ is a Lipschitz continuous function. **CSIR NET DEC 2014**

Then

(a) $y(x)=0$ if and only if $x\in\{0,1\}$ (b) y is bounded

(c) y is strictly increasing (d) $\frac{dy}{dx}$ is unbounded

Ans. **(b)**

Explanation :

Consider $f(y)=0$

by Lipschitz continuous function $\frac{dy}{dx}=f(y)\Rightarrow y=c\ [\because y(0)=0; y(1)=0]$

$\Rightarrow y(x)\equiv 0$ Hence, option (b) is correct.

14. Consider the *IVP* $\frac{dy}{dx}=x+y^2; y(0)=0$ the largest interval in which picard's method guaranties the unique solution |x| < α. Then α is

(a) $\alpha=\left(\frac{1}{2}\right)^{2/3}$ (b) $\alpha=(2)^{3/2}$ (c) $\alpha=\left(\frac{1}{2}\right)^{3/2}$ (d) $\alpha=(2)^{2/3}$

Ans. **(a)**

Explanation :

$\frac{dy}{dx}=x+y^2$ and y(0) = 0 $\Rightarrow x_0=y_0=0$

and $f(x, y) = x + y^2$

Let a rectangle $\{(x, y)\in R^2| \ |x - x_0| \le a, |y - y_0| \le b\}$
$a, b \in R^+$

$= \{(x, y)\in IR^2| \ |x|\le a, |y|\le b\}$

Then $m = \sup f(x, y) = a + b^2$

$$h = \min\left\{a, \frac{b}{m}\right\} = \min\left\{a, \frac{b}{a+b^2}\right\}$$

Let $\phi(b) = \dfrac{b}{a+b^2}$

$$\phi'(b) = 0 = \frac{a+b^2-b(2b)}{(a+b^2)^2} = \frac{a-b^2}{(a+b^2)^2} = 0 \quad \Rightarrow b = \pm\sqrt{a} \quad \Rightarrow b = \sqrt{a}$$

Then $h = \min\left\{a, \dfrac{\sqrt{a}}{a+a}\right\} = \min\left\{a, \dfrac{1}{2\sqrt{a}}\right\}$

$$a = \frac{1}{2\sqrt{a}} \Rightarrow a^{3/2} = \frac{1}{2} \qquad a = \left(\frac{1}{2}\right)^{2/3} \qquad h = \left(\frac{1}{2}\right)^{2/3}$$

Unique solution exist in $|x| < h$ where $h = \alpha = \left(\dfrac{1}{2}\right)^{2/3}$

UNSOLVED QUESTIONS

1. The initial value problem $\left(x^2 - x\right)\dfrac{dy}{dx} = (2x-1)y, y(x_0) = y_0$ has a unique solution if (x_0, y_0) equals **GATE-2002**

(a) (2, 1) (b) (1, 1) (c)(0, 0) (b) (0, 1)

2. The solution of $\dfrac{dy}{dx} = y^2, y(0) = 1$ does not exist for all **GATE-1995**

(a) $x \in (-\infty, 1)$ (b) $x \in [0, a]$ where $a > 1$ (c) $x \in (-\infty, \infty)$ (b) $x \in [1, a]$ where $a > 1$

3. Let $u(t)$ be a continuously differentiable function taking non-negative values for $t > 0$ and satisfying $u'(t) = 4u^{3/4}(t); u(0) = 0$.
Then **CSIR NET DEC 2015**

(a) $u(t) = 0$

(b) $u(t) = t^4$

(c) $u(t) = \begin{cases} 0 & for\ 0 < t < 1 \\ (t-1)^4 & for\ t \ge 1 \end{cases}$

(d) $u(t) = \begin{cases} 0 & for\ 0 < t < 10 \\ (t-10)^4 & for\ t \ge 10 \end{cases}$

4. The Differential equation $\frac{dy}{dx}=60(y^2)^{1/5}$, y(0) = 0

(a) unique solution (b) two solution (c) no solution
(d) infinite number of solution

5. Consider the initial value problem $y'(t)=f(t)y(t), y(0)=1$ where $f:R\to R$ is continuous.
Then this initial value problem has
(a) infinitely many solutions for some f (b) A unique solution in R
(c) No solution in R for some f

CSIR NET JUNE 2012

(d) A solution in an interval containing 0, but not on R for some f

6. Given, continuous function $f(x,y)=y^{2/3}$ on rectangle $|x|\le 1, |y|\le 1$, which of the following are true?
(a) This function satisfies Lipschitz condition on a rectangle.
(b) This function does not satisfy Lipschitz condition on a rectangle.
(c) $\left|\frac{\partial f}{\partial y}\right|$ is unbounded for y = 0 (d) $\left|\frac{\partial f}{\partial y}\right|$ is bounded for y = 0

7. Given, the initial value problem $\frac{dy}{dx}=y^2, y(1)=-1$, then
(a) there exist atleast one solution (b) there does not exist any solution
(c) the initial value problem has no unique solution
(d) the given problem possesses a unique solution when $|x-1|<\frac{1}{4}$

8. The largest value of c such that there exists a function h(x) for $-c<x<c$ that is solution of $\frac{dy}{dx}=1+y^2$ with h(0) = 0, is given by

(a) $\frac{\pi}{3}$ (b) $\frac{\pi}{2}$ (c) $\frac{\pi}{4}$ (d) π

Answer Key

1	2	3	4	5	6	7	8
a	c	a,b,c,d	d	b	b,c	a,d	b

Chapter 5

System of Ordinary Differential Equation

1. System of ODE

A pair of linear differential equations

$\frac{dy}{dx} = f(x, y)$ and $\frac{dy}{dx} = g(x, y)$ is called system of linear differential equation and it is denoted by Y` = AY Where A is Auxiliary matrix of this system

Example : Let f(D)y = 0 be a differential equation and Y` = AY be a system corresponding to given equation then

$C_A(x)$ = characteristic polynomial of A
= auxiliary equation of F(D)y = 0

Let $F(D)y = (D^2 + aD + b)\, y = 0$ and Y` = AY be an equivalent system of ODE then A is

(a) $A = \begin{bmatrix} -a & 1 \\ -b & 0 \end{bmatrix}$ (b) $\begin{bmatrix} 0 & b \\ 1 & -a \end{bmatrix}$ (c) $\begin{bmatrix} -a & b \\ 1 & 0 \end{bmatrix}$ (d) $\begin{bmatrix} 0 & -1 \\ b & -a \end{bmatrix}$

Solution : Characteristic polynomial of A = $m^2 + am + b$ then sum of eigen value = Tr(A) = -a

Product of eigen value = |A| = b

Option (b) and (c) is incorrect.

Option (a) Y` = AY $\begin{bmatrix} y_1' \\ y_2' \end{bmatrix} = \begin{bmatrix} -a & 1 \\ -b & 0 \end{bmatrix}\begin{bmatrix} y_1 \\ y_2 \end{bmatrix}$

$y_1' = -ay_1 + y_2$ & $y_2' = -by_1$ Then $y_1'' = -ay_1' + y_2'$ & $y_2'' = -by_1'$

Put $y_2' = -by_1$ in $y_1'' = -ay_1' + y_2' \Rightarrow y_1'' = -ay_1' - by_1 \Rightarrow y_1'' + ay_1' + by_1 = 0$

Option (a) is correct.

2. Method for solving of system of ODE

Let the system of ODE Y` = AY

If λ is eigen value of A and $C = (C_1, C_2)^T$ be eigen vector

Then $Y(x) = Ce^{\lambda x} = \begin{bmatrix} C_1 e^{\lambda_1 x} \\ C_2 e^{\lambda_2 x} \end{bmatrix}$

Case – 1 If $\lambda_1 \neq \lambda_2$

and $u_1 = (u_{11}, u_{12})^T$ $u_2 = (u_{21}, u_{22})^T$ are eigen vectors corresponding to λ_1 & λ_2.

Then $y_1(x) = u_1 e^{\lambda_1 x}$ & $y_2(x) = u_2 e^{\lambda_2 x}$ are two LI solution.

Then $y(x) = c_1 y_1(x) + c_2 y_2(x)$ is G.S.

Example : $Y' = AY$ where $A = \begin{bmatrix} 0 & 1 \\ 1 & 0 \end{bmatrix}$

Eigen value of A are 1, -1 then

Eigen vector corresponding to 1.

$$\begin{bmatrix} -1 & 1 \\ 1 & -1 \end{bmatrix}\begin{bmatrix} x \\ y \end{bmatrix} = \begin{bmatrix} 0 \\ 0 \end{bmatrix} \Rightarrow \quad x = y \quad \Rightarrow y_1(x) = \begin{bmatrix} 1 \\ 1 \end{bmatrix} e^x$$

$u_1 = (1, 1)^T$

and u_2 is eigen vector corresponding to -1

$$[A + I]\, x = 0 \Rightarrow x = -y \qquad u_2 = (1, -1)^T \Rightarrow y_2(x) = \begin{bmatrix} 1 \\ -1 \end{bmatrix} e^{-x}$$

General solution $\quad Y(x) = c_1 y_1 + c_2 y_2 \quad \begin{bmatrix} y_1 \\ y_2 \end{bmatrix} = C_1 \begin{bmatrix} e^x \\ e^x \end{bmatrix} + C_2 \begin{bmatrix} e^{-x} \\ -e^{-x} \end{bmatrix}$

Case – 2

Let $\lambda_1 = \lambda_2 = \lambda$ and G.M. of λ is 2.

We know that G.M. of λ = Number of LI eigen vector corresponding to λ.

$\Rightarrow$ Number of LI eigen vectors are 2.

Let u_1 & u_2 are eigen vector. $y_1(x) = u_1 e^{\lambda x}$ & $y_2 = u_2 e^{\lambda x}$

General solution $\quad y(x) = c_1 y_1 + c_2 y_2$

Case – 3

If $\quad \lambda_1 = \lambda_2 = \lambda \quad$ and G.M of λ is 1.

$\Rightarrow$ Number of LI eigen vectors is 1 and let u_1 be eigen vector

Then another eigen vector u_2.

$u_1 = (A - \lambda I)u_2 \qquad y_1(x) = u_1 e^{\lambda} \qquad y_2(x) = (u_1 x + u_2)\, e^{\lambda x}$

General solution $\quad y(x) = c_1 y_1 + c_2 y_2$

3. Class of Critical point

Consider system of the form $\dfrac{dx}{dt} = P(x, y); \dfrac{dy}{dt} = Q(x, y)$ a point (x_0, y_0) at which both $P(x_0, y_0) = 0$ and both $Q(x_0, y_0) = 0$ is called a critical point.

Consider linear system $\dfrac{dx}{dt} = ax + by = u \quad \dfrac{dy}{dt} = cx + dy = v \quad$ where $ad - bc \neq 0$

If $A = \begin{bmatrix} u_x & u_y \\ v_x & v_y \end{bmatrix}$, Suppose λ_1 and λ_2 are eigen value of A

λ_1 & λ_2 are eigen value of A	
(1) $\lambda_1 \neq \lambda_2$ are real with same sign	Node
(2) $\lambda_1 \neq \lambda_2$ and real with opposite sign	Saddle point

$(3)\ \lambda_1 = \lambda_2 = \lambda \in R$	Node
$(4)\ \lambda_1 = \alpha + i\beta$ & $\lambda_2 = \alpha - i\beta$ $\beta \neq 0$ (a) $\alpha \neq 0$ (b) $\alpha = 0$	 Spiral point Centre point

4. Stability of Critical point

1	$\lambda_1 \neq \lambda_2$ with same sign	(a) $\lambda_1 > 0$ & $\lambda_2 > 0$ (b) $\lambda_1 < 0$ & $\lambda_2 < 0$	Unstable Asymtotic stable
2.	$\lambda_1\lambda_2 < 0$ $\lambda_1, \lambda_2 \in R$		Unstable
3.	$\lambda_1 = \lambda_2 = \lambda = R$	(a) $\lambda > 0$ (b) $\lambda < 0$	Unstable Asymtotic stable
4.	$\lambda = \alpha \pm i\beta$		
5.	$\alpha \neq 0$	(i) $\alpha > 0$ (ii) $\alpha < 0$	Unstable Asymtotic stable
	$\alpha = 0$		Stable but not Asymtotic stable

Non linear system of ODE:

$x'(t) = ax + by + P(x,y) = u$ $\qquad$ $y'(t) = cx + dy + Q(x,y) = v$

Then Matrix will be $A = \begin{bmatrix} u_x & u_y \\ v_x & v_y \end{bmatrix}$

SOLVED QUESTIONS

1. Consider a system of ODE $x`(t) = Ax$, where $A = \begin{bmatrix} 5 & 4 \\ 1 & 2 \end{bmatrix}$, $x(0) = x_0$. Then

CSIR NET 2023

(a) $e^{-10t}|x(t)| \to 0$ as $t \to \infty$, $x_0 \neq 0$
(b) $e^{-t}|x(t)| \to \infty$ as $t \to \infty$, $x_0 \neq 0$
(c) $e^{-6t}|x(t)| \to \infty$ as $t \to \infty$, $x_0 \neq 0$
(d) None of these

Ans. (a)

Explanation :

Eigen value of A are 1 & 6, then $x(t) = Ae^{t} + Be^{6t}$

Using condition $x(0) = x_0 \neq 0 \Rightarrow \quad A + B \neq 0$

(a) $e^{-10t}|x(t)| = (Ae^{-9t} + Be^{-4t}) \to 0$ as $t \to 0$ $\qquad$ For any value of A & B

(b) $e^{-t}|x(t)| = (A + Be^{5t}) \to \infty$ as $t \to \infty$ for $B \neq 0$
But for $B = 0$ & $A \in R$ $\qquad$ Option (b) is incorrect.

(c) $e^{-6t}|x(t)| = (Ae^{-5t} + Be^{-t}) \to 0$ as $t \to \infty$ $\qquad$ For any value of A & B.

2. For $\lambda \in R$, consider the system of differential equations

$x_1' = x_1 + 2x_2 + 2x_3$
$x_2' = 2x_1 + x_3$
$x_3' = -x_1 + 2x_2 + \lambda x_3.$ **CSIR NET NOV 2020**

If $\vec{x}(t) = \vec{\alpha}te^{2t}$ (for some $\vec{\alpha}$) is a solution of the system then the value of λ is

(a) 2 (b) 4 (c) 6 (d) 1

Ans. (a)

Explanation :

Given that $A = \begin{bmatrix} 1 & 2 & 2 \\ 2 & 0 & 1 \\ -1 & 2 & \lambda \end{bmatrix}$

and te^{2t} is a solution of system. So 2 is eigen value of with multiplicity 2

Now, $Tr(A) = \lambda + 1 = 2 + 2 + k$; k is third eigen value $\Rightarrow k = \lambda - 3$

Again $|A| = 4 - 4\lambda = 4k \Rightarrow 4 - 4\lambda = 4(\lambda - 3) \Rightarrow \lambda = 2$

3. For the following system of ODE $\frac{dx}{dt} = x(3-2x-2y);\ \frac{dy}{dt} = y(2-2x-y)$,
the critical point (0, 2) is **CSIR NET DEC 2019**

(a) a stable spiral (b) an unstable spiral (c) a stable node (d) an unstable node

Ans. (c)

Explanation :

Let u = x(3 – 2x – 2y) and v = y(2 – 2x – y).

Then $A = \begin{bmatrix} u_x & u_y \\ v_x & v_y \end{bmatrix}_{(0,2)} = \begin{bmatrix} 3-4x-2y & -2x \\ -2y & 2-2x-2y \end{bmatrix}_{(0,2)} \Rightarrow A = \begin{bmatrix} -1 & 0 \\ -4 & -2 \end{bmatrix}$

Eigen value of auxiliary matrix are -1, -2.

This critical point is stable node.

4. Consider the system of differential equation

$\frac{dx}{dt} = 2x - 7y$ $\qquad \frac{dy}{dt} = 3x - 8y$

Then the critical point (0, 0) of the system is an **CSIR NET JUNE 2018**

(a) Asymtotic stable node (b) Unstable node
(c) Asymtotic stable spiral (d) Unstable spiral.

Ans. (a)

Explanation :

The system $\begin{bmatrix} x' \\ y' \end{bmatrix} = \begin{bmatrix} 2 & -7 \\ 3 & -8 \end{bmatrix}\begin{bmatrix} x \\ y \end{bmatrix}$ Where $A = \begin{bmatrix} 2 & -7 \\ 3 & -8 \end{bmatrix}$

Eigen value of A are -5, -1

Both are distinct with same sign then critical point node.

And both are negative $\Rightarrow$ Asymtotic stable.

5. The critical point (0, 0) for the system

$x'(t) = x - 2y + y^2 (\sin x)$

$y'(t) = 2x - 2y - 3y (\cos y^2)$ is a

CSIR NET DEC 2018

(a) Stable spiral point
(b) Unstable spiral point
(c) Saddle point
(d) Stable node

Ans. (c)

Explanation :

This system is not linear.

Let $u = x - 2y + y^2 \sin x$ and $v = 2x - 2y - 3y \cos y^2$

Then $A = \begin{bmatrix} u_x & u_y \\ v_x & v_y \end{bmatrix}_{(0,0)} \Rightarrow A = \begin{bmatrix} 1 & -2 \\ 2 & -5 \end{bmatrix}$

$C_A(x) = x^2 + 4x - 1 = 0$

Both are distinct and opposite sign. $\Rightarrow$ Saddle point.

6. Consider a system of first order DE $\begin{bmatrix} x'(t) \\ y'(t) \end{bmatrix} = \begin{bmatrix} x(t) + y(t) \\ -y(t) \end{bmatrix}$

The solution space is spanne by

CSIR NET DEC. 2017

(a) $\begin{bmatrix} 0 \\ e^{-t} \end{bmatrix}$ and $\begin{bmatrix} e^t \\ 0 \end{bmatrix}$
(b) $\begin{bmatrix} e^t \\ 0 \end{bmatrix}$ and $\begin{bmatrix} \cosh t \\ e^{-t} \end{bmatrix}$

(c) $\begin{bmatrix} e^{-t} \\ -2e^{-t} \end{bmatrix}$ and $\begin{bmatrix} \sinh t \\ e^{-t} \end{bmatrix}$
(d) $\begin{bmatrix} e^t \\ 0 \end{bmatrix}$ and $\begin{bmatrix} e^t - \frac{1}{2}e^{-t} \\ e^{-t} \end{bmatrix}$

Ans. (c), (d)

Explanation :

The system is $\begin{bmatrix} x' \\ y' \end{bmatrix} = \begin{bmatrix} 1 & 1 \\ 0 & -1 \end{bmatrix} \begin{bmatrix} x \\ y \end{bmatrix}$ Where $A = \begin{bmatrix} 1 & 1 \\ 0 & -1 \end{bmatrix}$

Eigen value of A are 1 & -1 (distinct) then find eigen vector corresponding to 1 & -1.

For $u_1 = 0$

$(A - I) = \begin{bmatrix} 0 & 1 \\ 0 & -2 \end{bmatrix} \begin{bmatrix} x \\ y \end{bmatrix} = \begin{bmatrix} 0 \\ 0 \end{bmatrix} \Rightarrow y = 0$ Then $u_1 = [1\ 0]^T$ $y_1(t) = \begin{bmatrix} 1 \\ 0 \end{bmatrix} e^t$

And eigen vector corresponding to $-1 (A + I) x = 0$

$\begin{bmatrix} 2 & 1 \\ 0 & 0 \end{bmatrix} \begin{bmatrix} x \\ y \end{bmatrix} = \begin{bmatrix} 0 \\ 0 \end{bmatrix} \Rightarrow 2x + y = 0$ $u_2 = \begin{bmatrix} 1 \\ -2 \end{bmatrix}$ Then $y_2(t) = \begin{bmatrix} 1 \\ -2 \end{bmatrix} e^{-t}$

General solution $y(t) = c_1 y_1 + c_2 y_2$

Then $y(t) = \begin{bmatrix} c_1 e^t + c_2 e^{-t} \\ -2 c_2 e^{-t} \end{bmatrix}$

Option (c) is correct.

Let $c_1 = 0$ & $c_2 = 1$ & $c_1 = 1/2$ & $c_2 = -1/2$

$$\begin{bmatrix} e^{-t} \\ -2e^{-t} \end{bmatrix} \begin{bmatrix} \frac{e^t - e^{-t}}{2} \\ e^{-t} \end{bmatrix} = \begin{bmatrix} \sinh t \\ e^{-t} \end{bmatrix}$$

Option (d) is correct.

$c_1 = 1$ & $c_2 = 0$ & $c_1 = 1$, $c_2 = -1/2$

$$\begin{bmatrix} e^t \\ 0 \end{bmatrix} \begin{bmatrix} e^t - \frac{1}{2}e^{-t} \\ e^{-t} \end{bmatrix}$$

Option (b) is incorrect,

$c_1 = 1$ & $c_2 = 0$ & $c_2 = -1/2$ & $c_1 = -1/2$

$$\begin{bmatrix} e^t \\ 0 \end{bmatrix} \begin{bmatrix} -\left[\frac{e^t + e^{-t}}{2}\right] \\ e^{-t} \end{bmatrix} = \begin{bmatrix} -\cosh t \\ e^{-t} \end{bmatrix}$$

Option (a) is incorrect.

$c_1 = 0$, $c_2 = 0$ $\begin{bmatrix} 0 \\ 0 \end{bmatrix}$

7. Solution of the system of equation $x_1' = x_1 + x_2$ and $x_2' = 2x_2$

(a) $c_1 e^t \begin{bmatrix} 1 \\ 1 \end{bmatrix} + c_2 e^{2t} \begin{bmatrix} 1 \\ 1 \end{bmatrix}$ (b) $c_1 e^t \begin{bmatrix} 1 \\ 0 \end{bmatrix} + c_2 e^{2t} \begin{bmatrix} 1 \\ 1 \end{bmatrix}$

(c) $c_1 e^t \begin{bmatrix} 0 \\ 0 \end{bmatrix} + c_2 e^{2t} \begin{bmatrix} 1 \\ 1 \end{bmatrix}$ (d) $c_1 e^t \begin{bmatrix} 1 \\ 0 \end{bmatrix} + c_2 e^{2t} \begin{bmatrix} 4 \\ 1 \end{bmatrix}$

Ans. (b)

Explanation :

It is equivalent to the form $X' = AX$, i.e., $\begin{bmatrix} x_1' \\ x_2' \end{bmatrix} = \begin{bmatrix} 1 & 1 \\ 0 & 2 \end{bmatrix} \begin{bmatrix} x_1 \\ x_2 \end{bmatrix}$

Consider matrix $A = \begin{bmatrix} 1 & 1 \\ 0 & 2 \end{bmatrix}$

The eigenvalues of A are 1, 2. (by solving $|A - \lambda I| = 0$)

Let $\lambda_1 = 1$, $\lambda_2 = 2$, let X_1 be its eigenvector

For $\lambda_1 = 1$, solving $(A-I)X_1 = 0$ $\begin{bmatrix} 0 & 1 \\ 0 & 1 \end{bmatrix} \begin{bmatrix} y_1 \\ y_2 \end{bmatrix} = \begin{bmatrix} 0 \\ 0 \end{bmatrix} \Rightarrow y_2 = 0$

Eigenvector is $X_1 = \begin{bmatrix} 1 \\ 0 \end{bmatrix}$. corresponding to $\lambda_1 = 1$

For $\lambda_2 = 2$, let X_2 be eigen vector.

Then (A – 2I) X2 = 0 $\begin{bmatrix} -1 & 1 \\ 0 & 0 \end{bmatrix}\begin{bmatrix} y_1 \\ y_2 \end{bmatrix} = \begin{bmatrix} 0 \\ 0 \end{bmatrix} \Rightarrow -y_1 + y_2 = 0 \Rightarrow y_1 = y_2$

Eigenvector is $X_2 = \begin{bmatrix} 1 \\ 1 \end{bmatrix}$ corresponding to eigen value $\lambda_2 = 2$

Hence, the solution is $X = c_1 e^t X_1 + c_2 e^{2t} X_2 = c_1 e^t \begin{bmatrix} 1 \\ 0 \end{bmatrix} + c_2 e^{2t} \begin{bmatrix} 1 \\ 1 \end{bmatrix}$

$\Rightarrow x_1 = c_1 e^t + c_2 e^{2t}$ *and* $x_2 = c_2 e^{2t}$ are required solutions.

8. Let $y(t) = \begin{pmatrix} y_1(t) \\ y_2(t) \end{pmatrix}$ satisfy $\frac{dy}{dt} = Ay; t > 0,\ y(0) = \begin{pmatrix} 0 \\ 1 \end{pmatrix}$ where A is a 2×2 constant matrix with real entries satisfying trace $A = 0$ and det $A > 0$. Then $y_1(t)$ and $y_2(t)$ both are **CSIR NET DEC 2012**

(a) Monotonically decreasing functions of t (b) Monotonically increasing functions of t

(c) Oscillating function of t (d) Constnat function of t

Ans. (c)

Explanation :

Given that $Tr(A) = 0$ and det $(A) > 0$

$\Rightarrow$ Roots of characteristic polynomial of matrix are purely imaginary.

Hence, $y_1(t)$ and $y_2(t)$ are functions of $\cos t$ and $\sin t$, which are oscillating on R.

Hence, option (c) is correct.

9. Let $(x(t), y(t))$ satisfy for $t > 0$ $\frac{dx}{dt} = -x + y, \frac{dy}{dt} = -y,$ $x(0) = y(0) = 1$

Then $x(t)$ is equal to

(a) $e^{-t} + ty(t)$ (b) $y(t)$ (c) $e^{-t}(1+t)$ (d) $-y(t)$

Ans. (a),(c)

Explanation :

The matrix corresponding to the given system is $A = \begin{bmatrix} -1 & 1 \\ 0 & -1 \end{bmatrix}$ and its eigenvalues are obtained by $\lambda^2 + 2\lambda + 1 = 0 \Rightarrow \lambda = -1, -1$ are eigen values

Let u_1 be eigen vector corresponding to -1.

Then (A + I)u_1 = 0 $\begin{bmatrix} 0 & 1 \\ 0 & 0 \end{bmatrix}\begin{bmatrix} x_1 \\ y_1 \end{bmatrix} = \begin{bmatrix} 0 \\ 0 \end{bmatrix}$ $u_1 = (x_1, y_1)^T = (1, 0)^T$

$y_1(t) = \begin{bmatrix} 1 \\ 0 \end{bmatrix} e^{-t}$ and u_2 be another eigen vector corresponding to -1 then (A + I) $u_2 = u_1$

$$\begin{bmatrix}0&1\\0&0\end{bmatrix}\begin{bmatrix}x_2\\y_2\end{bmatrix}=\begin{bmatrix}1\\0\end{bmatrix}$$ $y_2 = 1, x_2 = 1$ $$Y_2(t)=\left[\begin{bmatrix}1\\0\end{bmatrix}t+\begin{bmatrix}1\\1\end{bmatrix}\right]e^{-t}=\begin{bmatrix}1+t\\1\end{bmatrix}e^{-t}$$

Then general solution $Y(t)=\alpha\begin{bmatrix}1\\0\end{bmatrix}e^{-t}+\beta\begin{bmatrix}1+t\\1\end{bmatrix}e^{-t}$

$\Rightarrow x(t)=\alpha e^{-t}+\beta(1+t)e^{-t}$ and $y(t)=\beta e^{-t} \because y(0)=1\Rightarrow\beta=1$

and $x(0)=1\Rightarrow\alpha+1=1,\alpha=0\Rightarrow x(t)=(1+t)e^{-t}\quad y(t)=e^{-t}$

$\Rightarrow x(t)=e^{-t}+ty(t)$

Hence, option (a) and (c) are correct.

UNSOLVED QUESTIONS

1. Consider the system of linear differential equation $\frac{dx_1}{dt}=5x_1-2x_2\quad \frac{dx_2}{dt}=4x_1-x_2$

With the initially condition $x_1(0) = 0$, $x_2(0) = 1$. Then $\log_e(x_2(2) - x_1(2))$ is equal to

(a) 0 (b) 1 (c) 2 (d) 3

2. Consider initial value problem in R^2, $Y'(t)=AY+BY;Y(0)=Y_0$, where

$A=\begin{bmatrix}1&0\\-1&1\end{bmatrix},B=\begin{bmatrix}1&-1\\0&1\end{bmatrix}$. then $Y(t)$ is given by **CSIR NET JUNE 2012**

(a) $e^{tA}e^{tB}Y_0$ (b) $e^{tB}e^{tA}Y_0$ (c) $e^{t(A+B)}Y_0$ (d) $\epsilon^{-t(A+B)}Y_0$

3. Consider the system of ODE in R^2, $\frac{dY}{dt}=AY,Y(0)=\begin{pmatrix}0\\1\end{pmatrix},t>0$

where $A=\begin{bmatrix}1&0\\-1&1\end{bmatrix},b=\begin{bmatrix}1&-1\\0&1\end{bmatrix}$. Then $Y(t)$ is given by

(a) $y_1(t)$ and $y_2(t)$ are monotonically increasing for $t>0$

(b) $y_1(t)$ and $y_2(t)$ are monotonically increasing for $t>1$

(c) $y_1(t)$ and $y_2(t)$ are monotonically decreasing for $t>0$

(d) $y_1(t)$ and $y_2(t)$ are monotonically decreasing for $t>1$

4. Let $A=\begin{bmatrix}-2&1&0\\0&-2&1\\0&0&-2\end{bmatrix},x(t)=\begin{bmatrix}x_1(t)\\x_2(t)\\x_3(t)\end{bmatrix}$. and $|x(t)|=\left(x_1^2(t)+x_2^2(t)+x_3^2(t)\right)^{1/2}$.

CSIR NET JUNE 2016

Then any solution of the first order system of the ordibnary differential equation

$\left.\begin{matrix}x'(t)=Ax(t)\\x(0)=x_0\end{matrix}\right\}$ satisfies

(a) $\lim_{t\to\infty}|x(t)|=0$ (b) $\lim_{t\to\infty}|x(t)|=0$ (c) $\lim_{t\to\infty}|x(t)|=2$ (d) $\lim_{t\to\infty}|x(t)|=12$

5. Let $(x(t), y(t))$ satisfy the system of ODEs **CSIR NET DEC 2016**

$$\frac{dx}{dt} = -x + ty, \quad \frac{dy}{dt} = tx - y$$

If $(x_1(t), y_1(t))$ and $(x_2(t), y_2(t))$ are two solutions and $\Phi(t) = x_1(t)y_2(t) - x_2(t)y_1(t)$ then $\frac{d\Phi}{dt}$ is equal to

(a) -2Φ (b) 2Φ (c) $-\Phi$ (d) Φ

6. Consider the first order system of linear equations

$$\frac{dX}{dt} = AX; A = \begin{bmatrix} 3 & 2 \\ -2 & -1 \end{bmatrix}; x = \begin{bmatrix} x_1(t) \\ x_2(t) \end{bmatrix}. \text{Then}$$

(a) The coefficient matrix A has a repeated eigenvalues $\lambda = 1$

(b) There is only one linearly independent eigenvector $X_1 = \begin{bmatrix} 1 \\ -1 \end{bmatrix}$

(c) The general solution of the ODE is $(aX_1 + bX_2)e^t$, where a,b are arbitrary constants and $X_1 = \begin{bmatrix} 1 \\ -1 \end{bmatrix}, X_2 = \begin{bmatrix} t \\ \frac{1}{2} - t \end{bmatrix}$

(d) The vectors X_1 and X_2 in the option 3 given above are linearly independent.

7. Consider the system of ODE **CSIR NET JUNE 2012**

$$\frac{d}{dx}Y = AY, Y(0) = \begin{bmatrix} 2 \\ -1 \end{bmatrix} \text{ where } A = \begin{pmatrix} 1 & 2 \\ 0 & -1 \end{pmatrix} \text{ and } Y = \begin{bmatrix} y_1(x) \\ y_2(x) \end{bmatrix}. \text{ Then}$$

(a) $y_1(x) \to \infty$ and $y_2(x) \to 0$ as $x \to \infty$ (b) $y_1(x) \to 0$ and $y_2(x) \to 0$ as $x \to \infty$

(c) $y_1(x) \to \infty$ and $y_2(x) \to -\infty$ as $x \to -\infty$ (d) $y_1(x), y_2(x) \to -\infty$ as $x \to -\infty$

8. The system of ODE **CSIR NET DEC 2014**

$$\left.\begin{aligned} \frac{dx}{dt} &= (1 + x^2)y, t \in R \\ \frac{dy}{dt} &= -(1 + x^2)x, t \in R \\ (x(0), y(0)) &= (a, b) \end{aligned}\right\}$$

(a) Only if $(a, b) = (0,0)$ (b) For any $(a, b) \in R \times R$

(c) Such that $x^2(t) + y^2(t) = a^2 + b^2$ for all $t \in R$

(d) Such that $x^2(t) + y^2(t) \to \infty$ as $t \to \infty$ if $a > 0$ and $b > 0$

Answer Key

1	2	3	4	5	6	7	8
c	c	d	a	a	a,b,c,d	a,c	b,c

DIFFERENTIAL EQUATION

Chapter 6

Boundary Value Problem

1. Boundary Value Problem

The most general second order ordinary linear differential equation is $P(x)y'' + Q(x)y' + R(x)y = S(x)$

Now consider the following types of boundary condition at two point $x = a$ and $x = b$

(1) $a_0y(a) + a_1y'(a) = c_1$

$b_0y(b) + b_1y'(b) = c_2$

(2) $y(a) = y(b)$

$y'(a) = y'(b)$

(1) is called regular condition.

The bounded condition are said to be homogeneous if both c_1 & c_2 are zero.

(2) is called periodic condition.

Example : $y'' + y = \cos 2x$

$y'(0) = 0,\ y'(\pi) = 0$

2. Eigen value and eigen function

(i) **Homogeneous BVP :** If differential equation and all boundary condition are homogeneous then BVP is called homogeneous BVP.

A homogeneous BVP posses a trivial solution $y(x) = 0$

However, if a λ occurs either in DE or boundary condition then a non-trivial solution is possible and in such a case the value of λ are determined which are called eigen value and such non-trivial solution is called eigen function.

Adjoint of BVP

Let $a_0(x)y'' + a_1(x)y' + a_2(x)y = b$ ....(1)

be a differential equation where $a_0(x) \neq 0 \quad \forall x$

and a_0, a_1, a_2 are continuous.

Then adjoint $\dfrac{d^2}{dx^2}(a_0(x)y) - \dfrac{d}{dx}(a_1(x)y) + a_2(x)y = 0$

$a_0(x)y'' + 2a_0'(x)y' + a_0''(x)y - a_1(x)y' - a_1'(x)y + a_2(x)y = 0$

$a_0(x)y'' + [2a_0'(x) - a_1(x)]y' + [a_0''(x) - a_1'(x) + a_2(x)]y = 0$ ..... (2)

(2) is called adjoint of (1)

DIFFERENTIAL EQUATION

If (1) is self adjoint then (1) = (2)

Compare coefficient of (1) & (2) both

$\Rightarrow a_0`(x) = a_1(x)$

Self adjoint homogeneous BVP with homogeneous condition

1. The value of λ for which differential equation has non-trivial solution is called an eigen value and corresponding non-trivial solution is eigen function.
2. DE has infinite solution iff λ is an eigen value.
3. DE has unique solution iff λ is not an eigen value.
4. The eigen value are always real number.
5. The eigen vector corresponding to distinct eigen value are orthogonal.
6. If $\lambda_1 \neq \lambda_2$ are eigen value and y_1 & y_2 are corresponding eigen function or vector

$$\Rightarrow \quad \int_a^b w(x)y_1(x).y_2(x) = 0$$

Where w(x) is weight function usually we take w(x) = 1

Regular / Periodic BVP

Let A = {$\lambda \in R$ |y(x) is non-trivial solution corresponding to λ of BVP}

(1) A is always C'ably infinite.

(2) The element of A can be arranged into a strictly increasing sequence.

s.t. $\lambda_1 < \lambda_2 < \lambda_3 \ldots\ldots < \lambda_n \ldots \quad \Rightarrow \lim_{n\to\infty} \lambda_n = \infty$

(3) A always bounded below and unbounded above.

(4) Every elements of A are isolated point.

(5) A is nowhere dense

(6) In Regular BVP

dim $(E_{\lambda_i}) = 1$ ⇒ Only one LI eigen vector corresponding to each eigen value.

(7) In periodic BVP

dim$(E_{\lambda_0}) = 1$

dim $(E_{\lambda_i}) = 2$ i = 1, 2, 3……

Regular BVP (SLP)

Let $$\frac{d}{dx}\{p(x)y'\} + [q(x) + \lambda r(x)]y = 0 \quad \ldots..(1)$$

B.C. $\begin{matrix} a_0y(a)+a_1y'(a)=0 \\ b_0y(b)+b_1y'(b)=0 \end{matrix}$ ..... (2)

Where $p(x) > 0$ $\forall\ x \in [a, b]$

Result ⇒

If $a_0, a_1, b_0, b_1 \geq 0$

Then each $\lambda_n \geq 0$ ⇒ Each eigen values are non-negative

Singular BVP ⇒

(1) With (2) B.C. s.t.

$p(a) = 0 = p(b)$ and $p(x) > 0$ $\forall x \in (a, b)$

Example : $(1 - x^2)y'' - 2xy' + \lambda x^2 y = 0$

$x \in [-1, 1]$

Periodic Problem

(1) with B.C. $y(a) = y(b)$ $y'(a) = y'(b)$ $p(a) = p(b)$

$p(x) > 0$ $\forall\ x \in (a, b)$

SOLVED QUESTIONS

1. The following two – point boundary value problem

$$\begin{cases} y''(x)+\lambda y(x)=0\ for x\in(0,\pi) \\ y(0)=0 \\ y(\pi)=0 \end{cases}$$

has trivial solution y = 0. It also has a non- trivial solution for

(a) no values of $\lambda \in R$ (b) $\lambda = 1$ (c) $\lambda = n^2$ for all $n \in N, n > 1$ (d) $\lambda \leq 0$

CSIR NET JUNE 2020

Ans. **(b), (c)**

Explanation :

$$\begin{cases} y''(x)+\lambda y(x)=0\ for x\in(0,\pi) \\ y(0)=0 \\ y(\pi)=0 \end{cases}$$

Put $\lambda = 1$, $y(0) = 0$, $y(\pi) = 0$, Auxiliary equation is $m^2 + 1 = 0 \Rightarrow m = \pm i$

$y(x) = c_1 \cos x + c_2 \sin x$

$y(0) = 0 \Rightarrow c_1 = 0$, $y(x) = c_2 \sin x$ $y(\pi) = 0 \Rightarrow 0 = c_2 \sin \pi \Rightarrow 0 = 0$,

c_2 arbitrary. i.e. for $\lambda = 1$ the differential equation has non-trivial solution. Option (b) is correct and option (a) is false.

For $\lambda = -n^2$, $y''(x) - n^2y(x) = 0$. Auxiliary equation is $m^2 - n^2 = 0 \Rightarrow m = \pm n$, solution

$y(x) = c_1e^{nx} + c_2e^{-nx}$

DIFFERENTIAL EQUATION

$y(0) = 0 \Rightarrow c_1 + c_2 = 0$ $\quad$ $y(\pi) = 0 \Rightarrow c_1e^{n\pi} + c_2e^{-n\pi} = 0$

We get $c_1 = c_2 = 0$, trivial solution. Option (d) is false.

For $\lambda = n^2$, $y``(x) + n^2y(x) = 0$. Auxiliary equation is $m^2 + n^2 = 0 \Rightarrow m = \pm i\,n$

$y(x) = c_1 \cos x + c_2 \sin x$

$y(0) = 0 \Rightarrow c_1 = 0$

$y(\pi) = 0 \Rightarrow c_2$ arbitrary.

$y(x) = c_2 \sin x$, non-trivial solution. Option (c) is correct.

Option (b) and (c) are correct.

2. Consider the eigen value problem.

$((1 + x^4)y`)` + \lambda y = 0$ $\quad$ $x \in (0, 1)$ $\quad$ $y(0) = 0$ $\quad$ $y(1) + 2y`(1) = 0$

Then which of the following are TRUE? **CSIR NET DEC 2019**

(a) All eigen value are negative (b) All eigen value are positive

(c) $\exists$ some negative and some positive value (d) there are no eigen value

Ans. (b)

Explanation :

Given that

$((1 + x^4)y`)` + \lambda y = 0$ ...(1) $\quad$ $x \in (0, 1)$

Where $p(x) = (1 + x^4) > 0$ $\quad$ $\forall x \in (0, 1)$

B.C.

$y(0) = 0$ $\quad$ $y(1) + 2y`(1) = 0$

where $\left.\begin{matrix} a_0 = 1,\ a_1 = 0 \\ b_0 = 1, b_1 = 2 \end{matrix}\right\}$ all non-negative

This problem is Regular BVP.

By result each $\lambda_n \geq 0$

$\Rightarrow$ All eigen value are non-negative.

3. The positive value of λ for which the equation $y`` + \lambda^2y(x) = 0$ has non-trivial solution $y(0) = y(\pi)$ and $y`(0) = y`(\pi)$ are **CSIR NET JUNE 2019**

(a) $\lambda = \dfrac{2n+1}{n}$ (b) $\lambda = 2n$ (c) $\lambda = n$ (d) $\lambda = 2n - 1$

Ans. (b)

Explanation :

It is a periodic problem.

Given that $\quad y`` + \lambda^2 y = 0$

Auxiliary equation $\quad m^2 + \lambda^2 = 0 \quad \Rightarrow m = \pm i\lambda$

then $y(x) = c_1\cos\lambda x + c_2\sin\lambda x$ ....(1)

Given condition. $\quad y(0) = y(\pi)$

$c_1\cos 0 + c_2 \sin 0 = c_1 \cos \lambda\pi + c_2 \sin \lambda\pi$

$c_1 = c_1 \cos \lambda\pi \quad = c_1(1 - \cos \lambda\pi) = 0$ ....(2)

and $y`(0) = y`(\pi)$

$y`(x) = -c_1\lambda \sin\lambda x + c_2 \lambda \cos \lambda x$

$c_2\lambda = c_2\lambda \cos \lambda\pi \quad \Rightarrow \lambda c_2(1 - \cos \lambda\pi) = 0$

$\lambda \neq 0 \quad \Rightarrow c_2(1 - \cos \lambda\pi) = 0$ ..... (3)

From (2) and (3)

$$\begin{bmatrix} 1-\cos\lambda\pi & 0 \\ 0 & 1-\cos\lambda\pi \end{bmatrix}\begin{bmatrix} c_1 \\ c_2 \end{bmatrix} = \begin{bmatrix} 0 \\ 0 \end{bmatrix}$$

Where $A = \begin{bmatrix} 1-\cos\lambda\pi & 0 \\ 0 & 1-\cos\lambda\pi \end{bmatrix}$

Given that the equation has non-trivial solution.

$\Rightarrow |A| = 0 \quad = (1 - \cos \lambda\pi)^2 = 0 \quad \cos \lambda\pi = 1 \quad \lambda\pi = 2n\pi \quad \lambda = 2n$

4. Consider the SLP

$y`` + \lambda y = 0, \quad y(0) = 0$ and $y(\pi) = 0$

Which of the following are TRUE? **CSIR NET JUNE 2018**

(a) ∃ only c'ably many eigen value (b) ∃ uncountably many eigen value

(c) each eigen function corresponding to eigen value λ has exactly $|\sqrt{\lambda}|-1$ zeros in $(0, \pi)$

(d) each eigen function corresponding to eigen value λ has exactly zeros in $|\sqrt{\lambda}|$ $(0, \pi)$

Ans. (a), (c)

Explanation :

We know that eigen values are countably infinite

Option (a) is correct.

(1) Let $\lambda = 0$ then $y`` = 0 \Rightarrow y(x) = Ax + B$ Condition $y(0) = 0$ & $y(\pi) = 0$

Then $y = 0$ (trivial solution) $\Rightarrow \lambda = 0$ is not eigen value.

(2) Let $\lambda > 0 \quad \lambda = \mu^2$

Then $y`` + \mu^2 y = 0$

A equation $m^2 + \mu^2 = 0 \quad m = \pm \mu i$

$y(x) = c_1 \cos \mu x + c_2 \sin \mu x$

$y(0) = 0 \quad \Rightarrow c_1 = 0 \quad y(\pi) = 0$

then $y(\pi) = 0 = c_2 \sin \mu\pi$ if c_2 is zero then $y = 0$ (trivial) solution.

For non-trivial solution $c_2 \neq 0$

Then $c_2 \sin \mu\pi = 0 \quad \Rightarrow \quad \mu\pi = n\pi$

$\Rightarrow \quad \mu = n \quad [\lambda = \mu^2]$

$\Rightarrow \quad \sqrt{\lambda} = n$ where $n \in Z$ (countably infinite)

Then eigen function $y(x) = c_2 \sin \sqrt{\lambda} x$

Let $\sqrt{\lambda} = n = 1$

Then y(x) = c_2 sin x c_2 sin x has no zero in (0, π)

⇒ It has no zero in (0, π)

Option (c) is correct. $|\sqrt{\lambda}|-1$ zero ⇒ 1 – 1 = 0 (no zero)

Option (d) is incorrect. $|\sqrt{\lambda}|=1$ zeros

5. Consider the eigen value

$y'' + \lambda y = 0$ for $x \in (-1, 1)$

y(-1) = y(1)

y'(-1) = y'(1)

Which of the following are TRUE? **CSIR NET DEC 2018**

(a) All eigen value are strictly positive (b) All eigen value are non-negative

(c) Distinct eigen function are orthogonal $L^2[-1,1]$

(d) The sequence of eigen value is bounded above.

Ans. (b), (c)

Explanation :

In SLP

Let A be the set of eigen value then A is bounded below but not bounded above.

Option (d) is incorrect.

(1) λ = 0 ⇒ $y'' = 0 \Rightarrow y(x) = Ax + B$

Apply condition y(-1) = y(1)   A = 0

Then y(x) = B (non-trivial solution) ⇒ λ = 0 is an eigen value.

(2) $\lambda = \mu^2$

Then $y'' + \mu^2 y = 0$

$m^2 + \mu^2 = 0$

$m = \pm \mu i$

y(x) = A cos μx + B sin μx

Condition y(-1) = y(1)   = A cos μ - B sin μ = A cos μ + B sin μ

= 2B sin μ = 0 ..... (1)

and y'(x) = -Aμ sin μx + Bμ cos μx

then condition y'(-1) = y'(1)

= A μ sin μ + B μ cos μ = -Aμ sin μ + Bμ cos μ

= 2Aμ sin μ = 0 ..... (2)

For eliminating A & B

$$\begin{bmatrix} 0 & 2\sin\mu \\ 2\mu\sin\mu & 0 \end{bmatrix}\begin{bmatrix} A \\ B \end{bmatrix} = \begin{bmatrix} 0 \\ 0 \end{bmatrix}$$

This system has non-trivial solution if |A| = 0

$-2\mu \sin^2 \mu = 0$ ⇒ sin μ = 0

$\mu = n\pi$ $[\lambda = \mu^2]$ $\Rightarrow \lambda = n^2\pi^2$ $n = 0, \pm1, \pm2 \ldots.$

$n^2 = 0, 1, 2, \ldots.$

Eigen values are non-negative.

Option (b) is correct.

Option (a) is incorrect.

Two vector u & v are orthogonal

If $\int_{-1}^{1} w(x)u(x).v(x) = 0$

Where w(x) is weight function usually we take w(x) = 1

Then $\int_{-1}^{1} u(x).v(x) = 0$ …… (3)

Here $u(x) = A \cos n\pi x + B \sin n\pi x$

$y_1(x) = A \cos \pi x + B \sin \pi x$ [Take A = B = 1]

$y_2(x) = A \cos 2\pi x + B \sin 2\pi x$

then $\int_{-1}^{1} (\cos \pi x + \sin \pi x)(\cos 2\pi x + \sin 2\pi x)dx = 0$ $\Rightarrow$ both are orthogonal.

Option (c) is correct.

6. The set of real number λ for which the BVP $y'' + \lambda y = 0,$ $y(0) = 0,$ $y(\pi) = 0$ has non-trivial solution is **CSIR NET DEC 2017**

(a) $(-\infty, 0)$ (b) $\{\sqrt{n} \mid n \in Z^+\}$ (c) $\{n^2 \mid n \in Z^+\}$ (d) R

Ans. (c)

Explanation :

Let $\lambda = 0$

(1) Then $y'' = 0$ $\Rightarrow y = Ax + B$ $y(0) = 0 \Rightarrow B = 0$ $y = Ax$

and $y(\pi) = 0$ $\Rightarrow$ $A\pi = 0 \Rightarrow A = 0$ $\Rightarrow$ $y(x) = 0$ (trivial solution)

$\lambda = 0$ is not an eigen value.

(2) let $\lambda = \mu^2$ Then $y'' + \mu^2 y$ 0

$m^2 + \mu^2 = 0$ $\Rightarrow$ $m = \pm\mu i$

$y(x) = c_1 \cos \mu x + c_2 \sin \mu x$ use $y(0) = 0$ then we get $c_1 = 0$

and $y(\pi) = 0 = c_2 \sin \mu\pi$ for non-trivial solution $c_2 \neq 0$

$\Rightarrow \sin \mu\pi = 0$ $\Rightarrow \mu\pi = n\pi$ $\Rightarrow \mu = n$ $\Rightarrow \sqrt{\lambda} = n$ $\Rightarrow \lambda = n^2$

7. Consider the BVP $\frac{d^2y}{dx^2} = f(x)$ with y(0) = y(1) = y`(1). Where f is real valued continuous function on [0, 1] then which of the following are TRUE?

(a) The given BVP has a unique solution for every f

(b) doesn't unique solution for some f

(c) $y(x) = \int_0^x xt\, f(t)dt + \int_x^1 (t - x + xt) f(t)dt$ is a solution

(d) $y(x) = \int_0^x (x - t + xt) f(t)dt + \int_x^1 xt\, f(t)dt$ **CSIR NET DEC 2017**

Ans. (a), (c)

Explanation :

Let $y(x) = \int_0^x xtf(t)dt + \int_x^1 (t - x + xt) f(t)dt$ then it satisfy the condition

y(0) = y(1) = y`(1)

Let $y(0) = \int_0^0 0 + \int_0^1 (t) f(t)dt = \int_0^1 t\, f(t)dt$

$y(1) = \int_0^1 tf(t)dt + \int_1^1 (t - x + xt) f(t)dt = \int_0^1 t\, f(t)dt \Rightarrow$ y(0) = y(1) = y`(1)

And $y'(1) = \int_0^1 t\, f(t)dt$

Option (c) is correct.

Option (d) is incorrect.

Let $y(x) = \int_0^x (x - t + xt) f(t)dt + \int_1^x xt\, f(t)dt$

$y(0) = \int_0^0 (x - t + xt) f(t)dt + \int_1^0 0.t\, f(t)dt = 0 \quad y(1) = \int_0^1 f(t)\, dt$

Here y(0) ≠ y(1) for every f(t)

(a) and (b)

(1) Let f(x) = 0

Then y`` = 0 ⇒ y(x) = Ax + B

Then y(0) = y(1) = y`(1) = A(0) + B = A + B = A ⇒ A = B = 0 ⇒ y(x) = 0 (unique)

(2) Let f(x) = 1

Then y`` = 1 $y(x)=\frac{x^2}{2}+Ax+B$

Here y(0) = y(1) = y'(1) $=B=\frac{1}{2}+A+B=1+A$

Then $y(x)=\frac{x^2}{2}+\frac{x}{2}-\frac{1}{2}$ (unique) ⇒ We can easily find the value of A & B

⇒ unique solution. Option (a) is correct

8. For Strum – Lioville problem y`` + λy = 0 and y(0) = 0 = y(π)

(a) Eigen value are $\lambda = n^2$, n = 1, 2, ... (b) Eigen value are λ = -n, n = 1, ...

(c) y(x) = sin nx eigen function (d) y = cos nx eigen function

Ans. **(a), (c)**

Explanation :

Let $\lambda > 0 \Rightarrow \lambda = \mu^2$ Then $y'' + \mu^2 y = 0$

Root of auxiliary equation $m = \pm i\mu$ ⇒ $y(x) = c_1 \cos \mu x + c_2 \sin \mu x$

Given condition y(0) = 0 ⇒ $c_1 = 0$

$y(x) = c_2 \sin \mu x$

$y(\pi) = 0$ ⇒ $c_2 \sin \mu\pi = 0$

$c_2 \neq 0$ then $\mu\pi = n\pi$ ⇒ $\mu = n$ ⇒ $\lambda = n^2$

then $y(x) = c_2 \sin nx$

9. The value of least possible positive number λ such that the following problem has a non-trivial solution $u''(x) + \lambda u(x) = 0$ $0 < x < 1$

$u'(0) = u'(1) = 0$

(a) π^2 (b) $2\pi^2$ (c) 0 (d) $4\pi^2$

Ans. **(a)**

Explanation :

Let $\lambda > 0$ $\lambda = \mu^2$

Then $u'' + \mu^2 u = 0$ ⇒ $u(x) = c_1 \cos \mu x + c_2 \sin \mu x$

$u'(x) = -c_1 \mu \sin \mu x + c_2 \mu \cos \mu x$

$u'(0) = u'(1) = 0$ ⇒ $c_2 = 0$ ⇒ $u'(x) = c_1 \mu \sin \mu x$ $u'(1) = 0$

⇒ $c_1 \mu \sin \mu = 0$ $\sin \mu = 0$ ⇒ $\mu = n\pi$

& $\lambda = n^2\pi^2$ $n \in Z$ $\lambda = \pi^2, 4\pi^2, 9\pi^2$ ⇒ Least positive integer π^2

UNSOLVED QUESTIONS

1. Which of the followings is/are true about Sturm-Liouville Problem (SLP)?

(a) All eigen values of SLP are real and non negative

(b) Eigen functions corresponding to different eigen values are orthogonal with respect to weight function.

(c) SLP has always an eigen function

(d) For each eigen values of a SLP there exists only one linearly independent eigen function.

2. The characteristic values of the Sturm-Liouville problem

$$\frac{d^2y}{dx^2}+\lambda y=0; y(0)=0, y(\pi)-y'(\pi=0) \text{ are}$$

(a) $\lambda=\alpha_n^2$ where $\alpha_n(n=1,2,3...)$ are the positive roots of the equation $\alpha = \cot \pi\alpha$

(b) $\lambda=\alpha_n^2$ where $\alpha_n(n=1,2,3...)$ are roots of the equation $\alpha = \tan \pi\alpha$

(c) 0, 1 (d) negative real numbers

3. The Sturm – Liouville problem : $y''+\lambda^2 y=0, y'(0)=0, y'(\pi)=0$ has its eigenvectors given by

(a) $y=sin\left(n+\frac{1}{2}\right)x$ (b) $y=sin\,nx$

(c) $y=cos\left(n+\frac{1}{2}\right)x$ (d) $y=cos\,nx; where\, n=0,1,2,...$

4. The eigenvalues of the Strum Liouville system $y''+\lambda y=0, 0\le x\le\pi, y(0)=0, y'(\pi)=0$ are

(a) $\frac{n^2}{4}$ (b) $\frac{(2n-1)^2\pi^2}{4}$ (c) $\frac{(2n-1)^2}{4}$ (d) $\frac{n^2\pi^2}{4}$

5. Let n be a non-negative integer. The eigenvalues of the Sturm-Liouville problem $\frac{d^2y}{dx^2}+\lambda=0$, with boundary conditions $y(0)=y(2\pi), \frac{dy}{dx}(0)=\frac{dy}{dx}(2\pi)$, are

(a) n (b) $n^2\pi^2$ (c) $n\pi$ (d) n^2

6 Consider the boundary value problem (BVP) $\frac{d^2y}{dx^2}+\alpha y(x)=0,\ \alpha\in IR$

s.t. $y(0) = 0$, $y(\pi) = k$, k is non-zero real number. Then which of the following is true?

(a) for $\alpha = 1$, the BVP has infinitely many solution.

(b) for $\alpha = 1$, unique solution

(c) for $\alpha = -1$, $k < 0$ the BVP has a solution y(x) such that $y(x) > 0\ \forall\ x\in(0, \pi)$

(d) for $\alpha = -1$, $k > 0$ the BVP has a solution y(x) such that $y(x) > 0\ \forall\ x \in (0, \pi)$

Answer Key

1	2	3	4	5	6
b,d	b	d	c	d	d

Chapter 7

Introduction to Partial Differential Equations and Partial Differential Equations of First Order

1. Partial Differential Equation

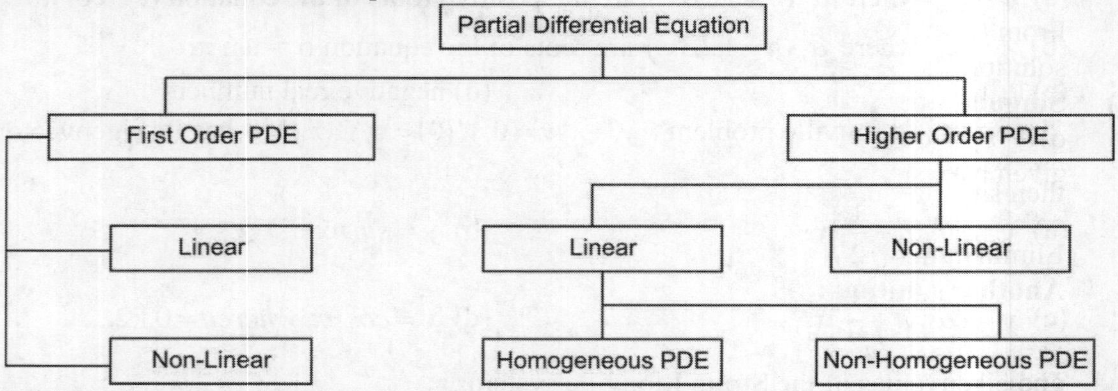

Definition : If a dependent variable is of function of two or more independent variables than an equation involving partial differential coefficients is called a partial differential equation.

Example : $\frac{\partial z}{\partial x}+\frac{\partial z}{\partial y}=xy$

(x & y both are independent variables)

Note : In $\frac{\partial z}{\partial x}$, z is dependent variable and x is independent variable.

1. **Formation of Partial differential Equations:** The partial differential equations can be formed either by elimination of arbitrary constants from a relation between x, y, z or by the elimination of arbitrary functions of variables.
2. **Solutions of Partial Differential Equations :** The solution f(x, y, z, a, b) = 0 of a first order partial differential equation with two arbitrary constants is called a complete solution or complete integral.

Type of solution or integral :

(i) **Complete Solution :** First order PDE is given by

$f(x, y, z, p, q) = 0$ ..... (1)

Let F(x, y, z a, b) = 0 is called the complete integral of equation (1) if

(a) it satisfy equation (1)

(b) Rank of the matrix $\begin{bmatrix} F_a & F_{xa} & F_{ya} \\ F_b & F_{xb} & F_{yb} \end{bmatrix}$ is 2

Note : Number of independent variable in PDE = Number of arbitrary constant in complete solution.

(ii) General Solution : Let given PDE is F(x, y, z, p, q) = 0 (non-linear) and z = f(x, y, z, a, b) be complete integral of equation (1)

If we take b = ϕ(a)

Then solution become z = η(x, y, z, a ϕ(a)) ..... (3)

Then find $\frac{\partial \eta}{\partial a} = 0$ ...... (4)

From (3) & (4) eliminating (a) then we obtain z = [x, y, z, a(x,y), ϕ[a(x,y)]] which is a solution.

(iii) Singular Integral : Let f(x, y, z, a, b) = 0 be the complete solution of PDE F(x, y, z, p, q) = 0

then singular integral is given by $f = 0, \frac{\partial f}{\partial a} = 0, \frac{\partial f}{\partial b} = 0$

Eliminating (a) & (b) then we get ϕ(x, y, z) = 0

Another solution :

f(x, y, z, p, q) = 0

and $\frac{\partial f}{\partial p} = 0 = \frac{\partial f}{\partial q}$

eleminating p & q we get η(x, y, z) = 0

Example : The singular integral of z = px + qy + p^2 + q^2 is

(a) $\frac{-(x^2+y^2)}{2}$ (b) $\frac{-(x^2+y^2)}{4}$ (c) $\frac{-(x^2+y^2)}{8}$ (d) $-(x^2+y^2)$

Solution : (b)

Given that z = px + qy + p^2 + q^2

$$\frac{\partial z}{\partial p} = x + 2p = 0 \Rightarrow p = -\frac{x}{2} \quad \frac{\partial z}{\partial q} = y + 2q = 0 \Rightarrow q = -\frac{y}{2}$$

Put value of p & q in given PDE.

$$z = -\frac{x^2}{2} - \frac{y^2}{2} + \frac{x^2}{4} + \frac{y^2}{4} \quad = -\frac{(x^2+y^2)}{4}$$

Types of Partial Differential Equations

(i) **Linear P.D.E.:** A first order PDE is said to be a linear equation if it is linear in p, q and z, i.e., if it is of the form P(x, y) p + Q(x, y) q = R (x, y) z + S(x, y).

For example: yp – xq = xyz + x.

(ii) **Semi-linear P.D.E.:** A first order PDE is said to be semi-linear equation if it is linear in p and q and the coefficients of p and q are the function of x and y only. i.e., it is of the form

P(x, y) p + Q(x, y) q = R (x, y, z)

For example: e^xp – y x q = xz^2

DIFFERENTIAL EQUATION

(iii) **Quasi linear P.D.E.:** A first order PDE is said to be a quasi linear equation if it is linear in p and q, i.e., if it is of the form

$P(x,y,z)p + Q(x,y,z)q = R(x,y,z)$

Example : $(x^2 + z^2)p - xyq = z^3x + y^2$

(iv) **Non-linear P.D.E.:** Partial differential equations of the form f(x,y,z,p,q) = 0 which do not come under the above three types are said to be non-linear equations.

Example. pq = z does not belong to any of the first three types. So, it is an non-linear first order p.d.e.

2. Linear Partial Differential Equations of First order

A differential equation involving first order partial derivatives p and q only is called a partial differential equation of the first order.

If p and q both occur in the first degree only and are not multiplied together, then it is called a linear partial differential equation of the first order

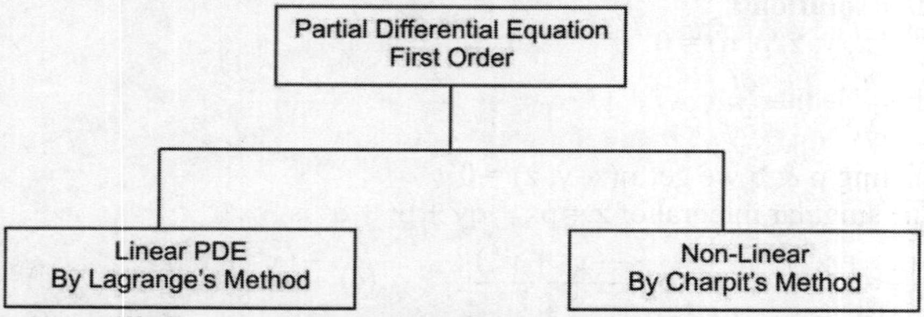

In this chapter we will discuss about Lagrange's method and we will discuss Charpit's method in next chapter.

Lagrange's Linear Equation

Lagrange's equation : If PDE of the form Pp + Qq = R where P, Q, R are functions of x, y, z such a PDE is known as Lagrange's equation **Example** xy p + yz q = zx

Solve Pp + Qq = R from Lagrange's method :

(1) Given Pp + Qq = R

(2) Write down the auxiliary equation $\frac{dx}{P} = \frac{dy}{Q} = \frac{dz}{R}$

(3) Then solve this following method

Type I for solving :

Let Pp + Qq = R be a given PDE and then Lagrange's equation is $\frac{dx}{P} = \frac{dy}{Q} = \frac{dz}{R}$ ... (1)

Take two pair of fractions of (1) and we get

$u(x, y, z) = c_1$ & $v(x, y, z) = c_2$ and general solution $\phi(u, v) = 0$

Example : yzp + zxy = xy

By Lagrange's method

$$\frac{dx}{yz}=\frac{dy}{zx}=\frac{dz}{xy}$$

From $\frac{dx}{yz}=\frac{dy}{zx}\Rightarrow xdx=ydy \underset{Integration}{\Rightarrow} x^2-y^2=c_1$

From $\frac{dy}{zx}=\frac{dz}{xy}\Rightarrow zdz=ydy\Rightarrow z^2-y^2=c_2$

From $\frac{dx}{yz}=\frac{dz}{xy}\Rightarrow zdz-xdx\Rightarrow z^2-x^2=c_3$

Then general solution

$\phi(x^2-y^2,z^2-y^2)=0 \quad ,\phi(x^2-y^2,z^2-x^2)=0 \;,\phi(z^2-x^2,x^2-y^2)=0$ all three are solutions of given PDE.

II. Type II for solving :

Let Lagrange's equation $\frac{dx}{P}=\frac{dy}{Q}=\frac{dz}{R}$

By taking two fractions we get u = c_1 then other v(x, y, z) is obtain by using of solution.

Example : p – q = log (x + y) then Lagrange's auxiliary equation is

$$\frac{dx}{1}=\frac{dy}{-1}=\frac{dz}{\log(x+y)}$$

from first two fractions $\frac{dx}{1}=\frac{dy}{-1}\Rightarrow x+y=c_1$ ...... (1)

Then (i), (iii) fractions

$\frac{dx}{1}=\frac{dz}{\log(x+y)}$ using first solution x + y = c_1 $\frac{dx}{1}=\frac{dz}{\log c_1}\Rightarrow(\log c_1)(x)=z+c_2$

$\Rightarrow$ x log c_1 – z = c_2 x log (x + y) – z = c_2 .... (2)

then (1) & (2) are solution.

Type III for solving :

Let $\frac{dx}{P}=\frac{dy}{Q}=\frac{dz}{R}$ ..... (1)

We choose S, T, U s.t. SP + TQ + RU = 0

Then from equation (1) $\frac{dx}{P}=\frac{dy}{Q}=\frac{dz}{R}=\frac{Sdx+Tdy+Udz}{SP+TQ+RU}$

Example : x (y – z)p + y(z – x)q = z (x – y)

By Lagrange's equation $\frac{dx}{xy-xz}=\frac{dy}{yz-yx}=\frac{dz}{zx-zy}$

Choose S = T = U = 1

Here $\dfrac{dx+dy+dz}{xy-xz+yz-yx+zx-zy=0}$

$\Rightarrow$ dx + dy + dz = 0

$x + y + z = c_1$ ..... (1)

Now $\dfrac{\frac{1}{x}dx+\frac{1}{y}dy+\frac{1}{z}dz}{y-z+z-x+x-y=0}$ $\left[Choose\ S=\frac{1}{x}, T=\frac{1}{y}, U=\frac{1}{z}\right]$

$\Rightarrow$ log x + log y + log z = $c_2 \Rightarrow$ log (x y z) = c_2 $\Rightarrow$ xyz = c_2 ..... (2)

Equation (1) & (2) are solutions

Type IV for solving

Let P_1, Q_1, R_1 be functions of x, y, z then $\dfrac{P_1dx+Q_1dy+R_1dz}{P_1P+Q_1Q+R_1R}$ ..... (3)

Here P_1dx $Q_1dy + R_1dz$ is exact differential of the denominator of (2)

Then solve it

Example : (y + z)p + (z + x)q = (x + y)

$$\Rightarrow \frac{dx}{y+z}=\frac{dy}{z+x}=\frac{dz}{x+y} \Rightarrow \frac{dx-dy}{y-x}=\frac{dy-dz}{z-y}=\frac{dx-dz}{z-x}$$

First two fraction $\dfrac{dx-dy}{y-x}=\dfrac{dy-dz}{z-y} \Rightarrow \dfrac{dx-dy}{+(x-y)}=\dfrac{dy-dz}{+(y-z)}$

$$\Rightarrow \log\left[\frac{x-y}{y-z}\right]=c_1 \Rightarrow \frac{x-y}{y-z}=c_1$$

$\Rightarrow$ log(x – y) – log(y – z) = c_1

Now second & third fraction $\dfrac{dy-dz}{+(y-z)}=\dfrac{dx-dz}{+(x-z)}$

$= \log\left[\dfrac{y-z}{x-z}\right]=c_2 \Rightarrow \dfrac{y-z}{x-z}=c_2$ Then $\phi\left(\dfrac{x-y}{y-z},\dfrac{y-z}{x-z}\right)=0$ is a general sclution.

3. Classification of solution of first order pde

Cauchy Problem : (Integral surface passing through a given curve)

Let Pp + Qq = R be PDE & curve x = x(t), y = y(t), z = z(t)

Then find the solution passing through this curve Such type of problem are called Cauchy problem.

(A) Uniqueness of Cauchy problem When R ≠ 0

(1) **Unique solution** : Given the PDE Pp + Qq = R with initial data curve $x_0(t)$, $y_0(t)$, $z_0(t)$

then the problem has unique solution if $\dfrac{P(x_0,y_0,z_0)}{\frac{dx_0}{dt}} \neq \dfrac{Q(x_0,y_0,z_0)}{\frac{dy_0}{dt}}$

(2) **No-solution :** $\dfrac{P(x_0,y_0,z_0)}{\frac{dx_0}{dt}} = \dfrac{Q(x_0,y_0,z_0)}{\frac{dy_0}{dt}} \neq \dfrac{R(x_0,y_0,z_0)}{\frac{dz_0}{dt}}$

(3) **Infinite solution :** $\dfrac{P(x_0,y_o,z_0)}{\frac{dx_0}{dt}} = \dfrac{Q(x_0,y_0,z_0)}{\frac{dy_0}{dt}} = \dfrac{R(x_0,y_0,z_0)}{\frac{dz_0}{dt}}$

(B) Uniqueness of Cauchy problem when R = 0

Given PDE Pp + Qq = 0, then two possibility arise (1) Unique solution (2) Infinite solution

(1) Unique solution $\dfrac{P(x_0,y_o,z_0)}{\frac{dx_0}{dt}} \neq \dfrac{Q(x_0,y_0,z_0)}{\frac{dy_0}{dt}}$

(2) Infinite solution $\dfrac{P(x_0,y_o,z_0)}{\frac{dx_0}{dt}} = \dfrac{Q(x_0,y_0,z_0)}{\frac{dy_0}{dt}}$

SOLVED QUESTIONS

1. Let u(x ,t) be the solution of Cauchy problem $uu_x + u_y = 0$, u(x, 0) = x. Then the value of u(2, 3)

CSIR NET 2023

(a) 1/2 (b) 1/3 (c) 3 (d) 2

Ans. (a)

Explanation :

Lagrange's auxiliary equation $\dfrac{dx}{u} = \dfrac{dy}{1} = \dfrac{du}{0}$

Here $u = c_1$ .... (1)

$uy - x = c_2$ .... (2)

Hence $u = \phi(uy - x)$

Using condition u(x, 0) = x

$\Rightarrow x = \phi(-x) \Rightarrow \phi(x) = -x \Rightarrow \phi(uy - x) = x - uy$

$\Rightarrow u = x - uy \Rightarrow u(1 + y) = x \Rightarrow u = \dfrac{x}{1+y} \Rightarrow u(2,3) = \dfrac{2}{4} = \dfrac{1}{2}$

2. Let u(x ,t) be the solution of Cauchy problem $u_x + u_y = e^u$; u(x, 0) = 1. Then which of the following is true **CSIR NET 2023**

(a) $u\left(\frac{1}{2e},\frac{1}{2e}\right)=1$ (b) $u_x\left(\frac{1}{2e},\frac{1}{2e}\right)=0$ (c) $u_y\left(\frac{1}{4e},\frac{1}{4e}\right)=\log 4$ (d) $u_y\left(0,\frac{1}{4e}\right)=\frac{4e}{3}$

Ans. (b),(d)

DIFFERENTIAL EQUATION

Explanation :

$\frac{dx}{1}=\frac{dy}{1}=\frac{du}{e^u}$ $\quad x-y=c_1$ & $y+e^{-u}=c_2$ then $y+e^{-u}=\phi(x-y)$ ..... (1)

Using u(x, 0) = 1, then (1) become $e^{-1}=\phi(x)$ $\quad y+e^{-u}=e^{-1}$ .... (2)

$\Rightarrow u(x,y)=\log\left(\frac{e}{1-ey}\right)$

(a) $u\left(\frac{1}{2e},\frac{1}{2e}\right)=\log(2e)$ (b) $u_x(x,y)=0$ (c) $u_y(x,y)=\left(\frac{e}{1-ey}\right)$

$\Rightarrow u_y\left(\frac{1}{4e},\frac{1}{4e}\right)=\frac{4e}{3}$

Similar option (c) and (d) are incorrect.

3. Which of the following expressions for u = u(x, t) are solutions of $u_t-e^{-t}u_x+u=0$ with u(x, 0) = x? **CSIR NET FEB 2022**

(a) $e^t(x+e^t-1)$ (b) $e^{-t}(x-e^t+1)$ (c) xe^t (d) xe^t

Ans. (b)

Explanation :

Consider the given PDE $\quad u_t-e^{-t}u_x+u=0$ .... (1)

Lagrange auxiliary equation are $\frac{dx}{-e^{-t}}=\frac{dt}{1}=\frac{du}{-u}$

Consider, $\quad \frac{dt}{1}=\frac{du}{-u}$ $\quad -t=\log u+\log c$

$uc_1=e^{-t}$ $\quad \Rightarrow \quad ue^t=c_1$ ..... (2)

Now, $\frac{dx}{-e^{-t}}=dt$ $\quad dx=-e^{-t}dt$

Integrating, we get $\quad x=e^{-t}+c_2$ $\quad \Rightarrow \quad x-e^{-t}=c_2$ ..... (3)

From equation (2) and (3) $\quad \phi(ue^t)=x-e^{-t}$ ...... (4)

Now using given condition $\quad u(x, 0) = x$

We get $\quad \phi(ue^t)=ue^t-1$

Equation (4) become $ue^t-1=x-e^{-t}$ $\quad u=e^{-t}(x-e^{-t}+1)$

Option (b) is the correct answer.

4. Let u(x, y) solve the Cauchy problem $\frac{\partial u}{\partial y}-x\frac{\partial u}{\partial x}+u-1=0$ where $-\infty<x<\infty$, $y\geq 0$ and u(x, 0) = sin x. Then u(0, 1) is equal to **CSIR NET FEB 2022**

(a) $1-\frac{1}{e}$ (b) $1+\frac{1}{e}$ (c) e (d) 1/e

Ans. (a)

Explanation :

Consider the given PDE $\frac{\partial u}{\partial y}-x\frac{\partial u}{\partial x}+u-1=0$ .... (1)

Lagrange's auxiliary equation are $\frac{dy}{1}=\frac{dx}{-x}=\frac{du}{1-u}$

Consider $\frac{dy}{1}=\frac{dx}{-x}$

Integrating we get $e^{-y}=xc_1$ $\Rightarrow$ $\frac{e^{-y}}{x}=c_1$ ..... (2)

Now, $\frac{dy}{1}=\frac{du}{1-u}$

Integrating we get $e^{-y}=(u-1)c_2$ $\Rightarrow$ $c_2=\frac{e^{-y}}{(u-1)}$ ..... (3)

$\Rightarrow$ $\phi(c_1)=c_2$ $\phi\left(\frac{e^{-y}}{x}\right)=\frac{e^{-y}}{u-1}$ ..... (4)

Now using given condition u(x, 0) = sin x

Equation (4) become $\phi\left(\frac{1}{x}\right)=\frac{1}{\sin x-1}$

Put above value in equation (4)

We get $\frac{1}{\sin xe^{y}-1}=\frac{e^{-y}}{u-1}$ $u-1=\frac{e^{-y}(\sin xe^{y}-1)}{1}$ $u=1+e^{-y}(\sin x\, e^{y}-1)$

Now $u(0,1)=1-e^{-1}=1-\frac{1}{e}$

Option (a) is the correct answer.

5. Which of the following is a solution to $u_x+x^2u_y=0$ with $u(x, 0)=e^x$?

CSIR NET NOV 2020

(a) e^x (b) $e^{(x^3+y)^{1/3}}$ (c) $e^{(x^3-y)^{1/3}}$ (d) $(x^2y+1)e^x$

Ans. (c)

Explanation :

The lagrange's auxiliary equation is

$\frac{dx}{1}=\frac{dy}{x^2}=\frac{du}{0}$ ..... (*) From $\frac{du}{0}\Rightarrow u=c_1$ ..... (i)

$\frac{dx}{1}=\frac{dy}{x^2}=x^2dx=dy\Rightarrow\frac{1}{3}x^3=y+c_2$ ..... (ii)

Given $u(x, 0)=e^x$ We get $c_1=e^t$ ..... (iii)

We get $\frac{1}{3}t^3 = 0 + c_2 \Rightarrow c_2 = \frac{1}{3}t^3$ ..... (iv)

$\Rightarrow \quad 3c_2 = t^3 \quad \Rightarrow \quad t = (3c_2)^{1/3}$

i.e. $u(x, y) = e^x$ so option (a) is correct.

From (ii) $\quad c_2 = \frac{1}{3}x^3 - y$

Now $\quad c_1 = e^{(3c_2)^{1/3}} \Rightarrow \quad u(x,y) = e^{\left(3\left(\frac{1}{3}x^3 - y\right)\right)^{1/3}} = e(x^3 - 3y)^{1/3}$

Option (c) is correct.

6. The general solution of the surfaces which are perpendicular to the family of surfaces $z^2 = kxy$, $k \in R$ is

(a) $\phi(x^2 - y^2, xz) = 0, \phi \in C^1(R^2)$ (b) $\phi(x^2 - y^2, x^2 + z^2) = 0, \phi \in C^1(R^2)$

(c) $\phi(x^2 - y^2, 2x^2 + z^2) = 0, \phi \in C^1(R^2)$ (d) $\phi(x^2 + y^2, 3x^2 - z^2) = 0, \phi \in C^1(R^2)$

Ans. (c) **CSIR NET JUNE 2020**

Explanation :

We need to find the general solution of the surfaces which are perpendicular to the family of surfaces $z^2 = kxy$, $k \in R$.

Result: The integral surface of Quisi – linear partial differential equation $f_x u_x + f_y u_y - f_u = 0$ is orthogonal to the given surface $f(x, y, z) = 0$.

Given surfaces is $z^2 = kxy$, $k \in R$, $f = \frac{z^2}{xy}$.

The integral curves are given by

$$\frac{dx}{\frac{\partial f}{\partial x}} = \frac{dy}{\frac{\partial f}{\partial y}} = \frac{dz}{\frac{\partial f}{\partial z}} \Rightarrow \frac{dx}{\frac{z^2}{-x^2y}} = \frac{dy}{\frac{z^2}{-xy^2}} = \frac{dz}{\frac{2z}{xy}}, \quad \Rightarrow \frac{-xdx}{z} = \frac{-ydy}{z} = \frac{dz}{2}$$

Taking pairs 1 and 2, we have, $\frac{-xdx}{z} = \frac{-ydy}{z} \quad \Rightarrow x\,dx - y\,dy = 0$

Integrating both sides we get $x^2 - y^2 = c_1$.

Taking pairs 1 and 3, we have, $\frac{-xdx}{z} = \frac{dz}{2} \quad \Rightarrow 2x\,dx + z\,dz = 0,$

Integrating both sides we get, $2x^2 + z^2 = c_2$.

Thus, the required surface is $\phi(c_1, c_2) = 0$

$\Rightarrow \phi(x^2 - y^2, 2x^2 + z^2) = 0$, hence Option (c) is correct.

7. The general solution of the equation $x\frac{\partial z}{\partial x} + y\frac{\partial z}{\partial y} = 0$ is

(a) $z = \phi\left(\frac{|x|}{|y|}\right), \phi \in C^1(R)$ (b) $z = \phi\left(\frac{x-1}{y}\right), \phi \in C^1(R)$

(c) $z = \phi\left(\frac{x+1}{y}\right), \phi \in C^1(R)$ (d) $z = \phi(|x| + |y|), \phi \in C^1(R)$

Ans. (a) **CSIR NET JUNE 2020**

Explanation :

$x\frac{\partial z}{\partial x} + y\frac{\partial z}{\partial y} = 0$ $\frac{dx}{x} = \frac{dy}{y} = \frac{dz}{0}$ $dz = 0 \Rightarrow z = c_1$ ... (1)

$\frac{dx}{x} = \frac{dy}{y} \Rightarrow \log|x| = \log|y| + \log c_2$ $\Rightarrow \log|x| - \log|y| = \log|c_2|$

$\Rightarrow \log\frac{|x|}{|y|} = \log|c_2| \Rightarrow \frac{|x|}{|y|} = c_2$ ... (2)

From (1) and (2), $c_1 = \phi(c_2) \Rightarrow z = \phi\left(\frac{|x|}{|y|}\right)$, $\phi \in c^1(R)$, Option (a) is correct.

8. The general solution of $(x + y)z\, z_x + (x - y)z\, z_y = x^2 + y^2$ is **CSIRNET JUNE 2019**

(a) $F(x^2 + y^2 + z^2, z^2 - xy) = 0$ (b) $F(x^2 - y^2 - z^2, z^2 - 2xy) = 0$

(c) $F(x + y + z, z - 2xy) = 0$ (d) $F(x^3 - y^3 - z^3, z - 2x^2y^2) = 0$

Ans. (b)

Explanation :

Lagrange's auxiliary equation $\frac{dx}{(x+y)z} = \frac{dy}{(x-y)z} = \frac{dz}{x^2+y^2}$

$\frac{xdx - ydy - zdz}{x^2z - xyz - xyz + y^2z - x^2z - zy^2} = 0$ $\Rightarrow xdx - ydy - zdz = 0$

$x^2 - y^2 - z^2 = c_1$

This factor is in (b) $\Rightarrow$ Only (b) option is correct.

9. The Cauchy problem $y\frac{\partial z}{\partial x} - x\frac{\partial z}{\partial y} = 0$ and $x_0(t) = \cos t$, $y_0(t) = \sin t$, $z_0(t) = 1$ has

(a) unique solution (b) no solution (c) infinite solution

(d) more than one but finite solution

Ans. (c) **CSIR NET DEC 2019**

Explanation :

In this PDE, $R = 0$

and $P = y$, $Q = -x$ then

$$\frac{P(x_0, y_o, z_0)}{\frac{dx_0}{dt}} = \frac{y_0}{\frac{dx_0}{dt}} = -\frac{\sin t}{\sin t} = -1$$

DIFFERENTIAL EQUATION

$$\frac{Q(x_0,y_o,z_0)}{\frac{dy_0}{dt}}=\frac{-x_0}{\frac{dy_0}{dt}}=\frac{-\cos t}{\cos t}=-1 \quad \Rightarrow -1=-1$$ Infinite solution.

10. The Cauchy problem $2u_x + 3u_y = 5$, $u = 1$ on the line $3x-2y = 0$ has
(a) unique solution (b) exactly two solution (c) infinitely solution
(d) no solution

Ans. (d) : **CSIR NET JUNE 2018**

Explanation :

Here P = 2, Q = 3, R = 5 $\neq 0$
and given that u = 1 on the line 3x – 2y = 0

$\Rightarrow$ 3x – 2y $\quad \Rightarrow x=\frac{2y}{3}$ put y = t

Then curve

$$x_0(t)=\frac{2t}{3}, y_0(t)=t, u_0(t)=1$$

Then

$$\frac{P(x_0,y_o,z_0)}{\frac{dx_0}{dt}}=\frac{3\times 2}{2}=3 \qquad \frac{Q(x_0,y_o,z_0)}{\frac{dy_0}{dt}}=\frac{3}{1}=3 \qquad \frac{R(x_0,y_o,z_0)}{\frac{dz_0}{dt}}=\frac{5}{0}=no-sol.$$

11. Consider the Lagrange equation $x^2\frac{\partial z}{\partial x}+y^2\frac{\partial z}{\partial y}=(x+y)z$. Then the general solution of the given equation is **CSIR NET DEC 2017**

(a) $f\left(\frac{xy}{z},\frac{x-y}{z}\right)=0$ for an arbitrary differentiable function f.

(b) $f\left(\frac{x-y}{z},\frac{1}{x}-\frac{1}{y}\right)=0$ for an arbitrary differentiable function f.

(c) $z=f\left(\frac{1}{x}-\frac{1}{y}\right)=0$ for an arbitrary differentiable function f

(d) $z=xyf\left(\frac{1}{x}-\frac{1}{y}\right)=0$ for an arbitrary differentiable function f

Ans. (a), (b), (d)

Explanation :

Given equation is $x^2\frac{\partial z}{\partial x}+y^2\frac{\partial z}{\partial y}=(x+y)z$

By Lagrange linear equation, $\frac{dx}{x^2}=\frac{dy}{y^2}=\frac{dz}{(x+y)z}$

DIFFERENTIAL EQUATION

$$\therefore \frac{dx-dy}{x^2-y^2}=\frac{dz}{(x+y)z}\Rightarrow\frac{dx-dy}{x-y}=\frac{dz}{z}\Rightarrow log(x-y)=log\,z+log\,c_1\Rightarrow\frac{x-y}{z}=c_1$$

and $\frac{dx}{x^2}=\frac{dy}{y^2}\Rightarrow\frac{-1}{x}=\frac{-1}{y}+c_2\Rightarrow\frac{1}{x}-\frac{1}{y}=c_2\;\therefore f\left(\frac{x-y}{z},\frac{1}{x}-\frac{1}{y}\right)=0$

Thus, option (b) is correct

$$\because x-y=zc_1\Rightarrow\frac{1}{x}-\frac{1}{y}=\frac{y-x}{xy}=\frac{-zc_1}{xy}$$

$\therefore \frac{1}{x}-\frac{1}{y}=c_2$ can be written as $\frac{z}{xy}=c_3\Rightarrow\frac{z}{xy}=f\left(\frac{1}{x}-\frac{1}{y}\right), ie., z=xyf\left(\frac{1}{x}-\frac{1}{y}\right)$

or $\quad \frac{xy}{z}=c_4$

$\therefore$ option (a) & (d) are correct.

Now for option (c), if $z=f\left(\frac{1}{x}-\frac{1}{y}\right)$, then $p=f'\left(\frac{1}{x}-\frac{1}{y}\right)\left(\frac{-1}{x^2}\right)$ $q=f'\left(\frac{1}{x}-\frac{1}{y}\right)\left(\frac{1}{y^2}\right)$

$\therefore$ By given equation $x^2p+y^2q=(x+y)z$

$$\Rightarrow x^2f'\left(\frac{1}{x}-\frac{1}{y}\right)\left(\frac{-1}{x^2}\right)+y^2f'\left(\frac{1}{x}-\frac{1}{y}\right)\left(\frac{1}{y^2}\right)=(x+y)z$$

$$\Rightarrow -f'\left(\frac{1}{x}-\frac{1}{y}\right)+f'\left(\frac{1}{x}-\frac{1}{y}\right)=(x+y)z\Rightarrow 0=(x+y)z$$

which is not true.

UNSOLVED QUESTIONS

1. Let a be a fixed real constant. Consider the first order partial differential equation $\frac{\partial u}{\partial t}+a\frac{\partial u}{\partial x}=0,\ x\in R,\ t>0$ with the initial data $u(x, 0) = u_0(x)$, $x\in R$, where u_0 is a continuously differentiable function. Consider the following two statements

S_1 : There exists a bounded function u_0 for which the solution u is unbounded.

S_2 : If u_0 vanishes outside a compact set then for each fixed $T > 0$ there exists a compact set $K_T \subset R$ such that u(x, T) vanishes for $x \notin K_T$. **CSIR UGC JUNE – 2018**

(a) S_1 is true, S_2 is false
(b) S_1 is true, S_2 is true
(c) S_1 is false, S_2 is true
(d) S_1 is false, S_2 is false

2. The solution of the partial differential equation $u_t-xu_x+1-u=0, x<R, t>0$ subject to $u(x,0)=g(x)$ is **CSIR NET JUNE 2017**

(a) $u(x,t)=1-e^{-t}(1-g(xe^t))$
(b) $u(x,t)=1+e^t(1-g(xe^t))$

DIFFERENTIAL EQUATION

(c) $u(x,t)=1-e^{-t}\left(1-g\left(xe^{-t}\right)\right)$ (d) $u(x,t)=e^{-t}\left(1-g\left(xe^{t}\right)\right)$

3. A general solution of PDE $uu_x + yu_y = x$ is of form **CSIR NET JUNE 2011**

(a) $f(u^2 - x^2, y/(x + u)) = 0$ (b) $u^2 = g(y/(x+u)) + x^2$

(c) $f(x^2 + y^2) = 0$ (d) $f(x + y) = 0$

4. For an arbitrary continuously differentiable function f, which of the following is a general solution of $z(px-qy)=y^2-x^2$ **CSIR NET JUNE 2015**

(a) $x^2+y^2+z^2=f(xy)$ (b) $(x+y)^2+z^2=f(xy)$

(c) $x^2+y^2+z^2=f(y-x)$ (d) $x^2+y^2+z^2=g\left((x+y)^2+z^2\right)$

5. The differential equation $x\frac{\partial u}{\partial x}+y\frac{\partial u}{\partial y}=2u$ satisfying the initial condition $y=xg(x)$, $u=f(x)$ with **CSIR NET DEC 2011**

(a) $f(x)=2x, g(x)=1,$ has no solution

(b) $f(x)=2x^2, g(x)=1,$ has infinite number of solutions

(c) $f(x)=x^3, g(x)=x,$ has a unique solution.

(d) $f(x)=x^4, g(x)=x,$ has a unique solution.

6. The Cauchy problem

$$\left.\begin{array}{l} y\frac{\partial u}{\partial x}-x\frac{\partial u}{\partial y}=0 \\ u=g \text{ on } \Gamma \end{array}\right\}$$

has a unique solution in a neighbourhood of Γ for every differentiable function $g:\Gamma\to R$ if **CSIR NET DEC 2016**

(a) $\Gamma=\{(x,0):x>0\}$ (b) $\Gamma=\{(x,y):x^2+y^2=1\}$

(c) $\Gamma=\{(x,y);x+y=1,x>1\}$ (d) $\Gamma=\{(x,y):y=x^2,x>0\}$

7. The IVP $z\frac{\partial z}{\partial x}+\frac{\partial z}{\partial y}=1$ with the initial curve data $x_0 = S$, $y_0 = S$, $z_0=\frac{S}{2}$ has unique solution if

(a) $0 \le S \le 2$ (b) $0 \le S < 2$ (c) $0 < S \le 2$ (d) For any S

8. Find the P.I. of $x(y - z)p + y(z - x)q = z(x - y)$ which represent a surface passing through $x = y = z$.

(a) $(x + y + z)^3 = 27xyz$ (b) $(x + y + z) = 27xyz$

(c) $(x + y + z)^2 = 27xyz$ (d) None of these

DIFFERENTIAL EQUATION

9. Let px(x + y) – qy(x + y) = -(x – y)(2x + 2y + z) be a PDE then its solution
(a) u = xy, v = (x + y)(x + y + z) (b) u = x/y, v = (x + y)(x + y + z)
(c) u = xy, v = (x + y)/(x + y + z) (d) u = x/y, v = (x + y + z) / (x + y)

10. Consider the PDE $\frac{\partial u}{\partial t}+u\frac{\partial u}{\partial x}=0$ satisfying the initial condition

u(x, 0) = α + βx if u(x, t) = 1 along the x = t + 1, then
(a) α = 1, β = 1 (b) α = 2, β = 0 (c) α = 0 = β (d) α = 0, β = 1

11. Consider $x^2p + y^2q = (x + y)z$

(a) It is quasi – linear equation (b) $x-y=zF\left(\frac{xy}{z}\right)$ is a solution

(c) $G\left(\frac{1}{x}-\frac{1}{y},\frac{x-y}{z}\right)=0$ (d) $xyF\left(\frac{1}{x}-\frac{1}{y}\right)$ is a solution

12. The differential equation of surface $z = e^{ax+by} f(ax – by)$ is

(a) $a\left(\frac{\partial z}{\partial x}\right)+b\left(\frac{\partial z}{\partial y}\right)=2abz$ (b) $b\left(\frac{\partial z}{\partial x}\right)+a\left(\frac{\partial z}{\partial y}\right)=2z$

(c) $\left(\frac{\partial z}{\partial x}\right)+\left(\frac{\partial z}{\partial y}\right)=2abz$ (d) $b\left(\frac{\partial z}{\partial x}\right)+a\left(\frac{\partial z}{\partial y}\right)=2abz$

13. The integral surface of the PDE p + q = 1 passing through the curve x = 0, $y^2 = z$ is
(a) $yz = (x + y)^2$ (b) $x^3 – 3xy^2 = z^2 – 2yz$ (c) $(y – x)^2 = z – x$ (d) None of these

14. The largest interval in which the PDE $\frac{\partial z}{\partial x}-z\frac{\partial z}{\partial y}+z=0$ with the initially data.

$x_0 = 0$, $y_0 = 2s$, $z_0 = -2s$ has unique solution.
(a) $-1 \le s \le 1$ (b) $-\infty \le s \le 0$ (c) $0 < s < \infty$ (d) $-\infty < s < \infty$

15. The integral surface of the partial differential equation $x\frac{\partial u}{\partial x}+y\frac{\partial u}{\partial y}=0$ satisfying the condition u(1, y) = y, is given by **GATE- 2005**

(a) $u(x,y)=\frac{y}{x}$ (b) $u(x,y)=\frac{2y}{x+1}$ (c) $u(x,y)=\frac{y}{2-x}$ (d) $u(x,y)=y+x-1$

16. The partial differential equation of the family of surfaces z = (x + y) + A (xy), is
(a) xp – yq = 0 (b) xp – yq = x – y (c) xp + yq = x + y (d) xp + yq = 0

17. Consider the partial differential equation $x\frac{\partial u}{\partial x} + yu\frac{\partial u}{\partial y} = -xy$ for x > 0 subject to u = 5 on xy = 1. Then **CSIR NET JUNE 2017**

(a) u (x, y) exists when xy ≤ 19 and u(x, y) = u (y, x) for x > 0, y > 0
(b) u (x, y) exists when xy ≥ 19 and u(x, y) = u (y, x) for x > 0, y > 0
(c) u (1, 11) = 3, u (13, -1) = 7
(d) u (1, -1) = 5, u (11, 1) = -5

18. General solution of the p.d.e. $(y+z)p+(z+x)q = x+y$ is/ae

(a) $\phi\left[(x-y)^2(x+y+z),\left(\frac{x-y}{y-z}\right)\right]=0$
(b) $\frac{x-y}{y-z} = f\left[(x-y)^2(x+y+z)\right]$
(c) $(x-y)^2(x+y+z) = F\left(\frac{x-y}{y-z}\right)$
(d) $\psi\left[(y-z)^2(x+y+z),\left(\frac{y-z}{z-x}\right)\right]=0$

19. Let u(x, y) be the solution of the Cauchy problem

$x^2\frac{\partial u}{\partial x} - y^2\frac{\partial u}{\partial y} = 0, u \to e^x \text{ as } y \to \infty$. Then u(1,1)

(a) -1
(b) 0
(c) 1
(d) e^2

20. The general solution of px + qy = 3z, is given by

(a) $f\left(\frac{z}{x^3},\frac{y}{3}\right)=0$
(b) $f\left(\frac{y}{x^3},\frac{z}{x}\right)=0$
(c) $f\left(\frac{x}{y^3},\frac{z}{y}\right)=0$
(d) $f\left(\frac{x}{y},\frac{z}{x^3}\right)=0$

21. The solution of Cauchy problem $u_l + uu_s = x$, u(x, 0) = 1 is u(x, t) =

(a) x tan ht + sec ht
(b) tan ht + sec ht
(c) $(x^2 + t^2)$ sin t
(d) none of these

22. Integral surface of $x^2p + y^2q = -z^2$ which passes through hyperbola xy= x + y, z = 1, is given by

(a) xy + 2yz + xz = 3xyz
(b) yz + 2xy + xz 3xyz
(c) xy + 2zx + xz = 3xyz
(d) none of these

Answer Key

1	2	3	4	5	6	7	8	9	10
d	b	a,b	a,b,d	All	a,c,d	b	a	a	d
11	12	13	14	15	16	17	18	19	20
All	d	c	c	a	b	a,c	All	d	d
21	22								
a	b								

Chapter 8

Non-Linear Partial Differential Equations of First Order

These equations will contain p and q with powers higher than unity and the products of p and q. A solution of such an equation containing as many arbitrary constants as there are independent variables, is called the complete integral. A particular integral is obtained by giving particular values to the constant.

1. Compatible Systems of First Order Equations

Two first order partial differential equations are said to be compatible, if they have a common solution. The compatibility condition for two partial differential equations

$f(x,y,z,p,q)=0$ ....(1) and

$g(x,y,z,p,q) = 0$ ..... (2)

is $\dfrac{\partial(f,g)}{\partial(x,p)}+p\dfrac{\partial(f,g)}{\partial(z,p)}+\dfrac{\partial(f,g)}{\partial(y,q)}+q\dfrac{\partial(f,g)}{\partial(z,q)}=0$

Example. Show that partial differential equations $p^2 + q^2 = 1$, $(p^2 + q^2)x = pz$ are compatible

Solution. Let $f = p^2 + q^2 - 1 = 0, g = p^2x + q^2x - pz = 0$

$$\frac{\partial(f,g)}{\partial(x,p)}=\begin{vmatrix}0 & 2p\\ p^2+q^2 & -z+2px\end{vmatrix}=-2p(p^2+q^2)$$

$$\frac{\partial(f,g)}{\partial(z,p)}=\begin{vmatrix}0 & 2p\\ -p & -z+2px\end{vmatrix}=2p^2 \ and \ \frac{\partial(f,g)}{\partial(y,q)}=\begin{vmatrix}0 & 2q\\ 0 & 2qx\end{vmatrix}=0$$

$$\frac{\partial(f,g)}{\partial(z,q)}=\begin{vmatrix}0 & 2q\\ -p & 2qx\end{vmatrix}=2pq$$

According to compatibility condition

$$\frac{\partial(f,g)}{\partial(x,p)}+p\frac{\partial(f,g)}{\partial(z,p)}+\frac{\partial(f,g)}{\partial(y,q)}+q\frac{\partial(f,g)}{\partial(z,q)}=0$$

$$-2p^3-2pq^2+2p^3+2pq^2=0$$

Hence PDEs are compatible

DIFFERENTIAL EQUATION

Charpit's method :

Let the given PDE f(x, y, z, p, q) = 0

and we know that dz = pdx + qdy then the Charpit's auxiliary equation

$$\frac{-dp}{f_x+pf_z}=\frac{-dq}{f_y+qf_z}=\frac{dz}{pf_p+qf_q}=\frac{dx}{f_p}=\frac{dy}{f_q}$$

Working Rule while using Charpit's method

Step – 1 : Transfer all term of the given equation to LHS.

Step – 2 : Write down the Charpit's auxiliary equation

Step – 3 : Put all value in Charpit's equation.

Step – 4 : Find the value of p & q.

Step – 5 : and put the value of p and q in dz = pdx + qdy

Example : Let the PDE $z = px + qy + p^2 + q^2$ then the complete integral is

(a) $z = ax + by + a^2 + b^2$ (b) z = a

(c) z = x (d) z = bx + ay

Solution. By Charpit's equation

Given that $z = px + qy + p^2 + q^2$ $\Rightarrow f = px + qy + p^2 + q^2 - z = 0$ ...... (1)

$f_p = x + 2p$; $f_x = p$, $f_z = -1$

$f_q = y + 2q$; $f_y = q$

then $\frac{dx}{x+2p}=\frac{dy}{y+2q}=\frac{-dp}{p-p}=\frac{-dq}{q-q}\Rightarrow\frac{dp}{0}\Rightarrow dp=0\Rightarrow p=a$

$\Rightarrow \frac{dq}{0}\Rightarrow dq=0\Rightarrow q=b$

Put value of p & q in (1) then $z = ax + by + a^2 + b^2$

Standard forms of Charpit's Method

Standard Form – 1 : Equation containing p and q only

Partial Differential equation is given by f(p, q) = 0 (x, y, z are absent)

Let p = a then q = ϕ(a) Put p & q in dz = pdx + qdy

dz = adx + ϕ(a) dy z = ax + ϕ(a) y + b

Note : Singular solution doesn't exist.

Example : Let f (p,q) = p + q – pq = 0, Let p = a then a + q – aq = 0

$$a=q(a-1)\Rightarrow q=\frac{a}{a-1}$$

DIFFERENTIAL EQUATION

Put the value of p & q in dz = pdx + qdy

Then $dz = adx + \frac{a}{a-1}dy \qquad z = ax + \frac{a}{a-1}y + b$

Standard Form – 2 : If Partial differential equation involving z, p & q only

Let f (z, p, q) = 0 .....(1)

Here $p = \frac{dz}{du} \;\&\; q = a\frac{dz}{du}$

Where u = x + ay Then put in (1) $f(z, \frac{dz}{du}, a\frac{dz}{du}) = 0$

And solve this problem

Standard Form – 3 : If the equation involving x, y, p & q.

Partial Differential is given by f(x, y, p, q) = 0 (z is absent)

We have to do

$\phi(x, p) = a \quad \Rightarrow \quad p = P(x, a)$

And $\phi(y, a) = a \quad \Rightarrow \quad q = Q(y, a)$

Then dz = P(x, a)dx + Q(y, a) dy

Solve this problem and get solution.

Note : Singular solution doesn't exist.

Standard Form – 4 :

Clairaut's equation :

A first order partial differential equation is said to be of Clairaut form if it can be written in the form z = px + qy+ f(p,q) then its solution is z = ax + by + f(a, b)

SOLVED QUESTIONS

1. Consider the PDE $z = x\frac{\partial z}{\partial x} + y\frac{\partial z}{\partial y} + \frac{\partial z}{\partial x}\frac{\partial z}{\partial y}$. Which of the following statements are true? **CSIR NET NOV 2020**

(a) The complete integral is z = xa + yb + ab, a, b arbitrary constants.

(b) The complete integral is $z = ax + by + \sqrt{a^2 + b^2}\; a, b$ arbitrary constants.

(c) The particular solution passing through x = 0 and $z = y^2$ is $\left(\frac{x}{y} - y\right)^2$.

(d) The particular solution passing through x = 0 and $z = y^2$ is $\left(\frac{x}{y} + y\right)^2$.

Ans. (a), (c)

Explanation :

$z = px + qy + pq$ so it is Clairaut's equation,

so the complete integral is $z = ax + by + ab$, a, b arbitrary constants.

So, option (a) correct but not (b).

Now, $z = ax + by + ab$ passing through $x = 0$ and $z = y^2$

$z = ax + by + ab$ ..... (1)

$x = 0, z = y^2$ let $y = t, z = t^2$

So, equation (1) becomes,

$t^2 = a.0 + b.t + ab \Rightarrow t^2 = bt + ab$ .... (2)

By t discriminant , we get $a = -\frac{b}{4}$ Putting $a = -\frac{b}{4}$ in equation (1),

$$\Rightarrow \quad z = -\frac{b}{4}x + by + \left(-\frac{b}{4}\right).b \quad \Rightarrow \quad z = -\frac{b}{4}x + by - \frac{b^2}{4} \quad .... (3)$$

$$\Rightarrow \quad b^2 + (x-4y)b + 4z = 0$$

Using 'b' discriminant , we get $\therefore \quad z = \left(\frac{x}{4} - y\right)^2$

So, option (c) is correct and (d) is wrong.

2. Consider the Partial differential equation :

$$x\left(\frac{\partial u}{\partial x}\right)^2 + y\left(\frac{\partial u}{\partial y}\right)^2 + (x+y)\frac{\partial u}{\partial x}\frac{\partial u}{\partial y} - u\left(\frac{\partial u}{\partial x} + \frac{\partial u}{\partial y}\right) + 1 = 0$$

Which of the following statements are correct? **CSIR NET NOV 2020**

(a) The general solution of the PDE can be expressed in the form $u(x,y) = ax + by + \frac{1}{a+b}$, where a and b arbitrary constants.

(b) The general solution of the PDE can be expressed in the form $u(x,y) = f(ax+by) + \frac{1}{a+b}$, where a and b arbitrary constant and f is an arbitrary function.

(c) The Charpit's equation are

$$\frac{dx}{p^2+pq} = \frac{dy}{q^2+pq} = \frac{dU}{p(p^2+pa)+q(p^2+pq)} = \frac{dp}{0} = \frac{dq}{0}$$

(d) The Charpit's equation are $\frac{dx}{2px+(x+y)q-u} = \frac{dy}{2qy+(x+y)p-u} = \frac{dP}{0} =$

$$\frac{du}{p(2px+(x+y)q-u)+q(2qy+(x+y)p-u)}$$

Ans. **(a), (d)**

Explanation :

$xp^2 + yq^2 + (x + y)pq - u(p + q) + 1 = 0 \Rightarrow xp^2 + xpq + yq^2 + ypq + 1 = u(p + q)$

$\Rightarrow \quad uq(q + p) + 1 = u(p + q) \quad \Rightarrow \quad u = px + qy + \frac{1}{p+q}$

This is of the form u = px + qy + f(p, q) i.e. Clairaut's form.

So, the general solution is $u(x,y) = ax + by + \frac{1}{a+b}$.

i.e. option (a) is correct and option (b) is wrong.

Charpit's auxiliary equations are

$$\frac{dp}{f_x + pf_u} = \frac{dq}{f_y + qf_u} = \frac{du}{-(pf_p + qf_q)} = \frac{dx}{-f_p} = \frac{dy}{-f_q}$$

$$\Rightarrow \frac{dp}{p(p+q) - p(p+q)} = \frac{dq}{q(p+q) - q(p+q)} = \frac{dx}{-2px - q(x+y) + u}$$

$$= \frac{du}{p(2px + (x+y)q - u) + q(2qy + (x+y)p - u)} = \frac{dy}{-2qy - (x+y)p + u}$$

So, option (d) is correct and option (c) is wrong

3. If a complex integral of the partial differential equation $x(p^2 + q^2) = zp; p = \frac{\partial z}{\partial x}, q = \frac{\partial z}{\partial y}$ passes through the curve x = 0, $z^2 = 4y$, then the envelope of this family passing through x = 1 and y = 1 has

(a) z = -2 (b) z = 2 (c) $z = \sqrt{2 + 2\sqrt{2}}$ (d) $z = -\sqrt{2 + 2\sqrt{2}}$

CSIR NET JUNE – 2017

Ans. (c), (d)

Explanation :

Given P.D.E. is $x(p^2 + q^2 \_ = zp$

Let $f(x,y,z,p,q) = x(p^2 + q^2) - zp$

By Charpit's equations $\frac{dx}{f_p} = \frac{dy}{f_q} = \frac{dz}{pf_p + qf_q} = \frac{dp}{-(pf_z + f_x)} = \frac{dq}{-(qf_z + f_y)}$

$$\Rightarrow \frac{dx}{2xp - z} = \frac{dy}{2xq} = \frac{dz}{2x(p^2 + q^2) - zp} = \frac{dp}{-(-p^2 + p^2 + q^2)} = \frac{dq}{-(-qp)}$$

$$\Rightarrow \frac{dp}{q^2} = \frac{dq}{-pq} \Rightarrow pdp + qdq = 0 \Rightarrow p^2 + q^2 = a^2$$

from (1), $p = \frac{a^2 x}{z}$ {using $p^2 + q^2 = a^2$} $\Rightarrow q = \frac{a}{z}.\sqrt{z^2 - a^2 x^2}$

DIFFERENTIAL EQUATION

Now $dz = pdx + qdy \Rightarrow dz = \frac{a^2x}{z}dx + \frac{a}{z}\sqrt{z^2 - a^2x^2}\,dy \Rightarrow \frac{zdz - a^2xdx}{\sqrt{z^2 - a^2x^2}} = ady$

By integrating, $z^2 - a^2x^2 = (ay+b)^2 \Rightarrow z^2 = a^2x^2 + (ay+b)^2$ is the complete solution

Now given conditions are x = 0, $z^2 = 4y$

It's Parametric form is x = 0, $y = t^2$, z = 2t

∴ By complete Integral,

$4t^2 = (at^2 + b)^2 \Rightarrow a^2t^4 + 2(ab-2)t^2 + b^2 = 0$ ..(2)

⇒(2) has equal roots if discriminate = 0 $\Rightarrow 4(ab-2)^2 - 4a^2b^2 = 0$

$\Rightarrow ab = 1 \Rightarrow b = \frac{1}{a}$ $\therefore z^2 = a^2x^2 + \left(ay + \frac{1}{a}\right)^2$ $\Rightarrow a^4(x^2+y^2) + a^2(2y - z^2) + 1 = 0$

which is quadratic in 'a' which also gives envelope of the complete Integral

Take discriminate = 0

$\Rightarrow (2y - z^2)^2 - 4(x^2+y^2) = 0$

Now at x = 1, y = 1 $\Rightarrow (2-z^2)^2 - 4(1+1) = 0 \Rightarrow (2-z^2)^2 = 8 \Rightarrow 2 - z^2 = \pm 2\sqrt{2}$

$\Rightarrow \pm\sqrt{2+2\sqrt{2}} = z$ ∴ option (c), (d) are correct.

4. Which of the following are complete integrals of the partial differential equation $pqx + yq^2 = 1$? **CSIR NET JUNE 2015**

(a) $z = \frac{x}{a} + \frac{ay}{x} + b$ (b) $z = \frac{x}{b} + \frac{ay}{x} + b$

(c) $z^2 = 4(ax+y) + b$ (d) $(z-b)^2 = 4(ax+y)$

Ans. (a),(d)

Explanation :

Option (a):

$x = \frac{x}{a} + \frac{ay}{x} + b \Rightarrow p = \frac{1}{a} - \frac{ay}{x^2}$ and $q = \frac{a}{x}$ Now, $pqx + yq^2 = 1 - \frac{a^2y}{x^2} + \frac{a^2y}{x^2} = 1$

Option (d): $(z-b)^2 = 4(ax+y)$ $2(z-b)p = 4a$ $\Rightarrow p = \frac{a}{(z-b)}$

and $2(z-b)q = 4$ $\Rightarrow q = \frac{2}{z-b}$

Now, $pqx + yq^2 = \frac{4ax+4y}{(z-b)^2} = \frac{(z-b)^2}{(z-b)^2} = 1$

Similarly it can be verified that (b) and (c) does not form the solution.

Hence, option (a) and (d) are correct.

5. Consider the Cauchy problem for the Eikonal equation $p^2+q^2=1;\ p\equiv\frac{\partial u}{\partial x}, q\equiv\frac{\partial u}{\partial y}$,

$u(x,y)=0$ on $x+y=1,(x,y)\in R^2$. Then **CSIR NET JUNE 2016**

(a) The Charpit's equations for the differential equation are

$$\frac{dx}{dt}=2p;\frac{dy}{dt}=2q;\frac{du}{dt}=2;\frac{dp}{dt}=-p;\frac{dq}{dt}=-q$$

(b) The Charpit's equations for the differential equation are

$$\frac{dx}{dt}=2p;\frac{dy}{dt}=2q;\frac{du}{dt}=2;\frac{dp}{dt}=0;\frac{dq}{dt}=0$$

(c) $u(1,\sqrt{2})=\sqrt{2}$ (d) $u(1,\sqrt{2})=1$

Ans. (b),(d)

Explanation :

From Charpit's method

$$\frac{dx}{f_p}=\frac{dy}{f_q}=\frac{dz}{pf_p+qf_p}=\frac{dp}{-(f_x+pf_z)}=\frac{dq}{-(f_y+qf_z)}=dt$$

$$\frac{dx}{dt}=2p,\frac{dy}{dt}=2q,\frac{dz}{dt}=2,\frac{dp}{dt}=0,\frac{dq}{dt}=0$$

For option (c) and (d)

From standard form f(p, q) = 0

Let p = a then $a^2+q^2=1\Rightarrow q=\pm\sqrt{1-a^2}$ then $du=adx\pm\sqrt{1-a^2}dy$.

$u=ax\pm\sqrt{1-a^2}y+b$ ......(1)

Given that u(x, 1 – x) = 0

From (1) $ax\pm\sqrt{1-a^2}(1-x)+b=0$

Special case: $ax+\sqrt{1-a^2}(1-x)+b=0\ \ (a-\sqrt{1-a^2})x+(b+\sqrt{1-a^2})=0$.

Compare both sides $a^2=1-a^2\Rightarrow a^2=\frac{1}{2}$ and $b+\sqrt{1-a^2}=0\Rightarrow b=\frac{-1}{\sqrt{2}}$.

Then $u(x,y)=\frac{x}{\sqrt{2}}+\frac{y}{\sqrt{2}}-\frac{1}{\sqrt{2}}$ From (1)

$\Rightarrow u(1,\sqrt{2})=1$ Option (d) is correct.

6. Find singular solution of $z=px+qy-2\sqrt{(pq)}$

(a) xy = 1 (b) x + y = 1 (c) x – y = 1 (d) none of these

Ans. (a)

Explanation :

The complete integral is $ax+by-2\sqrt{(ab)}$ ....(1)

For Singular Integral, Differentiating (1) partially w.r.t. a and b, we have

$$0 = x-\frac{2}{2\sqrt{(ab)}}b, i.e. x = \sqrt{\left(\frac{b}{a}\right)} \quad ....(2)$$

And $$0 = y-\frac{2}{2\sqrt{(ab)}}a \text{ i.e. } y = \sqrt{\left(\frac{a}{b}\right)} \quad ....(3)$$

Eliminating a, b, the singular solution is xy = 1

7. Let z = z(x, y) be a solution of p.q = 1 passing through (0, 0, 0) then z(0, 1) is –

(a) 0 (b) 1 (c) 2 (d) 4

Ans. (a)

Explanation :

Let p = a then pq = 1 ⇒ q = 1/a dz = pdz + qdy

$dz = adx+\frac{1}{a}dy$ $z = ax+\frac{y}{a}+b$

This is passing through (0, 0, 0) ⇒ b = 0 $z = ax+\frac{y}{a}$

$a^2x - az + y = 0$ a – discriminant $z^2 - 4xy = 0$ $z = 2\sqrt{xy}$ z(0, 1) = 0

8. The general solution of $\frac{\partial z}{\partial y} = 3\left(\frac{\partial z}{\partial x}\right)^2$ is

(a) z = ax + 3ay + c (b) $z = ax + 3a^2y + c$ (c) $z = ax^2 + 3ay + c$ (d) none of these

Ans. (b)

Explanation :

Given that $q = 3p^2$ Let p = a then $q = 3a^2$ dz = pdx + qdy

$dz = adx + 3a^2dy$ $\Rightarrow z = ax + 3a^2y + c$ Option (b) is correct

UNSOLVED QUESTIONS

1. Let $u=u(x,y)$ be the complete integral of the PDE $\frac{\partial u}{\partial x}.\frac{\partial u}{\partial y} = xy$ passing through the points $(0,0,1)$ and $\left(0,1,\frac{1}{2}\right)$ in the $x-y-u$ space. Then the value of $u(x,y)$ evaluated at $(-1,1)$ is **CSIR NET DEC 2011**

(a) 0 (b) 1 (c) 2 (d) 3

2. The complete integral of the partial differential equation

$xp^3q^2 + yp^2q^3 + (p^3+q^3) - zp^2q^2 = 0$ is not **GATE-1997**

(a) $ax+by+\left(ab^{-2}+ba^{-2}\right)$

(b) $ax-by+\left(ab^{-2}-ba^{-2}\right)$

(c) $-ax+by+\left(ba^{-2}-ab^{-2}\right)$

(d) $ax+by-a\left(b^{-3}+ba^{-2}\right)$

3. Complete solution of $p^2-q^2=1$ is given by

(a) $z=ax+\left(a^2+1\right)y+C$

(b) $z=aye^{-x}$

(c) $z=ax-\left(a^2-1\right)^{1/2}y+C$

(d) none of these

4. Let $x=x(s), y=y(s), u=u(s), s\in R$ be the characteristic curve of the PDE

$$\left(\frac{\partial u}{\partial x}\right)\left(\frac{\partial u}{\partial y}\right)-u=0$$

passing through a given curve $x=0, y=\tau, u=\tau^2, \tau\in R$

Then the characteristics are given by **CSIR NET JUNE 2014**

(a) $x=3\tau\left(e^{S}-1\right), y=\frac{\tau}{2}\left(e^{-s}+1\right), u=\tau^2e^{-2s}$

(b) $x=2\tau\left(e^{-S}-1\right), y=\tau\left(2e^{2s}-1\right), u=\frac{\tau^2}{2}\left(1+e^{-2s}\right)$

(c) $x=2\tau\left(e^{S}-1\right), y=\frac{\tau}{2}\left(e^{s}+1\right), u=\tau^2e^{2s}$

(d) $x=\tau\left(e^{-S}-1\right), y=-2\tau\left(e^{-s}-\frac{3}{2}\right), u=\tau^2\left(2e^{-2s}-1\right)$

5. The Charpit's equation for the PDE **CSIR NET DEC 2014**

$$up^2+q^2+x+y=0, p=\frac{\partial u}{\partial x}, q=\frac{\partial u}{\partial y}$$

are given by

(a) $\frac{dx}{-1-p^3}=\frac{dy}{-1-qp^2}=\frac{du}{2p^2u+2q^2}=\frac{dp}{2pu}=\frac{dq}{2q}$

(b) $\frac{dx}{2pu}=\frac{dy}{2}=\frac{du}{2p^2u+2q^2}=\frac{dp}{-1-p^3}=\frac{dq}{-1-qp^2}$

(c) $\frac{dx}{up^2}=\frac{dy}{q^2}=\frac{du}{0}=\frac{dp}{x}=\frac{dq}{y}$

(d) $\frac{dx}{2q}=\frac{dy}{2pu}=\frac{du}{x+y}=\frac{dp}{p^2}=\frac{dq}{qp^2}$

6. Let u be of the boundary value problem **CSIR NET JUNE 2012**

$$\left.\begin{array}{l} u''+\frac{1}{t}u'=f(t), t\in(0,1) \\ u'(0)=a, u(1)=b \end{array}\right\}$$

Define for $x^2+y^2 \le 1, v(x,y)=u\left(\sqrt{x^2+y^2}\right)$ and $g(x,y)=f\left(\sqrt{x^2+y^2}\right)$, then v is a solution of the PDE

$$\left.\begin{array}{l} v_{xx}+v_{yy}=g \ in \ \{(x,y): x^2+y^2<1\} \\ v(x,y)=0 \quad on \ \{(x,y): x^2+y^2=1\} \end{array}\right\} \text{ if}$$

(a) $a>0$ and $b>0$ (b) $a>0$ and $b=0$ (c) $a=0$ and $b=0$ (d) $a<0$ and $b=0$

7. Consider the first order PDE **CSIR NET JUNE 2013**

$p+q=pq$ where $p \equiv \frac{\partial z}{\partial x}, q \equiv \frac{\partial z}{\partial y}$. Then which of the following are correct?

(a) The Charpit's equation for the above PDE reduce to

$$\frac{dx}{1-q}=\frac{dy}{1-p}=\frac{dz}{-pq}=\frac{dp}{p+q}=\frac{dq}{0}$$

(b) A solution of the Charpit's equation is q=b, where b is a constant

(c) The corresponding value of p is $p=\frac{b}{b-1}$

(d) A solution of the equation is $z=\frac{b}{b-1}x+by+a,$ Where a and b are constants

8. Let $u=u(x,t)$ be the solution of the Cauchy problem **CSIR NET DEC 2014**

$$\frac{\partial u}{\partial t}+\left(\frac{\partial u}{\partial x}\right)^2=1 \ x\in R, t>0 \quad u(x,0)=-x^2 \quad x\in R$$ Then

(a) $u(x,t)$ exists for all $x\in R$ and $t>0$

(b) $|u(x,t)|\to\infty$ as $t\to t^*$ for some $t^*>0$ and $x\neq 0$

(c) $u(x,t)\le 0$ for all $x\in R$ and for all $t<1/4$

(d) $u(x,t)>0$ for some $x\in R$ and $0<t<1/4$

9. Which of the following are complete integrals of the partial differential equation $pqx+yq^2=1$?

(a) $z=\frac{x}{a}+\frac{ay}{x}+b$ (b) $z=\frac{x}{b}+\frac{ay}{x}+b$

(c) $z^2=4(ax+y)+b$ (d) $(z-b)^2=4(ax+y)$

Answer Key

1	2	3	4	5	6	7	8	9
a	d	c	c	b	c	b,c,d	b,d	a,d

Chapter 9

Homogeneous Linear Equations With Constant Coefficient and Classification of Partial Differential Equation

1. Homogeneous PDE with constant coefficient :

An equation of the form $\frac{\partial^n z}{\partial x^n}+k_1\frac{\partial^n z}{\partial x^{n-1}\partial y}+\ldots+k_n\frac{\partial^n z}{\partial y^n}=F(x,y)$ ...(i)

In which k_i's are constants, is called a homogeneous linear partial differential equation of the nth order with constant coefficients. It is called homogeneous because all terms contain derivatives of the same order. This can be written as , $\phi(D, D')\, z = F(x, y)$ Its solution consists of two parts.

(i) The Complementary Function (C.F.): The complementary function is the complete solution of the equation $\phi(D, D')z = 0$. It must contain n arbitrary functions where n is the order of the differential equation.

(ii) The particular Integral (P.I): The particular integral is a particular solution (free from arbitrary constant) of $\phi(D, D')z = F(x, y)$.

a. The complete solution of above differential equation is
z = C.F. + P.I.

Rules to write complementary function:

Consider the equation $\frac{\partial^2 z}{\partial x^2}+k_1\frac{\partial^2 z}{\partial x\partial y}+k_2\frac{\partial^2 z}{\partial y^2}=0$ ....(i)

which is symbolic form is $\left(D^2+k_1DD'+k_2D'^2\right)z=0$ ...(ii)

Form the (A.E.) $m^2 + k_1m + k_2 = 0$, by putting D = m and D' = 1 in (ii). Solve the (A.E.) and find its roots. If

(a) The roots of A.E. are different say m_1 and m_2 then
$z=\phi_1(y+m_1x)+\phi_2(y+m_2x)$ is the C.F.

(b) the roots of the A.E. are equal each equal to say, m_1 then $z=\phi_1(y+m_1x)+x\phi_2(y+m_1x)$ is the C.F. In general if the A.E. has r roots equal, then $z = \phi_1(y + mx) + x\phi_2(y+mx)+ \ldots + x^{r-1}\phi_r(y + mx)$

Example. Solve $\left(D^4-2D^3D'+2DD'^3-D'^4\right)z=0$

Solution. The auxiliary equation of the given equation is $m^4 - 2m^3 + 2m - 1 = 0$
or $(m + 1)(m - 1)^3 = 0$ so that m = -1, 1, 1, 1.
Hence, the general solution of the given equation is
$z=\phi_1(y-x)+\phi_2(y+x)+x\phi_3(y+x)+x^2\phi_4(y+x)$ where ϕ_1, ϕ_2, ϕ_3 and ϕ_4 are arbitrary functions.

Rules to obtain particular integral:

(i) When $F(x,y)=e^{ax+by}$ P.I. $=\frac{1}{\phi(D,D')}e^{ax+by}=\frac{1}{\phi(a,b)}e^{ax+by}$

(i.e. put D = a and D' = b) provided $\phi(a, b) \neq 0$

If $\phi(a, b) = 0$, we have the case of failure, in that case

$$P.I.=x.\frac{1}{\frac{\partial\phi}{\partial D}}e^{ax+by} \text{ or } y.\frac{1}{\frac{\partial\phi}{\partial D'}}e^{ax+by}$$

(ii) When F(x, y) = sin (ax + by)

$$\text{P.I.} =\frac{1}{\phi(D^2,DD',D'^2)}sin(ax+by)=\frac{1}{\phi(-a^2,-ab,-b^2)}sin(ax+by)$$

(i.e., put $D^2 = -a^2$, $DD' = -ab$, $D^2 = -b^2$), provided $\phi(-a^2,-ab,-b^2)\neq 0$

If $\phi(-a^2,-ab,-b^2)=0$, then it is a case of failure and we can repeat the process of (i). A similar rule holds when F(x, y) = cos (ax + by).

Example. Solve $(D^2 - 3DD' + 2D'^2)\, z = \cos(x + 2y)$

Solution. Its auxiliary equation is given by $m^2-3m+2=0$

$\therefore$ Its C.F. is $y=\phi_1(y+x)+\phi_2(y+2x)$

$$P.I.=\frac{1}{(D^2-3DD'+2D'^2)}\cos(x+2y)=\frac{1}{-1-3(-2)+2(-4)}\cos(x+2y)$$

$$=\frac{1}{-1+6-8}cos(x+2y)=\frac{-1}{3}cos(x+2y)\left\{\because D^2=-a^2,DD'=-ab,D'^2=-b^2\right\}$$

It's integral solution is $z=\phi_1(y+x)+\phi_2(y+2x)-\frac{1}{3}cos(x+2y)$

(iii) When $F(x, y) = x^p y^q$, where p, q are positive integers

$$P.I.=\frac{1}{\phi(D,D')}x^p y^q=[\phi(D,D')]^{-1}x^p y^q$$

If $p<q, expand\ [\phi(D,D')]^{-1}$ in powers of $\frac{D}{D'}$

If $q<p, expand\ [\phi(D,D')]^{-1}$ in powers of $\frac{D'}{D}$

Also, we have $\frac{1}{D}F(x,y)=\int_{y\,constant}F(x,y)dx$ and $\frac{1}{D'}F(x,y)=\int_{x\,constant}F(x,y)dy$

Example. Solve $(D^3-D'^3)z=x^3y^3$

Solution. Its auxiliary equation is $m^3 - 1 = 0$

$\therefore$ Its C.F. is $\phi_1(y+x)+\phi_2(y+\omega x)+\phi_3(y+\omega^2x), \phi_1,\phi_2,\phi_3$ being arbitrary functions, where ω and ω^2 are complex cube roots of unity.

$$\text{Now, } P.I. = \frac{1}{\left(D^3 - D'^3\right)} x^3 y^3 = \frac{1}{D^3\left[1-\left(D'^3 / D^3\right)\right]} x^3 y^3 = \frac{1}{D^3}\left(1-\frac{D'^3}{D^3}\right)^{-1} x^3 y^3$$

$$= \frac{1}{D^3}\left(1+\frac{D'^3}{D^3}+\ldots\right) x^3 y^3$$

$$= \frac{1}{D^3}\left(x^3 y^3 + \frac{1}{D^3} 6x^3\right) = \frac{1}{D^3\left(x^3 y^3 + 6x \frac{x^6}{4\times5\times6}\right)}$$

$$= (1/120) x^6 y^3 + (1/60480) x^9 .$$

Hence, the required general solution is z = C.F. + P.I.

$$\Rightarrow z = \phi_1(y+x) + \phi_2(y+\omega x) + \phi_3\left(y+\omega^2 x\right) + (1/120)x^6 y^3 + (1/60480)x^9$$

(iv) If $f(x,y) = e^{ax+by} V(x,y)$, where V(x,y) is a function of x and y

$$P.I. = \frac{1}{\phi(D,D')} e^{ax+by} . V = e^{ax+by} \frac{1}{\phi(D+a, D'+b)} . V$$

(v) A short method when f(x, y) is a function of ax + by

We may apply a shorter method to find the particular integral.

Working rule: To get the particular integral of an equation ϕ(D, D') = z = F(ax + by) where ϕ(D, D') is a homogeneous function of D, D' of degree n.

Put ax + by = t; then integrate ϕ(t), n times with respect to t. Put a for D and b for D' in ϕ(D, D') we get ϕ(a, b). Thus $P.I. = \frac{1}{\phi(a,b)} X\, n^{th}$ integral of ϕ(t) with respect to t, where t = ax + by.

In case of failure $\frac{1}{(bD - aD')^n} F(ax+by) = \frac{x^n}{n! b^n} F(ax+by)$

(vi) Where F(x, y) = Any function

Then $P.I. = \frac{1}{\phi(D,D')} F(x,y)$ Resolve $\frac{1}{\phi(D,D')}$ into partial fractions.

$$P.I. = \frac{1}{D - mD'} F(x,y)$$

$= \int F(x, c - mx) dx$, where c is to be replaced by y + mx after integration.

2. Non-Homogeneous Linear PDE

A linear PDE which is not homogeneous, i.e. all the derivatives are not of the same order, is called a non-homogenous linear partial differential equation.

F(D, D') z= f(x, y), where F(D, D') is non-homogeneous in D and D'.

F(D, D') is not always resolvable into linear factors as in homogeneous linear equations. Therefore we classify linear differential operators F(D, D') into two following types.

(i) F(D, D') cannot be resolved into linear factors, for example $D^2 - D'$.

(ii) F(D, D') can be expressed as product of linear factors of the form $(\alpha D + \beta D' + \gamma)$ where α, β, and γ are constants.

Method of finding C.F. of non-homogeneous linear PDE

Type – 1 :

Consider F(D, D') = f(x, y)

Where F(D, D') cannot be factorized into linear factors:

In such cases, we apply-trial method.

Consider the equation $(D - D'^2)\, z = 0$ .....(i)

Let a trial solution of (i) be $z = Ae^{hx + ky}$ .....(ii)

where, A, h and k are constants.

from (ii), $Dz = \dfrac{\partial z}{\partial x} = hAe^{hx+ky}$ and $D'^2 z = \dfrac{\partial^2 z}{\partial y^2} = k^2 Ae^{hx+ky}$

putting these in (i), we get $(h-k^2)Ae^{hx+ky} = 0$ or $h = k^2$ ....(iii)

Putting the value of h in (ii), a solution (which is also C.F.) of (i) is

$z = Ae^{k^2x+ky}$ ....(iv)

Since all values of k satisfy the given equation (i), a more general solution (which is also C.F.) is taken as $z = \sum Ae^{k^2x+ky}$

where A and k are arbitrary constants.

Example. solve $(2D^4 - 3D^2D' + D'^2)z = 0$

Solution: The given equation can be written as $(2D^2 - D')(D^2 - D')z = 0$ ....(1)

Consider $(2D^2 - D')z = 0$ .....(2)

Let $z = Ae^{hx+ky}$ be a trial solution of (1). Then, we have $D^2z = Ah^2e^{hx+ky}$ and $D'z = Ake^{hx+ky}$

Putting these values in (2), we gt $A(2h^2 - k)e^{hx+ky} = 0$, so that $2h^2 - k = 0$ or $k = 2h^2$

Hence, the most general solution of (2) is $z = \sum Ae^{hx+2h^2y}$ .....(3)

Next, consider $(D^2 - D')z = 0$ ....(4)

Let $z = A'e^{h'x+k'y}$ be a trial solution of (1).

Then, we have $D^2z = A'h'^2 e^{h'x+k'y}$ and $D'z = A'k'e^{h'x+k'y}$

Putting these values in (4), we get $A'(h'-k')e^{h'x+k'y} = 0$ so that $h' - k' = 0$ or $h' = k'$

Hence, the most general solution of (3) is $z = \sum A'e^{h'x+h'y}$ ....(5)

From (3) and (5) the most general solution of (1) is $\sum Ae^{hx+2h^2y} + \sum A'e^{h'x+h'y}$

A. A', h, h' being arbitrary constants.

Type – 2 :

1. Corresponding to each non repeated factor (D – mD' – γ) the part of C.F. is $e^{\gamma x}\phi(y+mx)$
2. If the factor (D – mD'- γ) repeats k times then the part of C.F. corresponding to it is $e^{\gamma x}\left[\phi_1(y+mx)+x\phi_2(y+mx)+\ldots+x^{k-1}\phi_k(y+mx)\right]$
3. If a factor $(\beta D'+\gamma)$ occurs only once hen C.F. corresponding to it is $e^{\frac{-\gamma}{\beta}y}\phi(\beta x)$. In case βD'+γ repeats k times the part of C.F. is $e^{\frac{-\gamma_1}{\beta}y}\left[\phi_1(\beta x)+x\phi_2(\beta x)+\ldots+x^{k-1}\phi_k(\beta x)\right]$

Method of find P.I. of non-homogeneous linear partial differential equation :

$$F(D,D')z=f(x,y) \qquad P.I.=\frac{1}{F(D,D')}f(x,y)$$

Case I : When $F(x,y)=e^{ax+by}$ and $F(a,b)\neq 0$

$$P.I.=\frac{1}{F(D,D')}e^{ax+by}=\frac{1}{F(a,b)}e^{ax+by}$$

Case – II. When f(x, y) = sin (ax + by) or cos (ax + by)

$$P.I.=\frac{1}{F(D,D')}\sin(ax+by)=\frac{1}{F(D^2,DD',D'^2,D,D')}\sin(ax+by)$$ Which can be evaluated further.

Case – III. When f(x, y) = $x^m y^n$, m and n being positive integers,

$$P.I.=\frac{1}{F(D,D')}x^m y^n=[F[D,D')]^{-1}x^m y^n$$

Case – IV. When f(x, y) = Ve^{ax+by}, where V is a function of x and y, then

$$P.I.=\frac{1}{F(D,D')}(Ve^{ax+by})=e^{ax+by}\frac{1}{F(D+a,D'+b)}V$$, which can be evaluated further.

3. Classification of Partial Differential Equation

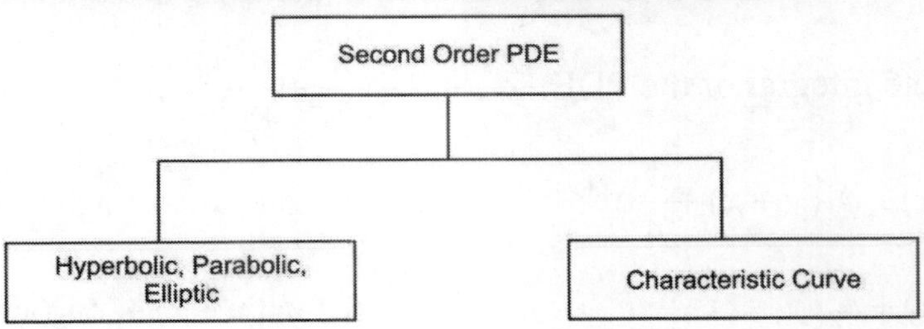

If PDE is in the form R u_{xx} + S u_{xy} + T u_{yy} + f(x, y, u, u_x, u_y) = 0

Or Rr + Ss + Tt + f(x, y, z, p, q) = 0 .... (1)

Then Equation (1) is

(a) Hyperbolic if $S^2 - 4RT > 0$

(b) Parabolic if $S^2 - 4RT = 0$

(c) Elliptic if $S^2 - 4RT < 0$

Characteristic Root's : Let $Rr + Ss + Tt + f(x, y, z, p, q) = 0$ be PDE then the roots of this equation $R\lambda^2 + S\lambda + T = 0$ ..... (2) are called characteristic roots.

Characterictic curve : If λ_1 & λ_2 are roots of (2) then $\frac{dy}{dx} + \lambda_1 = 0$ and $\frac{dy}{dx} + \lambda_2 = 0$ are called characteristic curve.

Case – 1 : If $S^2 - 4RT > 0$

i.e. equation (1) is hyperbolic then equation (2) has 2 distinct root. So we have two real family of characteristic curve.

Case – 2 : If $S^2 - 4RT < 0$ equation (1) is elliptic

Equation (2) has imaginary root.

So we have two imaginary family of curve.

Classification second order PDE with three independent variable.

Let $\sum_{i=1}^{3}\sum_{j=1}^{3} a_{ij}\frac{\partial^2 z}{\partial x_i \partial x_j} + \sum_{i=1}^{3} b_i \frac{\partial z}{\partial x_i} + cz + f = 0$ ....(1)

Where $a_{ij} = a_{ji}$ Then $A = \begin{bmatrix} a_{11} & a_{12} & a_{13} \\ a_{21} & a_{22} & a_{23} \\ a_{31} & a_{32} & a_{33} \end{bmatrix}$

(1) Equation (1) is hyperbolic if all eigen values of A are non-zero & have same sign except one of them.

(2) Equation (1) is parabolic if one of eigen value of A is zero.

(3) Equation (1) is elliptic if all eigen value of A are non-zero and same sign.

SOLVED QUESTIONS

1. The complete integral of the PDE $\frac{\partial^2 u}{\partial x^2} + 2\frac{\partial^2 u}{\partial x \partial y} + \frac{\partial^2 u}{\partial y^2} = xe^{x+y}$

(a) $\phi_1(y+x) + \phi_2(y+x) + \frac{1}{4}e^{x+y}$

(b) $\phi_1(y+x) + x\phi_2(y+x) + \frac{(x-1)}{4}e^{x+y}$

(c) $\phi_1(y-x) + \phi_2(y-x) + \frac{1}{4}e^{x+y}$

(d) $\phi_1(y-x) + x\phi_2(y-x) + \frac{(x-1)}{4}e^{x+y}$

Ans. (d) **CSIR NET DEC 2011**

Explanation :

Given that a non-homogeneous PDE with constant co-efficient.

$\Rightarrow$ Complete integral = C.F. + P.I.

For C.F.

$D^2 - 2DD` + D`^2 = 0$ Let D = m & D` = 1

$m^2 + 2m + 1 = 0$

$(m + 1)^2 = 0$

m = -1, -1

Then $u(x, y) = \phi_1(y - x) + x\phi_2(y - x)$

For P.I.

$$\frac{1}{D^2+2DD'+D'^2}xe^{x+y} = \frac{1}{(D+D')^2}xe^{x+y} \quad e^{x+y}\frac{1}{(D+D'+2)^2}x$$

$$\frac{e^{x+y}}{4}\left[1+\left(\frac{D+D'}{2}\right)\right]^{-2}x \quad \frac{e^{x+y}}{4}\left[1-2\left(\frac{D+D'}{2}\right)+......\right]x \quad \frac{e^{x+y}}{4}[x-1]$$

Complete integral

$$u(x,y) = \varphi_1(y-x) + x\varphi_2(y-x) + \frac{(x-1)}{4}e^{x+y}$$

2. $\frac{\partial^2 u}{\partial x^2} + 2\frac{\partial^2 u}{\partial x \partial y} + \frac{\partial^2 u}{\partial y^2} = x$ has

(a) Only one P.I. (b) A P.I. which is linear in x and y

(c) A P.I. which is quadratic polynomial (d) More than one P.I.

Ans. (d)

Explanation :

For P.I.

$$\frac{1}{(D+D')^2}x = \frac{1}{D^2\left[1+\frac{D'}{D}\right]^2}x = \frac{1}{D^2}\left[1-\frac{2D'}{D}+.....\right]x \qquad \frac{1}{D^2}[x-0] = \frac{x^3}{6}$$

Or another method $\frac{1}{D'^2}\left[1+\frac{D}{D'}\right]^{-2}x = \frac{1}{D'}\left[1-\frac{2D}{D'}+.....\right]x = \frac{1}{D'}[x-2y] = xy - y^2$

3. Let $P(x,y)$ be a particular integral of the partial differential equation

$\frac{\partial^2 z}{\partial x^2} - \frac{\partial z}{\partial y} = 2y - x^2$ Then $P(2,3)$ equals **CSIR NET DEC 2013**

(a) 2 (b) 8 (c) 12 (d) 10

Ans. (c)

Explanation :

DIFFERENTIAL EQUATION

$$P(x,y)=\frac{1}{D^2-D'}(2y-x^2)=\frac{1}{D^2}\left[1-\frac{D'}{D^2}\right]^{-1}(2y-x^2)=\frac{1}{D^2}\left[1+\frac{D'}{D^2}+...\right](2y-x^2)$$

$$=\frac{1}{D^2}[2y-x^2+x^2]=\frac{1}{D^2}[2y]=yx^2$$

$P(2,3)=3\times 4=12$ Hence, option (c) is correct.

4. Which of the following partial differential equations is NOT PARABOLIC for all x, y $\in$ R?

CSIR NET FEB 2022

(a) $x^2\frac{\partial^2 u}{\partial x\partial y}-2xy\frac{\partial u}{\partial y}+y^2=0$ (b) $x^2\frac{\partial^2 u}{\partial x^2}-2xy\frac{\partial^2 u}{\partial x\partial y}+y^2\frac{\partial^2 v}{\partial y^2}=0$

(c) $x^2\frac{\partial^2 u}{\partial x^2}+2xy\frac{\partial^2 u}{\partial x\partial y}+y^2\frac{\partial^2 v}{\partial y^2}=0$ (d) $x^2\frac{\partial^2 u}{\partial x^2}-2xy\frac{\partial^2 u}{\partial x\partial y}+y^2\frac{\partial^2 u}{\partial y^2}+x\frac{\partial u}{\partial x}+y\frac{\partial u}{\partial y}=0$

Ans. (a)

Explanation :

Consider the standard form of semi-linear PDE is

Rr + Ss + Tt + F(x,y,z,p,q) = 0 ..... (1)

For option (a)

Comparing equation (1) with $x^2\frac{\partial^2 u}{\partial x\partial y}-2xy\frac{\partial u}{\partial y}+y^2=0$. $\Rightarrow$ $S=x^2$

$S^2-4RT=x^2-0=x^2$

Which is parabolic only for x = 0 not for R – {0}

Option (a) is the correct answer.

For option (b) $x^2\frac{\partial^2 u}{\partial x^2}-2xy\frac{\partial^2 u}{\partial x\partial y}+u^2\frac{\partial^2 u}{\partial y^2}=0$

Comparing with equation (1)

$\Rightarrow$ $R=x^2$, $S=-2xy$, $T=y^2$ $\Rightarrow$ $S^2-4RT=4x^2y^2-4x^2y^2=0$

It is parabolic for all x, y $\in$ R

For option (c) $x^2\frac{\partial^2 u}{\partial x^2}+2xy\frac{\partial^2 u}{\partial x\partial y}+y^2\frac{\partial^2 u}{\partial y^2}=0$

Comparing with equation (1)

$R=x^2$, $S=2xy$, $T=y^2$ $=S^2-4RT=4x^2y^2-4x^2y^2=0$

It is parabolic for all x, y $\in$ R

For option (d)

$$x^2\frac{\partial^2 u}{\partial x^2}-2xy\frac{\partial^2 u}{\partial x\partial y}+y^2\frac{\partial^2 u}{\partial y^2}+x\frac{\partial u}{\partial x}+y\frac{\partial u}{\partial y}=0$$

Comparing with equation (1)

$R = x^2$, $S = -2xy$, $T = y^2$ $\quad = S^2 - 4RT = ux^2y^2 - ux^2y^2 = 0$

It is parabolic for all $x, y \in R$

Option (a) is correct answer.

5. Consider the partial differential equation

(i) $\frac{\partial^2 v}{\partial x^2}+2\frac{\partial^2 v}{\partial x\partial y}+(1-\text{sgn}(y))\frac{\partial^2 y}{\partial y^2}=0$ (ii) $y\frac{\partial^2 v}{\partial x^2}+x\frac{\partial^2 v}{\partial y^2}=0$

Which of the following statements are true? **CSIR NET NOV 2020**

(a) Equation (i) is parabolic for $y > 0$ and elliptic for $y < 0$

(b) Equation (i) is hyperbolic for $y > 0$ and elliptic for $y < 0$

(c) Equation (ii) is elliptic in I and III quadrant and hyperbolic in II and IV quadrant.

(d) Equation (ii) is hyperbolic in I and III quadrant and elliptic in II and IV quadrant.

Ans. (b), (c)

Explanation :

Comparing (i) with

$Ru_{xx} + Su_{xy} + Tu_{yy} + F(x,y,u,u_x,u_y) = 0$

Now $\text{sgn}(y)=\begin{cases}1 & if \quad y>0\\ 0 & if \quad y=0\\ -1 & if \quad y<0\end{cases}$

$S^2 - 4RT = \begin{cases}4-(4.1.0) & if \quad y>0\\ 4-(4.1.1) & if \quad y=0\\ 4-(4.1.2) & if \quad y>0\end{cases} = \begin{cases}4 & if \quad y>0\\ 0 & if \quad y=0\\ -4 & if \quad y<0\end{cases}$

So, hyperbolic for $y > 0$ and elliptic for $y < 0$

i.e. option (b) is correct and option (a) is wrong.

Now comparing (ii) with (*). we get $R = y$, $S = 0$, $T = x$

Now $S^2 - 4RT = 0 - (4. y. x) = -4xy$

So, $-4xy > 0$ if ($x < 0$ or $y > 0$) $-4xy<0$ if $\begin{pmatrix}x>0\,\&\,y>0\\ x<0\,\&\,y<0\end{pmatrix}$

So, $S^2 - 4RT < 0$ in I and III quadrant.

$S^2 - 4RT > 0$ in II and IV quadrant.

i.e. elliptic in I and III quadrant and hyperbolic in II and IV quadrant.

So, option (c) is correct and option (d) is wrong.

6. Consider the PDE

$$P(x,y)\frac{\partial^2 u}{\partial x^2}+e^{x^2}e^{y^2}\frac{\partial^2 u}{\partial x\partial y}+Q(x,y)\frac{\partial^2 u}{\partial y^2}+e^{2x}\frac{\partial u}{\partial x}+e^{y}\frac{\partial u}{\partial y}=0,$$

where P and Q are polynomials in two variables with real coefficients. Then which of the following is true for all choices of P and Q? **CSIR NET JUNE 2019**

(a) There exists R>0 such that the PDE is elliptic in $\{(x, y) \in R^2 : x^2 + y^2 > R\}$

(b) There exists R>0 such that the PDE is hyperbolic in $\{(x, y) \in R^2 : x^2 + y^2 > R\}$

(c) There exists R>0 such that the PDE is parabolic in $\{(x, y) \in R^2 : x^2 + y^2 > R\}$

(d) There exists R>0 such that the PDE is hyperbolic in $\{(x, y) \in R^2 : x^2 + y^2 < R\}$

Ans. **(b)**

Explanation :

In the given partial differential equation

$R=P(x,y), S=e^{x^2}e^{y^2}, T=Q(x,y)\quad \Rightarrow S^2-4RT=\left(e^{x^2+y^2}\right)^2-4PQ$

Let $P(x,y)=Q(x,y)=1 \quad \Rightarrow S^2-4RT=\left(e^{(x^2+y^2)}\right)^2-4$

For PDE to be hyperbolic, we need $S^2-4RT>0$

For PDE to be parabolic, we need $S^2-4RT=0$

For PDE to be elliptic, we need $S^2-4RT<0$

$e^{2(x^2+y^2)}-4=0 \quad \Rightarrow e^{2(x^2+y^2)}=2^2 \quad \Rightarrow \ln e^{2(x^2+y^2)}=\ln 2^2 \ \Rightarrow (x^2+y^2)=\ln 2$

If $(x^2+y^2)>\ln 2,$ then PDE is hyperbolic.

If $(x^2+y^2)<\ln 2,$ then PDE is elliptic

If $(x^2+y^2)=\ln 2,$ then PDE is parabolic.

Treating $R=\ln 2,$ we can clearly observe that option (b) is correct.

7. The PDE $x\frac{\partial^2 u}{\partial x^2}+y\frac{\partial^2 u}{\partial y^2}=0$ is **CSIR NET DEC 2016**

(a) Hyperbolic for $x>0, y<0$

(b) Elliptic for $x>0, y<0$

(c) Hyperbolic for $x>0, y>0$

(d) Elliptic for $x<0, y>0$

Ans. **(a)**

Explanation :

Comparing given equation with the standard form

$Rr+Ss+Tt+f(x,y,z,p,q)=0$ we get, $R=x, S=0, T=y \Rightarrow S^2-4RT=-4xy$

Option (a): If $x>0, y<0$ then $S^2-4RT>0 \Rightarrow$ The equation is hyperbolic.

Option (c): If $x>0, y>0$ then $S^2-4RT<0 \Rightarrow$ The equation is elliptic.

Option (d): If $x<0, y>0$ then $S^2-4RT>0$

$\Rightarrow$ The equation is hyperbolic. Hence, option (a) is correct.

8. The PDE $y^3u_{xx}-(x^2-1)u_{yy}=0$ is

(a) hyperbolic in $\{(x, y) \mid y < 0, |x| > 1\}$ (b) Parabolic in $\{(x, y) \mid y > 0, |x| > 1\}$

(c) Hyperbolic in $\{(x, y) \mid y > 0, |x| > 1\}$ (d) Hyperbolic for all pair (x, y)

Ans. **(c)**

Explanation :

$Rr + Ss + Tt = 0$ be any PDE

Where $r = u_{xx}$, $s = u_{xy}$, $t = u_{yy}$

and $S^2 - 4RT > 0$ hyperbolic

$S^2 - 4RT = 0$ parabolic

$S^2 - 4RT < 0$ elliptic

Here $+4y^3(x^2 - 1) = S^2 - 4RT$ Option (c) is correct.

Let $y > 0$ and $|x| > 1$ Then $4y^3 (x^2 - 1) > 0$ Hyperbolic

Option (a) is incorrect. Option (b) is incorrect. Option (d) is incorrect.

9. Which of the following are not characteristic curve of $y^2r - x^2t = 0$ are

(a) $x^2 + y^2 = c_1$ & $y^2 - x^2 = c_2$ (b) $xy = c_1$, $x - y = c_2$

(c) $xy = c_1$ & $y^2 - x^2 = c_2$ (d) $x^2+y^2=c_1\sqrt{2}$, $x - y = c_2$

Ans. (b), (c), (d)

Explanation :

Given that $R = y^2$, $S = 0$, $T = -x^2$ $R\lambda^2 + S\lambda + T = 0$ $y^2\lambda^2 - x^2 = 0$

$$\lambda=\pm\frac{x}{y}$$

Characteristic curve

$$\frac{dy}{dx}=\frac{x}{y} \quad \text{and} \quad \frac{dy}{dx}=\frac{-x}{y}$$

$x^2 - y^2 = c_1$ $x^2 + y^2 = c_2$

Option (b), (c) & (d) is correct.

Option (a) is incorrect.

10. For a second order PDE $(1-\sqrt{xy})u_{xx}+2u_{xy}+(1+\sqrt{xy})u_{yy}=0$. Which of the following are incorrect.

(a) hyperbolic in II and IV quadrant (b) elliptic in I and III quadrant

(c) hyperbolic in II and elliptic in IV quadrant (d) hyperbolic in I and III quadrant

Ans. **(a), (b), (c)**

Explanation :

Here $S = 2$, $R=1-\sqrt{xy}$ $T=1+\sqrt{xy}$ $S^2 - 4RT$

$4-4(1-\sqrt{xy})(1+\sqrt{xy})$ $4 - 4(1 - xy) = 4xy$

DIFFERENTIAL EQUATION

Option (a) is correct. In II quadrant. $x < 0, y > 0$ $S^2 - 4RT < 0 \Rightarrow$ elliptic
Option (b) is correct.
In I $x > 0, y > 0$ $\Rightarrow$ $S^2 - 4RT > 0 \Rightarrow$ hyperbolic
Option (c) is correct.
Option (d) is incorrect.

UNSOLVED QUESTIONS

1. The number of real characteristic curves of the PDE **CSIR NET JUNE 2011**
$(x^2+2y)u_{xx}+(y^3-y+x)u_{yy}+x^2(y-1)u_{xy}+3u_x+u=0$ passing through the point $x=1, y=1$ is

(a) 0 (b) 1 (c) 2 (d) 3

2. The second order partial differential equation

$$\frac{(x-y)^2}{4}\frac{\partial^2 u}{\partial x^2}+(x-y)\sin(x^2+y^2)\frac{\partial^2 u}{\partial x\partial y}+\cos^2(x^2+y^2)\frac{\partial^2 u}{\partial y^2}+(x-y)\frac{\partial u}{\partial x}$$

$$+\sin^2(x^2+y^2)\frac{\partial u}{\partial y}+u=0 \text{ is}$$ **CSIR NET DEC 2011**

(a) Elliptic in the region $\left\{(x,y): x\neq y, x^2+y^2<\frac{\pi}{6}\right\}$

(b) Hyperbolic in the region $\left\{(x,y): x\neq y, \frac{\pi}{4}<x^2+y^2<\frac{3\pi}{4}\right\}$

(c) Elliptic in the region $\left\{(x,y): x\neq y, \frac{\pi}{4}<x^2+y^2<\frac{3\pi}{4}\right\}$

(d) Hyperbolic in the region $\left\{(x,y): x\neq y, x^2+y^2<\frac{\pi}{4}\right\}$

3. The variable ξ&η which reduce the PDE $u_{xx} - x^2u_{yy} = 0$ into canonical form are

(a) $y^2+\frac{x}{2}, y^2-\frac{x}{2}$ (b) $y+\frac{x^2}{2}, y-\frac{x^2}{2}$ (c) $y + x^2, y - x^2$ (d) $y^2 + x, y^2 - x$

4. The PDE $y\frac{\partial^2 u}{\partial x^2}+x\frac{\partial^2 u}{\partial y^2}=0$ is hyperbolic in

(a) II & IV quadrant (b) I & II quadrant
(c) II & III quadrant (d) II & I & IV quadrant

5. Consider the following PDE?

I : $5y_{xx} + 6u_{yy} + 2u_{yy}$
II : $2u_{xx} + 6u_{yy} + 2u_{yy}$ Then

(a) I is elliptic (b) II is hyperbolic (c) I is parabolic (d) II is elliptic.

6. The partial differential equation for the family of surfaces $z = ce^{wt}\cos(wx)$, where c and w are arbitrary constants, is

(a) $z_{xx} + z_{tt} = 0$ (b) $z_{xx} - z_{tt} = 0$ (c) $z_{xt} + z_{tt} = 0$ (d) $z_{xt} + z_{xx} = 0$

7. The particular integral of the p.d.e. $(D^2 + 3DD' + 2D'^2)\, z = x + y$ cannot be

(a) $z = \frac{1}{2}yx^2 - x^3$ (b) $z = \frac{1}{2}yx^2 + \frac{x^3}{3}$ (c) $z = yx^2 - \frac{x^2}{3}$ (d) $z = \frac{1}{2}yx^2 - \frac{x^3}{3}$

8. The particular integral of the p.d.e. $\frac{\partial^2 z}{\partial x^2} + \frac{\partial^2 z}{\partial y^2} = \cos mx.\cos ny$ is not

(a) $\frac{1}{(m^2+n^2)}\cos mx.\cos ny$ (b) $\frac{1}{(m^2+n^2)}\cos .nx.\sin ny$

(c) $-\frac{1}{(m^2+n^2)}\cos mx.\cos ny$ (d) $\frac{-1}{(m^2+n^2)}\sin mx.\sin ny$

9. Complementary function of the p.d.e. (bD – aD' – c) (bD – aD'-c) z = 0 is not

(a) $z = e^{\frac{c}{a}x}\left[\phi_1(by+ax) + y\phi_2(by+ax)\right] if\ b \neq 0$

(b) $z = e^{\frac{c}{b}x}\left[\phi_1(by+ax) + x\phi_2(by+ax)\right] if\ b \neq 0$

(c) $z = e^{\frac{c}{b}x}\left[\psi_1(by+ax) + y\psi_2(by+ax)\right] if\ a \neq 0$

(d) none of these

10. General solution of the p.d.e. $\frac{\partial^2 u}{\partial x^2} + \frac{\partial^2 u}{\partial y^2} = 0$ is not of the form

(a) $u = f(x+iy) + g(x-iy)$ (b) $u = f(x-iy) + g(x-iy)$

(c) $u = f(x+iy) - g(x-iy)$ (d) $u = f(x-iy) - g(x-iy)$

11. The general solution of the partial differential equation $(D^2 - D'^2 - 2D - 2D')z = 0$, where $D = \frac{\partial}{\partial x}$ and $D' = \frac{\partial}{\partial y}$ is

(a) $f(y+x) + e^{2x}g(y-x)$ (b) $e^{2x}f(y+x) + g(y-x)$

(c) $e^{-2x}f(y+x) + g(y-x)$ (d) $f(y+x) + e^{-2}g(y-x)$

Answer Key

1	2	3	4	5	6	7	8	9	10	11
a	a,b	b	a	a,b	a	a,b,c	a,d	a,c,d	b,d	b

Chapter 10

Boundary Value Problem

1. Heat Equation

Homogeneous heat equation

Let $u_t = k\, u_{xx}$ $0 < x < l$ and $t > 0$ be heat equation

(1) $u(0, t) = u(l, t) = 0$ Boundary condition

(2) $u(x, 0) = f(x)$ Initial condition

Then the solution $u(x,t) = \sum D_n \sin\left(\frac{n\pi x}{l}\right) e^{-k\left(\frac{n^2\pi^2}{l^2}\right)t}$

Where $D_n = \frac{2}{l}\int_0^l f(x)\sin\left(\frac{n\pi x}{l}\right)dx$

Note :

(1) Homogeneous boundary condition $u(0, t) = 0 = u(l, t)$

(2) Non-homogeneous boundary condition $u(0, t) = C_1$ and $u(l, t) = C_2$

(3) Homogeneous heat equation $u_t - k\, u_{xx} = 0$

(4) Non-homogeneous heat equation $u_t - ku_{xx} = f(x, t)$

Non-homogeneous heat equation :

Let $u_t - ku_{xx} = f(x, t)$ (1) $u(0, t) = u(l, t) = 0$ (2) $u(x, 0) = \phi(x)$

Then the solution

$$u(x,t) = \sum_{n=1}^{\infty} D_n \sin\left(\frac{n\pi x}{l}\right) e^{-k\left(\frac{n^2\pi^2}{l^2}\right)t} + \sum_{n=1}^{\infty} \sin\left(\frac{n\pi x}{l}\right) \int_0^t e^{-k\left(\frac{n^2\pi^2}{l^2}\right)(t-s)} f_n(s)ds$$

Where $D_n = \frac{2}{l}\int_0^l \phi(x)\sin\frac{n\pi x}{l}dx$ & $f_n(S) = \frac{2}{l}\int_0^l f(x,t)\sin\left(\frac{n\pi x}{l}\right)ds$

Heat equation with non-homogeneous boundary condition

Let $u_t = ku_{xx}$ $0 < x < l$ and $t > 0$

(1) $u(0, t) = C_1$ & $u(l, t) = C_2$ (2) $u(x, 0) = f(x)$

Then we use a transformation $u(x, t) = v(x, t) + Ax + B$ ....(1)

where $v(x, 0) = g(x)$ and $v(0, t) = v(l, t) = 0$

From boundary condition find the value of A & B.

Then find the solution for v(x,t) as like a homogeneous heat equation.

The put v(x,t) in (1) and get final solution.

SOLVED QUESTIONS

1. Let u be the unique solution of $\frac{\partial u}{\partial t} = \frac{\partial^2 u}{\partial x^2}$ where (x, t) ∈ (0, 1) × (0, ∞), u(x, 0) = sin πx, x ∈ (0, 1) u(0, t) = u(1, t) = 0 t ∈ (0, ∞) Then which of the following are true?

(a) ∃ (x, t) ∈ (0, 1) × (0, ∞) s.t. u(x, t) = 0

(b) ∃ (x, t) ∈ (0, 1) × (0, ∞) s.t. $\frac{\partial u}{\partial t}(x,t) = 0$

(c) The function $e^t u(x, t)$ is bounded for (x, t) ∈ (0, 1) × (0, ∞)

(d) ∃ (x, t) ∈ (0, 1) ×(0, ∞) s.t. u(x, t) > 1

Ans. (c) **CSIR NET JUNE 2019**

This is homogeneous heat equation

$$u(x,t) = \sum D_n \sin\left(\frac{n\pi x}{l}\right) e^{-k\left(\frac{n^2\pi^2}{l^2}\right)t}$$

Here l = 1 & k = 1

then $u(x,t) = \sum D_n \sin n\pi x \, e^{-n^2\pi^2 t}$ ….. (1)

Given that u(x, 0) = sin πx

take t = 0 in (1)

u(x, 0) = sin πx = $\sum D_n \sin n\pi x$

Comparision both side

⇒ n = 1, $D_1 = 1$

n ≠ 1, $D_n = 0$

⇒ From (1)

$u(x,t) = \sin \pi x \, e^{-\pi^2 t}$ is solution.

Option (a) is incorrect.

We know that $e^{-\pi^2 t} \neq 0$ ∀ t ∈ (0, ∞)

and sinπx ≠ 0 ∀ x ∈ (0, 1)

⇒ u(x, t) ≠ 0 ∀(x, t) ∈ (0, 1) × (0, ∞)

Option (b) is incorrect.

$$\frac{\partial u}{\partial t} = -\pi^2 \sin \pi x \, e^{-\pi^2 t} \neq 0$$

∀ (x, t) ∈ (0, 1) × (0, ∞)

Option (c) is correct.

$e^t u(x, t) = \sin \pi x \, e^{(1-\pi^2)t}$

We have to check it is bounded on x ∈ (0, 1) & t ∈ (0, ∞)

Here sin πx is bounded in (0, 1)

and $e^{(1-\pi^2)t}$ is also bounded t ∈ (0, ∞) $1 - \pi^2 < 0$ $\Rightarrow \lim_{t\to\infty} e^{(1-\pi^2)t} = 0$

Option (d) is incorrect.

We know that

$e^{-\pi^2 t} < 1$ ∀ t ∈ (0, ∞)

sin πx < 1 ∀ x ∈(0, π)

⇒ $\sin \pi x . e^{-\pi^2 t} < 1$ ∀ (x, t) ∈ (0, 1) × (0, ∞)

2. If u(x, t) is a solution of $u_t = u_{xx}$ 0 < x < 1 t > 0

u(x, 0) = 1 + x + sin (πx) cos (πx)

u(0, t) = 1, u(1, t) = 2. Then **CSIR NET JUNE 2018**

(a) $u\left(\frac{1}{2},\frac{1}{4}\right)=\frac{3}{2}$ (b) $u\left(\frac{1}{4},\frac{3}{4}\right)=\frac{5}{4}+\frac{1}{2}e^{-3\pi^2}$ (c) $u\left(\frac{1}{2},\frac{1}{2}\right)=\frac{3}{2}$ (d) $u\left(\frac{1}{4},1\right)=\frac{5}{4}+\frac{1}{2}e^{-4\pi^2}$

Ans. **(a), (b), (c), (d)**

Explanation :

This is a question of non-homogeneous boundary condition. Let transformation

u(x, t) = v(x, t) + Ax + B ..... (1)

and v(0, t) = 0 = v (1, t) [homogeneous boundary condition]

then u(0, t) = v(0, t) + B = 1 = 0 + B ⇒ B = 1

and u(1, t) = v(1, t) + A + 1 = 2 = A + 1 ⇒ A = 1

Then from (1)

u(x, t) = v(x, t) + x + 1 .....(2)

Given that

u(x, 0) = 1 + x + sin (πx) cos(πx)

= v(x, 0) + x + 1⇒ v(x, 0) = sin (πx) cos(πx) ⇒ v(x, 0) = $\frac{1}{2}$ sin(2πx) .....(3)

Here new bounded condition v(0, t) = v(1, t) = 0

and initial condition $v(x,0)=\frac{1}{2}\sin 2\pi x$ then $v(x,t)=\sum D_n \sin n\pi x\, e^{-n^2\pi^2 t}$ .....(4)

where $l = 1 = k$ $v(x,0)=\sum D_n \sin(n\pi x)$

Compare (3) and (4)

If n = 2 then D_2 = ½ n ≠ 2 then D_n = 0 then $v(x,t)=\frac{1}{2}\sin 2\pi x\ xe^{-4\pi^2 t}$

this value put in (2) then $u(x,t)=1+x+\frac{1}{2}\sin 2\pi x\, e^{-4\pi^2 t}$

Option (a) is correct.

$$u\left(\frac{1}{2},\frac{1}{4}\right)=1+\frac{1}{2}+\frac{1}{2}\sin 2\pi\left(\frac{1}{2}\right)e^{-4\pi^2\times\frac{1}{4}}=\frac{3}{2}$$

Option (b) is correct. $u\left(\frac{1}{2},\frac{1}{2}\right)=1+\frac{1}{2}+\frac{1}{2}\sin 2\pi\left(\frac{1}{2}\right)e^{\frac{-4\pi^2}{2}}=\frac{3}{2}$

Option (c) is correct. $u\left(\frac{1}{4},\frac{3}{4}\right)=1+\frac{1}{4}+\frac{1}{2}\sin 2\pi\left(\frac{1}{4}\right)e^{-4\pi^2\times\frac{3}{4}}=\frac{5}{4}+\frac{1}{2}e^{-3\pi^2}$

Option (d) is correct. $u\left(\frac{1}{4},1\right)=1+\frac{1}{4}+\frac{1}{2}\sin 2\pi\left(\frac{1}{4}\right)e^{-4\pi^2(1)}=\frac{5}{4}+\frac{1}{2}e^{-4\pi^2}$

3. For the diffusion problem $u_t = u_{xx}$

$u(0, t) = u(\pi, t) = 0$ and $u(x, 0) = 3 \sin 2x$ then solution is given by

(a) $3e^{-t}\sin2x$ (b) $3e^{-9t}\sin2x$ (c) $3e^{-4t}\sin2x$ (d) $3e^{-2t}\sin2x$

Ans. (c)

Explanation :

We know that a solution of heat equation $u(x,t)=\sum D_n \sin\frac{n\pi x}{l}e^{-k\left(\frac{n^2\pi^2}{l^2}\right)t}$ .... (1)

Here $l = \pi$ & k = 1 $u(x,t)=\sum D_n \sin(nx)e^{-n^2t}$ ..... (2)

Initially condition $u(x, 0) = 3 \sin 2x$ From (2)

$u(x,0)=\sum D_n \sin(nx)=3\sin 2x$ n = 2, $D_2 = 3$ $D_n = 0$ $n \neq 2$

Then from (2) $u(x, t) = 3\sin2x\, e^{-4t}$

Option (c) is correct.

4. Let u be a solution of the heat equation

$$\left.\begin{aligned} &u_t-u_{xx}=0 \quad ; \quad 0<x<\pi \text{ and } t>0\\ &u(0,t)=u(\pi,t)=0 \quad ; \quad t>0\\ &u(x,0)=\sin x+\sin 2x \; ; \quad 0\le x<\pi\end{aligned}\right\} . \text{ Then}$$

(a) $u(x,t)\to 0$ as $t\to\infty$ for all $x\in(0,\pi)$

(b) $t^2u(x,t)\to 0$ as $t\to\infty$ for all $x\in(0,\pi)$

(c) $e^tu(x,t)$ is a bounded function for $x\in(0,\pi),t>0$

(d) $e^{2t}u(x,t)\to 0$ as $t\to\infty$ for all $x\in(0,\pi)$

Ans. (a),(b),(c) **CSIR NET JUNE 2012**

Explanation :

The solution of the problem is given by

$$u(x,t)=\sum_{n=1}^{\infty} D_n \sin\left(\frac{n\pi x}{l}\right)e^{-k\left(\frac{n^2\pi^2}{l^2}\right)t}$$

DIFFERENTIAL EQUATION

Here $l = \pi$ and k = 1 then $u(x,t) = \sum_{n=1}^{\infty} D_n \sin nx\, e^{-n^2 t}$ ... (1)

Put t = 0 in (1) $u(x,0) = \sum_{n=1}^{\infty} D_n \sin nx$

Given that u(x, 0) = sin x + sin 2x

Compare and we get

$D_1 = 1$ for n = 1, $D_2 = 1$ for n = 2 and $D_n = 0$ for all n except 1 & 2.

Then solution $u(x,t) = \sin x e^{-t} + \sin 2x e^{-4t}$

(a) As $t \to \infty, u(x,t) \to 0$ (b) $t^2 u(x,t) = \frac{t^2}{e^t}\sin x + \frac{t^2}{e^{4t}}\sin 2x \to 0$ as $t \to \infty$

(c) $e^t u(x,t) = \sin x + \sin 2x e^{-3t}$ (d) $e^{2t} u(x,t) = \sin x e^t + \sin 2x e^{-2t} \to \infty$ as $t \to \infty$

Hence, options (a),(b) and (c) are correct.

5. Let $u(x,t)$ satisfy the initial boundary value problem

$$\frac{\partial u}{\partial t} = \frac{\partial^2 u}{\partial x^2}; x \in (0,1), t > 0, \quad u(x,0) = \sin(\pi x); x \in [0,1], \quad u(0,t) = u(1,t) = 0, t > 0$$

Then for $x \in (0,1), u\left(x, \frac{1}{\pi^2}\right)$ is equal to **CSIR NET JUNE 2016**

(a) $e\sin(\pi x)$ (b) $e^{-1}\sin(\pi x)$ (c) $\sin(\pi x)$ (d) $\sin(\pi^{-1}x)$

Ans. (b)

Explanation :

The solution for the given heat problem is $u(x,t) = \sum_{n=1}^{\infty} D_n \sin\left(\frac{n\pi x}{l}\right) e^{-k(n^2\pi^2/l^2)t}$

where, $D_n = \frac{2}{l}\int_0^l \sin(\pi x)\sin(n\pi x)dx$

Here, $l = 1, k = 1 \Rightarrow D_1 = 1, D_i = 0; \forall i \neq 1$ $\quad \therefore u(x,t) = \sin(\pi x)e^{-\pi^2 t}$

$u\left(x, \frac{1}{\pi^2}\right) = e^{-1}\sin \pi x$

Hence, option (b) is correct.

6. If the initial value problem for partial differential equation $\frac{\partial u}{\partial t} + \frac{\partial^2 u}{\partial x^2} = 0;$ $u(x,0) = \sin(\pi x)$, has a solution of the form $u(x,t) = \phi(t)\sin(\pi x)$, then

(a) ϕ is always negative (b) ϕ is always positive

(c) ϕ is an increasing function (d) ϕ is a decreasing function

DIFFERENTIAL EQUATION

Ans. **(b),(c)** **CSIR NET DEC 2013**

Given IVP is $\frac{\partial u}{\partial t}+\frac{\partial^2 u}{\partial x^2}=0; u(x,0)=\sin(\pi x)$

Since, $u(x,t)=\phi(t)\sin(\pi x)$, $\Rightarrow \phi(0)=1$

Also, $\frac{\partial u}{\partial t}=\phi'(t)\sin \pi x$ and $\frac{\partial^2 u}{\partial x^2}=-\phi(t)\pi^2 \sin \pi x$

Since, $\frac{\partial u}{\partial t}+\frac{\partial^2 y}{\partial x^2}=0$ $\Rightarrow \phi'(t)\sin \pi x-\pi^2\phi(t)\sin \pi x=0$

$\Rightarrow \phi'(t)=\pi^2\phi(t)$ $\Rightarrow \phi(t)=Ae^{\pi^2 t}$, where A is a constant

Since, $\phi(0)=1 \Rightarrow A=1$ $\Rightarrow \phi(t)=e^{\pi^2 t}$, which is always positive and increasing.

Hence, option (b) and (c) are correct.

2. Wave Equation

2.1 Homogenous wave equation

Let $u_{tt} = c^2u_{xx}$ $0 < x < l$; $t > 0$

B.C. $u(0, t) = u(l, t) = 0$

I.C. $u(x, 0) = \phi(x)$; $u_t(x, 0) = g(x)$

Type – 1

$u(x,0) = \phi(x)$ and $u_t(x, 0) = 0$ then $u(x,t)=\sum D_n \sin\frac{n\pi x}{l}\cos\frac{n\pi ct}{l}$

Type – 2

$u(x, 0) = 0$ and $u_t(x, 0) = g(x)$

then $u(x,t)=\sum E_n \sin\frac{n\pi x}{l}\sin\frac{n\pi ct}{l}$

Non-homogeneous wave equation

Let $u_{tt} - c^2 u_{xx} = f(x, t)$ $0 < x < l$ & $t > 0$

B.C. $u(0, t) = u(l, t) = 0$

I.C. $u(x,0) = \phi(x)$, & $u_t(x, 0) = g(x)$

Then

$$u(x,t)=\sum_{n=1}^{\infty}\sin\frac{n\pi x}{l}\left[D_n\cos\frac{n\pi ct}{l}+E_n\sin\frac{n\pi ct}{l}\right]+\sum_{n=1}^{\infty}\sin\frac{n\pi x}{l}\int_0^t f_n(s)\sin(t-s)\frac{n\pi c}{l}ds$$

Where $D_n=\frac{2}{l}\int_0^l \phi(x)\sin\frac{n\pi x}{l}dx$ $E_n=\frac{2}{n\pi c}\int_c^l g(x)\sin\frac{n\pi x}{l}dx$

$$f_n(x,t)=\frac{2}{n\pi c}\int_0^l f(x,t)\sin\frac{n\pi x}{l}dx$$

DIFFERENTIAL EQUATION

Wave Equation for infinite string

Let $u_{tt} = c^2u_{xx}$ $\quad -\infty < x < \infty$

with $u(x, 0) = f(x)$

$u_t(x, 0) = g(x)$

then **D'Alembert solution** $u(x,t)=\frac{1}{2}[f(x+ct)+f(x-ct)]+\frac{1}{2c}\int_{x-ct}^{x+ct} g(\xi)d\xi$

7. Let $u_{tt} - u_{xx} = 0$; $0 < x < 2$, $t > 0$ s.t. B.C. $u(0, t) = u(2, t) = 0$

CSIR NET 2023

$u(x, 0) = \sin \pi x + \sin 2\pi x$ & $u_t(x, 0) = 0$, $0 \le x \le 2$

(a) $u(1, 1) = -1$ (b) $u\left(\frac{1}{2},1\right)=0$ (c) $u\left(\frac{1}{2},2\right)=1$ (d) $u_t\left(\frac{1}{2},\frac{1}{2}\right)=\pi$

Ans. **(c)**

Explanation :

We know that

$$u(x,t)=\sum_{n=1}^{\infty} D_n \sin\left(\frac{n\pi x}{l}\right)\cos\left(\frac{n\pi ct}{l}\right) = \sum_{n=1}^{\infty} D_n \sin\left(\frac{n\pi x}{2}\right)\cos\left(\frac{n\pi t}{2}\right)$$

Given that $u(x, 0) = \sin \pi x + \sin 2\pi x \Rightarrow$ $\quad \sin \pi x+\sin 2\pi x=\sum_{n=1}^{\infty} D_n \sin\left(\frac{n\pi x}{2}\right)$

Hence $D_2 = 1$ & $D_4 = 1$, then $u(x, t) = \sin \pi x \cos \pi t + \sin 2\pi x \cos 2\pi t$

(a) $u(1, 1) = 0$ (b) $u\left(\frac{1}{2},1\right)=-1$ (c) $u\left(\frac{1}{2},2\right)=1+0=1$

(d) $u_t(x, t) = -\pi \sin \pi x \sin \pi t - 2\pi \sin 2\pi x \sin 2\pi t$

$u_t\left(\frac{1}{2},\frac{1}{2}\right)=-\pi+0=-\pi$

8. Let $u(x,t)$ be the solution of the initial boundary value problem

$$\frac{\partial^2 u}{\partial t^2}=\frac{\partial^2 u}{\partial x^2}, 0<x<\infty, t>0 \quad u(x,0)=\cos\left(\frac{\pi x}{2}\right), 0\le x<\infty \quad \frac{\partial u}{\partial t}(x,0)=0, 0\le x<\infty$$

Then,

(a) The value of $u(2,2)=-1$ (b) The value of $u(2,2)=1$

(c) The value of $u\left(\frac{1}{2},\frac{1}{2}\right)=\frac{1}{\sqrt{2}}$ (d) The value of $u\left(\frac{1}{2},\frac{1}{2}\right)=\frac{1}{2}$

Ans. **(b),(d)**

Explanation :
The solution of the problem is given by

$$u(x,t)=\frac{1}{2}[f(x+ct)+f(x-ct)]+\frac{1}{2c}\int_{x-ct}^{x+ct}g(s)ds$$

Here, $f(x)=\cos\left(\frac{\pi x}{2}\right), g(x)=0$

$$\Rightarrow u(x,t)=\frac{1}{2}\left[\cos\frac{\pi}{2}(x+t)+\cos\frac{\pi}{2}(x-t)\right]$$

$$\Rightarrow u(2,2)=1 \text{ and } u\left(\frac{1}{2},\frac{1}{2}\right)=\frac{1}{2}$$

Hence, option (b) and (d) are correct.

9. Let u be the solution of $u_{tt}=u_{xx}$, $(x, t) \in R \times (0, \infty)$, $u(x, 0) = f(x)$, $u_t(x, 0) = g(x)$ where f, g are in $C^2(R)$ and satisfying the following conditions

(i) $f(x) = g(x) = 0$ for $x \le 0$

(ii) $0 < f(x) \le 1$ for $x > 0$

(iii) $g(x) > 0$ for $x > 0$

(iv) $\int_0^\infty g(x)dx < \infty$

Then which of the following are true?

(a) $u(x, t) = 0$ for all $x \le 0$ & $t > 0$

(b) u is bounded on $R \times (0, \infty)$

(c) $u(x, t) = 0$ whenever $x + t < 0$

(d) $u = 0$ for some (x,t) satisfy $x + t > 0$

CSIR NET JUNE 2019

Ans. (b), (c)

Explanation :

D'Alembert's solution for wave equation

$$u(x,t)=\frac{1}{2}[f(x+t)+f(x-t)]+\frac{1}{2c}\int_{x-t}^{x+t}g(\xi)d\xi$$

For c = 1 .... (1)

Option (a) is incorrect.

Let x = -1 and t = 2 Then $x + t = 1 > 0$ $x - t = -3 < 0$

$\Rightarrow$ By property $f(x - t) = 0$ but $f(x + t) \neq 0$

$\Rightarrow u(x, t)$ may not zero for $x \le 0$ & $t > 0$

Option (b) is correct.

We have to show that u(x, t) is bounded.

For $x \in R$ $t > 0$

Case-1 : If $x \le 0$

then $f(x) = 0 = g(x)$ $\Rightarrow u(x, t)$ is bounded.

Case – 2 : If $x > 0$

Then $\left.\begin{array}{l} 0 < f(x) \le 1 \\ \int_0^\infty g(x) < \infty \end{array}\right\}$ f(x) & g(x) are bounded.

$\le u(x, t)$ is bounded.

Option (c) is correct.

Let $x + t < 0$

Given that $t > 0 \Rightarrow x < 0 \quad \Rightarrow x - t < 0$ (also negative) $\Rightarrow$ from equation (1)
$u(x, t) = 0$
because

$$f(x + t) = 0 = g(x + t)$$
$$f(x - t) = 0 = g(x - t)$$

Option (d) is incorrect.
Let $x + t > 0$ then $f(x + t) > 0 \quad \Rightarrow u(x, t) \neq 0$

10. Consider the PDE $u_{tt} = c^2 u_{xx}$ $c > 0$
s.t. $u(x, 0) = f(x)$ and $u_t(x, 0) = g(x)$, then

(a) If $g(x) = 0$ then $u(x,t) = \frac{1}{2}[f(x-ct)+f(x+ct)]$
(b) If $f(x) = \sin x$ & $g(x) = 0$ then $u(x, t) = \sin x \cos t$
(c) If $g(x) = 1$ then $u(x, t) = f(x, t) + t$ (d) None of these

Ans. **(a), (b), (c)**

Explanation :

Option (a) is correct. If $g(x) = 0$ Then $u(x,t) = \frac{1}{2}[f(x-ct)+f(x+ct)]$

Option (b) is correct. If $f(x) = \sin x$ & $g(x) = 0$

Then $\frac{1}{2}[\sin(x-ct)+\sin(x+ct)]$

$= \frac{1}{2}[\sin x \cos ct - \cos x \sin ct + \sin x \cos ct + \cos x \sin ct]$ $= \sin x \cos ct$

Option (c) is correct.
If $g(x) = 1$

Then $u(x,t) = \frac{1}{2}[f(x-ct)+f(x+ct)] + \frac{1}{2c}\int_{x-ct}^{x+ct} d\xi$

$\frac{1}{2}[f(x-ct)+f(x+ct)] + \frac{1}{2c}[x+ct-x+ct)]$ $\frac{1}{2}[f(x-ct)+f(x+ct)]+t = f(x, t) + t$

11. Which of the following solution of PDE?

$$u_{tt} = u_{xx},\ x \in R, t > 0$$
$$u(x, 0) = x^2, x \in R$$
$$u_t(x, 0) = 0, x \in R$$

(a) $u(x, t) = x^2 + t^2$ (b) $u(x, t) = x^2 - t^2$ (c) $u(x, t) = 2x^2 + t^2$ (d) $u(x, t) = x^2 + 2t^2$

Ans. **(a)**

Explanation :

Using of d'Alembert's solution $u(x,t) = \frac{1}{2}[f(x+t)+f(x-t)] + \frac{1}{2}\int_{x-t}^{x+t} g(\xi)d\xi$

Here f(x) = x^2 and g(x) = 0 & c = 1 $\frac{1}{2}\left[(x+t)^2+(x-t)^2\right]-0=\frac{1}{2}[x^2+2t^2]=x^2+t^2$

$\Rightarrow u(x,t) = x^2 + t^2$ Option (a) is correct.

12. let u be the unique solution of $u_{tt} - u_{xx} = 0 \quad x \in R, t > 0$
u(x, 0) = f(x) and $u_t(x, 0) = 0$, $x \in R$, where f : R → R satisfies the relation f(x) = x(1 – x) $\forall x \in [0, 1]$ and f(x + 1) = f(x) $\forall x \in R$. Then $u\left(\frac{1}{2},\frac{5}{4}\right)$ is

(a) $\frac{1}{8}$ (b) $\frac{1}{16}$ (c) $\frac{3}{16}$ (d) $\frac{5}{16}$

Ans. (c) **CSIR NET JUNE 2018**

Explanation :

It is unbounded homogeneous wave equation

By D'Alembert solution

$$u(x,t)=\frac{1}{2}[f(x+ct)+f(x-ct)]+\frac{1}{2c}\int_{x-ct}^{x+ct}g(\xi)d\xi$$

Here c = 1

$$u(x,t)=\frac{1}{2}[f(x+t)+f(x-t)]+\frac{1}{2}\int_{x-t}^{x+t}g(\xi)d\xi$$

Then $u\left(\frac{1}{2},\frac{5}{4}\right)=\frac{1}{2}\left[f\left(\frac{1}{2}+\frac{5}{4}\right)+f\left(\frac{1}{2}-\frac{5}{4}\right)\right]+0\ [\because g(x)=0]=\frac{1}{2}\left[f\left(\frac{7}{4}\right)+f\left(\frac{-3}{4}\right)\right]$

Given that f(1 + x) = f(x) $\quad f\left(\frac{7}{4}\right)=f\left(1+\frac{3}{4}\right)=f\left(\frac{3}{4}\right)$

Put value

Then $\frac{1}{2}\left[f\left(\frac{3}{4}\right)+f\left(\frac{-3}{4}\right)\right]$ And $\quad f\left(\frac{-3}{4}\right)=f\left(\frac{-3}{4}+1\right)=f\left(\frac{1}{4}\right)$

Then $\frac{1}{2}\left[f\left(\frac{3}{4}\right)+f\left(\frac{1}{4}\right)\right]\ \frac{1}{2}\left(\frac{3}{4}\left(-\frac{3}{4}+1\right)+\frac{1}{4}\left(1-\frac{1}{4}\right)\right)=\frac{1}{2}\left[\frac{3}{4}\times\left(\frac{1}{4}\right)+\frac{1}{4}\left(\frac{3}{4}\right)\right]$

$=\frac{1}{2}\left(\frac{3}{8}\right)=\frac{3}{16}$

13. Let u(x, t) be the solution of the initial value problem
$u_{tt}-u_{xx}=0, u(x,0)=x^3, u_t(x,0)=\sin x$. Then u (π, π) is **CSIR NET DEC-2017**

(a) $4\pi^3$ (b) π^3 (c) 0 (d) 4

Ans. (a)

Explanation :

Given equation is $u_{tt} = u_{xx}$

$u(x, 0) = x^3$

$u_t(x, 0) = \sin x$

By D' Alembert Solution

$$u(x,t)=\frac{1}{2}\left[(x+t)^3+(x-t)^3\right]+\frac{1}{2}\int_{x-t}^{x+t}\sin s.ds$$

$$u(x,t)=\frac{1}{2}\left[(x+t)^3+(x-t)^3\right]+\frac{1}{2}\left[\cos(x-t)-\cos(x+t)\right]$$

Now $u(\pi,\pi)=\frac{1}{2}\left[(\pi+\pi)^3+(\pi-\pi)^3\right]+\frac{1}{2}\left[\cos 0-\cos 2\pi\right]$

$\therefore u(\pi,\pi)=4\pi^3$

$\therefore$ option (a) is correct.

Non-homogeneous wave equation for infinite string

Let $u_{tt} - c^2u_{xx} = F(x, t)$

$u(x, 0) = f(x)$ $u_t(x, 0) = g(x)$

then solution

$$u(x,t)=\frac{1}{2}\left[f(x+ct)+f(x-ct)\right]+\frac{1}{2c}\int_{x-ct}^{x+ct}g(s)ds+\frac{1}{2c}\int_0^t\int_{x-c(t-\theta)}^{x+c(t-\theta)}F(s,\theta)\,ds\,d\theta$$

14. Let u(x, t) be the function that satisfies the PDE $u_{xx} - u_{tt} = e^x + 6t$ $x \in \mathbb{R}, t > 0$ and the initial condition $u(x, 0) = \sin x$, $u_t(x, 0) = 0$ for every $x \in \mathbb{R}$. Then the values of $u\left(\frac{\pi}{2},\frac{\pi}{2}\right)$ is **CSIR NET DEC 2018**

(a) $e^{\pi/2}\left(1+\frac{1}{2}e^{\pi/2}\right)+\left(\frac{\pi^3+4}{8}\right)$

(b) $e^{\pi/2}\left(1+\frac{1}{2}e^{\pi/2}\right)+\left(\frac{\pi^3-4}{8}\right)$

(c) $e^{\pi/2}\left(1-\frac{1}{2}e^{\pi/2}\right)-\left(\frac{\pi^3+4}{8}\right)$

(d) $e^{\pi/2}\left(1-\frac{1}{2}e^{\pi/2}\right)-\left(\frac{\pi^3-4}{8}\right)$

Ans. **(c)**

Explanation :

Given that $u_{tt} - u_{xx} = -(e^x + 6t)$

By Alembert equation $u(x,t)=\frac{1}{2}\left[f(x+ct)+f(x-ct)\right]+\frac{1}{2c}\int_{x-ct}^{x+ct}g(\xi)\,d\xi$

$$+\frac{1}{2c}\int_{0}^{t}\int_{x-c(t-\theta)}^{x+c(t-\theta)}f(s,\theta)ds\,d\theta \qquad .....(1)$$

Here c = 1, f(x, t) = (e^x + 6t)

f(x) = sinx, g(x) = 0

Put value in (1)

$$u(x,t)=\frac{1}{2}[\sin(x+t)+\sin(x-t)]+\frac{1}{2}\int_{x-t}^{x+t}(0) \qquad \frac{1}{2}\int_{0}^{t}\int_{x-(t-\theta)}^{x+(t-\theta)}-(e^s+6\theta)ds\,d\theta$$

$$u\left(\frac{\pi}{2},\frac{\pi}{2}\right)=\frac{1}{2}[\sin(\pi)+\sin 0]-\frac{1}{2}\int_{0}^{\pi/2}\int_{\theta}^{\pi-\theta}(e^s+6\theta)ds\,d\theta$$

$$u\left(\frac{\pi}{2},\frac{\pi}{2}\right)=\frac{1}{2}[\sin(\pi)+\sin 0]-\frac{1}{2}\int_{0}^{\pi/2}\int_{\theta}^{\pi-\theta}(e^s+6\theta)\,ds\,d\theta$$

$$=-\frac{1}{2}\int_{0}^{\pi/2}\int_{\theta}^{\pi-\theta}e^s ds\,d\theta-\frac{1}{2}\int_{0}^{\pi/2}\int_{\theta}^{\pi-\theta}6\theta\,ds\,d\theta \qquad =\frac{1}{2}\int_{0}^{\pi/2}[e^s]_{\theta}^{\pi-\theta}d\theta-\frac{6}{2}\int_{0}^{\pi/2}[\theta s]_{\theta}^{\pi-\theta}d\theta$$

$$=\frac{-1}{2}\int_{0}^{\pi/2}[e^{\pi-\theta}-e^{\theta}]d\theta-3\int_{0}^{\pi/2}[\theta(\pi-\theta)-\theta^2]d\theta \qquad =\frac{-1}{2}\left[-e^{\pi-\theta}-e^{\theta}\right]_{0}^{\pi/2}-3\left[\frac{\pi\theta^2}{2}-\frac{\theta^3}{3}-\frac{\theta^3}{3}\right]_{0}^{\pi/2}$$

$$=\frac{+1}{2}\left[e^{\pi/2}+e^{\pi/2}-e^{\pi}-1\right]-3\left[\frac{\pi^3}{8}-\frac{2\pi^3}{24}\right] \qquad =+\frac{1}{2}\left[2e^{\pi/2}-e^{\pi}-1\right]-\frac{3\pi^3}{4}\left[\frac{1}{2}-\frac{1}{3}\right]$$

$$=+e^{\pi/2}-\frac{e^{\pi}}{2}-\frac{1}{2}-\frac{\pi^3}{8} \qquad =e^{\pi/2}\left[+1-\frac{1}{2}e^{\pi/2}\right]-\left(\frac{\pi^3+4}{8}\right)$$

15. Let u(x, t) be solution of $u_{tt}-u_{xx}=xt$ $-\infty<x<\infty$ & $t>0$

$u(x,0)=u_t(x,0)=0$, Then u(2, 3) is equal to **CSIR NET DEC 2019**

(a) 9 (b) 1 (c) 27 (d) 12

Ans. (a)

Explanation :

This is non-homogeneous wave equation for infinite string

$x\in(-\infty,\infty)$

$$u(x,t)=\frac{1}{2}\left[f(x+ct)+f(x-ct)\right]+\frac{1}{2c}\int_{x-ct}^{x+ct}g(s)ds+\frac{1}{2c}\int_{0}^{t}\int_{x-c(t-\theta)}^{x+c(t-\theta)}F(s,\theta)\,ds\,d\theta$$

here f(x) = 0 = g(x)

DIFFERENTIAL EQUATION

$$\Rightarrow u(x,t)=\frac{1}{2c}\int_{0}^{t}\int_{x-c(t-\theta)}^{x+c(t-\theta)} f(s,\theta)ds\,d\theta$$

here c = 1, F(x, t) = xt

then $u(x,t)=\frac{1}{2}\int_{0}^{t}\int_{x-(t-\theta)}^{x+(t-\theta)} s\theta\,ds\,d\theta$

$$u(2,3)=\frac{1}{2}\int_{0}^{3}\int_{-1+\theta}^{5-\theta} s\theta\,ds\,d\theta=\frac{1}{2}\int_{0}^{3}\theta\left[\frac{s^2}{2}\right]_{\theta-1}^{5-\theta}d\theta=\frac{1}{2}\times\frac{1}{2}\int_{0}^{3}\theta[(5-\theta)^2-(\theta-1)^2]d\theta$$

$$=\frac{1}{4}\int_{0}^{3}\theta(25+\theta^2-10\theta-\theta^2-1+2\theta)d\theta=\frac{1}{4}\int_{0}^{3}(24\theta-8\theta^2)d\theta=\frac{1}{4}\left[24\left(\frac{\theta^2}{2}\right)_0^3-8\left(\frac{\theta^3}{3}\right)_0^3\right]$$

$$=\frac{1}{4}\left[24\times\frac{9}{2}-8\times\frac{27}{3}\right]=\frac{1}{4}[108-72]=\frac{1}{4}[36]=9$$

16. Let $u(x,t)$ be the solution of the initial boundary value problem

$$\frac{\partial^2 u}{\partial t^2}=\frac{\partial^2 u}{\partial x^2},0<x<\infty,t>0 \qquad u(x,0)=\cos\left(\frac{\pi x}{2}\right),0\le x<\infty \qquad \frac{\partial u}{\partial t}(x,0)=0,0\le x<\infty$$

$\frac{\partial u}{\partial x}(0,t)=0,t\ge 0$ Then, **CSIR NET JUNE 2011**

(a) The value of $u(2,2)=-1$

(b) The value of $u(2,2)=1$

(c) The value of $u\left(\frac{1}{2},\frac{1}{2}\right)=\frac{1}{\sqrt{2}}$

(d) The value of $u\left(\frac{1}{2},\frac{1}{2}\right)=\frac{1}{2}$

Ans. (b),(d)

Explanation :

The solution of the problem is given by

$$u(x,t)=\frac{1}{2}[f(x+ct)+f(x-ct)]+\frac{1}{2c}\int_{x-ct}^{x+ct}g(s)ds$$

Here, $f(x)=\cos\left(\frac{\pi x}{2}\right),g(x)=0$

$$\Rightarrow u(x,t)=\frac{1}{2}\left[\cos\frac{\pi}{2}(x+t)+\cos\frac{\pi}{2}(x-t)\right]$$

$$\Rightarrow u(2,2)=1 \text{ and } u\left(\frac{1}{2},\frac{1}{2}\right)=\frac{1}{2}$$

Hence, option (b) and (d) are correct.

3. Another Method :

Solution of Partial Differential Equation by Laplace Transform

$L\left\{\frac{\partial^2 u}{\partial t^2}\right\} = s^2\overline{U} - su(x,0) - u_t(x,0)$	$L\left\{\frac{\partial^2 u}{\partial x^2}\right\} = \frac{d^2\overline{U}}{dx^2}$
$L\left\{\frac{\partial u}{\partial t}\right\} = s\overline{U} - su(x,0)$	$L\left\{\frac{\partial u}{\partial x}\right\} = \frac{d\overline{U}}{dx}$
$L\{u(x,t)\} = \overline{U}(x,s)$	

17. $\frac{\partial u}{\partial t} = 2\frac{\partial^2 u}{\partial x^2}$ given $u(0,t) = 0 = u(5,t)$ and $u(x,0) = 10\sin 4\pi x$

(a) $u(x,t) = 10\sin(4\pi x)e^{-32\pi^2 t}$ (b) $u(x,t) = 10\cos(4\pi x)e^{-32\pi^2 t}$

(c) $u(x,t) = 10\sin(4\pi x)e^{32\pi^2 t}$ (d) None of these

Ans. (a)

Explanation :

$L\left(\frac{\partial u}{\partial t}\right) = 2L\left(\frac{\partial^2 u}{\partial x^2}\right) \quad \Rightarrow s\overline{U} - u(x,0) = 2\frac{d^2\overline{U}}{dt^2}$

$\Rightarrow \frac{d^2\overline{U}}{dt^2} - \frac{s}{2}\overline{U} = \frac{-u(x,0)}{2} \Rightarrow \left(D^2 - \frac{s}{2}\right)\overline{U} = -5\sin 4\pi x \quad m^2 - \frac{s}{2} = 0 \Rightarrow m = \pm\sqrt{\frac{s}{2}}$

$\text{CF} \Rightarrow \overline{U} = c_1 e^{\left(\sqrt{s/2}\right)x} + c_1 e^{-\left(\sqrt{s/2}\right)x}$

$\text{PI} \Rightarrow \overline{U} = -\frac{5\sin 4\pi x}{\left(D^2 - s/2\right)} \Rightarrow \overline{U} = -\frac{5\sin 4\pi x}{\left(-16\pi^2 - s/2\right)}$

$\Rightarrow \overline{U} = \frac{10\sin 4\pi x}{\left(s + 32\pi^2\right)} \Rightarrow \overline{U} = CF + PI \quad \Rightarrow \overline{U} = c_1 e^{\left(\sqrt{s/2}\right)x} + c_1 e^{-\left(\sqrt{s/2}\right)x} + \frac{10\sin 4\pi x}{\left(s + 32\pi^2\right)}$

Apply the condition $u(0,t) = 0 = u(5,t) \Rightarrow c_1 = c_2 = 0$

$\Rightarrow \overline{U} = \frac{10\sin 4\pi x}{\left(s + 32\pi^2\right)} \Rightarrow L\{u(x,t)\} = \frac{10\sin 4\pi x}{\left(s + 32\pi^2\right)}$

$\Rightarrow u(x,t) = 10\sin 4\pi x L^{-1}\left[\frac{1}{\left(s + 32\pi^2\right)}\right] \Rightarrow u(x,t) = 10\sin(4\pi x).e^{-32\pi^2 t}$

18. $\frac{\partial u}{\partial t} = 3\frac{\partial^2 u}{\partial x^2}$ given $u\left(\frac{\pi}{2}, t\right) = 0 = \left(\frac{\partial u}{\partial x}\right)_{x=0}$ and $u(x,0) = 30\cos 5x$

(a) $30\cos 5x.e^{-75t}$ (b) $30\cos 5x.e^{+75t}$ (c) $10\cos 5x.e^{-75t}$ (d) $10\cos 5x.e^{+75t}$

Ans. (a)

DIFFERENTIAL EQUATION

Explanation :

$$L\left(\frac{\partial u}{\partial t}\right)=3L\left(\frac{\partial^2 u}{\partial x^2}\right)\Rightarrow s\overline{U}-u(x,0)=3\frac{d^2\overline{U}}{dt^2}$$

$$\Rightarrow\frac{d^2\overline{U}}{dt^2}-\frac{s}{3}\overline{U}=\frac{-u(x,0)}{3}\Rightarrow\left(D^2-\frac{s}{3}\right)\overline{U}=-10\cos 5x$$

$$m^2-\frac{s}{3}=0\Rightarrow m=\pm\sqrt{\frac{s}{3}}$$

$$\text{CF}\Rightarrow\overline{U}=c_1e^{\left(\sqrt{s/3}\right)t}+c_1e^{-\left(\sqrt{s/3}\right)t}$$

$$\text{PI}\Rightarrow\overline{U}=-\frac{10\cos 5x}{\left(D^2-s/3\right)}\Rightarrow\overline{U}=-\frac{10\cos 5x}{(-25-s/3)}$$

$$\Rightarrow\overline{U}=\frac{30\cos 5x}{(s+75)}$$

$$\Rightarrow\overline{U}=CF+PI$$

$$\Rightarrow\overline{U}=c_1e^{\left(\sqrt{s/3}\right)t}+c_1e^{-\left(\sqrt{s/3}\right)t}+\frac{30\cos 5x}{(s+75)}$$

Apply the condition $u\left(\frac{\pi}{2},t\right)=0=\left(\frac{\partial u}{\partial x}\right)_{x=0}\Rightarrow c_1=c_2=0$

$$\Rightarrow\overline{U}=\frac{30\cos 5x}{(s+75)}\Rightarrow L\{u(x,t)\}=\frac{30\cos 5x}{(s+75)}\Rightarrow u(x,t)=30\cos 5xL^{-1}\left[\frac{1}{(s+75)}\right]$$

$$\Rightarrow u(x,t)=30\cos 5x.e^{-75t}$$

UNSOLVED QUESTIONS

1. Let U(x, t) be the solution of $u_{tt}=u_{xx};0<x<1,t>0,u(x,0)=x(1-x)$, $u_t(x,0)=0$. Then $u\left(\frac{1}{2},\frac{1}{4}\right)$ is

(a) $\frac{3}{16}$ (b) $\frac{1}{4}$ (c) $\frac{3}{4}$ (d) $\frac{1}{16}$

2. The solution of the initial value problem $u_{tt}=4u_{xx},t>0,-\infty<x,\infty$. Satisfying the conditions u(x, 0) = x, $u_t(x, 0) = 0$, is

(a) x (b) $\frac{x^2}{2}$. (c) 2x (d) 2t

3. Let u be a solution of the initial value problem $\frac{\partial^2 u}{\partial t^2}-\frac{\partial^2 u}{\partial x^2}=0; u(x,0)=x^2, \frac{\partial u}{\partial t}(x,0)=0.$ Then u(0, 1) equals to

(a) 1 (b) -2 (c) 2 (d) $\frac{1}{2}$

4. Let U(x, t) be the solution of the initial value problem $\frac{\partial^2 u}{\partial t^2}-\frac{\partial^2 u}{\partial x^2}=0;$

$u(x,0)=\sin x; \frac{\partial u}{\partial t}(x,0)=1$. Then, $u(\pi,\pi/2)$ equals

(a) π/2 (b) 1 – (π/2) (c) 1 (d) 1 + π/ 2

5. Consider the wave equation $\frac{\partial^2 u}{\partial t^2}=4\frac{\partial^2 u}{\partial x^2}, 0<x<\pi, t>0$, with u(0, t) = u(π, t) = 0,

$u(x,0)=\sin x \text{ and } \frac{\partial u}{\partial t}=0 \text{ at } t=0. \text{ Then } u\left(\frac{\pi}{2},\frac{\pi}{2}\right)=$

(a) 2 (b) 1 (c) 0 (d) -1

6. Consider the diffusion equation $u_{xx} = u_t$ with 0 < x < π and t > 0, subject to the initial and boundary conditions ; u (x, 0) = 4 sin 2x for 0 < x < π and u (0, t) = u (π, t) for t > 0. Then, u(π/8, 1) is equal to :

(a) $4e^{-4}/\sqrt{2}$ (b) $4e^{-9}/\sqrt{2}$ (c) $4/e^2$ (d) $4/\sqrt{e}$

7. The diffusion equation $\frac{\partial^2 u}{\partial x^2}=\frac{\partial u}{\partial t}, u=u(x,t), u(0,t)=u(\pi,t)=0, u(x,0)=\cos x \sin 5x$ admits the solution

(a) $\frac{e^{-36t}}{2}\left[\sin 6x+e^{20t}\sin 4x\right]$ (b) $\frac{e^{-36t}}{2}\left[\sin 4x+e^{20t}\sin 6x\right]$

(c) $\frac{e^{-20}}{2}\left[\sin 3x+e^{15t}\sin 5x\right]$ (d) $\frac{e^{-36t}}{2}\left[\sin 5x+e^{20t}\sin x\right]$

Answer Key

1	2	3	4	5	6	7
a	a	a	a	d	b	a

DIFFERENTIAL EQUATION

Chapter 1

Calculus of Variation

1. Variational problem with fixed boundaries.

Let a curve y = y(x) where $y(x_0) = y_0$ & $y(x_1) = y_1$
Such that the given function F(x, y, y').

Then $I[y(x) = \int_{x_0}^{x_1} F(x,y,y')dx$ is called a functional.

Note :

(1) A problem in which investigate maxima or minima of a functional are known as variational problem.

(2) A function I[y(x)] is said to be linear if it satisfiesI[αy(x)] = αI[y(x)]

Euler's Equation :

For $\int_{x_1}^{x_2} F(x,y,y')dx$ extremal are given by $\frac{\partial F}{\partial y} - \frac{d}{dx}\left(\frac{\partial F}{\partial y'}\right) = 0$ ... (1)

Other form of Euler's equation

$$\frac{d}{dx}\left[F - y'\frac{\partial F}{\partial y'}\right] - \frac{\partial F}{\partial x} = 0 \quad \text{... (2)}$$

(i) Particular case of Euler's equation :

Case – 1 : If F is independent of x i.e. F = F(y, y')

Then from (2) $F - y'\frac{\partial F}{\partial y'}$ = Constant

Case – 2 : If F is independent of y' i.e. F = F(x, y) then Euler's equation

From (1) $\frac{\partial F}{\partial y} = 0$

Case – 3 : If F is independent of y i.e. F = F(x, y') then $\frac{\partial F}{\partial y'}$ = Constant

Case – 4 : If F is independent of x and y both i.e. F = F(y') then

$\frac{\partial^2 y}{\partial x^2} = 0 \quad \Rightarrow \quad y = c_1x + c_2$

(ii) When functional have a several dependent variable

i.e. $\int_{x_1}^{x_2} F(x, y_1, y_2, ..., y_n, y_1', y_2', y_n')dx$

Then Euler's equation $\frac{\partial F}{\partial y_1}-\frac{d}{dx}\left(\frac{\partial F}{\partial y_1'}\right)=0,\ \frac{\partial F}{\partial y_2}-\frac{d}{dx}\left(\frac{\partial F}{\partial y_2'}\right)=0......\frac{\partial F}{\partial y_n}-\frac{d}{dx}\left(\frac{\partial F}{\partial y_n'}\right)=0$

(iii) When functional involving higher order derivative

i.e. $\int_{x_1}^{x_2} F(x,y,y',y'',.....y^{(n)})dx$

Then Euler's equation $\frac{\partial F}{\partial y}-\frac{d}{dx}\left(\frac{\partial F}{\partial y'}\right)+\frac{d^2}{dx^2}\left(\frac{\partial F}{\partial y''}\right)+.....(-1)^n\frac{d^n}{dx^n}\left(\frac{\partial F}{\partial y^{(n)}}\right)$

(iv) When functional involve more than one independent variable

$\iint F(x,y,z,z_x,z_y)dx\,dy$

Then Euler's equation $\frac{\partial F}{\partial z}-\frac{d}{dx}\left(\frac{\partial F}{\partial z_x}\right)-\frac{d}{dy}\left(\frac{\partial F}{\partial z_y}\right)=0$

Isoperimetric Problem :

Let $I[y(x)]=\int_{x_0}^{x_1} f(x,y,y')dx$ be a functional and subject to constraint

$J[y(x)]=\int_{x_0}^{x_1} g(x,y,y')dx$ = constant

This type of problem are called a isoperimetric problem.

Working rule :

Consider F = f + λg Where λ is called Lagrange's multiplier.

Then the use of Euler's equation on F.

i.e. $\frac{\partial f}{\partial y}-\frac{d}{dx}\left(\frac{\partial f}{\partial y`}\right)=0$ .... (1)

Solve equation (1) and get solution

Example : Find the extremal of functional

$I=\int_0^{\pi}(y'^2-y^2)dx$ under the condition y(0) = 0; y(π) = 1

and subject to constraint $\int_0^{\pi} ydx=1$

Solution :

Let here f = y'² – y² and g = y Then F = f + λg = y'² – y² + λy

Euler's equation

$\frac{\partial F}{\partial y}-\frac{d}{dx}\left(\frac{\partial F}{\partial y'}\right)=0\ -2y+\lambda-\frac{d}{dx}(2y')=0\Rightarrow -2y+\lambda-2y''=0\ \Rightarrow y''+y=\frac{\lambda}{2}$ .....(1)

Solve this ODE

C.F. = $m^2 + 1 = 0$

$m = \pm i$

$y(x) = c_1 \cos x + c_2 \sin x$

P.I. $\dfrac{1}{D^2+1}\left(\dfrac{\lambda}{2}\right) = (1+D^2)^{-1}\dfrac{\lambda}{2} = (1-D^2+....)\dfrac{\lambda}{2} = \dfrac{\lambda}{2}$

y(x) = C.F. + P.I. $= c_1 \cos x + c_2 \sin x + \dfrac{\lambda}{2}$

Use condition

$y(0) = 0 \Rightarrow y(0) = c_1 + \dfrac{\lambda}{2} = 0 \qquad \Rightarrow c_1 = -\dfrac{\lambda}{2}$ ....(2)

& $y(\pi) = 1$

$\Rightarrow \quad y(\pi) = 1 = -c_1 + \dfrac{\lambda}{2} \qquad \Rightarrow c_1 = \dfrac{\lambda}{2} - 1$ ....(3)

Solving (2) & (3)

We get $\lambda = 1$ & $c_1 = -\dfrac{1}{2}$ Then $y(x) = -\dfrac{1}{2}\cos x + c_2 \sin x + \dfrac{1}{2}$ ..... (4)

Now we have to find the value c_2

Given that $\int_0^{\pi} y dx = 1 = \int_0^{\pi}\left(-\dfrac{1}{2}\cos x + c_2 \sin x + \dfrac{1}{2}\right)dx = 1 = \left[-\dfrac{1}{2}\sin x - c_2 \cos x + \dfrac{x}{2}\right]_0^{\pi} = 1$

$= c_2 + \dfrac{\pi}{2} + c_2 = 1 = 2c_2 = 1 - \dfrac{\pi}{2} \qquad c_2 = \dfrac{1}{2}\left(1 - \dfrac{\pi}{2}\right)$

Then from (4) $f(x) = -\dfrac{1}{2}\cos x + \left(\dfrac{1}{2} - \dfrac{\pi}{4}\right)\sin x + \dfrac{1}{2}$

2 Variational problem with moving boundaries :

Consider the functional $I[y(x) = \int_{x_1}^{x_2} F(x, y, y')dx$

Let us assume that one of the boundary point (x_1, y_1) is fixed while the other boundary point. (x_2, y_2) can moves on the curve $y = \phi(x)$.

Then $[F + (\phi' - y')F_{y'}]_{x=x_2} = 0$

This is called Transversality condition.

Note : If the boundary point (x_1, y_1) moves along a curve $y = \phi(x)$ and other point (x_2, y_2) moves along the curve $y = \psi(x)$, then

$[F + (\phi' - y')F_{y'}]_{x=x_1} = 0 \qquad [F + (\psi' - y')F_{y'}]_{x=x_2} = 0$

This gives required extremal of functional.

Orthogonality conditions

If one boundary point (x_2, y_2) moves on a curve $y = \phi(x)$

Then $\phi' y' = -1$

3. Sufficiency condition for an extremal is to be maximum or minimum

Legendre condition :

For achieving maximum or minimum on the extremal, central field is considered.

Considered functional $\int_{x_1}^{x_2} F(x,y,y')dx$

(a) If $\frac{\partial^2 F}{\partial y'\partial y'} > 0$ on C weak minima

(b) If $\frac{\partial^2 F}{\partial y'\partial y'} < 0$ on C weak maxima

(c) If $\frac{\partial^2 F}{\partial y'\partial y'} \geq 0$ on C strong minima.

(d) If $\frac{\partial^2 F}{\partial y'\partial y'} \leq 0$ on C strong maxima.

Weriestrass condition :

Consider, $I[y] = \int_a^b F(x,y,y')dx$ With $y(a) = y_a; y(b) = y_0$

The Werierstrass function is given by

$E(x,y,y',p) = F(x,y,y') - F(x,y,p) - (y'-p)F_p(x,y,p)$

Where, $p = \frac{dy}{dx}$ along the extremal of the field.

(1) For strong maximum is $E \leq 0$ $\forall$ y'.

(2) For weak maximum is $E \leq 0$ $\forall$ y' close to p.

(3) For strong minimum is $E \geq 0$ $\forall$y'

(4) For weak minimum is $E \geq 0$ $\forall$y' close to p.

4. Broken extremal and corner point

Let $J[y] = \int_{x_1}^{x_2} f(x,y,y')dx$ be a functional s. t. $y(x_1) = a$ and $y(x_2) = b$ where a and b are constant.

Step for extrem (broken)

(1) use the E-L equation and get extremal

(2) if this extremal doesn't satisfy this condition then we can construct a new extremal with a corner

point, i.e. $y(x) = \begin{cases} \phi(x) & x \in [x_1, x^*] \\ \psi(x) & x \in [x^*, x_2] \end{cases}$ where x^* is corner point.

(3) if this new function is not diff then this functional has broken extremal.

Example : $J[y]=\int_{-1}^{1} y^2(1-y')^2 \quad y(-1)=0\, and\, y(1)=1$

Solution : F is independent of x then $F-y'\frac{\partial F}{\partial x}=c_1$

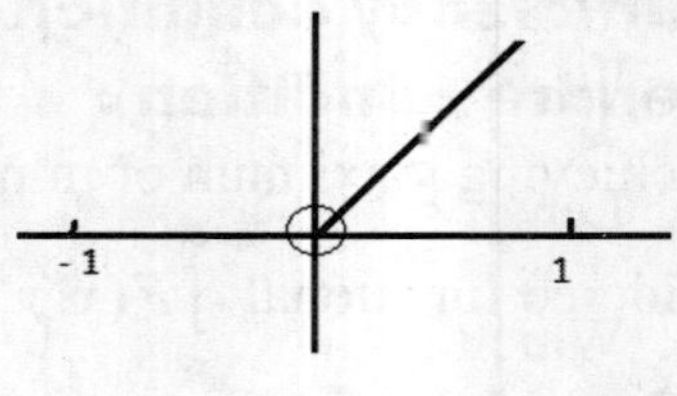

Using E-L equation

$$F-y'\frac{\partial F}{\partial x}=c_1 \Rightarrow y^2\left(1-y'^2\right)=c_1$$

take $c_1=0$

then either y = 0 or $y(x)=x+A$

Both solution not satisfy the condition

$\Rightarrow$ we can construct a new function with corner point.

Let $y(x)=\begin{cases}0 & -1\le x<0\\ x & 0\le x\le 1\end{cases}$

it satisfy the condition and it is broken external and x = 0 is a corner point.

SOLVED QUESTIONS

1. On what curve can the extremal of **CSIR NET SEPT. 2022**

$I[y(x)]=\int_{-1}^{0}\{12xy-(y')^2\}dx$, y(0) = 0, y(-1) = 1

(a) $y = x^2$ (b) $y = x^3$ (c) $y = -x^3$ (d) None of these

Ans. (c)

Explanation:

Given that $F(x, y, y') = 12xy - (y')^2$

By the Euler's equation $\frac{\partial f}{\partial y}-\frac{d}{dx}\left(\frac{\partial f}{\partial y'}\right)=0$ ...... (1)

$\Rightarrow \frac{\partial f}{\partial y}=12x$ and $\frac{\partial f}{\partial y'}=-2y'$ From (1) $12x+\frac{d}{dx}(2y')=0$

$12x + 2y'' = 0$ $\quad y'' = -6x$

Integration both side $y'=-\frac{6x^2}{2}+C$ $\quad y' = -3x^2 + C$

Again integration $y(x)=-\frac{3x^3}{3}+Cx+D$ $\quad y(x) = -x^3 + Cx + D$ ......(2)

Given condition y(0) = 0 $\quad$ = y(0) = 0 = D $\Rightarrow$ D = 0 $\quad$ &y(-1) = 1

$\Rightarrow$y(-1) = 1 = 1 + C + D $\quad$ C + D = 0

Put the value of D⇒ C = 0

Put value of C & D in (2)

$y(x) = -x^3$ ⇒ The extremal can be achieved only on the curve $y = -x^3$

2. Let $X = \{u \in C^1[0,1] : u(0) = u(1) = 0\}$. Let $I : X \to R$ be defined as $I(u) = \int_0^1 e^{-u'(t)^2} dt \; ; \forall u \in X$ **CSIR NET SEPT. 2022**

Let $M = \sup_{f\in X} I[f]$ & $m = \inf_{f \in X} I[f]$

Which of the following statements are true

(a) M = 1 , m = 0 (b) 1 = M > m > 0 (c) M is attained (d) m is attained

Ans. (a), (c)

Explanation:

Given that $F = e^{-u'(t)^2}$ Then Euler's equation is

u(t) = at + b Apply condition u(0) = u(1) = 0

we get u = 0 Now $I(u) = \int_0^1 e^0 dt = 1$

Check given extremal is maximum or minimum

Given that $F = e^{-u'(t)^2}$

$$\frac{\partial F}{\partial y'} = e^{-u'(t)^2}(-2u'(t)); \frac{\partial^2 F}{\partial y'^2} = e^{-u'(t)^2}(4u'(t)) - 2e^{-u'(t)^2}$$

For u = 0 , $\frac{\partial^2 F}{\partial y'^2} = -2 < 0$ ⇒ Maximum attained

3. Let x*(t) be the curve which minimizes the functional $J(x) = \int_0^1 [x^2(t) - x'^2(t)]dt$ satisfying x(0) = 0 & x(1) = 1. Then the value of $x*\left(\frac{1}{2}\right)$ is **CSIR NET JUNE 2019**

(a) $\frac{\sqrt{e}}{1+e}$ (b) $\frac{2\sqrt{e}}{1-e}$ (c) $\frac{\sqrt{e}}{1+2e}$ (d) $\frac{2\sqrt{e}}{1+2e}$

Ans. (a)

Explanation :

Given that $F(t, x, x') = x^2 - x'^2$

By Euler equation $\frac{\partial f}{\partial x} - \frac{d}{dt}\left(\frac{df}{dx'}\right) = 0 = 2x - \frac{d}{dt}(2x') = 0 \Rightarrow 2x - 2x'' = 0$

$x - x'' = 0 \quad \Rightarrow x'' - x = 0$

Auxilliary equation $(D^2 - 1)\,x = 0 \quad M^2 - 1 = 0$

$M = \pm 1 \Rightarrow x(t) = c_1e^t + c_2e^{-t}$ .....(1)

Use condition

COV, NA, I.E.

$x(0) = c^1 + c^2 = 0$ .....(1)

$x(1) = c_1 e + c_2 e^{-1} = 1$ .... (2)

Find c_1& c_2 from (1) and (2)

$$c_2 = \frac{1}{e^{-1} - e} \;\&\; c_1 = \frac{-1}{e^{-1} - e}$$

From (1) $x(t) = \frac{-e^t}{e^{-1} - e} + \frac{1}{e^{-1} - e} e^{-t} = \frac{1}{e^{-1} - e}\left[e^{-t} - e^t\right]$ $x^*\left(\frac{1}{2}\right) = \frac{1}{e^{-1} - e}\left[e^{-1/2} - e^{1/2}\right]$

$$\frac{1}{e^{-1} - e}\left[\frac{e^{-1/2} - e^{1/2}}{e^{-1/2} + e^{1/2}}\right]e^{-1/2} + e^{1/2} = \frac{1}{\frac{1}{\sqrt{e}} + \sqrt{e}} \quad x^*\left(\frac{1}{2}\right) = \frac{\frac{1}{1+e}}{\sqrt{e}} = \frac{\sqrt{e}}{1+e}$$

4. Consider the functional $J(y) = \int_0^1 \left[y'(x)^2 + (y'(x))^3\right] dx$ y(0) = 1 & y(1) = 2 then

CSIR NET JUNE 2019

(a) ∃an extremal $y \in C^1[0, 1] / C^2[0, 1]$ (b) ∃an extremal $y \in C[0, 1] / C^1[0, 1]$

(c) every extremal y belongs to $C^1[0, 1]$ (d) every extremal y belongs to $C^2[0, 1]$

Ans. (c), (d)

Explanation :

C[0, 1] = {The set of all continuous function}

$C^1[0, 1]$ = {The set of all continuous * 1st derivative exist function}

$C^2[0, 1]$ = {The set of all continuous & 2nd derivative exist function}

f is independent of x & y then extremal

$$\frac{d^2y}{dx^2} = 0 \quad \Rightarrow y(x) = Ax + B$$

Use condition y(0) = 1 & y(1) = 2 then y(x) = x + 1

This is continuous, 1st derivative & 2nd derivative exist.

Option (c), (d) is correct.

Option (a), (b) discarded.

5. The extremals of the functional $J[y] = \int_0^1 \left[720x^2y - (y'')^2\right] dx$, subject to

$y(0) = y'(0) = y(1) = 0, y'(1) = 6$, are

CSIR NET DEC 2018

(a) $x^6 + 2x^3 - 3x^2$ (b) $x^5 + 4x^4 - 5x^3$ (c) $x^5 + x^4 - 2x^3$ (d) $x^6 + 4x^3 - 6x^2$

Ans. (a)

COV, NA, I.E.

Explanation :

Euler's equation is $\frac{\partial F}{\partial y}-\frac{d}{dx}\left(\frac{\partial F}{\partial y'}\right)+\frac{d^2}{dx^2}\left(\frac{\partial F}{\partial y''}\right)$

Here

$$F(x,y,y',y'')=720x^2y-(y'')^2 \Rightarrow 720x^2-\frac{d}{dx}(0)+\frac{d^2}{dx^2}(-2y'') \Rightarrow 720x^2-2y''''=0$$

$$\Rightarrow y''''=360x^2 \Rightarrow y'''=120x^3+c_1 \Rightarrow y''=30x^4+c_1x+c_2 \Rightarrow y'=6x^5+c_1\frac{x^2}{2}+c_2x+c_3$$

$$\Rightarrow y(x)=x^6+c_1\frac{x^3}{6}+c_2\frac{x^2}{2}+c_3x+c_4$$

Also, $y(0)=0 \Rightarrow c_4=0$ and $y'(0)=0 \Rightarrow c_3=0$

Since, $y(1)=0 \Rightarrow 1+\frac{c_1}{6}+\frac{c_2}{2}=0 \Rightarrow c_1+3c_2=-6\ldots(1)$

Since, $y'(1)=6 \Rightarrow 6+\frac{c_1}{2}+c_2=6 \Rightarrow c_1+2c_2=0\ldots(1)$

Solving equation (1) and (2), we get $\Rightarrow c_2=-6$ and $c_1=12 \Rightarrow y(x)=x^6+2x^3-3x^2$

Hence, option (a) is correct.

6. Let y = y(x) $\in C^4[0, 1]$ be extremize function for the functional $I[y]=\int_0^1[(y'')^2-2y]dx$, y (0) = 0 = y(1). With Natural Boundary Condition is given by **CSIR NET DEC 2019**

(a) $\frac{x}{24}(x-1)^3$ (b) $\frac{x}{24}(x^3-2x^2+1)$ (c) $\frac{x^2}{24}(x-1)^2$ (d) $\frac{x}{24}(x^3+x^2-2)$

Ans. (b)

Explanation :

Euler equation $\frac{\partial f}{\partial y}-\frac{d}{dx}\left(\frac{\partial f}{\partial y'}\right)+\frac{d^2}{dx^2}\left(\frac{\partial f}{\partial y''}\right)=0=-2-0+\frac{d^2}{dx^2}(2y'')=0$

$= -2 + 2y^{iv} = 0$ $\quad y^{iv} = 1$

Integration $y''' = x + A$ $\quad y''=\frac{x^2}{2}+Ax+B$ …. Ist

$y'=\frac{x^3}{6}+\frac{Ax^2}{2}+Bx+C$ $\quad y(x)=\frac{x^4}{24}+\frac{Ax^3}{6}+\frac{Bx^2}{2}+Cx+D$ …..(1)

Using the condition

y(0) = 0 then D = 0 & y(1) = 0 then $\frac{1}{24}+\frac{A}{6}+\frac{B}{2}+C=0$

COV, NA, I.E.

$$\frac{1+4A+12B+24C}{24} = 0 = 4A + 12B + 24C + 1 = 0 \qquad \text{..... (2)}$$

Natural bounded condition

$$\frac{\partial f}{\partial y''} = 0$$

at the end point = 2y'' = 0

y'' (0) = 0 = y''(1)

Ist from

$$y'' = \frac{x^2}{2} + Ax + B$$

y'' (0) = 0 = B

$$\& \; y''(1) = 0 = \frac{1}{2} + A + B$$

$$\Rightarrow A = -\frac{1}{2}$$

Now A = -1/2 , B = 0, D = 0

From (2)

$$4\left(-\frac{1}{2}\right) + 12(0) + 24C + 1 = 0$$

= -2 + 24C + 1 = 0

24C = 1

$$C = \frac{1}{24}$$

Put in (1)

$$y(x) = \frac{x^4}{24} - \frac{x^3}{12} + \frac{x}{24}$$

$$= \frac{x}{24}[x^3 - 2x^2 + 1]$$

7. The extremal of the functional $J(y) = \int_0^1 y'^2(x)dx$ subject to y(0) = 0, y(1) = 1 and $\int_0^1 y(x)dx = 0$ is **CSIR NET 2018**

(a) $3x^2 - 2x$ (b) $8x^3 - 9x^2 + 2x$ (c) $\frac{5}{3}x^4 - \frac{2}{3}x$ (d) $\frac{-21}{2}x^5 + 10x^4 + 4x^3 - \frac{5}{2}x$

Ans. **(a)**

Explanation :

Given problem is an isoperimetric problem

∴ Functional for this problem H = f + λg, where f = y'^2(x) and g = y(x)

$$H = y'^2 + \lambda y$$

Now by Euler's equation

$$\frac{\partial H}{\partial y} - \frac{d}{dx}\left(\frac{\partial H}{\partial y'}\right) = 0 \Rightarrow \lambda - \frac{d}{dx}(2y') = 0 \Rightarrow \lambda - 2y'' = 0 \; y'' = \frac{\lambda}{2} \Rightarrow y = \frac{\lambda x^2}{4} + cx + d$$

which is a quadratic equation. Options (b), (c), (d) are clearly incorrect.

Hence, option (a) is correct.

COV, NA, I.E.

8. The extremal of the functional $J(y)=\int_0^1 y'^2(x)dx$ subject to y(0) = 0, y(1) = 1 and $\int_0^1 y(x)dx = 0$ is **CSIR NET JUNE 2018**

(a) $3x^2-2x$ (b) $8x^3-9x^2+2x$ (c) $\frac{5}{3}x^4-\frac{2}{3}x$ (d) $\frac{-21}{2}x^5+10x^4+4x^3-\frac{5}{2}x$

Ans. **(a)**

Explanation :

Given problem is an isoperimetric problem

∴ Functional for this problem H = f + λg, where f = y'²(x) and g = y(x)

$H = y'^2+\lambda y$

Now by Euler's equation

$$\Rightarrow \frac{\partial H}{\partial y}-\frac{d}{dx}\left(\frac{\partial H}{\partial y'}\right)=0 \Rightarrow \lambda-\frac{d}{dx}(2y')=0 \Rightarrow \lambda-2y''=0$$

$$\Rightarrow y''=\frac{\lambda}{2} \Rightarrow y=\frac{\lambda x^2}{4}+cx+d$$, which is a quadratic equation.

∴ Options (b), (c), (d) are clearly incorrect.

Hence, option (a) is correct.

9. Let y(x) be a piecewise continuously differentiable function on [0, 4]. Then the functional $J[y]=\int_0^4 (y'-1)^2(y'+1)^2 dx$ attains minimum if y = y(x) is

(a) $y=\frac{x}{2}, 0\le x\le 4$

(b) $y=\begin{cases}-x, & 0\le x\le 1\\ x-2, & 1\le x\le 4\end{cases}$

(c) $y=\begin{cases}2x, & 0\le x\le 2\\ -x+6 & 2\le x\le 4\end{cases}$

(d) $y=\begin{cases}x, & 0\le x\le 3\\ -x+6 & 3\le x\le 4\end{cases}$

Ans. **(a), (d)** **CSIR NET DEC 2016**

COV, NA, I.E.

Explanation :

Given functional is $J[y]=\int_0^4 (y'-1)^2(y'+1)^2\,dx$

We can check it by options From option (a),

$$y=\frac{x}{2}\Rightarrow y'=\frac{1}{2}\ J[y]=\int_0^4\left(\frac{1}{2}-1\right)^2\left(\frac{1}{2}+1\right)^2 dx=\frac{9}{16}(4)=\frac{9}{4}$$

$$J[y]=\int_0^1(-2)^2(0)^2\,dx+\int_1^4(0)^2(2)^2\,dx=0$$

From option (b),

$$J(y)=\int_0^2(2-1)^2(2+1)^2\,dx+\int_2^4(-1-1)^2(0)^2\,dx=9\times 2=18$$

From option (c),

$$J(y)=\int_0^3(1-1)^2(1+1)^2\,dx+\int_1^4(-1-1)^2(-1+1)^2\,dx=0$$

From option (d),

For minimum value, option (b) and (d) are correct.

10. The infimum of $\int_0^1 (u'(t))^2\,dt$ on the class of functions $\{u\in C^1[0,1]$, such that u(0) = 0 and $\max\limits_{[0,1]}|u|=1\}$ is equal to **CSIR NET JUNE 2017**

(a) 0 (b) 1/2 (c) 1 (d) 2

Ans. (c)

Explanation :

Given functional is $\int_0^1 (u'(t))^2\,dt$ on $\left\{u(0)=0,\max\limits_{x\in|0,1|}|u|=1 \text{ and } u\in c'[0,1]\right\}$

Given functional is independent of u, t

$\therefore$ From Euler's equation solution is given by u'' (t) = 0 u(t) = C_1t + C_2, as u(0) = 0

$\therefore$ $C_2 = 0$ Also u(1) = 1 as $\max\limits_{x\in|0,1|}|u|=1$

$\therefore$ $C_1 = 1$

$\therefore$ u = t and u' = (1)

$\therefore$ $\int_0^1 (1)^2\,dt=1$

The infimum of given functional is equal to 1.

11. Consider the functional $I(y(x))=\int_{x_0}^{x_1} f(x,y)\sqrt{1+y'^2}e^{\tan^{-1}y'}dx$, where $f(x,y) \neq 0$. Let the left end of the extremal be fixed at the point $A(x_0, y_0)$ and the right end $B(x_1, y_1)$ be movable along the curve $y = \psi(x)$. Then the extremal $y = y(x)$ intersects the curve $y = \psi(x)$ along which the boundary point $B(x_1, y_1)$ slides at an angle

(a) $\pi/3$ (b) $\pi/2$ (c) $\pi/4$ (d) $\pi/0$

Ans. (c) **CSIR NET JUNE 2017**

Explanation :

Given functional is

$$I(y(x))=\int_{x_0}^{x_1} f(x,y)\sqrt{1+y'^2}e^{\tan^{-1}y'}dx$$

, where $f(x, y) \neq 0$.

$\therefore$ the right end of the extremal is movable along the curve $y = \psi(x)$.

$\therefore$ By transversality condition on right end.

i.e.,

$$F+(\psi'-y')\frac{\partial F}{\partial y'}=0$$

where

$$F=f(x,y)\sqrt{1+y'^2}\,e\tan^{-1}y'$$

$\Rightarrow$

$$f(x,y)\sqrt{1+y'^2}e^{\tan^{-1}y'}+(\psi'-y')\left[f(x,y).\frac{y'}{\sqrt{1+y'^2}}e\tan^{-1}y'+f(x,y)\sqrt{1+y'^2}.e^{\tan^{-1}y}.\frac{1}{1+y'^2}\right]=0$$

$$f(x,y)[1+y'^2+(\psi'-y')(y'+1)]e^{\tan^{-1}y'}=0$$

$$1+y'^2+\psi'y'+\psi'-y'^2-y'=0$$

$\Rightarrow$

$$1+\psi'y'+\psi'-y'=0\Rightarrow\frac{y'-\psi'}{1+\psi'y'}=1$$

$\Rightarrow$

....(1)

$$\tan(\theta)=\frac{m_1-m_2}{1+m_1m_2}$$

$\therefore$

$$\theta=\frac{\pi}{4}$$

$\therefore$ from (1), clearly

Option (c) is correct.

12. Let $X=\{u\in C^1[0,1] \mid u(0)=u(1)=0\}$ and define $J:X\to R$ by $J(u)=\int_0^1 e^{-u'^2(x)}dx$. Then

(a) J does not attain its infimum (b) J attains its infimum at a unique $u \in X$

(c) J attains its infimum at exactly two elements $u \in X$.
(d) J attains its infimum at infinitely many $u \in X$

Ans. (a) **CSIR NET DEC 2017**

Explanation :

$$J(u) = \int_0^1 e^{-(u'(x))^2} dx$$

Given functional is

$$J(u) = \int_0^1 f(u')dx$$

$\therefore$ and f is a function of u' only

$\therefore$ extremal of above functional is straight line $\Rightarrow u(x) = ax + b$.

By applying given conditions $u(0) = u(1) = 0$; we get $u(x) = 0$, which is unique

But $f_{u'u'}$ at $u(x) = 0$ is negative which shows functional attains maxima at $u(x) = 0$

$\therefore$ J does not attain its minima.

Hence, only option (a) is correct.

UNSOLVED QUESTIONS

1. The variational problem of extremizing the functional $f(y(x)) = \int_0^{2\pi} \left[\left(\frac{dy}{dx}\right)^2 - y^2\right] dx$;

$y(0) = 1$, $y(2\pi) = 1$ has **CSIR NET JUNE 2011**

(a) a unique solution
(b) exactly two solutions
(c) an infinite number of solutions
(d) no solution

2. Consider the functional $I(y) = \int_{\frac{1}{2}}^{x_1} \sqrt{1 + y'^2}\, dx$ subject to the condition that the left end of the extremal is fixed at the point $\left(\frac{1}{2}, \frac{1}{4}\right)$ and the right end (x_1, y_1) be movable along the straight line $y = x - 5$. Let C_1 and C_2 be arbitrary constants. Then

CSIR NET DEC 2011

(a) the general solution of Euler's equation satisfies $2C_1 + 4C_2 = 1$
(b) the transversality condition that holds at the right end yields

$$\sqrt{(1 + C_1^2)} + (1 - C_1)\frac{C_1}{\sqrt{1 + C_1^2}} = 0$$

(c) the relation $C_1x_1 + C_2 = x_1 - 5$ is true
(d) the extremal is $y = -x + \frac{3}{4}$

3. An approximate solution $z = z_0(x, y)$ to the problem of extremizing the functional $I(z(x,y)) = \iint_D \left[\left(\frac{\partial z}{\partial x}\right)^2 + \left(\frac{\partial z}{\partial y}\right)^2 - 2z\right]dxdy$, where D is the square $-1 \le x \le 1$, $-1 \le y \le 1$ and $z = 0$ on the boundary of the square, is of the form **CSIR NET JUNE 2012**

(a) $z_0 = \sum_{i=1}^{n} \alpha_i \phi_i(x,y)$, where α_i are constants and functions ϕ_i are linearly independent in D.

(b) $z_0 = \alpha_1\phi_1(x,y) + \alpha_2\phi_2(x,y)$, where α_1 and α_2 are constants and ϕ_1 and ϕ_2 have continuous partial derivatives.

(c) $z_0 = \alpha\phi(x, y)$, where α is a constant and ϕ is continuous in D.

(d) $z_0 = (x^2 - 1)(y^2 - 1) / 16$

4. Let $J(u) = \int_0^1 \left[u_x^2 + 4\frac{u^2}{x^2}\right]xdx$, where u(x) is a smooth function defined on [0, 1] satisfying u(0) = 0 and u(1) = 1. which of the functions minimizes J?

CSIR NET DEC 2012

(a) $u(x) = x^2$ (b) $u(x) = \frac{1}{\sqrt{2}}x^2$ (c) $u(x) = \frac{1}{2}x^2$ (d) $u(x) = \frac{1}{4}x^2$

5. The extremal of $\int_1^2 \frac{x'^2}{t^3}dt; x(1) = 3, x(2) = 18$ (where $x' = \frac{dx}{dt}$) using Lagrange;s equation is given by which of the following? **CSIR NET JUNE 2013**

(a) $x = t^4 + 2$ (b) $x = \frac{15}{7}t^3 + \frac{6}{7}$ (c) $x = 5t^2 - 2$ (d) $x = 5t^3 + 3$

6. An extremal of the functional $J(y) = \int_a^b \sqrt{1 + |y'(x)|^2}\,dx$ **CSIR NET DEC 2013**

(a) is the straight line connecting (a, y(a)) and (b, y(b)).

(b) is a solution to the differential equation $y' = C\sqrt{1 + y'^2}$ for some constant C.

(c) is a solution to $y'' = 0$ (d) does not exist.

7. The curve extremizing the functional $I(y) = \int_1^2 \frac{\sqrt{1 + (y'(x))^2}}{x}dx, y(1) = 0, y(2) = 1$ is

CSIR NET JUNE 2014

(a) an ellipse (b) a parabola (c) a circle (d) a straight line

COV, NA, I.E.

8. Let u(x, y) be an extremal of the functional $f(u)=\iint_D\left[\frac{1}{2}u_x^{\ 2}+\frac{1}{2}u_y^{\ 2}+e^{xy}u\right]dxdy$, where D is the open unit disk in R^2. Then u satisfies **CSIR NET JUNE 2014**

(a) $u_{xx}+u_{yy}-e^{x+y}=0$ (b) $u_{xx}+u_{yy}=e^{xy}$ (c) $u_{xx}+u_{yy}=-e^{xy}$

(d) $\iint_D\left[u_{xx}+u_{xy}-e^{xy}\right]h(x,y)dxdy=0$ for every smooth h vanishing on the boundary of D.

9. Consider the functional $J(y)=y^2(1)+\int_0^1 y'^2(x)dx, y(0)=1$, where $y\in C^2([0,1])$. If y extremizes J, then **CSIR NET DEC 2014**

(a) $y(x)=1-\frac{1}{2}x^2$ (b) $y(x)=1-\frac{1}{2}x$ (c) $y(x)=1+\frac{1}{2}x$ (d) $y(x)=1+\frac{1}{2}x^2$

10. Let $y\in C^2([0,\pi])$ satisfying y(0) = y(π) = 0 and $\int_0^\pi y^2(x)dx=1$ extremize the functional $J(y)=\int_0^\pi (y'(x))^2dx; y'=\frac{dy}{dx}$. Then **CSIR NET DEC 2014**

(a) $y(x)=\sqrt{\frac{2}{\pi}}\sin x$ (b) $y(x)=-\sqrt{\frac{2}{\pi}}\sin x$ (c) $y(x)=\sqrt{\frac{2}{\pi}}\cos x$ (d) $y(x)=-\sqrt{\frac{2}{\pi}}\cos x$

11. The extremal of the functional $\int_0^\alpha (y'^2-y^2)dx$ that passes through (0, 0) and (α, 0) has a **CSIR NET JUNE 2015**

(a) weak minimum if a <π (b) strong minimum if a <π

(c) weak minimum if a >π (d) strong minimum if a >π

12. Let y = y(x) be the extremal of the functional $I[y(x)]=\int_{x_1}^{x_2}\sqrt{1+\left(\frac{dy}{dx}\right)^2}dx$, subject to the condition that the left end of the extremal moves along $y = x^2$, while the right end moves along x – y = 5. Then the **CSIR NET JUNE 2016**

(a) shortest distance between the parabola and the straight line is $\left(\frac{19\sqrt{2}}{8}\right)$

(b) slope of the extremal at (x, y) is $\left(-\frac{3}{2}\right)$

(c) point $\left(\frac{3}{4},0\right)$ lies on the extremal (d) extremal is orthogonal to the curve $y=\frac{x}{2}$

COV, NA, I.E.

13. The variational problem of extremizing the functional $I(y(x)) = \int_1^3 y(3x - y)dx$;

$y(3) = 4\frac{1}{2}, y(1) = 1$ has

(a)a unique solution (b)exactly two solutions
(c)an infinite number of solutions (d)no solution

14. Consider the functional $I(y) = \int_a^b F(y, y')dx; y' \equiv \frac{dy}{dx}; y(a), y(b) = y_2$, where $y \in C^2[a,b]$, F has second order continuous partial derivatives with respect to y, y' and y_1, y_2 are given real numbers. Let y = y(x) be an extremizing function for the functional I. Then, along the extremizing curve.

(a)F remains constant (b) $\frac{\partial F}{\partial y} = 0$ (c) $F - y\frac{\partial F}{\partial y'} =$ constant (d) $F - y'\frac{\partial F}{\partial y'} =$ constant

15. Consider the functional $J(y) = \int_a^b F(x, y, y')dx$, where $F(x, y, y') = y' + y$ for admissible functions y. Then J has

(a) no extremals (b) several extremals (c)$y(x) = e^{-x}$ as an extremal
(d) y(x) = constant as an extremal

16. Suppose any two points $P(x_0, y_0)$, $Q(x_1, y_1)$ and a function F(x, y, y') of three independent variables are given, where $y' \equiv \frac{dy}{dx}$. In order to find among all curves y = y(x) joining P and Q that one which furnishes for the definite integral $I(y) = \int_{x_0}^{x_1} F(x, y(x), y'(x))dx$ the smallest value, which of the following assumption suffices

(a) Function F is of class C^1
(b) Functional F is of class C^2 for all systems of values x,y(x) and y'(x) furnished by all the admissible functions
(c) Functional F is of class C^3 for all system of values x, y(x) and y'(x) furnished by all the admissible functions.
(d) It is enough to treat y(x) and F to be of class C^1 only with respect to their arguments.

17. Consider the functional $J = \int_a^b F(x, y, y')dx$, where $F(x, y, y') = (I + y^2) / y'^2$ for admissible functions y(x). Which of the following are extremals for J?

(a) $y(x) = A\sin(x)$ (b) $y(x) = A\sinh(x) + B\cosh(x)$
(c) $y(x) = A\sinh(Ax + B)$ (d) $y(x) = A\sin(x) + B\cos(x)$

COV, NA, I.E.

18. The functional $\int_0^1 (y'^2+4y^2+8ye^x)dy, y(0)=-\frac{4}{3}, y(1)=-\frac{4e}{3}$ possesses :

GATE-2012

(a) strong minima on $y=-\frac{1}{3}e^x$

(b) strong minima on $y=-\frac{4}{3}e^x$

(c) weak maxima on $y=-\frac{1}{3}e^x$

(d) weak maxima on $y=-\frac{1}{3}e^x$

19. The functional $\int_0^1 (y'^2+(y+2y')y''+kxyy'+y^2)dx$, with y(0) = 0, y(1) = 1, y'(0) = 2, y'(1) = 3 is path independent if k equals

GATE-2012

(a) 1 (b) 2 (c) 3 (d) 4

20. Suppose that among all continuously differentiable functions y(x), x ∈ R, with y(0) = 0 and $y(1)=\frac{1}{2}$, the function $y_0(x)$ minimizes the functional $\int_0^1 (e^{-(y'-x)}+(1+y)y')dx$.

Then $y_0\left(\frac{1}{2}\right)$ is equal to

GATE-2015

(a) 0 (b) 1/8 (c) 1/4 (d) 1/2

Answer Key

1	2	3	4	5	6	7	8	9	10
c	All	a,b,c	a	a	a,b,c	a	a,d	b	a,b
11	12	13	14	15	16	17	18	19	20
a,b	a,c	d	d	a	b	a,b,c,d	b	b	b

COV, NA, I.E.

Chapter 2 Numerical Analysis

In this section we will find the roots of equations by such type of method
There are two method : (1) Direct method (2) Iteration method

(1) Direct Method :

These method gives the exact root of equation.

Let $ax^2 + bx + c = 0$

Then $x = \dfrac{-b \pm \sqrt{b^2 - 4ac}}{2a}$

(2) Iteration Method :

These types of method are based on the ideal of successive approximation. It mean they starting with one or more initial approximation to the root and obtain a sequence of iterates (x_i) which in the limit converges to the root. It can be grouped into two types :

(a) Bracketing methods (b) Open end methods

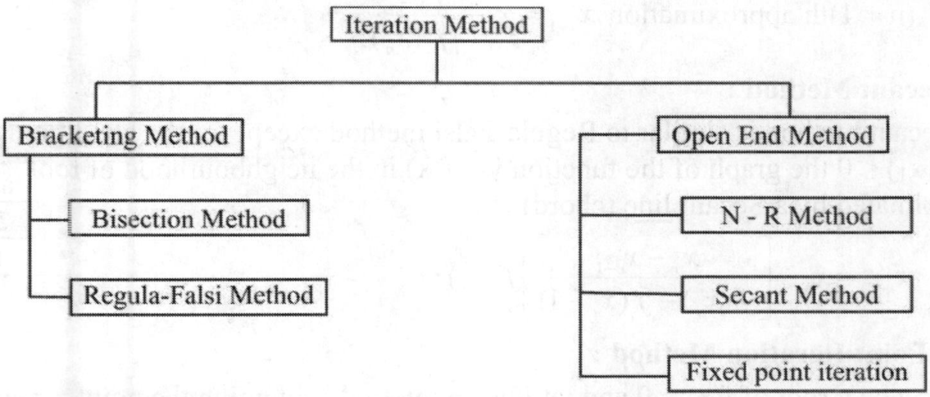

(1) Bisection Method :

Let f(x) be a continuous on $[x_1, x_2]$
and $f(x_1).f(x_2) < 0$ then $\exists$ atleast one point $x \in (x_1, x_2)$ s.t. $f(x) = 0$.

Working Rule :

(1) Let $f(x_1) < 0$ and $f(x_2) > 0$ Then $x_3 = \dfrac{x_1 + x_2}{2}$

If $f(x_3) = 0$ then x_3 is root of equation.
Otherwise root lie between either x_1 and x_3 or x_3 and x_2 according to sign of $f(x_3)$.

(2) And continue the process until the root is found.

(2) Regula Falsi Method :

Working Rule :

(1) Equation of line passing through (x_1, $f(x_1)$) and (x_2, $f(x_2)$) is

$$y - f(x_1) = \frac{f(x_2) - f(x_1)}{x_2 - x_1}(x - x_1)$$

(2) Intersection point between line and x-axis is x_3.

(3) If $f(x_3) = 0$ then x_3 is a root.

(4) If $f(x_3).f(x_2) < 0$ then root lie between (x_2) and (x_3)

(5) Continue this process until the root is found.

General Formula : $x_{n+1} = x_n - \dfrac{(x_n - x_{n-1})f(x_n)}{f(x_n) - f(x_n - 1)}, n \geq 2$

(3) Newton-Raphson Method :

One of the most widely used method of solving the equation is Newton's Method

n^{th} approximation is $x_n = x_{n-1} - \dfrac{f(x_{n-1})}{f'(x_{n-1})}$

& (n + 1)th approximation $x_{n+1} = x_n - \dfrac{f(x_n)}{f'(x_n)}$

(4) The Secant Method :

The Secant method is similar to Regula-Falsi method except for the condition $f(x_0).f(x_1) < 0$ the graph of the function y = f(x) in the neighbourhood of root is approximated by a Secant line (chord)

$$x_{n+1} = x_n - \left[\frac{x_n - x_{n-1}}{f(x_n) - f(x_n - 1)}\right] f(x_n)$$

(5) Fixed Point Iteration Method :

Let x = a be a root of f(x) = 0 and let I be an interval containing the point x = a. Let $\phi(x)$ is defined by the equation $x = \phi(x)$ which is equivalent to f(x) = 0 if $|\phi`(x)| < 1$ for $\forall\ x \in I$.

Then sequence of approximation $x_0, x_1, x_2 \ldots, x_n$ defined by $x_{n+1} = \phi(x_n)$ converges to root a .

Rate of convergence :

In numerical analysis the speed at which a convergent sequence approaches its limit is called the rate of convergences.

Table : Comparision of iterative methods

Method	Iterative formulae	Order of convergence	Reliability
Bisection	$x_{n+1} = \frac{(x_n + x_{n-1})}{2}$ Either $[x_{n-1}, x_{n+1}]$ or $[x_{n+1}, x_n]$ encloses a root	1	Guaranteed convergence
False Position	$x_{n+1} = \frac{x_{n-1}f(x_n) - x_n f(x_{n-1})}{f(x_n) - f(x_{n-1})}$	1	Guaranteed convergence
Newton Raphson	$x_{n+1} = x_n - \frac{f(x_n)}{f'(x_n)}$ If root is of multiplicity of 1	2	Convergence sensitive to initial value
	If root is of multiplicity of 2	1	Fast if starting point is near to the root
Secant	$x_{n+1} = x_n - \frac{(x_n - x_{n-1})f(x_n)}{f(x_n) - f(x_{n-1})}$	1.618	No guarantee of convergence, if not near the root, economical method
Iteration	$x_i = g(x_{i+1})$	1	Easy to program, no guarantee of convergence

Solution of system of linear algebraic equation

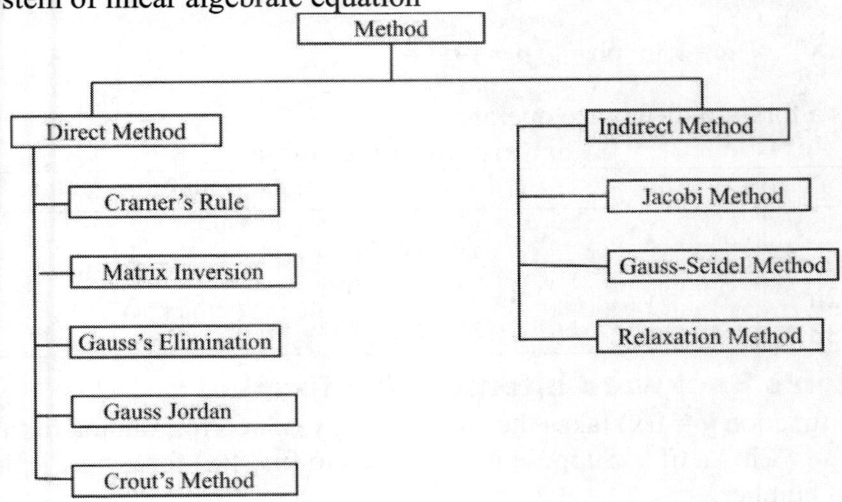

Calculus of Finite difference

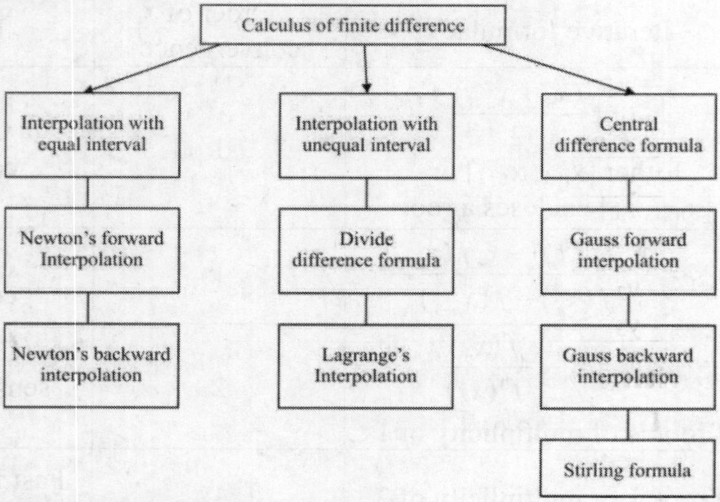

(1) Interpolation with equal interval.
(2) Interpolation with unequal interval.
(3) Central Differences

1. Interpolation with equal interval.

Newton's forward interpolation formula.

Let the function y = f(x) takes the value y_0, y_1, y_2 .....corresponding to the values x_0, $x_0 + h$, $x_0 + 2h$ ..... of x. Suppose it is required to find f(x) for $x = x_0 + ph$ where p is any real number.

Then $$y_p = y(x_0 + ph) = y_0 + p\Delta y_0 + \frac{p(p-1)}{2!}\Delta^2 y_0 + \frac{p(p-1)(p-2)}{3!}\Delta^3 y_0 + ...$$

Where $x = x_0 + ph \qquad p = \dfrac{x - x_0}{h}$

andΔ is a forward difference operator.

Forward difference table

x	y	Δy	$\Delta^2 y$	$\Delta^3 y$
$x_0 = x_0$	y_0			
$x_1 = x_0 + h$	y_1	$y_1 - y_0 = \Delta y_0$	$y_2 - y_0 = \Delta^2 y_0$	
$x_2 = x_0 + 2h$	y_2	$y_2 - y_1 = \Delta y_1$	$y_3 - y_1 = \Delta^2 y_1$	$y_3 - y_0 = \Delta^3 y_0$
$x_3 = x_0 + 3h$	y_3	$y_3 - y_2 = \Delta y_2$		

2. Newton's Backward interpolation formula :

Let the function y = f(x) takes the value y_0, y_1, y_2 .....corresponding to the values x_0, $x_0 + h$, $x_0 + 2h$ ..... of x. Suppose it is required to find f(x) for $x = x_n + ph$ where p is any real number.

Then $$y_p = y(x_n + ph) = y_n + p\nabla y_n + \frac{p(p+1)}{2!}\nabla^2 y_n + ...$$

Where $$p = \frac{x - x_n}{h}$$

Construct Backward difference table

x	y	∇y	$\nabla^2 y$	$\nabla^3 y$
$x_0 = x_0$ $x_1 = x_0 + h$ $x_2 = x_0 + 2h$ $x_3 = x_0 + 3h$	y_0 y_1 y_2 y_3	$y_1 - y_0 = \nabla y_0$ $y_2 - y_1 = \nabla y_1$ $y_3 - y_2 = \nabla y_2$	$y_2 - y_0 = \nabla^2 y_3$ $y_3 - y_1 = \nabla^2 y_3$	$y_3 - y_0 = \nabla^3 y_0$

3. Interpolation with unequal interval.

Newton's Divided Differences formula :

Let $f(x_0), f(x_1), \ldots, f(x_n)$ be the value of function f(x) corresponding to the unequal point $x_0, x_1, \ldots, x_n$

Then $f(x) = f(x_0) + (x - x_0)\, f(x_0, x_1) + (x - x_0)(x - x_1)\, f(x_0, x_1, x_2) + \ldots\ldots (x - x_0)(x - x_1)\ldots (x - x_{n-1})\, f(x_0, x_1, \ldots, x_n)$

Firstly construct divide difference table

x	f(x)	$\Delta f_1(x)$	$\Delta^2 f(x)$
x_0 x_1 x_2 …	$f(x_0)$ $f(x_1)$ $f(x_2)$ ….	$f(x_0, x_1) = \dfrac{f(x_1) - f(x_0)}{x_1 - x_0}$ $f(x_1, x_2) = \dfrac{f(x_2) - f(x_1)}{x_2 - x_1}$	$f(x_0, x_1, x_2) = \dfrac{f(x_2) - f(x_0)}{x_2 - x_0}$ …..

Lagrange's interpolation formula :

$$f(x) = \frac{(x - x_1)(x - x_2)\ldots.(x - x_n)}{(x_0 - x_1)(x_0 - x_2)\ldots..(x_0 - x_n)} f(x_0) + \frac{(x - x_1)(x - x_2)\ldots..(x - x_n)}{(x_1 - x_0)(x_1 - x_2)\ldots.(x_1 - x_n)} f(x_1) + \ldots\ldots$$

3. Central Difference interpolation :

Gauss's forward interpolation

$$y_p = y_0 + p\Delta y_0 + \frac{p(p-1)}{2!}\Delta^2 y_{-1} + \frac{(p+1)p(p-1)}{3!}\Delta^3 y_{-1} + \frac{(p+1)p(p-1)(p-2)}{4!}\Delta^4 y_{-2} + \ldots$$

Gauss's backward interpolation

$$y_p = y_0 + p\Delta y_0 + \frac{p(p+1)}{2!}\Delta^2 y_{-1} + \frac{(p+1)p(p-1)}{3!}\Delta^3 y_{-2} + \frac{(p+2)p(p+1)(p-1)}{4!}\Delta^4 y_{-2} + \dots$$

Stirling's Formula :

$$y_p = y_0 + p\left(\frac{\Delta y_0 + \Delta y_{-1}}{2}\right) + \frac{p^2}{2!}\Delta^2 y_{-1} + \frac{p(p^2-1)}{3!}\times \left(\frac{\Delta^3 y_{-1} + \Delta^3 y_{-2}}{2}\right) + \frac{p^2(p^2-1)}{4!}\Delta^4 y_{-2} + \dots$$

4. Numerical Integration :

Let y = f(x) be a function and $y_0, y_1, \dots, y_n$ are the value of function corresponding to $x_0, x_1, \dots, x_n$. Where $x_n = x_0 + nh$.

Trapezoidal rule : $\int_{x_0}^{x_0+nh} f(x)dx = \frac{h}{2}[(y_0 + y_n) + 2(y_1 + y_2 \dots y_{n-1})]$

Simpson's 1/3rd rule : $\int_{x_0}^{x_0+nh} f(x)dx = \frac{h}{3}\left[(y_0 + y_n) + 4\underbrace{(y_1 + y_3) + \dots)}_{\text{Odd term}} + 2\underbrace{(y_2 + y_{4\dots})}_{\text{Even term}}\right]$

Simpson's 3/8th rule : $\int_{x_0}^{x_0+nh} f(x)dx = \frac{3}{8}h\left[(y_0 + y_n) + 3(y_1 + y_2 + y_4 \dots) + 2\underbrace{(y_3 + y_6 + \dots\dots)}_{\text{Multiple of 3}}\right]$

Weddle's rule : $\int_{x_0}^{x_0+nh} f(x)dx = \frac{3}{10}h[(y_0 + 5y_1 + y_2 + 6y_3 + y_4 + 5y_5 + \dots]$

5. Numerical solution of ODE :

Let $\frac{dy}{dx} = f(x,y)$ and $y(x_0) = y_0$

Picard's theorem

Consider the first order equation $\frac{dy}{dx} = f(x,y)$ and $y(x_0) = y_0$.

Then by picard method first approximation $y_1 = y_0 + \int_{x_0}^{x} f(x, y_0)dx$ and second approximation $y_2 = y_1 + \int_{x_0}^{x} f(x, y_1)dx$.

Runga- Kutta Method :

Let $\frac{dy}{dx} = y' = f(x, y)$ and initially condition $y(x_0) = y_0$. We have to find out $y(x_0 + h)$

Solution :

Case-I. Second order Runga-Kutta method

$y(x_0 + h) = y_0 + k$

where $k = w_1k_1 + w_2k_2$

$k_1 = h\, f(x_0, y_0)$

$k_2 = h\, f(x_0 + \alpha h, y_0 + \beta k_1)$

$w_1 + w_2 = 1$, $w_2\alpha = ½$; $w_2\beta = ½$

Case-II. Third order Runge-Kutta method :

$y(x_0 + h) = y_0 + k$

$k = w_1k_1 + w_2k_2 + w_3k_3$

$k_1 = h\, f(x_0, y_0)$

$k_2 = h\, f(x_0 + \alpha h, y_0 + \beta k_1)$

$k_3 = h\, f(x_0 + \alpha_1 h, y_0 + \beta_1 k_1 + y_1 k_2)$

$w_1 + w_2 + w_3 = 1$, $w_2\alpha = ½$; $w_2\beta = ½$

SOLVED QUESTIONS

1. Fix a $\alpha \in (0,1)$. Consider the iteration defined by $x_{k+1} = \frac{1}{2}(x_k^2 + \alpha), k = 0,1,2....$ (*)

The above iteration has two distinct fixed points ρ_1 and ρ_2 such that $0 < \rho_1 < 1 < \rho_2$. which of the following are true **CSIR NET JUNE 2020**

(a) The iteration (*) is equivalent to the recurrence relation

$$x_{k+1} - \rho_1 = \frac{1}{2}(x_k + \rho_1)(x_k - \rho_1) \;,\; k = 0,1,2...$$

(b) The iteration (*) is equivalent to the recurrence relation

$$x_{k+1} - \rho_1 = \frac{1}{2}(x_k + \rho_2)(x_k - \rho_1) \;,\; k = 0,1,2...$$

(c) If $0 \le x_0 < \rho_2$ then $\lim_{k\to\infty} x_k = \rho_1$

(d) If $-\rho_2 < x_0 \le 0$ then $\lim_{k\to\infty} x_k = \rho_1$

Ans. (a) , (c) , (d)

Explanation :

We know that if any iteration $x_{k+1} = \phi(x_k)$ is fixed point iteration then $|\phi'(x_k)| < 1$ so , $|x_k| < 1$

(a) $x_{k+1} - \rho_1 = \frac{1}{2}(x_k + \rho_1)(x_k - \rho_1) \;,\; k = 0,1,2... \Rightarrow x_{k+1} = \rho_1 + \frac{1}{2}(x_k^2 + \rho_1^2)$, $k = 0,1,2...$

Here $\phi(x) = \rho_1 + \frac{1}{2}(x_k^2 + \rho_1^2) \Rightarrow |\phi'(x)| = |x_k| < 1$

Hence (a) is true

(b) $x_{k+1} - \rho_1 = \frac{1}{2}(x_k + \rho_2)(x_k - \rho_1) \;,\; k = 0,1,2...$

$$\Rightarrow x_{k+1} = \rho_1 + \frac{1}{2}(x_k^2 + x_k(\rho_2 - \rho_1) - \rho_1\rho_2) \;,\; k = 0,1,2...$$

COV, NA, I.E.

here$\phi(x) = \rho_1 + \frac{1}{2}(x_k{}^2 + x_k(\rho_2 - \rho_1) - \rho_1\rho_2) \Rightarrow |\phi'(x)| = |x_k + \left|\frac{(\rho_2 - \rho_1)}{2}\right|| < 1$

So , (b) is not true

(c) If $0 \le x_0 < \rho_2$ then $\lim_{k\to\infty} x_k = \rho_1 \Rightarrow \lim_{k\to\infty}|x_k| = |\rho_1| < 1$

(d) If $-\rho_2 < x_0 \le 0$ then $\lim_{k\to\infty} x_k = \rho_1 \Rightarrow \lim_{k\to\infty}|x_k| = |\rho_1| < 1$

So, (c) and (d) are true

2. Consider ordinary differential equation $\begin{cases} y`(x) + y(x) = 0 \\ y(0) = 1 \end{cases}$, x > 0 and following numerical scheme to solve ODE $\begin{cases} \frac{Y_{n+1} - Y_{n-1}}{2h} + Y_{n-1} = 0. & n \ge 1 \\ Y_0 = 1 & Y_1 = 1 \end{cases}$. If $0 < h < \frac{1}{2}$, then which of the following statement are true? **CSIR NET DEC. 2019**

(a) $Y_n \to \infty$ as $n \to \infty$

(b) $Y_n \to 0$ as $n \to \infty$

(c) Y_n is bounded

(d) $\max_{nh\in[0,T]} |y(nh) - Y_n| \to \infty$ as $T \to \infty$

Ans. (b), (c)

Explanation :

Given that $\frac{Y_{n+1} - Y_{n-1}}{2h} + Y_{n-1} = 0.$

$\Rightarrow Y_{n+1} - Y_{n-1} + 2hY_{n-1} = 0.$

$\Rightarrow Y_{n+1} = Y_{n-1} - 2hY_{n-1}$

$\Rightarrow Y_2 = Y_0 - 2hY_0 = 1 - 2h$

$Y_3 = Y_1 - 2hY_1 = (1 - 2h)$

$Y_4 = Y_2 - 2hY_2 = (1 - 2h)^2$

$Y_5 = Y_3 - 2hY_3 = (1 - 2h)^2$

$\vdots$

$\Rightarrow Y_{2n} = (1 - 2h)^n \; and \; Y_{2n+1} = (1 - 2h)^n$

Given that $0 < h < \frac{1}{2} \Rightarrow 0 < 1 - 2h < 1$

$\Rightarrow Y_n \to 0 \; as \; n \to \infty$

Option (b) is true

For (d)

The given differential equation is

$y`(x) + y(x) = 0$

$y(x) = c_1 e^{-x}$

$y(0) = 1 \Rightarrow c_1 = 1$

$y(x) = e^{-x}$

$|y(nh) - Y_n| \le |y(nh)| + |Y_n|$

$= |e^{-nh}| + |y_n|$

Here, $e^{-nh} \to 0$ as $n \Rightarrow \infty$

and $Y_n \to 0$ as $n \to \infty$

$\Rightarrow |y(nh) - Y_n| \to 0$ as $n \to \infty$

Opion (b) and (c) are correct.

3. The values of α, A, B, C for which quadrature formula **CSIR NET DEC. 2019**

$\int_{-1}^{1}(1-x)f(x)dx = Af(-\alpha)+Bf(0)+Cf(\alpha)$ is exact for polynomials of highest possible degree are

(a) $\alpha=\sqrt{\frac{3}{5}}, A=\frac{5}{9}+\frac{\sqrt{5}}{3\sqrt{3}}, B=\frac{8}{9}, C=\frac{5}{9}-\frac{\sqrt{5}}{3\sqrt{3}}$

(b) $\alpha=\sqrt{\frac{3}{5}}, A=\frac{5}{9}-\frac{\sqrt{5}}{3\sqrt{3}}, B=\frac{8}{9}, C=\frac{5}{9}+\frac{\sqrt{5}}{3\sqrt{3}}$

(c) $\alpha=\sqrt{\frac{3}{5}}, A=\frac{5}{9}\left(1-\frac{\sqrt{3}}{\sqrt{5}}\right), B=\frac{8}{9}, C=\frac{5}{9}\left(1+\frac{\sqrt{3}}{\sqrt{5}}\right)$

(d) $\alpha=\sqrt{\frac{3}{5}}, A=\frac{5}{9}\left(1+\frac{\sqrt{3}}{\sqrt{5}}\right), B=\frac{8}{9}, C=\frac{5}{9}\left(1-\frac{\sqrt{3}}{\sqrt{5}}\right)$

Ans. **(a), (d)**

Explanation :

From each option $\alpha=\sqrt{\frac{3}{5}}, B=\frac{8}{9}$ $\int_{-1}^{1}(1-x)f(x)dx = Af(-\alpha)+Bf(0)+Cf(\alpha)$

Let f(x) = 1 $\int_{-1}^{1}(1-x)dx = A+B+C$ A + B + C = 2 ..... (1)

Now, let f(x) = x $\int_{-1}^{1}f(x)dx=\int_{-1}^{1}(1-x)xdx=-A\alpha+C\alpha$

$\left[\frac{x^2}{2}-\frac{x^3}{3}\right]_{-1}^{1}=\alpha(-A+C)$ $-\frac{2}{3}=\sqrt{\frac{3}{5}}(-A+C)$ $\Rightarrow$ $A-C=\frac{2}{3}\sqrt{\frac{5}{3}}$ ..... (2)

Let f(x) = x² $\int_{-1}^{1}(1-x)x^2dx = A\alpha^2+C\alpha^2$ $\left[\frac{x^3}{3}-\frac{x^4}{4}\right]_{-1}^{1}=(A+C)\alpha^2$

$\left[\frac{1}{3}-\frac{1}{4}+\frac{1}{3}+\frac{1}{4}\right]=\frac{3}{5}(A+C)$ $\frac{2}{3}=\frac{3}{5}(A+C)$ $A+C=\frac{10}{9}$ ..... (3)

From (2) and (3) $A=\frac{5}{9}\left(1+\frac{\sqrt{3}}{\sqrt{5}}\right), C=\frac{5}{9}\left(1-\frac{\sqrt{3}}{\sqrt{5}}\right)$

Option (a) and (d) are correct.

COV, NA, I.E.

4. For f $\in$C[0, 1] and n > 1. Let $T(f)=\frac{1}{n}\left[\frac{1}{2}f(0)+\frac{1}{2}f(1)+\sum_{j=1}^{n-1}f\left(\frac{j}{n}\right)\right]$ be an approximation $I(f)=\int_0^1 f(x)dx$. For which of the following function f is T(f) = I(f)?

CSIR NET JUNE 2017

(a) 1 + sin 2πnx (b) 1 + cos 2πnx (c) $\sin^2 2\pi nx$ (d) $\cos^2 2\pi(n+1)x$

Ans. (a)

Explanation :

(a) f = 1 + sin 2nπx

$$I(f)=\int_0^1(1+\sin 2\pi n)dx=\left[x-\frac{\cos 2n\pi x}{2\pi x}\right]_0^1=1-\frac{1}{2n\pi}-0+\frac{1}{2n\pi}=1$$

$$T(f)=\frac{1}{n}\left[\frac{1}{2}(1)+\frac{1}{2}(1)+\left(1+\sin 2n\pi.\frac{1}{n}\right)+.....+\left(1+\sin 2\pi n.\frac{1}{n-1}\right)\right]$$

$$=\frac{1}{n}\left[\frac{1}{2}+\frac{1}{2}+n-1\right]=\frac{1}{n}(n)=1$$

T(f) = I(f) Option (a) is correct.

(b) f(x) = 1 + cos 2nπx

$$I(f)=\int_0^1(1+\cos 2n\pi x)dx=\left[x+\frac{\sin 2n\pi x}{2n\pi}\right]_0^1=1+0-0-0=1$$

$$T(f)=\frac{1}{n}\left[\frac{1}{2}.2+\frac{1}{2}.2+2n-2\right]=\frac{1}{n}(2n)=2$$

T(f) ≠ I(f) Option (b) is incorrect.

(c) $f(x) = \sin^2 2n\pi x$ $=\frac{1-\cos 4n\pi x}{2}$

$$I(f)=\frac{1}{2}\int_0^1(1-\cos 4n\pi x)dx=\frac{1}{2}\left[x-\frac{\sin 4n\pi x}{4\pi n}\right]_0^1=\frac{1}{2}[1-0-0+0]=\frac{1}{2}$$

$$T(f)=\frac{1}{n}\left[\frac{1}{2}.0+\frac{1}{2}.0+0\right]=0$$

T(f) ≠ I(f) Option (c) is incorrect.

(d) $f = \cos^2 2(n+1)\pi x$

$$=\frac{1+\cos 4(n+1)\pi x}{2}$$

$$I(f)=\frac{1}{2}\int_0^1(1+\cos 4(n+1)\pi x)dx=\frac{1}{2}\left[x+\frac{\sin 4(n+1)\pi x}{4(n+1)\pi}\right]_0^1=\frac{1}{2}[1+0]=\frac{1}{2}$$

$$T(f)=\frac{1}{n}\left[\frac{1}{2}.1+\frac{1}{2}.1+(n-1)\right]=\frac{1}{n}(n)=1$$

I(f) ≠ T(f) Option (d) is incorrect.

5. Consider the equation $x^2 + ax + b = 0$ which has two real roots α and β. Then which of the following iteration scheme converges when x_0 is chosen sufficient close to α?

CSIR NET JUNE 2019

(a) $x_{n+1}=-\frac{ax_n+b}{x_n}\ if\ |\alpha|>|\beta|$

(b) $x_{n+1}=-\frac{x_n^2+b}{a}\ if\ |x|>1$

(c) $x_{n+1}=-\frac{b}{x_n+a}\ if\ |\alpha|<|\beta|$

(d) $x_{n+1}=-\frac{x_n^2+b}{a};\ if\ 2|\alpha|<|\alpha+\beta|$

Ans. (a), (c), (d)

Explanation :

α&β are two roots ⇒ α + β = -a, αβ = b

(a) $g(x_n)=-\left(a+\frac{b}{x_n}\right)$ $g`(x_n)=\frac{b}{x_n^2}$ for converge to α

We need |g\`(α)| < 1 $|g`(\alpha)|=\left|\frac{b}{\alpha^2}\right|=\left|\frac{\alpha\beta}{\alpha^2}\right|=\left|\frac{\beta}{\alpha}\right|<1$

⇒ |β| < |α| Option (a) is correct.

(b) $g(x_n)=-\frac{x_n^2+b}{a}$ $g`(x_n)=-\frac{2x_n}{a}$

$|g`(\alpha)|=\left|-\frac{2\alpha}{a}\right|=\left|\frac{2\alpha}{\alpha+\beta}\right|<1$ ⇒ 2|α| < |α + β|

Option (b) is incorrect.

(c) $g(x_n)=-\frac{b}{x_n+a}$ $g`(x)=\frac{b}{(x+a)^2}$ $|g`(\alpha)|=\left|\frac{b}{(\alpha+a)^2}\right|=\left|\frac{\alpha\beta}{\beta^2}\right|=\left|\frac{\alpha}{\beta}\right|<1$

⇒ |α| < |β| Option (c) is correct.

(d) $g(x_n)=-\left(\frac{x_n^2}{a}+\frac{b}{a}\right)$ $g`(x_n)=\frac{-2x_n}{a}$ $|g`(x)|=\left|-\frac{2\alpha}{a}\right|=\left|\frac{2\alpha}{\alpha+\beta}\right|<1$

⇒ 2|α| < |α + β| Option (d) is correct.

6. Let f(x) be a polynomial of unknown degree taking the values

x	0	1	2	3
f(x)	2	7	13	16

COV, NA, I.E.

All the fourth divided differences are -1/6. Then the coefficient of x^3 is

CSIR NET DEC. 2018

(a) 1/3 (b) -2/3 (c) 16 (d) 1

Ans. (b)

Explanation :

By Newton's divided difference formula, we have

$f(x) = f(0) + (x - 1)\Delta f(0) + (x - 1)(x - 2)\Delta^2 f(0) + (x - 1)(x - 2)(x - 3)\Delta^3 f(0)$

x	f(x)	Δf(x)	Δ^2f(x)	Δ^3f(x)
0 1 2 3	2 7 13 16	$\frac{7-2}{1-0}=5$ $\frac{13-7}{2-1}=6$ $\frac{16-13}{3-2}=3$	$\frac{6-5}{2-0}=\frac{1}{2}$ $\frac{3-6}{3-1}=-\frac{3}{2}$	$\frac{-\frac{3}{2}-\frac{1}{2}}{3-0}=-\frac{2}{3}$

Hence option (b) is correct.

7. The iteration $x_{n+1} = \frac{1}{2}\left(x_n + \frac{2}{x_n}\right), n \geq 0$ for a given $x_0 \neq 0$ is an instance of

(a) fixed point iteration for $f(x) = x^2 - 2$

(b) Newton's method for $f(x) = x^2 - 2$

(c) fixed point iteration for $f(x) = \frac{x^2+2}{2x}$

(d) Newton's method for $f(x) = x^2 + 2$

Ans. (a), (c) **CSIR NET DEC 2015**

Explanation :

The given iteration is $x_{n+1} = \frac{1}{2}\left(x_n + \frac{2}{x_n}\right)$ ..... (1)

To find square root of 2, we have $f(x) = x^2 - 2$ and $f(x_n) = x_n^2 - 2$

From Newton Raphson Method we have $x_{n+1} = x_n - \frac{f(x_n)}{f'(x_n)}; f'(x_n) \neq 0$

$x_{n+1} = x_n - \frac{(x_n^2 - 2)}{2x_n} = \frac{2x_n^2 - x_n^2 + 2}{2x_n} = \frac{x_n^2 + 2}{2x_n}$ $x_{n+1} = \frac{1}{2}\left(x_n + \frac{2}{x_n}\right)$, which is same as (1)

∴ Option (b) is correct. In similar way we can check option (d) is incorrect.

For option (c),

$f(x) = \frac{x^2+2}{2x}$ $\therefore f'(x) = \frac{1}{2} - \frac{1}{x^2}$ Clearly, $|f'(x)| < 1\ \forall x$

∴ Fixed point iteration for $f(x) = \frac{x^2+2}{2x}$ is given by

COV, NA, I.E.

$$x_{n+1} = \frac{x_n^2 + 2}{2x_n} = \frac{1}{2}\left(x_n + \frac{2}{x_n}\right)$$ which is same as (1)

$\therefore$ Option (c) is correct.

In similar way, we can check option (a) is incorrect.

8. Let f(x) = $x^2 + 2x + 1$ and the derivative of f at x = 1 is approximated by using the central-difference formula $f'(1) \approx \frac{f(1+h) - f(1-h)}{2h}$ with $h = \frac{1}{2}$. Then the absolute value of the error in the approximation of f'(1) is equal to **CSIR NET JUNE 2016**

(a) 0 (b) 1/2 (c) 0 (d) 1/12

Ans. (c)

Explanation :

Given that, f(x) = $x^2 + 2x + 1$ $f'(1) \approx \frac{f(1+h) - f(1-h)}{2h}$ with $h = \frac{1}{2}$ i.e.,

$$f'(1) \approx \frac{f\left(\frac{3}{2}\right) - f\left(\frac{1}{2}\right)}{2.\frac{1}{2}} \approx f\left(\frac{3}{2}\right) - f\left(\frac{1}{2}\right) = \left(\frac{9}{4} + 2.\frac{3}{2} + 1\right) - \left(\frac{1}{4} + 2.\frac{1}{2} + 1\right) \approx \frac{9}{4} - \frac{1}{4} + 4 - 2 \approx 2 + 2 \approx 4$$

Also, f'(x) = 2x + 2 f'(1) = 4 $\therefore$ Error = 4 – 4 = 0

$\therefore$ Option (c) is correct.

Trick : In Runga-Kutta method

$y_{n+1} = y_n + ak_1 + bk_2$ where $k_1 = hf(x_n, y_n)$ and $k_2 = hf(x_n + \alpha h, y_n + \beta k_1)$ be a solution if and only iff a + b = 1, $\alpha = \beta$, $\alpha b = 1/2$.

9. Consider the Runge – Kutta method of the form $y_{n+1} = y_n + ak_1 + bk_2$ **CSIR NET JUNE 2016**

$k_1 = hf(x_n, y_n)$

$k_2 = hf(x_n + \alpha h, y_n + \beta k_1)$

$y'(x) = f(x, y(x))$, $y(x_0) = y_0$. Which of the following choices of a, b, α and β yield a second order method?

(a) $a = \frac{1}{2}, b = \frac{1}{2}, \alpha = 1, \beta = 1$ (b) $a = 1, b = 1, \alpha = \frac{1}{2}, \beta = \frac{1}{2}$

(c) $a = \frac{1}{4}, b = \frac{3}{4}, \alpha = \frac{2}{3}, \beta = \frac{2}{3}$ (d) $a = \frac{3}{4}, b = \frac{1}{4}, \alpha = 1, \beta = 1$

Ans. (a), (c)

Explanation :

In option (a) and (c) a + b = 1, $\alpha = \beta$, $\alpha b = 1/2$.

So, option (a) and (c) are true.

Trick : $x_{n+1} = \phi(x_n)$ be an iteration.
Let, e_{n+1} denotes error in $(n + 1)^{th}$ iteration, then error equation is given by

$$e_{n+1} = e_n\phi'(\xi) + \frac{e_n^2}{2!}\phi''(\xi) + ...$$

For m^{th} order convergence, $\phi'(\xi) = 0, \phi'' = 0, ..., \phi^{(m-1)} = 0$, but $\phi^m \neq 0$.

10. The values of α and β, such that $x_{n+1} = \alpha x_n\left(3 - \frac{x_n^2}{a}\right) + \beta x_n\left(1 + \frac{a}{x_n^2}\right)$ has 3rd convergence to $\sqrt{a}$, are

CSIR NET DEC 2016

(a) $\alpha = \frac{3}{8}, \beta = \frac{1}{8}$ (b) $\alpha = \frac{1}{8}, \beta = \frac{3}{8}$ (c) $\alpha = \frac{2}{8}, \beta = \frac{2}{8}$ (d) $\alpha = \frac{1}{4}, \beta = \frac{3}{4}$

Ans. (b)

Explanation :

Given iteration is $x_{n+1} = \alpha x_n\left(3 - \frac{x_n^2}{a}\right) + \beta x_n\left(1 + \frac{a}{x_n^2}\right)$ .... (1)

We know that fixed point iteration

$x_{n+1} = \phi(x_n)$ has 3rd order convergence if $\phi''(\xi) = 0$, where ξ is exact root of $f(x) = 0$

Comparing (1) with $x_{n+1} = \phi(x_n)$, we get

$$\phi(x_n) = \alpha x_n\left(3 - \frac{x_n^2}{a}\right) + \beta x_n\left(1 + \frac{a}{x_n^2}\right) = 3\alpha x_n - \frac{\alpha}{a}x^3{}_n + \beta x_n + \frac{\beta a}{x_n}$$

Here, $\xi = \sqrt{a}$ Now, $\phi'(x_n) = 3\alpha - \frac{3\alpha}{a}x_n^2 + \beta - \frac{\beta a}{x_n^2}$

$$\phi''(x_n) = -\frac{6\alpha}{a}x_n + \frac{2\beta a}{x_n^3} \qquad \therefore \phi''(\sqrt{a}) = 0 \Rightarrow \left(\frac{-6\alpha}{a}\right)\sqrt{a} + \left(\frac{2\beta}{a^3}\right)a = 0 \Rightarrow \beta = 3\alpha$$

Only, option (b) satisfies it.

$\therefore$ Option (b) is correct

11. The iterative method $x_{n+1} = g(x_n)$ for the solution of $x^2 - x - 2 = 0$ converges quadratically in a neighbourhood of the root $x = 2$ if $g(x)$ equals

(a) $x^2 - 2$ (b) $(x - 2)^2 - 6$ (c) $1 + \frac{2}{x}$ (d) $\frac{x^2 + 2}{2x - 1}$

Ans. (d)

CSIR NET DEC 2017

Explanation :

Given equation is $f(x) = x^2 - x - 2 = 0$

Result 1 : let $f(x) = 0$ is written as $x = \phi(x)$. Then the sequence of approximations $x_{n+1} = \phi(x_n)$ converges $\forall n$

if $|\phi'(x)| < 1 \ \forall \ x \in I$, where I is interval containing root $x = \xi$ of $f(x) = 0$

Result 2 : Let, e_{n+1} denotes error in $(n + 1)^{th}$ iteration, then error equation is given by

$$e_{n+1} = e_n\phi'(\xi) + \frac{e_n^2}{2!}\phi''(\xi) + ...$$

For second order convergence, $\phi'(\xi) = 0$

We can easily verify that only $\phi(x) = \frac{x^2+2}{2x-1}$ satisfies $\phi'(2) = 0$ and $|\phi'(x)| < 1 \ \forall \ x \in$ I, where I is interval containing '2'.

$\therefore$ Option (d) is correct.

12. For $f \in C[0, 1]$ and $n > 1$, let $T(f) = \frac{1}{n}\left[\frac{1}{2}f(0) + \frac{1}{2}f(1) + \sum_{j-1}^{n-1} f\left(\frac{j}{n}\right)\right]$ be an approximation of the integral $I(f) = \int_0^1 f(x)dx$. For which of the following function f is T(f) = I(f) ? **CSIR NET DEC 2017**

(a) $1 + \sin 2\pi nx$ (b) $1 + \cos 2\pi nx$ (c) $\sin^2 2\pi nx$ (d) $\cos^2 2\pi(n + 1)x$

Ans. **(a)**

Explanation :

Given, $T(f) = \frac{1}{n}\left[\frac{1}{2}f(0) + \frac{1}{2}f(1) + \sum_{j=1}^{n=1} f\left(\frac{j}{n}\right)\right]$ *and* $I(f) = \int_0^1 f(x)dx$

For option (a),

$f(x) = 1 + \sin 2\pi nx$

Then, $T(f) = \frac{1}{n}\left[\frac{1}{2} + \frac{1}{2} + \left(1 + \sin 2\pi n.\frac{1}{n}\right) + ... + \left(1 + \sin 2\pi n.\left(\frac{n-1}{n}\right)\right)\right]$

$\Rightarrow$ $T(f) = \frac{1}{n}[n] = 1$ *and* $I(f) = \int_0^1 [1 + \sin(2n\pi x)]dx = 1$

$\therefore$ T(f) = I(f)

$\therefore$ Option (a) is correct, For option (b), $f(x) = 1 + \cos(2n\ \pi x)$

$$T(f) = \frac{1}{n}\left[\frac{1}{2}(2) + \frac{1}{2}(2) + 2n - 2\right] \Rightarrow T(f) = \frac{2n}{n} = 2 \text{ and } I(f) = \int_0^1 1 + \cos(2n\pi x)dx = 1$$

$\therefore$ $T(f) \neq I(f)$

Thus, option (b) is incorrect. For option (c).

$$f(x) = \sin^2(2\pi nx) = \frac{1-\cos(4n\pi x)}{2} \text{ and } T(f) = \frac{1}{n}\left[\frac{1}{2}(0) + \frac{1}{2}(0) + 0\right] = 0$$

COV, NA, I.E.

$$I(f)=\int_0^1 \frac{1-\cos(4n\pi x)}{2}dx \neq 0 \Rightarrow T(f)\neq I(f)$$

$\therefore$ Option (d) is incorrect.

13. A Runge-Kutta method for numerical solving the initial value ordinary differential equation $\frac{dy}{dx} = y' = f(x,y)$; $y(x_0) = y_0$ is given by **CSIR NET JUNE 2013**

$y(x + h) = y(x) + w_1F_1 + w_2F_2$

$F_1 = h\, f(x, y)$, $F_2 = h\, f(x + \alpha h, y + \beta F_1)$

Which of the following are correct sets of the value for these constants.

(a) $w_1 = \frac{1}{2}, w_2 = \frac{1}{2}, \alpha = 1, \beta = 1$ (b) $w_2 = 1, w_1 = 2, \alpha = \frac{1}{2}, \beta = \frac{1}{2}$

(c) $w_1 = \frac{1}{3}, w_2 = \frac{2}{3}, \alpha = \frac{3}{4}, \beta = \frac{3}{4}$ (d) $w_1 = \frac{3}{4}, w_2 = \frac{1}{4}, \alpha = 2, \beta = 2$

Ans. **(c), (d)**

Explanation :

It is third order Runge-Kutta method

We know about condition $w_1 + w_2 = 1$

$$w_2\alpha = w_2\beta = \frac{1}{2}$$

Option (a) is correct.

Option (b) is incorrect.

$w_1 + w_2 = 3$

Option (c) is correct.

$$w_2\alpha = \frac{2}{3}\times\frac{3}{4} = \frac{1}{2}$$

$$w_2\beta = \frac{2}{3}\times\frac{3}{4} = \frac{1}{2}$$

Option (d) is correct.

$$w_1 + w_2 = \frac{3}{4} + \frac{1}{4} = 1$$

$$w_2\alpha = \frac{1}{4}\times 2 = \frac{1}{2}$$

$$w_2\beta = \frac{1}{4}\times 2 = \frac{1}{2}$$

UNSOLVED QUESTIONS

1. Let $f(x) = ax + b$ for $a, b \in R$. Then the iteration $x_{n+1} = f(x_n)$ starting from any given x_0 for $n \geq 0$ converges **CSIR NET JUNE 2014**

(a) for all $a \in R$ (b) for no $a \in R$ (c) for $a \in [0, 1]$ (d) only for $a = 0$

2. Consider the iteration $x_{n+1} = \frac{1}{2}\left(x_n + \frac{2}{x_n}\right)$, $n \geq 0$ for given $x_0 \neq 0$. Then

CSIR NET JUNE 2014

(a) x_n converges to $\sqrt{2}$ with rate of convergence 1.

(b) x_n converges to $\sqrt{2}$ with rate of convergence 2.
(c) The given iteration is the fixed point iteration for $f(x) = x^2 - 2$.
(d) The given iteration is the Newton's method for $f(x) = x^2 - 2$.

3. Let $f : R \to R$ be a smooth function with non-vanishing derivative. The Newton's method for finding a root of $f(x) = 0$ is the same as **CSIR NET DEC 2014**
(a) fixed point iteration for the map $g(x) = x - f(x)/f'(x)$
(b) Forward Euler method with unit step length for the differential equation $\frac{dy}{dt} + \frac{f(y)}{f'(y)} = 0$
(c) fixed point iteration for $g(x) = x + f(x)$ (d) fixed point iteration for $g(x) = x - f(x)$

4. Consider the iteration function for Newton's method $g(x) = x - \frac{f(x)}{f'(x)}$ and its application to find (approximate) square root of 2, starting with $x_0 = 2$. Consider the first and the second iterates x_1 and x_2, respectively; then **CSIR NET DEC 2012**
(a) $1.5 < x_1 \le 2$ (b) $1.5 \le x_1 < 2$ (c) $x_1 \le 1.5$; $x_2 \le 1.5$ (d) $x_1 = 1.5$; $x_2 < I$

5. For the computation of $\sqrt{x+1} - 1$ at $x = 1.2345678 \times 10^{-5}$ using a machine which keeps 8 significant digits, which of the following equivalent expression would be best to use **CSIR NET DEC 2011**
(a) $\sqrt{x+1} - 1$ (b) $\left(1 - \frac{1}{\sqrt{x-1}}\right)\sqrt{x+1}$ (c) $\frac{x}{\sqrt{x+1}+1}$ (d) $\frac{1}{2}x - \frac{1}{8}x^2 + \frac{1}{16}x^3 \ldots$

6. Consider the interpolation data given below : **CSIR NET DEC 2011**

x	1	1/2	3
y	3	-10	2

The interpolating polynomial corresponding to this data is given by
(a) $p(x) = -3\left(x - \frac{1}{2}\right)(x-3) - 8(x-1)(x-3) + \frac{2}{5}(x-1)\left(x - \frac{1}{2}\right)$
(b) $q(x) = 3 + 26(x-1) + \frac{-53}{5}(x-1)\left(x - \frac{1}{2}\right)$
(c) $r(x) = \frac{-53}{5}x^2 + \frac{419}{10}x + \frac{-283}{10}$ (d) p(x) q(x) + r(x)

7. Given that an upper triangular matrix (UTM) is invertible if and only if all its diagonal elements are different from zero, consider the linear system **CSIR NET JUNE 2012**
$2x_1 + 3x_2 - x_3 = 5$
$4x_1 + 4x_2 - 3x_3 = 3$
$-2x_1 + 3x_2 - x_3 = 1$
Then system is

COV, NA, I.E.

(a) can be transformed into an UTM but is not invertible because the diagonal entries of the UTM are not different from zero.
(b) is invertible though cannot be transformed into a UTM
(c) can be transformed into an UTM because above diagonal entries are all different from zero.
(d) can be transformed into a UTM and the solution of the UTM is the solution of (a).

8. Let f be a continuous map from the interval [0,1] into itself and consider the iteration $x_{n+1} = f(x_0)$. Which of the following maps will yield a fixed point for f?

CSIR NET DEC 2012

(a) $f(x) = x^2/4$ (b) $f(x) = x^2/8$ (c) $f(x) = x^2/16$ (d) $f(x) = x^2/3$

9. If the points $x_1, x_2, \ldots, x_n$ are distinct, then for arbitrary real values $y_1, y_2, \ldots, y_n$ the degree of the unique interpolating polynomial p(x) such that $p(x_1) = y_1 (1 \le i \le n)$ is
(a) n (b) n – 1 (c) ≤ n – 1 (d) ≤ n

10. Consider a sufficiently smooth function f(x). A formula for estimating its derivative is given by $\frac{df}{dx} = \frac{1}{4h}[f(x+2h) - f(x-2h)]$ + error term where h > 0. Let $f^{(n)}$ denote the nth derivative of f and let ξ be a point between x – 2h and x + 2h. Which of the following expressions for the error term are correct? **CSIR NET JUNE 2013**

(a) $\frac{-f^{(2)}(\xi)h^2}{2}$ (b) $\frac{-f^{(3)}(\xi)h^2}{3}$ (c) $-f^{(1)}(\xi)h$ (d) $\frac{-f^{(4)}(\xi)h^4}{12}$

11. A Runge-Kutta method for numerically solving the initial – value ordinary differential equation $y' = f(x, y)$; $y(x_0) = y_0$ is given by (for h small) $y(x + h) = y(x) + w_1F_1(x, y) + w_2F_2(x, y)$
$F_1(x, y) = hf(x, y)$
$F_2(x, y) = hf(x + \alpha h, y + \beta F_1)$.
The objective is to determine the constants w_1, w_2, α and β such that the above formula is accurate to order 2 (that is, the error term is $O(h^3)$). Which of the following are correct sets of values for these constants? **CSIR NET JUNE 2013**

(a) $w_1 = \frac{1}{2}, w_2 = \frac{1}{2}; \alpha = 1, \beta = 1$ (b) $w_1 = 2, w_2 = 1; \alpha = \frac{1}{2}, \beta = \frac{1}{2}$

(c) $w_1 = \frac{1}{3}, w_2 = \frac{2}{3}; \alpha = \frac{3}{4}, \beta = \frac{3}{4}$ (d) $w_1 = \frac{3}{4}, w_2 = \frac{1}{4}; \alpha = 2, \beta = 2$

12. Let $f(x) = e^x$ be approximated by Taylor's polynomial of degree n at the point x = ½ and the entire interval [0, 1]. If the absolute error in this approximation does not exceed 10^{-2}, then the value of n should be taken as? **CSIR NET DEC 2013**
(a) 0 (b) 1 (c) 2 (d) 3

13. Let $f \in C^3([x_{-1}, x_1])$ where $x^{-1} = x_0 - h$, $x_1 = x_0 + h$ with h > 0, $f(x_0) = f_0$, $f(x_2) = f_1$ for j = -1, 1 and $f'(x_0) = f'_0$. Then for some $\xi \in (x_{-1}, x_1)$ we have **CSIR NET DEC 2013**

(a) $f'_0 = \frac{f_1 - f_0}{h} - \frac{h^2}{2} f''(\xi)$ (b) $f'_0 = \frac{f_1 - f_{-1}}{h} - \frac{h^3}{3} f''(\xi)$

(c) $f'_0 = \frac{f_1 - f_{-1}}{2h} - \frac{h^2}{6} f''(\xi)$ (d) $f'_0 = \frac{f_1 - f_{-1}}{2h} - \frac{h^3}{6} f''(\xi)$

14. Let f(x) = ax + 100 for a ∈ R. Then the iteration $x^{n+1} = f(x_n)$ for n ≥ 0 and $x_0 = 0$ converges for **CSIR NET DEC 2015**

(a) a = 5 (b) a = 1 (c) a = 0.1 (d) a = 10

15. The order of linear multi step method **CSIR NET DEC 2016**

$u_{j+1} = (1-a)u_j + au_{j-1} + \frac{h}{4}\{(a+3)u'_{j+1} + (3a+1)u'_{j-1}\}$ for solving u' = f(x, u) is

(a) 2 if a = -1 (b) 2 if a = -2 (c) 3 if a = -1 (d) 3 if a = -2

16. The forward difference operator is defined as $\Delta U_n = U_{n+1} - U_n$. Then which of the following difference equations has an unbounded general solution?

CSIR NET JUNE 2018

(a) $\Delta^2 U_n - 3\Delta U_n + 2U_n = 0$ (b) $\Delta^2 U_n + \Delta U_n + \frac{1}{4} U_n = 0$

(c) $\Delta^2 U_n - 2\Delta U_n + 2U_n = 0$ (d) $\Delta^2 U_{n+1} - \frac{1}{3}\Delta^2 U_n = 0$

17. Let $f(x) = \sqrt{x+3}$ for x ≥ -3. Consider the iteration $x_{n+1} = f(x_n)$, $x_0 = 0$; n ≥ 0. The possible limits of the iteration are **CSIR NET DEC 2015**

(a) -1 (b) 3 (c) 0 (d) $\sqrt{3+\sqrt{3+\sqrt{3+....}}}$

18. The values of a, b, c such that $\int_0^h f(x)dx = h\left\{af(0) + bf\left(\frac{h}{3}\right) + cf(h)\right\}$ is exact for polynomials f of degree as high as possible are **CSIR NET JUNE 2018**

(a) $a = 0, b = \frac{3}{4}, c = \frac{1}{4}$ (b) $a = \frac{3}{4}, b = \frac{2}{4}, c = \frac{1}{4}$

(c) $a = \frac{-2}{4}, b = \frac{3}{4}, c = \frac{1}{4}$ (d) $a = 0, b = \frac{1}{4}, c = \frac{3}{4}$

Answer Key

1	2	3	4	5	6	7	8	9	10
c	b,d	a,b	a, c	d	a, b, c	c, d	a,b,c,d	c	b
11	12	13	14	15	16	17	18		
a,c,d	d	c	c	b,c	a,c,d	d	a		

COV, NA, I.E.

Chapter 3

Integral Equation

1. Integral Equation :

An equation of the form $g(x).y(x) = f(x) + \lambda \int_a k(x,t)y(t)dt$ where g(x), f(x), k(x,t) are given functions and λ is a constant is linear integral equation.

Linear and non-linear integral equations :

Definition : An integral equation is called linear if only linear operation are perform.
If L is a integral equation
Then, $L[C_1y_1(x) + C_2y_2(x) = C_1L[y_1(x)] + C_2L[Y_2(x)]$
An integral equation is not a linear then it is called non-linear integral equation.

The most general form of Integral Equation :

$$g(x).y(x) = f(x) + \lambda \int_a k(x,t)y(t)dt \quad .....(1)$$

Where the upper limit may be either variable x or fixed, λ is non-zero real or complex.
Then k(x,t) is known as the kernel of integral equation.

1. If g(x) = 0 then equation (1) is called 1st kind integral equation.
2. If g(x) ≠ 0 then equation (1) is called 2[nd]kind integral equation.

Fredholm integral equation :

$$g(x).y(x) = f(x) + \lambda \int_a^b k(x,t)y(t)dt$$ if both limit are constant.

Voltera integral equation :

$$g(x).y(x) = f(x) + \lambda \int_a^x k(x,t)y(t)dt$$ if upper limit is variable.

Homogeneous integral equation : If f(x) = 0 then equation (1) is called homogeneous integral equation.

Singular integral equation : When one or both limit of integration become infinite or when kernel become infinite at one or more point within then range of integration.

$$y(x) = f(x) + \lambda \int_{-\infty}^{\infty} k(x,t)y(t)dt$$

Type of Kernal :

1. **Symmetric kernel→** if k(x,t) = k(t, x)
   Example. (i) k(x,t) = sin(x + t)
   (ii) $k(x,t) = e^{x^2+y^2}$

2. **Separable or degenerate kernel→** if $k(x,t)=\sum_{i=1}^{n} g_i(x)h_i(t)$

Example. sin(x + t) = sin x cos t + cos x sin t

Integral equation of convolution type : k(x,t) = k(x – t)

Leibnitz's rule of differential under integral sign

$$\frac{d}{dx}\int_{G(x)}^{H(x)} F(x.t)dt=\int_{G(x)}^{H(x)}\frac{\partial F}{\partial x}dt+F[x,H(x)]\frac{dH}{dx}-F[x,G(x)]\frac{dG}{dx}$$

Case. If G and H are constant then $\frac{d}{dx}\int_{G}^{H} F(x,t)dt=\int_{G}^{H}\frac{\partial F}{\partial x}dt$

Note : $\underbrace{\int_a^x\int_a^x\int_a^x ....\int_a^x}_{n-times} y(t)\ dtdt.....dt=\int_a^x\frac{(x-t)^{n-1}}{(n-1)!}y(t)dt$

Example :- Show that y(x) = 3 is a solution of integral equation $x^3=\int_0^x (x-t)^2 y(t)dt$

Solution : Given that y(t) = 3 $\quad 3\int_0^x (x-t)^2\,dt=-3\int_0^x\frac{(x-t)^3}{3}\Big]_0^x=+x^3$

2. **Conversion of differential equation into integral equation.**

A boundary value problem of form y'' + c_1(x)y'+ c_2(x)y = c_3(x), y(a) = y_0 and y(b) = y_1 can be transferred into Fredholm integral equation.

A initial value problem of form y'' + c_1(x)y' + c_2(x)y = c_3(x), y(a) = y_0, y'(a) = y_1 can be transformed into Volterra integral equation.

Example : The initial value problem

$$\frac{d^2y}{dx^2}+y=0;x>0$$
$$y(0)=1,y'(0)=0,$$

In equaivalent to the Volterra integral equation

(a) $y(x)=1+\int_0^x (t-x)y(t)dt$ (b) $y(x)=1+\int_0^x (t+x)y(t)dt$

(c) $y(x)=1+\int_0^x xty(t)dt$ (d) $y(x)=1+\int_0^x (x-t)y(t)dt$

Ans. **(a)**

Explanation :

Given that $y''=-y$

COV, NA, I.E.

Integrating (1) both sides, we get

$$y` = -\int_0^x y(t)dt + c_1$$

Integrating again, we get

$$y'' = -\int_0^x\int_0^x y(t)dt + c_1x + c_2$$

$$\Rightarrow y(x) = -\int_0^x (x-t)y(t)dt + c_1x + c_2$$

$$\Rightarrow y(0) = c_2 = 1$$

Since, $y'(x) = -\int_0^x y(t)dt + c_1$

$$y'(0) = c_1 = 0$$

$$\Rightarrow y(x) = -\int_0^x (x-t)y(t) + 1$$

$$\Rightarrow y(x) = 1 + \int_0^x (t-x)y(t)dt$$

Hence, option (a) is correct.

3. Conversion of integral equation into ordinary differential equation (by using Leibnitz formula)

Example. The initial value problem corresponding to integral equation

$y(x) = 1 + \int_0^x y(t)dt$ is **GATE-2001**

(a) y' – y = 0, y(0) = 1 (b) y' + y = 0, y(0) = 0

(c) y' – y = 0, y(0) = 0 (d) y' + y = 0, y(0) = 1

Ans. **(a)**

Explanation :

Given that

$$y(x) = 1 + \int_0^x y(t)dt \quad .....(1)$$

Differential both side w.r.t. x

$$y' = 0 + \frac{d}{dx}\int_0^x y(t)dt \quad \text{[By using Leibnitz's rule]}$$

$$y' = \int_0^x \frac{\partial}{\partial x} y(t)dt + y(x)\frac{dx}{dx} - y(0)\frac{d.0}{dx} \quad y' = \int_0^x \frac{\partial y(t)}{\partial x}dt + y(x)$$

y' – y = 0

By equation (1) put x = 0

$$y(0) = 1 + \int_0^0 y(t)dt$$

y(0) = 1

COV, NA, I.E.

SOLVED QUESTIONS

1. Consider continuous solution f of the following integral equation in [0, 1]

$$f^2(t)=1+2\int_0^t f(s)ds,\ \forall t\in[0,1]$$

Which of the following statements are true? **CSIR NET NOV 2020**

(a) There is no solution
(b) There is exactly one solution
(c) There are exactly two solution
(d) There are more than two solution

Ans. (c)

Explanation :

Option (c) is correct.

2. A characteristic number and the corresponding eigen function of the homogeneous Fredholm integer equation with kernel $K(x,t)=\begin{cases} x(t-1), & 0\le x\le t \\ t(x-1), & t\le x\le 1\end{cases}$ are

(a) $\lambda=-\pi^2,\varphi(x)=\sin\pi x$
(b) $\lambda=-2\pi^2,\varphi(x)=\sin 2\pi x$
(c) $\lambda=-3\pi^2,\varphi(x)=\sin 3\pi x$
(d) $\lambda=-4\pi^2,\varphi(x)=\sin 2\pi x$

Ans. (a),(d) **CSIR NET 2018**

Explanation :

Let $\varphi(x)=\lambda\int_0^1 k(x,t)\varphi(t)dt$ be a homogeneous Fredholm integral equation

$$\varphi(x)=\lambda\left[\int_0^x t(x-1)\varphi(t)dt+\int_x^1 x(t-1)\varphi(t)dt\right]$$

$$\varphi'(x)=\lambda\int_0^x t\varphi(t)dt+\lambda x(x-1)\varphi(x)+\lambda\int_x^1(t-1)\varphi(t)dt-\lambda x(x-1)\varphi(x)$$

$$\varphi'(x)=\lambda\int_0^x t\varphi(t)dt+\lambda\int_x^1(t-1)\varphi(t)dt \qquad \varphi''(x)=\lambda x\varphi(x)-\lambda(x-1)\varphi(x)$$

$$\varphi''(x)=\lambda\varphi \qquad ....(1)$$

Also, $\varphi(0)=\varphi(1)=0$

Solving the boundary value problem (1)

Case – 1 : If $\lambda=0$ $\qquad \varphi''(x)=0\Rightarrow\varphi(x)=c_1x+c_2$

Since $\varphi(0)=\varphi(1)=0\Rightarrow\varphi(x)=0$ (Trivial solution)

$\Rightarrow\lambda=0$ is not characteristic number.

Case: 2 If $\lambda>0$, then $\lambda=\mu^2,\mu\neq 0$

$$\varphi''(x)-\mu^2\varphi(x)=0\Rightarrow\varphi(x)=c_1e^{\mu x}+c_2e^{-\mu x}$$

COV, NA, I.E.

$\varphi(0)=c_1+c_2=0 \Rightarrow c_1=-c_2$

$\varphi(1)=c_1e^{\mu}+c_2e^{-\mu}=0$

$\Rightarrow c_1\left(e^{\mu}-e^{-\mu}\right)=0 \Rightarrow c_1\left(e^{2\mu}-1\right)=0$

Since $\mu \neq 0$

Therefore, $c_1=-c_2=0$, $\lambda>0$

is not characteristic number.

$\Rightarrow \varphi(x)=c_1 \cos \mu x+c_2 \sin \mu x$

$\Rightarrow \varphi(0)=c_1=0$

and $\varphi(1)=c_1 \cos \mu+c_2 \sin \mu=0$

$\Rightarrow c_2 \sin \mu=0$

If $c_2=0 \Rightarrow \varphi(x)=0$

If $c_2 \neq 0 \Rightarrow \mu=n\pi$

$\Rightarrow \varphi(x)=c_2 \sin \mu x=c_2 \sin n\pi x$

and $\lambda=-\mu^2=-n^2\pi^2$

For $n=1, c_2=1$

$\lambda=-\pi^2, \varphi(x)=\sin 2\pi x$

And For $n=3, c_2=1$

$\lambda=-9\pi^2, \phi(x)=\sin 3\pi x$

Hence, options (a) and (d) are correct

3. Which are the characteristic numbers and the corresponding eigen-functions for the Fredholm homogeneous equation whose kernel is

$$K(x,t)=\begin{cases}(x+1)t, & 0 \le x \le t \\ (t+1)x, & t \le x \le 1\end{cases}$$

CSIR NET 2016

(a) $1, e^x$ (b) $-\pi^2, \pi \sin \pi x+\cos \pi x$ (c) $-4\pi^2, \pi \sin \pi x+\pi \cos 2\pi x$

(d) $-\pi^2, \pi \cos \pi x+\sin \pi x$

Ans. (a),(d)

Explanation :

$$\phi(x)=\lambda\int_0^1 K(x,t)\phi(t)dt \quad \phi(x)=\lambda\left[\int_0^x (t+1)x\phi(t)dt+\int_x^1 t(x+1)\phi(t)dt\right]$$

$$\phi'(x)=\lambda\int_0^x (t+1)\phi(t)dt+\lambda x(x+1)\phi(x)+\lambda\int_x^1 t\phi(t)dt-\lambda x(x+1)\phi(x)$$

$\phi''(x)=\lambda(x+1)\phi(x)-\lambda x\phi(x)$

$\phi''(x)=\lambda\phi(x) \Rightarrow \phi''(x)=\lambda\phi(x)$

$\phi''(x)-\lambda\phi(x)=0$

Also $\phi(0)=\phi'(0)$ and $\phi(1)=\phi'(1)$

Case – I : If $\lambda=0$

$\phi(x)=c_1(x)+c_2 \Rightarrow \phi(0)=\phi'(0) \Rightarrow c_2=c_1$

and $\phi(1)=\phi'(1) \Rightarrow c_1+c_2=c_1 \Rightarrow c_2=c_1=0$

$\Rightarrow \phi(x)=0$

Case – II: If $\lambda=\mu^2, \mu\neq 0$

$$\phi''(x)-\mu^2\phi(x)=0$$

$\Rightarrow \phi(x)=c_1e^{\mu x}+c_2 2^{-\mu x}$

$\because \phi(0)=\phi'(0)\Rightarrow c_1+c_2=\mu c_1-\mu c_2$

$\Rightarrow c_1(1-\mu)+c_2(1+\mu)=0 \ldots(1)$

Also $\phi(1)=\phi'(1)$

$\Rightarrow c_1e^{\mu}+c_2e^{-\mu}=\mu c_1e^{\mu}-\mu c_2e^{-\mu}$

$\Rightarrow c_1e^{\mu}(1-\mu)+c_2e^{-\mu}(1+\mu)=0 \ldots(2)$

From equation (1) $\Rightarrow c_2=\dfrac{(\mu-1)}{(\mu+1)}c_1$

Put this value in (2), we get

$c_1e^{\mu}(1-\mu)+(\mu-1)c_1e^{-\mu}=0$

$c_1e^{\mu}\left[(1-\mu)+(\mu-1)e^{-2\mu}\right]=0$

$e^{\mu}\neq 0$ and $c_1\neq 0\Rightarrow \mu=1\Rightarrow c_2=0$

$\Rightarrow \lambda=1$ and $\phi(x)=c_1e^x$

Choose $c_1=1\Rightarrow \phi(x)=e^x$

Case – III: If $\lambda=-\mu^2, \mu\neq 0$

$\phi''(x)+\mu^2\phi(x)=0$

$\Rightarrow \phi(x)=c_1\cos\mu x+c_2\sin\mu x$

$\Rightarrow \phi'(x)=-\mu c_1\sin\mu x+\mu c_2\cos\mu x$

$\Rightarrow \phi(0)=\phi'(0)\Rightarrow c_1=\mu c_2$

$\Rightarrow c_1-\mu c_2=0$ and $\phi(1)=\phi'(1)$

$\Rightarrow c_1\cos\mu+c_2\sin\mu=-\mu c_1\sin\mu+\mu c_2\cos\mu$

$\Rightarrow (c_1-\mu c_2)\cos\mu+(c_2+\mu c_1)\sin\mu=0$

$\Rightarrow (c_2+\mu c_1)\sin\mu=0$

$\Rightarrow \mu=n\pi$, because if $c_2=-\mu c_1$

Then again $\phi(x)=0\Rightarrow \lambda=-n^2\pi^2$

$\Rightarrow \phi(x)=\mu c_2\cos\mu x+c_2\sin\mu x$

Let $c_2=1$

$\Rightarrow \phi(x)=\mu c_2\cos\mu x+c_2\sin\mu x$

Let $c_2=1\Rightarrow \phi(x)=n\pi\cos n\pi x+c_2\sin\mu x$

For $n=1\Rightarrow \phi(x)=\pi\cos\pi x+\sin\pi x$

and $\lambda=-\pi^2$ For $n=2$

$\lambda=-4\pi^2$ and $\phi(x)=-2\pi\cos 2\pi x+\sin 2\pi x$

Hence, options (a) and (d) are correct.

4. The value of λ for which the following equation has a non-trivial solution $\phi(x)=\lambda\int_0^{\pi}K(x,t)\phi(t)dt, 0\leq x\leq\pi$ where $K(x,t)=\begin{cases}\sin x\cos t, & 0\leq x\leq t\\ \cos x\sin t, & t\leq x\leq\pi\end{cases}$ are

CSIR NET 2018

(a) $\left(n+\dfrac{1}{2}\right)^2-1, n\in N$ (b) $n^2-1, n\in N$

(c) $\dfrac{1}{2}(n+1)^2-1, n\in N$ (d) $\dfrac{1}{2}(2n+1)^2-1, n\in N$

Ans. (a)

Explanation :

$$\phi(x)=\lambda\int_0^x\cos x\sin t d(t)dt+\lambda\int_x^{\pi}\sin x\cos t\phi(t)dt \ldots(1)$$

By using Leibnitz's Rule, Differential w.r.t. x

COV, NA, I.E.

$$\phi'(x)=-\lambda\int_0^x \sin x\sin t\ \phi(t)dt+\lambda\cos x\sin x\phi(x)+\lambda\int_x^\pi \cos x\cos t\,\phi(t)+0-\lambda\sin x\cos x\phi(x)$$

$$\Rightarrow \phi'(x)=-\lambda\int_0^x \sin x.\sin t\ \phi(t)dt+\lambda\int_x^\pi \cos x.\cos t\phi(t)dt\ .....(2)$$

$$\Rightarrow \phi''(x)=-\lambda\int_0^x \cos x\sin t\ \phi(t)dt-\lambda\sin x.\sin x\ \phi(x)-\lambda\int_x^\pi \sin x\cos t\phi(t)dt-\lambda\cos x\cos x\ \phi($$

$$\Rightarrow \phi''(x)=-\lambda\int_0^x \cos x\sin t\ \phi(t)dt+\lambda\int_x^\pi \sin x\cos t\ \phi(t)dt-\left[\lambda\sin^2 x\phi(x)+\lambda\cos^2 x\phi(x)\right]$$

Put the value of $\phi(x)$ from (1)

$\Rightarrow \phi''(x) - \phi(x) - \lambda\phi(x)$

$\phi''(x)+(\lambda+1)\phi(x)=0$ …. (3)

Put x = 0 in (1), Then $\phi(0) = 0$

Put $x = \pi$ in (2) then $\phi'(\pi) = 0$

Auxilliary equation of (3)

$m^2 + (\lambda + 1) = 0,\ m^2 = -(1 + \lambda)$

$m=\pm\sqrt{1+\lambda}\ i$

$\phi(x)=c_1\sin\left(\sqrt{1+\lambda}\right)x+c_2\cos\left(\sqrt{1+\lambda}\right)x$

By using of condition

$\phi(0) = 0 = c_2,\quad \phi(x)=c_1\sin(\sqrt{1+\lambda})x$

$$\phi'(x)=\frac{c_1}{2\sqrt{1+\lambda}}\cos(\sqrt{1+\lambda})x$$

$$\phi'(\pi)=0=\frac{c_1}{2\sqrt{1+\lambda}}\cos(\sqrt{1+\lambda})\pi$$

$$\sqrt{1+\lambda}\pi=(2n+1)\frac{\pi}{2}$$

$$1+\lambda=\left(n+\frac{1}{2}\right)^2\quad \lambda=\left(n+\frac{1}{2}\right)^2-1$$

4. Fredhold I.E. of second kind with separable kernel :

Solving Fredholm integral equations

Separable kernel : $k(x,t)=\sum_{i=1}^n f_i(x)g_i(t)$, then Fredholm integral equation reduces to

$$y(x)=f(x)+\lambda\int_a^b \sum f_i(x)g_i(t)y(t)dt\ or\ y(x)=f(x)+\lambda\sum f_i(x)\int_a^b g_i(t)y(t)dt$$

$$\text{Let } \int_a^b g_i(t)y(t)dt=c_i \qquad i=1,2,\ldots..,n$$

$$\therefore\ y(x)=f(x)+\lambda\sum_{i=1}^n c_i f_i(x),\qquad i=1,2,\ldots.,n$$

Find c_i and solve.

5. For the Fredholm integral equation $\lambda\int_0^1 e^s e^t y(t)dt$ which of the following statements are true?

CSIR NET NOV 2020

(a) It has a non trivial solution satisfying $\int_0^1 e^t y(t)dt = 0$.

(b) Only the trivial solution satisfies $\int_0^1 e^t y(t) = 0$. (c) It has non-trivial solution for all $\lambda \neq 0$

(d) It has non-trivial solution if $\lambda = \frac{2}{e^2-1}$ and $\int_0^1 e^t y(t)dt \neq 0$.

Ans. **(b), (d)**

Explanation :

$$y(s) = \lambda\int_0^1 e^s e^t y(t)dt = \lambda e^s \int_0^1 e^t y(t)dt = \lambda e^s c_1$$

Where $c_1 = \int_0^1 e^t y(t)dt \Rightarrow \quad c_1 = \int_0^1 e^t \lambda e^t c_1 dt = \lambda c_1 \int_0^1 e^{2t} dt = \lambda c_1 \left.\frac{e^{2t}}{2}\right|_0^1 = \lambda c_1 \frac{e^2-1}{2}$

$\Rightarrow \quad c_1 = c_1.\lambda\frac{e^2-1}{2} \Rightarrow \quad \lambda = \frac{2}{e^2-1} \quad$ If $\int_0^1 e^t y(t)dt = 0 \Rightarrow c_1 = 0$

Then y(x) = 0 is the only solution which is the trivial solution.

So, we concluded y(x) = 0 satisfy $\int_0^1 e^t y(t)dt = 0$ i.e. option (b) is correct.

And $\lambda = \frac{2}{e^2-1}$ and $\int_0^1 e^t y(x)dt \neq 0$ then we get non-trivial solution i.e. Option (d) is correct.

From (3) y(x) = 0 which is a zero solution i.e. no characteristic number.

6. Consider the integral equation $\phi(x) - e/2\int_{-1}^1 xe^t \phi(t)dt = f(x)$ then

CSIR NET JUNE 2019

(a) ∃a continuous function f : [-1, 1] → (0, ∞) for which solution exist
(b) ∃continuous function f : [-1, 1] → [-∞, 0] for which solution exist
(c) $f(x) = e^{-x}(1 - 3x^2)$, a solution exist. (d) $f(x) = e^{-x}(x + x^3 + x^5)$ a solution exist.
Ans. **(c),(d)**

COV, NA, I.E.

Explanation :

Given that $\phi(x) = f(x) + \frac{e}{2}\int_{-1}^{1} xe^t\phi(t)dt$

$$\phi(x) = f(x) + \frac{e}{2}x\int_{-1}^{1} e^t\ \phi(t)dt \qquad ....(1)$$

$$\phi(x) = f(x) + \frac{ex}{2}.C \qquad .....(2)$$

where $C = \int_{-1}^{1} e^t\phi(t)dt \qquad ..... (3)$

So $\boldsymbol{\phi(t) = f(t) + \frac{et}{2}C}$

After solve we get

$$\Rightarrow \int_{-1}^{1} e^t f(t)dt = 0$$

We got a relation.

Option (a) is incorrect.

Let f(t) > 0 in [a, b] then $\int_a^b f(t)dt > 0$

Given that f : [-1, 1] → (0, ∞)

i.e. Range of f is contain in R*

i.e. f(x) > 0 and e^t> 0

$\int_{-1}^{1} e^t f(t)dt > 0$ (never zero)

Option (b) is incorrect.

Result : Let f(x) < 0 in [a, b]

then $\int_a^b f(t)dt < 0$

here we know that e^t> 0 and

f : [-1, 1] → (-∞, 0)

i.e. f(x) < 0

then

$$e^t f(t) < 0 \quad \Rightarrow \quad \int_{-1}^{1} e^t f(t) < 0$$

(non-zero)

Option (c) is correct.

Given that $f(x) = e^{-x}(1 - 3x^2)$

$$\int_{-1}^{1} e^t[e^{-t}(1-3t^2)]dt = \int_{-1}^{1}(1-3t^2)dt$$

$$= [t - t^3]_{-1}^{1} = 0$$

Option (d) is correct.

Given that $f(x) = e^{-x}(x + x^3 + x^5)$

$$\int_{-1}^{1} e^t e^{-t}(t + t^3 + t^5)dt$$

$$= \left[\frac{t^2}{2} + \frac{t^4}{4} + \frac{t^6}{6}\right]_{-1}^{1}$$

$$= \left[\frac{1}{2} + \frac{1}{4} + \frac{1}{6}\right] - \left[\frac{1}{2} + \frac{1}{4} + \frac{1}{6}\right] = 0$$

7. Let φ be the solution of the integral equation $\frac{1}{2}\phi(x) = x^2 + \int_0^1 e^{x-y}\phi(y)dy$. Then

(a) $\phi(0) = 20e^{-1} - 8$ (b) $\phi(0) = 20e - 8$ (c) $\phi(1) = 22 - 8e$ (d) $\phi(1) = 22 - 8e^{-1}$

Ans. (b) **CSIR NET 2017**

Explanation :

Given that $\phi(x) = 2x^2 + 2\int_0^1 e^{x-y}\phi(y)dy \qquad ...... (1)$

$$\phi(x) = 2x^2 + 2e^x\int_0^1 e^{-y}\phi(y)dy$$

Let $C=\int_0^1 e^{-y}\phi(y)dy$ .....(2)

$\phi(x)=2x^2+2e^x C$ ..... (3)

$\phi(y) = 2y^2 + 2e^yC$ then by (2)

$C=\int_0^1 e^{-y}\left[2y^2+2e^y C\right]dy$

$C=2\int_0^1 y^2e^{-y}dy+2\int_0^1 Cdy$

$C=2\left[\left\{-y^2e^{-y}\right\}_0^1+2\int_0^1 ye^{-y}dy\right]+2[cy]_0^1$

$C=2\left[-e^{-1}+2\left[-ye^{-y}\right]_0^1-2\left[e^{-y}\right]_0^1\right]+2C$

$C=2\left[-2e^{-1}-2e^{-1}-2e^{-1}+2\right]+2C$

$C=2\left[-5e^{-1}+2\right]+2C$

$C=-10e^{-1}+4+2C$

$C=10e^{-1}-4$

By using (3) $\phi(x) = 2x^2 + 2e^x [10e^{-1} - 4]$ Put x = 0

$\phi(0) = 20e^{-1} - 8$ Put x = 1

$\phi(1) = 2 + 2e[10e^{-1} - 4]$ $= 2 + 20 + 8e$

$\phi(1) = 22 - 8e$

8. The integral equation $\phi(x)=1+\frac{2}{\pi}\int_0^{\pi}(\cos^2 x)\phi(t)dt$ has **CSIR NET DEC 2017**

(a) no solution
(b) unique solution
(c) more than one but infinitely many solution
(d) infinitely many solution

Ans. (a)

Explanation :

Let $C=\int_0^{\pi}\phi(t)dt$ ......(1)

Then $\phi(x)=1+\frac{2\cos^2 x}{\pi}C \Rightarrow \phi(t)=1+\frac{2C}{\pi}\cos^2 t$

Put value in (1)

Then $C=\int_0^{\pi}\left[1+\frac{2C}{\pi}\cos^2 t\right]dt$ $C=[t]_0^{\pi}+\frac{2C}{\pi}\int_0^{\pi}\cos^2 t\ dt$ $C=\pi+\frac{2C}{\pi}\int_0^{\pi}\left(\frac{1+\cos 2t}{2}\right)dt$

$C=\pi+\frac{2C}{2\pi}[t]_0^{\pi}+\frac{C}{\pi}\left[\frac{\sin 2t}{2}\right]_0^{\pi}$ $C=\pi+\frac{C}{\pi}[\pi]+\frac{C}{2\pi}[\sin 2\pi-0]$ $C = \pi + C + 0$

$\Rightarrow \pi = 0 \Rightarrow \phi(x)$ has no solution.

COV, NA, I.E.

9. For the homogeneous Fredholm equation $y(x)=\lambda\int_0^{\pi}\sin(x+\varsigma)y(\zeta)d\zeta$ the eigenvalue λ and the corresponding eigen function $y(x)$, involving arbitrary constants A and B, are **CSIR NET 2013**

(a) $\lambda=\frac{2}{\pi}, y(x)=A(\sin x-\cos x)$ (b) $\lambda=\frac{-2}{\pi}, y(x)=B(\sin x+\cos x)$

(c) $\lambda=\frac{-2}{\pi}, y(x)=B(\sin x-\cos x)$ (d) $\lambda=\frac{2}{\pi}, y(x)=A(\sin x+\cos x)$

Ans. (c),(d)

Explanation :

$$y(x)=\lambda\int_0^{\pi}\sin(x+\varsigma)y(\varsigma)d\varsigma \qquad y(x)=\lambda\int_0^{\pi}(\sin x\cos\varsigma+\sin\varsigma\cos x)y(\varsigma)d\varsigma$$

Let $C_1=\lambda\int_0^{\pi}\cos\zeta y(\zeta)d\zeta$ .....(1) $\quad C_2=\lambda\int_0^{\pi}\sin\zeta y(\zeta)d\zeta$ .....(2)

From (1) and (2), we get $\quad y(x)=C_1\sin x+C_2\cos x$ .....(3)

From (1) and (3), we get $\quad C_1=\lambda\int_0^{\pi}\cos\zeta(C_1\sin\zeta+C_2\cos\zeta)a\zeta$

$$C_1=\lambda\frac{C_1}{2}\int_0^{\pi}\sin 2\zeta d\zeta+\lambda\frac{C_2}{2}\int_0^{\pi}(\cos 2\zeta+1)d\zeta \quad C_2=\lambda\frac{C_1}{2}(0)+\lambda\frac{C_2}{2}(\pi)$$

$$C_1-\frac{\lambda\pi}{2}C_2=0 \quad ...(4)$$

From (2) and (3), we get $\quad C_2=\lambda\int_0^{\pi}\sin\zeta(C_1\sin\zeta+C_2\cos\zeta)d\zeta$

$$C_2=\lambda\frac{C_1}{2}\int_0^{\pi}(1-\cos 2\varsigma)d\varsigma+\lambda\frac{C_2}{2}\int_0^{\pi}\sin 2\varsigma d\varsigma \quad C_2=\lambda\frac{C_1}{2}\pi+\lambda\frac{C_2}{2}(0)$$

$$\frac{\lambda\pi}{2}C_1-C_2=0 \quad .....(5)$$

From equation (4) and (5), we get $\begin{bmatrix}1 & -\frac{\lambda\pi}{2}\\ \frac{\lambda\pi}{2} & -1\end{bmatrix}\begin{bmatrix}C_1\\ C_2\end{bmatrix}=0$

For non-zero solution of above system of equations, determinant of co-efficient matrix must be zero i.e.

$$\begin{vmatrix} 1 & -\frac{\lambda\pi}{2} \\ \frac{\lambda\pi}{2} & -1 \end{vmatrix} = 0 \qquad \Rightarrow \frac{\lambda^2\pi^2}{4} - 1 = 0 \Rightarrow \lambda = \pm\frac{2}{\pi}$$

If $\lambda = -\frac{2}{\pi}$ then $C_1 = -C_2$ $\quad \Rightarrow y(x) = B(\sin x - \cos x)$

If $\lambda = \frac{2}{\pi}$ then $C_1 = C_2$ $\quad \Rightarrow y(x) = A(\sin x + \cos x)$

Hence, option (c) and (d) are correct.

10. The integral equation $y(x) = \lambda\int_0^1 (3x-2)t\ y(t)$ has **CSIR NET JUNE 2015**

(a) only one characteristic number (b) two characteristic number
(c) more than two character number (d) no characteristic number

Ans. (d)

Explanation :

$$y(x) = \lambda(3x-2)\int_0^1 t\ y(t)dt \qquad .....(1)$$

Let $\quad \boldsymbol{C = \int_0^1 t\ y(t)dt}$ $\qquad ....(2)$

Then (1) become

y(x) = λ(3x – 2)C ..... (3)

y(t) = λC(3t – 2)

From (2)

$$C = \lambda C\int_0^1 t(3t-2)dt$$

$$C = \lambda C\int_0^1 (3t^2 - 2t)dt$$

$$C = \lambda C[t^3 - t^2]_0^1$$

C = λC(0)

C = 0

5. Solution of Fredhold integral equation of second kind with separable kernel.

Consider a Fredhold integral equation $y(x) = f(x) = \lambda\int_a^b k(x,t)y(t)dt$ ....(1)

Since k(x, t) is separable $\quad k(x,t) = \sum_{i=1}^n f_i(x)g_i(t)$ ..... (2)

From (1) $y(x) = f(x) = \lambda\int_a^b [\sum f_i(x).g_i(t)]y(t)dt \quad y(x) = f(x) + \lambda\sum_{ni=1}^n f_i(x)\int_a^b g_i(t)y(t)dt$ .....(3)

Let $\quad \boldsymbol{\int_a^b g_i(t)y(t)dt = C_i}$ ..... (4)

Using (4), (3) reduces to

COV, NA, I.E.

$$y(x) = f(x) + \lambda \sum_{j=1}^{n} C_j f_j(x) \qquad \text{...... (5)}$$

Re-writing $y(t) = f(t) + \lambda \sum_{j=1}^{n} C_j f_j(t)$ ...... (6)

Substituting the value of y(x) &y(t) in (3)

$$\boldsymbol{f(x) + \lambda \sum_{i=1}^{n} C_i f_i(x) = f(x) + \lambda \sum_{i=1}^{n} f_i(x) \int_{C_i}^{b} g_i(t) \left\{ f(t) + \lambda \sum_{j=1}^{n} C_j f_j(t) \right\} dt}$$

$$\Rightarrow \boldsymbol{\sum_{i=1}^{n} C_i f_i(x) = \sum_{i=1}^{n} f_i(x) \int_{C_i}^{b} g_i(t) \left\{ f(t) + \lambda \sum_{j=1}^{n} C_j f_j(t) \right\} dt}$$

$$\Rightarrow \sum C_i f_i(x) = \sum_{i=1}^{n} f_i(x) \left\{ \int_a^b g_i(t) f(t) dt + \lambda \sum_{j=1}^{n} C_j \int_a^b g_i(t) f_j(t) dt \right\} \qquad \text{... (7)}$$

We now introduce the useful notation

$$\boldsymbol{\int_a^b g_i(t) f(t) dt = \beta_i} \qquad \& \qquad \boldsymbol{\int_a^b g_i(t) f_j(t) dt = \alpha_{ij}} \qquad \text{... (8)}$$

Then (7) may be simplified as

$$\sum_{i=1}^{n} C_i f_i(x) = \sum_{i=1}^{n} f_i(x) \left\{ \beta_i + \lambda \sum_{j=1}^{n} \alpha_{ij} C_j \right\} \qquad \Rightarrow \sum f_i(x) \left\{ C_i - \beta_i - \lambda \sum_{j=1}^{n} \alpha_{ij} C_j \right\} = 0$$

In integral equation function $f_i(x)$ are L.I.

So $C_i - \beta_i - \lambda \sum_{j=1}^{n} \alpha_{ij} C_j = 0 \quad C_i - \lambda \sum_{j=1}^{n} \alpha_{ij} C_j = \beta_i$

Taking i = 1 $\quad C_i - \lambda \sum_{j=1}^{n} \alpha_{ij} C_j = \beta_i$

Or $C_1 - \lambda(\alpha_{11} C_1 + \alpha_{12} C_2 \text{.......} + \alpha_{1n} C_n) = \beta_1$

$\Rightarrow (1 - \lambda\alpha_{11}) C_1 - \lambda\alpha_{12} C_2 \text{.....} = \lambda\alpha_{1n} C_n = \beta_1$ ....(β_1)

Taking i = 2

$\boldsymbol{-\lambda\alpha_{21} C_1 + (1 - \lambda\alpha_{22}) C_2 \text{.......} - \lambda\alpha_{2n} C_n = \beta_2}$ .... (β_2)

Taking i = n

$-\lambda\alpha_{n_1} C_{n_1} - \lambda\alpha_{n_2} C_{n_2} \text{.......} + (1 - \lambda\alpha_{nn}) C_n = \beta_n$ .... (β_n)

$$D(\lambda) = \begin{vmatrix} 1 - \lambda\alpha_{11} & -\lambda\alpha_{12} & \cdots\cdots & -\lambda\alpha_1 n \\ -\lambda\alpha_{21} & 1 - \lambda\alpha_{22} & \cdots\cdots & -\lambda\alpha_{2n} \\ \vdots & & & \\ -\lambda\alpha_{n_1} & -\lambda\alpha_{n2} & \cdots\cdots & 1 - \lambda\alpha_{nn} \end{vmatrix} \qquad \text{...... (9)}$$

Situation 1st: When atleast one right member of the system (B_1), (B_2)...... (B_n) is non – zero

1. If D(λ) ≠ 0, then unique non-zero solution.
2. D(λ) = 0, then either no solution or infinite solution.

Situation 2nd : When f(x) = 0, then (B) $\beta_i = 0$

Hemog. System (B_1) ...... (B_n)

1. D(λ) ≠ 0, zero + unique solution
2. D(λ) = 0, infinite solution.

Situation 3rd : When f(x) ≠ 0 but

$\int_a^b g_n(x) f(x) dx = 0$ i.e. all $g_1(x)$ ...... $g_n(x)$ are orthogonal to f(x) &β_1, β_2,...... β_n zero.

Homogeneous System

1. D(λ) ≠ 0 unique solution, zero solution
2. D(λ) = 0, Infinite solution

11. Consider the integral equation $\phi(x) = \cos 7x + \lambda \int_0^\pi [\cos x \cos t - 2\sin x \sin t] \phi(t) dt$.

Which of the following are true?

CSIR NET 2018

(a) For every λ∈ R, a solution exist. (b) ∃λ∈R s.t. no solution

(c) ∃λ∈R s.t. more than one but finitely many solution

(d) ∃λ∈R s.t. infinitely solution

Ans. (a), (d) :

Explanation :

When k(x, t) is separable the integral equation

$$y(x) = f(x) + \lambda \sum_{i=1}^{n} \int_a^b [f_i(x) g_i(t)] y(t) dt$$

Then $f(x) = \cos 7x$ $\quad f_1(x) = \cos x, g_1(t) = \cos t$

$f_2(x) = -2\sin x, g_2(t) = \sin t$

Using (8) $\quad \beta_1 = \int_0^\pi g_1(x) f(x) dx = \int_0^\pi \cos x \cos 7x dx = 0$

$[\because \int_0^\pi \cos mx . \cos nx dx = 0 \text{ if } m \neq n]$ $\quad \beta_2 = \int_0^\pi \sin x \cos 7x dx = \frac{1}{2}\int_0^\pi 2\sin x \cos 7x dx$

$= \frac{1}{2}\int_0^\pi \{\sin 4x - \sin 3x\} dx$ $\quad \frac{1}{2}\left[-\frac{\cos 4x}{4}\right]_0^\pi - \frac{1}{2}\left[-\frac{\cos 3x}{3}\right]_0^\pi = 0$

COV, NA, I.E.

$$\alpha_{11}=\int_0^{\pi} g_1(x)f_1(x)dx=\int_0^{\pi}\cos^2 xdx=\frac{1}{2}\int_0^{\pi}[1+\cos 2x]dx=\frac{1}{2}\left[x+\frac{\sin 2x}{2}\right]_0^{\pi}=\frac{\pi}{2}$$

$$\alpha_{12}=\int_0^{\pi} g_1(x)f_2(x)=-\int_0^{\pi}2\sin x\cos xdx=-\int_0^{\pi}\sin 2xdx+\left[\frac{\cos 2x}{2}\right]_0^{\pi}=\left[\frac{1}{2}-\frac{1}{2}\right]=0$$

$$\alpha_{21}=\int_0^{\pi} g_2.f_2dx=\frac{1}{2}\int_0^{\pi}2\sin x\cos xdx=0$$

$$\alpha_{22}=\int_0^{\pi} g_2.f_2dx=-2\int_0^{\pi}\sin^2 xdx=-2\int_0^{\pi}\left(\frac{1-\cos 2x}{2}\right)dx-\left[x-\frac{\sin 2x}{2}\right]_0^{\pi}=-\pi \text{ By (9)}$$

$$D(\lambda)=\begin{vmatrix}1-\lambda\frac{\pi}{2} & 0\\ 0 & 1+\lambda\pi\end{vmatrix}=(1-\lambda\pi/2)(1+\lambda\pi)\text{For }\lambda=\frac{2}{\pi},-\frac{1}{\pi}$$

$D(\lambda)=0$

By situation III :-β_1, β_2 are zero.

&$D(\lambda)=0$ then infinite solution. If $D(\lambda)\neq 0$ then unique zero solution.

Option (a) is correct. For every $\lambda\in R$

Either $D(\lambda)=0$ or $D(\lambda)\neq 0$ then solution always exist.

Option (d) is correct.

$\exists\lambda=\frac{2}{\pi}\ \&-\frac{1}{\pi}\in R$ system has infinite solution.

i.e. integral equation has infinite solution.

12. The integral equation $\phi(x)=\frac{2}{\pi}\int_0^{\pi}\cos(x+t)\phi(t)dt$ has infinitely many solution if

CSIR NET DEC 2016

(a) f(x) = cosx (b) f(x) = cos 3x (c) f(x) = sin x (d) f(x) = sin3x

Ans. ( )

Explanation :

Given that

$$k(x,t)=\underset{f_1}{\cos x}\underset{g_1}{\cos t}-\underset{f_2}{\sin x}\underset{g_2}{\sin t}$$

Find $\alpha_{11}=\int_0^{\pi} g_1f_1dx=\int_0^{\pi}\cos^2 xdx=\frac{1}{2}\int_0^{\pi}(1+\cos 2x)dx=\frac{1}{2}\left[x-\frac{\sin 2x}{2}\right]_0^{\pi}=\frac{1}{2}[\pi]$

$$\alpha_{12}=-\int_0^{\pi}\cos x\sin xdx=-\frac{1}{2}\int_0^{\pi}2\sin x\cos xdx$$

$$= -\frac{1}{2}\int_0^{\pi} \sin 2x dx = +\frac{1}{2}\left[\frac{\cos 2x}{2}\right]_0^{\pi} = \frac{1}{2}\left[\frac{1}{2}-\frac{1}{2}\right] = 0$$

$$\alpha_{21} = \int_0^{\pi} g_2(x) f_1(x) dx = \int_0^{\pi} \sin x \cos x dx = 0$$

$$\alpha_{22} = \int_0^{\pi} g_2(x).f_2(x) dx = -\int_0^{\pi} \sin^2 x dx = -\int_0^{\pi}\frac{1-\cos 2x}{2} dx = -\frac{1}{2}\left[x - \frac{\cos 2x}{2}\right]_0^{\pi}$$

$$= -\frac{1}{2}\left[\pi - \frac{1}{2} + \frac{1}{2}\right] = -\frac{\pi}{2}$$

Option (a)

Now $\beta_1 = \int_0^{\pi} \cos x.\cos x dx = \int_0^{\pi} \cos^2 x dx = \frac{\pi}{2}$

$$\beta_2 = \int_0^{\pi} g_2(x) f(x) dx = \int_0^{\pi} \sin x \cos x dx = 0$$

and $D(\lambda) = \begin{vmatrix} 1-\lambda\alpha_{11} & -\lambda\alpha_{12} \\ -\lambda\alpha_{21} & 1-\lambda\alpha_{22} \end{vmatrix}$

$$D(\lambda) = \begin{vmatrix} 1-\frac{2}{\pi}\left(\frac{\pi}{2}\right) & -\frac{2}{\pi}(0) \\ -\frac{2}{\pi}(0) & 1-\frac{2}{\pi}\left(-\frac{\pi}{2}\right) \end{vmatrix} = \begin{vmatrix} 0 & 0 \\ 0 & 2 \end{vmatrix}$$

$D(\lambda) = 0$

By situation (1) it has no solution

Option (B)

$$\beta_1 = \int_0^{\pi} \cos x.\cos 3x dx = 0$$

$$\beta_2 = \frac{1}{2}\int_0^{\pi} 2\sin x.\cos 3x dx = \frac{1}{2}\int_0^{\pi} (\sin 4x - \sin 2x) dx$$

$$= \frac{1}{2}\left[-\frac{\cos 4x}{4} + \frac{\cos 2x}{2}\right]_0^{\pi} = \frac{1}{2}\left[-\frac{1}{4}+\frac{1}{2}+\frac{1}{4}-\frac{1}{2}\right] = 0$$

Let AX = b be a system of equation t hen f(A) = f(A/b) then system has a solution

Here $\begin{bmatrix} 0 & 0 \\ 0 & 2 \end{bmatrix}\begin{bmatrix} c_1 \\ c_2 \end{bmatrix} = \begin{bmatrix} 0 \\ 0 \end{bmatrix}$

$$f\begin{bmatrix} 0 & 0 \\ 0 & 2 \end{bmatrix} = 1 \quad \& \quad f[A/b] = 1$$

So system has infinite solution

Option (c)

$$\beta_1 = \int_0^\pi \cos x.\sin x dx = 0 \qquad \beta_2 = \int_0^\pi \sin^2 x dx = \frac{\pi}{2}$$

It has also infinite solution.

Option (d) $\beta_1 = \frac{1}{2}\int_0^\pi 2\cos x \sin 3x dx = \frac{1}{2}\int_0^\pi [\sin 4x + \sin 2x] dx$

$$= \frac{1}{2}\left[-\frac{\cos 4x}{4} - \frac{\cos 2x}{2}\right]_0^\pi = \text{-1} \qquad \frac{1}{2}\left[-\frac{1}{4} - \frac{1}{2} + \frac{1}{4} + \frac{1}{2}\right] = 0$$

$\& \ \beta_2 = \int_0^\pi \sin x.\sin 3x.dx$ may be zero or non-zero.

Note : System of equation

$$\left.\begin{aligned} (1-\lambda\alpha_{11})c_1 - \lambda\alpha_{12}c_2 = \beta_1 \\ -\lambda\alpha_{21} + (1-\lambda\alpha_{22})c_2 = \beta_2 \end{aligned}\right\} \text{be a system}$$

$$\Rightarrow \begin{bmatrix} 1-\lambda_{11} & -\lambda\alpha_{12} \\ -\lambda\alpha_{21} & 1-\lambda\alpha_{22} \end{bmatrix}\begin{bmatrix} c_1 \\ c_2 \end{bmatrix} = \begin{bmatrix} \beta_1 \\ \beta_2 \end{bmatrix}$$

If (1) f(A) = f(A : b) then system has solution

f(A) $\neq$ f(A : b) then no solution.

13. The integral equation, involving a parameter $\boldsymbol{\lambda}$, $\boldsymbol{\phi(x) = \cos 2x + \lambda\int_0^\pi \cos(x+\zeta)d\zeta}$ has

(a) A unique solution if $\lambda = 1$, and an inifinite number of solution if $\lambda = \frac{2}{\pi}$

(b) A unique solution if $\lambda = -1$, and an inifinite number of solution if $\lambda = -\frac{2}{\pi}$

(c) A unique solution if $\lambda \neq \frac{2}{\pi}$

(d) No solution if $\lambda = \pm\frac{2}{\pi}$

CSIR NET 2011

Ans. **(a),(b)**

Explanation :

$$\phi(x) = \cos 2x + \lambda\int_0^\pi \cos(x+\zeta)d\zeta$$

COV, NA, I.E.

$$\phi(x)=\cos 2x+\lambda\int_0^{\pi}\underbrace{\cos x}_{f_1}\underbrace{\cos\zeta}_{g_1}+\underbrace{(-\sin x)}_{f_2}\underbrace{(\sin\zeta)}_{g_2}.d\zeta$$

$$\phi(x)=\cos 2x+\lambda\cos x.C_1+\lambda\sin x.C_2$$

Find $\alpha_{11}=\int_0^{\pi}g_1f_1dx=\int_0^{\pi}\cos^2 xdx=\frac{1}{2}\int_0^{\pi}(1+\cos 2x)dx$

$$=\frac{1}{2}\left[x-\frac{\sin 2x}{2}\right]_0^{\pi}\quad=\frac{1}{2}[\pi]$$

$$\alpha_{12}=-\int_0^{\pi}\cos x\sin xdx=-\frac{1}{2}\int_0^{\pi}2\sin x\cos xdx$$

$$=-\frac{1}{2}\int_0^{\pi}\sin 2xdx=+\frac{1}{2}\left[\frac{\cos 2x}{2}\right]_0^{\pi}=\frac{1}{2}\left[\frac{1}{2}-\frac{1}{2}\right]=0$$

$$\alpha_{21}=\int_0^{\pi}g_2(x)f_1(x)dx=\int_0^{\pi}\sin x\cos xdx=0$$

$$\alpha_{22}=\int_0^{\pi}g_2(x).f_2(x)dx=-\int_0^{\pi}\sin^2 xdx$$

$$=-\int_0^{\pi}\frac{1-\cos 2x}{2}dx=-\frac{1}{2}\left[x-\frac{\cos 2x}{2}\right]_0^{\pi}=-\frac{1}{2}\left[\pi-\frac{1}{2}+\frac{1}{2}\right]=-\frac{\pi}{2}$$

Now $\beta_1=\int_0^{\pi}g_1(x)f(x)dx=\int_0^{\pi}\cos x.\cos xdx=\int_0^{\pi}\cos^2 xdx=\frac{\pi}{2}$

$$\beta_2=\int_0^{\pi}g_2(x)f(x)dx=\int_0^{\pi}\sin x\cos xdx=0$$

and $D(\lambda)=\begin{vmatrix}1-\lambda\alpha_{11} & -\lambda\alpha_{12}\\ -\lambda\alpha_{21} & 1-\lambda\alpha_{22}\end{vmatrix}=\begin{vmatrix}1-\lambda\frac{\pi}{2} & 0\\ 0 & 1+\lambda\frac{\pi}{2}\end{vmatrix}$

$$\begin{bmatrix}1-\lambda\frac{\pi}{2} & 0\\ 0 & 1+\lambda\frac{\pi}{2}\end{bmatrix}\begin{bmatrix}C_1\\ C_2\end{bmatrix}=\begin{bmatrix}\beta_1\\ \beta_2\end{bmatrix}$$

COV, NA, I.E.

Option (a) If $\lambda = 1$

System of equation becomes,

$$\begin{bmatrix} 1-\frac{\pi}{2} & 0 \\ 0 & 1+\frac{\pi}{2} \end{bmatrix}\begin{bmatrix} c_1 \\ c_2 \end{bmatrix} = \begin{bmatrix} 0 \\ -\frac{4}{3} \end{bmatrix}$$

$$\Rightarrow c_1 = 0 \Rightarrow c_2 = \frac{-8}{3(2+\pi)}$$

Unique solution

$$\phi(x) = \cos 2x - \frac{8}{3(2+\pi)}\sin x$$

Option (a) is correct and Option (d) is wong

Now option (b)

If $\lambda = -1$

System of equation becomes,

$$\begin{bmatrix} 1+\frac{\pi}{2} & 0 \\ 0 & 1-\frac{\pi}{2} \end{bmatrix}\begin{bmatrix} c_1 \\ c_2 \end{bmatrix} = \begin{bmatrix} 0 \\ -\frac{4}{3} \end{bmatrix}$$

$$\Rightarrow c_1 = 0$$

$$\Rightarrow c_2 = \frac{-8}{3(2-\pi)}$$

Unique solution

$$\phi(x) = \cos 2x + \frac{8}{3(2-\pi)}\sin x$$

If $\lambda = -\frac{2}{\pi}$

System of equation becomes,

$$\begin{bmatrix} 2 & 0 \\ 0 & 0 \end{bmatrix}\begin{bmatrix} c_1 \\ c_2 \end{bmatrix} = \begin{bmatrix} 0 \\ -\frac{4}{3} \end{bmatrix}$$

$\Rightarrow c_1 = 0$ and c_2 is arbitrary

$\Rightarrow$ Infinity solution

$$\phi(x) = \cos 2x + c_2\left(\frac{-2}{\pi}\right)\sin x$$

So, option (b) is correct

And option (c) is wrong.

Answer option (a) and (b).

6. Method of successive approximation

IteratedKernal :

1. Fredhold integral equation : $y(x) = f(x) + \lambda\int_a^b k(x,t)y(t)dt$

Then iterated kernel kn(x, t), n = 1, 2, 3, …..are define as follow

$k_1(x, t) = k(x, t)$

$$k_2(x,t) = \int_a^b k(x,z).k_1(z,t)dz \quad k_n(x,t) = \int_a^b k(x,z).k_{n-1}(z,t)dz$$

2. Voltera integral equation :

Then iterated kernel $k_n(x, t)$

$$k_1(x, t) = k(x, t) \qquad k_n(x,t) = \int_t^x k(x,z).k_{n-1}(z,t)dz$$

Resolvent kernel or reciprocal kernel :

Let $k_m(x, t)$ be iterated kernel then $R(x,t;\lambda) = \sum_{m=1}^{\infty}\lambda^{m-1}k_m(x,t)$ is called resolvent kernel.

Trick : Let $y(x) = f(x) + \lambda\int_a^x k(,t)y(t)dt$ be voltera integral equation and if k(x, t) = k₁(x, t) =

$k_1(z, t) .k(x, z) = k_2(x, t)$
then resolvent kernel $R(x, t, \lambda) = k_1 e^{\lambda(x-t)}$

14. y(x) be the solution of integral equation $y(x) = x - \int_0^x xt^2 y(t)dt \quad x > 0$. Then the value of function y(x) at $x = \sqrt{2}$ is **CSIR NET JUNE 2017**

(a) $\frac{1}{\sqrt{2e}}$ (b) $\frac{e}{2}$ (c) $\frac{\sqrt{2}}{e^2}$ (d) $\frac{\sqrt{2}}{e}$

Ans. (d)

Explanation :

Given that $k(x, t) = xt^2 = k_1(x, t)$

$$k_2 = \int_t^x xz^2 zt^2 dz = xt^2\left[\frac{z^4}{4}\right]_t^x = \frac{xt^2}{4}\left[x^4 - t^4\right]$$

$$k_3 = \int_t^x xz^2.\frac{zt^2}{4}\left[z^4 - t^4\right]dz = \int_t^x \frac{xt^2}{4}z^7 dz - \int_t^x \frac{xt^6}{4}z^3 dz = \frac{xt^2}{4}\int_t^x z^7 dz - \frac{xt^6}{4}\int_t^x z^3 dz$$

$$= \frac{xt^2}{4}\left[\frac{x^8}{8} - \frac{t^8}{8}\right] - \frac{xt^6}{4}\left[\frac{x^4}{4} - \frac{t^4}{4}\right] = \frac{xt^2}{32}\left[x^4 - t^4\right]\left[x^4 + t^4\right] - \frac{xt^6}{4}\left[\frac{x^4 - t^4}{4}\right]$$

$$= \frac{xt^2}{16}\left[x^4 - t^4\right]\left\{\frac{x^4}{2} + \frac{t^4}{2} - t^4\right\} = \frac{xt^2}{16}\frac{(x^4 - t^4)^2}{2!}$$

So $$k_m(x,t) = \sum_{m=1}^{\infty}\frac{xt^2}{4^{m-1}(m-1)!}(x^4 - t^4)^{m-1}$$

Resolvent kernel $$R(x,t;-1) = \sum_{m=1}^{\infty}(-1)^{m-1}\frac{xt^2}{4^{m-1}(m-1)!}(x^4 - t^4)^{m-1}$$

$$= \frac{xt^2}{1} - \frac{xt^2}{4}(x^4 - t^4) + \frac{xt^2}{4^2\ 2!}(x^4 - t^4)^2 + ...$$

COV, NA, I.E.

$$xt^2\left[1-\left\{\frac{x^4-t^4}{4}\right\}+\frac{\left\{\frac{x^4-t^4}{4}\right\}^2}{2!}+\ldots\right] \quad xt^2 e^{-\left(\frac{x^4-t^4}{4}\right)} \quad \left[1-x+\frac{x^2}{2!}-\frac{x^3}{3!}+\ldots\right]=e^{-x}$$

Then solution of integral equation $y(x)=x-\int_0^x R(x,t;-1)t\,dt = x-\int_0^x xt^3 e^{-\left(\frac{x^4-t^4}{4}\right)}dt$

Let $\frac{x^4-t^4}{4}=V \quad -\frac{4t^3}{4}dt=dv \Rightarrow -t^3 dt = dv$

After integration we get $\Rightarrow y(x)=xe^{-\frac{x^4}{4}} \quad y(\sqrt{2})=\sqrt{2}e^{-1}=\frac{\sqrt{2}}{e}$

15. Consider a non-zero, real valued polynomial function $p(x) = a_0 + a_1x + a_2x^2$ of degree atmost 2. Let $y = y(x)$ be a solution of integral equation. **CSIR NET DEC 2017**

$y(x)=p(x)+\int_0^x \sin(x-t)y(t)dt$ Which of the following are correct?

(a) y(x) is a polynomial of degree ≤ 2
(b) y(x) is a polynomial of degree ≤ 4
(c) If $a_1 \neq 0$ & $a_0 + 2a_2 = 0$ then $y'(0) = 0$
(d) If $a_1 \neq 0$ & $a_0 + 2a_2 = 0$ then $y''(0) = 0$

Ans. (b), (d)

Explanation :

Given that $y(x)=p(x)+\int_0^x \sin(x-t)y(t)dt$ …… (1)

Differential both side w.r.t. x & using Leibnitz's rule

$$y'(x)=p'(x)+\int_0^x \cos(x-t)y(t)dt$$

Again differentiate

$$y''(x)=p''(x)+\int_0^x -\sin(x-t)y(t)dt+\cos(x-x)y(x)$$

$$y''=p''-\int_0^x \sin(x-t)y(t)dt+y(x)$$

Using (1) $y''(x) = p''(x) - [y(x) - p(x)] + y(x)$
$y''(x) = p''(x) - y(x) + p(x) + y(x)$
$y''(x) = p''(x) + p(x)$ ….. (2)

Given that $p(x) = a_0 + a_1x + a_2x^2$

Then $p' = a_1 + 2a_2x$ $\quad p''(x) = 2a_2$

COV, NA, I.E.

Put value in (2)

$y''(x) = 2a_2 + a_0 + a_1x + a_2x^2$

Integration both side and we get

$$y(x) = \frac{(a_0 + 2a_2)x^2}{2} + \frac{a_1x^3}{6} + \frac{a_2x^4}{12} + c_1x + c_2$$

Option (b) and (d) are correct.

16. For the integral equation $y(x) = 1 + x^3 + \int_0^x k(x,t)y(t)dt$ with k(x, t) = 2^{x-t} then iterated kernel k_3(x, t) is **CSIR NET 2015**

(a) $2^{x-t}(x-t)^2$ (b) $2^{x-t}(x-t)^3$ (c) $2^{x-t-1}(x-t)^2$ (d) $2^{x-t-1}(x-t)^3$

Ans. (c)

Explanation :

Find the resolvent kernel for voltera integral equation.

(a) $k(x,t) = e^{x^2-t^2}$

Solution :

$k(x,t) = e^{x^2-t^2} = k_1(x,t)$

$k_2(x,t) = e^{x^2-z^2}e^{z^2-t^2} = e^{x^2-t^2}$

Then $k_1e^{\lambda(x-t)}$

$= e^{x^2-t^2}e^{\lambda(x-t)}$ is resolvent kernel.

(b) $k(x,t) = \dfrac{1+x^2}{1+t^2}$

Solution :

$$k = \frac{1+x^2}{1+t^2} = k_1(x,t)$$

$$k(x,z).k_1(z,t) = \frac{1+x^2}{1+z^2} \times \frac{1+z^2}{1+t^2} = \frac{1+x^2}{1+t^2}$$

Resolvent kernel $\dfrac{1+x^2}{1+t^2}e^{\lambda(x-t)}$

17. Resolvent kernel of voltera integral equation with kernel $k(x,t) = \dfrac{2+\cos x}{2+\cos t}$.

(a) $ke^{\lambda(x+t)}$ (b) $ke^{-\lambda(x-t)}$ (c) $ke^{\lambda(x-t)}$ (d) $ke^{-\lambda(x+t)}$

Ans. (c)

Explanation :

$$k(x,t) = k_1(x,t) = \frac{2+\cos x}{2+\cos t}$$

$$k_2(x,t) = k(x,z).k_1(z,t) = \frac{2+\cos x}{2+\cos z} \times \frac{2+\cos z}{2+\cos t} = \frac{2+\cos x}{2+\cos t} = k(x,t)$$

Then $R(x,t;\lambda) = \dfrac{2+\cos x}{2+\cos t}e^{\lambda(x-t)}$

18. For the linear integral equation $\phi(x) = x + \int_0^{1/2} \phi(\xi)d\xi$, the resolvent kernel $R(x,\xi;1)$ is

CSIR NET JUNE 2012

(a) $\frac{1}{2}$ (b) 2 (c) $\frac{3}{2}$ (d) 4

Ans. (b)

Explanation :

Given, linear integral equation is $\phi(x) = x + \int_0^{1/2} \phi(\varepsilon)d\varepsilon$ and $K_1(x,t) = 1$

$$K_2(x,t) = \int_0^{1/2} K_1(x,z)K_1(z,t)dz = \frac{1}{2} \qquad K_3(x,t) = \int_0^{1/2} K_2(x,z)K_1(z,t)dz = \left(\frac{1}{2}\right)^2$$

and so on $K_{n+1}(x,t) = \left(\frac{1}{2}\right)^n$

Here, $R(x,\varepsilon:\lambda) = \sum_{n=1}^{\infty}\left(\frac{1}{2}\right)^{n-1}\lambda^{n-1} = \sum_{n=0}^{\infty}\left(\frac{1}{2}\right)^n\lambda^n = \frac{1}{1-\frac{\lambda}{2}}; \left(\frac{\lambda}{2}\right) < 1 \qquad R(x,\varepsilon:1) = 2$

Hence, option (b) is correct.

7. Solving Voltera integral equation :

(a) **By Resolvent kernel :** Resolvent kernel of a non-homogeneous volterra integral equation of second kind $\varphi(x) = F(x) + \lambda\int_a^x k(x,\xi)\varphi(\xi)d\xi$ has a unique solution of the form

$$\varphi(x) = F(x) + \lambda\int_a^x R(x,t;\lambda)F(t)dt, \text{ where } R(x,t;\lambda) = \sum_{n=1}^{\infty}\lambda^{n-1}k_n(x,t), \text{ where}$$

$$k_n(x,t) = \int_t^x k(x,\xi)k_{n-1}(\xi,t)d\xi$$

.(b) **Laplace transform :** Laplace transform provides a useful method for solving Volterra's integral equation in which kernel k(x,t) is function of (t – x) alone.

(c) By convert into differential equation.

Solution of Voltera integral equation of 1st Kind :

We know that general form of Voltera integral equation is $g(x)y(x) = f(x) + \lambda\int_a^x k(x,t)y(t)dt$

If g(x) = 0 then it is called 1st kind of integral equation. i.e. $f(x)+\lambda\int_a^x k(x,t)y(t)dt=0$

19. Consider the integral equation $\int_0^x (x-t)u(t)dt = x; x \ge 0$. For continuous functions u defined on [0, ∞). The equation has **CSIR NET FEB 2022**

(a) A unique bounded solution (b) No solution

(c) A unique solution u such that $|u(x)| \le C(1+|x|)$ for some constant C.

(d) More than one solution u such that $|u(x)| \le C(1+|x|)$ for some constant C.

Ans. (b)

Explanation :

Consider the given integral equation

$$\int_0^x (x-t)u(t)dt = x; x \ge 0 \quad \text{..... (1)}$$

Differentiating equation (1) w.r.t x with the help of Leibnitz rule

$$\int_0^x u(t)dt = 1 \quad \text{..... (2)}$$

Again apply Leibnitz rule, ⇒ u(x) = 0 ⇒ u(t) = 0

Equation (2) become, 0 = 1

Which is contradiction.

So, it has no solution.

Option (b) is the correct answer.

20. The solution of the Fredholm integral equation $y(s) = s + 2\int_0^1 (st^2+s^2t)y(t)\,dt$ is

(a) $y(s) = -(50s + 40s^2)$ (b) $y(s) = (30s + 15s^2)$

(c) $y(s) = -(30s + 40s^2)$ (d) $y(s) = (60s + 50s^2)$

Ans. (c) **CSIR NET JUNE 2020**

Explanation :

$y(s) = s + 2\int_0^1 (st^2 + s^2t)y(t)\,dt$

We have $f(s) = s$, $\alpha_{ij} = \int_0^1 g_i(t)\,f_j(t)\,dt$ $\beta_i = \int_0^1 g_i(t)\,f(t)\,dt$ $c_i = \int_0^1 g_i(t)\,y(t)\,dt$

Here, $K(s, t) = st^2 + s^2t = f_1g_1 + f_2g_2$ $c_1 = \int_0^1 t^2y(t)\,dt$ $c_2 = \int_0^1 t\,y(t)\,dt$

So, $y(s) = s + 2sc_1 + 2s^2c_2$

For finding c_1 and c_2, $\begin{bmatrix} 1-\lambda\alpha_{11} & -\lambda\alpha_{12} \\ -\lambda\alpha_{21} & 1-\lambda\alpha_{22} \end{bmatrix}\begin{bmatrix} c_1 \\ c_2 \end{bmatrix} = \begin{bmatrix} \beta_1 \\ \beta_2 \end{bmatrix}$ ... (1)

$\alpha_{11} = \int_0^1 t^3dt = ¼$ $\alpha_{12} = \int_0^1 t^4dt = 1/5$ $\alpha_{21} = \int_0^1 t^2dt = 1/3$ $\alpha_{22} = \int_0^1 t^3dt = 1/4$

COV, NA, I.E.

$\beta_1 = \int_0^1 t^2 dt = 1/4 \qquad \beta_1 = \int_0^1 t\,dt = 1/3 \qquad (1) \Rightarrow \begin{bmatrix} 1/2 & -2/5 \\ -2/3 & 1/2 \end{bmatrix} \begin{bmatrix} c_1 \\ c_2 \end{bmatrix} = \begin{bmatrix} 1/4 \\ 1/3 \end{bmatrix} \ldots$

We get $c_1 = -31/2$, $c_2 = -20$.

$y(s) = s + 2sc_1 + 2s^2c_2 = s + 2s\ -\dfrac{31}{2} + 2s^2\,(-20) = -30s - 40s^2,$

Option (c) is correct.

21. Let ϕ be the solution of $\phi(x) = 1 - 2x - 4x^2 + \int_0^x \left[3 + 6(x-t) - 4(x-t)^2\right]\phi(t)\,dt$

Then $\phi(1)$ equal to **CSIR NET DEC 2019**

(a) e^{-1} (b) e^{-2} (c) e (d) e^2

Ans. (c)

Explanation :

This is a Voltera integral equation.

Differential both side & use Leibnitz formula.

$\phi'(x) = -2 - 3x + \int_0^x [6 - 8(x-t)]\phi(t)\,dt + 3\phi(x)$ Differential again both side

$\phi(x) = -8 + \int_0^x -8\phi(t)\,dt + 6\phi(x) + 3\phi'(x) \qquad \phi'''(x) = -8\,\phi(x) + 6\,\phi'(x) - 3\phi''(x)$

$\Rightarrow \quad \phi'''(x) + 8\phi(x) - 6\phi'(x) - 3\phi''(x) = 0$ Continued

1. $\phi(0) = 1$
2. $\phi'(0) = -2 + 3\phi(0) = -2 + 3 = 1$
3. $\phi''(0) = -8 + 6\phi(0) + 3\phi'(0)$
   $= 8 - 6 = -14 \quad -8 + 9 = 1$

Auxiliary equation of … (4)

$m^3 - 3m^2 - 6m + 8 = 0$

$\Rightarrow (m-1)(m^2 - 2m - 8) = 0$

$\Rightarrow m = 1, 4, -2 \qquad m = \dfrac{2 \pm \sqrt{4+32}}{2} \qquad \dfrac{2 \pm 6}{2}$

$\phi(x) = c_1e^x + c_2e^{4x} + c_3e^{-2x}$ …..(4*)

Use of condition

$\phi(0) = c_1 + c_2 + c_3 = 1$ ….(5)

$\phi'(x) = c_1e^x + 4c_2e^{4x} - 2c^3e^{-2x}$

$\phi'(0) = c_1 + 4c_2 - 2c_3 = 1$ …..(6)

$\phi''(x) = c_1e^x + 16\,c_2e^{4x} + 4\,c_3e^{-2x}$

$\phi''(0) = c_1 + 16c_2 + 4c_3 = 1$ ….. (7)

Solve (5), (6) & (7) & get c_1, c_2 & c_3

$$\begin{bmatrix}1&1&1&1\\1&4&-2&1\\1&16&4&1\end{bmatrix}\sim\begin{bmatrix}1&1&1&1\\0&3&-3&0\\0&15&3&0\end{bmatrix}\qquad\begin{bmatrix}1&1&1&1\\0&3&-3&0\\0&0&18&0\end{bmatrix}$$

$\Rightarrow$ $18c_3 = 0$

$3c_2 - 3c_3 = 0$ & $c_1 + c_2 + c_3 = 1$

$C_3 = 0 = C_2$ & $C_1 = 1$

Equation – (4)

$\phi(x) = e^x$

$\phi(1) = e$

22. If y is a solution of $y(x)-\int_0^x (x-t)y(t)dt = 1$, then which of the following is true?

(a) y is bounded but not periodic in R

(b) y is periodic in R

(c) $\int_R y(x)dx < \infty$

(d) $\int_R \frac{dx}{y(x)} < \infty$

Ans. **(d)** **CSIR NET 2019**

Explanation :

$$y(x) = 1 + \int_0^x (x-t)y(t)dt$$

$$\Rightarrow y(0) = 0$$

Differentiatting using Leibnitz Rule, we get

$$y'(x) = \int_0^x y(t)dt$$

$$\Rightarrow y'(0) = 0$$

Again differentiating using Leibnitz Rule, we get

$$y'(x) = \int_0^x y(t)dt$$

$$\Rightarrow y'(0) = 0$$

Again differentiating using Leibnitz Rule, we get

$$y''(x) = y(x) \quad \Rightarrow y''(x) - y(x) = 0$$

$$\Rightarrow y(x) = c_1 e^x + c_2 e^{-x}$$

Using $y(0)=1$ and $y'(0)=0$

We get $c_1 = \frac{1}{2}$ and $c_2 = \frac{1}{2}$

$$\Rightarrow y(x) = \frac{e^x + e^{-x}}{2} = \cosh x$$

which is neither bounded nor periodic in R.

Also,

$$\int_R y(x)dx = \int_{-\infty}^{\infty} \cosh x\,dx = [\sinh x]_{-\infty}^{\infty} \not< \infty$$

and $$\int_{-\infty}^{\infty} \frac{1}{y(x)}dx = \int_{-\infty}^{\infty} \frac{1}{\cosh x}dx$$

$$= \int_{-\infty}^{\infty} \operatorname{sech} x\,dx = \left[2\tan^{-1}(e^x)\right]_{-\infty}^{\infty}$$

$$= 2\pi < \infty$$

Hence, opion (d) is correct.

COV, NA, I.E.

23. If ϕ is the solution of $\int_0^x (1-x^2+t^2)\phi(t)dt = \frac{x^2}{2}$ then $\phi(\sqrt{2})$ is equals to

(a) $\sqrt{2}e^{\sqrt{2}}$ (b) $\sqrt{2}e^2$ (c) $\sqrt{2}e^{2\sqrt{2}}$ (d) $2e^4$

Ans. (b) **CSIR NET DEC 2018**

Explanation :

It is 1st kind voltera integral equation.

So differentiate both side & using Leibnitz rule

$$x = \int_0^x -2x\phi(t)dt + \phi(x)\frac{dx}{dx} \qquad \phi(x) = x + 2\int_0^x \phi(t)dt \qquad \text{..... (1)}$$

Find resolvent kernel k (x, t) = x = k1(x, t)

$$k_2(x,t) = \int_t^x xzdz = \frac{x}{2}\left[z^2\right]_t^x = \frac{x}{2}\left[x^2 - t^2\right]$$

$$k_3 = \int_t^x x\left\{\frac{z^3}{2} - \frac{zt^2}{2}\right\}dz = \frac{x}{2}\left[\frac{z^4}{4}\right]_t^x - \frac{xt^2}{2}\left[\frac{z^2}{2}\right]_t^x = \frac{x}{8}\left[x^4 - t^4\right] - \frac{xt^2}{4}\left[x^2 - t^2\right]$$

$$= \frac{x}{8}\left[x^2 - t^2\right]\left[x^2 + t^2\right] - \frac{xt^2}{4}\left[x^2 - t^2\right] = \frac{x}{4}(x^2 - t^2)\left[\frac{x^2}{2} + \frac{t^2}{2} - t^2\right]$$

$$= \frac{x}{4}(x^2 - t^2)\left(\frac{x^2 - t^2}{2}\right) = \frac{x}{2^2}\frac{(x^2 - t^2)^2}{2!} \quad \text{i.e. iterated kernel}$$

$$k_m(x,t) = \frac{x}{2^{m-1}}\frac{(x^2 - t^2)^{m-1}}{m-1}$$

Then resolvent kernel $R(x,t,z) = \sum_{m=1}^{\infty} 2^{m-1}\frac{x}{2^{m-1}}\frac{\left(x^2 - t^2\right)^{m-1}}{(m-1)!} = x\sum_{m=1}^{\infty}\frac{(x^2 - t^2)^{m-1}}{(m-1)!}$

$= xe^{(x^2 - t^2)}$

Then solution of voltera integral equation $\phi(x) = x + 2\int_0^x xe^{(x^2 - t^2)}t\, dt$

Let $x^2 - t^2 = u$

$-2t\, dt = du$

$tdt = -\frac{du}{2}$

After integration we get

$\phi(x) = xe^{x^2}$

Then $\phi(\sqrt{2}) = \sqrt{2}e^2$

Solution of Volterra integral equation of 2nd kind

24. If ϕ is a solution of $\phi(x) = 1 - 2x - 4x^2 + \int_0^x [3 + 6(x-t) - 4(x-t)^2]\phi(t)dt$ then $\phi(\log 2)$ is equal to **CSIR NET DEC 2018**

(a) 2 (b) 4 (c) 6 (d) 8

Ans. (a)

Explanation :

Differentiate both side of integral equation and using Lebnitz rule.

$$\phi'(x) = -2 - 8x + \int_0^x [6 - 8(x-t)\phi(t) + 3\phi(x) \quad .....(2)$$

Again $\boldsymbol{\phi''(x) = -8t\int_0^x - 8\phi(t)dt + 6\phi(x) + 3\phi'(x)}$ .... (3)

$$\phi''x) = -8 - 8\int_0^x \phi(t)dt + 6\phi(x) + 3\phi'(x)$$

Then again

$$\phi'''(x) = -8\phi(x) + 6\phi'(x) + 3\phi''(x) \quad ... (4)$$

$$\phi''' - 3\phi'' - 6\phi' + 8\phi = 0 \quad ... (5)$$

This is a differential equation of 3rd order.

Solve this integral equation

Auxiliarry E

$m^3 - 3m^2 - 6m + 8 = 0$

$(m - 1)(m^2 - 2m - 8) = 0$

$m = 1, -2, 4$

then $\phi(x) = Ae^x + Be^{-2x} + Ce^{4x}$ ...(6)

Now we have to find condition

$\phi(0) = 1$ $\phi'(0) = -1 + 3\phi(0) = -2 + 3 = 1$

& $\phi''(0) = 1$

Differentiate of (6)

$\phi'(x) = Ae^x - 2Be^{-2x} + 4Ce^{4x}$ .....(7)

$\phi''(x) = Ae^x + 4Be^{-2x} + 16Ce^{4x}$ ..... (8)

Using condition in (6), (7) & (8)

Then $A + B + C = 1$ ... (9)

$A - 2B + 4C = 1$ ... (10)

$A + 4B + 16C = 1$ ... (11)

Solve (9), (10) & (11)

and get value of A, B, C

$A = 1, B = 0, C = 0$

Then by (6)

$\phi(x) = e^x$

$\phi(\log 2) = e^{\log 2} = 2$

25. Let $\phi(x)$ be the solution of $\int_0^x e^{x-t}\phi(t)dt = x$ then $\phi(1)$ equals **CSIR NET JUNE 2017**

(a) -1 (b) 0 (d) 1 (d) 2

Ans. (b)

Explanation :

Step – 1 : Firstly we have to do derivative both side of integral equation

$$1 = \int_0^x e^{x-t}\phi(t)dt + e^{x-x}\phi(x) \qquad 1 = \int_0^x e^{x-t}\phi(t)dt + \phi(x)$$

[Using Leibnitz rule] $\phi(x) = 1 - \int_0^x e^{x-t}\phi(t)dt$

Now, 1st kind convert into 2nd kind.

Step – 2 : Find resolvent kernel

$k(x,t) = e^{x-t} = k_1(x,t)$

$k_2(x,t) = k(x,z).k_1(z,t) = e^{x-z}e^{z-t} = e^{x-t} = k(x,t)$

then resolvent kernel

$R(x,t,\lambda) = ke^{\lambda(x-t)}$

$R(x,t,-1) = e^{x-t}e^{-(x-t)} = 1$

Step – 3 :

Solution of Voltera integral equation

$$\phi(x) = f(x) - \int_0^x R(x,t,-1)f(t)dt \qquad \phi(x) = f(x)\int_0^x f(t)dt \qquad \phi(x) = 1 - \int_0^x dt$$

$\phi(x) = 1 - [t]_0^x = 1 - x$ Then $\phi(1) = 0$

Laplace Transform Method to Solve Volterra Integral Equation

The integral equation $u(x) = f(x) + \int_0^x k(x-t)u(t)dt$ in which the kernel $k(x-t)$ is a function of the difference $x-t$ only , is known as integral equation of the convolution type and can be written as

$u(x) = f(x) + k(x)*u(x)$

Taking Laplace transformation of both sides $\Rightarrow L[u(x)] = L[f(x)] + L[k(x)].L[u(x)]$

$$\Rightarrow [1 - L\{k(x)\}]L[u(x)] = L[f(x)] \Rightarrow L[u(x)] = \frac{L[f(x)]}{1 - L\{k(x)\}}$$

$$\Rightarrow u(x) = L^{-1}\left\{\frac{L[f(x)]}{1-L[k(x)]}\right\}$$ Which is required solution.

Example 1: Solve $y(x) = 1 - 2x - 4x^2 + \int_0^x [3 + 6(x-t) - 4(x-t)^2] y(t)dt$

Solution:-Taking Laplace transform on either sides we get

$$L[y(x)] = L[1] - 2L[x] - 4L[x^2] + L[3 + 6x - 4x^2] * L[y(x)]$$

$$L[y(x)] = L[1] - 2L[x] - 4L[x^2] + [3L\{1\} + 6L\{x\} - 4L\{x^2\}]L[y(x)]$$

$$\Rightarrow L[y(x)] = \frac{1}{s} - \frac{2}{s^2} - \frac{4\times 2!}{s^3} + \left[\frac{3}{s} + \frac{6}{s^2} - \frac{4\times 2!}{s^3}\right] L[y(x)]$$

$$\Rightarrow \left(1 - \frac{3}{s} - \frac{6}{s^2} + \frac{8}{s^3}\right) L[y(x)] = \frac{1}{s} - \frac{2}{s^2} - \frac{8}{s^3}$$

$$\Rightarrow L[y(x)] = \frac{s^2 - 2s - 8}{s^3 - 3s^2 - 6s + 8} = \frac{(s-4)(s+2)}{(s-4)(s+2)(s-1)}$$

$$\Rightarrow L[y(s)] = \frac{1}{s-1} \Rightarrow y(x) = L^{-1}\left[\frac{1}{s-1}\right] = e^x$$

$$\Rightarrow y(x) = e^x$$

26. Which of the following function is solution of the volterra type equation?

$y(x) = x + \int_0^x \sin(x-t) y(t)dt$

(a) $x + \frac{x^3}{3}$ (b) $x - \frac{x^3}{3}$ (c) $x + \frac{x^3}{6}$ (d) $x - \frac{x^3}{6}$

Ans. (c)

Explanation :

Let $y(x) = f(x) + k(x) * y(x)$ $\Rightarrow y(x) = x + [\sin x] * y(x)$

Taking laplace transform on both sides we get

$$\Rightarrow L[y(x)] = L[x] + L[\sin x] L[y(x)] \qquad \Rightarrow L[y(x)] = \frac{1}{s^2} + \frac{1}{1+s^2} . L[y(x)]$$

$$\Rightarrow \left[1 - \frac{1}{s^2+1}\right] L[y(x)] = \frac{1}{s^2} \Rightarrow \left[\frac{s^2}{s^2+1}\right] L[y(x)] = \frac{1}{s^2}$$

$$\Rightarrow y(x) = L^{-1}\left[\frac{s^2+1}{s^4}\right] = L^{-1}\left[\frac{1}{s^2}\right] + L^{-1}\left[\frac{1}{s^4}\right] \qquad \Rightarrow y(x) = \frac{x}{1!} + \frac{x^3}{3!} \qquad \Rightarrow y(x) = x + \frac{x^3}{6}$$

27. Let $\phi(x)$ be the solution of $\int_0^x e^{x-t}\phi(t)dt = x,\ x > 0.$ then $\phi(1)$ equals

CSIR NET 2017

(a) -1 (b) 0 (c) 1 (d) 2

Ans. (b)

Explanation :

Given integral equation is $\int_0^x e^{x-t}\phi(t)dt = x,\ x>0 \Rightarrow e^x * \phi(x) = x$

Taking Laplace on both sides, we get $L(e^x).L[\phi(x)] = L(x)$

[using convolution theorem of Laplace Transformation]

$$\Rightarrow \frac{1}{s-1}.L[\phi(x)] = \frac{1}{s^2} \Rightarrow L[\phi(x)] = \frac{s-1}{s^2} = \frac{1}{s} - \frac{1}{s^2}$$

Now applying Laplace inverse on both sides

$$\Rightarrow \phi(x) = L^{-1}\left(\frac{1}{s}\right) - L^{-1}\left(\frac{1}{s^2}\right) \qquad \begin{array}{l} \Rightarrow \phi(x) = 1 - x \\ \therefore \phi(1) = 1 - 1 = 0 \end{array}$$

Hence option (b) is correct.

28. Consider a non –zero real-valued polynomial function $p(x) = a_0 + a_1x + a_2x^2$ of degree at most 2. Let $y = y(x)$ be a solution of the integral equation $y = p(x) + \int_0^x \sin(x-t)y(t)dt$

Which of the following statement are necessarily correct? **CSIR NET 2017**

(a) $y(x)$ is a polynomial function of degree ≤ 2

(b) $y(x)$ is a polynomial function of degree ≤ 4

(c) if $a_1 \ne 0$ *and* $a_0 + 2a_2 = 0$, *then* $y'(0) = 0$

(d) if $a_1 \ne 0$ *and* $a_0 + 2a_2 = 0$, *then* $y''(0) = 0$

Ans. (b), (d)

Explanation :

Given integral equation is $y = p(x) + \int_0^x \sin(x-t)y(t)dt$, where

$p(x) = a_0 + a_1x + a_2x^2$ of degree at most 2.

$$\therefore y = a_0 + a_1x + a_2x^2 + \int_0^x \sin(x-t)y(t)dt$$

Let $y(x) = p(x) + k(x) * y(x)$

$$\Rightarrow y(x) = a_0 + a_1x + a_2x^2 + [\sin x]y(x)$$

Taking laplace transform on both sides we get

$$\Rightarrow L[y(x)] = a_0L[1] + a_1L[x] + a_2L[x^2] + L[\sin x]L[y(x)]$$

$$\Rightarrow \bar{y} = \frac{a_0}{s} + \frac{a_1}{s^2} + \frac{2a_2}{s^3} + \left(\frac{1}{s^2+1}\right)\bar{y}, \text{ where } \bar{y} = L[y(x)]$$

$$\Rightarrow \bar{y}\left[1 - \frac{1}{s^2+1}\right] = \frac{a_0s^2 + a_1s + 2a_2}{s^3}$$

COV, NA, I.E.

$$\Rightarrow \bar{y}=\left(\frac{a_0s^2+a_1s+2a_2}{s^3}\right)\times\frac{s^2+1}{s^2} \qquad \Rightarrow \bar{y}=\left(\frac{a_0s^4+a_1s^3+2a_2s^2+a_0s^2+a_1s+2a_2}{s^5}\right)$$

$$\Rightarrow \bar{y}=\frac{a_0}{s}+\frac{a_1}{s^2}+\frac{2a_2}{s^3}+\frac{a_0}{s^3}+\frac{a_1}{s^4}+\frac{2a_2}{s^5} \Rightarrow y=a_0+a_1x+a_2x^2+a_0\frac{x^2}{2}+a_1\frac{x^3}{6}+a_2\frac{x^4}{12}$$

$$\Rightarrow y=a_0+a_1x+\frac{x^2}{2}(2a_2+a_0)+a_1\frac{x^3}{6}+a_2\frac{x^4}{12}$$ which is a polynomial of degree ≤ 4

Option (b) is correct. Using Condition option (d) is also correct

29. Let ϕ satisfy **CSIR NET 2016**

$\phi(x)=f(x)+\int_0^x \sin(x-t)\phi(t)dt.$ then ϕ is given by

(a) $\phi(x)=f(x)+\int_0^x (x-t)f(t)dt.$ (b) $\phi(x)=f(x)-\int_0^x (x-t)f(t)dt.$

(c) $\phi(x)=f(x)-\int_0^x \cos(x-t)f(t)dt.$ (d) $\phi(x)=f(x)-\int_0^x \sin(x-t)f(t)dt.$

Ans. (a)

Explanation :

Given integral equation is $\phi(x)=f(x)+\int_0^x \sin(x-t)\phi(t)dt.$

Take $f(x)=1$ $\therefore \phi(x)=1+\int_0^x \sin(x-t)\phi(t)dt.$

Now, take Laplace transformation on both sides,

$$\bar{\phi}=\frac{1}{s}+L[\sin x*\phi]\Rightarrow \bar{\phi}=\frac{1}{s}+\frac{1}{s^2+1}\bar{\phi}\Rightarrow\left(1-\frac{1}{s^2+1}\right)\bar{\phi}=\frac{1}{s}\Rightarrow\bar{\phi}=\frac{1}{s}.\frac{s^2+1}{s^2}=\frac{s^2+1}{s^3}$$

Now from options, from option (a), $\phi(x)=1+\int_0^x (x-t)dt$

Taking Laplace transformation $\bar{\phi}=\frac{1}{s}+L[x*1]\Rightarrow\bar{\phi}=\frac{1}{s}+\frac{1}{s^2}.\frac{1}{s}=\frac{1}{s}+\frac{1}{s^3}$

Now, similarly for other options, From option (b) , $\bar{\phi}=\frac{1}{s}-\frac{1}{s^3}$

From option (c) , $\bar{\phi}=\frac{1}{s}-\frac{1}{s^2+1}$ From option (d) , $\bar{\phi}=\frac{1}{s}-\frac{1}{s(s^2+1)}$

So option (b), (c) and (d) are incorrect. Hence, option (a) is correct.

30. Consider the integral equation $y(x)=x^3+\int_0^x \sin(x-t)y(t)dt$ $x\in[0,\pi]$ then the value of $y(1)$ is **CSIR NET 2016**

(a) 19/20 (b) 1 (c) 17/20 (d) 21/20

Ans. (d)

COV, NA, I.E.

Explanation :

$y(x)=x^3+\int_0^x \sin(x-t)y(t)dt,\ x\in[0,\pi]$..(i)

Solving the given integral equation by Laplace method

$\because \int_0^x \sin(x-t)y(t)dt.$ is convolute of $y(t)*\sin x$

Taking Laplace on both sides of the given equation (1)

$L[y(x)]=L(x^3)+L\left[\int_0^x \sin(x-t)y(t)dt\right]\quad L[y(x)]=L(x^3)+L[\sin x*y(x)]$

$\bar{y}(s)=\frac{3!}{s^4}+L[\sin x*y(x)]\quad \left[\because L(t^n)=\frac{n!}{s^{n+1}}\right]$

$\bar{y}(s)=\frac{3!}{s^4}+\frac{1}{s^2+1}.\bar{y}(s)\Rightarrow\left[1-\frac{1}{s^2+1}\right]\bar{y}(s)=\frac{3!}{s^4}\Rightarrow\bar{y}(s)=\frac{3!}{s^4}\times\frac{s^2+1}{s^2}\Rightarrow\bar{y}(s)=3!\left[\frac{1}{s^2}+\frac{1}{s^4}\right].\frac{1}{s^2}$

Taking inverse Laplace on both sides

$\therefore L^{-1}[\bar{y}(s)]=3!\left[L^{-1}\left(\frac{1}{s^4}\right)+L^{-1}\left(\frac{1}{s^6}\right)\right]\Rightarrow y(t)=3!\times\left[\frac{t^3}{3!}+\frac{t^5}{5!}\right]\quad\left[\because L^{-1}\left[\frac{1}{s^{n+1}}\right]=\frac{t^n}{n!}\right]$

$\Rightarrow y(t)=3!\times\left[\frac{t^3}{3!}+\frac{t^5}{5\times4\times3!}\right]\Rightarrow y(t)=\left[t^3+\frac{t^5}{20}\right]\Rightarrow y(1)=\left[1+\frac{1}{20}\right]=\frac{21}{20}$

$\therefore$ Option (d) is correct.

31. The curve $y=y(x)$, passing through the point $(\sqrt{3},1)$ and defined by the following property (Volterraintegral equation of the first kind) $\int_0^y \frac{f(v)dv}{\sqrt{y-v}}=4\sqrt{y}$, where

$f(y)=\sqrt{1+\frac{1}{y'^2}}$ is the part of a **CSIR NET 2016**

(a)Straight line (b) Circle (c)Parabola (d) Cycloid.

Ans. (a)

Explanation :

$\because \int_0^y \frac{f(v)}{(y-v)^{1/2}}dv=4y^{1/2}$

Taking Laplace on both sides we get,

$L\left[\int_0^y \frac{f(v)}{(y-v)^{1/2}}dv\right]=4L[y^{1/2}]\quad\Rightarrow L\left[f(y)*\frac{1}{y^{1/2}}\right]=4L[y^{1/2}]$

$\bar{f}(s).\frac{\Gamma(1/2)}{s^{1/2}}=4.\frac{\Gamma(3/2)}{s^{3/2}}$ {using convolution theorem}

COV, NA, I.E.

$$\Rightarrow \bar{f}(s).\Gamma(1/2)=\frac{4}{s}.1/2.\Gamma(1/2)\Rightarrow \bar{f}(s)=\frac{2}{s} \quad \Rightarrow f(x)=L^{-1}\left[\frac{2}{s}\right]\Rightarrow f(x)=2$$

Which is constant function.

Now, $f(y)=\sqrt{1+\frac{1}{y'^2}}\Rightarrow 2=\sqrt{1+\frac{1}{y'^2}}\Rightarrow 3=\frac{1}{y'^2}$

$\Rightarrow y'=\pm\frac{1}{\sqrt{3}}\Rightarrow y=\pm\frac{1}{\sqrt{3}}x+d$ which is a straight line. So option (a) is correct.

UNSOLVED QUESTIONS

1. Eigen value of integeral equation $y(x)=\lambda\int_0^1 e^x e^t y(t)dt$. **CSIR NET DEC 2012**

(a) $\frac{2}{e^2+1}$ (b) $\frac{e^2+1}{2}$ (c) $\frac{2}{e^2-1}$ (d) $\frac{e^2-1}{2}$

2. The integral equation $y(x)=\int_0^x (x-t)y(t)dt - x\int_0^1 (1-t)y(t)dt$ is **GATE-2013**

(a) y'' – y = 0, y(0) = y(1) (b) y'' – y = 0, y(0) = 0, y'(0) = 0
(c) y'' + y = 0, y(0) = 0, y(1) = 0 (d) y'' + y = 0, y(0) = 0 = y'(0)

3. The value of λ for which the integral equation $y(x)=\lambda\int_0^1 (6x-t)y(t)dt$ has a non-trivial solution are given by roots of the equation. **GATE**

(a) $(3\lambda - 1)(2+\lambda) - \lambda^2 = 0$ (b) $(3\lambda - 1)(2+\lambda) + 2 = 0$
(c) $(3\lambda - 1)(2+\lambda) - 4\lambda^2 = 0$ (d) $(3\lambda - 1)(2+\lambda) + \lambda^3 = 0$

4. For the Volterra type linear integral equation $\phi(x)=x+2\int_0^x e^{x-\zeta}\phi(\zeta)d\zeta$, the resolvent kernel $R(x,\zeta;2)$ of the kernel $e^{x-\zeta}$ is

(a) $(x-\zeta)^2 e^{2(x-\zeta)}$ (b) $(x-\zeta)e^{x-\zeta}$ (c) $e^{3(x-\zeta)}$ (d) $e^{(x-\zeta)}$

5. For the linear integral equation $\phi(x)=x+\int_0^{1/2}\phi(\xi)d\xi$, the resolvent kernel $R(x,\xi;1)$ is **CSIR NET JUNE 2012**

(a) 1/2 (b) 2 (c) 3/2 (d) 4

6. The resolvent kernel for the integral equation $u(x)=F(x)+\int_{\log 2}^{x} e^{(t-x)}u(t)dt$, is

(a) $\cos(x-t)$ (b) 1 (c) e^{t-x} (d) $e^{2(t-x)}$

COV, NA, I.E.

7. The general solution of the differential equation $\frac{d^2y}{dx^2}+y=f(x), x\in(-\infty,\infty)$, where f is a continuous, real valued function on $(-\infty,\infty)$, is (where A,B,C and k are arbitrary constants)

(a) $y(x)=A\cos x+B\sin x+\int_0^x f(t)\sin(x-t)dt$ (b) $y(x)=\cos(x+k)+C\int_0^x f(t)\sin(x-t)dt$

(c) $y(x)=A\cos x+B\sin x+\int_0^x f(x-t)\sin tdt$ (d) $y(x)=A\cos x+B\sin x+\int_0^x f(x+t)\cos tdt$

8. The integral equation $y(x)=x-\int_1^x xy(t)dt; y\in C^1[1,\infty)$ has the solution

(a) $y=x(1-\ln x)$ (b) $y=xe^{x-\frac{1}{2}}(x-1)+x$ (c) $y=xe^{(1-x^2)/2}$ (d) $y=x-x(e^{x^2}-e)$

9. Let $\lambda>0$ and $F(x)=1-e^{-\lambda x}$ for $x>0$. Then for $t>0$, $\int_0^\infty e^{-tx}dF(x)$ equals

(a) $\frac{\lambda}{\lambda+1}$ (b) $\frac{\lambda}{\lambda-t}$ (c) 0 (d) ∞

10. The integral equation, involving a parameter λ, $\phi(x)=\cos x+\lambda\int_0^\pi \cos(x+\zeta)\phi(\zeta)d\zeta$ has

(a) a unique solution if $\lambda=1$, and an infinite number of solution if $\lambda=\frac{2}{\pi}$

(b)a unique solution if $\lambda=-1$, and an infinite number of solution if $\lambda=-\frac{2}{\pi}$

(c)a unique solution if $\lambda\neq\frac{2}{\pi}$ (d) no solution if $\lambda=\pm\frac{2}{\pi}$

11. The integral equation $y(x)=1+\lambda\int_0^{\pi/2}\cos(x-t)y(t)dt$ has

(a)a unique solution for $\lambda\neq 4/(\pi+2)$ (b) a unique solution for $\lambda\neq 4/(\pi-2)$

(c)no solution for $\lambda=4/(\pi+2)$, but the corresponding homogeneous equation has non-trival solutions

(d)no solution for $\lambda=4/(\pi-2)$, but the corresponding homogeneous equation has non-trival Solutions

12. For the integral equation $u(x)=f(x)+\lambda\int_a^b K(x,t)u(t)dt$ to have a continuous solution in the Interval $a\leq x\leq b$, which of the following assumptions are necessary?

COV, NA, I.E.

(a) $K(x,t)\not\equiv 0$, is real and continuous in the region $a\le x\le b, a\le t\le b$ with $|K(x,t)|\le M$

(b) $f(x)\not\equiv 0$, is real and continuous in the interval $a\le x\le b$

(c) λ is a constant (d) $|\lambda|<\dfrac{1}{M(b-a)}$

13. The integral equation $\phi(x)=\lambda\int_0^\pi K(x,\zeta)\phi(\zeta)d\zeta$, where λ is a parameter and

$K(x,\zeta)=\begin{cases}\cos x\sin\zeta \;\; for\; 0\le x<\zeta \\ \cos\zeta\,\sin x \;\; for\; 0\le x<\zeta\end{cases}$ leads to a boundary value problem

$\phi''(x)-f(\lambda)\phi(x)=0,\ \phi(\pi)=0, \phi(0)=0$, where $f(\lambda)$ is known. Then the boundary value problem has

(a) a unique solution when $f(\lambda)=0$ (b) Infinite number of solutions when $f(\lambda)>0$

(c) no solution when $f(\lambda)<0$ (d) a unique solution when $\lambda>1$

14. For the homogeneous Fredholm equation $y(x)=\lambda\int_0^\pi \sin(x+\zeta)y(\zeta)d\zeta$, the eigenvalues λ and the corresponding eigenfunction $y(x)$, involving arbitrary constants A and B, are

(a) $\lambda=\dfrac{2}{\pi}, y(x)=A(\sin x-\cos x)$ (b) $\lambda=-\dfrac{2}{\pi}, y(x)=B(\sin x+\cos x)$

(c) $\lambda=-\dfrac{2}{\pi}, y(x)=B(\sin x-\cos x)$ (d) $\lambda=\dfrac{2}{\pi}, y(x)=A(\sin x+\cos x)$

15. The integral equation $\int_a^x K(x,y)\phi(y)dy=f(x)$ with $K(x,x)\ne 0$, for all x can be transformed to $\phi(x)+\int_a^x G(x,y)\phi(y)dy=g(x)$, where for all x,y

(a) $K(x,x)=1, g=f'$ and $G(x,y)=\dfrac{\partial K}{\partial x}(x,y)$ (b) $K(x,x)=1, g=f$ and $G(x,y)=0$

(c) $G(x,y)=\dfrac{1}{K(x,x)}\dfrac{\partial K}{\partial x}(x,y)$ and $g(x)=\dfrac{f'(x)}{K(x,x)}$ (d) $G(x,y)=1$ and $g(x)=f(x)$

16. Consider the integral equation $y(x)=x^3+\int_0^x \sin(x-t)y(t)dt\;\; x\in[0,\pi]$ then the value of $y(1)$ is

(a) 19/20 (b) 1 (c) 17/20 (d) 21/20

17. Let $u \in c^2$ $[0,1]$ satisfy for some $\lambda \neq 0$ and $a \neq 0$, $u(x)+\frac{\lambda}{2}\int_0^1 |x-s| u(s)ds = ax+b$

Then 'u' also satisfies

(a) $\frac{d^2u}{dx^2}+\lambda u=0$ (b) $\frac{d^2u}{dx^2}-\lambda u=0$ (c) $\frac{du}{dx}-\frac{\lambda}{2}\int_0^1 \frac{x-s}{|x-s|}u(s)ds=a$ (d) $\frac{du}{dx}+\frac{\lambda}{2}\int_0^1 \frac{x-s}{|x-s|}u(s)ds=a$

18. For the integral equation $y(x)=1+x^3+\int_0^x K(x,t)y(t)dt$ with kernel $K(x,t)=2^{x-t}$, the iterated kernel $K_3(x,t)$ is

(a) $2^{x-t}(x-t)^2$ (b) $2^{x-t}(x-t)^3$ (c) $2^{x-t-1}(x-t)^2$ (d) $2^{x-t-1}(x-t)^3$

19. Let $y:[0,\infty)\to[0,\infty)$ be a continuously differentiable function satisfying

$y(t)=y(0)+\int_0^t y(s)ds \quad for \quad t\geq 0$. Then

(a) $y^2(t)=y^2(0)+\int_0^t y^2(s)ds$ (b) $y^2(t)=y^2(0)+2\int_0^t y^2(s)ds$

(c) $y^2(t)=y^2(0)+\int_0^t y(s)ds$ (d) $y^2(t)=y^2(0)+\left(\int_0^t y(s)ds\right)^2+2y(0)\int_0^t y(s)ds$

20. Which of the following are the characteristics numbers and the corresponding eigen Functions for the Fredholm homogeneous equation whose kernel is

$$K(x,t)=\begin{cases}(x+1)t, & 0\leq x\leq t\\ (t+1)x, & t\leq x\leq 1\end{cases}?$$

(a) $1, e^x$ (b) $-\pi^2, \pi\sin(\pi x)+\cos(\pi x)$

(c) $-4\pi^2, \pi\sin(\pi x)+\pi\cos(2\pi x)$ (d) $-\pi^2, \pi\cos(\pi x)+\sin(\pi x)$

21. Let $y(x)$ be the solution of the integral equation $y(x)=x-\int_0^x xt^2 y(t)dt,\ x>0$ Then the value of the function $y(x)$ at $x=\sqrt{2}$ is equal to

(a) $\frac{1}{\sqrt{2e}}$ (b) $\frac{e}{2}$ (c) $\frac{\sqrt{2}}{e^2}$ (d) $\frac{\sqrt{2}}{e}$

22. The solution of Volterra integral equation $y(x)=1+x+\int_0^x (x-t)y(t)dt$ is

(a) $y=e^{-x}$ (b) $y=e^x$ (c) $y=e^{x^2}$ (d) $y=e^{-x^2}$

Answer Key

1	2	3	4	5	6	7	8	9	10
a	c	a	c	b	b	a	c	a	b
11	12	13	14	15	16	17	18	19	20
c	a,c,d	a,d	c,d	a,c	d	a,d	c	b,d	a,d
21	22								
d	b								

MODEL TEST PAPER – 1

PART – B

1. Given a system of equations :

$$x + 2y + z = b_1$$
$$5x + y + 3z = b_2$$

Which of the following is true regarding its solution?
(a) The system has a unique solution for any given b_1 and b_2.
(b) The system will have infinitely many solutions for any given b_1 and b_2.
(c) Whether or not a solution exists depends on the given b_1 and b_2.
(d) The system would have no solution for any values of b_1 and b_2.

2. The eigenvectors of the matrix $\begin{bmatrix} 1 & 2 \\ 0 & 2 \end{bmatrix}$ are written in the form $\begin{bmatrix} 1 \\ a \end{bmatrix}$ and $\begin{bmatrix} 1 \\ b \end{bmatrix}$.
What is a + b?
(a) 0 (b) 1/2 (c) 1 (d) 2

3. Given a function $f(x, y) = 4x^2 + 6y^2 - 8x - 4y + 8$.
The optimal value of f(x, y)
(a) is a minimum equal to 10/3 (b) is a maximum equal to 10/3
(c) is a minimum equal to 8/3 (d) is a maximum equal to 8/3

4. Which one of the following graphs describes $f(x) = e^{-x}(x^2 + x + 1)$?

(a) 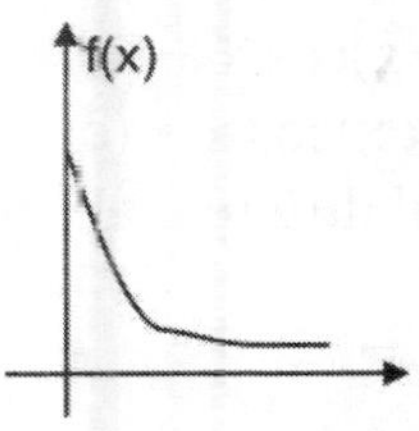

(b) 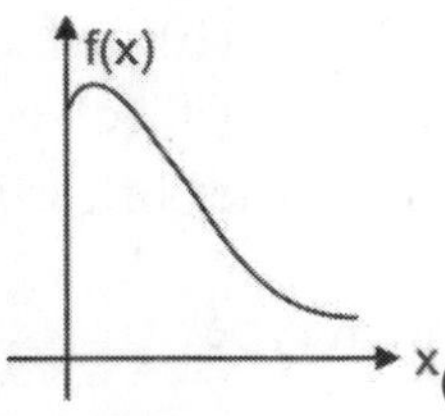

(c) 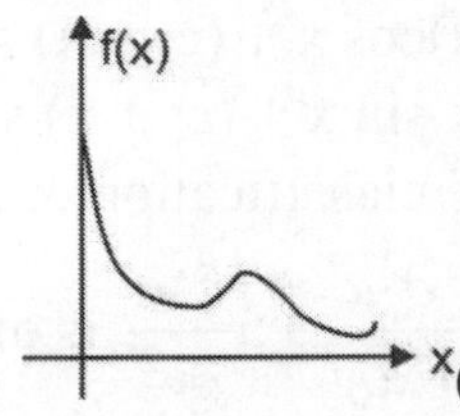

(d) 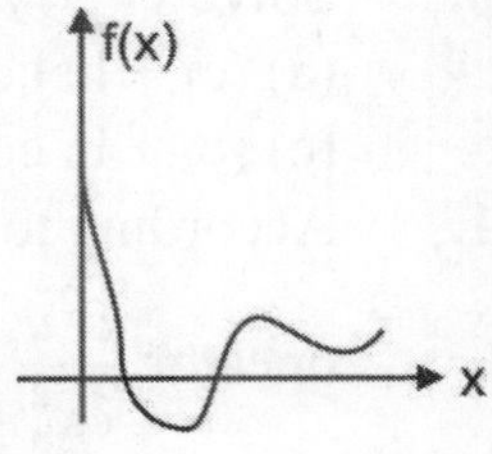

5. All the values of the multi-valued complex function 1^i, where $i = \sqrt{-1}$, are
(a) purely imaginary (b) real and non-negative
(c) on the unit circle (d) equal in real and imaginary parts

6. Consider likely applicability of Cauchy's integral theorem to evaluate the following integral counter clockwise around the unit circle c.

$$I - \oint_c \sec z dz,$$

z being a complex variable. The value I will be
(a) I = 0 : singularities set = ϕ

(b) I = 0 : singularities set = $\left\{\pm\frac{2n+1}{2}\pi; n = 0,1,2,....\right\}$

(c) I = π/2 : singularities set = {±nπ; n = 0, 1, 2, ...}

(d) none of above

7. Let G = {z ∈ C | z^n = 1 for some positive integer n}. Then under multiplication of complex numbers,

(a) G is a group of finite order.

(b) G is a group of infinite order, but every element of G has finite order.

(c) G is a cyclic group. (d) None of the above

8. Which of the following statements is false?

(a) The polynomial $x^2 + x + 1$ is irreducible in Z/2Z [x].

(b) The polynomial $x^2 - 2$ is irreducible in Q[x].

(c) The polynomial $x^2 + 1$ is reducible in Z/5Z[x].

(d) The polynomial $x^2 + 1$ is reducible in Z/7Z[x].

9. Consider the following difference equation $x(ydx + xdy)\cos\frac{y}{x} = y(xdy - ydx)\sin\frac{y}{x}$.

Which of the following is the solution of the above equation (c is an arbitrary constant)?

(a) $\frac{x}{y}\cos\frac{y}{x} = C$ (b) $\frac{x}{y}\sin\frac{y}{x} = C$ (c) $xy\cos\frac{y}{x} = C$ (d) $xy\sin\frac{y}{x} = C$

10. Solve y`` + y = sec x.

(a) (c_1 + ln cos x)cos x + (c_2 + x) sin x (b)(c_1 + ln tan x)cos x + (c_2 + x) sin x

(c) (c_1 + ln cos x)sin x + (c_2 + x) sin x (d) (c_1 + ln cot x)cos x + (c_2 + x) sin x

11. According to the classification of second order linear partial differentia operators, the opeator $\frac{\partial^2 u}{\partial x^2} - 4\frac{\partial^2 u}{\partial x \partial y} + 5\frac{\partial^2 u}{\partial y^2}$ is of which type?

(a) elliptic (b) parabolic (c) hyperbolic (d) none of these

12. Let u(x, t) be the solution of the following initial value problem :

$\frac{\partial^2 u}{\partial t^2} = \frac{\partial^2 u}{\partial x^2}, t > 0, x \in R$ $u(x,0) = u_0(x), x \in R$ $\frac{\partial u}{\partial t}(x,0) = 0, x \in R$

If $u_0(x)$ vanishes outside the interval [-1, 1], find the interval outside which u(x, t) vanishes, which t > 1.

(a) [-2t, t] (b) [-(t + 1), (t + 1)] (c) [-t/2, t/2]

(d) [-(2t + 1), (2t + 1)]

13. Simpson's rule gives the exact value of $\int_0^1 p(t)dt$ for every polynomial of degree less than or equal to

(a) 1 (b) 2 (c) 3 (d) 4

14. The functional $\int_0^1 [y'^2 + (y+2y')y'' + kxyy' + y^2]dx$, y(0) = 0, y(1) = 1, y`(0) = 2, y`(1) = 3 is path independent, if k equals

(a) 1 (b) 2 (c) 3 (d) 4

15. The eigen value λ of the following Fredhold integral equation $y(x) = \lambda \int_0^1 x^2 t\, y(t)dt$ is

(a) -2 (b) 2 (c) 4 (d) -4

PART – C

16. Let V be the real vector space of all polynomials in a single variable and with real coefficient, of degree at most 3. Let V* be its dual space. Let $t_1 = 1$, $t_2 = 2$, $t_3 = 3$, $t_4 = 4$. Which of the following sets of functionals {f, $1 \le i \le 4$} form a basis for V*?

(a) For $1 \le i \le 4$, and for $p \in V$, $f_i(p) = p(t_i)$

(b) For all $p \in V$, $f_i(p) = p(t_i)$ for i = 1, 2, $f_3(p) = p`(t_1)$ and $f_4(p) = p`(t_2)$

(c) For all $p \in V$, $f_i(p) = p(t_i)$ for $1 \le i \le 3$ and $f_4(p) = \int_1^2 p'(t)dt$.

(d) None of these

17. Let $A \in GL_n(R)$ have integer entries. Let $b \in R^n$ be a (column) vector, also with integer entries. Which of the following statements are true?

(a) If Ax = b, then the entries of x also integers. (b) If Ax = b, then the entries of x are rational.

(c) The matrix A^{-1} has integer entries if, and only if, det(A) = ±1. (d) None of these

18. Let $x_0 = 0$. For $n \ge 0$, define $x_{n+1} = x_n^2 + \frac{1}{4}$. Pick out the true statements :

(a) The sequence $\{x_n\}$ is bounded (b) The sequence $\{x_n\}$ is monctonic

(c) The sequence $\{x_n\}$ is convergent (d) None of these

19. Let $f \in C[a, b]$. Assume that $\min_{x\in[a,b]} f(x) = m > 0$ and $M = \max_{x\in[a,b]} f(x)$. Which of the following inequalities are true?

(a) $\frac{1}{M}\int_a^b f(x)dx + m\int_a^b \frac{1}{f(x)}dx \geq 2\sqrt{\frac{m}{M}}(b-a)$ (b) $\int_a^b f(x)dx + \int_a^b \frac{1}{f(x)}dx \geq (b-a)^2$

(c) $\int_a^b f(x)dx + \int_a^b \frac{1}{f(x)}dx \leq (b-a)^2$ (d) None of these

20. Let f : C → C be analytic except for a simple pole at z = 0 and let g : C → C be analytic. Then, the value of $\frac{\operatorname{Res}_{z\to 0}\{f(z)g(z)\}}{\operatorname{Res}_{z\to 0} f(z)}$ is

(a) g(0) (b) g`(0) (c) $\lim_{z\to 0} z\, f(z)$ (d) $\lim_{z\to 0} z\, f(z)g(z)$

21. If D is the open unit disk in C and f : C → D is analytic with f(10) = 1/2, then f(10 + i) is not

(a) $\frac{1+i}{2}$ (b) $\frac{1-i}{2}$ (c) $\frac{1}{2}$ (d) $\frac{i}{2}$

22. Which one of the following groups is not simple?

(a) S_3 (b) GL(2, R) (c) $Z_2 \times Z_2$ (d) A_5

23. Consider the algebraic extension $E = Q(\sqrt{2}, \sqrt{3}, \sqrt{5})$ of the field Q of rational numbers. Then, [E : Q], i.e., degree of E over Q, is

(a) 3 (b) 4 (c) 7 (d) 8

24. Let y_1 and y_2 be two linearly independent solutions of y`` + (sin x)y = 0, 0 ≤ x ≤ 1. Let g(x) = W(y_1, y_2) (x) be the Wronskian of y_1 and y_2. Then,

(a) g` > 0 on [0, 1] (b) g` < 0 on [0, 1]

(c) g` vanishes at only one point of [0, 1] (d) g` vanishes at all point of [0, 1]

25. The partial differential equation $x\frac{\partial^2 u}{\partial x^2} + 2xy\frac{\partial^2 u}{\partial x \partial y} + y\frac{\partial^2 u}{\partial y^2} + x\frac{\partial u}{\partial y} + y\frac{\partial u}{\partial x} = 0$ is not

(a) elliptic in the region x < 0, y < 0, xy > 1 (b) elliptic in the region x > 0, y > 0, xy > 1

(c) parabolic in the region x < 0, y < 0, xy > 1 (d) hyperbolic in the region x < 0, y < 0, xy > 1

Answer Key

(Paper – 1)

1	2	3	4	5	6	7	8	9	10	11	12	13
b	b	a	b	b	a	c	d	c	a	a	b	c
14	**15**	**16**	**17**	**18**	**19**	**20**	**21**	**22**	**23**	**24**	**25**	
b	c	a,b	b,c	a,b,c	a,b	a	a,b,d	a,b,c	d	c	a,b,c	

MODEL TEST PAPER – 2

PART – B

1. Let u be the real $n \times 1$ vector satisfying $u^T u = 1$. Define $A = I - 2uu^T$ where I is nth order identity matrix. Which of the following are true?
(a) A is singular (b) $A^2 = A$ (c) Tr(A) = n-2 (d) $A^2 \neq I$

2. Let $T : R^4 \to R^4$ be the linear map satisfying $T(e_1) = e_2$, $T(e_2) = e_3$, $T(e_3) = 0$, $T(e_4) = e_3$
Where $\{e_1, e_2, e_3, e_4\}$ the standard basis of R^4. Then
(a) T is idempotent (b) T is invertible (c) $\rho(T) = 3$ (d) T is nilpotent

3. The set $A = \{\cos n \mid n \in IN\}$ is –
(a) Dense in [-1, 1] (b) Dense in [0, 2] (c) no-where dense in IR
(d) uncountable

4. Let $f : (0, 2) \to IR$ be defined by $f(x) = \begin{cases} x^2 & x \in Q \\ 2x-1 & x \in Q^C \end{cases}$. Then
(a) f is differentiable at one point (b) f is differentiable at two point
(c) f is not differentiable at any point of (0, 2)
(d) f is differentiable at every point of (0, 2)

5. $f(z) = \begin{cases} \dfrac{z^5}{|z|^4} & if \quad z \neq 0 \\ 0 & if \quad z = 0 \end{cases}$. Then
(a) f(z) is satisfy C-R equation at z=0 (b) f is differentiable at z = 0
(c) f is analytic at z = 0 (d) f is not continuous at z = 0

6. Let $f : C \to C$ be a complex function given by $f(z) = u + iv$, $v = 3xy^2$ then
(a) f can't be analytic for any choice of u
(b) f is analytic for any choice of u
(c) f is analytic for suitable choice of u
(d) v is not differentiable of x and y

7. Which of the following condition on a group G implies that G is abelian?
(a) The order of G is p^3 for some prime p (b) Every subgroup of G is cyclic
(c) Every subgroup of G is normal
(d) The functin $f : G \to G$ defined by $f(x) = x^{-1}$ $\forall x \in G$ is homomorphism

8. Let G_1 and G_2 be two group s.t. $O(G_1)$ and $O(G_2)$ are relatively prime. If $f : G_1 \to G_2$ is homomorphism then

(a) f is trivial (b) f is onto (c) f is 1 – 1 (d) f is an isomorphism

9. Let $R = \left\{ \begin{bmatrix} a & b \\ c & d \end{bmatrix} : a, b, c, d \in Z \right\}$ and $I = \left\{ \begin{bmatrix} a & b \\ c & d \end{bmatrix} : a, b, c, d \in 3Z \right\}$ then

(a) I be subring but not ideal (b) I be ideal of R s.t. $\left| \frac{R}{I} \right| = 81$

(c) I be ideal of R s.t. $\left| \frac{R}{I} \right| = 64$ (d) $\frac{R}{I}$ is I.D. with card. 64

10. For the diffusion problem $u_t = u_{xx}$

$u(0, t) = u(\pi, t) = 0$ and $u(x, 0) = 3 \sin 2x$ then solution is given by

(a) $3e^{-t} \sin 2x$ (b) $3e^{-9t} \sin 2x$ (c) $3e^{-4t} \sin 2x$ (d) $3e^{-2t} \sin 2x$

11. Find the P.I. of $x(y - z)p + y(z - x)q = z(x - y)$ which represent a surface passing through $x = y = z$.

(a) $(x + y + z)^3 = 27\, xyz$ (b) $(x + y + z) = 27\, xyz$

(c) $(x + y + z)^2 = 27\, xyz$ (d) None of these

12. Consider the IVP $dy = (1 + y^2)dx$; $y(0) = 0$. Then the global interval is

(a) $-\infty < x < \infty$ (b) $0 < x \le \frac{\pi}{2}$ (c) $-\frac{\pi}{2} < x < \frac{\pi}{2}$ (d) $-\frac{\pi}{2} \le x \le \frac{\pi}{2}$

13. If $g(x, y)dx + (x + y)dy = 0$ is an exact differential equation and if $g(x, 0) = x^2$ then the general solution of DE is

(a) $2x^3 + 2xy + y^2 = c$ (b) $2x^3 + 6xy + 3y^2 = c$

(c) $2x + 2xy + y^2 = c$ (d) $x^2 + 2xy + y^2 = c$

14. Which of the following curve can be the functional

$I[y(x)] = \int_0^1 \{y'^2 + 12xy\}dx$ $y(0) = 0,\ y(1) = 1$ be extrimized.

(a) $y = x^2$ (b) $y = x^3$ (c) $y = x$ (d) $y = x^4$

15. The integral equation corresponding to initial value problem.

$y'' + y = 0;\ y(0) = 0,\ y'(0) = 1$ is

(a) $u(x) = x + \int_0^1 (t - x)u(t)dt$ (b) $u(x) = x + \int_0^x (t - x)u(t)dt$

(c) $u(x) = \sin x + \int_0^x (t - x)u(t)dt$ (d) $u(x) = x(1 + x) + \int_0^1 (t - x)u(t)dt$

PART – C

16. Let T and V be linear operator on R^2 define by $T(x_1, x_2) = (x_2, x_1)$ and $V(x_1, x_2) = (x_1, 0)$ then

(a) $TV(x_1, x_2) = VT(x_1, x_2)$ (b) $T^2(x_1, x_2) = T(x_1, x_2)$

(c) $V^2(x_1, x_2) = V(x_1, x_2)$ (d) $T^2(x_1, x_2) = (x_1, x_2)$

17. If w = {p(A) | p(x) is a polynomial in x with real coefficient}, $A = \begin{bmatrix} 1 & 0 & 1 \\ 0 & 1 & 1 \\ 1 & 1 & 0 \end{bmatrix}$ then

(a) w is a subspace of $M_3(IR)$ of dim 3 (b) A is diagonalizable

(c) $w = \{x \in IR^3 \mid (A+I)x = 0\}$ is of one dim. (d) $C_A(x) = m_A(x)$

18. Let $A = \{f \mid f : R \to R\}$ and $|f(x)| \le |x|$

$f(x + y) = f(x) + f(y) \quad \forall x, y \in IR$ then which of the following are incorrect?

(a) $\forall$ f $\in$ A; f is uniformly continuous

(b) $\exists$ f $\in$ A; f doesn't have any fixed point

(c) $\exists$ f $\in$ A; f is not Lipschitz function (d) $\exists$ f $\in$ A; f is not continuous

19. Suppose f is a continuous function on R s.t. $\lim_{x\to\infty} f(x)$ & $\lim_{x\to-\infty} f(x)$ exist then

(a) f is uniformly continuous (b) f may not uniformly continuous

(c) f is bounded (d) f is always differentiable

20. Which of the following are correct?

(a) Every rational function is meromorphic

(b) Every meromorphic function is rational

(c) A meromorphic function can have finite number of poles (d) None of these

21. Let C is a simple closed curve and z_0 is any constant then

(a) $\oint_C dz = 0$ (b) $\oint_C z\,dz = 0$ (c) $\oint_C \overline{(z - z_0)}\,dz = 0$

(d) $\oint_C (z - z_0)dz = 0$

22. Let G be a finite group and p be +ve integer and p | O(G) then choose the incorrect option

(a) G has a subgroup of order p

(b) G has a subgroup of order except p = 1 or p = O(G)

(c) G has a subgroup of order if order p is odd.

(d) G has a subgroup of order if order p is prime.

MODEL TEST PAPERS

23. For a second order PDE $(1-\sqrt{xy})u_{xx} + 2u_{xy} + (1+\sqrt{xy})u_{yy} = 0$. Which of the following are incorrect.
(a) hyperbolic in II and IV quadrant
(b) elliptic in I and III quadrant
(c) hyper in II and elliptic in IV quadrant
(d) hyperbolic in I and III quadrant

24. Consider the differential equation
$x^3y'' + x^2y' - xy = 0$. If y_1 and y_2 are two L.I. solution of DE then choose incorrect.
(a) $w(y_1, y_2)$ is continuous on R.
(b) D.E. has unique solution $y(x)$ s.t. $y(0) = 0$, $y'(0) = 1$
(c) D.E. has no solution s.t. $y(0) = 0 = y'(0)$
(d) has unique solution if $y(1) = 0 = y'(1)$

25. The value of α for which the I.E. $u(x) = \alpha\int_0^1 e^{x-t}u(t)dt$ has non-trivial solution.
(a) -2 (b) -1 (c) 1 (d) 2

Answer Key

(Paper – 2)

1	2	3	4	5	6	7	8	9	10	11	12	13
c	d	a	a	a	a	d	a	b	c	a	c	b
14	15	16	17	18	19	20	21	22	23	24	25	
b	b	c,d	b	b,c,d	a,c	a,b,c	a,b,d	a,b,c	a,b,c	a,c	c	

MODEL TEST PAPER – 3

PART – B

1. Let $A=\begin{bmatrix}1&0&6\\3&1&0\\4&0&6\end{bmatrix}\in GL(3,Z_7)$, then number of eigen vector corresponding to an eigen value $\lambda = 1$

(a) 6 (b) 13 (c) 48 (d) 1

2. Let ξ be the primitive third root of unity $A=\begin{bmatrix}\xi^{-1}&0&0\\0&1&0\\0&0&\xi\end{bmatrix}$. Define for a vector V = (V_1, V_2, V_3) $\in$ IR^3.

Define $|V|_A=\sqrt{|VAV^T|}$. where V^T is transpose of V. If w=(-1,1,1) then $|w|_A$ equals to

(a) 0 (b) 1 (c) -1 (d) 2

3. If $<a_n>$ be a sequence s.t. $\overline{\lim} < a_n >= a$ and $\underline{\lim}\ a_n = b$ then which of the following is TRUE?

1. $<a_n>$ is oscillate finitely if a = b
2. $<a_n>$ oscillate infinitely if a = b
3. $<a_n>$ convergent if a = b

(a) 1 and 2 only (b) 1 and 3 only (c) 1 only (d) None of these

4. Let A = $\{x^2 : 0 < x < 2\}$ and B = $\{\sin x . 0 < x < \pi/2\}$

(a) There is 1 – 1 and onto function from A to B
(b) There is no 1 – 1 but onto function from A to B
(c) There is no onto but 1 – 1 function from A to B
(d) None of these

5. Value of the integral $I=\int\frac{e^z}{e^z-1}dz$, where $|z| = 3\pi$ will be

(a) $6\pi i$ (b) $2\pi i$ (c) $4\pi i$ (d) 0

6. $I=\oint_C\frac{dz}{z^2}$ where the contour C is the ellipse $(x-2)^2+\frac{1}{4}(y-5)^2=1$ then I is

MODEL TEST PAPERS

(a) $\frac{5}{2}\pi$ (b) $\frac{5}{2}\pi i$ (c) $\frac{5}{4}\pi i$ (d) None of these

7. The set $M_n(R)$ of all n × n matrices under matrix multiplication is
(a) Abelian group (b) Non-abelian group (c) Non-cyclic abelian group (d) None of these

8. Consider an ideal $I = <x^4 + 1>$ of Q[x]. Then
(a) I is maximal ideal of Q[x] (b) I is not maximal ideal of Q[x]
(c) $\frac{Q[x]}{<x^4+1>}$ is not I.D. (d) $\frac{Q[x]}{<x^4+1>} \cong Z$

9. $xy'' + 2y' + xe^x y = 0$. Let y_1 and y_2 are L.I. solution if $w(x) = y_1(x)\, y_2'(x) - y_2(x)\, y_1'(x)$ with
w(1) = 2. Find w(5) =
(a) $\frac{2}{13}$ (b) $\frac{2}{19}$ (c) $\frac{2}{25}$ (d) None of these

10. Consider the *IVP* $\frac{dy}{dx} = x + y^2; y(0) = 0$ the largest interval in which picard's method guaranties the unique solution $|x| < \alpha$. Then α is
(a) $\alpha = \left(\frac{1}{2}\right)^{2/3}$ (b) $\alpha = (2)^{3/2}$ (c) $\alpha = \left(\frac{1}{2}\right)^{3/2}$ (d) $\alpha = (2)^{2/3}$

11. The IVP $z\frac{\partial z}{\partial x} + \frac{\partial z}{\partial y} = 1$ with the initial curve data $x_0 = S$, $y_0 = S$, $z_0 = \frac{S}{2}$ has unique solution if
(a) $0 \le S \le 2$ (b) $0 \le S < 2$ (c) $0 < S \le 2$ (d) For any S

12. Let z = z(x, y) be a solution of p.q = 1 passing through (0, 0, 0) then z(0, 1) is –
(a) 0 (b) 1 (c) 2 (d) 4

13. The resolvent kernel R(x, t; λ) for the Voltera integeral equation $\phi(x) = x + \lambda\int_a^x \phi(t)dt$ is
(a) $e^{\lambda(x+t)}$ (b) $e^{\lambda(x-t)}$ (c) $\lambda e^{(x+t)}$ (d) $e^{\lambda xt}$

14. The extremal of the variational problem. $\int_0^5 \left(9y^2 + y'^2 - 3x\right)dx$ with the boundary condition P(0, 0) & Q(5, 0)
(a) $y(x) = e^x - 1$ (b) $y(x) = 0$ (c) $y(x) = x$ (d) $y(x) = e^{-x}$

15. Consider the function for $x \ge -2$ and the iteration $x_{n+1} = f(x_n)$; $n \ge 0$ for $x_0 = 1$. What are the possible limits of the iteration?

(a) $\sqrt{2+\sqrt{2+\sqrt{2+.....}}}$ (b) -1 (c) 5 (d) 1

PART – C

16. Let T : $P_2(IR) \to P_2(IR)$ the linear operator define by $T[f(x)] = f(1) + f(0)x + [f'(0) + f''(0)]x^2$

(a) T is diagonalizable (b) T is invertible

(c) $\exists$ an eigen value with A.M. = 2 (d) $\rho(T) = 2$

17. Consider $A = \begin{bmatrix} 2 & 0 & 0 & 3 \\ 0 & 2 & 3 & 0 \\ 0 & 3 & 2 & 0 \\ 3 & 0 & 0 & 2 \end{bmatrix}$, define a linear operator T : $M_4(R) \to M_4(R)$ s.t. T(X) = AX then

(a) $\lambda = 2$ is eigen value of T with G.M. 8 (b) $\lambda = 5$ is eigen value of T with G.M. 8

(c) T is diagonalizable over R

(d) $W_T = \{\text{span } \{T^k \mid k = 0, 1, 2, ...\}$ is of two dimensional.

18. The function f(x) = 1 - |1 – x| on R is

(a) Not continuous (b) Not differentiable but continuous

(c) Differential but not continuous (d) Differentiable at one point

19. Select the correct statement.

(a) If f is continuous at x = 0 then x f(x) is derivable at x = 0

(b) if f is differentiable function on [a, b] then $\exists$ exactly one zero of f' between two zero of f.

(c) If f [a, b] $\to$ R s.t. f(a) = f(b) = 0 and $f(x) \neq 0 \ \forall x \in (a, b)$ then f'(a), f'(b) must have opposite sign.

(d) If f is differentiable function then f Rolle's theorem can be deduced from Lagrange's mean value theorem.

20. Let $f(z) = \dfrac{1}{e^{1/z} - 1}$ then

(a) f has infinitely many poles (b) each pole of f is simple

(c) f has isolated singularity (d) f' has non-isolated singularity

21. Consider the ring Z[x] & I = {f(x) $\in$ Z[x] | f(0) is divisible by 2} then –

(a) I be maximal & prime ideal both (b) $\left|\dfrac{R}{I}\right|$ = prime

MODEL TEST PAPERS

(c) $\frac{R}{I}$ is field with characteristic '0' (d) all of the above

22. Let G_1 and G_2 be two group then

(a) G_1 & G_2 cyclic $\Rightarrow G_1 \times G_2$ is cyclic (b) $Z(G_1 \times G_2) = Z(G_1) \times Z(G_2)$

(c) $Z(G_1 \times G_2) = Z(G_1) \times Z(G_2)$ if G_1 & G_2 are finite

(d) $Z(G_1 \times G_2) = Z(G_1) \times Z(G_2)$ only if G_1 & G_2 are infinite.

23. $\frac{dy}{dx} + \left(\frac{2x+1}{x}\right)y = e^{-2x}$; be the solution of DE then which of the following are incorrect.

(a) $y \to \infty$ as $x \to \infty$ (b) $y \to -\infty$ as $x \to \infty$

(c) $y \to 0$ as $x \to \infty$ (d) y is unbounded.

24. Which of the following are not characteristic curve of $y^2r - x^2t = 0$ are

(a) $x^2 + y^2 = c_1$ & $y^2 - x^2 = c_2$ (b) $xy = c_1$, $x - y = c_2$

(c) $xy = c_1$ & $y^2 - x^2 = c_2$ (d) $x^2 + y^2 = c_1\sqrt{2}$, $x - y = c_2$

25. A characteristic number and the corresponding eigen function of the homogeneous Fredholm integral equation with kernel $K(x,t) = \begin{cases} x(t-1), & 0 \le x \le t \\ t(x-1), & t \le x \le 1 \end{cases}$ are

(a) $\lambda = -\pi^2, \varphi(x) = \sin \pi x$ (b) $\lambda = -2\pi^2, \varphi(x) = \sin 2\pi x$

(c) $\lambda = -3\pi^2, \varphi(x) = \sin 3\pi x$ (d) $\lambda = -4\pi^2, \varphi(x) = \sin 2\pi x$

Answer Key

(Paper – 3)

1	2	3	4	5	6	7	8	9	10	11	12	13
a	a	d	a	a	d	d	a	c	a	b	a	b
14	**15**	**16**	**17**	**18**	**19**	**20**	**21**	**22**	**23**	**24**	**25**	
b	a	a,b	b,c	b	c,d	a,b,c,d	a,b	b,c	a,b	b,c,d	a,d	

MODEL TEST PAPER – 4

PART – B

1. The application of Gram-Schmidt process of orthonormalization to $u_1=(1,1,0), u_2=(1,0,0), u_3=(1,1,1)$ yields

(A) $\frac{1}{\sqrt{2}}(1,1,0),(1,0,0),(0,0,1)$ (B) $\frac{1}{\sqrt{2}}(1,1,0),\frac{1}{\sqrt{2}}(1,-1,0),\frac{1}{\sqrt{2}}(1,1,1)$

(C) $(0,1,0),(1,0,0),(0,0,1)$ (D) $\frac{1}{\sqrt{2}}(1,1,0),\frac{1}{\sqrt{2}}(1,-1,0),(0,0,1)$

2. Consider the basis $\{u_1,u_2,u_3\}$ of R^3, where $u_1=(1,0,0), u_2=(1,1,0), u_3=(1,1,1)$. Let $\{f_1,f_2,f_3\}$ be the dual basis of $\{u_1,u_2,u_3\}$ and f be the linear functional defined by $f(a,b,c)=a+b+c;(a,b,c)\in R^3$ if $f=\alpha_1 f_1+\alpha_2 f_2+\alpha_3 f_3$, then $\{\alpha_1,\alpha_2,\alpha_3\}$ is

(A) (1, 2, 3) (B) (1, 3, 2) (C) (2, 3, 1) (D) (3, 2, 1)

3. Suppose that f is real valued, bounded on [a, b] and f^3 is Riemann integrable on [a, b]. Then

(A) f^2 is Riemann integrable (B) f^2 is not Riemann integrable

(C) $\int_0^1 f(x)dx=\lim_{c\to 0}\int_c^1 f(x)dx$ (D) f^2 is discontinuous

4. The sequence $f_n=\begin{cases} n^2x & , 0\le x\le 1/n \\ -n^2x+2n & , 1/n\le x\le 2/n \\ 0 & , 2/n\le x\le 1 \end{cases}$

(A) Converges uniformly (B) The point-wise limit is f(x) = x

(C) $\lim_{n\to\infty}\int_0^1 f_n(x)dx=\int_0^1 f(x)dx$ (D) $\lim_{n\to\infty}\int_0^1 f_n(x)dx\ne\int_0^1 f(x)dx$

5. The function $f(z)=|z|^2+i\bar{z}+1$ is differentiable at

(A) i (B) 1 (C) $-i$ (D) no point in C

6. The radius of convergence of the power series $\sum_{n=0}^{\infty}4^{(-1)^n n}z^{2n}$ is

(A) 1/2 (B) e (C) 1 (D) - 1/2

7. Let H_1, H_2 be two distinct subgroups of a finite group G, each of order 2. Let H be smallest subgroup containing H_1, and H_2. Then the order of H is
(A) always 2 (B) always 4 (C) always 8 (D) none of the above

8. Let F be a field of 8 elements and
$A=\{x\in F | x^7=1 \text{ and } x^k\neq 1 \text{ for all natural numbers } k<7\}$.
Then the numbers of elements in A is
(A) 1 (B) 2 (C) 3 (D) 6

9. If $y=\sum_{m=0}^{\infty} c_m x^{r+m}$ is assumed to be a solution of the differential equation
$x^2y''-xy'-3(1+x^2)y=0$, then the value of r are
(A) 1 and 3 (B) –1 and 3 (C) 1 and – 3 (D) –1 and –3

10. The solution to the initial value problem
$\frac{d^2y}{dt^2}+2\frac{dy}{dt}+5y=3e^{-t}\sin t, y(0)=0 \text{ and } \frac{dy}{dt}(0)=3$, is
(A) $y(t)=e^t(\sin t+\sin 2t)$ (B) $y(t)=e^{-t}(\sin t+\sin 2t)$
(C) $y(t)=3e^t\sin t$ (D) $y(t)=3e^{-t}\sin t$

11. The resolvent kernel R(x, t, λ) for the Volterra integral equation
$\varphi(x)=x+\lambda\int_a^x \varphi(s)\,ds$ is
(A) $e^{\lambda(x+t)}$ (B) $e^{\lambda(x-t)}$ (C) $\lambda e^{\lambda(x+t)}$ (D) $e^{\lambda xt}$

12. The extremals of the functional $J(y)=\int_1^2 \frac{\sqrt{1+y'^2}}{x}dx, y(1)=0, y(2)=1$ is
(A) $x^2+y^2=2$ (B) $x^2+(y-2)^2=5$ (C) $x^2-y^2=4$ (D) $x^2+(y-2)^2=4$

13. Using Taylor expansion, which of the following is true?
(for some ξ lying between x_0 and x_0+h)
(A) $f'(x_0)=\frac{f(x_0+h)-f(x_0)}{h}-\frac{h}{2}f''(\xi)$ (B) $f'(x_0)=\frac{f(x_0+h)-f(x_0)}{h}-\frac{h^2}{2}f'(\xi)$
(C) $f'(x_0)=\frac{f(x_0+h)+f(x_0)}{h}+\frac{h^2}{2}f''(\xi)$ (D $f'(x_0)=\frac{f(x_0-h)+f(x_0)}{h}-\frac{h^2}{2}f''(\xi)$

14. If g(x, y)dx + (x + y)dy = 0 is an exact differential equation and if g(x, 0) = x^2 then the general solution of DE is
(a) $2x^3 + 2xy + y^2 = c$ (b) $2x^3 + 6xy + 3y^2 = c$

(c) $2x + 2xy + y^2 = c$ (d) $x^2 + 2xy + y^2 = c$

15. For the diffusion problem $u_t = u_{xx}$

$u(0, t) = u(\pi, t) = 0$ and $u(x, 0) = 3 \sin 2x$ then solution is given by

(a) $3e^{-t} \sin 2x$ (b) $3e^{-9t} \sin 2x$ (c) $3e^{-4t} \sin 2x$ (d) $3e^{-2t} \sin 2x$

PART – C

16. Suppose that $B \in R^{n\times n}$ is a symmetric matrix and that the characteristics polynomial of B given by $t^2\left(t^2-1\right)^3\left(t-2\right)^2$

(A) The value of n is 10 (B) Trace(B) =4 and dat(B) = 0

(C) The eigenvalue of B are 0, 1, - 1, 2 (D) The value of n is 7

17. Let V be the set of 2×2 matrix $\begin{bmatrix} a_{11} & a_{12} \\ a_{21} & a_{22} \end{bmatrix}$ with complex entries such that $a_{11}+a_{22}=0$

Let W be the set of matrices in V with $a_{12}+\overline{a_{21}}=0$.

Then under usual matrix addition and scalar multiplication, which of following is (are) true?

(A) V is a vector space over R (B) W is a vector space over C

(C) V is a vector space over C (D) W is a vector space over R

18. Let $\{a_n\}$ be a sequence of real numbers such that the series $\sum_{n=1}^{\infty}|a_n|^2$ is convergent. Which of the following statement are true?

(A) The series $\sum_{n=1}^{\infty}\frac{a_n}{n}$ is convergent

(B) The series $\sum_{n=1}^{\infty}|a_n|^p$ is convergent for all $2<p<\infty$

(C) The series $\sum_{n=1}^{\infty}|a_n|^p$ is convergent for all $1<p<2$ (D) None of these

19. Which of the following limits does not exists

(A) $\lim\limits_{(x,y)\to(1,1)}\frac{e^x \ln y}{x^2+2y^2}$ (B) $\lim\limits_{(x,y,z)\to(0,0,0)}\frac{x^2y-z^2}{x^2+y^2+z^2}$ (C) $\lim\limits_{(x,y)\to(0,0)}\frac{x\sin y}{x^2+2y^2}$

(D) $\lim\limits_{(x,y)\to(0,0)}\frac{x^2\sin^2 y}{x^2+2y^2}$

20. Which of the following sets are open

(A) $0<|z-1|<2$ (B) $|\mathrm{Im}\, z|<1$ (C) $|z+3|<2$ (D) $\mathrm{Re}\, z=1$

21. The Laurent series for $f(z)=\frac{e^z}{\sin z}$ valid in some punctured disk centered at zero is

(A) $f(z)=z^{-1}+1+\frac{2}{3}z+...$ (B) $f(z)=1+z+\frac{z^2}{2!}+...$

(C) $f(z)=\frac{1}{z}+\frac{1}{z^2}+\frac{1}{z^3}+...$ (D) $f(z)=z^{-1}+z+\frac{z^2}{2!}+...$

22. Which of the following sets is/are subgroups of the given group G?

(A) $G=\langle C^*,\cdot\rangle, H=\{Z\in C;|z|=1\}$ (B) $G=\langle C^*,\cdot\rangle, H=\{Z\in C^*;|z|\le 1\}$

(C) $G=\langle GL_n(\mathrm{R}),\cdot\rangle, H=\{A\in GL_n(\mathrm{R}); A=A^T\}$

(D) $G=\langle Z_{100},+\rangle, H=\{n\in Z_{100};10/n\}$

23. Which of the following rings are integral domains?

(A) $R[x]$, the ring of all polynomials in one variable with real coefficients

(B) $M_n(R)$

(C) The ring of complex analytic functions defined on the unit disc of the complex plane (with point-wise addition and multiplication as the ring operations)

(D) None of these

24. Consider the differential equation $y_1'=y_1-y_1^2; y_2'=y_2$ Then

(A) The eigenvalues are $\lambda_1=-1,\lambda_2=1$ and the fixed point is a saddle.

(B) The are negative and the fixed point is an unstable node

(C) the eigenvalues are $\lambda_1=\lambda_2=1$ and the fixed point is an unstable node

(D) None of these

25. Consider the integral $J(x)=\int_0^1\left((y')^2-\frac{\pi^2y^2}{4}\right)dx$, $y(0)=0\ \&\ y(1)=1$ Then the solution is

(A) $y(x)=\cos\left(\pi x/2\right)$ (B) $y(x)=\sin\left(\pi x/2\right)$ (C) $y(x)=\tan\left(\pi x/2\right)$ (D) $y(x)=\cot\left(\pi x/2\right)$

Answer Key

(Paper – 4)

1	2	3	4	5	6	7	8	9	10	11	12	13
d	a	a	b	c	a	b	d	d	d	b	b	a
14	**15**	**16**	**17**	**18**	**19**	**20**	**21**	**22**	**23**	**24**	**25**	
b	c	a,b,c	a,b,c	a,b	b,c	a,b,c	a	a,d	a,c	a,b	b	

MODEL TEST PAPER – 5

PART – B

1. Let $M = \begin{bmatrix} 1 & 1 & 0 \\ 0 & 1 & 1 \\ 0 & 0 & 1 \end{bmatrix}$ and $e^M = Id + M + \frac{1}{2!}M^2 + \frac{1}{3!}M^3 + \ldots\ldots$

If $e^M = \left[b_{ij}\right]$, then $\frac{1}{e}\sum_{i=1}^{3}\sum_{j=1}^{3} b_{ij}$ is equal to

(A) 5 (B) 5.5 (C) 0 (D) 1

2. $U = \{(x,y,z) | x = y = z\}$, $V = \{(x,y,z) | x = 0\}$. Then what is U+V?

(A) xy plane (B) yz plane (C) zx plane (D) R^3

3. $\lim_{(x,y)\to(0,0)} \frac{\sin x \sin y}{x^2 + y^2}$ is

(A) 0 (B) 1 (C) 2 (D) does not exist

4. Let f be a one to one function on the closed interval [0, 1] to the set of real numbers R, then

(A) f must be onto (B) range of f must contain a rational number

(C) range of f must contain an irrational number

(D) range of f must contain both rational and irrational numbers

5. If $f(z)$ is that branch of $\log z$ for which $0 \le \theta_p < 2\pi$. The sum of the residues of $\frac{f(z)}{z^2+1}$ at its poles is

(A) π/2 (B) 2π (C) -π/2 (D) 0

6. Let $u(x,y) = x^3 + ax^2y + bxy^2 + 2y^3$ be a harmonic function and $v(x,y)$ its harmonic conjugate. If $v(0,0) = 1$, then $|a + b + v(1,1)|$ is equal to

(A) 10 (B) -1 (C) 9 (D) 0

7. If (a, 7), then $a^{12} - 1$ is divisible by k, where k is

(A) 3 (B) 5 (C) 7 (D) 8

8. Which one of the following is true?

(A) The characteristics of the ring 6Z is 6 (B) The ring 6Z has a zero divisor

(C) The characteristics of the ring (Z / 6Z) × 6Z is zero

(D) The ring 6Z × 6Z is an integral domain

9. Eigen value for BVP $X''(x)+\lambda X(x)=0;\ X(0)=0,\ X(\pi)+X'(\pi)=0$ satisfy

(A) $\lambda+\tan\lambda\pi=0$ (B) $\sqrt{\lambda}+\tan\lambda\pi=0$

(C) $\sqrt{\lambda}+\tan\sqrt{\lambda}\pi=0$ (D) $\lambda+\tan\sqrt{\lambda}\pi=0$

10. Which of the following pair of functions is a linearly independent pair of solution of $y''+9y=0$?

(A) cos 3x, sin 3x – cos 5x (B) $\sin 3x+\cos 3x,\ 3\sin x-4\sin^3 x$

(C) $\sin 3x, \sin 3x\cos 3x$ (D) $\sin 3x+\cos 3x,\ \sin 3x\cos 3x$

11. The eigenvalue of the homogeneous Fredholm integral equation

$g(s)=\lambda\int_0^1 e^s e^t g(t)\,dt$ is

(A) $\lambda=2(e^2+1)$ (B) $\lambda=2(e^2-1)$ (C) $\lambda=2/(e^2-1)$ (D) $\lambda=2/(e^2+1)$

12. Consider the functional $J(y)=\int_0^{\pi/4} y'(x)^2\cos^2 x\,dx$

A = $\left\{y : y\in c'\left[0,\frac{\pi}{4}\right], y(0)=1, y\left(\frac{\pi}{4}\right)=2\right\}$. Then extrema is

(A) $y(x)=\tan x+x$ (B) $y(x)=\tan x+1$ (C) $y(x)=\sec^2 x$ (D) $y(x)=x^2+x$

13. Let f : [0, 4] → R be a thrice continuously differentiable function. Then the value of $f[1,2,3,4]$ is

(A) $\frac{f''(\xi)}{3}$ for some $\xi\in(0,4)$ (B) $\frac{f''(\xi)}{6}$ for some $\xi\in(0,4)$

(C) $\frac{f'''(\xi)}{3}$ for some $\xi\in(0,4)$ (D) $\frac{f'''(\xi)}{6}$ for some $\xi\in(0,4)$

14. The PDE $y^3u_{xx}-(x^2-1)u_{yy}=0$ is

(a) Hyperbolic in $\{(x, y) \mid y<0, |x|>1\}$

(b) Parabolic in $\{(x, y) \mid y>0, |x|>1\}$

(c) Hyperbolic in $\{(x, y) \mid y>0, |x|>1\}$ (d) Hyperbolic for all pair (x, y)

15. Let px(x + y) – qy(x + y) = -(x – y)(2x + 2y + z) be a PDE then its solution

(a) u = xy v = (x + y)(x + y + z)

(b) u = x/y v = (x + y)(x + y + z)

(c) u = xy v = (x + y)/(x + y + z)

(d) u = x/y v = (x + y + z) / (x + y)

PART – C

16. For any vector u in a finite dimensional Euclidean space X the Housholder transformation is defined by $H_u = I - 2\frac{u}{\|u\|_2}\frac{u^*}{\|u\|_2}$. Then

(A) $H_u^2 = I$ (B) H_u is unitary (C) $H_u x = x$ if $x \perp u$ (D) H_u is not unitary

17. Consider the matrices $A = \begin{bmatrix} 2 & 2 & 1 \\ 0 & 2 & -1 \\ 0 & 0 & 3 \end{bmatrix}$ and $B = \begin{bmatrix} 2 & 1 & 0 \\ 0 & 2 & 0 \\ 0 & 0 & 3 \end{bmatrix}$. Then

(A) A and B are similar over the field of rational numbers Q
(B) A is diagonalizable over the field of rational numbers Q
(C) B is the Jordan canonical form of A
(D) The minimal polynomial and the characteristic polynomial of A are the same

18. Let A and B be subset of R^n. Define $A + B = \{x + y | x \in A, y \in B\}$.

Pick cut the true statement
(A) if A and B are closed sets, then A + B is a closed set
(B) if A is an open set and if B is a closed set, then A + B is an open set
(C) if A and B are compact sets, then so is A + B (D) None of these

19. Pick out the convergent series:

(A) $\sum_{n=1}^{\infty} \frac{(n+1)^n}{n^{n+\frac{3}{2}}}$ (B) $1 + \frac{1}{2^2} + \frac{2^2}{3^3} + \frac{3^3}{4^4} + \ldots$ (C) $\sum_{n=1}^{\infty} \sqrt{\frac{1+4^n}{1+5^n}}$ (D) None of these

20. Let H = {z = x + iy ∈ c : y > 0} be the upper half plane and D = {z ∈ c : |z| < 1} be the open unit disc. Suppose that f is Mobius transformation, which maps at conformally onto D. Suppose that $f(2i) = 0$. Pick each correct statement from below.

(A) f has a simple pole at $z = -2i$ (B) f satisfies $f(i)\overline{f(-i)} = 1$

(C) f has an essential singularity at $z = -2i$ (D) $|f(2+2i)| = \frac{1}{\sqrt{5}}$

21. Let f be an entire function such that $\lim_{|z|\to\infty} |f(z)| = \infty$. Then

(A) $f\left(\frac{1}{z}\right)$ has an essential singularity at 0 (B) f cannot be a polynomial

(C) f has finitely many zeros (D) $f\left(\frac{1}{z}\right)$ has a pole at 0

22. Suppose that $f : \frac{Z}{50Z} \to \frac{Z}{15Z}$ is a group homomorphism with $f(7)=6$. Then

(A) $f(n) = 3n \bmod 15$ (B) image of f is $\{0, 3, 6, 9, 12\}$

(C) kernel of f is 3x = 15q for some $q \in Z$ and x = 5q

(D) $f^{-1}(3) = \{1,6,11,16,21,26,31,36,41,46\}$

23. Which of the following quotient rings are fields?

(A) $F_3[x]/\langle(X^2+X+1)\rangle$, where F_3 is the finite field with three elements

(B) $Q[x]/\langle(X-3)\rangle$ (C) $Q[x]/\langle(X^2+X+1)\rangle$

(D) $F_2[x]/\langle(X^2+X+1)\rangle$, where F_2 is the finite field with two elements

24. The solution of $\frac{dy}{dx} = y^2, y(0) = 1$ does not exist for all

(A) $x \in (-\infty, 1)$ (B) $x \in [0, a]\, where\, a > 1$

(C) $x \in (-\infty, \infty)$ (D) $x \in [1, a]\, where\, a > 1$

25. Given the differential equation y"(x) - 3y'(x) + 2y(x) = 5 sin x, y(0) = 1, y'(0) = 2

(A) Equivalent equation is volterra Integral equation of first kind

(B) Equivalent equation is volterra Integral equation of the second kind

(C) The equivalent Integer equation is $y(x) + \int_0^x [2(x-t)-3]y(t)dt = 1 - 5\sin x$

(D) y(x) = 1 - 5 sin x.

Answer Key

(Paper – 5)

1	2	3	4	5	6	7	8	9	10	11	12	13
b	d	d	c	c	a	c	c	c	b	d	b	d
14	**15**	**16**	**17**	**18**	**19**	**20**	**21**	**22**	**23**	**24**	**25**	
c	a	a,b,c	a,c,d	b,c	a,c	a,b,d	c,d	a,b,c,d	c,d	b,c,d	b,c	

MODEL TEST PAPER – 6

PART – B

1. Let C be and real matrix. Let W be the vector space spanned by $\{I, C, C^2, ...C^{2n}\}$. Then dimension of the vector space W is

(A) 2n (B) atmost n (C) n^2 (D) None of these

2. Let V be the real vector space of all polynomials in one variable with real coefficients and having degree at most 20. Define the subspaces

$$W_1 = \left\{p \in V : p(1) = 0, p\left(\frac{1}{2}\right) = 0, p(5) = 0, p(7) = 0\right\},$$

$$W_2 = \left\{p \in V : p\left(\frac{1}{2}\right) = 0, p(3) = 0, p(4) = 0, p(7) = 0\right\}.$$ Then the dimension of $W_1 \cap W_2$ is

(A) 21 (B) 6 (C) 15 (D) 27

3. The sequence $\sqrt{7}, \sqrt{7+\sqrt{7}}, \sqrt{7+\sqrt{7+\sqrt{7}}}, ...$ converges to

(A) $\frac{1+\sqrt{33}}{2}$ (B) $\frac{1+\sqrt{32}}{2}$ (C) $\frac{1+\sqrt{30}}{2}$ (D) $\frac{1+\sqrt{29}}{2}$

4. For a positive real number p, let $(f_n : n = 1, 2, ...)$ be sequence of functions defined on [0, 1] by

$$f_n(x) = \begin{cases} n^{p+1}, & \text{if } 0 \le x \le \frac{1}{n} \\ \frac{1}{x^p}, & \text{if } \frac{1}{n} < x \le 1 \end{cases}$$, let $f(x) = \lim_{n\to\infty} f_n(x), x \in [0,1]$. Then on [0, 1]

(A) f is Riemann integrable

(B) the improper integral $\int_0^1 f(x)\,dx$ converge for $p \ge 1$

(C) the improper integral $\int_0^1 f(x)\,dx$ converge for p < 1 (D) f_n converges uniformly

5. If the power series $\sum_{n=0}^{\infty} a_n (z+3-i)^n$ converges at $5i$ and diverges at $-3i$, then the power series

(A) converges at $-2+5i$ and diverges at $2-3i$ (B) converges $2-3i$ at and diverges at $-2+5i$

(C) converges at both $2-3i$ and $-2+5i$ (D) diverges at both $2-3i$ and $-2+5i$

6. The value of $\frac{i}{4-\pi} \int_{|z|=4} \frac{dz}{z \cos z}$ is equal to

(A) 2 (B) π (C) 0 (D) $-\frac{4}{\pi}$

7. Let G be a group, suppose $O(G) = p^2 q$, where p and q are distinct prime numbers satisfying

$q \neq 1$ mod p. Which of the following is always true?

(A) G has more than one p-Sylow subgroup (B) G has a normal p-Sylow subgroup

(C) The number of q-Sylow subgroup of G is divisible by P

(D) G has unique q-Sylow subgroup

8. Let F be a field of 8 elements and

$A = \{x \in F | x^7 = 1 \, and \, x^k \neq 1$ for all natural numbers $k > 7\}$. Then the number of elements in A is

(A) 1 (B) 2 (C) 3 (D) 6

9. The initial value problem $(x^2 - x)\frac{dy}{dx} = (2x-1)y,\ y(x_0) = y_0$ has a unique solution if $f(x_0, y_0)$ equals

(A) (2, 1) (B) (1, 1) (C) (0, 0) (D) (0, 1)

10. Consider the integral equation $\frac{36}{55} x^{\frac{11}{6}} = \int_0^x \frac{1}{(x-t)^{1/6}} y(t)\, dt$, then the solution is

(A) $y(x) = x$ (B) $y(x) = e^x$ (C) $y(x) = \cos x$ (D) $y(x) = \sin x$

11. Given boundary value problem $y'' - y + x = 0,\ (0 \leq x \leq 1), y(0) = y(1) = 0$, then

(A) The approximate solution is $\bar{y}(x) = \frac{5}{22} x(1-x)$

(B) The solution is $y(x) = \sin x \cos x$ (C) The solution is $y(x) = \tan x \cot x$

(D) The solution is $y(x) = c_1 \sin x + c_2$

12. The local truncation error of $w_{i+1} - 4w_i - 3w_{i-1} - 2hf(t_{i-1}, w_{i-1})$

(A) $0(h^2)$ (B) $0(h^3)$ (C) $0(h^4)$ (D) $0(h)$

13. The complete integral of the partial differential equation $xp^3q^2 + yp^2q^3 + (p^3 + q^3) - zp^2q^2 = 0$ is

(A) $ax + by + (ab^{-2} + ba^{-2})$ (B) $ax - by + (ab^{-2} - ba^{-2})$

(C) $-ax + by + (ba^{-2} - ab^{-2})$ (D) $ax + by - a(b^{-3} + ba^{-2})$

14. Let $R = \left\{ \begin{bmatrix} a & b \\ c & d \end{bmatrix} : a,b,c,d \in Z \right\}$ and $I = \left\{ \begin{bmatrix} a & b \\ c & d \end{bmatrix} : a,b,c,d \in 3Z \right\}$ then

(a) I be subring but not ideal (b) I be ideal of R s.t. $\left|\frac{R}{I}\right| = 81$

(c) I be ideal of R s.t. $\left|\frac{R}{I}\right| = 64$ (d) $\frac{R}{I}$ is I.D. with card. 64

15. Which of the following curve can be the functional $I[y(x)] = \int_0^1 \{y'^2 + 12xy\}dx$

$y(0) = 0, \quad y(1) = 1$ be extrimized.

(a) $y = x^2$ (b) $y = x^3$ (c) $y = x$ (d) $y = x^4$

PART – C

16. Let $T: V \to V$ be a Linear operator on a finite dimensional vector space over the field F. Suppose $S: V \to V$ is a linear operator that commutes with T. Then which of the following subspace of V are T invariant

(A) $E_\lambda(s)$ (for any eigen value λ of S) (B) R(S)

(C) $N(\{S - cI_v\}^2)$ for all $c \in F$ (D) $n(T) \oplus n(T^2)$

17. Let T be a linear operator on $M_{n\times n}(R)$ defined by $T(B) = B'$. Then

(A) ± 1 are the only eigenvalue of T

(B) The eigenvectors corresponding to 1 and -1 are symmetric and skew symmetric matrices

(C) The order basis $\alpha = \left\{ \begin{pmatrix} 1 & 0 \\ 0 & 1 \end{pmatrix}, \begin{pmatrix} 0 & 0 \\ 0 & 1 \end{pmatrix}, \begin{pmatrix} 0 & 1 \\ 1 & 0 \end{pmatrix}, \begin{pmatrix} 0 & 1 \\ -1 & 0 \end{pmatrix} \right\}$ for $M_{2\times 2}(R)$ such that $[T]_\alpha$ is a diagonal matrix

(D) (n – 1) is the only eigenvalue of T

18. Let $f : R \to [0, \infty)$ be continuous function such that $g(x)=(f(x))^2$ is uniformly continuous.

Which of the following statements is always true?

(A) f is bounded
(B) f may not be uniformly continuous
(C) f is uniformly continuous
(D) f is unbounded

19. $\lim_{n\to\infty}\int_0^1 s_n(x)\,dx = 1$

(A) By dominated convergence theorem
(B) By Fatou's lemma
(C) By the fact that $\{s_n\}$ converges uniformly on [0, 1]
(D) By the fact that $\{s_n\}$ converges point-wise on [0, 1]

20. The fixed point of $f(z)=\dfrac{az+b}{cz+d}$

(A) $\dfrac{(a-d)+\sqrt{(d-a)^2+4bc}}{2c}$
(B) $\dfrac{(a-d)-\sqrt{(d-a)^2+4bc}}{2c}$
(C) $\dfrac{(a+d)-\sqrt{d^2+4bc}}{2}$
(D) $\dfrac{a\pm\sqrt{d^2+c}}{2a}$

21. For the function $g(z)=\left(\dfrac{z+1}{z-1}\right)^2$. Then

(A) The image of the extended real axis under g is $[0,\infty)$
(B) g maps the half plane Re z > 0 onto the region $|z|>1$ outside the unit circle
(C) g maps the half plane z > 0 onto the region $|z|<1$ inside the unit circle
(D) g maps the extended imaginary axis onto the unit circle.

22. Let G be a group. Which of the following statements are true?

(A) Let H and K be subgroup of G of order 3 and 5 respectively. Then $H\cap K=\{e\}$, where e is the identity element of G.
(B) If G is an abelian group of odd order, then $\varphi(x)=x^2$ is an automorphism of G
(C) If G has exactly one element of order 2, then this element belongs to the center of G
(D) None of these

23. Pick out the true statements:

(A) Let R be a commutative ring with unity. Let M be an ideal such that every element of R not in M is a unit. Then R/M is a field

(B) Let R be as above and let M be an ideal such that R/M is an integral domain. Then M is a prime ideal.

(C) Let R = C[0, 1] be the ring of real valued continuous functions on [0, 1] with respect to point-wise addition and point-wise multiplication. Let $M = \{f \in R | f(0) = (1) = 0\}$, then M is a maximal ideal

24. Consider the integral equation $g(s) = f(s) + \int_0^s e^{s-t} g(t)\,dt$. Then

(A) $\Gamma(s-t) = e^{2s+2t}$

(B) $g(s) = f(s) + \int_0^s e^{2s-2t} f(t)\,dt$

(C) $\Gamma(s-t) = e^{2s-2t}$

(D) $g(s) = f(s) + \int_0^s e^{2s+2t} f(t)\,dt$

25. Let the variational problem $\int_0^5 (9y^2 + y'^2 - 3x)dx$ with the boundary y(0) = 0 & y(5) = 0. Choose incorrect.

(a) $y(x) = e^x - 1$ (b) $y(x) = x$ (c) $y(x) = 0$ (d) $y(x) = e^{-x}$

Answer Key

(Paper – 6)

1	2	3	4	5	6	7	8	9	10	11	12	13
b	c	d	c	a	a	b	d	a	a	a	a	d
14	15	16	17	18	19	20	21	22	23	24	25	
b	b	a,b,c	a,b,c	c	a,b	a,b	a,b,d	a,b,c	a,b	b,c	a,b,d	

MODEL TEST PAPER – 7

PART – B

1. The signature of the quadratic form defined by the formula $Q(x,y,z)=x^2+z^2+4xy+4yz$ is

(A) (2, 1) (B) (2, 2) (C) (4, 2) (D) (0, 1)

2. Let $P_2(R)$ be the vector space of polynomials in x of degree at most 2 with real coefficients. Let $M_2(R)$ be the vector space of 2×2 real matrices. If the linear transformation $T : P_2(R) \to M_2(R)$ is defined $T(f)=\begin{bmatrix} f(0)-f(2) & 0 \\ 0 & f(1) \end{bmatrix}$. Then

(A) T is one-one but not onto (B) T is onto but not one-one

(C) Range (T) = span $\left\{\begin{bmatrix} 0 & 0 \\ 0 & 1 \end{bmatrix}, \begin{bmatrix} -2 & 0 \\ 0 & 1 \end{bmatrix}\right\}$ (D) Null (T) = span $\{x^2-2x, 1-x\}$

3. Let $S = \{x \in R : x^6 - x^5 \le 100\}$ and $T=\{x^2-2x : x\in(0,\infty)\}$. The set $S\cap T$ is

(A) Closed and bounded in R (B) Closed but not bounded in R

(C) Bounded but not closed in R (D) neither closed nor bounded in R

4. Let $\{a_n\}$ be a sequence of real numbers. Which of the following is true?

(A) If $\sum a_n$ converges, then so does $\sum a_n^4$ (B) If $\sum |a_n|$ converges, then so does $\sum a_n^2$

(C) If $\sum a_n$ diverges, then so does $\sum a_n^3$ (D) If $\sum |a_n|$ diverges, then so does $\sum a_n^2$

5. Let f be an entire function on C such that $|f(z)| \le 100 \log|z|$ for each z with $|z|\ge 2$. If $f(i)=2i$, then $f(1)$ must be

(A) 2 (B) 2i (C) I

(D) cannot be determined by given data

6. The coefficient of $(z-\pi)^2$ in the Taylor series expansion of

$$f(z)=\begin{cases}\dfrac{\sin z}{z-\pi} & if \quad z\neq\pi \\ -1 & if \quad z=\pi\end{cases} \text{ around } \pi \text{ is}$$

(A) ½ (B) -1/2 (C) 1/6 (D) -1/6

7. How many isomorphism classes of associative rings (with identity) are there with 35 elements

(A) 0 (B) 35 (C) 1

(D) cannot be determined by given data

8. Let $y_1(x)=x^3$ and $y_2(x)=x^2|x|$ for $x \in R$ Consider the following statements:

(P) : $y_1(x)$ and $y_2(x)$ are L.I. solution of $x^2\dfrac{d^2y}{dx^2}-4x\dfrac{dy}{dx}+6y=0$ on R

(Q): The Wronskian $y_1(x)\dfrac{dy_2(x)}{dx}-y_2(x)\dfrac{dy_1(x)}{dx}=0$ for all $x \in R$

Which of the following statements hold true?

(A) Both P and Q (B) Only P (C) Only Q (D) Neither P nor Q

9. Consider the Volterra integral equation $u(x)=\sin x+\left(e-e^{\cos x}\right)-\int_0^x e^{\cos t}u(t)dt$ Then the solution is

(A) u(x) = sinx (B) u(x) = cosx (C) u(x) = tanx (D) u(x) = cotx

10. Consider the functional $\int_0^{\frac{\pi}{2}}\left(q(x)^2+q'(x)^2-2q(x)\sin x\right)dx$. Then the stationary point is

(A) $q(x)=c_1e^x+c_2e^{-x}+\frac{1}{2}\sin x$ (B) $q(x)=c_1e^{-x}+c_2e^{2x}+\frac{1}{2}\cos x$

(C) $q(x)=c_1\cos x+c_2\sin x+\frac{1}{2}\sec x$ (D) $q(x)=c_1\cos x+c_2\sin x+\frac{1}{2}\operatorname{cosec} x$

11. Let f: [0, 2] → R be twice continuously differentiable function. If $\int_0^2 f(x)dx\approx 2f(1)$, then the error in the approximation is

(A) $\dfrac{f'(\xi)}{12}$ for some $\xi\in(0,2)$ (B) $\dfrac{f'(\xi)}{2}$ for some $\xi\in(0,2)$

(C) $\dfrac{f''(\xi)}{3}$ for some $\xi\in(0,2)$ (D) $\dfrac{f''(\xi)}{6}$ for some $\xi\in(0,2)$

12. Let $f(x) = x^3 + 2x^2 + 1$ and $g(x) = 2x^2 + x + 2$. Then over Z_3 -
(a) f(x) & g(x) are irreducible. (b) f(x) is irreducible but g is not
(c) g(x) is irreducible but f is not (d) Neither f nor g is irreducible.

13. Which of the following is not maximal ideal of (Z, +, .)
(a) (2Z, +, .) (b) (7Z, +, .) (c) (11Z, +, .) (d) (27Z, +, .)

14. The integral equation corresponding to initial value problem.
$y'' + y = 0; y(0) = 0, y'(0) = 1$ is
(a) $u(x) = x + \int_0^1 (t-x)u(t)dt$ (b) $u(x) = x + \int_0^x (t-x)u(t)dt$
(c) $u(x) = \sin x + \int_0^x (t-x)u(t)dt$ (d) $u(x) = x(1+x) + \int_0^1 (t-x)u(t)dt$

15. The solution of : $\frac{d^2y}{dx^2} - 4y = x^2$ is
(a) $c_1e^x + c_2e^{-x} - \frac{1}{4}\left[x^2 + \frac{1}{2}\right]$ (b) $c_1e^{2x} + c_2e^{-x} + \frac{1}{4}\left[x^2 + \frac{1}{2}\right]$.
(c) $c_1e^{2x} + c_2e^{-2x} + \frac{1}{4}\left[x^2 + \frac{1}{2}\right]$ (d) $c_1e^{2x} + c_2e^{-2x} - \frac{1}{4}\left[x^2 + \frac{1}{2}\right]$

PART – C

16. Let X be a proper closed subset of [0, 1]. Which of the following statements are always true?
(a) the set X is countable
(b) there exists $x \in X$ such that $X \setminus \{x\}$ is closed
(c) the set X contains an open interval (d) none of the above

17. Let $f(x) = \frac{e^{-\frac{1}{x}}}{x}$, where $x \in (0,1)$. Then on $(0,1)$
(a) f is uniformly continuous (b) f is continuous but not uniformly continuous
(c) f is bounded (d) f is not continuous

18. Let V be a finite dimensional inner product space and let $B = (\alpha_1, \alpha_2 \cdots \alpha_n)$ be an (ordered)
orthogonal basis for V. Let T be a linear operator on V and let A be the matrix of T in the order basis B

(a) $A_{kj} = \left(\frac{T\alpha_j}{\alpha_k}\right)$ (b) $\alpha = \sum_{k=1}^{n}(\alpha/\alpha_k)\alpha_k$ (c) $\alpha = \sum_{k=1}^{n}(\alpha/\alpha_k)$ (d) $T_{\alpha j} = \sum_{k=1}^{n}(T\alpha_j/\alpha_k)\alpha_k$

19. Let V be R^3 and U = R^2. Then consider the map $T: V \to U$ defined by T(a, b, c) = (0, a + b + c)

(a) T is not a linear transformation (b) dim(V) = 3

(c) Basis of ker (T) is $\begin{pmatrix}-1\\1\\0\end{pmatrix}, \begin{pmatrix}-1\\0\\1\end{pmatrix}$ (d) dim (Im (T)) = 1

20. When C is the unit circle $|z| = 1$ in either direction. Then

(a) $\int_c \frac{z^3}{z-3}\,dz = 0$ (b) $\int_c \tan z\,dz = 2\pi i$ (c) $\int_c \log(z+2)\,dz = 0$ (d) $\int_c \tan z\,dz = 0$

21. Let $\Omega = \{z \in C : \text{Im}(z) > 0\}$ and let C be a smooth curve lying in Ω with initial point $-1+2i$ and final point $1+2i$. The value of $\int_C \frac{1+2z}{1+z}\,dz$ is

(a) $4 - \frac{1}{2}\ln 2 + i\frac{\pi}{4}$ (b) $-4 + \frac{1}{2}\ln 2 + i\frac{\pi}{4}$ (c) $4 + \frac{1}{2}\ln 2 - i\frac{\pi}{4}$ (d) $4 - \frac{1}{2}\ln 2 + i\frac{\pi}{2}$

22. Which of the following primes satisfy the congruence $(a)^{24} \equiv 6a + 2 \pmod{13}$?

(a) 41 (b) 47 (c) 67 (d) 83

23. Let y(x) be a piecewise continuously differentiable function on [0,4]. Then the functional $J[y] = \int_0^4 (y'-1)^2 (y'+1)^2\,dx$ attains minimum if y = y(x) is

(A) $y = \frac{x}{2}$ $0 \le x \le 4$ (B) $y = \begin{cases} -x & 0 \le x \le 1 \\ x-2 & 1 \le x \le 4 \end{cases}$

(C) $y = \begin{cases} 2x & 0 \le x \le 2 \\ -x+6 & 2 \le x \le 4 \end{cases}$ (D) $y = \begin{cases} x & 0 \le x \le 3 \\ -x+6 & 3 \le x \le 4 \end{cases}$

24. The solution of the Integral equation $\phi(x) = X + \int_0^x (\xi - x)\phi(\xi)d\xi$ is

(A) sin x (B) cos x (C) sec x (D) None of these

25. The solution of the Fredholm integral equation $y(x) = \sin x + \sin x \int_0^{\pi/2} \cos t \; y(t)dt, n \geq 0$

(A) $y(x) = \sec^2 x$ (B) $y(x) = \cos x$ (C) $y(x) = \cos 2x$ (D) $y(x) = 2\sin x$

Answer Key

(Paper – 7)

1	2	3	4	5	6	7	8	9	10	11	12	13
a	c	a	b	b	c	c	a	a	a	c	b	d
14	15	16	17	18	19	20	21	22	23	24	25	
b	d	d	a	a,b,d	b,c,d	a,c,d	a	a,c	b,d	a	d	

MODEL TEST PAPERS